A selection of bounds of the standard normal distribution and of the χ^2 distribution (1 DF) for the one sided and for the two sided test

P	z		χ^2 For one degree of freedom	
	One sided	Two sided	One sided	Two sided
0.001	3.090	3.291	9.550	10.828
0.01	2.326	2.576	5.412	6.635
0.05	1.645	1.960	2.706	3.841
0.10	1.282	1.645	1.642	2.706
0.20	0.842	1.282	0.708	1.642
0.50	0	0.674	0	0.455

The Greek alphabet

Greek letter		Name	Greek letter		Name
A	α	Alpha	N	ν	Nu
B	β	Beta	Ξ	ξ	Xi
Γ	γ	Gamma	O	o	Omicron
Δ	δ	Delta	Π	π	Pi
E	ε	Epsilon	P	ρ	Rho
Z	ζ	Zeta	Σ	$\sigma\ \varsigma$	Sigma
H	η	Eta	T	τ	Tau
Θ	ϑ	Theta	Y	υ	Upsilon
I	ι	Iota	Φ	φ	Phi
K	κ	Kappa	X	χ	Chi
Λ	λ	Lambda	Ψ	ψ	Psi
M	μ	Mu	Ω	ω	Omega

Bounds (critical values) for the t-, χ^2-, and F-distributions for $P = 0.95$ ($\alpha = 0.05$)

DF ν	t One sided	t Two sided	χ^2	F upper bounds $\nu_1=1$	2	3	4	5	6	7	8	9	10	12	15	20	24	30	50	100	200	∞	ν_2
1	6.31	12.71	3.84	161	200	216	225	230	234	237	239	241	242	244	246	248	249	250	252	253	254	254	1
2	2.92	4.30	5.99	18.51	19.00	19.16	19.25	19.30	19.33	19.35	19.37	19.39	19.40	19.41	19.43	19.44	19.45	19.46	19.47	19.49	19.49	19.50	2
3	2.35	3.18	7.81	10.13	9.55	9.28	9.12	9.01	8.94	8.89	8.85	8.81	8.79	8.74	8.70	8.66	8.64	8.62	8.58	8.55	8.54	8.53	3
4	2.13	2.78	9.49	7.71	6.94	6.59	6.39	6.26	6.16	6.09	6.04	6.00	5.96	5.91	5.86	5.80	5.77	5.75	5.70	5.66	5.65	5.63	4
5	2.02	2.57	11.07	6.61	5.79	5.41	5.19	5.05	4.95	4.88	4.82	4.77	4.74	4.68	4.62	4.56	4.53	4.50	4.44	4.41	4.38	4.37	5
6	1.94	2.45	12.59	5.99	5.14	4.76	4.53	4.39	4.28	4.21	4.15	4.10	4.06	4.00	3.94	3.87	3.84	3.81	3.75	3.71	3.69	3.67	6
7	1.89	2.36	14.07	5.59	4.74	4.35	4.12	3.97	3.87	3.79	3.73	3.68	3.64	3.57	3.51	3.44	3.41	3.38	3.32	3.27	3.25	3.23	7
8	1.86	2.31	15.51	5.32	4.46	4.07	3.84	3.69	3.58	3.50	3.44	3.39	3.35	3.28	3.22	3.15	3.12	3.08	3.02	2.97	2.96	2.93	8
9	1.83	2.26	16.92	5.12	4.26	3.86	3.63	3.48	3.37	3.29	3.23	3.18	3.14	3.07	3.01	2.93	2.90	2.86	2.80	2.76	2.73	2.71	9
10	1.81	2.23	18.31	4.96	4.10	3.71	3.48	3.33	3.22	3.14	3.07	3.02	2.98	2.91	2.85	2.77	2.74	2.70	2.64	2.59	2.56	2.54	10
11	1.80	2.20	19.68	4.84	3.98	3.59	3.36	3.20	3.09	3.01	2.95	2.90	2.85	2.79	2.72	2.65	2.61	2.57	2.51	2.46	2.42	2.40	11
12	1.78	2.18	21.03	4.75	3.89	3.49	3.26	3.11	3.00	2.91	2.85	2.80	2.75	2.69	2.62	2.54	2.50	2.47	2.40	2.35	2.32	2.30	12
13	1.77	2.16	22.36	4.67	3.81	3.41	3.18	3.03	2.92	2.83	2.77	2.71	2.67	2.60	2.53	2.46	2.42	2.38	2.31	2.26	2.24	2.21	13
14	1.76	2.14	23.68	4.60	3.74	3.34	3.11	2.96	2.85	2.76	2.70	2.65	2.60	2.53	2.46	2.39	2.35	2.31	2.24	2.19	2.16	2.13	14
15	1.75	2.13	25.00	4.54	3.68	3.29	3.06	2.90	2.79	2.71	2.64	2.59	2.54	2.48	2.40	2.33	2.29	2.25	2.18	2.12	2.10	2.07	15
16	1.75	2.12	26.30	4.49	3.63	3.24	3.01	2.85	2.74	2.66	2.59	2.54	2.49	2.42	2.35	2.28	2.24	2.19	2.12	2.07	2.04	2.01	16
17	1.74	2.11	27.59	4.45	3.59	3.20	2.96	2.81	2.70	2.61	2.55	2.49	2.45	2.38	2.31	2.23	2.19	2.15	2.08	2.02	1.99	1.96	17
18	1.73	2.10	28.87	4.41	3.55	3.16	2.93	2.77	2.66	2.58	2.51	2.46	2.41	2.34	2.27	2.19	2.15	2.11	2.04	1.98	1.95	1.92	18
19	1.73	2.09	30.14	4.38	3.52	3.13	2.90	2.74	2.63	2.54	2.48	2.42	2.38	2.31	2.23	2.15	2.11	2.07	2.00	1.94	1.91	1.88	19
20	1.72	2.09	31.41	4.35	3.49	3.10	2.87	2.71	2.60	2.51	2.45	2.39	2.35	2.28	2.20	2.12	2.08	2.04	1.97	1.91	1.87	1.84	20
21	1.72	2.08	32.67	4.32	3.47	3.07	2.84	2.68	2.57	2.49	2.42	2.37	2.32	2.25	2.18	2.09	2.05	2.01	1.94	1.88	1.84	1.81	21
22	1.72	2.07	33.92	4.30	3.44	3.05	2.82	2.66	2.55	2.46	2.40	2.34	2.30	2.23	2.15	2.07	2.03	1.98	1.91	1.85	1.81	1.78	22
23	1.71	2.07	35.17	4.28	3.42	3.03	2.80	2.64	2.53	2.44	2.37	2.32	2.27	2.20	2.13	2.04	2.00	1.96	1.88	1.82	1.79	1.76	23
24	1.71	2.06	36.42	4.26	3.40	3.01	2.78	2.62	2.51	2.42	2.36	2.30	2.25	2.18	2.11	2.02	1.98	1.94	1.86	1.80	1.76	1.73	24
25	1.71	2.06	37.65	4.24	3.39	2.99	2.76	2.60	2.49	2.40	2.34	2.28	2.24	2.16	2.09	2.00	1.96	1.92	1.84	1.78	1.74	1.71	25
26	1.71	2.06	38.89	4.23	3.37	2.98	2.74	2.59	2.47	2.39	2.32	2.27	2.22	2.15	2.07	1.99	1.95	1.90	1.82	1.76	1.72	1.69	26
27	1.70	2.05	40.11	4.21	3.35	2.96	2.73	2.57	2.46	2.37	2.31	2.25	2.20	2.13	2.06	1.97	1.93	1.88	1.81	1.74	1.71	1.67	27
28	1.70	2.05	41.34	4.20	3.34	2.95	2.71	2.56	2.45	2.36	2.29	2.24	2.19	2.12	2.04	1.96	1.91	1.87	1.79	1.73	1.69	1.65	28
29	1.70	2.05	42.56	4.18	3.33	2.93	2.70	2.55	2.43	2.35	2.28	2.22	2.18	2.10	2.03	1.94	1.90	1.85	1.77	1.71	1.68	1.64	29
30	1.70	2.04	43.77	4.17	3.32	2.92	2.69	2.53	2.42	2.33	2.27	2.21	2.16	2.09	2.01	1.93	1.89	1.84	1.76	1.70	1.66	1.62	30
34	1.69	2.03	48.60	4.13	3.28	2.88	2.65	2.49	2.38	2.29	2.23	2.17	2.12	2.05	1.97	1.89	1.84	1.80	1.71	1.65	1.61	1.57	34
40	1.68	2.02	55.76	4.08	3.23	2.84	2.61	2.45	2.34	2.25	2.18	2.12	2.08	2.00	1.92	1.84	1.79	1.74	1.66	1.59	1.55	1.51	40
44	1.68	2.02	60.48	4.06	3.21	2.82	2.58	2.43	2.31	2.23	2.16	2.10	2.05	1.98	1.90	1.81	1.76	1.72	1.63	1.56	1.52	1.48	44
50	1.68	2.01	67.50	4.03	3.18	2.79	2.56	2.40	2.29	2.20	2.13	2.07	2.03	1.95	1.87	1.78	1.74	1.69	1.60	1.52	1.48	1.44	50
60	1.67	2.00	79.08	4.00	3.15	2.76	2.53	2.37	2.25	2.17	2.10	2.04	1.99	1.92	1.84	1.75	1.70	1.65	1.56	1.48	1.44	1.39	60
70	1.67	1.99	90.53	3.98	3.13	2.74	2.50	2.35	2.23	2.14	2.07	2.02	1.97	1.89	1.81	1.72	1.67	1.62	1.53	1.45	1.40	1.35	70
80	1.66	1.99	101.88	3.96	3.11	2.72	2.49	2.33	2.21	2.13	2.06	2.00	1.95	1.88	1.79	1.70	1.65	1.60	1.51	1.43	1.38	1.32	80
90	1.66	1.99	113.15	3.95	3.10	2.71	2.47	2.32	2.20	2.11	2.04	1.99	1.94	1.86	1.78	1.69	1.64	1.59	1.49	1.41	1.36	1.30	90
100	1.66	1.98	124.34	3.94	3.09	2.70	2.46	2.31	2.19	2.10	2.03	1.97	1.93	1.85	1.77	1.68	1.63	1.57	1.48	1.39	1.34	1.28	100
150	1.66	1.98	179.58	3.90	3.06	2.66	2.43	2.27	2.16	2.07	2.00	1.94	1.89	1.82	1.73	1.64	1.59	1.53	1.44	1.34	1.29	1.22	150
200	1.65	1.97	233.99	3.89	3.04	2.65	2.42	2.26	2.14	2.06	1.98	1.93	1.88	1.80	1.72	1.62	1.57	1.52	1.41	1.32	1.26	1.19	200
∞	1.64	1.96	∞	3.84	3.00	2.60	2.37	2.21	2.10	2.01	1.94	1.88	1.83	1.75	1.67	1.57	1.52	1.46	1.35	1.24	1.17	1.00	∞

Springer Series in Statistics

Advisors:
D. Brillinger, S. Fienberg, J. Gani, J. Hartigan
J. Kiefer, K. Krickeberg

Important Statistical Tables

An index of statistical tables follows the Contents

Four place common logarithms	12, 13
Four place antilogarithms	14, 15
Random numbers	52

Standard normal distribution	Top of front end-paper, 62, 217
t-distribution	136, 137
χ^2-distribution (DF = 1: Top of front end-paper, 349),	140, 141
F-distribution	144–150

Binomial coefficients	158
Factorials	160
$2n \ln n$	353–358

Tolerance limits (normal distribution)	282
Distribution-free tolerance limits	284

Confidence interval:
Median	317, 318
Lambda (Poisson distribution)	344, 345
π (Binomial distribution)	703

Sample sizes:
Counting	36, 181, 218, 283, 284, 337, 338, 350
Measurement	250, 262, 274, 283, 284

Correlation:
Spearman's ρ	398, 399
Correlation coefficient	425
Finding $\hat{z}$ in terms of r and conversely	428

Cochran's test	497
Departure from normality test	253
Friedman's test	551
H-test	305
Hartley's test	496
Run test	376, 377
Kolmogoroff–Smirnoff goodness of fit test	331
Link–Wallace test	543, 544
Nemenyi comparisons	546, 547
Page test	554
Siegel–Tukey test	287
Standardized mean square successive differences test	375
Studentized range	535, 536
U-test	297–301
Sign test	317, 318
Wilcoxon paired differences test	313
Wilcoxon–Wilcox comparisons	556, 557

Lothar Sachs

Applied Statistics
A Handbook of Techniques

Second Edition

Translated by Zenon Reynarowych

With 59 Figures

Springer-Verlag
New York Berlin Heidelberg Tokyo

Lothar Sachs
Abteilung Medizinische Statistik
  und Dokumentation im Klinikum
  der Universität Kiel
D-2300 Kiel 1
Federal Republic of Germany

Translator
Zenon Reynarowych
Courant Institute of
  Mathematical Sciences
New York University
New York, NY 10012
U.S.A.

AMS Classification: 62-XX

Library of Congress Cataloging in Publication Data
Sachs, Lothar.
  Applied statistics.
  (Springer series in statistics)
  Translation of: Angewandte Statistik.
  Includes bibliographies and indexes.
  1. Mathematical statistics.   I. Title.
II. Series.
QA276.S213   1984   519.5   84-10538

Title of the Original German Edition: *Angewandte Statistik*, 5 Auflage, Springer-Verlag, Berlin Heidelberg New York, 1978.

© 1982, 1984 by Springer-Verlag New York Inc.
Softcover reprint of the hardcover 1st Edition 1984

All rights reserved. No part of this book may be translated or reproduced in any form without written permission from Springer-Verlag, 175 Fifth Avenue, New York, New York 10010, U.S.A.

Typeset by Composition House Ltd., Salisbury, England.

9 8 7 6 5 4 3 2 1

ISBN-13: 978-1-4612-9755-0   e-ISBN: 978-1-4612-5246-7
DOI: 10.1007/978-1-4612-5246-7

To my wife

CONTENTS

Index of Statistical Tables	xiii
Preface to the Second English Edition	xvii
Preface to the First English Edition	xviii
From the Prefaces to Previous Editions	xix
Permissions and Acknowledgments	xxiii
Selected Symbols	xxv
Introduction	1
Introduction to Statistics	3

0 Preliminaries 7

0.1	Mathematical Abbreviations	7
0.2	Arithmetical Operations	8
0.3	Computational Aids	19
0.4	Rounding Off	20
0.5	Computations with Inaccurate Numbers	21

1 Statistical Decision Techniques 23

1.1	What Is Statistics? Statistics and the Scientific Method	23
1.2	Elements of Computational Probability	26
▶1.2.1	Statistical probability	26
▶1.2.2	The addition theorem of probability theory	28
▶1.2.3	Conditional probability and statistical independence	31
1.2.4	Bayes's theorem	38
▶1.2.5	The random variable	43
1.2.6	The distribution function and the probability function	44

1.3	The Path to the Normal Distribution		47
	▶1.3.1 The population and the sample		47
	▶1.3.2 The generation of random samples		49
	▶1.3.3 A frequency distribution		53
	▶1.3.4 Bell-shaped curves and the normal distribution		57
	▶1.3.5 Deviations from the normal distribution		65
	▶1.3.6 Parameters of unimodal distributions		66
	▶1.3.7 The probability plot		81
	1.3.8 Additional statistics for the characterization of a one dimensional frequency distribution		85
	1.3.9 The lognormal distribution		107
1.4	The Road to the Statistical Test		112
	1.4.1 The confidence coefficient		112
	1.4.2 Null hypotheses and alternative hypotheses		114
	1.4.3 Risk I and risk II		117
	1.4.4 The significance level and the hypotheses are, if possible, to be specified before collecting the data		120
	1.4.5 The statistical test		120
	1.4.6 One sided and two sided tests		124
	1.4.7 The power of a test		125
	1.4.8 Distribution-free procedures		130
	1.4.9 Decision principles		133
1.5	Three Important Families of Test Distributions		134
	1.5.1 The Student's t-distribution		135
	1.5.2 The χ^2 distribution		139
	1.5.3 The F-distribution		143
1.6	Discrete Distributions		155
	1.6.1 The binomial coefficient		155
	▶1.6.2 The binomial distribution		162
	1.6.3 The hypergeometric distribution		171
	1.6.4 The Poisson distribution		175
	▶1.6.5 The Thorndike nomogram		183
	1.6.6 Comparison of means of Poisson distributions		186
	1.6.7 The dispersion index		189
	1.6.8 The multinomial coefficient		192
	1.6.9 The multinomial distribution		193

2 Statistical Methods in Medicine and Technology 195

2.1	Medical Statistics		195
	2.1.1 Critique of the source material		196
	2.1.2 The reliability of laboratory methods		197
	2.1.3 How to get unbiased information and how to investigate associations		202
	2.1.4 Retrospective and prospective comparisons		206
	2.1.5 The therapeutic comparison		210
	2.1.6 The choice of appropriate sample sizes for the clinical trial		214
2.2	Sequential Test Plans		219
2.3	Evaluation of Biologically Active Substances Based on Dosage-Dichotomous Effect Curves		224
2.4	Statistics in Engineering		228
	2.4.1 Quality control in industry		228
	2.4.2 Life span and reliability of manufactured products		233

2.5	Operations Research	238
	2.5.1 Linear programming	239
	2.5.2 Game theory and the war game	239
	2.5.3 The Monte Carlo method and computer simulation	241

3 The Comparison of Independent Data Samples — 245

3.1	The Confidence Interval of the Mean and of the Median	246
	▶ 3.1.1 Confidence interval for the mean	247
	▶ 3.1.2 Estimation of sample sizes	249
	3.1.3 The mean absolute deviation	251
	3.1.4 Confidence interval for the median	254
▶ 3.2	Comparison of an Empirical Mean with the Mean of a Normally Distributed Population	255
▶ 3.3	Comparison of an Empirical Variance with Its Parameter	258
3.4	Confidence Interval for the Variance and for the Coefficient of Variation	259
3.5	Comparison of Two Empirically Determined Variances of Normally Distributed Populations	260
	3.5.1 Small to medium sample size	260
	3.5.2 Medium to large sample size	263
	3.5.3 Large to very large sample size ($n_1, n_2 \gtrsim 100$)	264
▶ 3.6	Comparison of Two Empirical Means of Normally Distributed Populations	264
	3.6.1 Unknown but equal variances	264
	3.6.2 Unknown, possibly unequal variances	271
3.7	Quick Tests Which Assume Nearly Normally Distributed Data	275
	3.7.1 The comparison of the dispersions of two small samples according to Pillai and Buenaventura	275
	3.7.2 The comparison of the means of two small samples according to Lord	276
	3.7.3 Comparison of the means of several samples of equal size according to Dixon	277
3.8	The Problem of Outliers and Some Tables Useful in Setting Tolerance Limits	279
3.9	Distribution-Free Procedures for the Comparison of Independent Samples	285
	3.9.1 The rank dispersion test of Siegel and Tukey	286
	3.9.2 The comparison of two independent samples: Tukey's quick and compact test	289
	3.9.3 The comparison of two independent samples according to Kolmogoroff and Smirnoff	291
	▶ 3.9.4 Comparison of two independent samples: The U-test of Wilcoxon, Mann, and Whitney	293
	3.9.5 The comparison of several independent samples: The H-test of Kruskal and Wallis	303

4 Further Test Procedures — 307

4.1	Reduction of Sampling Errors by Pairing Observations: Paired Samples	307

4.2	Observations Arranged in Pairs		309
	4.2.1 The t-test for data arranged in pairs		309
	4.2.2 The Wilcoxon matched pair signed-rank test		312
	4.2.3 The maximum test for pair differences		315
	4.2.4 The sign test of Dixon and Mood		316
4.3	The χ^2 Goodness of Fit Test		320
	4.3.1 Comparing observed frequencies with their expectations		321
	4.3.2 Comparison of an empirical distribution with the uniform distribution		322
	4.3.3 Comparison of an empirical distribution with the normal distribution		322
	4.3.4 Comparison of an empirical distribution with the Poisson distribution		329
4.4	The Kolmogoroff–Smirnoff Goodness of Fit Test		330
4.5	The Frequency of Events		333
	4.5.1 Confidence limits of an observed frequency for binomially distributed population. The comparison of a relative frequency with the underlying parameter		333
	4.5.2 Clopper and Pearson's quick estimation of the confidence intervals of a relative frequency		340
	4.5.3 Estimation of the minimum size of a sample with counted data		341
	4.5.4 The confidence interval for rare events		343
	4.5.5 Comparison of two frequencies; testing whether they stand in a certain ratio		345
4.6	The Evaluation of Fourfold Tables		346
	4.6.1 The comparison of two percentages—the analysis of fourfold tables		346
	4.6.2 Repeated application of the fourfold χ^2 test		360
	4.6.3 The sign test modified by McNemar		363
	4.6.4 The additive property of χ^2		366
	4.6.5 The combination of fourfold tables		367
	4.6.6 The Pearson contingency coefficient		369
	4.6.7 The exact Fisher test of independence, as well as an approximation for the comparison of two binomially distributed populations (based on very small samples)		370
4.7	Testing the Randomness of a Sequence of Dichotomous Data or of Measured Data		373
	4.7.1 The mean square successive difference		373
	4.7.2 The run test for testing whether a sequence of dichotomous data or of measured data is random		375
	4.7.3 The phase frequency test of Wallis and Moore		378
4.8	The S_3 Sign Test of Cox and Stuart for Detection of a Monotone Trend		379

5 Measures of Association: Correlation and Regression 382

5.1	Preliminary Remarks and Survey	382
	5.1.1 The Bartlett procedure	390
	5.1.2 The Kerrich procedure	392
5.2	Hypotheses on Causation Must Come from Outside, Not from Statistics	393

5.3	Distribution-Free Measures of Association		395
	▶ 5.3.1 The Spearman rank correlation coefficient		396
	5.3.2 Quadrant correlation		403
	5.3.3 The corner test of Olmstead and Tukey		405
5.4	Estimation Procedures		406
	5.4.1 Estimation of the correlation coefficient		406
	▶ 5.4.2 Estimation of the regression line		408
	5.4.3 The estimation of some standard deviations		413
	5.4.4 Estimation of the correlation coefficients and the regression lines from a correlation table		419
	5.4.5 Confidence limits of correlation coefficients		423
5.5	Test Procedures		424
	5.5.1 Testing for the presence of correlation and some comparisons		424
	5.5.2 Further applications of the $\hat{z}$-transformation		429
	▶ 5.5.3 Testing the linearity of a regression		433
	▶ 5.5.4 Testing the regression coefficient against zero		437
	5.5.5 Testing the difference between an estimated and a hypothetical regression coefficient		438
	5.5.6 Testing the difference between an estimated and a hypothetical axis intercept		439
	5.5.7 Confidence limits for the regression coefficient, for the axis intercept, and for the residual variance		439
	▶ 5.5.8 Comparing two regression coefficients and testing the equality of more than two regression lines		440
	▶ 5.5.9 Confidence interval for the regression line		443
5.6	Nonlinear Regression		447
5.7	Some Linearizing Transformations		453
▶ 5.8	Partial and Multiple Correlations and Regressions		456

6 The Analysis of $k \times 2$ and Other Two Way Tables — 462

6.1	Comparison of Several Samples of Dichotomous Data and the Analysis of a $k \times 2$ Two Way Table	462
	6.1.1 $k \times 2$ tables: The binomial homogeneity test	462
	6.1.2 Comparison of two independent empirical distributions of frequency data	467
	6.1.3 Partitioning the degrees of freedom of a $k \times 2$ table	468
	6.1.4 Testing a $k \times 2$ table for trend: The share of linear regression in the overall variation	472
6.2	The Analysis of $r \times c$ Contingency and Homogeneity Tables	474
	▶ 6.2.1 Testing for independence or homogeneity	474
	6.2.2 Testing the strength of the relation between two categorically itemized characteristics. The comparison of several contingency tables with respect to the strength of the relation by means of the corrected contingency coefficient of Pawlik	482
	6.2.3 Testing for trend: The component due to linear regression in the overall variation. The comparison of regression coefficients of corresponding two way tables	484
	6.2.4 Testing square tables for symmetry	488

▶ 6.2.5		Application of the minimum discrimination information statistic in testing two way tables for independence or homogeneity	490

7   Analysis of Variance Techniques 494

▶ 7.1   Preliminary Discussion and Survey   494
7.2   Testing the Equality of Several Variances   495
    7.2.1   Testing the equality of several variances of equally large groups of samples   495
    7.2.2   Testing the equality of several variances according to Cochran   497
  ▶ 7.2.3   Testing the equality of the variances of several samples of the same or different sizes according to Bartlett   498
7.3   One Way Analysis of Variance   501
    7.3.1   Comparison of several means by analysis of variance   501
    7.3.2   Assessment of linear contrasts according to Scheffé, and related topics   509
    7.3.3   Transformations   515
7.4   Two Way and Three Way Analysis of Variance   518
    7.4.1   Analysis of variance for $2ab$ observations   518
  ▶ 7.4.2   Multiple comparison of means according to Scheffé, according to Student, Newman and Keuls, and according to Tukey   533
  ▶ 7.4.3   Two way analysis of variance with a single observation per cell. A model without interaction   537
7.5   Rapid Tests of Analysis of Variance   542
    7.5.1   Rapid test of analysis of variance and multiple comparisons of means according to Link and Wallace   542
    7.5.2   Distribution-free multiple comparisons of independent samples according to Nemenyi: Pairwise comparisons of all possible pairs of treatments   546
7.6   Rank Analysis of Variance for Several Correlated Samples   549
  ▶ 7.6.1   The Friedman test: Double partitioning with a single observation per cell   549
    7.6.2   Multiple comparisons of correlated samples according to Wilcoxon and Wilcox: Pairwise comparisons of several treatments which are repeated under a number of different conditions or in a number of different classes of subjects   555
▶ 7.7   Principles of Experimental Design   558

Bibliography and General References   568

Exercises   642

Solutions to the Exercises   650

Few Names and Some Page Numbers   657

Subject Index   667

INDEX OF STATISTICAL TABLES

Four place common logarithms   12
Four place antilogarithms   14
Probabilities of at least one success in independent trials with specified success probability   36
Random numbers   52
The z-test: The area under the standard normal density curve from z to ∞ for $0 \leq z \leq 4.1$ 62
Values of the standard normal distribution   63
Ordinates of the standard normal curve   79
Probability P that a coin tossed n times always falls on the same side, a model for a random event   115
Two sided and upper percentage points of the Student distribution   136
Significance bounds of the χ^2-distribution   140
5%, 1%, and 0.1% bounds of the χ^2-distribution   141
Three place natural logarithms for selected arguments   142
Upper significance bounds of the F-distribution for $P = 0.10$, $P = 0.05$, $P = 0.025$, $P = 0.01$, $P = 0.005$, and $P = 0.001$ 144
Binomial coefficients   158
Factorials and their common logarithms   160
Binomial probabilities for $n \leq 10$ and for various values of p 167
Values of $e^{-\lambda}$ for the Poisson distribution   177
Poisson distribution for small parameter λ and no, one, and more than one event   178
Poisson distribution for selected values of λ 179

Number of "rare" events to be expected in random samples of prescribed size with given probability of occurrence and a given confidence coefficient $S = 95\%$ 182

Selected bounds of the standard normal distribution for the one and two sided tests   217

Optimal group size for testing in groups   218

Upper limits for the range chart (R-chart)   230

Sample size needed to estimate a standard deviation with given relative error and given confidence coefficient   250

Factors for determining the 95% confidence limits about the mean in terms of the mean absolute deviation   253

Number of observations needed to compare two variances using the F test, ($\alpha = 0.05$; $\beta = 0.01$, $\beta = 0.05$, $\beta = 0.1$, $\beta = 0.5$)   262

Angular transformation: the values $x = \arcsin \sqrt{p}$ 269

The size of samples needed to compare two means using the t-test with given level of significance, power, and deviation   274

Upper significance bounds of the F'-distribution based on the range   276

Bounds for the comparison of two independent data sequences of the same size with regard to their behavior in the central portion of the distribution, according to Lord   277

Significance bounds for testing arithmetic means and extreme values according to Dixon   278

Upper significance bounds of the standardized extreme deviation   281

Tolerance factors for the normal distribution   282

Sample sizes for two sided nonparametric tolerance limits   283

Nonparametric tolerance limits   284

Critical rank sums for the rank dispersion test of Siegel and Tukey   287

Critical values for comparing two independent samples according to Kolmogoroff and Smirnoff   292

Critical values of U for the Wilcoxon–Mann–Whitney test according to Milton   296

Levels of significance for the H-test of Kruskal and Wallis   305

Critical values for the Wilcoxon signed-rank test   313

Bounds for the sign test   317

Lower and upper percentages of the standardized 3rd and 4th moments, $\sqrt{b_1}$ and b_2, for tests for departure from normality   326

Critical limits of the quotient R/s 328

Critical values of D for the Kolmogoroff–Smirnoff goodness of fit test   331

Estimation of the confidence interval of an observed frequency according to Cochran   336

One sided confidence limits for complete failure or complete success   337

Critical 5% differences for the comparison of two percentages   338

Confidence intervals for the mean of a Poisson distribution   344

χ^2 table for a single degree of freedom   349

Sample sizes for the comparison of two percentages   350

$2n \ln n$ values for $n = 0$ to $n = 2009$ 353

$2n \ln n$ values for $n = \frac{1}{2}$ to $n = 299\frac{1}{2}$ 358

Supplementary table for computing large values of $2n \ln n$ 359

Critical bounds for the quotients of the dispersion of the mean squared successive differences and the variance   375

Critical values for the run test   376

Significance of the Spearman rank correlation coefficient   398

Upper and lower critical scores of a quadrant for the assessment of quadrant correlation   404

Bounds for the corner test   405

Testing the correlation coefficient for significance against zero   425

Converting the correlation coefficient into $\dot{z} = \frac{1}{2} \ln [(1 + r)/(1 - r)]$ 428

Normal equations of important functional equations   452

Linearizing transformations   454

Upper bounds of the Bonferroni χ^2-statistics   479

Values of the maximal contingency coefficient for $r = 2$ to $r = 10$ 483

Distribution of F_{max} according to Hartley for the testing of several variances for homogeneity   496

Significance bounds for the Cochran test   497

Factors for estimating the standard deviation of the population from the range of the sample   507

Factors for estimating a confidence interval of the range   508

Upper significance bounds of the Studentized range distribution, with $P = 0.05$ and $P = 0.01$ 535

Critical values for the test of Link and Wallace, with $P = 0.05$ and $P = 0.01$ 542

Critical differences for the one way classification: Comparison of all possible pairs of treatments according to Nemenyi, with $P = 0.10$, $P = 0.05$, and $P = 0.01$ 546

Bounds for the Friedman test   551

Some 5% and 1% bounds for the Page test   554

Critical differences for the two way classification: Comparison of all possible pairs of treatments according to Wilcoxon and Wilcox, with $P = 0.10$, $P = 0.05$, and $P = 0.01$ 556

The most important designs for tests on different levels of a factor or of several factors   563

Selected 95% confidence intervals for π (binomial distribution)   703

PREFACE TO THE SECOND ENGLISH EDITION

This new edition aims, as did the first edition, to give an impression, an account, and a survey of the very different aspects of applied statistics—a vast field, rapidly developing away from the perfection the user expects. The text has been improved with insertions and corrections, the subject index enlarged, and the references updated and expanded. I have tried to help the newcomer to the field of statistical methods by citing books and older papers, often easily accessible, less mathematical, and more readable for the non-statistician than newer papers. Some of the latter are, however, included and are attached in a concise form to the cited papers by a short "[see also . . .]".

I am grateful to the many readers who helped me with this revision by asking questions and offering advice. I took suggestions for changes seriously but did not always follow them. Any further comments are heartily welcome. I am particularly grateful to the staff of Springer-Verlag New York.

Klausdorf/Schwentine   LOTHAR SACHS

PREFACE TO THE FIRST ENGLISH EDITION

An English translation now joins the Russian and Spanish versions. It is based on the newly revised fifth edition of the German version of the book. The original edition has become very popular as a learning and reference source with easy to follow recipes and cross references for scientists in fields such as engineering, chemistry and the life sciences. Little mathematical background is required of the reader and some important topics, like the logarithm, are dealt with in the preliminaries preceding chapter one. The usefulness of the book as a reference is enhanced by a number of convenient tables and by references to other tables and methods, both in the text and in the bibliography. The English edition contains more material than the German original. I am most grateful to all who have in conversations, letters or reviews suggested improvements in or criticized earlier editions. Comments and suggestions will continue to be welcome. We are especially grateful to Mrs. Dorothy Aeppli of St. Paul, Minnesota, for providing numerous valuable comments during the preparation of the English manuscript. The author and the translator are responsible for any remaining faults and imperfections. I welcome any suggestions for improvement.

My greatest personal gratitude goes to the translator, Mr. Zenon Reynarowych, whose skills have done much to clarify the text, and to Springer-Verlag.

Klausdorf   LOTHAR SACHS

FROM THE PREFACES TO PREVIOUS EDITIONS

FIRST EDITION (November, 1967)

"This cannot be due merely to chance," thought the London physician Arbuthnott some 250 years ago when he observed that in birth registers issued annually over an 80 year period, male births always outnumbered female births. Based on a sample of this size, his inference was quite reliable. He could in each case write a plus sign after the number of male births (which was greater than the number of female births) and thus set up a sign test. With large samples, a two-thirds majority of one particular sign is sufficient. When samples are small, a $\frac{4}{5}$ or even a $\frac{9}{10}$ majority is needed to reliably detect a difference.

Our own time is characterized by the rapid development of probability and mathematical statistics and their application in science, technology, economics and politics.

This book was written at the suggestion of Prof. Dr. H. J. Staemmler, presently the medical superintendent of the municipal women's hospital in Ludwigshafen am Rhein. I am greatly indebted to him for his generous assistance. Professor W. Wetzel, director of the Statistics Seminar at the University of Kiel; Brunhilde Memmer of the Economics Seminar library at the University of Kiel; Dr. E. Weber of the Department of Agriculture Variations Statistics Section at the University of Kiel; and Dr. J. Neumann and Dr. M. Reichel of the local University Library, all helped me in finding the appropriate literature. Let me not fail to thank for their valuable assistance those who helped to compose the manuscript, especially Mrs. W. Schröder, Kiel, and Miss Christa Diercks, Kiel, as well as the medical laboratory technician F. Niklewitz, who prepared the diagrams. I am indebted to Prof. S. Koller, director of the Institute of Medical Statistics and Documentation at Mainz University, and especially to Professor E. Walter, director of the Institute of Medical Statistics and Documentation at the University of Freiburg im Breisgau, for many stimulating discussions.

Mr. J. Schimmler and Dr. K. Fuchs assisted in reading the proofs. I thank them sincerely.

I also wish to thank the many authors, editors, and publishers who permitted reproduction of the various tables and figures without reservation. I am particularly indebted to the executor of the literary estate of the late Sir Ronald A. Fisher, F.R.S., Cambridge, Professor Frank Yates (Rothamsted), and to Oliver and Boyd, Ltd., Edinburgh, for permission to reproduce Table II 1, Table III, Table IV, Table V, and Table VII 1 from their book "Statistical Tables for Biological, Agricultural and Medical Research"; Professor O. L. Davies, Alderley Park, and the publisher, Oliver and Boyd, Ltd., Edinburgh, for permission to reproduce a part of Table H from the book "The Design and Analysis of Industrial Experiments;" the publisher, C. Griffin and Co., Ltd. London, as well as the authors, Professor M. G. Kendall and Professor M. H. Quenouille, for permission to reproduce Tables 4a and 4b from the book "The Advanced Theory of Statistics," Vol. II, by Kendall and Stuart, and the figures on pp. 28 and 29 as well as Table 6 from the booklet "Rapid Statistical Calculations" by Quenouille; Professors E. S. Pearson and H. O. Hartley, editors of the "Biometrika Tables for Staticians, Vol. 1, 2nd ed., Cambridge 1958, for permission to adopt concise versions of Tables 18, 24, and 31. I also wish to thank Mrs. Marjorie Mitchell, the McGraw-Hill Book Company, New York, and Professor W. J. Dixon for permission to reproduce Tables A-12c and A-29 (Copyright April 13, 1965, March 1, 1966, and April 21, 1966) from the book "Introduction to Statistical Analysis" by W. J. Dixon and F. J. Massey Jr., as well as Professor C. Eisenhart for permission to use the table of tolerance factors for the normal distribution from "Techniques of Statistical Analysis," edited by C. Eisenhart, W. M. Hastay, and W. A. Wallis. I am grateful to Professor F. Wilcoxon, Lederle Laboratories (a division of American Cyanamid Company), Pearl River, for permission to reproduce Tables 2, 3, and 5 from "Some Rapid Approximate Statistical Procedures" by F. Wilcoxon and Roberta A. Wilcox. Professor W. Wetzel, Berlin-Dahlem, and the people at de Gruyter-Verlag, Berlin W 35, I thank for the permission to use the table on p. 31 in "Elementary Statistical Tables" by W. Wetzel. Special thanks are due Professor K. Diem of the editorial staff of Documenta Geigy, Basel, for his kind permission to use an improved table of the upper significance bounds of the Studentized range, which was prepared for the 7th edition of the "Scientific Tables." I am grateful to the people at Springer-Verlag for their kind cooperation.

SECOND AND THIRD EDITIONS

Some sections have been expanded and revised and others completely rewritten, in particular the sections on the fundamental operations of arithmetic extraction of roots, the basic tasks of statistics, computation of the standard deviation and variance, risk I and II, tests of $\sigma = \sigma_0$ with μ known and unknown, tests of $\pi_1 = \pi_2$, use of the arc since transformation and of $\pi_1 - \pi_2 = d_0$, the fourfold χ^2-test, sample sizes required for this test when risk I and risk II are given, the U-test, the H-test, the confidence interval of the median, the Sperman rank correlation, point bivariate and multiple correlation, linear regression on two independent variables, multivariate methods, experimental design and models for the analysis of variance. The following tables were supplemented or completely revised: the critical values for the standard normal distribution, the t- and the χ^2-distribution, Hartley's F_{max}, Wilcoxon's R for pairwise differences, the values of $e^{-\lambda}$ and arc sin $\sqrt{p}$, the table for the $\dot{z}$-transformation of the coefficient of correlation and the bounds for the test of $\rho = 0$ in the one and two sided problem. The bibliography was completely overhauled. Besides corrections, numerous simplifications, and improved formulations, the third edition also incorporates updated material. Moreover, some of the statistical tables have been expanded (Tables 69a, 80, 84, 98, and 99, and unnumbered tables in Sections 4.5.1 and 5.3.3). The bibliographical references have been completely revised. The author index is a newly added feature. Almost all suggestions resulting from the first and second editions are thereby realized.

FOURTH EDITION (June, 1973)

This revised edition, with a more appropriate title, is written both as an introductory and follow-up text for reading and study and as a reference book with a collection of formulas and tables, numerous cross-references, an extensive bibliography, an author index, and a detailed subject index. Moreover, it contains a wealth of refinements, primarily simplifications, and statements made more precise. Large portions of the text and bibliography have been altered in accordance with the latest findings, replaced by a revised expanded version, or newly inserted; this is also true of the tables (the index facing the title page, as well as Tables 13, 14, 28, 43, 48, 56, 65, 75, 84, 183, and 185, and the unnumbered tables in Sections 1.2.3, 1.6.4, and 3.9.1, and on the reverse side of the next to last sheet). Further changes appear in the second, newly revised edition of my book "Statistical Methods. A Primer for Practitioners in Science, Medicine, Engineering, Economics, Psychology, and Sociology," which can serve as a handy companion volume for quick orientation. Both volumes benefited from the suggestions of the many who offered constructive criticisms—engineers in particular. It will be of interest to medical students that I have covered the material necessary for medical school exams in biomathematics, medical statistics and documentation. I wish to thank Professor Erna Weber and Akademie-Verlag, Berlin, as well as the author, Dr. J. Michaelis, for permission to reproduce Tables 2 and 3 from the paper "Threshold value of the Friedman test," Biometrische Zeitschrift 13 (1971), 122. Special thanks are due to the people at Springer-Verlag for their complying with the author's every request. I am also grateful for all comments and suggestions.

FIFTH EDITION (July, 1978)

This new edition gave me the opportunity to introduce simplifications and supplementary material and to formulate the problems and solutions more precisely. I am grateful to Professor Clyde Y. Kramer for permission to reproduce from his book (A First Course in Methods of Multivariate Analysis, Virginia Polytechnic Institute and State University, Blacksburg, Virginia, 1972) the upper bounds of the Bonferroni χ^2-statistics (Appendix D, pp. 326–351), which were calculated by G. B. Beus and D. R. Jensen in September, 1967. Special thanks are due the people at Springer-Verlag for their complying with the author's every request. I welcome all criticisms and suggestions for improvement.

LOTHAR SACHS

PERMISSIONS AND ACKNOWLEDGMENTS

Author and publisher are grateful for the permission to reproduce copyrighted and published material from a variety of sources. Specific thanks go to:

The executor of the literary estate of the late Sir Ronald A. Fisher, Professor Frank Yates, and to Oliver and Boyd, Ltd., for permission to reproduce Table II 1, Table III, Table IV, Table V and Table VII 1 from their book "Statistical Tables for Biological, Agricultural and Medical Research."

Professor, M. G. Kendall and Professor M. H. Quenouille and the publisher, C. F. Griffin and Co., Ltd. London, for permission to reproduce Table 4a and 4b from the book "The Advanced Theory of Statistics."

Professor W. J. Dixon and the McGraw-Hill Book Company, New York, for permission to reproduce Table A-12c and Table A-29 from the book "Introduction to Statistical Analysis."

Professor C. Eisenhart for permission to use the table of tolerance factors for the normal distribution from "Techniques of Statistical Analysis."

Professor F. Wilcoxon, Lederle Laboratories (a division of American Cyanamid Company), Pearl River, for permission to reproduce Tables 2, 3, and 5 from "Some Rapid Approximate Statistical Procedures."

Professor W. Wetzel, Berlin-Dahlem, and de Gruyter-Verlag, Berlin, for permission to use the table on p. 31 in "Elementary Statistical Tables."

Professor K. Diem of Documenta Geigy, Basel, for his permission to use an improved table of upper significance bounds of the Studentized range, which was prepared for the 7th edition of the "Scientific Tables."

Professor E. S. Pearson and the Biometrika Trustees for permission to reproduce percentiles of the standardized 3rd and 4th moments, $\sqrt{b_1}$ and

b_2, Tables 34B and C from Biometrika Tables for Statisticians, Volume I and to the editors of the Biometrical Journal for permission to reproduce additional data on $\sqrt{b_1}$ and b_2.

Professor Clyde Y. Kramer for permission to reproduce from his book "A First Course in Methods of Multivariate Analysis" the upper bounds of the Bonferroni statistics (Appendix D, pp. 326–351), which were calculated by G. B. Beus and D. R. Jensen in September, 1967.

Dr. J. Michaelis, Professor Erna Weber and Akademie-Verlag, Berlin, for permission to reproduce Tables 2 and 3 from the paper "Threshold values of the Friedman Test."

Professor J. H. Zar and the Permissions Editor of Prentice-Hall, Inc., Englewood Cliffs, N.J. for permission to reproduce critical values of Spearman's r_S from J. H. Zar, "Biostatistical Analysis," Table D.24.

SELECTED SYMBOLS*

$>$	is greater than	7
$\geq$	is greater than or equal to	7
$\approx$	or $\simeq$, is approximately equal to	7
$\neq$	is not equal to	7
$\sum$	(Sigma) Summation sign; $\sum x$ means "add up all the x's"	9
e	Base of the natural logarithms, the constant 2.71828	19
P	Probability	26
E	Event	28
X	Random variable, a quantity which may take any one of a specified set of values; any special or particular value or realization is termed x (e.g. $X =$ height and $x_{\text{Ralph}} = 173$ cm); if for every real number x the probability $P(X \leq x)$ exists, then X is called a random variable [thus $X, Y, Z,$ denote random variables and $x, y, z,$ particular values taken on by them]; in this book we nearly always use only x	43
∞	Infinity	45
π	(pi) Relative frequency in a population	48
μ	(mu) Arithmetic mean of a population	48
σ	(sigma) Standard deviation of a population	48
$\hat{p}$	Relative frequency in the sample [π is estimated by $\hat{p}$; estimated values are often written with a caret (^)]	48
$\bar{x}$	(x bar or overbar) Arithmetic mean (of the variable X) in a sample	48

* Explanation of selected symbols in the order in which they appear.

s	Standard deviation of the sample: the square of the standard deviation, s^2, is called the sample variance; σ^2 is the variance of a population	48
n	Sample size	50
N	Size of the population	50
k	Number of classes (or groups)	53
z	Standard normal variable, test statistic of the z-test; the z-test is the application of the standardized normal distribution to test hypotheses on large samples. For the standard normal distribution, that is, a normal distribution with mean 0 and variance 1 [$N(0;1)$ for short], we use the following notation: 1. For the ordinates: $f(z)$, e.g., $f(2.0) = 0.054$ or 0.0539910. 2. For the cumulative distribution function: $F(z)$, e.g., $F(2.0) = P(Z \leq 2.0) = 0.977$ or 0.9772499; $F(2.0)$ is the cumulative probability, or the integral, of the normal probability function from $-\infty$ up to $z = 2.0$.	61
f	Frequency, cell entry	73
V	Coefficient of variation	77
$\tilde{x}$	(x-tilde) Median (of the variable X) in a sample	91
$s_{\bar{x}}$	(s sub x-bar) Standard error of the arithmetic mean in a sample	94
$s_{\tilde{x}}$	(s sub x-tilde) Standard error of the median in a sample	95
R	Range = distance between extreme sample values	97
S	Confidence coefficient ($S = 1 - \alpha$)	112
α	(alpha) Level of significance, Risk I, the small probability of rejecting a valid null hypothesis	112, 115, 118
β	(beta) Risk II, the probability of retaining an invalid null hypothesis	118
H_0	Null hypothesis	116
H_A	Alternative hypothesis or alternate hypothesis	117
z_α	Critical value of a z-test: z_α is the upper α-percentile point (value of the abscissa) of the standard normal distribution. For such a test or tests using other critical values, the so-called P-value gives the probability of a sample result, provided the null hypothesis is true [thus this value, if it is very low, does not always denote the size of a real difference]	122
$\hat{t}$	Test statistic of the t-test; the t-test, e.g., checks the equality of two means in terms of the t-distribution or Student distribution (the distribution law of not large samples from normal distributions)	135
v	(nu) or DF, the degrees of freedom (of a distribution)	135
$t_{v;\alpha}$	Critical value for the t-test, subscripts denoting degrees of freedom (v) and percentile point (α) of the t_v distribution	137

Selected Symbols   xxvii

χ^2 (chi-square) Test statistic of the χ^2-test; the χ^2-test, e.g., checks the difference between an observed and a theoretical frequency distribution, $\chi^2_{\nu;\alpha}$ is the critical value for the χ^2-test   **139**

$\hat{F}$ Variance ratio, the test statistic of the F-test; the F-test checks the difference between two variances in terms of the F-distribution (a theoretical distribution of quotients of variances); $F_{\nu_1;\nu_2;\alpha}$ is the critical value for the F-test   **143**

$_xC_n$ or $\binom{n}{x}$, Binomial coefficient: the number of combinations of n elements taken x at a time   **155**

!   Factorial sign ($n!$ is read "n factorial"); the number of arrangements of n objects in a sequence is $n! = n(n-1)(n-2) \times \cdots \times 3 \times 2 \times 1$ **155**

λ (lambda) Parameter, being both mean and variance of the Poisson distribution, a discrete distribution useful in studying failure data   **175**

CI   Confidence interval, range for an unknown parameter, a random interval having the property that the probability is $1 - \alpha$ (e.g., $1 - 0.05 = 0.95$) that the random interval will contain the true unknown parameter; e.g., 95% CI   **248**

MD   Mean deviation (of the mean) $= (1/n)\sum |x - \bar{x}|$ **251**

Q Sum of squares about the mean [e.g., $Q_x = \sum_{i=1}^{i=n} (x_i - \bar{x})^2 = \sum (x - \bar{x})^2$]   **264**

U Test statistic of the Wilcoxon–Mann–Whitney test: comparing two independent samples   **293**

H Test statistic of the Kruskal-Wallis test: comparing several independent samples   **303**

O Observed frequency, occupancy number   **321**

E Expected frequency, expected number   **321**

a,b,c,d Frequencies (cell entries) of a fourfold table   **346**

ρ (rho) Correlation coefficient of the population: $-1 \leq \rho \leq 1$ **383**

r Correlation coefficient of a sample: $-1 \leq r \leq 1$ **384**

β (beta) Regression coefficient of the population (e.g., β_{yx})   **384**

b Regression coefficient or slope of a sample; gives the direction of the regression line; of the two subscripts commonly used, as for instance in b_{yx}, the second indicates the variable from which the first is predicted   **386**

r_S Spearman's rank correlation coefficient of a sample: $-1 \leq r_S \leq 1$ **395**

$s_{y \cdot x}$ Standard error in estimating Y from X (of a sample)   **414**

s_a	Standard error of the intercept	**415**
s_b	Standard error of the slope	**415**
$\dot{z}$	Normalizing transform of the correlation coefficient (note the dot above the z) **427**	
E_{yx}	Correlation ratio (of a sample) of y over x: important for testing the linearity of a regression **436**	
$r_{12.3}$	Partial correlation coefficient	**456**
$R_{1.23}$	Multiple correlation coefficient	**458**
SS	Sum of squares, e.g., $\sum(x - \bar{x})^2$	**501**
MS	Mean square: the sample variance $s^2 = \sum(x - \bar{x})^2/(n - 1)$ is a mean square, since a sum of squares is divided by its associated $(n - 1)$ degrees of freedom **502**	
MSE	Mean square for error, error mean square; measures the unexplained variability of a set of data and serves as an estimate of the inherent random variation of the experiment; it is an unbiased estimate of the experimental error variance **503**	
LSD	The least significant difference between two means **512**	
SSA	Factor A sum of squares, that part of the total variation due to differences between the means of the a levels of factor A; a factor is a series of related treatments or related classifications **522**	
MSA	Factor A mean squares: $MSA = SSA/(1 - a)$, mean square due to the main effect of factor A **522**	
SSAB	AB-interaction sum of squares, measures the estimated interaction for the ab treatments; there are ab interaction terms **522**	
MSAB	AB-interaction mean squares: $MSAB = SSAB/[(a - 1)(b - 1)]$ **522**	
SSE	Error sum of squares **522**	
χ_R^2	Test statistic of the Friedman rank analysis of variance **550**	

INTRODUCTION

This *outline of statistics as an aid in decision making* will introduce a reader with limited mathematical background to the most important modern statistical methods. This is a revised and enlarged version, with major extensions and additions, of my "Angewandte Statistik" (5th ed.), which has proved useful for research workers and for consulting statisticians.

Applied statistics is at the same time a collection of applicable statistical methods and the application of these methods to measured and/or counted observations. Abstract mathematical concepts and derivations are avoided. Special emphasis is placed on the basic principles of statistical formulation, and on the explanation of the conditions under which a certain formula or a certain test is valid. Preference is given to consideration of the analysis of small sized samples and of distribution-free methods. As a *text and reference* this book is written for non-mathematicians, in particular **for technicians, engineers, executives, students, physicians as well as researchers in other disciplines**. It gives any mathematician interested in the practical uses of statistics a general account of the subject.

Practical application is the main theme; thus an essential part of the book consists in the 440 fully worked-out numerical examples, some of which are very simple; the 57 exercises with solutions; a number of different *computational aids*; and an extensive bibliography and a very detailed index. In particular, a collection of 232 mathematical and mathematical-statistical tables serves to enable and to simplify the computations.

Now a few words as to its *structure*: After some preliminary mathematical remarks, techniques of statistical decision are considered in Chapter 1. Chapter 2 gives an introduction to the fields of medical statistics, sequential analysis, bioassay, statistics in industry, and operations research. Data samples and frequency samples are compared in Chapters 3 and 4. The three subsequent chapters deal with more advanced schemes: analysis of associations: correlation and regression, analysis of contingency tables, and analysis of variance. A comprehensive general and specialized bibliography, a collection of exercises, and a subject and author index make up the remainder of the book.

A *survey* of the most important statistical techniques is furnished by sections marked by an arrow ▶: 1.1, 1.2.1–3, 1.2.5, 1.3.1–7, 1.4, 1.5, 1.6.1–2, 1.6.4–6, 3.1.1–2, 3.1.4, 3.2–3, 3.5, 3.6, 3.9.4–5, 4.1, 4.2.1–2, 4.3, 4.3.1–3, 4.5.1–3, 4.6.1, 4.6.7, 5.1–2, 5.3.1, 5.4.1–3, 5.4.5, 5.5.1, 5.5.3–4, 5.5.8–9, 5.8, 6.1.1, 6.1.4, 6.2.1, 6.2.5, 7.1, 7.2.3, 7.3–4, 7.6–7.

A more casual approach consists in first examining the material on the inner sides of the front and back covers, and then going over the Introduction to Statistics and Sections 1.1, 1.2.1, 1.3.2–4, 1.3.6.2–3, 1.3.6.6, 1.3.8.3–4, 1.3.9, 1.4.1–8, 1.5, 3 (introduction), 3.1.1, 3.2, 3.6, 3.8, 3.9 (introduction), 3.9.4, 4.1, 4.2.1–2, 4.5.1, 4.6.1 (through Table 83), 5.1–2, 5.3 (introduction and 5.3.1), 6.2.1 [through (6.4)], 7.1, 7.2.1, 7.3.1, and 7.7.

As the author found some difficulty in deciding on the order of presentation—in a few instances references to subsequent chapters could not be entirely avoided—and as the presentation had to be concise, the beginner is advised to read the book at least twice. It is only then that the various interrelationships will be grasped and thus the most important prerequisite for the comprehension of statistics acquired. *Numerous examples*—some very simple—have been included in the text for better comprehension of the material and as applications of the methods presented. The use of such examples—which, in a certain sense, amounts to playing with numbers—is frequently more instructive and a greater stimulus to a playful, experimental follow-up than treating actual data (frequently involving excessive numerical computation), which is usually of interest only to specialists. It is recommended that the reader independently work out some of the examples as an exercise and also solve some of the problems.

The numerous cross references appearing throughout the text point out various *interconnections*. A serendipitous experience is possible, i.e., on setting out in search of something, one finds something else of even greater consequence.

My greatest personal gratitude goes to the translator, Mr. Z. Reynarowych, whose skill has done much to clarify the text.

INTRODUCTION TO STATISTICS

> Scientists and artists have in common their desire to comprehend the external world and to reduce its apparent complexity, even chaos, to some kind of ordered representation: Scientific work involves the representation of disorder in an orderly manner.

Statistics is the art and science of data: of generating or gathering, describing, analyzing, summarizing, and interpreting data to discover new knowledge.

> Basic tasks of statistics: To describe, assess, and pass judgement. To draw inferences concerning the underlying population.

Each of us, like a hypochondriac or like one who only imagines himself to be well, has at some time failed to recognize existing relationships or distinctions or else has imagined relationships or distributions where none existed. In everyday life we recognize a similarity or a difference with the help of factual knowledge and what is called instinctive understanding. The scientist discovers certain new phenomena, dependences, trends, or a variety of effects upon which he bases a working hypothesis; he then must check them against the simpler hypothesis that the effects observed are conditioned solely by chance. *The problem of whether observed phenomena can be regarded as strictly random or as typical is resolved by analytical statistics*, which thus becomes a method characteristic of modern science.

With the help of statistical methods one can **respond to questions** and **examine the validity of statements**. For example, on the one hand: How many persons should one poll before an election to get a rough idea of the outcome? Does a weekly two-hour period at school devoted to sports contribute to the strengthening of the heart and circulatory system? Which of several toothpastes is to be recommended as a decay preventative? How does the quality of steel depend on its composition? Or on the other hand: The new saleslady has increased the daily turnover by $1000. The 60% characteristic survival rate for a certain disease is raised to 90% by treatment A. The effects of fertilizers K_1, K_2, and K_3 on oats are indistinguishable.

When observations are made to obtain numerical values that are typical (representative) of the situation under study, the values so obtained are called *data*. They are important in evaluating hypotheses and in discovering new knowledge.

Statistical methods are concerned with data from our environment, with its gathering and processing: describing, evaluating, and interpreting; the **aim is to prepare for decision making**. "Statistics" was in the 18th century the "science of diagnosing the condition of various nations," where data were also gathered on the overall population, the military, business, and industry. This led to the development of **descriptive statistics**, whose task is to describe conditions and events in terms of observed data; use is made of tables, graphs, ratios, indices, and typical parameters such as location statistics (e.g. the arithmetic mean) and dispersion statistics (e.g. the variance).

ANALYTICAL STATISTICS deduces from the data general laws, whose validity extends beyond the field of observation. It developed from "political arithmetic" whose primary function is to estimate the sex ratio, fertility, age structure, and mortality of the population from baptismal, marriage, and death registers. Analytical statistics, also termed **mathematical** or **inductive** statistics, is based on **probability theory**, which builds mathematical models that encompass random or stochastic experiments. Examples of stochastic experiments are: rolling a die, the various games of chance and lotteries, the sex of a newborn, daytime temperatures, the yield of a harvest, the operating lifetime of a light bulb, the position of the dial of a measuring instrument during a trial—in a word, every observation and every trial in which the results are affected by random variation or measurement error. **The data themselves are here, as a rule, of lesser interest than the primary population in which the data originated.** For instance, the probability of rolling a 6 with a fair die or of guessing correctly six numbers in a lottery or the proportion of male births in the United States in 1978 is more interesting than the particular outcomes of some trial. In many problems involving replicable experiences, one cannot observe the set of all possible experiences or observations—the so-called population—but only an appropriately selected portion of it. In order to rate a wine, the wine taster siphons off a small sample from a large barrel. This **sample** then provides information on the frequency and composition of the relevant properties of the population considered, which cannot be studied as a whole, either for fundamental reasons or because of the cost or amount of time required.

We assume we are dealing with **RANDOM SAMPLES** in which every element of the population has the same chance of being included. If the population contains distinct subpopulations, then a stratified random sample is chosen. A meaningful and representative portion of a cake shipment would be neither a layer nor the filling nor the trimmings but rather a piece of cake. Better yet would be layer, filling and trimmings samples taken from several cakes.

In bingo, random samples are obtained with the help of a mechanical device. More usually, random samples are obtained by employing a table of random numbers: the elements are numbered and an element is regarded as chosen as soon as its number appears in the table. Samples taken by a random procedure have the advantage that the statistical parameters derived from them, when compared with those of the population, generally exhibit only the unavoidable statistical errors. These can be estimated, since they do not distort the result—with multiple replications, random errors cancel out on the average. On the other hand, in procedures without random choice there can also arise methodical or systematic errors, regarding whose magnitude nothing can be said as a rule. We place particular emphasis on estimating the random error and on testing whether observed phenomena are also characteristic for the populations rather than being chance results. This is the so-called **testing of hypotheses on the population**.

On translating a problem into hypotheses that can be statistically tested, care must be taken to choose and define characteristics that are significant and appropriate to the problem and readily measurable, to specify and keep constant the test conditions, and also to employ cost-optimal sample or test plans. We focus our attention on those parts of the setup which seem to us important and, using them, try to construct as a *model* a new, easy to survey, compound with an appropriate degree of [concreteness and] abstractness. **Models are important aids in decision making.**

The scientific approach is to devise a strategy aimed at finding general laws which, with the help of assertions that are open to testing and rejection (falsification), are developed into a logicomathematically structured *theory*. An approximate description of ascertainable reality is thereby obtained. This approximate description can be revised and refined further.

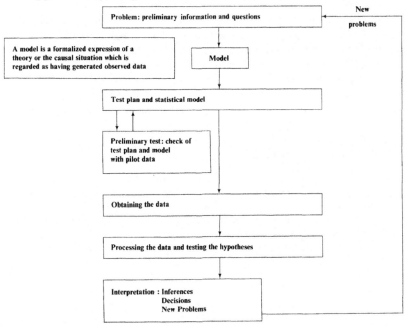

Typical of the scientific method is the *cycle process* or *iteration cycle*: conjectures (ideas) → plan (see also "Scientific Investigation" end of Section 7.7) → observations → analysis → results → new conjectures (new ideas) → ...; contradictions and incompatibilities are eliminated in the process (cf. also above) and the models and theories improved. *That theory is better which allows us to say more and make better predictions.*

It is important to keep the following in mind: Assumptions regarding the structure of the underlying model and the corresponding statistical model are made on the basis of the question particular to the problem. After testing the compatibility of the observations with the statistical model, characteristic quantities for the statistical description of a population, the so-called *parameters*, are established with a given confidence coefficient, and hypotheses on the parameters are tested. **Probabilistic statements** result in both cases. The task of statistics is thus to find and develop models appropriate to the question and the data, and using them, **to extract any pertinent information concealed in the data**—i.e., *statistics provides models for information reduction*.

These as well as other methods form the nucleus of a **data analysis** designed for the *skillful gathering and critical evaluation of data*, as is necessary in many branches of industry, politics, science and technology. *Data analysis* is the systematic search for fruitful information specific to phenomena, structures, and processes, utilizing data and employing *graphical, mathematical*, and especially *statistical techniques* with or without the concept of probability: to display and summarize the data to make them more comprehensible to the human mind, thus uncovering structures and detecting new features.

There is less concern here with reducing data to probabilities and obtaining significant results, which could in fact be meaningless or unimportant. What counts is the practical relevance rather than the statistical significance. An evaluation of results, only possible by a person with a thorough knowledge of the specific field and the observations under consideration, depends on many factors, such as the significance of the particular problem in question, compatibility with other results, or the predictions which they allow to be made. This evidence can hardly be evaluated statistically. Moreover, the data affect us in many ways that go beyond an evaluation. They give us comprehension, insight, suggestions, and surprising ideas.

Especially useful are the books written by Tukey (1977, cited in [1] on p. 570), Chambers and coworkers (1983, cited in [8:1] on page 579) and Hoaglin and coworkers (1983, cited on p. 582) as well as books and papers on diagrams and graphical techniques (e.g., Bachi 1968, Bertin 1967, Cox 1978, Dickinson 1974, Ehrenberg 1978, Fienberg 1979, Fisher 1983, King 1971, Lockwood 1969, Sachs 1977, Schmid and Schmid 1979, Spear 1969, and Wainer and Thissen 1981, all cited in [8:1]).

0 PRELIMINARIES

The following is a review of some **elementary mathematical concepts** which, with few exceptions, are an indispensable part of the background at the intermediate level. These concepts are more than adequate for the understanding of the problems considered in the text.

0.1 MATHEMATICAL ABBREVIATIONS

The language of mathematics employs symbols, e.g., letters or other marks, in order to present the content of a given statement precisely and concisely. Numbers are generally represented by lowercase Latin letters $(a, b, c, d, \ldots)$

Table 1 Some mathematical relations

Relation	Meaning	Example
$a = b$	a is equal to b	$8 = 12 - 4$
$a < b$	a is less than b	$4 < 5$
$a > b$	a is greater than b	$6 > 5$
$a \leq b$	a is less than or equal to b	profit a is at most \$b
$a \geq b$	a is greater than or equal to b	profit a is at least \$b
$a \simeq b$	a is roughly equal to,	$109.8 \simeq 110$
$a \approx b$	approximately equal to b	$109.8 \approx 110$
$a \neq b$	a is not equal to b	$4 \neq 6$

For "x greater than a and less than or equal to b" we write $a < x \leq b$.
For "x is much greater than a" we write $x \gg a$.
The inequality $a > b$ implies that $-a < -b$ and (for $b > 0$) $1/a < 1/b$.

7

or, if a large collection of distinct numbers is involved, by $a_1, a_2, a_3, \ldots, a_n$. Some other important symbols are listed in Table 1.

0.2 ARITHMETICAL OPERATIONS

A working knowledge of the 4 **basic arithmetical operations**—addition, subtraction, multiplication, and division—is assumed. An *arithmetical operation* is a prescription whereby to every pair of numbers a unique new number, e.g. the sum, is assigned.

1. **Addition:** Summand + summand = sum [$5 + 8 = 13$].

> **A survey of the relations among the four basic arithmetical operations**
>
> Computation means determining a new number from two or more given numbers. Each of the four standard **arithmetical symbols** $(+, -, \cdot, :)$ represents an operation:
>
> | | |
> |---|---|
> | + | plus, addition sign |
> | − | minus, subtraction sign |
> | · | times, multiplication sign |
> | : | divided by, division sign |
>
> The result of each computation should first be estimated, then worked out twice by performing the inverse operation and checked. For example $4.8 + 16.1$ equals approximately 21, exactly 20.9; check $20.9 - 4.8 = 16.1$; and $15.6 : 3$ equals approximately 5, exactly 5.2; check $5.2 \cdot 3 = 15.6$. The four basic arithmetical operations are subject to two rules:
>
> 1. **Dot operations (multiplication and division) precede dash operations (addition and subtraction).**
>
>   Examples $2 + 3 \cdot 8 = 2 + 24 = 26$,
> $6 \cdot 2 + 8 : 4 = 12 + 2 = 14$.
>
>   The positive integers ($+1, +2, +3, + \ldots$), zero, and the negative integers ($-1, -2, -3, - \ldots$) together form the integers, which have the collective property that every subtraction problem has an integer solution, (e.g., $8 - 12 = -4$). The following somewhat loosely formulated **sign rules** apply to "dot" operations:

0.2 Arithmetical Operations

> $+ \cdot + = +$ **Like signs**
> $+ : + = +$ **yield plus**
> $- \cdot - = +$ $(-8) : (-2) = +4 = 4$
> $- : - = +$
>
> operation symbol
>
> $+ \cdot - = -$ **Unlike signs**
> $+ : - = -$ **yield minus**
> $- \cdot + = -$ $(-8) : (+2) = -4$
> $- : + = -$
>
> signs

The size of a real number a (see Section 1.2.5 below) is independent of the sign, is called its **absolute value**, and is written $|a|$, e.g., $|-4| = |+4| = 4$.

2. **Expressions enclosed in parentheses like $(3 + 4)$ are worked out first.** If curly brackets $\{\ \}$ enclose brackets $[\ \ ]$ and parentheses $\{[(\ )]\}$, one begins with the innermost. In front of parentheses or brackets the multiplication sign is usually omitted, e.g.,

$$4(3 + 9) = 4(12) = 4 \cdot 12 = 48.$$

Division is often represented by a fraction, e.g.,

$$\frac{3}{4} = 3/4 = 3 : 4 = 0.75,$$

$$4[12 - (8 \cdot 2 + 18)] = 4[12 - (16 + 18)] = 4(-22) = -88,$$

$$12\left[\frac{(9-3)}{2} - 1\right] = 12\left[\frac{6}{2} - 1\right] = 12(3 - 1) = 12(2) = 24.$$

The symbol

$$z = \sum_{i=1}^{n} x_i$$

is introduced to indicate the sum of all the values $x_1, x_2, \ldots, x_n$. $\sum$ is an oversize Greek capital letter sigma, the sign for "sum of." This operation is to be read: z is the sum of all the numbers x_i from $i = 1$ to $i = n$. The subscript, or index, of the first quantity to be added is written below the summation sign, while the index of the last quantity goes above it. Generally the summation will run from the index 1 to the index n. The following ways of writing the sum from x_1 to x_n are equivalent:

$$x_1 + x_2 + x_3 + \cdots + x_n = \sum_{i=1}^{i=n} x_i = \sum_{i=1}^{n} x_i = \sum_i x_i = \sum x.$$

In evaluating expressions like $\sum_{i=1}^{n}(3 + 2x_i + x_i^2) = 3n + 2\sum_{i=1}^{n} x_i + \sum_{i=1}^{n} x_i^2$ involving constant values (k), here 3 and 2, the following three properties of the sum are used:

1. $\sum_{i=1}^{n}(x_i + y_i) = (x_1 + y_1) + (x_2 + y_2) + \cdots = (x_1 + x_2 + \cdots) + (y_1 + y_2 + \cdots)$

$$= \sum_{i=1}^{n} x_i + \sum_{i=1}^{n} y_i$$

2. $\sum_{i=1}^{n} kx_i = kx_1 + kx_2 + \cdots = k\sum_{i=1}^{n} x_i$

3. $\sum_{i=1}^{n}(k + x_i) = (k + x_1) + (k + x_2) + \cdots = nk + \sum_{i=1}^{n} x_i.$

2. Subtraction: Minuend − subtrahend = resulting difference [$13 - 8 = 5$].

3. Multiplication: Factor × factor = resulting product [$2 \times 3 = 6$].

In this book the product of two numbers will seldom be denoted by the symbol × between the two factors, since there could be confusion with the letter x. Multiplication will generally be indicated by an elevated dot or else the factors will simply be written side by side, for example $5 \cdot 6$ or pq. The expression (1.23)(4.56) or $1.23 \cdot 4.56$ is written in Germany as 1,23 · 4,56, in England and Canada as 1·23 . 4·56 or 1·23 × 4·56. A comma in Germany is used to represent the decimal point (e.g., 5837,43 instead of 5,837.43).

4. Division: dividend/divisor = resulting quotient [$\frac{6}{3} = 2$] (divisor $\neq 0$).

5. Raising to a power: A product of like factors a is a power a^n, read "a to the n" or "nth power of a." Here a is the base and n the exponent of the power ($a^1 = a$):

$$\text{base}^{\text{exponent}} = \text{power} \qquad 2 \cdot 2 \cdot 2 = 2^3 = 8.$$

The second powers a^2 are called **squares**, since a^2 gives the area of a square of side a, so that a^2 is also read "a squared." The third powers are called **cubes**; a^3 gives the volume of a cube of edge a. Of particular importance are the **powers of ten**. They are used in estimation, to provide a means of checking the order of magnitude, and in writing very small and very large numbers clearly and concisely: $1{,}000 = 10 \cdot 10 \cdot 10 = 10^3$, $1{,}000{,}000 = 10^6$. We will

0.2 Arithmetical Operations

return to this in Section 0.3 ($10^3 - 10^2$ does not equal 10^1; it equals $900 = 9 \cdot 10^2$ instead). First several power laws with examples (m and n are natural numbers):

$a^m \cdot a^n = a^{m+n}$	$2^4 \cdot 2^3 = 2^{4+3} = 2^7 = 128,$
$a^m : a^n = a^{m-n}$	$2^4 : 2^3 = 2^{4-3} = 2^1 = 2,$
$a^n \cdot b^n = (ab)^n$	$6^2 \cdot 3^2 = 6 \cdot 6 \cdot 3 \cdot 3 = 6 \cdot 3 \cdot 6 \cdot 3 = (6 \cdot 3)^2 = 18^2 = 324,$
$a^m : b^m = \left(\dfrac{a}{b}\right)^m$	(the reader should construct an example),
$(a^m)^n = a^{m \cdot n} = (a^n)^m$	$(5^2)^3 = 5^2 \cdot 5^2 \cdot 5^2 = 5^{2 \cdot 3} = 5^6 = 15{,}625,$
$a^{-n} = \dfrac{1}{a^n}$	$10^{-3} = \dfrac{1}{10^3} = \dfrac{1}{1{,}000} = 0.001,$
$a^0 = 1$ for $a \neq 0$	$\dfrac{a^5}{a^5} = a^{5-5} = a^0 = 1$ (cf. also: $0^a = 0$ for $a > 0$).

These power laws also hold when m, n are not integers; that is, if $a \neq 0$, the given power laws also hold for fractional exponents ($m = p/q$, $n = r/s$).

6. Extraction of roots: Another notation for $a^{1/n}$ is $\sqrt[n]{a^1} = \sqrt[n]{a}$. It is called the nth root of a. For $n = 2$ (square root) one writes $\sqrt{a}$ for short. $\sqrt[n]{a}$ is the number which when raised to the nth power yields the radicand a: $[\sqrt[n]{a}]^n = a$. The following is the usual terminology:

$$^{\text{index}}\sqrt{\text{radicand}} = \text{root}.$$

One extracts roots (the symbol $\sqrt{}$ is a stylized r from the latin radix = root) with the help of the electronic calculator. We give several formulas and examples for calculation with roots:

$\sqrt[n]{a} \cdot \sqrt[n]{b} = \sqrt[n]{ab}$	$\dfrac{\sqrt[n]{a}}{\sqrt[n]{b}} = \sqrt[n]{\dfrac{a}{b}}$	$a^{m/n} = \sqrt[n]{a^m}$	$[\sqrt[n]{a}]^m = \sqrt[n]{a^m}$	$\sqrt[m]{\sqrt[n]{a}} = \sqrt[m \cdot n]{a}$

$$\sqrt{50} = \sqrt{25 \cdot 2} = 5\sqrt{2}, \quad \dfrac{\sqrt{50}}{\sqrt{2}} = \sqrt{\dfrac{50}{2}} = \sqrt{25} = 5, \quad \sqrt[4]{3^{12}} = 3^{12/4} = 3^3 = 27.$$

7. Calculation with logarithms: Logarithms are exponents. If a is a positive number and y an arbitrary number (>0), then there is a uniquely defined number x for which $a^x = y$. This number x, called the logarithm of y to base a, is written

$$x = {}^a\!\log y \quad \text{or} \quad \log_a y.$$

Since $a^0 = 1$, we have $\log_a 1 = 0$.

Table 2 Four place common logarithms

x	log x										Differences								
	0	1	2	3	4	5	6	7	8	9	1	2	3	4	5	6	7	8	9
100	0000	0004	0009	0013	0017	0022	0026	0030	0035	0039	0	1	1	2	2	3	3	3	4
101	0043	0048	0052	0056	0060	0065	0069	0073	0077	0082	0	1	1	2	2	3	3	3	4
102	0086	0090	0095	0099	0103	0107	0111	0116	0120	0124	0	1	1	2	2	3	3	3	4
103	0128	0133	0137	0141	0145	0149	0154	0158	0162	0166	0	1	1	2	2	3	3	3	4
104	0170	0175	0179	0183	0187	0191	0195	0199	0204	0208	0	1	1	2	2	2	3	3	4
105	0212	0216	0220	0224	0228	0233	0237	0241	0245	0249	0	1	1	2	2	2	3	3	4
106	0253	0257	0261	0265	0269	0273	0278	0282	0286	0290	0	1	1	2	2	2	3	3	4
107	0294	0298	0302	0306	0310	0314	0318	0322	0326	0330	0	1	1	2	2	2	3	3	4
108	0334	0338	0342	0346	0350	0354	0358	0362	0366	0370	0	1	1	2	2	2	3	3	4
109	0374	0378	0382	0386	0390	0394	0398	0402	0406	0410	0	1	1	2	2	2	3	3	4
10	0000	0043	0086	0128	0170	0212	0253	0294	0334	0374	4	8	12	17	21	25	29	33	37
11	0414	0453	0492	0531	0569	0607	0645	0682	0719	0755	4	8	11	15	19	23	26	30	34
12	0792	0828	0864	0899	0934	0969	1004	1038	1072	1106	3	7	10	14	17	21	24	28	31
13	1139	1173	1206	1239	1271	1303	1335	1367	1399	1430	3	6	10	13	16	19	23	26	29
14	1461	1492	1523	1553	1584	1614	1644	1673	1703	1732	3	6	9	12	15	18	21	24	27
15	1761	1790	1818	1847	1875	1903	1931	1959	1987	2014	3	6	8	11	14	17	20	22	25
16	2041	2068	2095	2122	2148	2175	2201	2227	2253	2279	3	5	8	11	13	16	18	21	24
17	2304	2330	2355	2380	2405	2430	2455	2480	2504	2529	2	5	7	10	12	15	17	20	22
18	2553	2577	2601	2625	2648	2672	2695	2718	2742	2765	2	5	7	9	12	14	16	19	21
19	2788	2810	2833	2856	2878	2900	2923	2945	2967	2989	2	4	7	9	11	13	16	18	20
20	3010	3032	3054	3075	3096	3118	3139	3160	3181	3201	2	4	6	8	11	13	15	17	19
21	3222	3243	3263	3284	3304	3324	3345	3365	3385	3404	2	4	6	8	10	12	14	16	18
22	3424	3444	3464	3483	3502	3522	3541	3560	3579	3598	2	4	6	8	10	12	14	15	17
23	3617	3636	3655	3674	3692	3711	3729	3747	3766	3784	2	4	6	7	9	11	13	15	17
24	3802	3820	3838	3856	3874	3892	3909	3927	3945	3962	2	4	5	7	9	11	12	14	16
25	3979	3997	4014	4031	4048	4065	4082	4099	4116	4133	2	3	5	7	9	10	12	14	15
26	4150	4166	4183	4200	4216	4232	4249	4265	4281	4298	2	3	5	7	8	10	11	13	15
27	4314	4330	4346	4362	4378	4393	4409	4425	4440	4456	2	3	5	6	8	9	11	13	14
28	4472	4487	4502	4518	4533	4548	4564	4579	4594	4609	2	3	5	6	8	9	11	12	14
29	4624	4639	4654	4669	4683	4698	4713	4728	4742	4757	1	3	4	6	7	9	10	12	13
30	4771	4786	4800	4814	4829	4843	4857	4871	4886	4900	1	3	4	6	7	9	10	11	13
31	4914	4928	4942	4955	4969	4983	4997	5011	5024	5038	1	3	4	6	7	8	10	11	12
32	5051	5065	5079	5092	5105	5119	5132	5145	5159	5172	1	3	4	5	7	8	9	11	12
33	5185	5198	5211	5224	5237	5250	5263	5276	5289	5302	1	3	4	5	6	8	9	10	12
34	5315	5328	5340	5353	5366	5378	5391	5403	5416	5428	1	3	4	5	6	8	9	10	11
35	5441	5453	5465	5478	5490	5502	5514	5527	5539	5551	1	2	4	5	6	7	9	10	11
36	5563	5575	5587	5599	5611	5623	5635	5647	5658	5670	1	2	4	5	6	7	8	10	11
37	5682	5694	5705	5717	5729	5740	5752	5763	5775	5786	1	2	3	5	6	7	8	9	10
38	5798	5809	5821	5832	5843	5855	5866	5877	5888	5899	1	2	3	5	6	7	8	9	10
39	5911	5922	5933	5944	5955	5966	5977	5988	5999	6010	1	2	3	4	5	7	8	9	10
40	6021	6031	6042	6053	6064	6075	6085	6096	6107	6117	1	2	3	4	5	6	8	9	10
41	6128	6138	6149	6160	6170	6180	6191	6201	6212	6222	1	2	3	4	5	6	7	8	9
42	6232	6243	6253	6263	6274	6284	6294	6304	6314	6325	1	2	3	4	5	6	7	8	9
43	6335	6345	6355	6365	6375	6385	6395	6405	6415	6425	1	2	3	4	5	6	7	8	9
44	6435	6444	6454	6464	6474	6484	6493	6503	6513	6522	1	2	3	4	5	6	7	8	9
45	6532	6542	6551	6561	6571	6580	6590	6599	6609	6618	1	2	3	4	5	6	7	8	9
46	6628	6637	6646	6656	6665	6675	6684	6693	6702	6712	1	2	3	4	5	6	7	7	8
47	6721	6730	6739	6749	6758	6767	6776	6785	6794	6803	1	2	3	4	5	5	6	7	8
48	6812	6821	6830	6839	6848	6857	6866	6875	6884	6893	1	2	3	4	4	5	6	7	8
49	6902	6911	6920	6928	6937	6946	6955	6964	6972	6981	1	2	3	4	4	5	6	7	8
	0	1	2	3	4	5	6	7	8	9	1	2	3	4	5	6	7	8	9

Example: log 1.234 = 0.0899 + 0.0014 = 0.0913.

0.2 Arithmetical Operations

Table 2 Four place common logarithms (*continued*)

x	log x										Differences								
	0	1	2	3	4	5	6	7	8	9	1	2	3	4	5	6	7	8	9
50	6990	6998	7007	7016	7024	7033	7042	7050	7059	7067	1	2	3	3	4	5	6	7	8
51	7076	7084	7093	7101	7110	7118	7126	7135	7143	7152	1	2	3	3	4	5	6	7	8
52	7160	7168	7177	7185	7193	7202	7210	7218	7226	7235	2	3	3	4	5	6	7	7	7
53	7243	7251	7259	7267	7275	7284	7292	7300	7308	7316	1	2	2	3	4	5	6	6	7
54	7324	7332	7340	7348	7356	7364	7372	7380	7388	7396	1	2	2	3	4	5	6	6	7
55	7404	7412	7419	7427	7435	7443	7451	7459	7466	7474	1	2	2	3	4	5	5	6	7
56	7482	7490	7497	7505	7513	7520	7528	7536	7543	7551	1	2	2	3	4	5	5	6	7
57	7559	7566	7574	7582	7589	7597	7604	7612	7619	7627	1	2	2	3	4	5	5	6	7
58	7634	7642	7649	7657	7664	7672	7679	7686	7694	7701	1	1	2	3	4	4	5	6	7
59	7709	7716	7723	7731	7738	7745	7752	7760	7767	7774	1	1	2	3	4	4	5	6	7
60	7782	7789	7796	7803	7810	7818	7825	7832	7839	7846	1	1	2	3	4	4	5	6	6
61	7853	7860	7868	7875	7882	7889	7896	7903	7910	7917	1	1	2	3	4	4	5	6	6
62	7924	7931	7938	7945	7952	7959	7966	7973	7980	7987	1	1	2	3	3	4	5	6	6
63	7993	8000	8007	8014	8021	8028	8035	8041	8048	8055	1	1	2	3	3	4	5	5	6
64	8062	8069	8075	8082	8089	8096	8102	8109	8116	8122	1	1	2	3	3	4	5	5	6
65	8129	8136	8142	8149	8156	8162	8169	8176	8182	8189	1	1	2	3	3	4	5	5	6
66	8195	8202	8209	8215	8222	8228	8235	8241	8248	8254	1	1	2	3	3	4	5	5	6
67	8261	8267	8274	8280	8287	8293	8299	8306	8312	8319	1	1	2	3	3	4	5	5	6
68	8325	8331	8338	8344	8351	8357	8363	8370	8376	8382	1	1	2	3	3	4	4	5	6
69	8388	8395	8401	8407	8414	8420	8426	8432	8439	8445	1	1	2	2	3	4	4	5	6
70	8451	8457	8463	8470	8476	8482	8488	8494	8500	8506	1	1	2	2	3	4	4	5	6
71	8513	8519	8525	8531	8537	8543	8549	8555	8561	8567	1	1	2	2	3	4	4	5	5
72	8573	8579	8585	8591	8597	8603	8609	8615	8621	8627	1	1	2	2	3	4	4	5	5
73	8633	8639	8645	8651	8657	8663	8669	8675	8681	8686	1	1	2	2	3	4	4	5	5
74	8692	8698	8704	8710	8716	8722	8727	8733	8739	8745	1	1	2	2	3	3	4	5	5
75	8751	8756	8762	8768	8774	8779	8785	8791	8797	8802	1	1	2	2	3	3	4	5	5
76	8808	8814	8820	8825	8831	8837	8842	8848	8854	8859	1	1	2	2	3	3	4	5	5
77	8865	8871	8876	8882	8887	8893	8899	8904	8910	8915	1	1	2	2	3	3	4	4	5
78	8921	8927	8932	8938	8943	8949	8954	8960	8965	8971	1	1	2	2	3	3	4	4	5
79	8976	8982	8987	8993	8998	9004	9009	9015	9020	9025	1	1	2	2	3	3	4	4	5
80	9031	9036	9042	9047	9053	9058	9063	9069	9074	9079	1	1	2	2	3	3	4	4	5
81	9085	9090	9096	9101	9106	9112	9117	9122	9128	9133	1	1	2	2	3	3	4	4	5
82	9138	9143	9149	9154	9159	9165	9170	9175	9180	9186	1	1	2	2	3	3	4	4	5
83	9191	9196	9201	9206	9212	9217	9222	9227	9232	9238	1	1	2	2	3	3	4	4	5
84	9243	9248	9253	9258	9263	9269	9274	9279	9284	9289	1	1	2	2	3	3	4	4	5
85	9294	9299	9304	9309	9315	9320	9325	9330	9335	9340	1	1	2	2	3	3	4	4	5
86	9345	9350	9355	9360	9365	9370	9375	9380	9385	9390	1	1	2	2	3	3	4	4	5
87	9395	9400	9405	9410	9415	9420	9425	9430	9435	9440	0	1	1	2	2	3	3	4	4
88	9445	9450	9455	9460	9465	9469	9474	9479	9484	9489	0	1	1	2	2	3	3	4	4
89	9494	9499	9504	9509	9513	9518	9523	9528	9533	9538	0	1	1	2	2	3	3	4	4
90	9542	9547	9552	9557	9562	9566	9571	9576	9581	9586	0	1	1	2	2	3	3	4	4
91	9590	9595	9600	9605	9609	9614	9619	9624	9628	9633	0	1	1	2	2	3	3	4	4
92	9638	9643	9647	9652	9657	9661	9666	9671	9675	9680	0	1	1	2	2	3	3	4	4
93	9685	9689	9694	9699	9703	9708	9713	9717	9722	9727	0	1	1	2	2	3	3	4	4
94	9731	9736	9741	9745	9750	9754	9759	9763	9768	9773	0	1	1	2	2	3	3	4	4
95	9777	9782	9786	9791	9795	9800	9805	9809	9814	9818	0	1	1	2	2	3	3	4	4
96	9823	9827	9832	9836	9841	9845	9850	9854	9859	9863	0	1	1	2	2	3	3	4	4
97	9868	9872	9877	9881	9886	9890	9894	9899	9903	9908	0	1	1	2	2	3	3	4	4
98	9912	9917	9921	9926	9930	9934	9939	9943	9948	9952	0	1	1	2	2	3	3	4	4
99	9956	9961	9965	9969	9974	9978	9983	9987	9991	9996	0	1	1	2	2	3	3	3	4
	0	1	2	3	4	5	6	7	8	9	1	2	3	4	5	6	7	8	9

This table can be used to determine natural logarithms and values of e^x: $\ln x = 2.3026 \log x$; for $x = 1.23$ we have $\ln 1.23 = 2.3026 \cdot 0.0899 = 0.207$; $e^x = 10^{x \log e} = 10^{0.4343x}$; for $x = 0.207$ we have $e^{0.207} = 10^{0.4343 \cdot 0.207} = 10^{0.0899} = 1.23$.

Table 3 Four place antilogarithms

log x	x										Differences								
	0	1	2	3	4	5	6	7	8	9	1	2	3	4	5	6	7	8	9
.00	1000	1002	1005	1007	1009	1012	1014	1016	1019	1021	0	0	1	1	1	1	2	2	2
.01	1023	1026	1028	1030	1033	1035	1038	1040	1042	1045	0	0	1	1	1	1	2	2	2
.02	1047	1050	1052	1054	1057	1059	1062	1064	1067	1069	0	0	1	1	1	1	2	2	2
.03	1072	1074	1076	1079	1081	1084	1086	1089	1091	1094	0	0	1	1	1	1	2	2	2
.04	1096	1099	1102	1104	1107	1109	1112	1114	1117	1119	0	1	1	1	1	2	2	2	2
.05	1122	1125	1126	1130	1132	1135	1138	1140	1143	1146	0	1	1	1	1	2	2	2	2
.06	1148	1151	1153	1156	1159	1161	1164	1167	1169	1172	0	1	1	1	1	2	2	2	2
.07	1175	1178	1180	1183	1186	1189	1191	1194	1197	1199	0	1	1	1	1	2	2	2	2
.08	1202	1205	1208	1211	1213	1216	1219	1222	1225	1227	0	1	1	1	1	2	2	2	3
.09	1230	1233	1236	1239	1242	1245	1247	1250	1253	1256	0	1	1	1	1	2	2	2	3
.10	1259	1262	1265	1268	1271	1274	1276	1279	1282	1285	0	1	1	1	2	2	2	2	3
.11	1288	1291	1294	1297	1300	1303	1306	1309	1312	1315	0	1	1	1	2	2	2	2	3
.12	1318	1321	1324	1327	1330	1334	1337	1340	1343	1346	0	1	1	1	2	2	2	2	3
.13	1349	1352	1355	1358	1361	1365	1368	1371	1374	1377	0	1	1	1	2	2	2	3	3
.14	1380	1384	1387	1390	1393	1396	1400	1403	1406	1409	0	1	1	1	2	2	2	3	3
.15	1413	1416	1419	1422	1426	1429	1432	1435	1439	1442	0	1	1	1	2	2	2	3	3
.16	1445	1449	1452	1455	1459	1462	1466	1469	1472	1476	0	1	1	1	2	2	2	3	3
.17	1479	1483	1486	1489	1493	1496	1500	1503	1507	1510	0	1	1	1	2	2	2	3	3
.18	1514	1517	1521	1524	1528	1531	1535	1538	1542	1545	0	1	1	1	2	2	2	3	3
.19	1549	1552	1556	1560	1563	1567	1570	1574	1578	1581	0	1	1	1	2	2	3	3	3
.20	1585	1589	1592	1596	1600	1603	1607	1611	1614	1618	0	1	1	1	2	2	3	3	3
.21	1622	1626	1629	1633	1637	1641	1644	1648	1652	1656	0	1	1	2	2	2	3	3	3
.22	1660	1663	1667	1671	1675	1679	1683	1687	1690	1694	0	1	1	2	2	2	3	3	3
.23	1698	1702	1706	1710	1714	1718	1722	1726	1730	1734	0	1	1	2	2	2	3	3	4
.24	1738	1742	1746	1750	1754	1758	1762	1766	1770	1774	0	1	1	2	2	2	3	3	4
.25	1778	1782	1786	1791	1795	1799	1803	1807	1811	1816	0	1	1	2	2	2	3	3	4
.26	1820	1824	1828	1832	1837	1841	1845	1849	1454	1858	0	1	1	2	2	3	3	3	4
.27	1862	1866	1871	1875	1879	1884	1888	1892	1897	1901	0	1	1	2	2	3	3	3	4
.28	1905	1910	1914	1919	1923	1928	1932	1936	1941	1945	0	1	1	2	2	3	3	4	4
.29	1950	1954	1959	1963	1968	1972	1977	1982	1986	1991	0	1	1	2	2	3	3	4	4
.30	1995	2000	2004	2009	2014	2018	2023	2028	2032	2037	0	1	1	2	2	3	3	4	4
.31	2042	2046	2051	2056	2061	2065	2070	2075	2080	2084	0	1	1	2	2	3	3	4	4
.32	2089	2094	2099	2104	2109	2113	2118	2123	2128	2133	0	1	1	2	2	3	3	4	4
.33	2138	2143	2148	2153	2158	2163	2168	2173	2178	2183	0	1	1	2	2	3	3	4	4
.34	2188	2193	2198	2203	2208	2213	2218	2223	2228	2234	1	1	2	2	3	3	4	4	5
.35	2239	2244	2249	2254	2259	2265	2270	2275	2280	2286	1	1	2	2	3	3	4	4	5
.36	2291	2296	2301	2307	2312	2317	2323	2328	2333	2339	1	1	2	2	3	3	4	4	5
.37	2344	2350	2355	2360	2366	2371	2377	2382	2388	2393	1	1	2	2	3	3	4	4	5
.38	2399	2404	2410	2415	2421	2427	2432	2438	2443	2449	1	1	2	2	3	3	4	4	5
.39	2455	2460	2466	2472	2477	2483	2489	2495	2500	2506	1	1	2	2	3	3	4	5	5
.40	2512	2518	2523	2529	2535	2541	2547	2553	2559	2564	1	1	2	2	3	4	4	5	5
.41	2570	2576	2582	2588	2594	2600	2606	2612	2618	2624	1	1	2	2	3	4	4	5	5
.42	2630	2636	2642	2649	2655	2661	2667	2673	2679	2685	1	1	2	3	3	4	4	5	6
.43	2692	2698	2704	2710	2716	2723	2729	2735	2742	2748	1	1	2	3	3	4	4	5	6
.44	2754	2761	2767	2773	2780	2786	2793	2799	2805	2812	1	1	2	3	3	4	4	5	6
.45	2818	2825	2831	2838	2844	2851	2858	2864	2871	2877	1	1	2	3	3	4	5	5	6
.46	2884	2891	2897	2904	2911	2917	2924	2931	2938	2944	1	1	2	3	3	4	5	5	6
.47	2951	2958	2965	2972	2979	2985	2992	2999	3006	3013	1	1	2	3	3	4	5	5	6
.48	3020	3027	3034	3041	3048	3055	3062	3069	3076	3083	1	1	2	3	4	4	5	6	6
.49	3090	3097	3105	3112	3119	3126	3133	3141	3148	3155	1	1	2	3	4	4	5	6	6
	0	1	2	3	4	5	6	7	8	9	1	2	3	4	5	6	7	8	9

Example: antilog 0.0913 = 1.233 + 0.001 = 1.234.

Table 3 Four place antilogarithms (*continued*)

log x	x										Differences								
	0	1	2	3	4	5	6	7	8	9	1	2	3	4	5	6	7	8	9
.50	3162	3170	3177	3184	3192	3199	3206	3214	3221	3228	1	1	2	3	4	4	5	6	7
.51	3236	3243	3251	3258	3266	3273	3281	3289	3296	3304	1	2	2	3	4	5	5	6	7
.52	3311	3319	3327	3334	3342	3350	3357	3365	3373	3381	1	2	2	3	4	5	5	6	7
.53	3388	3396	3404	3412	3420	3428	3436	3443	3451	3459	1	2	2	3	4	5	6	6	7
.54	3467	3475	3483	3491	3499	3508	3516	3524	3532	3540	1	2	2	3	4	5	6	6	7
.55	3548	3556	3565	3573	3581	3589	3597	3606	3614	3622	1	2	2	3	4	5	6	7	7
.56	3631	3639	3648	3656	3664	3673	3681	3690	3698	3707	1	2	3	3	4	5	6	7	8
.57	3715	3724	3733	3741	3750	3758	3767	3776	3784	3793	1	2	3	3	4	5	6	7	8
.58	3802	3811	3819	3828	3837	3846	3855	3864	3873	3882	1	2	3	4	4	5	6	7	8
.59	3890	3899	3908	3917	3926	3936	3945	3954	3963	3972	1	2	3	4	5	5	6	7	8
.60	3981	3990	3999	4009	4018	4027	4036	4046	4055	4064	1	2	3	4	5	6	6	7	8
.61	4074	4083	4093	4102	4111	4121	4130	4140	4150	4159	1	2	3	4	5	6	7	8	9
.62	4169	4178	4188	4198	4207	4217	4227	4236	4246	4256	1	2	3	4	5	6	7	8	9
.63	4266	4276	4285	4295	4305	4315	4325	4335	4345	4355	1	2	3	4	5	6	7	8	9
.64	4365	4375	4385	4395	4406	4416	4426	4436	4446	4457	1	2	3	4	5	6	7	8	9
.65	4467	4477	4487	4498	4508	4519	4529	4539	4550	4560	1	2	3	4	5	6	7	8	9
.66	4571	4581	4592	4603	4613	4624	4634	4645	4656	4667	1	2	3	4	5	6	7	9	10
.67	4677	4688	4699	4710	4721	4732	4742	4753	4764	4775	1	2	3	4	5	7	8	9	10
.68	4786	4797	4808	4819	4831	4842	4853	4864	4875	4887	1	2	3	4	6	7	8	9	10
.69	4898	4909	4920	4932	4943	4955	4966	4977	4989	5000	1	2	3	5	6	7	8	9	10
.70	5012	5023	5035	5047	5058	5070	5082	5093	5105	5117	1	2	4	5	6	7	8	9	11
.71	5129	5140	5152	5164	5176	5188	5200	5212	5224	5236	1	2	4	5	6	7	8	10	11
.72	5248	5260	5272	5284	5297	5309	5321	5333	5346	5358	1	2	4	5	6	7	9	10	11
.73	5370	5383	5395	5408	5420	5433	5445	5458	5470	5483	1	3	4	5	6	8	9	10	11
.74	5495	5508	5521	5534	5546	5559	5572	5585	5598	5610	1	3	4	5	6	8	9	10	12
.75	5623	5636	5649	5662	5675	5689	5702	5715	5728	5741	1	3	4	5	7	8	9	10	12
.76	5754	5768	5781	5794	5808	5821	5834	5848	5861	5875	1	3	4	5	7	8	9	11	12
.77	5888	5902	5916	5929	5943	5957	5970	5984	5998	6012	1	3	4	5	7	8	10	11	12
.78	6026	6039	6053	6067	6081	6095	6109	6124	6138	6152	1	3	4	6	7	8	10	11	13
.79	6166	6180	6194	6209	6223	6237	6252	6266	6281	6295	1	3	4	6	7	9	10	11	13
.80	6310	6324	6339	6353	6368	6383	6397	6412	6427	6442	1	3	4	6	7	9	10	12	13
.81	6457	6471	6486	6501	6516	6531	6546	6561	6577	6592	2	3	5	6	8	9	11	12	14
.82	6607	6622	6637	6653	6668	6683	6699	6714	6730	6745	2	3	5	6	8	9	11	12	14
.83	6761	6776	6792	6808	6823	6839	6855	6871	6887	6902	2	3	5	6	8	9	11	13	14
.84	6918	6934	6950	6966	6982	6998	7015	7031	7047	7063	2	3	5	6	8	10	11	13	15
.85	7079	7096	7112	7129	7145	7161	7178	7194	7211	7228	2	3	5	7	8	10	12	13	15
.86	7244	7261	7278	7295	7311	7328	7345	7362	7379	7396	2	3	5	7	8	10	12	13	15
.87	7413	7430	7447	7464	7482	7499	7516	7534	7551	7568	2	3	5	7	9	10	12	14	16
.88	7586	7603	7621	7638	7656	7674	7691	7709	7727	7745	2	4	5	7	9	11	12	14	16
.89	7762	7780	7798	7816	7834	7852	7870	7889	7907	7925	2	4	5	7	9	11	13	14	16
.90	7943	7962	7980	7998	8017	8035	8054	8072	8091	8110	2	4	6	7	9	11	13	15	17
.91	8128	8147	8166	8185	8204	8222	8241	8260	8279	8299	2	4	6	8	9	11	13	15	17
.92	8318	8337	8356	8375	8395	8414	8433	8453	8472	8492	2	4	6	8	10	12	14	15	17
.93	8511	8531	8551	8570	8590	8610	8630	8650	8670	8690	2	4	6	8	10	12	14	16	18
.94	8710	8730	8750	8770	8790	8810	8831	8851	8872	8892	2	4	6	8	10	12	14	16	18
.95	8913	8933	8954	8974	8995	9016	9036	9057	9078	9099	2	4	6	8	10	12	15	17	19
.96	9120	9141	9161	9183	9204	9226	9247	9268	9290	9311	2	4	6	8	11	13	15	17	19
.97	9333	9354	9376	9397	9419	9441	9462	9484	9506	9528	2	4	7	9	11	13	15	17	20
.98	9550	9572	9594	9616	9638	9661	9683	9705	9727	9750	2	4	7	9	11	13	16	18	20
.99	9772	9795	9817	9840	9863	9886	9908	9931	9954	9977	2	5	7	9	11	14	16	18	20
	0	1	2	3	4	5	6	7	8	9	1	2	3	4	5	6	7	8	9

The number y is called the **numerus** of the logarithm to base a. Ordinarily logarithms to base 10, written $^{10}\log x$ or $\log_{10} x$ or simply $\log x$, are used. Other systems of logarithms will be mentioned at the end of this section. For $a = 10$ and $y = 3$ we have, using logarithms to base 10 (Briggs, decadic or common logarithms), $x = 0.4771$ and $10^{0.4771} = 3$. Other examples involving four place logarithms:

$$
\begin{aligned}
5 &= 10^{0.6990} & \text{or} && \log 5 &= 0.6990, \\
1 &= 10^{0} & \text{or} && \log 1 &= 0, \\
10 &= 10^{1} & \text{or} && \log 10 &= 1, \\
1000 &= 10^{3} & \text{or} && \log 1000 &= 3, \\
0.01 &= 10^{-2} & \text{or} && \log 0.01 &= -2.
\end{aligned}
$$

Since logarithms are exponents, the power laws apply, e.g.,

$$2 \cdot 4 = 10^{0.3010} \cdot 10^{0.6021} = 10^{0.3010+0.6021} = 10^{0.9031} = 8.$$

Taking the logarithm of a product of numbers reduces to the addition of the corresponding logarithms. Similarly taking the logarithm of a quotient becomes subtraction, taking the logarithm of a power becomes multiplication, taking the logarithm of a root becomes division — in general,

$$
\left.\begin{aligned}
\log(ab) &= \log a + \log b \\
\log \frac{a}{b} &= \log a - \log b
\end{aligned}\right\} (a > 0, b > 0),
$$

$$
\left.\begin{aligned}
\log a^n &= n \log a \\
\log \sqrt[n]{a} = \log a^{1/n} &= \frac{1}{n} \log a
\end{aligned}\right\} (a > 0, n = \text{decimal}),
$$

$$
\log \frac{1}{c} \left\{\begin{aligned} &= \log 1 - \log c = 0 - \log c = \\ &= \log c^{-1} = (-1)\log c = \end{aligned}\right\} - \log c.
$$

In the general statement $a = 10^{\log a}$, a is the numerus or **antilogarithm** and $\log a$ is the common logarithm of a, which decomposes into two parts: common logarithm = mantissa ± characteristic, e.g.,

Numerus			M C C M
$\log 210.0 = \log(2.1 \cdot 10^2)$	$= \log 2.1 + \log 10^2$	$= 0.3222 + 2 = 2.3222$
$\log 21.0 = \log(2.1 \cdot 10^1)$	$= \log 2.1 + \log 10^1$	$= 0.3222 + 1 = 1.3222$
$\log 2.1 = \log(2.1 \cdot 10^0)$	$= \log 2.1 + \log 10^0$	$= 0.3222 + 0 = 0.3222$
$\log 0.21 = \log(2.1 \cdot 10^{-1})$	$= \log 2.1 + \log 10^{-1}$	$= 0.3222 - 1.$

0.2 Arithmetical Operations

The sequence of digits following the decimal point of the logarithm (namely 3222) is called the **mantissa** (M). The mantissas are found in the logarithm table (Table 2), which could be more appropriately referred to as the mantissa table. We content ourselves with four place mantissas. Note that a mantissa is always nonnegative and less than 1. The largest integer which is less than or equal to the logarithm (in the examples, 2, 1, 0, -1) is called the **characteristic** (C). As in the four examples, the numerus is written in the following power of ten form (usually called scientific notation):

$$\text{Numerus} = \begin{bmatrix} \text{Sequence of digits in the numerus} \\ \text{with a decimal point after the} \\ \text{first nonzero digit} \end{bmatrix} \cdot 10^C.$$

EXAMPLE. Find the logarithm of:

(a) $0.000021 = 2.1 \cdot 10^{-5}$; $\log(2.1 \cdot 10^{-5}) = 0.3222 - 5$ [see Table 2].
(b) $987{,}000 = 9.87 \cdot 10^5$; $\log(9.87 \cdot 10^5) = 0.9943 + 5 = 5.9943$ [see Table 2].
(c) $3.37 = 3.37 \cdot 10^0$; $\log(3.37 \cdot 10^0) = 0.5276 + 0 = 0.5276$.

When working with logarithms the result must be written with M and C displayed. If in the process of finding a root a **negative characteristic** appears, this characteristic must always be brought into a form that is divisible by the index of the radical:

EXAMPLE. Calculate $\sqrt[3]{0.643}$

$\log 0.643 = 0.8082 - 1 = 2.8082 - 3$,

$\log \sqrt[3]{0.643} = \log 0.643^{1/3} = \tfrac{1}{3}(2.8082 - 3) = 0.93607 - 1$,
$\sqrt[3]{0.643} = 0.8631$ $\Big\}$ [see Table 3].

Now for the **inverse operation** of finding the antilogarithm. After the computations have been carried out in terms of logarithms the numerus corresponding to the result has to be determined. This is done with the help of the antilogarithm table in exactly the same way as the logarithm of a given number is found in the logarithm table. The logarithm has to be written in the proper form, with a positive mantissa M and an integer characteristic C, e.g.,

$$\log x = -5.7310 = (-5.7310 + 6) - 6 = 0.2690 - 6.$$

So also, $\log 1/x = (1 - \log x) - 1$; e.g., $\log \tfrac{1}{3} = (1 - 0.4771) - 1 = 0.5229 - 1$. The mantissa without characteristic determines the sought-after sequence of digits of the antilog with a decimal point after the first nonzero digit. The characteristic C, whether positive or negative, specifies the power:

$$\begin{bmatrix} \text{Sequence of digits in the numerus} \\ \text{with a decimal point after the} \\ \text{first nonzero digit} \end{bmatrix} \cdot 10^C = \text{Numerus}.$$

EXAMPLE. Find the antilog of:

(a) $\log x = 0.2690 - 6; x = 1.858 \cdot 10^{-6}$.
(b) $\log x = 0.0899 - 1; x = 1.23 \cdot 10^{-1}$.
(c) $\log x = 0.5276; x = 3.37$.
(d) $\log x = 5.9943; x = 9.87 \cdot 10^5$.

We summarize. **Every calculation with logarithms involves five steps:**

1. Formulating the problem.
2. Converting to logarithmic notation.
3. Recording the characteristic and determining the mantissa from the logarithm table.
4. Carrying out the logarithmic calculations.
5. Determining the antilog with the help of the antilogarithm table—the characteristic fixing the location of the decimal point.

If, as often happens, an antilogarithm table is unavailable, the numerus can of course be found with the help of the logarithm table. The procedure is simply the reverse of that used in determining the logarithms.

EXAMPLE. Calculate

$$\sqrt[6]{\frac{89.49^{3.5} \cdot \sqrt{0.006006}}{0.001009^2 \cdot 3{,}601{,}000^{4.2}}}.$$

We set

$$\sqrt[6]{\frac{(8.949 \cdot 10)^{3.5} \cdot \sqrt{6.006 \cdot 10^{-3}}}{(1.009 \cdot 10^{-3})^2 \cdot (3.601 \cdot 10^6)^{4.2}}} = x,$$

and using $\log x = \frac{1}{6} \cdot (\{\log(\text{numerator})\} - \{\log(\text{denominator})\})$, i.e.,

$$\log x = \tfrac{1}{6} \cdot (\{3.5 \cdot \log(8.949 \cdot 10) + \tfrac{1}{2} \cdot \log(6.006 \cdot 10^{-3})\} \\ - \{2 \cdot \log(1.009 \cdot 10^{-3}) + 4.2 \cdot \log(3.601 \cdot 10^6)\}),$$

we find from Table 4 that

$$\log x = \tfrac{1}{6} \cdot (\{5.7206\} - \{21.5447\}) = \tfrac{1}{6} \cdot (\{23.7206 - 18\} - \{21.5447\}),$$

$\log x = \frac{1}{6} \cdot (2.1759 - 18) = 0.36265 - 3$, and the desired value $x = 2.305 \cdot 10^{-3}$.

0.3 Computational Aids

Table 4

Numerus	Logarithm	Factor	Logarithm
$8.949 \cdot 10^1$	$0.9518 + 1$	3.5	6.8313
$6.006 \cdot 10^{-3}$	$0.7786 - 3$	0.5	$0.8893 - 2$
	$= 1.7786 - 4$		
Numerator			5.7206
$1.009 \cdot 10^{-3}$	$0.0039 - 3$	2	$0.0078 - 6$
$3.601 \cdot 10^6$	$0.5564 + 0$	4.2	27.5369
Denominator			21.5447

The so-called **natural logarithms** (ln) (cf. Table 29 and Table 36) have as base the constant

$$e \approx 2.718281828459 \cdots$$

which is the limit of the sequence

$$e = 1 + \frac{1}{1} + \frac{1}{1 \cdot 2} + \frac{1}{1 \cdot 2 \cdot 3} + \frac{1}{1 \cdot 2 \cdot 3 \cdot 4} + \cdots.$$

The conversion formulas with rounded-off coefficients are

$$\ln x = \ln 10 \cdot \log x \simeq 2.302585 \cdot \log x,$$
$$\log x = \log e \cdot \ln x \simeq 0.4342945 \cdot \ln x.$$

[Note that $\ln 1 = 0$, $\ln e = 1$, $\ln 10^k \simeq k \cdot 2.302585$; note also that $\ln e^x = x$, $e^{\ln x} = x$, and especially $a^x = e^{x \ln a}$ $(a > 0)$.] The symbols "$^e\log x$" and "$\log_e x$" are also used in place of "$\ln x$."

The logarithm to base 2 (logarithmus dualis) written as ld or lb [binary, consisting of two units]), can be obtained by the formulas

$$\text{ld } x = \frac{\log x}{\log 2} \simeq 3.321928 \cdot \log x,$$
$$\text{ld } x = \frac{\ln x}{\ln 2} \simeq 1.442695 \cdot \ln x$$

[e.g., ld 5 = 2.322 = 3.322 · 0.699 = 1.443 · 1.609].

or from a table, (e.g., Alluisi 1965).

0.3 COMPUTATIONAL AIDS

It is convenient to employ an electronic pocket calculator, an electronic calculator with printout or, better yet, a programmable calculator. For extensive calculations—large amounts of data and/or multi-variate methods

—it becomes necessary to make use of a computer. Programs are available for nearly all routines [see pages 572 and 573]. For calculations arising in statistical analysis one needs, in addition, a collection of **numerical tables** [see pages xiii–xvi].

The following procedure is recommended for every calculation:

> 1. **Arranging a computational setup:** Determine in full detail all steps involved in the computation. An extensive computation should be so well thought through and prepared for that a technician can carry it out. Clearly arranged computational schemes which include all the numerical computations and in which the computation proceeds systematically also lessen the chances of error.
> 2. **Use paper on one side only**; write all numbers clearly; leave **wide margins** for the rough work; avoid duplication; cross out any incorrect number and write the correct value above it.
> 3. **Use rough estimates to avoid misplacing the decimal point;** check your computation! **Each arithmetical operation should be preceded** or followed **by a rough estimate**, so that at least the location of the decimal point in the result is determined with confidence. **Scientific notation** is recommended:
>
> $$\frac{0.00904}{0.167} = \frac{9.04 \cdot 10^{-3}}{1.67 \cdot 10^{-1}} \simeq 5 \cdot 10^{-2};$$
>
> more precisely, to 3 decimal places: $5.413 \cdot 10^{-2}$.
> 4. To double check, the problem should, if possible, be solved by **still another method**. It is sometimes advantageous for two coworkers to carry out the computations independently and then to compare the results.
> 5. The recommendations and the computational checks mentioned in the text should be replaced by optimal versions and adapted to the computational aids at one's disposal.
>
> **Use formulas with care:** make sure that you really understand what the formula is about, that it really does apply to your particular case, and finally that you really have not made an arithmetical error.

0.4 ROUNDING OFF

If the quantities 14.6, 13.8, 19.3, 83.5, and 14.5 are to be rounded off to the nearest integer, the first three present no difficulty; they become 15, 14, and 19. For the last two quantities, one might choose the numbers 83 or 84 and 14 or 15 respectively. It turns out to be expedient to round off to the nearest **even** number, so that 83.5 goes over into 84 and 14.5 into 14. Here zero is

Table 5 Significant digits

Result	Number of significant digits	Limits of the error range	Greatest error $=\dfrac{0.5e}{R}(100)$ ($\pm\%$)
4	1	3.5–4.5	12.5
4.4	2	4.35–4.45	1.14
4.44	3	4.435–4.445	0.113

treated as an even number. The more values rounded off in this way and then summed, the more the roundoff errors cancel out.

Also important is the notion of **significant digits**. The significant digits of a number are the sequence of digits in the number without regard to a decimal point, if present, and, for numbers less than 1, without regard to the zeros preceding or immediately following the decimal. Table 5 compares three results of rounding off, the number of significant figures in the expressions and the accuracy inherent in each: by the corresponding error bounds as well as by the maximum rounding error. This clearly implies the following: If a method is used with which there is associated an error of at least 8% in the size, then it is misleading to state a result with more than two significant digits. If two numbers, each with x accurate or significant digits, are multiplied together, then at most $x - 1$ digits of the product can be regarded as reliable. A corresponding statement applies to division.

EXAMPLE. Compute the area of a rectangle with sides of measured length 38.22 cm and 16.49 cm. To write the result as $38.22 \cdot 16.49 = 630.2478$ cm^2 would be incorrect, since the area can take on any value between $38.216 \cdot 16.486 = 630.02898$ and $38.224 \cdot 16.494 = 630.4666$. This range is characterized by 630.2 cm^2 $\pm$ 0.3 cm^2. The result can be stated with only three significant figures (630 cm^2).

0.5 COMPUTATIONS WITH INACCURATE NUMBERS

If inaccurate numbers are tied together by arithmetical operations, then the so-called **propagation of error** can be estimated. Two parallel calculations can be carried out, one with error bounds which lead to the minimum value for the result and the other with error bounds which lead to the maximum.

EXAMPLE

30 ± 3 Range: from 27 to 33
20 ± 1 Range: from 19 to 21.

1. **Addition:** the actual sum of the two numbers lies between $27 + 19 = 46$ and $33 + 21 = 54$. The relative error of the sum equals

$$\frac{54 - 46}{54 + 46} = \frac{8}{100} = 0.08;$$

it lies within the $\pm 8\%$ limits.

2. **Subtraction:** The actual difference lies between $27 - 21 = 6$ and $33 - 19 = 14$ (subtraction "crossover," i.e., the maximal value of one number is subtracted from the minimal value of the other number, the minimal value of one number is subtracted from the maximal value of the other number). The relative error of the difference equals

$$\frac{14 - 6}{14 + 6} = \frac{8}{20} = 0.40, \qquad \pm \ 40\%.$$

3. **Multiplication:** The actual product lies somewhere between the limits $27 \cdot 19 = 513$ and $33 \cdot 21 = 693$. The relative error of the product equals

$$\frac{513 - 30 \cdot 20}{30 \cdot 20} = \frac{513 - 600}{600} = \frac{-87}{600} = -0.145 = -14.5\%,$$

$$\frac{693 - 30 \cdot 20}{30 \cdot 20} = \frac{693 - 600}{600} = \frac{93}{600} = 0.155 \quad = +15.5\%.$$

4. **Division:** The actual quotient lies between $\frac{27}{21} = 1.286$ and $\frac{33}{19} = 1.737$ (division "crossover"). The relative error of the quotient is found to be

$$\frac{1.286 - 30/20}{30/20} = -\frac{0.214}{1.500} = -0.143 = -14.3\%,$$

$$\frac{1.737 - 30/20}{30/20} = \frac{0.237}{1.500} = 0.158 \quad = +15.8\%.$$

Of all the basic arithmetic operations on inaccurate numbers **subtraction** is particularly risky, the final error being substantially higher than for the other arithmetic operations.

1 STATISTICAL DECISION TECHNIQUES

> The beginner should on first reading confine himself to sections indicated by an arrow ▶, **paying particular attention to the examples**, disregarding for the time being whatever he finds difficult to grasp, the remarks, the fine print, and the bibliography.

1.1 WHAT IS STATISTICS? STATISTICS AND THE SCIENTIFIC METHOD

Empirical science does not consist of a series of nonrecurring isolated events or characteristics relating to a particular individual or entity, but rather of **reproducible experiences**, a collection of events—regarded as of the same kind—about which information is sought.

In the year 1847, when Semmelweis introduced hygienic measures at the obstetrical clinic in Vienna in spite of opposition by his colleagues, he did not know of the bacteriological nature of childbed fever. He could also not prove directly that his experiments were successful, since even after the introduction of hygiene, women still died of childbed fever at his clinic. The maternal mortality decreased however from 10.7% (1840–1846) to 5.2% (1847) to 1.3% (1848) and since Semmelweis's calculations were based on a large number of women about to give birth (21,120; 3,375; 3,556) (Lesky 1964), it was **concluded** that hygienic measures should continue to be applied.

Statistical methods are necessary wherever results cannot be reproduced exactly and arbitrarily often. The sources of this nonreproducibility lie in **uncontrolled** and **uncontrollable** external influences, in the disparity among the test objects, in the variability of the material under observation, and in the test and observation conditions. In sequences of observations, these sources lead to **"dispersion"** of quantitatively recorded characteristics—(usually less for investigations in the natural sciences than for those in the social sciences). Since as a consequence of this dispersion any particular value will hardly ever be reproduced exactly, definite and unambiguous conclusions have to be deferred. The dispersion thus leads to an **uncertainty**, which frequently allows decisions but not exact inferences to be made. This is the starting point of a definition of statistics as an aid in decision making, which goes back to Abraham Wald (1902–1950): **Statistics is a combination of methods which permit us to make reasonable optimal decisions in cases of uncertainty.**

Descriptive statistics contents itself with the investigation and description of a whole population. Modern **inductive or analytic statistics** studies, in contrast, only some portion, which should be characteristic or representative for the population or aggregate in whose properties we are interested. **Conclusions about the population** are thus drawn from observations carried out on some part of it, i.e., one proceeds **inductively**. In this situation it is essential that the part of the population to be tested—the sample—be chosen **randomly**, let us say according to a lottery procedure. We call a sampling random if every possible combination of sample elements from the population has the same chance of being chosen. **Random samples are important because they alone permit us to draw conclusions about the population**. Overall surveys are frequently either not possible at all or else possible only with great expenditure of time and money.

Research means testing hypotheses and/or getting new insights (in particular the extension of factual knowledge; c.f., also the Introduction). Four levels can be distinguished:

1. Description of the problem and definitions. Observations are made.
2. Analysis: essential elements are abstracted to form the basis of a hypothesis or theory.
3. Solution I of problem: The hypothesis or theory is developed to where new conclusions can be stated and/or results predicted. Formulation of new (partial) problems.
4. New data are gathered to verify the predictions arrived at from the theory: observations II.

The whole sequence of steps then starts all over again. If the hypothesis is confirmed, then the test conditions are sharpened by more precisely wording and generalizing the predictions until finally some deviation is found, making it necessary to refine the theory. If any results contradicting the hypothesis are found, a new hypothesis that agrees with a larger number of factual observations is formulated. The final truth is entirely unknown to a science based on empirical data. The failure of all attempts to disprove a certain hypothesis will increase our confidence in it; this, however, does not furnish a conclusive proof that the hypothesis is always valid: **hypotheses can only be tested, they can never be proved**. Empirical tests are attempts at negation.

Statistics can intervene at every step of the (iterated) sequence described above:
1. In the choice of the observations (sampling theory).
2. In the presentation and summary of observations (descriptive statistics).
3. In the estimation of parameters (estimation theory).
4. In the formulation and verification of the hypotheses (test theory).

Statistical inference enables us to draw, from the sample, conclusions on the whole corresponding population (e.g. when we estimate election results from known particular results for a selected constituency)—**general statements which are valid beyond the observed aggregate**. In all empirical sciences, it makes possible the **assessing of empirical data** and the **verification of scientific theories** through confrontation of results derived from probability theoretical models—idealizations of special experimental situations—with empirical data; the probabilistic statements, which are of course the only kind here possible, then offer the practitioner indispensable information on which to base his decisions.

In **estimation theory** one is faced with deciding how, from a given sample, the greatest amount of information regarding the characteristic features of the corresponding parent population can be extracted. In **test theory** the problem is one of deciding whether the sample was drawn from a certain (specified) population. Modern statistics is concerned with designing experiments which are capable of efficiently answering the question asked (cf. also Section 7.7), and then carrying out and evaluating experiments and surveys.

STATISTICS is the science of obtaining, summarizing, analyzing and making inferences from both counted and measured observations, termed data. It deals with designing experiments and surveys in order to obtain main characteristics of the observations, especially kind and magnitude of variation and type of dependencies in both experimental and survey data. The defined total set of all possible observations, about which information is desired, is termed **population**. Commonly available is at best a representative part of the population, termed a **sample**, which may give us a tentative incomplete view of the unknown population.

Accordingly the science of statistics deals with:
1. **presenting and summarizing data** in tabular and graphic form to understand the nature of the data and to facilitate the detection of unexpected characteristics,
2. **estimating unknown constants** associated with the population, termed parameters, providing various measures of the accuracy and precision of these estimates,
3. **testing hypotheses** about populations.

Detecting different sources of error, giving estimates of uncertainty and, sometimes, trying to salvage experimental results are other activities of the statistician.

A discussion of the **philosophical roots of statistics** and of its position among the sciences is provided by Hotelling (1958) (cf. also Gini 1958; Tukey 1960, 1962, 1972; Popper 1963, 1966; Stegmüller 1972 and Bradley 1982 [8:7]; **common fallacies** (cf., Hamblin 1970) are pointed out by Campbell (1974) (cf. also Koller 1964 [8:2a] and Sachs 1977 [8:2a]). On statistical evidence [and the law] see Fienberg and Straf (1982 [8:2a]).

1.2 ELEMENTS OF COMPUTATIONAL PROBABILITY

The uncertainty of the decisions can be quantitatively expressed through the **theory of probability**. In other words: probability theoretic notions lead to the realization of optimal decision procedures. Hence we turn our attention for the present to the notion of "probability."

▶ 1.2.1 Statistical probability

We know in everyday life of various sorts of statements in which the word "probably" (range of significance: presumably to certainly) appears:

1. George probably has a successful marriage.
2. The president's handling of the crisis was probably correct.
3. The probability of rolling a "1" is $\frac{1}{6}$.
4. The probability of a twin birth is $\frac{1}{86}$.

The last two statements are closely related to the **notion of relative frequency**. It is assumed that in tossing the die each side turns up equally often on the average, so we expect that with frequent repetition the relative frequency with which 1 comes up will tend to $\frac{1}{6}$. The fourth statement originated from some relative frequency. It had been observed during the last few years that the relative frequency of twin births is 1:86; hence it can be assumed that a future birth will be a twin birth with probability equal to this relative frequency. In the first two statements however, no such relation to relative frequency exists. We wish, in the following, to consider only probabilities which can be interpreted as relative frequencies. With **frequent repetition**, these relative frequencies generally exhibit **remarkable stability**. This notion of probability is based historically on the well-known ratio

$$\frac{\text{number of favorable events}}{\text{number of possible events}} \quad (1.1)$$

1.2 Elements of Computational Probability

—the **definition of probability** due to Jakob Bernoulli (1654–1705) and Laplace (1749–1827). It is here tacitly assumed that all possible events are, as with the tossing of a die, equally probable. Every probability P is thus a number between zero and one:

$$\boxed{0 \leq P \leq 1.} \tag{1.2}$$

An impossible outcome has probability zero, while a sure outcome probability one. In everyday life these probabilities are multiplied by 100 and expressed as percentages ($0\% \leq P \leq 100\%$). The probability of rolling a 4 with a perfect die is $\frac{1}{6}$ because all six faces have the same chance of coming up. The six faces of a perfect die are assigned the same probabilities.

The definition of probability according to Bernoulli and Laplace obviously makes sense only when all possible cases are equally probable, statistically **symmetric**. It proves to be correct for the usual implements of games of chance (coins, dice, playing cards, and roulette wheels). They possess a certain **physical symmetry** which implies statistical symmetry. Statistical symmetry is however an unconditional requirement of this definition of probability. The question here is of an **a priori** probability, which can also be referred to as mathematical probability. An unfair die is not physically symmetric; therefore statistical symmetry can no longer be assumed, and the probability of a specified outcome in a toss of a die cannot be computed. The only way to determine the probability of a particular outcome consists in a very large number of tosses. Taking into account the information gained from the trial, we get in this case the **a posteriori probability** or the **statistical probability**. The distinction between mathematical and statistical probability concerns only the way the probability values are obtained. Probabilities are also stated in terms of odds, as in the following examples:

1. Buffon's experiments with coins. Here the odds are 2048 to 1996, whence $P = 2048/(2048 + 1996) = 0.5064$ (subjective probability). These numbers were obtained by Buffon (1787) in 4044 tosses of a coin. The value of P compares well with the probability of $p = 0.500$ for a fair coin to land heads up.
2. Wolf's experiments with dice. Here the odds are 3407 to 16,593, whence $P = 3407/(3407 + 16,593) = 0.17035$ (subjective probability). R. Wolf (1851) conducted an experiment in which a die was tossed 20,000 times. In 3407 tosses the face with one dot was up (in 2916 tosses four dots showed: $P = 0.146$). The mathematical probability of this outcome (for a fair die) is $p = \frac{1}{6} = 0.167$.

Another example of such a probability has 9 to 12 as odds, i.e., $P = 9/(9+12) = 0.429$ (subjective probability); this P approximates the probability that out of 12 fencing matches, three consecutive matches are won ($P = 1815/4096 = 0.443$; Hamlet: V, 2 [cf., Spinchorn 1970]).

The particularly important axiomatic definition of probability (Section 1.2.2) originated with A. N. Kolmogorov (1933), who connected the notion of probability with modern set theory, measure theory, and functional analysis (cf. Van der Waerden 1951) and thereby created the theoretical counterpart to empirical relative frequency (cf. also Hemelrijk 1968, Rasch 1969, and Barnett 1982).

▶ 1.2.2 The addition theorem of probability theory

The collection of possible outcomes of a survey or an experiment forms the so-called **space of elementary events**, S. One can now pose the question whether or not the outcome of an experiment falls in a particular region of the space of elementary events. The random outcomes can thus be characterized by subsets of the space of elementary events.

The space of elementary events which corresponds to a single tossing of a die consists of 6 points, which we number from 1 to 6. The space is thus finite. On the other hand, assume that in a game of Monopoly you land in jail. According to the rules you cannot move unless you toss a 6. Let an event consist of the number of times the die has to be tossed before a 6 comes up. Then, even in this simple situation, the space of elementary events is infinite, because every positive integer is a possible outcome (Walter 1966). If we are dealing with a characteristic of a continuous nature, such as the size of an object or the amount of rainfall, we can represent the events (outcomes) by points on the real axis. The space of elementary events then includes, for example, all the points in some interval.

Any subset of the space of elementary events is called an **event** and is denoted by a Latin capital letter, usually E or A. Let us emphasize that the whole space of elementary events, S, is also an event, called the sure or **certain event**. In the example involving the single toss of a die, $S = \{1, 2, 3, 4, 5, 6\}$ is the event that any number comes up.

If E_1 and E_2 are events, it is frequently of interest to know whether a measurement lies either in E_1 or in E_2 (or possibly in both). This event is characterized by the subset $E_1 \cup E_2$ of the space of elementary events that consists of all points lying in E_1 or E_2 (or in both). The "or conjunction," the logical sum $E_1 \cup E_2$ (also written $E_1 + E_2$)—read "E_1 union E_2"—is realized when at least one of the events E_1 or E_2 occurs. The symbol $\cup$ is reminiscent of the letter u (for Latin vel = or, in a nonexclusive sense).

EXAMPLE. $E_1 = \{2, 4\}$, $E_2 = \{1, 2\}$, $E_1 \cup E_2 = \{1, 2, 4\}$. This set characterizes the event "E_1 or E_2 or both."

Analogously one could ask whether a measurement lies in both E_1 and E_2. This event is characterized by the set of points in the space of elementary

1.2 $P(E)$—the Chance of the Event E Occurring

events, each of which lies in E_1 as well as in E_2. This set is denoted by $E_1 \cap E_2$. The "as-well-as conjunction"—the logical product $E_1 \cap E$, (also written $E_1 E_2$), read: "E_1 intersction E_2"—is realized when E_1 as well as E_2 occurs.

EXAMPLE. $E_1 \cap E_2 = \{2, 4\} \cap \{1, 2\} = \{2\}$.

If it so happens that E_1 and E_2 have no points in common, we say that the events E_1 and E_2 are **mutually exclusive**. The operation $E_1 \cap E_2$ then yields the so-called **empty set**, which contains no points. To the empty set $\emptyset$ there corresponds the **impossible event**. Since no measured values can possibly lie in the empty set, no measurement can fall in $\emptyset$. For any event E there is an event $\bar{E}$, consisting of those points in the sample space that do not lie in E. The set $\bar{E}$, read "not E", is called the event **complementary** to E or the logical complement.

If, for example, E is the event that in a toss of a die an even number comes up, then $E = \{2, 4, 6\}$ and $\bar{E} = \{1, 3, 5\}$. We have (1.3) and (1.4).

$$E \cup \bar{E} = S \quad \text{(sure or certain event)} \tag{1.3}$$

$$E \cap \bar{E} = \emptyset \quad \text{(impossible event)} \tag{1.4}$$

The diagrams in Figure 1A illustrate these relations. By (1.2) the probability $P(E)$ that as a result of a measurement the measured value x lies in E, is a number between zero and one. We shall assume that to every event E some probability $P(E)$ is assigned which will enable us to make statistical assertions. This assignment however is not arbitrary, but must adhere to the following rules (the axioms of probability theory):

I. Every event carries a probability, a number between zero and one:

$$0 \leq P(E) \leq 1 \quad \textbf{non-negativity.} \tag{1.5}$$

II. The certain event has probability one:

$$P(S) = 1 \quad \textbf{standardization.} \tag{1.6}$$

III. The probability that out of a collection of pairwise disjoint events ($E_i \cap E_j = \emptyset$ for $i \neq j$; i.e., every two distinct events exclude each other) one of the events occurs ("either or probability"), equals the sum of the probabilities of the events in the collection (**addition rule for mutually exclusive events**):

$$P(E_1 \cup E_2 \cup \ldots) = P(E_1) + P(E_2) + \ldots \quad \textbf{additivity.} \tag{1.7}$$

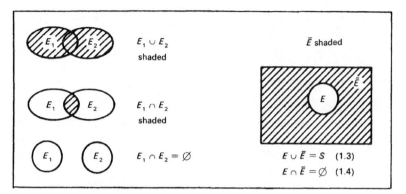

Figure 1A Euler's circles or Venn diagrams.

Axiom II could be written $\sum_i P(E_i) = 1$.

Simple version of III: $P(E_1 \cup E_2) = P(E_1) + P(E_2)$ if $E_1 \cap E_2 = \emptyset$. On combining this with (1.3) we get $1 = P(S) = P(E \cup \bar{E}) = P(E) + P(\bar{E})$, i.e.,

$$P(E) = 1 - P(\bar{E}). \tag{1.8}$$

EXAMPLE (Illustrating axiom III). The probability that on a single toss of an unbiased die either a 3 or 4 occurs, comes to $\frac{1}{6} + \frac{1}{6} = \frac{1}{3}$. Thus, in a series of tosses, we can expect a 3 or 4 to come up in 33% of the cases.

The probability that out of two events E_1 and E_2 that are not mutually exclusive, at least one occurs, is given by

$$P(E_1 \cup E_2) = P(E_1) + P(E_2) - P(E_1 \cap E_2). \tag{1.9}$$

The Venn diagram (Fig. 1A) shows that if we simply add $P(E_1)$ and $P(E_2)$, the "as well as probability" $P(E_1 \cap E_2)$ is counted twice. This is the **addition rule** or **addition theorem for arbitrary events which are not mutually exclusive**. (For three arbitrary events (1.9) extends (see Figure 1B) to

$$P(A \cup B \cup C) = P(A) + P(B) + P(C) - P(A \cap B) - P(A \cap C) - P(B \cap C) + P(A \cap B \cap C).$$

EXAMPLES

1. A card is drawn from a deck of 52 cards and one wishes to know the probability that the card was either an ace or a diamond. These conditions are not mutually exclusive. The probability of drawing an ace is $P(E_1) = \frac{4}{52}$, of drawing a diamond is $P(E_2) = \frac{13}{52}$ and of drawing an ace of diamonds is

1.2 Probability—the Mathematics of Uncertainty

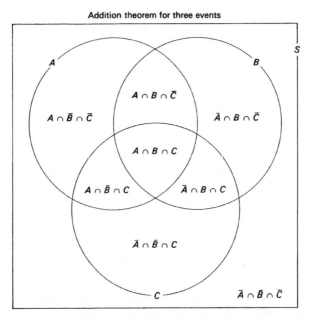

Area of $A + B + C = 3$ "circles" $- 3$ "ellipses" $+$ "triangle"
$P(A \cup B \cup C) = P(A) + P(B) + P(C) - P(A \cap B) - P(A \cap C) - P(B \cap C) + P(A \cap B \cap C)$

Figure 1B Venn diagram. **Comment:** Combining all elementary events in A, B, C ("circles"), the pairwise common events ("ellipses") have been counted twice, so we remove them once; but in doing this we removed the elementary events of $A \cap B \cap C$ ("triangle" in the middle) once too often, and so we have to add it.

$P(E_1 \cap E_2) = \frac{1}{52}$; thus we have $P(E_1 \cup E_2) = P(E_1) + P(E_2) - P(E_1 \cap E_2) = \frac{4}{52} + \frac{13}{52} - \frac{1}{52} = \frac{16}{52} = 0.308$.

2. Suppose the probability that it will rain is $P(E_1) = 0.70$, that it will snow is $P(E_2) = 0.35$, and that both events occur simultaneously is $P(E_1 \cap E_2) = 0.15$. Then the probability that it will rain, snow, or both is $P(E_1 \cup E_2) = P(E_1 \text{ or } E_2 \text{ or both}) = 0.70 + 0.35 - 0.15 = 0.90$ (cf. also Table 7, example 4).

▶ 1.2.3 Conditional probability and statistical independence

Two companies manufacture 70% and 30% respectively of the light bulbs on the market. On the average, 83 out of every 100 bulbs from the first company last the standard number of hours, while only 63 out of every 100 bulbs from the second company do so. Thus, out of every 100 light bulbs that reach the consumer, an average of $77 = [(0.83)(70) + (0.63)(30)]$ will have standard lifetimes; in other words the probability of buying a lightbulb with standard lifetime is 0.77. Now let us assume we have learned that the light bulbs a certain store carries were all manufactured by the first company. Then the probability of purchasing a light bulb which has a standard lifetime will be $\frac{83}{100} = 0.83$. The unconditional probability of buying a standard bulb is 0.77, while the **conditional probability**—conditioned on the knowledge that it was made by the first company—equals 0.83.

Two dice, when thrown in two separate places, lead to independent results. That events are **independent** means they do not mutually interact and are not jointly influenced by other events.

Assuming we toss a number of consecutive sixes with an unbiased die, the chance of getting any more sixes does not become negligible. It remains a constant $\frac{1}{6}$ for every toss. There is no need for the results of later tosses to balance off the preceding ones. An unbiased die, as well as the independence of individual tosses, is of course assumed, i.e., no preceding toss influences a subsequent one—the die is, for example, not deformed by the previous toss.

1. The probability of the event E_2, given the condition or assumption that the event E_1 has already occurred, [written $P(E_1|E_2)$], is called the conditional probability (see Figure 2A)

$$P(E_2|E_1) = \frac{P(E_1 \cap E_2)}{P(E_1)}, \qquad (1.10)$$

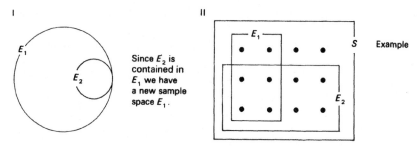

Figure 2A   Venn diagrams.

1.2 Probability—the Mathematics of Uncertainty

which is of course defined only for $P(E_1) \neq 0$; we have analogously

$$P(E_1|E_2) = \frac{P(E_1 \cap E_2)}{P(E_2)}, \qquad (1.10a)$$

for $P(E_2) \neq 0$. This leads to the **multiplication rule** or **multiplication theorem** for the simultaneous occurrence of E_1 and E_2:

$$P(E_1 \cap E_2) = P(E_1)P(E_2|E_1) = P(E_2)P(E_1|E_2) = P(E_2 \cap E_1). \qquad (1.11)$$

(1.11) gives the joint probability of E_1 and E_2, whether they are independent or not.

2. Two events are called **stochastically independent** ("stochastic" means: associated with random experiments and probabilities [cf., Sections 1.4.5, 3.2]) if

$$P(E_2|E_1) = P(E_2). \qquad (1.12)$$

In this case we have also

$$P(E_1|E_2) = P(E_1). \qquad (1.12a)$$

3. If E_1 and E_2 are stochastically independent, then so are (1) $\bar{E}_1$ and E_2, (2) E_1 and $\bar{E}_2$, or $P(E_2|E_1) = P(E_2|\bar{E}_1) = P(E_2)$ and $P(E_1|E_2) = P(E_1|\bar{E}_2) = P(E_1)$. Since there are more men (M) suffering from gout

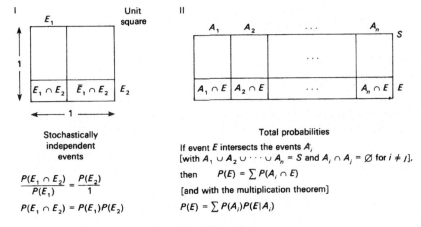

Figure 2B Venn diagrams.

than women (W), we have $P(G|M) > P(G|W)$. The **definition of stochastic independence** (see Figure 2BI) is a consequence of (1.11) and (1.12):

$$P(E_1 \cap E_2) = P(E_1) \cdot P(E_2). \tag{1.13}$$

EXAMPLE. Experiment: One die tossed twice or die I and die II tossed together; all tosses are stochastically independent. Sample space {1-1, 1-2, ..., 1-6; 2-1, ..., 5-6, 6-6}.
With

E_1 = {first toss [or die I] even}

  = {2, 4, 6} or three out of six {1, ..., 6} possible events, and

E_2 = {second toss [or die II] 2 or less}

  = {1, 2} or two out of six {1, ..., 6} possible events,

we have $P(E_1 \cap E_2) = P(E_1)P(E_2) = (\frac{3}{6})(\frac{2}{6}) = \frac{1}{6}$ or {2-1, 2-2; 4-1, 4-2; 6-1, 6-2} six of the 36 possible pairs.

For a composite event resulting from n mutually stochastically independent experiments with the outcomes E_i, $i = 1, 2, \ldots, n$, we find

$$P(E_1 \cap E_2 \cap \cdots \cap E_n) = P(E_1)P(E_2)\cdots P(E_n). \tag{1.14}$$

Writing this another way:

$P(E_1 \cap E_2 \cap \cdots \cap E_n)$
$= P(E_1)P(E_2|E_1)P(E_3|E_1 \cap E_2)\cdots P(E_n|E_1 \cap E_2 \cap \cdots \cap E_{n-1}).$

EXAMPLES

1. How large is the probability of getting three sixes simultaneously when three unbiased dice are tossed? $P = (\frac{1}{6})(\frac{1}{6})(\frac{1}{6}) = \frac{1}{216}$. In a long sequence of trials all three dice would show a six simultaneously in only one out of 216 tosses on the average (cf. also Table 7, Examples 1 and 2).

2. An unbiased die is tossed four times. What is the probability of getting a six at least once? Replace "a six at least once" by "no sixes." The probability of not getting a six with a single toss is $\frac{5}{6}$, with four tosses, it equals $(\frac{5}{6})^4$. Thus the probability of obtaining at least one six with four tosses is $1 - (\frac{5}{6})^4 = 0.518$, or a little larger than $\frac{1}{2}$. This predicts a profitable outcome for anyone who has patience, money, and an honest die and bets

1.2 Elements of Computational Probability

on the appearance of a six in four tosses. In the same way, for the case where a pair of dice is being tossed, one can ask the question of how many tosses are needed to render betting on a double six worthwhile. The probability of not getting a double six with one roll of the two dice is $\frac{35}{36}$, since 36 equals the number of possible outcomes 1-1, 1-2, ..., 6-6 of the roll. The probability of obtaining a double six at least once in a sequence of n tosses is again given by $P = 1 - (\frac{35}{36})^n$. P should be > 0.5, that is, $(\frac{35}{36})^n < 0.5$, so $n \log \frac{35}{36} < \log 0.5$ and hence $n > 24.6$. This last inequality follows from setting $n \log \frac{35}{36} = \log 0.5$ and solving for n:

$$n = \frac{\log 0.5}{\log(\frac{35}{36})} = \frac{0.6990 - 1}{\log 35 - \log 36} = \frac{9.6990 - 10}{1.5441 - 1.5563} = \frac{-0.3010}{-0.0122} = 24.6.$$

One would thus bet on the appearance of a double six if at least 25 tosses were allowed; the probability of tossing a double six is then greater than 50%.

The Chevalier de Méré acquired a substantial amount of money by betting that he would get at least one six in a sequence of four tosses of a die, and then lost it by betting that he would get at least one double six in a sequence of 24 tosses with two dice: $1 - (\frac{35}{36})^{24} = 0.491 < 0.5$.

> The exchange of letters between Pierre de Fermat (1601–1665) and Blaise Pascal (1623–1662), which had been requested by Chevalier de Méré in order to solve the problem mentioned above, established in 1654 a foundation for probability theory which was later developed by Jakob Bernoulli (1654–1705) into a mathematical theory of probability (Westergaard 1932, David 1963, King and Read 1963, Freudenthal and Steiner 1966, Pearson and Kendall 1970, Kruskal and Tanur 1978 (cited on p. 570), Pearson 1978; cf. end of Section 7.7) (cf. also pages 27, 59, 64, 123, and 567).

3. A certain bachelor insists that the girl of his dreams have a Grecian nose, Titian red hair, and a thorough knowledge of statistics. The corresponding probabilities are taken to be 0.01, 0.01 and 0.00001. Then the probability that the first young lady met (or any that is randomly chosen) exhibits the aforementioned properties is $P = (0.01)(0.01)(0.00001) = 0.000000001$ or exactly one in a billion. It is of course assumed that the three characteristics are independent of each other.

4. Three guns can shoot at the same airplane independently of one another. Each gun has a probability of $\frac{1}{10}$ to score a hit under the given conditions. What is the probability that an airplane is hit? In other words, the probability of at least one resulting hit is sought. Now the probability

Table 6

p	0.01						0.02						0.05			
n	1	5	10	15	30	50	2	5	10	15	30	50	2	5	10	15
P	0.010	0.049	0.096	0.140	0.260	0.395	0.040	0.096	0.183	0.261	0.455	0.636	0.098	0.226	0.401	0.537
p	0.10				0.20				0.30		0.50		0.75		0.90	
n	2	5	10	15	5	10	15	30	5	10	5	10	2	5	2	3
P	0.190	0.410	0.651	0.794	0.672	0.893	0.965	0.999	0.832	0.972	0.969	0.999	0.937	0.999	0.990	0.999

that no airplane is hit is $(\frac{9}{10})^3$. Thus the probability of at least one resulting hit is given by

$$P = 1 - \left(\frac{9}{10}\right)^3 = 1 - \frac{729}{1,000} = \frac{271}{1,000} = 0.271 = 27.1\%$$

$$\left(\text{cf. } P = 1 - \left[\frac{9}{10}\right]^{28} = 94.77\% \quad \text{or} \quad P = 1 - \left[\frac{1}{2}\right]^5 = 96.88\%\right).$$

Rule: The probability P of at least one successful result (hit) in n independent trials, given probability p for success in each trial, is given by

$$\boxed{P = 1 - (1 - p)^n \leqq np}$$

(cf. also p. 218). We list several examples in Table 6.

5. Four cards are drawn from a deck. What is the probability (a) that four aces turn up, and (b) that they all exhibit the same value? The probability of drawing an ace from a deck of cards is $\frac{4}{52} = \frac{1}{13}$. If the drawn card is replaced before the next card is picked, then the probability of obtaining two aces in two consecutive draws equals $(\frac{1}{13})(\frac{1}{13}) = \frac{1}{169}$. If the card drawn is not replaced, the probability comes to $(\frac{1}{13})(\frac{3}{51}) = \frac{1}{221}$. With replacement, the probability of a particular outcome is constant; without replacement it varies from draw to draw. Thus we have

$$\text{for (a):} \quad P = \frac{4}{52} \cdot \frac{3}{51} \cdot \frac{2}{50} \cdot \frac{1}{49} = \frac{24}{6,497,400} = \frac{1}{270,725} \simeq 3.7 \cdot 10^{-6},$$

$$\text{for (b):} \quad P = 13 \cdot \frac{4}{52} \cdot \frac{3}{51} \cdot \frac{2}{50} \cdot \frac{1}{49} = \frac{312}{6,497,400} = \frac{1}{20,825} \simeq 4.8 \cdot 10^{-5}.$$

6. 24 persons are chosen at random. How large is the probability that at least two persons have their birthday on the same day? It equals $P = 0.538$. It is assumed that the 365 days in a year are all equally likely as birthdays. We are interested in the event $\bar{E}$, "no 2 (from among n) persons have their birthday on the same day." For $\bar{E}$ there are then 365^n possible and

$$(365)(364)\cdots(365 - n + 1)$$

1.2 Elements of Computational Probability   37

favorable cases; i.e., the probability that in a group of 24 persons at least 2 persons have their birthday on the same day is equal to

$$P = P(E) = 1 - P(\bar{E}) = 1 - \frac{(365)(364)\cdots(342)}{365^{24}} = 0.5383.$$

In other words, a wager that out of 24 persons at least 2 celebrate their birthday on the same day would be profitable if repeated a large number of times, since out of 100 such wagers only 46 would be lost whereas 54 would be won. We have here ignored February 29; moreover we have not allowed for the fact that births are more frequent in certain months. The first lowers the probability while the last raises it.

For $n = 23$ we find $P = 0.507$; for $n = 30$, $P = 0.706$; and for $n = 50$, $P = 0.970$. Naus (1968) gives a table for the probability that two out of n persons ($n \leq 35$) have their birthdays within d days of each other ($d \leq 30$) [example: (1) $n = 7$, $d = 7$, $P = 0.550$; (2) $n = 7$, $d = 21$, $P = 0.950$; (3) $n = 15$, $d = 10$, $P = 0.999$] (cf. also Gehan 1968, Faulkner 1969, and Glick 1970).

Examples of conditional probability

1. An urn contains 15 red and 5 black balls. We let E_1 represent the drawing of a red ball, E_2 the drawing of a black ball. How large is the probability of obtaining first a red and then a black ball in two consecutive draws? The probability of drawing the red ball is $P(E_1) = \frac{15}{20} = \frac{3}{4}$. Without replacing the ball, another drawing is made. The probability of drawing a black ball, a red ball having been removed, is $P(E_2|E_1) = \frac{5}{19} \approx 0.26$. The probability of drawing a red and a black ball in two drawings without replacement is $P(E_1)P(E_2|E_1) = (\frac{3}{4})(\frac{5}{19}) = \frac{15}{76} \simeq 0.20$.

2. On the average ten percent of a population is, in a given period of time, stricken by a certain illness [$P(E_1) = 0.10$]. Of those stricken, 8% die as a rule [$P(E_2|E_1) = 0.08$]. The probability for this to occur, $P = 0.08$, is a conditional probability (condition: falling ill). The probability that a member of the population in question, in a given interval of time, contracts the illness and thereafter dies from this illness is thus

$$P(E_1 \cap E_2) = P(E_1)P(E_2|E_1) = (0.1)(0.08) = 0.008 = 0.8\%.$$

In medical terms, this would be stated: the morbidity of this illness is 10%, the lethality 8%, and the mortality rate 0.8%; that is, mortality = (morbidity) (lethality).

(p. 195)

Let us go even further. Suppose another disease infects 20% of the population (E_1); of these, 30% succumb to this disease in a certain interval of time (E_2); finally, 5% of those who have fallen ill die. The mortality is then given by $P(E_1 \cap E_2 \cap E)$ = $P(E_1)P(E_2|E_1)P(E_3|E_2) = (0.20)(0.30)(0.05) = 0.003 = 0.3\%$. No information about morbidity conditions (or about their age gradation) can be gained from clinical statistics without making reference to the population, since in the region served by the clinic, the group of people that could also have been afflicted by this illness (persons endangered) is usually unknown.

Table 7 This short survey table lists several probability formulas involving the independent events E_1 and E_2 with probabilities $P(E_1)$ and $P(E_2)$

Event	Probability	Example $P(E_1) = 0.10;\ P(E_2) = 0.01$
Both	$P(E_1) \cdot P(E_2)$	$P = 0.001$
Not both	$1 - P(E_1) \cdot P(E_2)$	$P = 0.999$
Either E_1 or E_2, not both	$P(E_1) + P(E_2) - 2\,P(E_1) \cdot P(E_2)$	$P = 0.108$
Either E_1 or E_2, or both	$P(E_1) + P(E_2) - P(E_1) \cdot P(E_2)$	$P = 0.109$
Neither E_1 nor E_2	$1 - P(E_1) - P(E_2) + P(E_1) \cdot P(E_2)$	$P = 0.891$
Both or neither	$(1 - P(E_1)) \cdot (1 - P(E_2)) + P(E_1) \cdot P(E_2)$	$P = 0.892$
E_1 but not E_2	$P(E_1) \cdot (1 - P(E_2))$	$P = 0.099$

Since one can speak of the probability of any event only under precisely specified conditions, every probability is, strictly speaking, a **conditional probability**. An unconditional probability cannot exist in the true sense of the word.

1.2.4 Bayes's theorem

Suppose $A_1, A_2, \ldots, A_n$ are mutually exclusive events. Let the union of all A_i be the certain event. Bayes's theorem is then as follows (see Figure 3): Assume that a random event E with $P(E) > 0$, which can occur only in combination with an event A_i, has already occurred. Then the probability that an event A_k occurs [Thomas Bayes: 1702–1761] is given by

$$P(A_k|E) = \frac{P(A_k \cap E)}{P(E)} = \frac{P(A_k)P(E|A_k)}{P(A_1)P(E|A_1) + \cdots + P(A_n)P(E|A_n)}$$

$$P(A_k|E) = \frac{P(A_k)P(E|A_k)}{\sum_{i=1}^{n} P(A_i)P(E|A_i)}. \tag{1.15}$$

Proof. The denominator equals $P(E)$, multiplication of (1.15) with $P(E)$ gives (1.11) $P(E)P(A_k|E) = P(A_k)P(E|A_k) = P(A_k \cap E)$. The theorem of total probabilities is given in Figure 2B, in Figure 3I, and in the summary at the end of this section.

1.2 Elements of Computational Probability

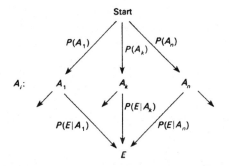

I. The probability of arriving at E is $P(E) = \sum_{i=1}^{n} P(A_i)P(E|A_i)$
II. Assume I arrives at E by way of A_k, then this probability is

$$P(A_k|E) = \frac{P(A_k)P(E|A_k)}{\sum_{i=1}^{n} P(A_i)P(E|A_i)}$$

Figure 3 Theorem of total probabilities (I) and Bayes's theorem (II).

Examples (see Table 8)

1. Two machines at some firm generate 10% and 90% respectively of the total production of a certain item. Assume the probability that the first machine (M_1) produces a reject is 0.01 and the probability that the second machine (M_2) does so is 0.05. What is the probability that an item randomly chosen from a day's output of daily production originated at M_1, given that the item is a reject? Let E be the event that an item is a reject, A_1 the event that it was produced by M_1, and A_2, that it can be traced to M_2, i.e., $P(M_1|\text{a reject}) = P(A_1|E)$:

$$P(A_1|E) = \frac{P(A_1) \cdot P(E|A_1)}{P(A_1) \cdot P(E|A_1) + P(A_2) \cdot P(E|A_2)},$$

$$P(A_1|E) = \frac{0.10 \cdot 0.01}{0.10 \cdot 0.01 + 0.90 \cdot 0.05} = \frac{1}{46} \simeq 0.022.$$

2. We assume there are two urns available. The probability of choosing urn I is $\frac{1}{10}$; for urn II it is then $\frac{9}{10}$. We suppose further that the urns contain black and white balls: in urn I 70% of the balls are black, in urn II 40% are black. What is the probability that a black ball drawn blindfolded came from urn I? Let E be the event that the ball is black, A_1 be the event that it is drawn from urn I, and A_2 be the event that it comes from urn II.

$$P(\text{from urn I}|\text{black}) = \frac{0.10 \cdot 0.70}{0.10 \cdot 0.70 + 0.90 \cdot 0.40} = 0.163.$$

This means that after many trials it is justified to conclude that in 16.3% of all cases in which a black ball is drawn, urn I was the source.

3. Let us assume that a chest x-ray, meant to uncover tuberculosis, properly diagnoses 90% of those afflicted with tuberculosis, i.e., 10% of those suffering from tuberculosis remain undetected in the process; for tuberculosis-free persons, the diagnosis is accurate in 99% of the cases, i.e., 1% of the tuberculosis-free persons are improperly diagnosed as being carriers of tuberculosis. Suppose, out of a large population within which the incidence of tuberculosis is 0.1%, a person is x-rayed and alleged to be afflicted with tuberculosis. What is the probability that this person has tuberculosis? Let E = the event that the x-ray gave a positive result, A_1 = the event that the person is afflicted with tuberculosis, and A_2 = the event that he is free of tuberculosis:

$$P(\text{afflicted with TB}|\text{pos. x-ray indication}) = \frac{0.001 \cdot 0.9}{0.001 \cdot 0.9 + 0.999 \cdot 0.01}$$

$$= 0.0826,$$

i.e., we find that of those diagnosed by x-ray as suffering from tuberculosis, only slightly over 8% are indeed so afflicted.

In a sequence of x-ray examinations one has to allow on the average for 30% incorrect negative results and 2% incorrect positive results (Garland 1959).

4. Four secretaries employed by an office file 40, 10, 30, and 20% of the documents. The probabilities that errors will be made in the process are 0.01, 0.04, 0.06, and 0.10. What is the probability that a misfiled document was misfiled by the third secretary?

$$P(\text{secretary No. 3}|\text{document misfiled})$$

$$= \frac{0.30 \cdot 0.06}{0.40 \cdot 0.01 + 0.10 \cdot 0.04 + 0.30 \cdot 0.06 + 0.20 \cdot 0.10}$$

$$= \frac{0.018}{0.046} = 0.391 \simeq 39\%.$$

Slightly over 39% of all misfiled documents! As an exercise, this computation should be carried out for each secretary, and the total result presented as a graph of the sort appearing as Table 8.

Bayesian methods require a prior distribution for the parameters. They then offer the possibility of incorporating prior information about the parameters and also of adding further information when it arrives. This is very important if **optimal decision making** is at stake. The choice of the prior distribution may cause trouble.

1.2 Elements of Computational Probability

Table 8 Summary of the first three examples illustrating Bayes's theorem: tree diagram with the associated "path weights" on the right

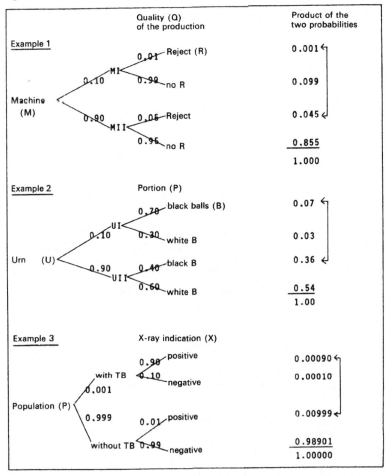

On the right we have $P(E_1 \cap E_2) = P(E_1)P(E_2|E_1)$ in all three cases. For Example 1, 0.001 = (0.10)(0.01) etc. The products joined by an arrow bracket enter Bayes's formula [Example 1: 0.001/(0.001 + 0.045) = 1/46].

More particulars on the Bayes theorem and on Bayesian methods can be found in Barnard (1967), Cornfield (1967, 1969), Schmitt (1969), de Groot (1970), Maritz (1970), Winkler (1972), Barnett (1982), Box and Tiao (1973), and Novik (1975) (cf. also Isaacs et al. [1974] in Section [2], and Martz and Waller [1982], and Tillman et al. [1982], both in Section [8:2d] of the bibliography).

The following list provides a summary of important formulas.

Probability

1. Axioms:

   I. $P(E) \geq 0$

   II. $P(S) = 1$

   III. If $E_1 \cap E_2 = \emptyset$, then $P(E_1 \cup E_2) = P(E_1) + P(E_2)$.

2. Complement (Co) and Partition (Pa):

   Co: If $E \cup \bar{E} = S$ and $E \cap \bar{E} = \emptyset$, then $P(\bar{E}) = 1 - P(E)$

   Pa: If subsets E_i form a partition of S, then $\sum_i P(E_i) = 1$.

3. Addition theorems:

   I. $P(E_1 \cup E_2) = P(E_1) + P(E_2) - P(E_1 \cap E_2)$

   II. $P(A \cup B \cup C) = P(A) + P(B) + P(C) - P(A \cap B)$
   $\qquad - P(A \cap C) - P(B \cap C) + P(A \cap B \cap C)$.

4. Definition of conditional probability. If $P(E_1) > 0$, then

   $$P(E_2|E_1) = \frac{P(E_1 \cap E_2)}{P(E_1)} = \frac{P(E_2 \cap E_1)}{P(E_1)}.$$

   Rewritten as multiplication theorem for arbitrary events

   $$P(E_1 \cap E_2) = P(E_1)P(E_2|E_1)$$
   $$P(A \cap B \cap C) = P(A)P(B|A)P(C|A \cap B),$$

   and so on.

5. Definitions of stochastical independence. If E_1 and E_2 are stochastically independent, then

   I. $P(E_2|E_1) = P(E_2|\bar{E}_1) = P(E_2)$ with $P(E_1) > 0$ and

   $\qquad P(E_1|E_2) = P(E_1|\bar{E}_2) = P(E_1)$ with $P(E_2) > 0$

   II. $P(E_1 \cap E_2) = P(E_1)P(E_2) > 0$.

   The 3 events A, B, C are (mutually) stochastically independent, if

   $P(A \cap B \cap C) = P(A)P(B)P(C), \qquad P(A \cap B) = P(A)P(B),$
   $P(A \cap C) = P(A)P(C), \quad$ and $P(B \cap C) = P(B)P(C).$

6. Theorem on total probabilities: If an arbitrary event E intersects the mutually exclusive and collectively exhaustive events A_i, then

   $$P(E) = \sum_i P(A_i \cap E) = \sum_i P(A_i)P(E|A_i).$$

7. **Bayes's theorem:** If n events A_i form a partition of the sample space and if event E can only occur in combination with one of the n events A_i, then for any event A_k, where k is an integer between 1 and n,

$$P(A_k|E) = \frac{P(A_k \cap E)}{P(E)} = \frac{P(A_k)P(E|A_k)}{\sum_i P(A_i)P(E|A_i)}.$$

▶ 1.2.5 The random variable

An event depending on random influences is called a stochastic event. A **random variable** maps the sample space into the real line. A random variable is a rule that associates to each possible outcome of an experiment a corresponding real number. In some cases the elementary outcomes are real numbers and hence are themselves random variables (e.g., the lifetime of a light bulb). In others, the outcomes have to be coded: e.g., for a toss of a coin the elementary outcomes are heads up (H), tails up (T). Then $X(T) = -1$, $X(H) = +1$ is a random variable; and $X(T) = a$, $X(H) = b$, a, b real numbers, $a \neq b$, is a random variable for the same sample space. If an experiment is performed in which the random variable X takes on a value x, then x **is called a realization** of X. The range of X is the set of **all** possible realizations of the random variable; the **sample** is an n-fold realization. The values of X are **real numbers**. By this we mean values which can be represented by integer $(2, -4)$, rational $(\frac{5}{12}, -\frac{31}{53})$, or irrational $(\sqrt{2}, \log 3, \pi, e)$ numbers. The probability of the event that X takes on any value in the interval from a to b is written $P(a < X < b)$. Accordingly $P(-\infty < X < \infty)$ is the certain event, because all the realizations of X lie on the real line. What is the probability that X assumes any value greater than c, $P(X > c)$? Since $P(X > c) + P(X \leq c) = 1$, it follows that for arbitrary real c

$$P(X > c) = 1 - P(X \leq c). \tag{1.16}$$

EXAMPLE. If X is the number that comes up when a fair die is rolled, then $P(X = 6)$ equals $\frac{1}{6}$, and

$$P(5 < X < 6) = 0, \quad P(5 \leq X < 6) = \tfrac{1}{6},$$
$$P(1 \leq X \leq 6) = 1, \quad P(5 < X \leq 6) = \tfrac{1}{6},$$
$$P(X > 1) = 1 - P(X \leq 1) = 1 - \tfrac{1}{6} = \tfrac{5}{6}.$$

Section 1.2.6 can be omitted in the first reading, since the material discussed is somewhat more complex and will not be assumed in the sequel.

1.2.6 The distribution function and the probability function

The probability distribution of a random variable specifies the probability with which the values of the variable will be realized. The probability distribution of the random variable X is uniquely defined by the **distribution function** [alternative terms: cumulative distribution function, cumulative frequency function, cumulative probability function]

$$F(x) = P(X \leqq x). \tag{1.17}$$

It specifies the probability that the random variable X assumes a value less than or equal to x. F is thus defined for all real numbers x and increases monotonically from 0 to 1. $F(x)$ is also referred to as the **cumulative frequency distribution**. The sequence $X_1, X_2, \ldots, X_n$ is a **random sample** of size n if each X has the same distribution and the n X's are stochastically independent.

EXAMPLE. The distribution function of the die experiment will serve as an example. The random variable is the number that comes up. The probability of each particular number that can turn up is $\frac{1}{6}$. $F(x)$ takes on the following values:

x	$x < 1$	$1 \leq x < 2$	$2 \leq x < 3$	$3 \leq x < 4$
$F(x)$	0	$\frac{1}{6}$	$\frac{1}{6} + \frac{1}{6} = \frac{1}{3}$	$\frac{1}{6} + \frac{1}{3} = \frac{1}{2}$

$4 \leq x < 5$	$5 \leq x < 6$	$x \geq 6$
$\frac{1}{6} + \frac{1}{2} = \frac{2}{3}$	$\frac{1}{6} + \frac{2}{3} = \frac{5}{6}$	$\frac{1}{6} + \frac{5}{6} = 1$

A so-called **step function** is obtained. It is constant over intervals which do not contain any values the random variable X can assume and jumps at the values x the random variable does assume. The size of the jump corresponds to the probability with which this value is realized. In our example this is $\frac{1}{6}$. One can plot this directly [Abscissa: x, the integers from 0 to 7; ordinate: $P(X \leq x)$, divided up in sixths from 0 to 1].

A random variable which assumes only finitely or countably many values as in the experiment with dice, is called a **discrete** random variable.

There is another way of describing the probability distribution of a random variable. As an example, it suffices in the die experiment to specify the probabilities with which the numbers that come up are rolled $[P(X = x_i) = \frac{1}{6}]$. In the case of discrete random variables we may consider the probability $f(x_i)$ associated with a value x_i a function of the point x_i. This function is called the **probability function** or **frequency function**. For

1.2 Elements of Computational Probability

discrete random variables the distribution function is found by simply summing up the probabilities $f(x_i)$. For continuous random variables, e.g. those whose values come about from measurements of length, weight, or velocity, one obtains the distribution function by integrating the co-called **probability density function**. In this way one likewise determines uniquely the distribution function. The probability function (or the probability density) and the distribution function are related in the following way:

---

1. For a discrete random variable

$$X: F(x) = \sum_{x_i \leq x} f(x_i); \qquad (1.18)$$

   $f(x_i)$ is the probability function.

2. For a continuous random variable

$$X: F(x) = \int_{-\infty}^{x} f(t) \, dt; \qquad (1.19)$$

   $f(t)$ is the probability density (∞ = infinity).

Note that $F(x)$ is a non-decreasing function with $F(-\infty) = 0$ and $F(\infty) = 1$.

As to the graphical meaning of the probability density function, one can say that for very small intervals dt the probability that X falls in the interval $(t, t + dt)$ is given approximately by the differential $f(t) \, dt$, which is also called a **probability element**:

$$\boxed{f(t) \, dt \simeq P(t < X \leq t + dt).} \qquad (1.20)$$

We have

$$\boxed{\int_{-\infty}^{+\infty} f(t) \, dt = 1} \qquad (1.21)$$

and, in particular,

$$\boxed{P(a < X \leq b) = F(b) - F(a) = \int_{a}^{b} f(t) \, dt.} \qquad (1.22)$$

The probability of the event $a < X \leq b$ is equal to the area under the probability density curve between $x = a$ and $x = b$ when the total area is equal to 1 [and the random variable is continuous].

We can now also define the discrete and the continuous random variable:

1. A random variable which can assume only finitely or countably many values is called **discrete**. We have called these values jump points. The distribution function associated with the random variable X has at most countably many jump points (points of discontinuity of the distribution function).

2. A random variable X is called **continuous** if the associated distribution function (1.17) can be written in the integral form (1.19). The values which the continuous variable X can assume form a **continuum**.

While the probability P of a particular event is usually meaningful in the case of a discrete distribution, the same cannot be said in the case of a continuous distribution (e.g., the probability that an egg weighs 50.00123 g); here probabilities of the sort where we say a variable X is $<a$ or $\geq a$ are of interest. For a continuous random variable, $P(X \leq x) = P(X < x)$ for all x. This is equivalent to stating that for every x the event $\{X = x\}$ has probability zero and that $P(a \leq X \leq b) = P(a < X < b)$. Since this book is aimed at practical application, we shall henceforth usually cease to distinguish between the (name of the) random variable X and their realizations x, the values that the random variable can assume [real numbers x assigned by the random variable X], and use x throughout.

Five important remarks

1. The mean μ or expected value $E(X) = \mu$ is given, for (a) discrete and (b) continuous random variables, by (a) $E(X) = \sum_i x_i P(x_i)$, (b) $E(X) = \int_{-\infty}^{\infty} x f(x)\, dx$, assuming the sum or integral is absolutely convergent ($\int |x| f(x)\, dx < \infty$).
2. For random variables with finite expected values, $E(X_1 + X_2) = E(X_1) + E(X_2)$ holds and, if the random variables are independent, then $E(X_1 X_2) = E(X_1) E(X_2)$ and $E(cX + k) = c E(X) + k = c\mu_x + k$ with c and k constant.
3. The expected value of the square of the deviation, $E[(X - \mu)^2] = E(X^2) - \mu^2$, is called the variance of X and is written $\mathrm{Var}(X)$ or σ^2; σ is called the standard deviation. Note that $\mathrm{Var}(cX + k) = c^2\, \mathrm{Var}(X)$.
4. For independent random variables: (1) The variance is additive: $\mathrm{Var}(X_1 + X_2) = \mathrm{Var}(X_1) + \mathrm{Var}(X_2)$. (2) Given n independent, identically distributed random variables and the mean $\bar{X} = (1/n)\sum_{i=1}^{n} X_i$, the variance of the sum is $\mathrm{Var}(\sum_{i=1}^{n} X_i) = n\, \mathrm{Var}(X)$ and the variance of the mean $\mathrm{Var}(\bar{X}) = (1/n^2) \sum_{i=1}^{n} \mathrm{Var}(X_i) = (1/n^2) n\, \mathrm{Var}(X) = (1/n)\, \mathrm{Var}(X) = (\sigma^2/n) = \sigma_{\bar{x}}^2$, which tends to zero with increasing n (cf. $\sigma_{\bar{x}}^2 = E\{[\bar{X} - E(\bar{X})]^2\}$).
5. For n independent, identically distributed random variables, with mean μ and finite variance σ^2, $(\bar{X} - \mu)\sqrt{n}/\sigma = (\bar{X} - \mu)/\sqrt{\sigma^2/n} = (\bar{X} - \mu)/\sigma_{\bar{x}}$ tends with increasing n to the standard normal distribution (central limit theorem).

The question of "stochastic independence of random variables" cannot be delved into further without presenting the accompanying theory; thus, a mere mention of the notion of independence in probabilistic calculations (Section 1.2.3) and a reference to theoretically oriented texts (cf. [1]) must suffice.

1.3 THE PATH TO THE NORMAL DISTRIBUTION

▶ 1.3.1 The population and the sample

Coins, dice, and cards are the implements of games of chance. Since every experiment which is affected by random influences and every random measurement can be represented approximately by an **urn model**, one can, instead of flipping an ideal coin, draw balls from an urn which contains exactly two completely identical balls, one of which is marked with an H and the other with a T (heads and tails). Instead of rolling a fair die we can draw balls from an urn containing exactly six balls, each ball distinguished by being marked with a 1, 2, 3, 4, 5, or 6. Instead of drawing a card from a deck we can draw balls from an urn containing exactly 52 numbered balls.

Several elementary observations are made in the following with regard to the urn model. We call the numbers 0, 1, 2, ... which index the balls **attributes**, and the attributes drawn from the urn **events**. Attributes can thus also be thought of as possible events "stored" in the urn. Attributes are fixed properties of statistical elements; these are also referred to as bearers of attributes, units of observation, or experimental units. It is the task of mathematical statistics to make inferences based on one or more samples from an urn, regarding the composition of the contents (the population) of this urn. These inferences are **probabilistic in nature**. Basic to statistical inference is the **replicability of the sample** (cf. Introduction).

The 52 balls form the **population** (cf., pages 5, 25). If the contents of the urn are thoroughly mixed, then every element of the population, that is, every ball, has the same chance of being drawn. We are referring to the random character of the sample, or **random sample** for short. The number of elements drawn—from 1 to a maximum of 51 balls—is called the sample size. The totality of possible samples forms the **sample space**. The relative frequency of the playing card attributes in the population is the **probability** of these attributes being drawn: for a ball corresponding to a single card it equals $\frac{1}{52}$, for the balls corresponding to the four kings it is $\frac{4}{52} = \frac{1}{13}$, for the balls corresponding to the spades it is $\frac{13}{52} = \frac{1}{4}$, and for the balls corresponding to the all black cards it comes to $\frac{26}{52} = \frac{1}{2}$.

In contrast to this, the relative frequency of the attributes in the sample is an **estimate of the probability** of these attributes. The more pronounced the "randomness" of the sample and the larger the sample, the better is the estimate. The **observations** are assumed to be independent. In finite populations, independence is obtained if after every single drawing the drawn element is returned to the population, which is then mixed again: this is the **urn model of sampling with replacement**. The number of samples can thus be regarded as **infinitely large**, an important concept in mathematical statistics.

> If the element drawn from a finite population is not replaced, we have an **urn model without replacement**: the composition of the residual population changes constantly. Every observation thus depends on the preceding ones. We are speaking of transmission of probability or probability linkage. Some models of this sort are presented in terms of so-called **Markov chains** (A. A. Markov, 1856–1922): Every observation depends only on one or on a finite number of observations directly preceding it. More detailed discussion of these and other classes of sequences of **random variables in time which are not assumed independent** can be found in the references in [8:1a] of the bibliography. Such **stochastic processes** are of considerable mathematical interest. Stochastic processes are at the foundation of many processes, theories, and models of the physics of small and elementary particles (Brownian motion of molecules, diffusion, quantum jumps of atoms, radioactive disintegration), of **demographic evolution** (the birth, death, and migration processes); of carcinogenesis and the development of cancer; of the spreading of epidemics; of the behavior of complex electronic equipment (while in operation, breakdown, repair); of queuing problems (theater ticket booths); and of the **prognosis models** for managerial problems. The theory of queues is also referred to as **service theory**: arriving units pass a **service location** where **queues** appear due to random fluctuations. Customers and sellers, ships and docks, patients and physicians exemplify the multitude of real life situations that can be treated as **service systems** (Saaty 1966).

We again turn to the urn model of sampling with replacement. The distribution of probabilities among the different attributes will be called the **probability distribution** or simply the distribution. Characteristic quantities of distributions will be called characteristics. **Characteristics** such as relative frequency, mean, or standard deviation, which refer to the population, are called **parameters**. Numerical values computed from samples are called **estimates or statistics**. Parameters will usually be denoted by Greek letters (Table 9 with the Greek alphabet is on the inner side of the front cover), and estimates by Latin letters. Thus the **symbols** for relative frequency, mean, and standard deviation relative to the population are π (pi), μ (mu), and σ (sigma); relative to the sample, they are $\hat{p}$, $\bar{x}$, and s. An object on which a measurement or observation may be made is termed a unit or element. The elements that form a population are almost always distinct from one another. Even if the differences are not initially "real," they are nevertheless introduced by the measurement. This difference within the population leads to the variation between samples, groups which have been chosen from the population (cf., also what was said in the Introduction and Section 1.1). In order to be able to make statements concerning the population, a sample is needed that is as similar as possible to the population, i.e., that is

1.3 The Path to the Normal Distribution

representative of the population. In such a sample every element of the population has the same chance of appearing in the sample.

By the **law of large numbers**, for a given population, the difference between the whole population and a sample (independent random variable assumed) decreases with increasing **sample size**; more precisely: $\overline{X}_n$ tends stochastically to μ as $n \to \infty$. This is called the **weak law of large numbers**, and states that $|\overline{X}_n - \mu|$ is usually small for n large, though for certain n it might be large; according to the so-called **strong law of large numbers**, however, the probability of this event is extremely small (cf., Section 1.3.6.1).

Beyond a certain sample size the **sampling error** becomes so small that a further increase in the size of the sample would no longer justify the additional expenditure.

Random samples are portions of a population from which they are drawn by a **random process**; they are representative of the population. A portion of a population can also be regarded as a representative sample if the partitioning or selection principle which determines the portion is in fact not random but is *independent* of the attributes under study.

Samples selected by some chance mechanism are known as **probability samples** if every item in the population has a known probability of being in the sample. In particular, if each item in the population has an equal chance of occurring in the sample, then the sample is known as a random sample. A **representative sample** is a probability sample arising ideally from perfect mixing in a population like a thimbleful of a mixture of miscible fluids or by some form of probability sampling (cf. Example A of the following section) to get (in enforced absence of selective forces) a mirror or miniature of the population.

One must be very cautious when generalizing on the basis of "samples which are obtained directly" and which cannot be regarded as random samples. Occasionally a generalization is possible through arbitrary augmentation of the available sample to an assumed imaginary population which will differ more or less from the population of interest, depending on the problem we are interested in.

▶ 1.3.2 The generation of random samples

The **lottery procedure** provides a method of generating authentic **random samples**. Suppose, for example, that from a population of 652 persons, two samples (I and II) of 16 elements each are to be chosen. Take 652 slips of paper, of which 16 are each marked with a I, and another 16 are each marked with a II; the other 620 slips remain blank. Now letting the 652 persons draw lots, we obtain the samples called for.

Tasks of this sort can be carried out more simply with the help of a table of random numbers; in Table 10 such numbers are recorded in groups of five digits. Suppose 16 random numbers less than 653 are needed. One reads the numbers from left to right in groups of three and records only those three digit numbers which are less than 653. As starting point for our search we might choose a point in the table that we mark blindfolded; assume it is the first digit of the third column of the sixth row from the bottom (first group of five digits is 17893). Then the sixteen numbers we seek will be 178, 317, 607, 436, 147, 601, 578 etc.

If from a population consisting of N elements a sample of n elements is to be chosen, the following procedure can be followed:

1. Assign to the N elements of the population the integers 1 through N. If $N = 600$, the individual elements are numbered from 001 to 600, each element being represented by a single three digit number.
2. Choose an arbitrary digit in the table as the starting point and read off the following digits, in groups of three if the population is a three digit number. (If the population is a z digit number, then groups are formed of z digits.)
3. If the number read off from the table is less than or equal to N, the population element so marked gets included in the random sample consisting of n elements. If the number read off is larger than N or if the corresponding element is already included in the sample, then this number will be disregarded and the process repeated until the n elements of the random sample are chosen.

Here are two further examples of using random digit tables to get special random samples.

(A) We require samples from the categories A, B, C, D with probabilities 0.60; 0.20; 0.16; 0.04 respectively, their sum being 1, and consider two successive digits as one of the 100 two-digit numbers 00, 01, ..., 99. Each has probability $\frac{1}{100}$ of occurring. Using the following correspondence

Random Number	00–59	60–79	80–95	96–99
Category	A	B	C	D

and obtaining the random digits 14, 93, 03, 65, ... from a table, for instance, we get the sample A, C, A, B, ...

(B) A doctor designing the comparison of a new treatment with the standard or old one has $2n$ patients available, grouped into n pairs, each consisting of two patients who have the disease in a similar state of advancement and who are similar in certain important factors such as age, sex, etc. One patient in each **matched pair** receives the new treatment, the other receives the old treatment. According to the "randomly selected" random digits the patient in the first column (I) is allocated to the new treatment if the digit is

1.3 The Path to the Normal Distribution

Table 9

	Names of matched pairs		Random digit	New treatment			
No.	I	II	0, 1, 2, 3, 4 5, 6, 7, 8, 9	Patient I Patient II			
1	K.H.	E.R.	No.	1	2	3	...
2	M.J.	L.S.					
3	U.S.	R.H.	New treatment Old treatment	E.R. K.H.	M.J. L.S.	R.H. U.S.	... ...
.	.	.					

0, 1, 2, 3, 4; otherwise it goes to the patient from II. The random digits are 6, 2, 9, ... Therefore in the first three pairs the patients E.R., M.J. and R. H. are to be given the new treatment.

One of the oldest methods of generating random numbers, more correctly termed **pseudorandom numbers**, which goes back to von Neumann, is the "middle-square" method. An s-digit number (s even) is squared, and the middle s digits of the $2s$-digit square are chosen [in case of a $(2s - 1)$-digit square, write it as a $2s$-digit square by putting a zero in front of it]. This number is in turn squared, etc.; the s-digit numbers then form sequences of pseudorandom numbers. As good random numbers there are also the nonperiodic decimal expansions of particular irrational numbers like $\sqrt{2}$, $\sqrt{3}$, $\pi = 3.141592653589793238462643...$, and most of the logarithms.

More on the meaning, generation, and examination of random numbers (cf., also Section 2.5.3) can be found in the survey article by Teichroew (1965); cf., also Good (1969), and the papers and books cited in [8:2g], e.g., Sowey (1978). The equally important random permutations (Moses and Oakford 1963, Plackett 1968) are here mentioned only briefly, (e.g., Sachs 1984).

Predictions

Unreliable forecasts of variables needed for long range plans, say in forestry or politics, are familiar to everyone. Since the future seems more uncertain than ever these days, its study (futurology)—questions of what could be, what is likely to be, and what shall be—is of increasing interest. Let us look briefly into several aspects of prediction.

There is a well-known and frequently used method of deducing facts about the population based on samples, as applied during elections, in official statistics, and in market and opinion studies, wherein the portion n_i/n of the elements with the attribute A_i, as determined from the sample, when multiplied by the total number of elements N in the population, yields the estimated value $\hat{N}_i = (n_i/n) \cdot N$. This is about the way a computer, being fed

Table 10 Random numbers (cf. Section 1.3.7)

44983	33834	54280	67850	96025	96117	00768	14821	69029	25453	48798	15486
89494	34431	44890	59892	79682	20308	82510	53609	13258	89631	80497	49167
54430	52632	94126	95597	48338	67645	44676	14730	22642	21919	21050	87791
96999	42104	34377	63309	82181	00278	28209	95629	75818	09043	48564	87355
87947	09427	32380	43636	58578	07761	28456	46570	11623	50417	37763	30136
30238	46126	85306	37114	22718	50584	92291	56575	24075	43889	40909	18741
22938	13073	32066	43098	75738	94910	15403	89151	73322	18370	90586	46115
89182	27750	63314	87302	49472	24885	79506	60638	07132	00908	92035	75518
16187	03303	40287	52435	23926	92544	54099	31497	06853	22864	72620	74169
21526	07401	30925	46148	20138	33874	56715	38424	38273	11361	15203	64912
42907	95158	27146	37012	43361	03173	97911	71313	44256	66609	42504	76799
21479	48265	01674	47274	56350	37512	14883	99673	62298	33948	32456	28675
90076	70233	76730	25043	16686	54737	57431	01786	20803	69465	37970	05673
93202	25355	93941	84434	22384	13240	93617	51549	28532	57150	77261	62643
46059	72208	90475	10341	39703	83224	37858	61657	04184	15597	29448	01922
38220	13972	86115	17196	24569	26820	66299	39960	02489	53079	72789	22562
82618	85756	51156	74037	12501	94162	42006	16135	82797	31296	93268	10104
07896	74085	59886	03051	78702	13402	74318	10870	72107	11550	61175	33345
95241	84360	13960	95736	43637	60399	19080	60261	11207	73065	48286	57057
53849	26578	39954	86726	91039	13884	25376	36880	02564	96978	62332	77321
72967	53031	47906	99501	27753	69946	66875	25601	30038	78786	65197	65283
87910	89260	66444	15979	83469	76952	50065	72802	70630	87336	16385	32784
10482	34277	40177	01081	57788	08612	39886	42234	04905	83274	22459	75032
68034	98561	46747	30655	41878	93610	51745	41771	61398	98154	61644	12405
80277	92450	60888	18689	45966	25837	70906	60733	11765	09293	70076	40751
59896	78185	60268	03650	36814	88460	34049	09111	64205	77930	32391	69076
78369	04163	77673	73342	78915	20537	06126	27222	17378	59359	00055	66780
23015	54261	95020	77705	81682	96907	37411	93548	87546	07687	47338	12240
55171	85448	12545	75992	08790	88992	69756	18960	85182	02245	11566	52527
58095	62204	69319	00672	96037	78680	98734	83719	40702	79038	68639	63329
19700	98193	37600	70617	58959	45486	58338	84563	62071	17799	96994	41635
12666	87597	23190	26243	36690	75829	71060	32257	15699	02654	83110	44278
66685	05344	71633	68536	18786	28575	08455	79261	49705	31491	25318	52586
72590	47283	45445	35611	98354	53804	45747	62026	13032	14048	16304	11959
30286	06434	50229	09070	44848	09996	77753	05018	92605	10316	07351	78020
87494	95585	25547	53500	45047	08406	66984	63390	48093	02366	05407	08325
32301	25923	76556	13274	39776	97027	56919	17792	09214	53781	90102	25774
70711	37921	54989	17828	60976	57662	61757	93272	09887	34196	98251	52453
36086	05468	41631	95632	78154	38634	47463	37514	24437	01316	04770	06534
37403	42231	17073	49097	54147	03656	14735	06370	18703	90858	55130	40869
41022	76893	29200	82747	97297	74420	18783	93471	89055	56413	77817	10655
70978	57385	70532	46978	87390	53319	90155	03154	20301	47831	86786	11284
19207	41684	20288	19783	82215	35810	39852	43795	21530	96315	55657	76473
50172	23114	28745	12249	35844	63265	26451	06986	08707	99251	06260	74779
43112	94833	72864	58785	53473	06308	56778	30474	57277	23425	27092	47759
64031	41740	69680	69373	73674	97914	77989	47280	71804	74587	70563	77813
92357	38870	73784	95662	83923	90790	49474	11901	30322	80254	99608	17019
79945	42580	86605	97758	08206	54199	41327	01170	21745	71318	07978	35440
48030	05125	70866	72154	86385	39490	57482	32921	33795	43155	30432	48384
80016	81500	48061	25583	74101	87573	01556	89184	64830	16779	35724	82103
34265	65728	89776	04006	06089	84076	12445	47416	83620	49151	97420	23689
82534	76335	21108	42302	79496	21054	80132	67719	72662	58360	57384	65406
72055	61146	82780	89411	53131	57879	39099	42715	24830	60045	23250	39847
26999	96294	20431	30114	23035	30380	76272	60343	57573	42492	47962	21439
01628	47335	17893	53176	07436	14799	78197	48601	97557	83918	20530	61565
66322	27390	73834	73494	21527	93579	20949	85666	25102	64733	93872	72698
96239	18521	67354	41883	58939	36222	43935	36272	47817	90287	91434	86453
10497	83617	39176	45062	63903	33862	14903	38996	60027	41702	78189	28598
69712	33438	85908	58620	50646	47857	96024	58568	67614	44370	40276	85964
51375	42451	76889	68096	80657	91046	95340	70209	23825	46031	45306	64476

1.3 The Path to the Normal Distribution 53

a few scattered results on the evening of election day, comes up with estimates of the election results (cf. Bruckmann 1966).

Long range predictions, or rather estimates of demographic evolution, energy requirements, developments in the labor market, etc. are generally made by means of trend analyses, less frequently (and subject to substantially greater bias and risk of false conclusions) by way of analogy (and intuition). Among the less well-known sources of error is the fact that a reasonable, generally acknowledged prediction can itself set events in motion which again influence the foretold events and thus the predicted trend ("forecast feedback"). The fear in 1955, which was confined to the USA, that there would be too few scientists in the years 1965-1970, proved groundless. The number of students increased by leaps and bounds (probably as a result of the gloomy prognosis). This example suggests the possible effect of predictions that are taken seriously (cf., Wold 1967, Polak 1970; also Theil 1966, Wagle 1966, Montgomery 1968, Cetron 1969).

If there is little or no reliable information at one's disposal, one can, after first viewing the potential developments, resort to interrogating a **panel of experts**. This is done by thinking the problem over thoroughly then submitting a carefully planned questionnaire to the experts. Potential prejudices, very subjective and exceptional opinions, can be eliminated to a great extent by "feeding back" to each participant the answers provided by all the others, so that each can once again reconsider his views ("feedback"). After running through several clarifications of this sort a common opinion is formed which may outweigh the individual views ("Delphi technique" see Martino 1970, as well as Linstone and Turoff 1975).

▶ 1.3.3 A frequency distribution

Statistics consist, in general, of measured or observed values of **continuous** (measurable) quantities (volume, time) or **discrete** (countable) quantities (number of children). **In addition to these quantitative attributes there are also qualitative attributes** (cf. also Section 1.4.8): **alternative attributes** (available, unavailable; gender), **dichotomous attributes** (synthetic alternative, e.g., stature: ≤ 175 cm, > 175 cm), **categorically joined attributes** (arbitrary sequences, e.g. occupations, eye colors, licence plate numbers [if several representations are possible at the same time, e.g., active hobbies, memberships, childhood illnesses, then we are dealing with coordinative attributes]), and **orderable** or **ordinal** attributes (natural sequences, e.g. rank sequences, gradings, pain intensities: $0, +, ++$).

The many results obtained in a survey are best tabulated and graphed. As an example, the classification of 200 infants according to the lengths of their bodies (range: 41-60 cm) leads to Table 11 with 7 classes [by a rule of thumb due to Sturges (1926), one would have as class number the

Table 11

Size class, in cm	Frequency	
	Absolute	Relative, in %
40 but less than 43	2	1.00
43 but less than 46	7	3.50
46 but less than 49	40	20.00
49 but less than 52	87	43.50
52 but less than 55	58	29.00
55 but less than 58	5	2.50
58 but less than 61	1	0.50
Total	200	100

number $k \simeq 1 + 3.32 \log n$, i.e., $1 + 3.32 \log 200 = 1.3 + (3.32)(2.30) = 8.6$, so that k could be chosen to equal either 8 or 9].

In this example the measurements were partitioned into an odd number of classes, thus creating a middle class. To prevent ambiguity the lower class limit was included in the class, while the upper class limit was not. For example, a child 52 cm tall has been included in the class of at least 52 cm but less than 55 cm. For the class interval "at least a but less than b" one writes $a \leq x < b$ (cf., Table 1, Section 0.1).

Compilation of data

The problem of how the statistical source material (primary statistical survey) was gathered—by written questioning (questionnaire), oral questioning (interview), or observation—will not be dealt with here. We only note that while it is almost impossible to eliminate deliberately or unwittingly false answers to questioning, in contrast errors in observations can usually be detected. Our material in the above example—the physical size of newborn infants—is compiled at every maternity clinic. Since these data were not compiled through a particular survey and serve statistical purpose only secondarily, they can be spoken of as secondary statistical material. To **prepare** statistical material, one uses listing procedures, point diagram procedures, or the filing procedures. For **listing procedures** a counting list is needed. Every measured child gets a (vertical) slash in the corresponding class on the list. More than four slashes per class are arranged in groups of five to facilitate counting.

Instead of the counting list one can also use millimeter paper or other squared paper, laying out the scale on the horizontal and plotting the individual measurements as points above the corresponding values. When the elements are all plotted on the point diagram, the class limits can be marked off by vertical lines. In this graphical procedure of partitioning into classes, points that lie on the boundary line are distributed between the adjacent classes.

1.3 The Path to the Normal Distribution

For punched card files and other files, especially in survey projects using raw data from interviews or questionnaires, Sonquist and Dunkelberg (1977) give an excellent overview with details for data collection and **data management operations** and provide examples and useful **checklists**.

More on planning an investigation (1), survey sampling (2), errors (3), data processing (4), and writing the report (5) may be found on pages 565/566 (1); 197, 245, 246 and 614/615 [8:3a] (2); 26 above, 67 above, 195/210, 393/395 (3); 572/573 [4] (4); 566 below (5). For experiments see Chapter 7 and especially Hahn (1977, cited in [8:7b] on page 640). For thinking with models see Saaty and Alexander (1981) [cf., also Box et al. 1978, cited in [8:1] on page 569].

Now back to our compilation on the newborn. On the right, beside the class decomposition in Table 11, it is indicated how many cases (the absolute frequency, the occupation number) or what fractions thereof (the relative frequency) fall into the individual classes.

A set of all the various values that individual observations may have and the frequency of their occurrence is called a **frequency distribution**. If plotted in the form of rectangles whose bases are equal to the class width and whose areas are proportional to the absolute or relative frequencies we have a **histogram** (cf., also p. 107). It gives an idea of the shape of an empirical distribution. If the abscissa is used for time intervals, e.g., produced cars per 3 years, a histogram results.

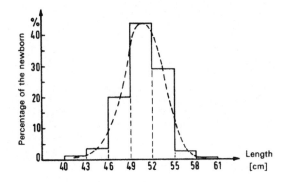

Figure 4 Frequency distribution of Table 11.

Table 12

Length in cm below	Cumulative frequency	
	absolute	per cent
43	2	1.00
46	9	4.50
49	49	24.50
52	136	68.00
55	194	97.00
58	199	99.50
61	200	100

The histogram is given in Figure 4. Here the percentage of the newborn is represented by the area of the rectangle drawn above the class width. Connecting the midpoints, we obtain a polygonal path. The finer the partition, the better the approximation by a curve. Not infrequently these curves are bellshaped and somewhat unsymmetric.

If the number of newborn whose body length is less than 49 cm is of interest, then $2 + 7 + 40 = 49$ infants or $1.00\% + 3.50\% + 20\% = 24.50\%$ can be read off from the table. If this calculation is carried out for different upper class limits, we get the cumulative table (Table 12) corresponding to the frequency distribution. The stepwise summation of frequencies yields the so-called **cumulative frequency distribution**; if we plot the cumulative frequency distribution (y-axis) against upper class limits (x-axis) and connect the points so determined by straight lines, we obtain a polygonal path. On refining the partitioning into classes, it can be well approximated by a monotonically nondecreasing, frequently S-shaped, curve (Figure 5) (cf., e.g., Sachs 1984, pp. 23–26).

The cumulative frequency distribution allows us to estimate how many elements are less than x cm in length, or what percentage of elements is smaller than x. **Cumulative frequency curves can be transformed into straight lines by a distortion of the ordinate scale.** A straight equalization line is drawn through the 50% point of the S-curve; for certain percentage values the points of the S-curve are then projected vertically onto the equalization line and the projected points transferred horizontally to the new ordinate axis.

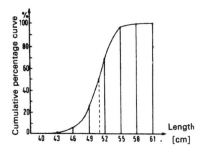

Figure 5 Percentage cumulative frequency curve or cumulative percentage curve of the body length of the newborn; the (not expressed in %) relative cumulative frequency curve is the empirical distribution function, ranging from 0 to 1.

1.3 The Path to the Normal Distribution

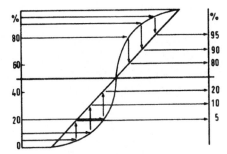

Figure 6 Flattening out the cumulative frequency curve into a straight line.

If the bell-shaped curve (and hence also the cumulative percentage curve derived from it) is **symmetric**, then all points $(50 \pm p)\%$ are situated symmetrically with respect to the 50% point of the equalization line (Figure 6; cf., also Figure 15, Section 1.3.7).

▶ 1.3.4 Bell-shaped curves and the normal distribution

Quantities which are essentially based on a **counting process**, which, by their very nature, can assume only integral values, form **discrete** frequency distributions, i.e., the associated stochastic variable can take on only integral values. The number of children born to a woman or the number of rejects in an output are examples for this situation. However, we wish in the following to study **continuous** random variables instead, that is, variables that are essentially based on a measuring process and that can take on every value, at least in a certain interval. Examples of this are the weight of a person, the size of his body (body length), and his age (time). Finely graduated discrete quantities like income can in practice be treated as continuous quantities. On the other hand, a continuous characteristic is often partitioned into classes, as when newborns are grouped according to length, which thereby becomes a discrete quantity.

If we keep in mind that every measurement consists basically in a comparison and that every measured value lies within some interval or on its boundary, then "ungrouped data" are in fact data that become classified in the course of measurement. The rougher the measuring process, the more evident this grouping effect becomes. **"Grouped data" in the usual sense are actually classified twice; first when they are measured, second when they are prepared for evaluation.** The classification induced by the "defective" measurement, not being due to the random variables, is generally neglected. There are thus no random variables that can assume, in the strong sense, every value in an interval, although in many cases such variables represent an appropriate idealization.

If one constructs a frequency distribution for a continuous quantity on the basis of observed values, it generally exhibits a more or less characteristic, frequently quite symmetric, bell-shaped form. In particular, the results of repeated measurements—say the length of a match or the girth of a child's head—often display this form.

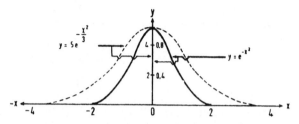

Figure 7 Bell-shaped curves.

A typical bell-shaped curve is given by the equation $y = e^{-x^2}$. More generally such curves are represented by

$$y = ae^{-bx^2} \tag{1.23}$$

(with $a, b > 0$). In Figure 7 curves are shown with $a = b = 1$ and with $a = 5$ and $b = \frac{1}{3}$: increasing a causes y to increase (for fixed x), and the curve rises proportionally; a reduction of b produces a flattening of the curve.

Many frequency distributions can be represented approximately by curves of this sort with appropriately chosen a and b. In particular, the distribution of a random measuring error or random error for repeated measurements (n large) of physical quantities exhibits a particular symmetric bell shape, with the typical maximum, the curve falling off on both sides, and large deviations from the measured value being extraordinarily rare. This distribution will be referred to as the **error law** or **normal distribution**. (Here the word "normal" has no connotations of "ideal" or "frequently found.") Before we delve into it further, let us give a short outline of its general significance. Quetelet (1796–1874) found that the body lengths of soldiers of an age group apparently follow a normal distribution. To him it was the distribution of the error that nature made in the reproduction of the ideal average man. The Quetelet school, which regarded the error law of de Moivre (1667–1754), Laplace (1749–1827), and Gauss (1777–1855) as a kind of natural law, also spoke of "homme moyen" with his "mean inclination toward suicide," "mean inclination toward crime," and so on. The number of rays in the tail fins of flounder is practically normally distributed. However, the majority of the unimodal distributions that we encounter in our environment deviate somewhat from a normal distribution or follow it only roughly.

1.3 The Path to the Normal Distribution

The normal distribution should properly be referred to as de Moivre's distribution. De Moivre discovered it and recognized its privileged position (Freudenthal and Steiner 1966 [see also Sheynin 1979]).

The primary significance of the de Moivre distribution lies in the fact that a **sum of many independent, arbitrarily distributed random variables is approximately normally distributed** and that the larger their number, the better the approximation. This statement is called the central limit theorem. It is on the basis of this theorem that very many sampling distributions can, for sufficiently large sample size, be approximated by this distribution and that for the corresponding test procedures the tabulated limits of the normal distribution suffice.

The normal distribution is a mathematical model with many favorable statistical properties and can be viewed as a **basic tool of mathematical statistics**. Its fundamental significance is based on the fact that random variables observed in nature can often be interpreted as superpositions of many individual, mutually more or less independent, influences, and thus as sums of many individual mutually independent random variables. One can easily produce an example: let dry sand run through a funnel into the space between two parallel vertically placed glass walls; an approximately normal distribution will appear on the glass panes. The occurrence of a de Moivre distribution is thus to be expected if the variables of the distribution considered are determined by the simultaneous effects of many mutually independent and equally influential factors, if the observed elements were randomly chosen and if a very large number of measurements or observations are available.

We now examine this distribution more closely (Figure 8). The ordinate y, which represents the height of the curve for every point on the x-scale, is the so-called **probability** density of the respective x-value. The probability density has its maximum at the mean, decreasing exponentially.

The probability density of the normal distribution is given by

$$y = f(x) = f(x|\mu, \sigma) = \frac{1}{\sigma\sqrt{2\pi}} e^{-\frac{1}{2}[(x-\mu)/\sigma]^2} \qquad (1.24)$$

$(-\infty < x < \infty, \quad -\infty < \mu < \infty, \quad \sigma > 0)$.

The symbol ∞ denotes infinity.

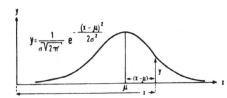

Figure 8   A normal curve.

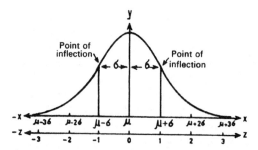

Figure 9 Normal distribution with standard deviation and inflection points. Relation between X and Z (transformation from the variable x to the standard normal variable Z): $Z = (X - \mu)/\sigma$.

Here x is an arbitrary value on the abscissa, y the corresponding ordinate value [y is a function of x: $y = f(x)$], σ the standard deviation of the distribution, and μ the mean of the distribution; π and e are mathematical constants with the approximate values $\pi = 3.141593$ and $e = 2.718282$. The right side of this formula involves both parameters μ and σ, the variable x, and both constants.

As indicated by the formula (1.24), the normal distribution is **fully** characterized by the parameters μ and σ. The mean μ fixes the location of the distribution along the x-axis. The standard deviation σ determines the shape of the curve (cf. Figure 9): the larger σ is, the flatter is the curve (the wider is the curve and the lower is the maximum).

Further properties of the normal distribution:

1. The curve is symmetric about the line $x = \mu$: it is symmetric with respect to μ. The values $x' = \mu - a$ and $x'' = \mu + a$ have equal density and thus the same value y.
2. The **maximum** of the curve is $y_{\max} = 1/(\sigma\sqrt{2\pi})$, and for $\sigma = 1$ it has the value $0.398942 \simeq 0.4$ (cf. Table 20). y tends to zero for very large positive x ($x \to \infty$) and very large negative x ($x \to -\infty$): the x-axis plays the role of an asymptote. Very extreme deviations from the mean μ exhibit so tiny a probability that the expression "**almost impossible**" seems appropriate.
3. The standard deviation of the normal distribution is given by the **abscissa of the inflection point** (Figure 9). The ordinate of the inflection point lies at approximately $0.6y_{\max}$. About $\frac{2}{3}$ of all observations lie between $\mu - \sigma$ and $\mu + \sigma$.
4. For large samples, approximately 90% of all observations lie between -1.645σ and $+1.645\sigma$. The limits -0.674σ and $+0.674\sigma$ are referred to as **probable deviations**; 50% of all observations lie in this interval [cf. the remarks on page 325 above].

Since μ and σ in the formula for the probability density of the normal distribution can assume arbitrary values (the deviation being subject to the condition $\sigma > 0$), infinitely many normally distributed collections with different distributions are possible. Setting $(X - \mu)/\sigma = Z$ in (1.24), where X depends on the scale, and Z is dimensionless, we obtain the unique **standardized normal distribution with mean zero and standard deviation one** [i.e.,

1.3 The Path to the Normal Distribution

since $f(x)\,dx = f(z)\,dz$, (1.24) goes over into (1.25a)]. This is plotted in Figure 10.

The normal distribution is usually abbreviated to $N(\mu, \sigma)$ or $N(\mu, \sigma^2)$, and the standard normal distribution correspondingly to $N(0, 1)$:

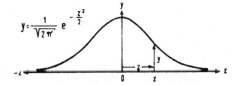

Figure 10 The standard normal curve.

The **standardized normal distribution**—y is here a function of the **standard normal variable** Z—is then defined by the probability density (cf. Section 1.3.6.7, Table 20, p. 79):

$$y = f(z) = \frac{1}{\sqrt{2\pi}} e^{-z^2/2} \simeq 0.3989 e^{-z^2/2} \simeq 0.4(0.6)^{z^2}, \quad -\infty < z < \infty,$$

(1.25abc)

with the distribution function $F(z) = P(Z \leq z)$

$$F(z) = \frac{1}{\sqrt{2\pi}} \int_{-\infty}^{z} e^{-v^2/2}\,dv \qquad (1.26)$$

[cf., Table 13, which lists $P = 1 - F(z)$ for $0 \leq z \leq 5.2$ or $P = P(Z \geq z) = 1 - F(z|0;1) = 1 - P(Z \leq z) = 1 - P(Z \leq (x - \mu)/\sigma) = 1 - P(X \leq x) = 1 - F(x|\mu; \sigma)$].

For every value of z one can read off from Table 13 the probability corresponding to the event that the random variable Z takes on values greater than z (examples in Section 1.3.6.7).

Two facts are important:

1. The total probability under the standard normal curve is one: this is why the equation for the normal distribution involves the constants $a = 1/\sqrt{2\pi}$ for $b = \frac{1}{2}$ (cf. $y = ae^{-bz^2}$).
2. The standard normal distribution is symmetric.

Table 13 indicates the "right tail" probabilities, namely the probabilities for z to be exceeded [$P(Z \geq z)$; see Figure 11]. For example, to the value $z = 0.00$ corresponds the probability $P = 0.5$, i.e., to the right of the mean lies half the area under the curve; for $z = 1.53$ we get $P = 0.0630 = 6.3\%$, i.e., to the right of $z = 1.53$ there lies 6.3% of the total area. The (cumulative) distribution function is $P(Z \leq z) = F(z)$, e.g., $F(1.53) = P(Z \leq 1.53) = 1 - 0.0630 = 0.937$. $F(1.53)$ is the cumulative probability,

Table 13 Area under the standard normal distribution curve from z to ∞ for the values $0 \leq z \leq 5.2$, i.e., the probability that the standard normal variable Z takes on values $\geq z$ [symbolically $P(Z \geq z)$] (taken from Fisher and Yates (1963), p. 45). Example: $P(Z \geq 1.96) = 0.025$

P-values for the one sided *z*-test

For the two sided *z*-test the tabulated *P*-values have to be doubled

z	0.00	0.01	0.02	0.03	0.04	0.05	0.06	0.07	0.08	0.09
0.0	0.5000	0.4960	0.4920	0.4880	0.4840	0.4801	0.4761	0.4721	0.4681	0.4641
0.1	0.4602	0.4562	0.4522	0.4483	0.4443	0.4404	0.4364	0.4325	0.4286	0.4247
0.2	0.4207	0.4168	0.4129	0.4090	0.4052	0.4013	0.3974	0.3936	0.3897	0.3859
0.3	0.3821	0.3783	0.3745	0.3707	0.3669	0.3632	0.3594	0.3557	0.3520	0.3483
0.4	0.3446	0.3409	0.3372	0.3336	0.3300	0.3264	0.3228	0.3192	0.3156	0.3121
0.5	0.3085	0.3050	0.3015	0.2981	0.2946	0.2912	0.2877	0.2843	0.2810	0.2776
0.6	0.2743	0.2709	0.2676	0.2643	0.2611	0.2578	0.2546	0.2514	0.2483	0.2451
0.7	0.2420	0.2389	0.2358	0.2327	0.2296	0.2266	0.2236	0.2206	0.2177	0.2148
0.8	0.2119	0.2090	0.2061	0.2033	0.2005	0.1977	0.1949	0.1922	0.1894	0.1867
0.9	0.1841	0.1814	0.1788	0.1762	0.1736	0.1711	0.1685	0.1660	0.1635	0.1611
1.0	0.1587	0.1562	0.1539	0.1515	0.1492	0.1469	0.1446	0.1423	0.1401	0.1379
1.1	0.1357	0.1335	0.1314	0.1292	0.1271	0.1251	0.1230	0.1210	0.1190	0.1170
1.2	0.1151	0.1131	0.1112	0.1093	0.1075	0.1056	0.1038	0.1020	0.1003	0.0985
1.3	0.0968	0.0951	0.0934	0.0918	0.0901	0.0885	0.0869	0.0853	0.0838	0.0823
1.4	0.0808	0.0793	0.0778	0.0764	0.0749	0.0735	0.0721	0.0708	0.0694	0.0681
1.5	0.0668	0.0655	0.0643	0.0630	0.0618	0.0606	0.0594	0.0582	0.0571	0.0559
1.6	0.0548	0.0537	0.0526	0.0516	0.0505	0.0495	0.0485	0.0475	0.0465	0.0455
1.7	0.0446	0.0436	0.0427	0.0418	0.0409	0.0401	0.0392	0.0384	0.0375	0.0367
1.8	0.0359	0.0351	0.0344	0.0336	0.0329	0.0322	0.0314	0.0307	0.0301	0.0294
1.9	0.0287	0.0281	0.0274	0.0268	0.0262	0.0256	0.0250	0.0244	0.0239	0.0233
2.0	0.02275	0.02222	0.02169	0.02118	0.02068	0.02018	0.01970	0.01923	0.01876	0.01831
2.1	0.01786	0.01743	0.01700	0.01659	0.01618	0.01578	0.01539	0.01500	0.01463	0.01426
2.2	0.01390	0.01355	0.01321	0.01287	0.01255	0.01222	0.01191	0.01160	0.01130	0.01101
2.3	0.01072	0.01044	0.01017	0.00990	0.00964	0.00939	0.00914	0.00889	0.00866	0.00842
2.4	0.00820	0.00798	0.00776	0.00755	0.00734	0.00714	0.00695	0.00676	0.00657	0.00639
2.5	0.00621	0.00604	0.00587	0.00570	0.00554	0.00539	0.00523	0.00508	0.00494	0.00480
2.6	0.00466	0.00453	0.00440	0.00427	0.00415	0.00402	0.00391	0.00379	0.00368	0.00357
2.7	0.00347	0.00336	0.00326	0.00317	0.00307	0.00298	0.00289	0.00280	0.00272	0.00264
2.8	0.00256	0.00248	0.00240	0.00233	0.00226	0.00219	0.00212	0.00205	0.00199	0.00193
2.9	0.00187	0.00181	0.00175	0.00169	0.00164	0.00159	0.00154	0.00149	0.00144	0.00139

z	P	z	P	z	P	z	P	z	P
0.25	0.4012937	2.9	0.0018658	3.5	0.0002326	4.1	$207 \cdot 10^{-7}$	4.7	$13 \cdot 10^{-7}$
0.5	0.3085375	3.0	**0.0013499**	3.6	0.0001591	4.2	$133 \cdot 10^{-7}$	4.8	$8 \cdot 10^{-7}$
1.0	**0.1586553**	3.1	0.0009676	3.7	0.0001078	4.3	$82 \cdot 10^{-7}$	4.9	$5 \cdot 10^{-7}$
1.5	0.0668072	3.2	0.0006871	3.8	$723 \cdot 10^{-7}$	4.4	$54 \cdot 10^{-7}$	5.0	$3 \cdot 10^{-7}$
2.0	**0.0227501**	3.3	0.0004834	3.9	$481 \cdot 10^{-7}$	4.5	$34 \cdot 10^{-7}$	5.1	$2 \cdot 10^{-7}$
2.5	0.0062097	3.4	0.0003369	4.0	$317 \cdot 10^{-7}$	4.6	$21 \cdot 10^{-7}$	5.2	$1 \cdot 10^{-7}$

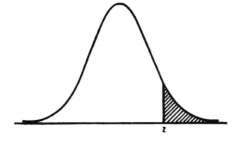

Figure 11 The portion A (shaded) of the area lying to the right of a certain value z. The portion of the area lying to the left of z is equal to $1 - A$, where A represents the values in Table 13 determined by z.

1.3 The Path to the Normal Distribution

Table 14 Values of the standard normal distribution (cf., also Table 43, Section 2.1.6) two sided: $P(|Z| \geq z)$, one sided: $P(Z \geq z)$

z	P two sided	P one sided	z	P two sided	P one sided
0.67448975	0.5	0.25	3.48075640	0.0005	0.00025
0.84162123	0.4	0.2	3.71901649	0.0002	0.0001
1.03643339	0.3	0.15	3.89059189	0.0001	0.00005
1.28155157	0.2	0.1	4.26489079	0.00002	0.00001
1.64485363	0.1	0.05	4.41717341	0.00001	0.000005
1.95996398	0.05	0.025	4.75342431	$2 \cdot 10^{-6}$	$1 \cdot 10^{-6}$
2.32634787	0.02	0.01	4.89163848	$1 \cdot 10^{-6}$	$5 \cdot 10^{-7}$
2.57582930	0.01	0.005	5.19933758	$2 \cdot 10^{-7}$	$1 \cdot 10^{-7}$
2.80703377	0.005	0.0025	5.32672389	$1 \cdot 10^{-7}$	$5 \cdot 10^{-8}$
3.09023231	0.002	0.001	5.73072887	$1 \cdot 10^{-8}$	$5 \cdot 10^{-9}$
3.29052673	0.001	0.0005	6.10941020	$1 \cdot 10^{-9}$	$5 \cdot 10^{-10}$

or the integral, of the normal probability function from $-\infty$ up to $z = 1.53$. Table 13 is supplemented by Table 14 and by Table 43 (Section 2.1.6).

The probability $P(Z \geq z)$ is easily approximated by

$$\tfrac{1}{2}[1 - \sqrt{1 - e^{-2z^2/\pi}}].$$

EXAMPLE

$$P(Z \geq 1) \approx \tfrac{1}{2}[1 - \sqrt{1 - 2.7183^{-2(1)^2/3.142}}]$$

$$\approx \tfrac{1}{2}[1 - \sqrt{1 - 0.529}]$$

$$\approx 0.157 \quad \text{(exact value: 0.159)}.$$

Better approximations to $P(Z \geq z)$ are given in Page (1977).

NORMAL DISTRIBUTION CURVE WITH THE SAME AREA AS A GIVEN HISTOGRAM
Fitting a normal curve to a histogram of absolute frequencies is easily done with the help of Table 20, n, $\bar{x}$ and s of the sample and

$$\hat{y} = \frac{bn}{s}\left[\frac{1}{\sqrt{2\pi}} e^{-((x-\bar{x})/s)^2/2}\right] = \frac{bn}{s} f(z)$$

with $z = (x - \bar{x})/s$ and class width b; $f(z)$ is found from the table and $\hat{y}$ is the height of the curve for a histogram of total area bn.

In the analysis of sampling results, reference is frequently made to the following regions:

$\mu \pm 1.96\sigma$	or	$z = 1.96$	with	95% of the total area,
$\mu \pm 2.58\sigma$	or	$z = 2.58$	with	99% of the total area,
$\mu \pm 3.29\sigma$	or	$z = 3.29$	with	99.9% of the total area,

$\mu \pm 1\sigma$	or	$z = \pm 1$	with	68.27% of the total area
$\mu \pm 2\sigma$	or	$z = \pm 2$	with	95.45% of the total area
$\mu \pm 3\sigma$	or	$z = \pm 3$	with	99.73% of the total area.

(Fig. 12)

A deviation of more than σ from the mean is to be expected about once in every three trials, a deviation of more than 2σ only about once in every 22 trials, and a **deviation of more than 3σ only about once in every 370 trials;** in other words, the probability that a value of x differs in absolute value from the mean by more than 3σ is **substantially less than** 0.01 (see Figure 12):

$$P(|X - \mu| > 3\sigma) = 0.0027.$$

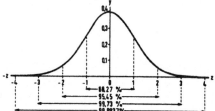

Figure 12 Portions of area of the standard normal distribution.

Because of this property of the normal distribution, the co-called **three sigma rule** used to be frequently applied; the probability that the absolute difference between an (at least approximately) normally distributed variable and its mean is greater than 3σ, is less than 0.3%.

For **arbitrary distributions,** the inequality of Bienaymé (1853) and Chebyshev (1874) holds: The probability that the absolute difference between the variable and its mean is greater than 3σ (in general: $\geq k\sigma$), is less than $1/3^2$ (in general, $\leq 1/k^2$) and hence less than 0.11:

$$P(|X - \mu| \geq 3\sigma) \leq \tfrac{1}{9} = 0.1111; \tag{1.27a}$$

in general,

$$P(|X - \mu| \geq k\sigma) \leq \frac{1}{k^2} \quad \text{with } k > 0, \tag{1.27}$$

i.e., in order to attain the 5% threshold one must specify 4.47σ, since $1/4.47^2$ is approximately equal to 0.05.

For **symmetric unimodal distributions, the sharper inequality due to Gauss** (1821 [see, e.g., Sheynin 1979, pp. 41/42]) **applies**:

$$P(|X - \mu| \geq k\sigma) \leq \frac{4}{9k^2} \quad \text{with } k > 0, \tag{1.28}$$

and thus the probability for

$$\boxed{P(|X - \mu| \geq 3\sigma) \leq \frac{4}{9 \cdot 9} = 0.0494}\qquad(1.28a)$$

comes to about 5%. More detailed discussions on inequalities of this sort can be found in Mallows (1956) and Savage (1961).

▶ 1.3.5 Deviations from the normal distribution

Certain attributes of objects which originated under similar conditions are sometimes approximately normally distributed. On the other hand, many distributions exhibit strong deviations from the normal distribution. Our populations, in contrast with the normal distribution, **are mostly finite, seldom consist of a continuum of values**, and **FREQUENTLY** have **asymmetric**—sometimes even multimodal—frequency distributions.

Deviations from the normal distribution may be caused by the use of an inappropriate scale. Surface areas and weights of organisms are ordinarily not normally distributed, but are rather instances of squares and cubes of normally distributed variables. In such cases the use of a **transformation** is indicated. For surface areas, volumes, and small frequencies, the square root and the cube root transformations respectively are appropriate; random variables with distributions that are flat on the right and bounded by zero on the left are frequently transformed by the logarithm into approximately normally distributed variables. Percentages are normalized by the angular transformation. More on this can be found in Sections 1.3.9, 3.6.1, and 7.3.3.

If the deviation from a normal distribution cannot be accounted for by the scale used, the **sampling technique** should be more fully investigated. If a sample contains only the largest individual values, which are intentionally or unintentionally favored, no normal distribution can be expected. **Sample heterogeneity**, in terms of e.g., age or kind, manifests itself similarly: more than one peak is obtained. Several methods for verifying the homogeneity of a sample, in other words, for controlling the deviation from the normal distribution, will be discussed later (Section 1.3.7 as well as 3.8 and 4.3.3).

If we suspect that a population exhibits considerable deviation from the normal distribution, particularly in the tails (Charles P. Winsor has pointed out that many empirical distributions are nearly normally distributed only in their central regions), then to improve the normality of the sample it can be expedient to do without the smallest and largest observations, i.e., to neglect a certain number of extreme observations at both ends of the distribution ($\leq 5\%$ of all values). Through such cutting (cf., Section 3.8) the variance is greatly reduced but the estimate of the mean is improved (McLaughlin and Tukey 1961, Tukey 1962, Gebhardt 1966). More on **ROBUST STATISTICS** [see pages 123 and 253/254] is provided by Huber (1972, 1981), Wainer (1976), R. V. Hogg (1979, The American Statistician **33**, 108–115), Hampel (1980), David (1981 [8 : 1b]), Box et al. (1983) and Hoaglin et al. (1983).

Graphical methods for determining $\bar{x}$, s, and s^2 of a trimmed normal distribution are given by Nelson (1967) (cf., also Cohen 1957, 1961, as well as Sarhan and Greenberg 1962).

▶ 1.3.6 Parameters of unimodal distributions

1.3.6.1 Estimates of parameters

Observed values of a random variable X, e.g., height of 18-year old men, are denoted by $x_1 = 172$ cm, $x_2 = 175$ cm, $x_3 = 169$ cm, or generally by $x_1, x_2, \ldots, x_n$; n denotes the sample size. The sample average $\bar{x} = (1/n)\sum x$ is an observed value of the random variable $\bar{X} = (1/n)\sum X$.

A summary value calculated from a sample of observations, usually but not necessarily to estimate some population parameter, is called a **statistic**. In short, a value computed entirely from the sample is called a statistic. The statistic being used as a strategy or recipe to estimate the parameter is called an **estimator of the parameter**. A specific value of the sample statistic, computed from a particular set of data, preferably from a random sample, is called an **estimate of the parameter**. So $\bar{X}$ is the estimator of the parameter μ and $\bar{x}$ is a corresponding estimate, for instance $\bar{x} = 173$ cm.

Estimators such as $\bar{X}$ should, if possible, satisfy the four following conditions:

1. They must be **unbiased**—i.e., if the experiment is repeated very often, the average of all possible values of $\bar{X}$ must converge to the true value. An estimator is said to be unbiased if its expected value is equal to the population quantity being estimated.

    If this is not the case, the estimator is biased (e.g., Section 1.3.6.3, Remark 3). A bias can be caused by the experiment, through contaminated or unstable solvent; unreliable equipment; poor calibration; through "instrument drift" errors in recording the data, in calculations, and in interpretation; or nonrandomness of a sample. Errors of this sort are referred to as **systematic errors**: They cause the estimate to be always too large or always too small. The size of systematic errors can be estimated only on the basis of specialized knowledge of the origin of the given values. They can be prevented only by careful planning of the experiments or surveys. If systematic errors are present, nothing can be said concerning the true value; this differs greatly from the situation when random errors are present (cf., also Sections 2.1.2, 2.1.4, and 3.1, Parrat 1961, Anderson 1963, Sheynin 1969, Szameitat and Deininger 1969, Campbell 1974, Fraser 1980 and Strecker 1980 [8:3a]).

2. They must be in agreement or **consistent**—i.e., as the sample size tends to infinity the estimator approaches the parameter (limit property of consistency).
3. They must be **efficient**—i.e., they must have the smallest possible deviation (or variance) for samples of equal size. Suppose an infinite number of samples of size n is drawn from a population and the variance is determined for an eligible statistic, one which fulfills conditions 1 and 2. Then this condition means that the statistic is to be chosen whose variance about the mean or expected value of the statistic is least. As a rule, the standard deviation of an estimate decreases absolutely and relatively to the expected value with increasing sample size. It can be shown that the sample mean is the most efficient estimator of μ. As a result the sample mean is called a minimum variance unbiased estimator of μ.
4. They must be **sufficient**—i.e., no statistic of the same kind may provide further information about the parameter to be estimated. This condition means that the statistic contains all the information that the sample furnishes with respect to the parameter in question. For a normal distribution with known variance σ^2 and unknown mean μ, the sample mean $\bar{X}$ is a sufficient estimator of the population mean μ.

The notions consistent, efficient, and sufficient go back to R. A. Fisher (1925).

An extensive methodology of estimation has been developed for **estimating the parameters** from sample values. Of particular importance is the **maximum likelihood method** (R. A. Fisher): It is the universal method for optimal estimation of unknown parameters. It is applicable only if the type of the distribution function of the variables is known; the maximum likelihood estimate of the unknown parameter is the parameter value that maximizes the probability that the given sample would occur (see Norden 1972, 1973). This method of constructing point estimates for parameters is closely related to the important **method of least squares** (C. F. Gauss; c.f., Section 5.1), concerning which Harter (1974, 1975) provides a survey.

Weak and strong law of large numbers

Consider as given n measurements or observations, conceivable as n independent identically distributed random variables with parameter μ, and the sample mean $\bar{X}$. Then $\bar{X}$ is the estimator of μ. The weak law of large numbers states that with increasing measurements or observations n ($n \to \infty$) the absolute difference $|\bar{X}_n - \mu|$ is ultimately small; but not every value is small; it might be that for some n it is large, although such cases will only occur infrequently. The strong law of large numbers says that the probability of such an event is extremely small. In other words:

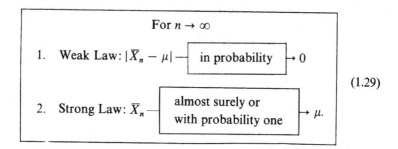

(1.29)

If a sample of n independent values is available, the **sample cumulative distribution function** $\hat{F}_n(x)$ is the proportion of the sample values which are less than or equal to x

$$\hat{F}_n(x) = \frac{n_{\leq x}}{n}.$$ (DF 1)

This empirical cumulative distribution function estimates the cumulative distribution function $F(x)$ of the population. For n large ($n \to \infty$) and fixed x the absolute difference

$$|\hat{F}_n(x) - F(x)|$$ (DF 2)

tends to zero, with probability one. This **theorem of Glivenko and Cantelli** indicates that **for n great** empirical distributions are practically identical with the pertinent theoretical distributions (cf. Sections 393 and 44).

The laws of large number (qualitative convergence statements) (1) imply that parameters can be estimated to any degree of accuracy given a sufficiently large sample, and (2) justify the Monte Carlo method.

1.3.6.2 The arithmetic mean and the standard deviation

The mean and the standard deviation are characteristic values for the Gaussian (or normal) distribution. They give the position of the average (mean) value of a sequence of measurements and the deviation (variation, variance, dispersion) of the individual values about the mean value respectively. Moreover, Chebyshev's inequality (1.27) shows that, even for other distributions, the standard deviation can serve as a general measure of dispersion. Analogous remarks apply to the mean value.

1.3 The Path to the Normal Distribution

Definitions

The **arithmetic mean** $\bar{x}$ (x bar) is the sum of all the observations divided by the number of observations:

$$\bar{x} = \frac{1}{n}(x_1 + x_2 + \ldots + x_n) = \frac{\sum x}{n}. \qquad (1.30)$$

The **standard deviation** is practically equal to the square root of the mean value of the squared deviations:

$$s = \sqrt{\frac{\sum (x - \bar{x})^2}{n - 1}}. \qquad (1.31)$$

s is usually computed according to (1.31a,b) in Section 1.3.6.3.

The expression "practically" here refers to the fact that inside the square root the denominator is not n, as is the case for a mean value, but rather $n - 1$. The square of the standard deviation is called the **variance**:

$$s^2 = \frac{\sum (x - \bar{x})^2}{n - 1}. \qquad (1.32)$$

s^2 is usually computed according to (1.32a) in Section 1.3.6.3.

If the mean value μ of the population is known, the quantity

$$s_0^2 = \frac{\sum (x - \mu)^2}{n}. \qquad (1.33)$$

is used, in place of s^2, as an estimate for σ^2 (cf., also end of section 1.2.6, item 3).

1.3.6.3 Computation of the mean value and standard deviation when the sample size is small

If a small number of values is involved or if a calculator is available, the mean value is calculated according to (1.30), the standard deviation (the positive value of $\sqrt{s^2}$) according to (1.31a) or (1.31b):

$$s = \sqrt{\frac{\sum x^2 - \frac{(\sum x)^2}{n}}{n - 1}}, \qquad s = \sqrt{\frac{n \sum x^2 - (\sum x)^2}{n(n - 1)}}. \qquad (1.31\ a), (1.31\ b)$$

EXAMPLE. Calculate $\bar{x}$ and s for the values 27, 22, 24 and 26 ($n = 4$).

$$\bar{x} = \frac{\sum x}{n} = \frac{99}{4} = 24.75;$$

$$s = \sqrt{\frac{\sum x^2 - \frac{(\sum x)^2}{n}}{n-1}} = \sqrt{\frac{2465 - \frac{99^2}{4}}{4-1}} = \sqrt{4.917} = 2.22,$$

$$s = \sqrt{\frac{n \sum x^2 - (\sum x)^2}{n(n-1)}} = \sqrt{\frac{4 \cdot 2465 - 99^2}{4(4-1)}} = \sqrt{4.917} = 2.22.$$

Applications of the arithmetic mean

1. **Tables of arithmetic means** should list next to the sample size (n) the corresponding standard deviation (s), in accordance with the table headings | Group | n | s | $\bar{x}$ |. With **random samples** from normally distributed populations one presents in a fifth column the 95% confidence interval (CI) for μ (cf., Sections 1.4.1 and 3.1.1; cf., also Sections 1.8.3 and 3.1.4: in general, either the mean $\bar{x}$ or the median $\tilde{x}$ is chosen and with random samples the corresponding 95% CI is stated). Occasionally the relative variation coefficient $V_r = s/(\bar{x}\sqrt{n})$ with $0 \leq V_r \leq 1$ (Section 1.3.6.6) and the extreme values (x_{min}, x_{max}) (or else the range, Section 1.3.8.5) can also be found in these tables.

2. **For the comparison of two means according to Student** (t-test; cf. Sections 3.5.3 and 3.6.2) it is more expedient to calculate the variances than the standard deviation, since these are needed for testing the inequality of variances (Section 3.5) as well as for the t-test. One simply omits the taking of roots in the formulas (1.31a,b); for example: $s = \sqrt{4.917}$ or $s^2 = 4.917$, i.e.,

$$s^2 = \frac{\sum x^2 - (\sum x)^2/n}{n-1}. \tag{1.32a}$$

Note that

$$\sum (x - \bar{x})^2 = \sum (x^2 - 2x\bar{x} + \bar{x}^2)$$

$$= \sum x^2 - 2\bar{x} \sum x + n\bar{x}^2 \quad \text{(cf. Section 0.2)}$$

$$= \sum x^2 - \frac{2(\sum x)^2}{n} + \frac{n(\sum x)^2}{n^2}$$

$$= \sum x^2 - \frac{(\sum x)^2}{n}.$$

The use "$n - 1$" in (1.32a) gives us an unbiased estimator of σ^2 and therefore an unbiased estimate of s^2.

1.3 The Path to the Normal Distribution

Remarks

1. Instead of using (1.32a) one could also estimate the variance using the formula $s^2 = \{1/[2n(n-1)]\} \sum_i \sum_j (x_i - x_j)^2$. Other interesting measures of dispersion are: $\{1/[n(n-1)]\} \sum_i \sum_j |x_i - x_j|$, $(1/n) \sum_i |x_i - \bar{x}|$ (cf., Sections 3.1.3 and 3.8), and $(1/n) \sum_i |x_i - \tilde{x}|$, where $\tilde{x}$ is the median.

2. In extensive samples the standard deviation can be quickly estimated as one-third the difference between the means of the largest sixth and the smallest sixth of the observations (Prescott 1968); cf., also D'Agostino 1970).

3. While the estimate of σ^2 by s^2 is unbiased, s is a biased estimate of σ. This bias is generally neglected. For a normally distributed population, a factor depending only on the sample size (e.g., Bolch 1968) turns s into an unbiased estimate of σ (e.g., 1.0854 for $n = 4$, i.e., $\sigma = 1.0854s$). For sample sizes that are not too small ($n \gtrsim 10$), this factor, which is approximately equal to $\{1 + 1/[4(n-1)]\}$, tends rapidly to one (e.g., 1.00866 for $n = 30$). For further details see Brugger (1969) and Stephenson (1970).

4. Taking an additional value x_z into consideration, if $\bar{x}$ and s^2 were computed for n observations, we have for the present $n + 1$ observations $\bar{x}_{n+1} = (x_z + n\bar{x})/(n + 1)$ and $s^2_{n+1} = (n + 1)(\bar{x}_{n+1} - \bar{x})^2 + (n - 1)s^2/n$.

5. It is characteristic of $\bar{x}$ that $\sum (x_i - \bar{x}) = 0$ and that $\sum (x_i - \bar{x})^2 \leq \sum (x_i - x)^2$ for every x; the median $\tilde{x}$ (cf., Section 1.3.8.3) has, on the other hand, the property that $\sum_i |x_i - \tilde{x}| \leq \sum_i |x_i - x|$ for every x; i.e., $\sum_i (x_i - \bar{x})^2$ and $\sum_i |x_i - \tilde{x}|$ are minima in the respective cases.

With multidigit individual values: To simplify the computation a provisional mean value d is chosen so as to make the difference $x - d$ as small as possible or positive throughout. Then we have

$$\boxed{\bar{x} = d + \frac{\sum(x - d)}{n}}, \tag{1.34}$$

$$\boxed{s = \sqrt{\frac{\sum (x - d)^2 - n(\bar{x} - d)^2}{n - 1}}}. \tag{1.35}$$

EXAMPLE. See Table 15. According to (1.34) and (1.35),

$$\bar{x} = d + \frac{\sum(x - d)}{n} = 11.26 + \frac{0.05}{5} = 11.27,$$

$$s = \sqrt{\frac{\sum (x - d)^2 - n(\bar{x} - d)^2}{n - 1}},$$

$$s = \sqrt{\frac{0.0931 - 5(11.27 - 11.26)^2}{5 - 1}} = \sqrt{0.02315} = 0.152.$$

In problems of this sort the decimal point can be removed through multiplication by an appropriate power of ten: In the present case we would

Table 15

x	x - 11.26	(x - 11.26)²
11.27	0.01	0.0001
11.36	0.10	0.0100
11.09	-0.17	0.0289
11.16	-0.10	0.0100
11.47	0.21	0.0441
	0.05	0.0931

form x^* ("x star") $= 100x$ and, as described, using the x^*-values, obtain $\bar{x}^* = 1{,}127$ and $s^* = 15.2$. These results again yield

$$\bar{x} = \frac{\bar{x}^*}{100} = 11.27 \quad \text{and} \quad s = \frac{s^*}{100} = 0.152.$$

The appearance of large numbers can be avoided in calculations of this sort by going a step further. By the encoding procedure the original values x are converted or transformed into the simplest possible numbers x^* by appropriate choice of k_1 and k_2, where k_1 introduces a change of scale and k_2 produces a shift of the origin (thereby generating a linear transformation):

$$\boxed{x = k_1 x^* + k_2,} \quad \text{i.e.} \quad \boxed{x^* = \frac{1}{k_1}(x - k_2).} \quad (1.36)$$
$$(1.36a)$$

From the parameters $\bar{x}^*$ and s^* or s^{*2}, calculated in the usual manner, the desired parameters are obtained directly:

$$\boxed{\bar{x} = k_1 x^* + k_2,} \qquad (1.37)$$

$$\boxed{s^2 = k_1^2 s^{*2}.} \qquad (1.38)$$

It is recommended that the example be once again independently worked out with $k_1 = 0.01$, $k_2 = 11.26$, i.e., with $x^* = 100(x - 11.26)$.

1.3.6.4 Computation of the mean value and standard deviation for large sample sizes: Grouping the individual values into classes [use perhaps the remark on page 81 as a control]

The sum of the ten numbers $\{2, 2, 2, 2; 3; 4, 4, 4, 4, 4\}$, namely 31, can just as well be written $(4)(2) + (1)(3) + (5)(4)$; the mean value of this sequence can then also be obtained according to

$$\bar{x} = \frac{(4)(2) + (1)(3) + (5)(4)}{4 + 1 + 5} = 3.1.$$

We have in this way partitioned the values of a sample into three classes (strata, groups). The frequencies 4, 1, and 5 assign different weights to the values 2, 3, and 4. Thus 3.1 can also be described as a weighted arithmetic mean. We shall return to this later (Section 1.3.6.5).

1.3 The Path to the Normal Distribution

In order to better survey extensive collections of numbers and to be able to more easily determine their characteristic statistics such as the mean value and standard deviation, one frequently combines into classes the values ordered according to magnitude. It is here expedient to maintain a **constant class width** b; you may use $b \approx \sqrt{x_{max} - x_{min}}$. Moreover, numbers as simple as possible (numbers with few digits) should be chosen as the **class midpoints**. The **number of classes** lies generally between 6 (for around 25–30 observations) and 25 (for around 10,000 or more values); cf., Sections 1.3.3 and 1.3.8.6.

The k classes are then occupied by the frequency values or frequencies $f_1, f_2, \ldots, f_k$ ($n = \sum_{i=1}^{k} f_i = \sum f$). A preliminary average value d is chosen, which often falls in the class that contains the largest number of values.

I The multiplication procedure

The individual classes are then numbered: d receives the index $z = 0$, the classes with means smaller than d get the indices $z = -1, -2 \ldots$ in descending order, those larger get $z = 1, 2, \ldots$ in ascending order. Then we have

$$\bar{x} = d + \frac{b}{n}\sum fz,$$

$$s = b\sqrt{\frac{\sum fz^2 - (\sum fz)^2/n}{n-1}},$$

(1.39)

$$s^2 = b^2 \left[\frac{n\sum fz^2 - (\sum fz)^2}{n(n-1)}\right],$$

(1.40)

with d = assumed average (midpoint of the class with index $z = 0$,
b = class width [classes given by the intervals $(d + b(z - \frac{1}{2}), d + b(z + \frac{1}{2}))$ — left endpoint excluded, right endpoint included]
n = number of values,
f = frequency within a class,
x = midpoint of a class ($x = d + bz$),
z = normed distance or index of the class with midpoint x:

$$z = (x - d)/b.$$

Table 16

	CM	f	z	fz	fz²
	13	1	-3	-3	9
	17	4	-2	-8	16
	21	6	-1	-6	6
d =	25	7	0	0	0
	29	5	1	5	5
	33	5	2	10	20
	37	2	3	6	18
	∑	30	–	4	74

CM = class midpoints
b = 4

An example is shown in Table 16, with $b = 4$, CM = class mean = x; we have

$$\bar{x} = d + \frac{b}{n}\sum fz = 25 + \frac{4}{30}\cdot 4 = 25.53$$

$$s = b\sqrt{\left(\frac{\sum fz^2 - (\sum fz)^2/n}{n-1}\right)} = 4\sqrt{\left(\frac{74 - 4^2/30}{30-1}\right)} = 6.37.$$

Control: One makes use of the identities

$$\boxed{\sum f(z+1) = \sum fz + \sum f = \sum fz + n,} \quad (1.41)$$

$$\sum f(z+1)^2 = \sum f(z^2 + 2z + 1),$$
$$\sum f(z+1)^2 = \sum fz^2 + 2\sum fz + \sum f,$$
$$\boxed{\sum f(z+1)^2 = \sum fz^2 + 2\sum fz + n,} \quad (1.42)$$

Table 17

z + 1	f	f(z + 1)	f(z + 1)²
-2	1	-2	4
-1	4	-4	4
0	6	0	0
1	7	7	7
2	5	10	20
3	5	15	45
4	2	8	32
n = ∑f = 30		∑f(z + 1) = 34	∑f(z + 1)² = 112

and notes the corresponding distributions. An example is shown in Table 17.
Control for the mean:

$$\sum f(z+1) = 34 \quad \text{(from Table 17),}$$
$$\sum fz + n = 4 + 30 = 34 \quad \text{(from Table 16).}$$

Control for the standard deviation:

$$\sum f(z+1)^2 = 112 \quad \text{(from Table 17),}$$
$$\sum fz^2 + 2\sum fz + n = 74 + (2)(4) + 30 = 112 \quad \text{(from Table 16).}$$

The multiplication procedure is particularly appropriate if a second computation based on data from which outliers have been removed (cf., Section 3.8) or based on an augmented data set becomes necessary, or if moments of higher order (cf., Section 1.3.8.7) are to be computed.

II The summation procedure (cf. Table 18)

The summation procedure consists of a stepwise summation of the frequencies from the top and the bottom of the table toward the class containing the preassigned average value d (column 3). The values so obtained are again sequentially added, starting with the top and the bottom of column

1.3 The Path to the Normal Distribution

3 and proceeding to classes adjacent to the class containing d (column 4). The resulting sums are denoted by δ_1 and δ_2 (Greek delta 1 and 2). The values obtained are once again added, starting with the top and the bottom of column 4 and working toward the classes adjacent to the class containing d

Table 18

	CM	f	S_1	S_2		S_3	
	13	1	1	1		1	
	17	4	5	6		7	
	21	6	11	17 = δ_1	24 = ε_1
d =	25	7					
	29	5	12	21 = δ_2	32 = ε_2
	33	5	7	9		11	
	37	2	2	2		2	
	n = 30						

(column 5). We represent the sums so obtained by ε_1 and ε_2 (Greek epsilon 1 and 2). Then on setting

$$\frac{\delta_2 - \delta_1}{n} = c$$

we have

$$\boxed{\bar{x} = d + b \cdot c,} \tag{1.43}$$

$$\boxed{s = b\sqrt{\frac{2(\varepsilon_1 + \varepsilon_2) - (\delta_1 + \delta_2) - nc^2}{n-1}},} \tag{1.44}$$

$$\boxed{s^2 = b^2 \left[\frac{2(\varepsilon_1 + \varepsilon_2) - (\delta_1 + \delta_2) - nc^2}{n-1}\right],}$$

where d is the chosen average value; b is the class width; n is the number of values; and $\delta_1, \delta_2, \varepsilon_1, \varepsilon_2$ denote the special sums defined above. We give a last example in Table 18 (CM = class mean):

$$c = \frac{\delta_2 - \delta_1}{n} = \frac{21 - 17}{30} = 0.133,$$

$$\bar{x} = d + bc = 25 + 4 \cdot 0.133 = 25.53,$$

$$s = b\sqrt{\frac{2(\varepsilon_1 + \varepsilon_2) - (\delta_1 + \delta_2) - nc^2}{n-1}},$$

$$s = 4\sqrt{\frac{2(24 + 32) - (17 + 21) - 30 \cdot 0.133^2}{30 - 1}},$$

$$s = 4\sqrt{2.533},$$

$$s = 6.37.$$

The standard deviation computed from grouped data is in general somewhat larger than when computed from ungrouped data, and in fact—within a small range—increases with the class width b; thus it is wise to choose b not too large (cf., Section 1.3.8.5):

$$b \leq s/2 \tag{1.45}$$

if possible. In our examples we used a coarser partition into classes. Moreover, Sheppard has proposed that the variance, when calculated from a frequency distribution partitioned into strata or classes, be corrected by subtracting $b^2/12$:

$$s^2_{\text{corr}} = s^2 - b^2/12. \tag{1.46}$$

This correction need only be applied if $n > 1{,}000$ with a coarse partition into classes, i.e., if the number of classes $k < 20$. Corrected variances must not be used in statistical tests.

1.3.6.5 The combined arithmetic mean, the combined variance, and the weighted arithmetic mean

If several samples of sizes $n_1, n_2, \ldots, n_k$, mean values $\bar{x}_1, \bar{x}_2, \ldots, \bar{x}_k$, and squares of standard deviations $s_1^2, s_2^2, \ldots, s_k^2$ are combined to form a single sequence of size $n = n_1 + n_2 + \cdots + n_k$, then the arithmetic mean of the combined sample is the **combined arithmetic mean**, $\bar{x}_{\text{comb}}$:

$$\bar{x}_{\text{comb}} = \frac{n_1 \cdot \bar{x}_1 + n_2 \cdot \bar{x}_2 + \cdots + n_k \cdot \bar{x}_k}{n}, \tag{1.47}$$

and the **standard deviation** s_{in} **within the samples** is

$$s_{\text{in}} = \sqrt{\frac{s_1^2(n_1 - 1) + s_2^2(n_2 - 1) + \cdots s_k^2(n_k - 1)}{n - k}}. \tag{1.48}$$

EXAMPLE

$$n_1 = 8, \quad \bar{x}_1 = 9, \quad (s_1 = 2) \quad s_1^2 = 4,$$
$$n_2 = 10, \quad \bar{x}_2 = 7, \quad (s_2 = 1) \quad s_2^2 = 1,$$
$$n_3 = 6, \quad \bar{x}_3 = 8, \quad (s_3 = 2) \quad s_3^2 = 4,$$

$$\bar{x} = \frac{8 \cdot 9 + 10 \cdot 7 + 6 \cdot 8}{24} = 7.917,$$

$$s_{\text{in}} = \sqrt{\frac{4(8-1) + 1(10-1) + 4(6-1)}{24 - 3}} = 1.648.$$

1.3 The Path to the Normal Distribution

The **variance** of the x-variable **in the combined sample** is calculated by the formula

$$s_{\text{comb}}^2 = \frac{1}{n-1}\left[\sum_i (n_i - 1)s_i^2 + \sum_i n_i(\bar{x}_i - \bar{x})^2\right], \quad (1.48a)$$

in our example

$$s_{\text{comb}}^2 = \tfrac{1}{23}[(7\cdot 4 + 9\cdot 1 + 5\cdot 4) + (8\cdot 1.083^2 + 10\cdot 0.917^2 + 6\cdot 0.083^2)]$$
$$= 3.254.$$

The weighted arithmetic mean: Unequal precision of individual measurements can be taken into account by the use of different weights w_i ($i = 1, 2, \ldots$; $w_i = 0.1$ or 0.01, etc., with $\sum w_i = 1$). The weighted arithmetic mean is found according to $\bar{x} = (\sum w_i x_i)/\sum w_i$ or, more appropriately, by choosing a convenient auxiliary value a and working with the translated variable $z_i = x_i - a$.

EXAMPLE

x_i	w_i	$x_i - a = z_i$ ($a = 137.8$)	$w_i z_i$
138.2	1	0.4	0.4
137.9	2	0.1	0.2
137.8	1	0.0	0.0

$\sum w_i = 4,$ $\sum w_i z_i = 0.6,$

$$\bar{x} = a + \frac{\sum w_i z_i}{\sum w_i}, \quad (1.49)$$

$$\bar{x} = 137.8 + \frac{0.6}{4} = 137.95.$$

For index numbers, see Mudgett (1951), Snyder (1955), Crowe (1965), and Craig (1969).

1.3.6.6 The coefficient of variation

Suppose that x can assume positive values only. The ratio of the standard deviation to the mean value is called the **coefficient of variation** (K. Pearson 1895): or, occasionally, the **coefficient of variability**, and denoted by V:

$$V = \frac{s}{\bar{x}} \quad \text{for all } x > 0. \quad (150)$$

The coefficient of variation is equal to the standard deviation when the mean value equals one. In other words the coefficient of variation is a dimensionless relative measure of dispersion with the mean value as unit. Since its maximum is $\sqrt{n}$ (Martin and Gray 1971), one can also readily specify the **relative coefficient** of variation V_r, expressed as a percentage, which can take on values between 0% and 100%:

$$V_r[\text{in }\%] = \frac{s/\bar{x}}{\sqrt{n}} 100 \quad \text{for all } x > 0. \tag{1.50a}$$

The coefficient of variation is useful in particular for **comparison of samples** of some population types (e.g., when mean and variance vary together).

EXAMPLE. For $n = 50$, $s = 4$, and $\bar{x} = 20$ we get from (1.50) and (1.50a) that

$$V = \frac{4}{20} = 0.20 \quad \text{and} \quad V_r = \frac{4/20}{\sqrt{50}} 100 = 2.83\% \quad \text{or} \quad V_r = 3\%.$$

1.3.6.7 Examples involving the normal distribution (for Section 1.3.4)

1. With the help of the ordinates of the normal distribution (Table 20), the normal curve can readily be sketched. For a **quick plotting of the normal curve** the values in Table 19 can be utilized. To abscissa values of $\pm 3.5\sigma$ corresponds the ordinate $\frac{1}{400} y_{\max}$, so for x-values larger than 3.5σ or smaller than -3.5σ the curve practically coincides with the x-axis because, e.g., to a maximum ordinate of 40 cm, there correspond, at the points $z = \pm 3.5\sigma$, 1 mm long ordinates.

2. Let the lengths of a collection of objects be normally distributed with $\mu = 80$ cm and $\sigma = 8$ cm. (a) What percentage of the objects fall between 66 and 94 cm? (b) In what interval does the "middle" 95% of the lengths fall?

For (a): The interval 80 ± 14 cm can also be written $80 \pm \frac{14}{8}\sigma = 80 \pm 1.75\sigma$. Table 13 gives for $z = 1.75$ a probability ($P = 0.0401$) of 4%. The percentage of the objects lying between $z = -1.75$ and $z = +1.75$ is to be determined. Since 4% lie above $z = 1.75$ and another 4% below $z = 1.75$

Table 19

Abscissa	0	$\pm 0.5\sigma$	$\pm 1.0\sigma$	$\pm 2.0\sigma$	$\pm 3.0\sigma$
Ordinate	$y_{\max}$	$\frac{7}{8} y_{\max}$	$\frac{5}{8} y_{\max}$	$\frac{1}{8} y_{\max}$	$\frac{1}{80} y_{\max}$

1.3 The Path to the Normal Distribution

Table 20 Ordinates of the standard normal curve: $f(z) = (1/\sqrt{2\pi}) e^{-z^2/2}$

z	0.00	0.01	0.02	0.03	0.04	0.05	0.06	0.07	0.08	0.09
0.0	0.3989	0.3989	0.3989	0.3988	0.3986	0.3984	0.3982	0.3980	0.3977	0.3973
0.1	0.3970	0.3965	0.3961	0.3956	0.3951	0.3945	0.3939	0.3932	0.3925	0.3918
0.2	0.3910	0.3902	0.3894	0.3885	0.3876	0.3867	0.3857	0.3847	0.3836	0.3825
0.3	0.3814	0.3802	0.3790	0.3778	0.3765	0.3752	0.3739	0.3725	0.3712	0.3697
0.4	0.3683	0.3668	0.3653	0.3637	0.3621	0.3605	0.3589	0.3572	0.3555	0.3538
0.5	0.3521	0.3503	0.3485	0.3467	0.3448	0.3429	0.3410	0.3391	0.3372	0.3352
0.6	0.3332	0.3312	0.3292	0.3271	0.3251	0.3230	0.3209	0.3187	0.3166	0.3144
0.7	0.3123	0.3101	0.3079	0.3056	0.3034	0.3011	0.2989	0.2966	0.2943	0.2920
0.8	0.2897	0.2874	0.2850	0.2827	0.2803	0.2780	0.2756	0.2732	0.2709	0.2685
0.9	0.2661	0.2637	0.2613	0.2589	0.2565	0.2541	0.2516	0.2492	0.2468	0.2444
1.0	0.2420	0.2396	0.2371	0.2347	0.2323	0.2299	0.2275	0.2251	0.2227	0.2203
1.1	0.2179	0.2155	0.2131	0.2107	0.2083	0.2059	0.2036	0.2012	0.1989	0.1965
1.2	0.1942	0.1919	0.1895	0.1872	0.1849	0.1826	0.1804	0.1781	0.1758	0.1736
1.3	0.1714	0.1691	0.1669	0.1647	0.1626	0.1604	0.1582	0.1561	0.1539	0.1518
1.4	0.1497	0.1476	0.1456	0.1435	0.1415	0.1394	0.1374	0.1354	0.1334	0.1315
1.5	0.1295	0.1276	0.1257	0.1238	0.1219	0.1200	0.1182	0.1163	0.1145	0.1127
1.6	0.1109	0.1092	0.1074	0.1057	0.1040	0.1023	0.1006	0.0989	0.0973	0.0957
1.7	0.0940	0.0925	0.0909	0.0893	0.0878	0.0863	0.0848	0.0833	0.0818	0.0804
1.8	0.0790	0.0775	0.0761	0.0748	0.0734	0.0721	0.0707	0.0694	0.0681	0.0669
1.9	0.0656	0.0644	0.0632	0.0620	0.0608	0.0596	0.0584	0.0573	0.0562	0.0551
2.0	0.0540	0.0529	0.0519	0.0508	0.0498	0.0488	0.0478	0.0468	0.0459	0.0449
2.1	0.0440	0.0431	0.0422	0.0413	0.0404	0.0396	0.0387	0.0379	0.0371	0.0363
2.2	0.0355	0.0347	0.0339	0.0332	0.0325	0.0317	0.0310	0.0303	0.0297	0.0290
2.3	0.0283	0.0277	0.0270	0.0264	0.0258	0.0252	0.0246	0.0241	0.0235	0.0229
2.4	0.0224	0.0219	0.0213	0.0208	0.0203	0.0198	0.0194	0.0189	0.0184	0.0180
2.5	0.0175	0.0171	0.0167	0.0163	0.0158	0.0154	0.0151	0.0147	0.0143	0.0139
2.6	0.0136	0.0132	0.0129	0.0126	0.0122	0.0119	0.0116	0.0113	0.0110	0.0107
2.7	0.0104	0.0101	0.0099	0.0096	0.0093	0.0091	0.0088	0.0086	0.0084	0.0081
2.8	0.0079	0.0077	0.0075	0.0073	0.0071	0.0069	0.0067	0.0065	0.0063	0.0061
2.9	0.0060	0.0058	0.0056	0.0055	0.0053	0.0051	0.0050	0.0048	0.0047	0.0046
3.0	0.0044	0.0043	0.0042	0.0040	0.0039	0.0038	0.0037	0.0036	0.0035	0.0034
3.1	0.0033	0.0032	0.0031	0.0030	0.0029	0.0028	0.0027	0.0026	0.0025	0.0025
3.2	0.0024	0.0023	0.0022	0.0022	0.0021	0.0020	0.0020	0.0019	0.0018	0.0018
3.3	0.0017	0.0017	0.0016	0.0016	0.0015	0.0015	0.0014	0.0014	0.0013	0.0013
3.4	0.0012	0.0012	0.0012	0.0011	0.0011	0.0010	0.0010	0.0010	0.0009	0.0009
3.5	0.0009	0.0008	0.0008	0.0008	0.0008	0.0007	0.0007	0.0007	0.0007	0.0006
3.6	0.0006	0.0006	0.0006	0.0005	0.0005	0.0005	0.0005	0.0005	0.0005	0.0004
3.7	0.0004	0.0004	0.0004	0.0004	0.0004	0.0004	0.0003	0.0003	0.0003	0.0003
3.8	0.0003	0.0003	0.0003	0.0003	0.0003	0.0002	0.0002	0.0002	0.0002	0.0002
3.9	0.0002	0.0002	0.0002	0.0002	0.0002	0.0002	0.0002	0.0002	0.0001	0.0001
4.0	0.0001	0.0001	0.0001	0.0001	0.0001	0.0001	0.0001	0.0001	0.0001	0.0001
z	0.00	0.01	0.02	0.03	0.04	0.05	0.06	0.07	0.08	0.09

Example: $f(1.0) = 0.242 = f(-1.0)$.
This table gives values of the standard normal density function.

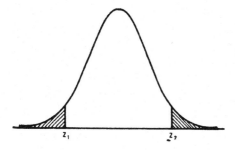

Figure 13 Standard normal distribution: The shaded portion of the area lies to the left of z_1 (negative value) and to the right of z_2 (positive value). In the figure we have $|z_1| = |z_2|$. Table 13 in Section 1.3.4 lists the portion of the area to the right of z_2 and, by symmetry, also to the left of the negative values $z_1 = z_2$, where we use $|z_2|$ in the table.

(cf., Figure 13 with $z_1 = -1.75$ and $z_2 = +1.75$), it follows that $100 - (4 + 4) = 92\%$ of the objects lie between the two boundaries, i.e., between the 66 and 94 cm lengths.

For (b): The text following Figure 11 and Table 14 in Section 1.3.4 indicate (for $z = 1.96$) that 95% of the objects lie in the interval 80 cm $\pm$ (1.96)(8) cm, i.e., between 64.32 cm and 95.68 cm.

3. Let some quantity be normally distributed with $\mu = 100$ and $\sigma = 10$. We are interested in determining the portion, in percent, (a) above $x = 115$, (b) between $x = 90$ and $x = 115$, and (c) below $x = 90$. First the values x are to be transformed to standard units: $\hat{z} = (x - \mu)/\sigma$.

For (a): $x = 115$, $\hat{z} = (115-110)/10 = 1.5$. The portion sought is determined from Table 13, for $\hat{z} = 1.5$, to be 0.0668 or 7%.

For (b): $x = 90$, $\hat{z} = (90-100)/10 = -1.0$; for $x = 115$ we just obtained $\hat{z} = 1.5$. We are to find the area A under the normal curve bounded by $\hat{z} = -1.0$ and $\hat{z} = 1.5$. Thus we must add the quantities:

(Area betw. $z = -1.0$ and $z = 0$) + (Area betw. $z = 0$ and $z = 1.5$).

Since the first area is by symmetry equal to the A_1 bounded by $z = 0$ and $z = +1.0$, the area A is given by $(A_1$ betw. $z = 0$ and $z = 1) + (A_1$ betw. $z = 0$ and $z = 1.5)$. Table 13 gives the probabilities of the right hand tail of the standard normal distribution. We know that the total probability is 1, that the distribution is symmetric with respect to $z = 0$, and that the probability integral (area) is additive. Thus the areas A_1 and A_2 can be written as differences: $A_1 = P(z > 0) - P(z > 1)$, $A_2 = P(z > 0) - P(z > 1.5)$ whence $A = (0.5 - 0.1587) + (0.5 - 0.0668) = 0.7745$ (cf., Figure 14).

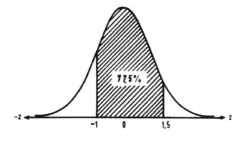

Figure 14

1.3 The Path to the Normal Distribution

For (c): For $x = 90$ a value of $\hat{z} = -1.0$ was just found. But by symmetry the area beyond $z = +1.0$ is the same as the area we want: 0.1587 or 16%.

Let's check the computations (a), (b), (c): $0.0668 + 0.7745 + 0.1587 = 1.000$.

4. For the normal distribution $\mu = 150$ and $\sigma = 10$ the value below which 6% of the probability mass lies is to be specified; moreover $P(130 < X < 160) = P(130 \leq X \leq 160)$ (cf., Section 1.2.6) is to be determined. The equation $(x - 150)/10 = -1.555$ implies $x = 134.45$. For $P(130 < X < 160)$ we can write

$$P\left(\frac{130 - 150}{10} < \frac{x - 150}{10} < \frac{160 - 150}{10}\right) = P(-2 < Z < 1)$$
$$= 1 - (0.0228 + 0.1587) = 0.8185.$$

5. In a normal distribution $N(11; 2)$ with $\mu = 11$ and $\sigma = 2$ find the probability for the interval, area under the curve, from $x = 10$ to $x = 14$ or $P(10 \leq X \leq 14)$. By putting $z_1 = (x_1 - \mu)/\sigma = (10 - 11)/2 = -0.5$ and $z_2 = (x_2 - \mu)/\sigma = (14 - 11)/2 = 1.5$ we have $P(10 \leq X \leq 14) = P(-0.5 \leq Z \leq 1.5)$ and with $P(-0.5 \leq Z \leq 0) = P(0 \leq Z \leq 0.5)$, from symmetry,

$$P(10 \leq X \leq 14) = [0.5 - P(Z \geq 0.5)] + [0.5 - P(Z \geq 1.5)]$$
$$= [0.5 - 0.3085] + [0.5 - 0.0668]$$
$$= 0.1915 + 0.4332 = 0.6247.$$

Remark: Quick estimation of $\bar{x}$ and s by means of a random sample from a normally distributed population. Using two arbitrary values $(W_l; W_u)$ one detaches from a sample a lower and an upper end of the distribution containing $\gtrsim 20$ values each, determines their relative frequencies p_l and p_u, and reads off the corresponding z_l and z_u from Table 13. Then one has $s \simeq (W_u - W_l)/(z_u + z_l)$ and $\bar{x} \simeq W_l + z_l s = W_u - z_u s$.

▶ 1.3.7 The probability plot

Graphical methods are useful in statistics: for description of data, for their screening, analysis, cross-examining, selection, reduction, presentation and their summary, and for uncovering distributional peculiarities. Moreover **probability plots** provide insight into the possible inappropriateness of certain assumptions of the statistical model. More on this can be found in King (1971) [cf., Wilk and Gnanadesikan (1968), Sachs (1977), Cox (1978), Fienberg (1979), D. Stirling (1982) [The Statistician **31**, 211–220] and Fisher (1983)].

A plot on standard probability paper may be helpful in deciding whether the sample at hand might come from a population with a normal distribution. In addition, the mean and the standard deviation can be read from the graph. The scale of the ordinate is given by the cumulative distribution function of the normal distribution, the abscissa carries a linear scale [logarithmic scale, if we are interested not in a normally distributed variable but in a variable whose logarithm is normally distributed, $Z = \log X, Z \sim N(\mu, \sigma^2)$] (cf., Figure 15). The graph of the observed values against the corresponding cumulative empirical distribution will, in the case of a sample from a normally distributed population, be approximately a straight line.

The ordinate values 0% and 100% are not included in the probability plot. The percentage frequencies with these values are thus disregarded in the graphical presentation.

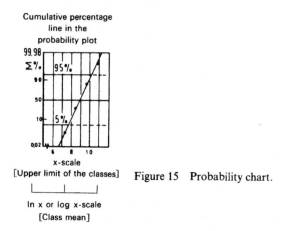

Figure 15   Probability chart.

Write $z = (x - \mu)/\sigma = x/\sigma - \mu/\sigma = (-\mu/\sigma) + (1/\sigma)x$, a straight line, with the points $F(\mu) = 0.5$ and $F(\mu + \sigma) = 0.8413$. The ordinate scale includes the corresponding percentages (50%, 84.13%) of the distribution function of the standard normal distribution (e.g., 15.87% of the distribution lies below $z = -1$):

y	0%	10%	15.87%	...	50%	...	84.13%	90%	100%
z	$-\infty$	-1.28	-1	...	0	...	$+1$	$+1.28$	$+\infty$

From the empirical frequency distribution one computes the cumulative sum distribution in percent and plots these values in the chart. It must here be kept in mind that class **limits** are specified on the abscissa. How straight the line is can be judged from the behavior of the curve between about 10% and 90%. To obtain the parameters of the sample, the point of intersection of the straight line drawn with the horizontal line through the 50% point

1.3 The Path to the Normal Distribution

of the ordinate is marked on the abscissa. This x-coordinate of the point of intersection is the graphically estimated mean ($\bar{x}_g$). Furthermore, the 16% and 84% horizontal lines are brought into intersection with the straight line drawn. These points of intersection are marked on the x-axis and one reads off $\bar{x}_g + s_g$ and $\bar{x}_g - s_g$. By subtracting the second from the first one gets $2s_g$ and hence the standard deviation. This simple calculation frequently determines the mean ($\bar{x}_g$) and standard deviation (s_g). The **cumulative sum line of the normal distribution**, also known as the **Hazen line**, can be obtained by proceeding in the opposite direction, starting with the following characteristic values:

$$\text{At} \quad x = \mu, \quad y = 50\%;$$
$$x = \mu + \sigma, \quad y \simeq 84\%;$$
$$x = \mu - \sigma, \quad y \simeq 16\%.$$

While a probability plot gives us an idea of the normality of a distribution, it is an inadequate method for precise analysis, since the weights of the individual classes manifest themselves indistinctly; moreover, only a poor assessment can be made of whether or not the deviations from the theoretical straight line remain within the domain of randomness (see also Section 4.3.3). The lower part of Figure 15 anticipates the important lognormal distribution (Section 1.3.9). Further discussion can be found in King (1971) (cf., also Mahalanobis 1960, and especially Wilk and Gnanadesikan 1968, as well as Sachs 1977).

Many empirical distributions are **mixtures of distributions**. Even if a sample looks homogeneous and is e.g. nearly normally distributed, we cannot ascertain that it was drawn from a population with a pure distribution function. Not infrequently a distribution that seems to be normal proves to be a mixture. Decompositions are then possible (Preston 1953, Weichselberger 1961, Ageno and Frontali 1963, Bhattacharya 1967, Harris 1968, Day 1969, Herold 1971).

The homogeneity of the material studied cannot be ascertained in principle. **Only the existence of inhomogeneities can be established**. Inhomogeneity does not indicate the material is useless; rather it requires consideration of the inhomogeneity in the estimate, mostly through the **formation of subgroups**.

Remark: Uniform or rectangular distributions

When a fair die is thrown, any one of the sides with 1, 2, 3, 4, 5, or 6 pips can come up. This gives a theoretical distribution in which the integers 1 through 6 have the *same* probability of $\frac{1}{6}$, i.e., $P(x) = \frac{1}{6}$ for $x = 1, 2, \ldots, 6$. If, as in the example, the possible events E can be assigned the numbers x, with the individual probabilities $P(x)$ corresponding to the relative frequencies, then we have quite generally for the **parameters of theoretical distributions** the relations (cf., end of Section 1.2.6)

$$\boxed{\mu = \sum x P(x)} \qquad (1.51)$$

and the so-called **translation law**

$$\sigma^2 = \sum x^2 P(x) - \mu^2, \qquad (1.52)$$

e.g., $\mu = 1 \cdot \frac{1}{6} + 2 \cdot \frac{1}{6} + \cdots + 6 \cdot \frac{1}{6} = 3.5$ and $\sigma^2 = 1 \cdot \frac{1}{6} + 4 \cdot \frac{1}{6} + \cdots + 36 \cdot \frac{1}{6} - 3.5^2 = 2.917$.

The **discrete uniform distribution** is defined by

$$P(x) = 1/n \quad \text{for} \quad 1 \leq x \leq n \qquad (1.53)$$

with mean μ and variance σ^2:

$$\mu = \frac{n+1}{2}, \qquad (1.54)$$

$$\sigma^2 = \frac{n^2 - 1}{12}. \qquad (1.55)$$

The values for our example are found directly ($n = 6$):

$$\mu = \frac{6+1}{2} = 3.5 \quad \text{and} \quad \sigma^2 = \frac{6^2 - 1}{12} = 2.917.$$

The uniform distribution comes up, for example, when rounding errors are considered. Here we have

$$P(x) = \tfrac{1}{10} \quad \text{for } x = -0.4, \ldots, +0.5.$$

The parameters are $\mu = 0.05$ and $\sigma^2 = 0.287$. The random numbers mentioned in the Introduction and Section 1.3.2 are realizations of a discrete uniform distribution of the numbers 0 to 9. By reading off three digits at a time (Table 10, Section 1.3.2) we get uniformly distributed random numbers from 0 to 999.

The probability density of the **continuous uniform or rectangular distribution** over the interval $[a, b]$ is given by

$$y = f(x) = \begin{cases} 1/(b-a) & \text{for } a \leq x \leq b, \\ 0 & \text{for } x < a \text{ or } x > b, \end{cases} \qquad (1.56)$$

i.e., $f(x)$ is constant on $a \leq x \leq b$ and vanishes outside this interval. The mean and variance are given respectively by

$$\mu = \frac{a+b}{2} \qquad (1.57)$$

1.3 The Path to the Normal Distribution

and

$$\sigma^2 = \frac{b - a^2}{12}. \tag{1.58}$$

The continuous uniform distribution is useful to a certain extent in applied statistics: for example, when any arbitrary value in some class of values is equally probable; as another example, to approximate relatively small regions of arbitrary continuous distributions. Thus, for example, the normally distributed variable X is nearly uniformly distributed in the region

$$\mu - \frac{\sigma}{3} < X < \mu + \frac{\sigma}{3}. \tag{1.59}$$

Rider (1951) gives a test for examining whether two samples from uniform populations actually come from the same population. The test is based on the quotient of their ranges; the paper also contains critical bounds at the 5% level.

1.3.8 Additional statistics for the characterization of a one dimensional frequency distribution

The tools for characterizing one dimensional frequency distributions are:

1. **Location statistics:** statistics for the location of a distribution (arithmetic, geometric, and harmonic mean; mode [and relative modes]; median and other quantiles).
2. **Dispersion statistics:** statistics that characterize the variability of the distribution (variance, standard deviation, range, coefficient of variation, interdecile region).
3. **Shape statistics:** statistics that characterize the deviation of a distribution from the normal distribution (simple skewness and kurtosis statistics, as well as the moment coefficients a_3 and a_4).

1.3.8.1 The geometric mean

Let $x_1, x_2, \ldots, x_n$ be positive numbers. The nth root of the product of all these numbers is called the geometric mean $\bar{x}_G$:

$$\bar{x}_G = \sqrt[n]{x_1 \cdot x_2 \cdot x_3 \cdots x_n} \quad \text{with } x_i > 0. \tag{1.60}$$

It may be evaluated with the help of logarithms (1.61) [cf., also (1.66)]:

$$\log \bar{x}_G = \frac{1}{n}(\log x_1 + \log x_2 + \log x_3 + \cdots + \log x_n) = \frac{1}{n} \sum_{i=1}^{n} \log x_i.$$

$$\tag{1.61}$$

We note that the logarithm of the geometric mean equals the arithmetic mean of the logarithms. The overall mean of several (say k), geometric means, based on sequences with $n_1, n_2, \ldots, n_k$ terms, is a weighted geometric mean:

$$\log \bar{x}_G = \frac{n_1 \log \bar{x}_{G1} + n_2 \log \bar{x}_{G2} + \cdots + n_k \log \bar{x}_{Gk}}{n_1 + n_2 + \cdots + n_k}. \quad (1.62)$$

The geometric mean is used, first of all, when an average of rates is to be calculated in which the changes involved are given for time intervals of equal length (cf., Example 1). It is applied if a variable changes at a rate that is roughly proportional to the variable itself. This is the case with various kinds of **growth**. The average increase of population with time, and the number of patients or the costs of treatment in a clinic, are familiar examples. One can get a rough idea of whether one is dealing with a velocity that is changing proportionately by plotting the data on semilogarithmic graph paper (ordinate, scaled logarithmically in units of the variable under consideration; abscissa, scaled linearly in units of time). With a velocity varying proportionately, the resulting graph must be approximately a straight line. $\bar{x}_G$ is then the **mean growth rate** (cf., Examples 2 and 3).

The geometric mean is also used when a sample contains a few elements with x-values that are much larger than most. These influence the geometric mean less than the arithmetic mean, so that the geometric mean is more appropriate as a typical value.

Examples

1. An employee gets raises in his salary of 6%, 10%, and 12% in three consecutive years. The percentage increase is in each case calculated with respect to the salary received the previous year. We wish to find the average raise in salary.

The geometric mean of 1.06, 1.10 and 1.12 is to be determined:

$$\log 1.06 = 0.0253$$
$$\log 1.10 = 0.0414$$
$$\log 1.12 = 0.0492$$
$$\overline{}$$
$$\sum \log x_i = 0.1159$$
$$\tfrac{1}{3} \sum \log x_i = 0.03863 = \log \bar{x}_G$$
$$\bar{x}_G = 1.093.$$

Thus the salary is raised 9.3% on the average.

2. The number of bacteria in a certain culture grows in three days from 100 to 500 per unit of culture. We are asked to find the average daily increase, expressed in percentages.

1.3 The Path to the Normal Distribution

Denote this quantity by x; then the number of bacteria is after the

1st day: $\qquad 100 + 100x = 100(1 + x)$,

2nd day: $\quad 100(1 + x) + 100(1 + x)x = 100(1 + x)^2$,

3rd day: $100(1 + x)^2 + 100(1 + x)^2 x = 100(1 + x)^3$.

This last expression must equal 500, i.e.,

$$100(1 + x)^3 = 500, \quad (1 + x)^3 = 5, \quad 1 + x = \sqrt[3]{5}.$$

With the help of logarithms we find $\sqrt[3]{5} = 1.710$, so that $x = 0.710 = 71.0\%$.

In general, beginning with a quantity M having a **constant growth rate** r per unit time, we have after n units of time the quantity

$$\boxed{B = M(1 + r)^n.}$$

3. Suppose a sum of 4 million dollars (M) grows in $n = 4$ years to 5 million dollars (B). We are asked to find the average annual growth.

If an initial capital of M (dollars) grows after n years to B (dollars), the geometric mean r of the growth rates for the n years is given by

$$B = M(1 + r)^n, \quad \text{or} \quad r = \sqrt[n]{\frac{B}{M}} - 1.$$

Introducing the given values, $n = 4$, $B = 5$, $M = 4$, we find

$$r = \sqrt[4]{\frac{5,000,000}{4,000,000}} - 1, \quad r = \sqrt[4]{\frac{5}{4}} - 1,$$

and setting $\sqrt[4]{\frac{5}{4}} = x$, so that $\log x = \frac{1}{4} \log \frac{5}{4} = \frac{1}{4}(\log 5 - \log 4) = 0.0217$, we thus get $x = 1.052$ and $r = 1.052 - 1 = 0.052$. Hence the average growth rate comes to 5.2% annually.

REMARK. The number of years required for an amount of **capital to double** can be well approximated by a formula due to Troughton (1968): $n = (70/p) + 0.3$, so that $p = 70/(n - 0.3)$; e.g., if $p = 5\%$, $n = (70/5) + 0.3 = 14.3$. [The exact calculation is $(1 + 0.05)^n = 2$; $n = (\log 2)/(\log 1.05) = 14.2$.]

Exponential Growth Function. If d is the doubling period, r the relative growth rate per year, and the growth equation is $y = ke^{rt}$ with k constant [since $\ln e = 1$ it can be written $\ln y = \ln k + rt$ (cf., Sections 5.6, 5.7)], then $d = (\ln 2)/r = 0.693/r$. If the relative growth rate is $r = 0.07$ per year or 7%, a doubling takes place in $d = 0.693/0.07 \approx 10$ years.

The **critical time** t_{cr} in years that it takes for a quantity Q to increase from its present value Q_0 to a critical or limiting value Q_{cr}, assuming exponential growth with constant rate r in % per year, is $t_{cr} = (230/r)\log(Q_{cr}/Q_0) = (100/r)(2.3)\log(Q_{cr}/Q_0)$. For instance: $Q_{cr}/Q_0 = 25$; $r = 7\%$; $t_{cr} = (230/7)\log 25 = (32.8571)1.3979 = 45.9$ or 46 years.

1.3.8.2 The harmonic mean

Let $x_1, x_2, \ldots, x_n$ be all positive (or all negative) values. The reciprocal value of the arithmetic mean of the reciprocals of these quantities is called the harmonic mean $\bar{x}_H$:

$$\bar{x}_H = \frac{n}{\dfrac{1}{x_1} + \dfrac{1}{x_2} + \cdots + \dfrac{1}{x_n}} = \frac{n}{\sum_{i=1}^{n} \dfrac{1}{x_i}} \quad \text{with} \quad x_i \neq 0 \tag{1.63}$$

[cf., also (1.67)]. In applications it is frequently necessary to assign weights w_i to the individual quantities x_i. The weighted harmonic mean (cf., Example 3) is given by

$$\bar{x}_H = \frac{w_1 + w_2 + \cdots + w_n}{\dfrac{w_1}{x_1} + \dfrac{w_2}{x_2} + \cdots + \dfrac{w_n}{x_n}} = \frac{\sum_{i=1}^{n} w_i}{\sum_{i=1}^{n} \left(\dfrac{w_i}{x_i}\right)}. \tag{1.64}$$

The combined harmonic mean is

$$\bar{x}_H = \frac{n_1 + n_2 + \cdots + n_k}{\dfrac{n_1}{\bar{x}_{H_1}} + \dfrac{n_2}{\bar{x}_{H_2}} + \cdots + \dfrac{n_k}{\bar{x}_{H_k}}}. \tag{1.65}$$

The harmonic mean is called for if the observations of what we wish to express with the arithmetic mean are given in inverse proportion: if the observations involve some kind of reciprocality (for example, if velocities are stated in hours per kilometer instead of kilometers per hour). It is used, in particular, **if the mean velocity is to be computed from different velocities over stated portions of a road** (Example 2) or **if the mean density of gases, liquids, particles, etc. is to be** calculated from the corresponding densities under various conditions. It is also used as a mean lifetime.

Examples

1. In three different stores a certain item sells at the following **prices**: 10 for $1, 5 for $1, and 8 for $1. What is the average number of **units of the item per dollar**?

$$\bar{x}_H = \frac{3}{\frac{1}{10} + \frac{1}{5} + \frac{1}{8}} = \frac{3}{\frac{17}{40}} = \frac{120}{17} = 7.06 \simeq 7.1.$$

1.3 The Path to the Normal Distribution

Check:

$$
\begin{aligned}
1 \text{ unit} &= \$0.100 \\
1 \text{ unit} &= \$0.200 \\
1 \text{ unit} &= \$0.125 \\
\overline{3 \text{ units} = \$0.425:} \quad 1 \text{ unit} &= \frac{\$0.425}{3} = \$0.1417,
\end{aligned}
$$

so that $1.000/0.1417 = 7.06$, which agrees with the above result of 7.1 units per dollar.

2. The classical use for the harmonic mean is a determination of the **average velocity**. Suppose one travels from C to B with an average velocity of 30 km/hr. For the return trip from B to C one uses the same streets, traveling at an average velocity of 60 km/hr. The average velocity for the whole round trip (A_R) is found to be

$$A_R = \frac{2}{\frac{1}{30} + \frac{1}{60}} = 40 \text{ km/hr}.$$

Note: Assuming the distance CB is 60 km, the trip from C to B would take $(60 \text{ km})/(30 \text{ km/hr}) = 2$ hours and the trip from B to C $(60 \text{ km})/(60 \text{ km/hr}) = 1$ hr, so that $A_R = $ (total distance)/(total time) $= (120 \text{ km})/(3 \text{ hr}) = 40$ km/hr.

3. For a certain manufacturing process, the so-called **unit item time** in minutes per item has been determined for $n = 5$ workers. The average time per unit item for the group of 5 workers is to be calculated, given that four of them work for 8 hours and the fifth for 4 hours.

The data are displayed in Table 21. The average unit item time comes to 1.06 minutes/unit.

Table 21

Working time w_i (in minutes)	Unit item time x_i (in minutes/unit)	Output w_i/x_i (in units)
480	0.8	480/0.8 = 600
480	1.0	480/1.0 = 480
480	1.2	480/1.2 = 400
480	1.2	480/1.2 = 400
240	1.5	240/1.5 = 160
$\Sigma w_i = 2{,}160$		$\Sigma(w_i/x_i) = 2{,}040$

$$\bar{x}_H = \frac{\sum w_i}{\sum (w_i/x_i)} = \frac{2{,}160}{2{,}040} = 1.059.$$

If observations are grouped into c classes, with class means x_i and frequencies f_i (where $\sum_{i=1}^{c} f_i = n$), we have

$$\bar{x}_G = \sqrt[n]{x_1^{f_1} \cdot x_2^{f_2} \cdot \ldots \cdot x_c^{f_c}} \qquad \log \bar{x}_G = \frac{1}{n} \sum_{i=1}^{c} f_i \log x_i \quad \text{with } x_i > 0,$$

(1.66)

$$\frac{1}{\bar{x}_H} = \frac{1}{n} \sum_{i=1}^{c} \frac{f_i}{x_i} \qquad \bar{x}_H = \frac{n}{\sum_{i=1}^{c} \frac{f_i}{x_i}} \quad \text{with } x_i \neq 0.$$

(1.67)

The three mean values are related in the following way:

$$\bar{x}_H \leq \bar{x}_G \leq \bar{x}.$$

(1.68)

Equality holds only if the x's are identical, $x_1 = x_2 = \cdots = x_n$. For $n = 2$ the means satisfy the equation

$$\frac{\bar{x}}{\bar{x}_G} = \frac{\bar{x}_G}{\bar{x}_H} \quad \text{or} \quad \bar{x}\bar{x}_H = \bar{x}_G^2.$$

(1.69)

1.3.8.3 Median and mode

A unimodal distribution is said to be **skewed** if considerably more probability mass lies on one side of the mean than on the other. A frequently cited example of a distribution with **skewness** is the frequency distribution of incomes in a country. The bulk of those employed in the U.S. earn less than $1,700 a month; the rest have high to very high income. The arithmetic mean would be much too high to be taken as the average income, in other words, the mean value lies too far to the right. A more informative quantity is in this case the **median** ($\tilde{x}$), which is the value that divides the distribution into halves. An estimate of the median is that value in the sequence of individual values, ordered according to size, which divides the sequence in half. It is important to note that the median is not influenced by the extreme values, whereas the arithmetic mean is rather sensitive to them. Further details are given in Smith (1958) as well as in Rusch and Deixler (1962). Since most of those employed realize an income which is "below average," the median income is smaller than the arithmetic mean of the incomes. The peak of the curve near the left end of the distribution (the mode) gives a still more appropriate value if the bulk of those employed is the object of our studies.

1.3 The Path to the Normal Distribution

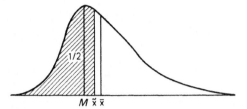

Figure 16 Positive skewness with mode (M), median ($\tilde{x}$) and mean ($\bar{x}$); the median divides the sample distribution into two equal parts.

In Figure 16 $\bar{x}$ lies to the right of $\tilde{x}$, so that the arithmetic mean is greater than the median, or $\bar{x} - \tilde{x}$ is positive; hence one also refers to the distribution as positively skewed. A simpler description is that distributions with positive skewness are "left steep" and exhibit an excessively large positive tail.

For unimodal distributions the **mode** (cf., Figure 16), is the **most frequent sample value** (absolute mode, approximated by $3\tilde{x} - 2\bar{x}$), while for multimodal distributions **relative** modes also appear, these being **values which occur more frequently than do their neighboring values**, in other words, **the relative maxima of the probability density** (cf., also Dalenius 1965). For multimodal distributions (cf., Figure 17) the modes are appropriate mean values. Examples of bi- and trimodal distributions are the colors of certain flowers and butterflies.

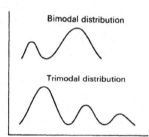

Figure 17 Distributions with more than one mode.

Estimation of the median

If the sequence consists of an odd number of values, then the median is the middle value of the values ordered by magnitude, while if n is even, then there are two middle values, $\tilde{x}_1$ and $\tilde{x}_2$, in which case the median (or better, pseudomedian) is given by $\tilde{x} = \frac{1}{2}(\tilde{x}_1 + \tilde{x}_2)$ (cf., also Remark 4 in Section 1.3.6.3 [and the Remark at the end of Section 1.3.8.6]).

If a sequence of specific values grouped into classes is given, the median is estimated by linear interpolation according to

$$\boxed{\tilde{x} = \tilde{L} + b\left(\frac{n/2 - (\sum f)_{\tilde{L}}}{f_{\text{Median}}}\right),} \qquad (1.70)$$

Table 22

Class	Class mean x_i	Frequency f_i
5 but less than 7	6	4
7 but less than 9	8	8
9 but less than 11	10	11
11 but less than 13	12	7
13 but less than 15	14	5
15 but less than 17	16	3
17 but less than 19	18	2
		n = 40

where $\tilde{L}$ = lower class limit of the median class; b = class width; n = number of values; $(\sum f)_{\tilde{L}}$ = sum of the frequency values of all classes below the median class; f_{Median} = number of values in the median class.

In the example in Table 22, since the median must lie between the 20th and 21st values, and since $4 + 8 = 12$, whereas $4 + 8 + 11 = 23$, it is clear that the median must lie in the 3rd class.

$$\tilde{x} = \tilde{L} + b\left(\frac{n/2 - (\sum f)_{\tilde{L}}}{f_{\text{Median}}}\right) = 9 + 2\left(\frac{40/2 - 12}{11}\right) = 10.45.$$

For the **median $\tilde{\mu}$ of a population** with random variable X (cf. Section 1.2.6) we have the inequalities $P(X < \tilde{\mu}) \leq 0.5$ and $P(X > \tilde{\mu}) \leq 0.5$. For a continuous population the median is defined by $F(\tilde{\mu}) = 0.5$.

REMARK. A **quantile** ξ_p (Greek xi) (also called a fractile, percentile) is a location measure (or parameter) defined by $P(X \leq \xi_p) \geq p$, $P(X \geq \xi_p) \geq 1 - p$ (cf., Section 1.2.6). The value ξ_p of a continuous distribution thus has the property that the probability of a smaller value is precisely equal to p, and that of a larger value, to $1 - p$. For a discrete distribution "precisely" is to be replaced by "at most". Particular cases of quantiles at $p = \frac{1}{2}, \frac{1}{4}, \frac{3}{4}, q/10$ ($q = 1, 2, \ldots, 9$), $r/100$ ($r = 1, 2, \ldots, 99$) are referred to as the median, lower **quartile** or Q_1 (cf., Section 1.3.8.7), upper quartile or Q_3, or qth **decile** (in Sections 1.3.3.6–7 called $DZ_1, \ldots, DZ_9$), and rth **percentile** or rth centile, respectively. For ungrouped samples, e.g., the value with the order number $(n + 1)p/100$ is an estimate x_p of the pth percentile ξ_p (x_p is the sample pth percentile; e.g. the 80th percentile for $n = 125$ values in ascending order is the $(125 + 1)80/100 = 100.8 = 101$st value. For grouped samples, the quantiles are computed according to (1.70) with $n/2$ replaced by $in/4$ ($i = 1, 2, 3$; quartile), $jn/10$ ($j = 1, 2, \ldots, 9$; decile), $kn/10$ ($k = 1, 2, \ldots, 99$; percentile), and the median and median class by the desired quantile and its class. The corresponding parameters are ξ_p. For discrete distributions a quantile cannot always be specified. Certain selected quantiles of the more important distribution functions, which play as upper tail probabilities a particular role in test theory, are tabulated in terms of $1 - p = \alpha$ (e.g., Tables 27 and 28, Section 1.5.2) or $1 - p = P$ (e.g., Table 30a, Section 1.5.3), rather than in terms of p as in the above definitions.

Rough estimate of the mode

Strictly speaking, the mode is the value of the variable that corresponds to the maximum of the ideal curve that best fits the sample distribution. Its determination is therefore difficult. For most practical purposes a satisfactory estimate of the mode is

$$M = L + b\left(\frac{f_u - f_{u-1}}{2 \cdot f_u - f_{u-1} - f_{u+1}}\right), \qquad (1.71)$$

where L = lower class limit of the most heavily occupied class; b = class width; f_u = number of values in the most heavily occupied class; f_{u-1}, f_{u+1} = numbers of values in the two neighboring classes.

EXAMPLE. We use the distribution of the last example:

$$M = L + b\left(\frac{f_u - f_{u-1}}{2 \cdot f_u - f_{u-1} - f_{u+1}}\right) = 9 + 2\left(\frac{11 - 8}{2 \cdot 11 - 8 - 7}\right) = 9.86.$$

Here M is the maximum of an approximating parabola that passes through the three points (x_{u-1}, f_{u-1}), (x_u, f_u), and (x_{u+1}, f_{u+1}). The corresponding arithmetic mean lies somewhat higher ($\bar{x} = 10.90$). For unimodal distributions with positive skewness, as in the present case (cf., Fig. 18), the inequality $\bar{x} > \tilde{x} > M$ holds. This is easy to remember because the sequence mean, median, mode is in alphabetical order.

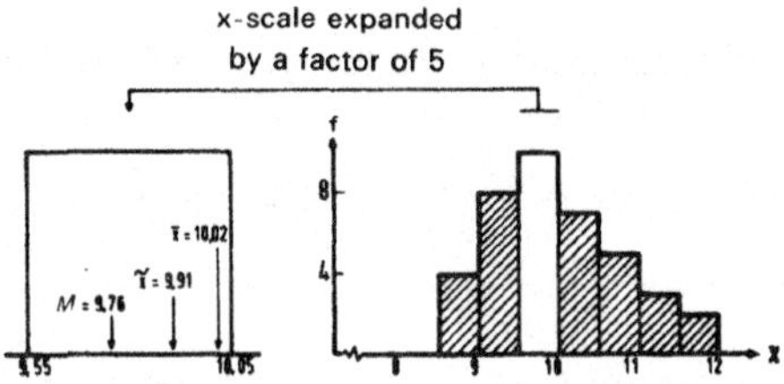

Figure 18 Positively skewed left steep frequency distribution.

For continuous unimodal symmetric distributions the mode, median, and mean coincide. With skewed distributions the median and mean can still coincide. This of course holds also for U-shaped distributions, characterized by the two modes and a minimum ($x_{\min}$) lying between them. Examples of distributions of this type are influenza mortality as a function of age (since it is greatest for infants and for the elderly) and cloudiness in the northern latitudes (since days on which the sky is, on the average, half covered are rare, while clear days and days on which the sky is overcast are quite common); cf., Yasukawa (1926).

1.3.8.4 The standard error of the arithmetic mean and of the median

Assuming independent random variables, we know that with increasing sample size suitable statistics of the sample tend to the parameters of the parent population; in particular the mean $\bar{X}$ determined from the sample tends to μ (cf., end of Section 1.3.1).

How much can $\bar{x}$ deviate from μ? The smaller the standard deviation σ of the population and the larger the sample size n, the smaller will be the deviation. Since the mean $\bar{X}$ is again a random variable, it also has a probability distribution. The (theoretical) standard deviation of the mean $\bar{X}$ of n random variables $X_1, \ldots, X_n$, all of which are independent and identically distributed (cf., remark 4(2) at end of Section 1.2.6: $\sigma_{\bar{x}} = \sqrt{\sigma_{\bar{x}}^2}$) is determined for $N = \infty$ (i.e., for random samples with replacement or with $N \gg n$; cf., Section 3.1.1) by the formula

$$\sigma_{\bar{x}} = \frac{\sigma}{\sqrt{n}}, \qquad (1.72)$$

where σ is the standard deviation of the x_i. As an estimate for $\sigma_{\bar{x}}$, the so-called **standard error of the arithmetic mean**, one has ($N = \infty$, cf. Section 3.1.1)

$$s_{\bar{x}} = \frac{s}{\sqrt{n}} = \sqrt{\frac{\sum (x - \bar{x})^2}{n(n-1)}} = \sqrt{\frac{\sum x^2 - (\sum x)^2/n}{n(n-1)}}. \qquad (1.73)$$

For observations with unequal weight w we have

$$s_{\bar{x}} = \sqrt{\frac{\sum w(x - \bar{x})^2}{(n-1)\sum w}} \quad \text{with} \quad \bar{x} = \frac{\sum wx}{\sum w}.$$

The physicist regards s as the mean error of a single measurement and $s_{\bar{x}}$ as the mean error of the mean. A halving of this error requires a quadrupling of the sample size: $(s/\sqrt{n})/2 = s/\sqrt{4n}$. For a **normal distribution** the standard error of the median (see box on page 95) has the value

$$\sqrt{\frac{\pi}{2}} \cdot \frac{\sigma}{\sqrt{n}} \quad \text{with} \quad \sqrt{\frac{\pi}{2}} \simeq 1.253; \qquad (1.74)$$

thus $\bar{x}$ is a more precise mean than $\tilde{x}$ (cf., also below).

1.3 The Path to the Normal Distribution

The **reliability of an estimate** used to be indicated by the standard deviation; in the case of the mean and summarizing some data, the result is written in the form

$$\boxed{\bar{x} \pm s_{\bar{x}}} \tag{1.75}$$

provided the observations come from a normal distribution or the deviations from a normal distribution are not weighty and a low degree of generalization is intended. If a **higher degree of generalization** is intended, then the confidence interval for μ is preferred, provided some requirements are met (cf. Sections 1.4.1 and 3.1.1). In the case of $\bar{x} = 49.36$, $s_{\bar{x}} = 0.1228$, we would write (1.75) as 49.4 ± 0.1. (Carrying more decimals would not make sense, because an "error" of 0.12 renders the second decimal of the estimate questionable). Frequently the percentage error was also stated. For our example it comes to

$$\boxed{\pm \frac{s_{\bar{x}} \cdot 100}{\bar{x}}} = \pm \frac{0.2 \cdot 100}{49.4} = \pm 0.4\%. \tag{1.76}$$

> If the observations are **not normally distributed** and some data are summarized, then the median $\tilde{x}$ with its standard error $s_{\tilde{x}}$ is stated: $\tilde{x} \pm s_{\tilde{x}}$. Arranging the observations in ascending order, the **standard error of the median** is estimated by $[1/3.4641]$ {[the value of the $(n/2 + \sqrt{3n}/2)$th observation] $-$ [the value of the $(n/2 - \sqrt{3n}/2)$th observation]}, with both values rounded up to the next whole number. If the observations are a **random sample**, it is better to generalize in giving the **confidence interval for the median** of the population (Sections 3.1.4 and 4.2.4).

EXAMPLE FOR $\tilde{x} \pm s_{\tilde{x}}$. x_i: 18, 50, 10, 39, 12 ($n = 5$).

We arrange the observations in ascending order of size from smallest to largest: 10, 12, 18, 39, 50; $\tilde{x} = 18$; $s_{\tilde{x}} = (a - b)/3.4641$ with

$$a = \left(\frac{n}{2} + \frac{\sqrt{3n}}{2}\right) = \left(\frac{5}{2} + \frac{\sqrt{15}}{2}\right) = 2.5 + 1.9 = \text{5th observation}$$

$$b = \left(\frac{n}{2} - \frac{\sqrt{3n}}{2}\right) = \left(\frac{5}{2} - \frac{\sqrt{15}}{2}\right) = 2.5 - 1.9 = \text{1st observation}$$

$$s_{\tilde{x}} = [\text{(value of the 5th observation)} - \text{(value of the 1st observation)}]/3.4641$$

$$= (50 - 10)/3.4641 = 11.55 \text{ or } 12.$$

Result: 18 ± 12.

Sums, differences, and quotients, together with their corresponding standard errors of means of independent samples, have the form (Fenner 1931):

Addition:

$$\bar{x}_1 + \bar{x}_2 \pm \sqrt{s_{\bar{x}_1}^2 + s_{\bar{x}_2}^2}$$

$$\bar{x}_1 + \bar{x}_2 + \bar{x}_3 \pm \sqrt{s_{\bar{x}_1}^2 + s_{\bar{x}_2}^2 + s_{\bar{x}_3}^2} \qquad (1.77)$$

Subtraction:

$$\bar{x}_1 - \bar{x}_2 \pm \sqrt{s_{\bar{x}_1}^2 + s_{\bar{x}_2}^2} \qquad (1.78)$$

Multiplication:

$$\bar{x}_1 \bar{x}_2 \pm \sqrt{\bar{x}_1^2 s_{\bar{x}_2}^2 + \bar{x}_2^2 s_{\bar{x}_1}^2}$$

$$\bar{x}_1 \bar{x}_2 \bar{x}_3 \pm \sqrt{\bar{x}_1^2 \bar{x}_2^2 s_{\bar{x}_3}^2 + \bar{x}_1^2 \bar{x}_3^2 s_{\bar{x}_2}^2 + \bar{x}_2^2 \bar{x}_3^2 s_{\bar{x}_1}^2} \qquad (1.79)$$

Division:

$$\frac{\bar{x}_1}{\bar{x}_2} \pm \frac{1}{\bar{x}_2^2} \sqrt{\bar{x}_1^2 s_{\bar{x}_2}^2 + \bar{x}_2^2 s_{x_1}^2} \qquad (1.80)$$

With stochastic dependence ($\rho \neq 0$)—between, not within the samples, one has

Addition:

$$\bar{x}_1 + \bar{x}_2 \pm \sqrt{s_{\bar{x}_1}^2 + s_{\bar{x}_2}^2 + 2r s_{\bar{x}_1} s_{\bar{x}_2}} \qquad (1.77a)$$

Subtraction:

$$\bar{x}_1 - \bar{x}_2 \pm \sqrt{s_{\bar{x}_1}^2 + s_{\bar{x}_2}^2 - 2r s_{\bar{x}_1} s_{\bar{x}_2}} \qquad (1.78a)$$

where r is an estimate of ρ. The corresponding relations for multiplication and division are quite complicated and hold only for large n.

Let us mention here the frequently applied **power product law of propagation of errors**. Suppose we are given the functional relation

$$h = k x^a y^b z^c \ldots \qquad (1.81)$$

(with known constants $k, a, b, c, \ldots$ and variables $x, y, z, \ldots$), and that we are interested in an estimate of the mean $\bar{h}$ and the mean relative error $s_{\bar{h}}/\bar{h}$. The observations $x_i, y_i, z_i, \ldots$ are assumed independent. We need the means $\bar{x}, \bar{y}, \bar{z}, \ldots$ and the corresponding standard deviations. Then the mean relative error is given by

$$s_{\bar{h}}/\bar{h} = \sqrt{(a \cdot s_{\bar{x}}/\bar{x})^2 + (b \cdot s_{\bar{y}}/\bar{y})^2 + (c \cdot s_{\bar{z}}/\bar{z})^2 + \ldots} \qquad (1.82)$$

More on this may be found in Parratt (1961), Barry (1978, cited on page 200), and Strecker (1980, cited on page 615).

1.3.8.5 The range

The simplest of all the elementary dispersion measures is the **range** R, which is the difference between the largest and the smallest value in a sample:

$$R = x_{\max} - x_{\min}. \qquad (1.83)$$

If the sample consists of only two values, specifying the range exhausts the information on the dispersion in the sample. As the size of the sample increases, the range becomes a less and less reliable measure for the dispersion in the sample since the range, determined by the two extreme values only, contains no information about the location of the intermediate values (cf., also end of Section 7.3.1).

Remarks concerning the range

1. If you have to determine standard deviations often, it is worthwhile to familiarize yourself with a method presented by Huddleston (1956). The author proceeds from systematically trimmed ranges which, when divided by appropriate factors, represent good estimates of s; tables and examples can be found in the original work (cf., also Harter 1968).

2. If n' mutually independent pairs of observations are given, then the ranges can be used in estimating the standard deviation

$$\hat{s} = \sqrt{\frac{\sum R^2}{2n'}}. \qquad (1.84)$$

The caret on the s indicates it is an estimate.

3. If several samples of size n, with $n \leq 13$, are taken, then the standard deviation can be roughly estimated from the mean range ($\bar{R}$):

$$\hat{s} = \frac{1}{d_n} \bar{R}. \qquad (1.85)$$

This formula involves $1/d_n$, a proportionality factor that depends on the size of the samples and that presupposes normal distribution. This factor can be found in Table 156. We will return to this later (end of Section 7.3.1).

4. A rule of thumb, due to Sturges (1926), for determining a suitable class width b of a frequency distribution is based on the range and size of the sample:

$$b \simeq \frac{R}{1 + 3.32 \log n}. \qquad (1.86)$$

For the distribution given in Section 1.3.3 (Table 11) we get $b = 2.4$; we had chosen $b = 3$.

5. The formula

$$\frac{R}{2}\sqrt{\frac{n}{n-1}} \geq s \qquad (1.87)$$

allows an estimate of the maximum standard deviation in terms of the range (Guterman 1962). The deviation of an empirical standard deviation from the upper limit can serve as a measure of the accuracy of the estimate. For the three values 3, 1, 5 with $s = 2$ we have

$$s < \frac{4}{2}\sqrt{\frac{3}{3-1}} = 2.45.$$

Equation (1.87) provides a means of obtaining a rough estimate of the standard deviation if only the range is known and nothing can be said concerning the shape of the distribution.

6. Rough estimate of the standard deviation based on the extreme values of hypothetical samples of a very large size: Assume the underlying distribution of the values is approximately normal. Then a rough estimate of the standard deviation of the population is given by

$$\hat{s} \simeq \frac{R}{6} \qquad (1.88)$$

because in a **normal distribution** the range 6σ is known to encompass 99.7% of all values. For the **triangular distribution**, we have $R/4.9 \lesssim \hat{s} \lesssim R/4.2$ ($\triangleright$: $\hat{s} \simeq R/4.2$; $\triangle$s: $\hat{s} \simeq R/4.9$; $\triangleleft$: $s \simeq R/4.2$)—which can be thought of as the basic forms of the positively skewed, symmetric, and negatively skewed distributions. For the **uniform or rectangular distribution** ($\square$) we have $\hat{s} \simeq R/3.5$, and for the **U-shaped distribution** $\hat{s} \simeq R/2$. As an example for the latter, we consider the sequence 3, 3, 3, 3, 10, 17, 17, 17, 17, which has an approximately U-shaped distribution. The standard deviation is

$$s = \sqrt{\frac{8 \cdot 7^2}{9-1}} = 7 \quad \text{whereas} \quad \hat{s} = \frac{17-3}{2} = 7.$$

The reader should examine other samples.

7. A certain peculiarity of the range is worth mentioning: Regardless of the distribution of the original population, the distributions of many statistics tend with increasing n to a normal distribution (by the central limit theorem, $\bar{X}_n$ is asymptotically normally distributed); this is not true for the distribution of the range. The distribution of the estimator S^2, estimated by s^2, does tend (very slowly) to a normal distribution as n increases.

1.3.8.6 The interdecile range

Suppose the data, ordered according to magnitude, are partitioned by nine values into ten groups of equal size. These values are called deciles and are denoted by $DZ_1, DZ_2, \ldots, DZ_9$. The first, second, ..., ninth decile is obtained by counting off $n/10, 2n/10, \ldots, 9n/10$ data points. The kth decile can be defined as the value that corresponds to a certain region on the scale of an

1.3 The Path to the Normal Distribution

symmetric distribution

left-steep distribution

U-shaped distribution

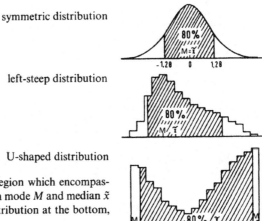

Figure 19 The interdecile region which encompasses 80% of a distribution with mode M and median $\tilde{x}$ (except for the U-shaped distribution at the bottom, which exhibits two modes).

interval-wise constructed frequency distribution in such a way that exactly $10k\%$ of the cases lie below this value. Let us note that in accordance with this the 5th decile, the point below which 5 tenths of the observations lie, is in fact the median.

A **dispersion statistic** that, in contrast to the full range, depends very little on the extreme values, nevertheless involves the overwhelming majority of the cases, and **exhibits very slight fluctuation from sample to sample** is the **interdecile range** I_{80}, which encompasses 80% of a sample distribution:

$$I_{80} = DZ_9 - DZ_1. \tag{1.89}$$

Deciles are interpolated linearly according to the formula (1.70) where, in place of $n/2$, $0.1n$ or $0.9n$ appears, $\tilde{L}$ is replaced by the lower class limit of the decile class, $(\sum f)_{\tilde{L}}$ by the sum of the frequency values of all the classes below the decile class (class containing the particular decile) and f_{Median} by the frequency value of the decile class. For the example in Section 1.3.8.3 one gets accordingly

$$DZ_1 = 5 + 2\frac{4-0}{4} = 7, \qquad DZ_9 = 15 + 2\frac{36-35}{3} = 15.67,$$

the interdecile range is $I_{80} = 15.67 - 7 = 8.67$.

We can also get DZ_1 directly as the lower class limit of the 2nd class by counting off $n/10 = 40/10 = 4$ values. DZ_9 must follow the $9n/10 = (9)(40)/10 = 36$th value. 35 values are distributed among classes 1–5. Thus we shall also need $36 - 35 = 1$ value from class 6, which contains 3 values. We multiply the number $\frac{1}{3}$ by the class width, obtaining thereby the correction term, which, when added to the lower class limit or class 6, gives the decile.

Two other dispersion statistics, the mean deviation of the mean and the median deviation will be introduced in Section 3.1.3.

A rough estimate of the mean and standard deviation for nearly normally distributed values, based on the first, fifth, and ninth deciles, is given by

$$\bar{x} \simeq 0.33(DZ_1 + \tilde{x} + DZ_9),\quad (1.90)$$

$$s \simeq 0.39(DZ_9 - DZ_1).\quad (1.91)$$

For our example (cf., Section 1.3.8.3) we find according to (1.90) and (1.91) that $\tilde{x} \simeq 0.33(7 + 10.45 + 15.67) = 10.93$, $s \simeq 0.39(15.67 - 7) = 3.38$. On comparing with $\bar{x} = 10.90$ and $s = 3.24$, we see that the quick estimates (cf., also the end of Section 1.3.6.7) are useful. For normally distributed samples the agreement is better (a good check on the method). If the samples are not normally distributed, quick estimates under circumstances similar to those given in the example can represent a better estimate of the parameters of interest than the standard estimates $\bar{x}$ and s.

REMARK. As a parameter of the central tendency or location of the distribution, in addition to the interquartile range $I_{50} = Q_3 - Q_1$ (for a normal distribution, the region $\tilde{x} \pm I_{50}/2$ includes the exact central 50% of the observations; see Section 1.3.8.7), we also have the two sided quartile-weighted median $\tilde{\tilde{x}} = (Q_1 + 2\tilde{x} + Q_3)/4$; $\tilde{\tilde{x}}$ is yet remarkably robust and frequently more informative than $\tilde{x}$, especially with odd and skewed distributions.

1.3.8.7 Skewness and kurtosis

With regard to possible deviations from the normal distribution, one singles out two distinct types (cf., Figure 20):

1. One of the two tails is lengthened, and the distribution becomes skewed: if the left part of the curve is lengthened, one speaks of negative skewness; if the right part is lengthened, positive skewness. In other words, if the principal part of a distribution is concentrated on the left side of the distribution (left-steep), it is said to have a positive skewness.

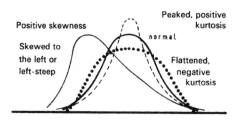

Figure 20 Deviations from the symmetric bell-shaped curve (normal distribution).

1.3 The Path to the Normal Distribution

2. The maximum lies higher or lower than that of the normal distribution. If it is higher, with the variances equal, so that the shape of the curve (the bell) is more peaked, the coefficient of kurtosis will be positive (i.e., scantily occupied flanks, with a surplus of values near the mean and in the tails of the distribution). With a negative kurtosis the maximum is lower, the bell is more squat, and the distribution is flatter than the normal distribution.

Skewness and **kurtosis** can be determined exactly from the moments. The following measures of skewness occasionally prove satisfactory: Of importance is

$$\boxed{\text{Skewness I} = \frac{3(\bar{x} - \tilde{x})}{s}} \tag{1.92}$$

with the rarely attained limits -3 and $+3$. If the arithmetic mean lies above the median, as in Figure 18, a positive skewness index arises. Another useful measure of skewness, the 1-9 decile coefficient of skewness, is based on the median and interdecile range:

$$\boxed{\text{Skewness II} = \frac{(DZ_9 - \tilde{x}) - (\tilde{x} - DZ_1)}{(DZ_9 - \tilde{x}) + (\tilde{x} - DZ_1)} = \frac{DZ_9 + DZ_1 - 2\tilde{x}}{DZ_9 - DZ_1}} \tag{1.93}$$

and varies between -1 and $+1$.

Remark on the quartiles. There exist 3 values which partition a frequency distribution into 4 equal parts. The central value is the median; the other two are designated the lower or first quartile and the upper or third quartile, i.e., the first quartile Q_1 is the value that lies at the end of the first quarter of the sequence of measured values, ordered by size; Q_3 is the value at the end of the third quarter of the sequence (cf., Section 1.3.8.3).

If we replace DZ_1 and DZ_9 by Q_1 and Q_3 in (1.93), thus emphasizing the central part of the distribution, we find a third measure for the skewness (range -1 to $+1$):

$$\boxed{\text{Skewness III} = \frac{(Q_3 - \tilde{x}) - (\tilde{x} - Q_1)}{(Q_3 - \tilde{x}) + (\tilde{x} - Q_1)} = \frac{Q_3 + Q_1 - 2\tilde{x}}{Q_3 - Q_1}.} \tag{1.94}$$

In a symmetric distribution, all three skewness coefficients vanish.

A simple measure for the coefficient of kurtosis based on quartiles and deciles is

$$\boxed{\text{Kurtosis} = \frac{Q_3 - Q_1}{2(DZ_9 - DZ_1)}.} \tag{1.95}$$

For the normal distribution it has the value 0.263.

If the difference between the mean and the mode is at least twice the corresponding standard error,

$$\bar{x} - M \geq 2[\sqrt{3s/2n}], \tag{1.96}$$

then the underlying distribution cannot be considered symmetric. For the data of Table 22 we have

$$10.90 - 9.86 = 1.04 > 0.697 = 2\left[\sqrt{\frac{3 \cdot 3.24}{2 \cdot 40}}\right];$$

thus the coefficient of skewness should be evaluated as a further characteristic of the underlying, unknown distribution.

EXAMPLE. We use the values of the last example:

$$\text{Skewness I} = \frac{3(10.90 - 10.45)}{3.24} = 0.417,$$

$$\text{Skewness II} = \frac{15.67 + 7.00 - 2 \cdot 10.45}{15.67 - 7.00} = 0.204,$$

$$\text{Skewness III} = \frac{13.00 + 8.50 - 2 \cdot 10.45}{13.00 - 8.50} = 0.133,$$

cf.,

$$Q_1 = 7 + 2\left(\frac{10-4}{8}\right) = 8.5 \qquad Q_3 = 13 + 2\left(\frac{30-30}{5}\right) = 13$$

(by (1.70) with $n/4$ or $3n/4$ in place of $n/2$, etc.), so

$$\text{Kurtosis} = \frac{13.00 - 8.50}{2(15.67 - 7.00)} = 0.260.$$

This distribution exhibits the kurtosis of a normal distribution and positive skewness.

Important measures of skewness and kurtosis in a population are the third and fourth moments about the mean, i.e., the average values of $(x - \mu)^3$ and $(x - \mu)^4$ over the whole population. To render these measures scale invariant they are divided by σ^3 and σ^4 respectively. The coefficient of skewness is $\alpha_3 = E(X - \mu)^3/\sigma^3$, and the coefficient of kurtosis

$$\alpha_4 = [E(X - \mu)^4/\sigma^4] - 3.$$

$$a_3 = \frac{\sum f_i(x_i - \bar{x})^3}{n \cdot s^3}, \tag{1.97}$$

1.3 The Path to the Normal Distribution

$$a_4 = \frac{\sum f_i(x_i - \bar{x})^4}{n \cdot s^4} - 3. \tag{1.98}$$

Note that s (1.31, 1.35, 1.40, 1.44) is here defined with the denominator "n," and not with "$n-1$". For a symmetric distribution $\alpha_3 = 0$; for the normal distribution $\alpha_4 = 0$. If α_3 is positive, we have positive skewness; if negative, negative skewness. A distribution with a high peak—steeper than the normal distribution—exhibits a positive value of α_4; a distribution that is flatter than the normal distribution exhibits a negative value of α_4. The kurtosis measures peakedness combined with tailedness and corresponding depletion of the flanks, and hence is strongly negative for a bimodal curve (Finucan 1964, cf., also Chissom 1970 and Darlington 1970). The rectangular distribution with distinct flanks therefore also has a negative kurtosis ($\alpha_4 = -1.2$). This is true as well for every triangular distribution ($\alpha_4 = -0.6$), which exhibits more fully developed flanks than a normal distribution with the same variance but has no tails.

But first another remark on moments. Quantities of the form

$$\frac{\sum f_i(x_i - \bar{x})^r}{n} = m_r \tag{1.99}$$

are called the **rth sample moments** (m_r). For $r = 2$, (1.99) yields approximately the sample variance. Both moment coefficients can be written more concisely as

$$a_3 = m_3/s^3 \quad \text{and} \quad a_4 = m_4/s^4 - 3. \tag{1.97a} \quad (1.98a)$$

If the class width does not equal one ($b \neq 1$), our definition becomes

$$m_r = \frac{\sum f_i \left(\frac{x_i - \bar{x}}{b}\right)^r}{n}. \tag{1.100}$$

In order to facilitate the calculation it is customary to relate the power moments not to the arithmetic mean but rather to some arbitrary origin, say to the number d, which identifies the most populous class of a frequency distribution. We are already acquainted with this method (multiplication procedure, cf. Section 1.3.6.4). The moments so obtained are denoted by m'_r to distinguish them from the moments m_r. Writing again $(x - d)/b = z$, we get the **first through fourth order moments of the sample** (cf., Table 23) from the formulas:

$$\text{1st order moment} \quad \boxed{m'_1 = \frac{\sum f_i \cdot z_i}{n}} = \frac{18}{40} = 0.45, \quad (1.101)$$

$$\text{2nd order moment} \quad \boxed{m'_2 = \frac{\sum f_i \cdot z_i^2}{n}} = \frac{110}{40} = 2.75, \quad (1.102)$$

$$\text{3rd order moment} \quad \boxed{m'_3 = \frac{\sum f_i \cdot z_i^3}{n}} = \frac{216}{40} = 5.40, \quad (1.103)$$

$$\text{4th order moment} \quad \boxed{m'_4 = \frac{\sum f_i \cdot z_i^4}{n}} = \frac{914}{40} = 22.85. \quad (1.104)$$

Table 23 includes an additional column with the products $f_i(z_i + 1)^4$, which will be used to test the computations. The column sums can be readily checked with the help of the relation

$$\boxed{\sum f_i(z_i+1)^4 = \sum f_i + 4\sum f_i z_i + 6\sum f_i z_i^2 + 4\sum f_i z_i^3 + \sum f_i z_i^4:} \quad (1.105)$$

$2550 = 40 + 72 + 660 + 864 + 914$. The above also provides us with estimates of the following parameters:

1. The mean

$$\boxed{\bar{x} = d + bm'_1,} \quad (1.106)$$

$$\bar{x} = 9.8 + 0.5 \cdot 0.45 = 10.025.$$

Table 23

x_i	f_i	z_i	$f_i z_i$	$f_i z_i^2$	$f_i z_i^3$	$f_i z_i^4$	$f_i(z_i+1)^4$
8.8	4	−2	−8	16	−32	64	4
9.3	8	−1	−8	8	−8	8	0
d = 9.8	11	0	0	0	0	0	11
10.3	7	1	7	7	7	7	112
10.8	5	2	10	20	40	80	405
11.3	3	3	9	27	81	243	768
11.8	2	4	8	32	128	512	1,250
	40		18	110	216	914	2,550

1.3 The Path to the Normal Distribution

2. The variance

$$s^2 = b^2(m_2' - m_1'^2),\qquad(1.107)$$

$$s^2 = 0.5^2(2.75 - 0.45^2) = 0.637.$$

3. The skewness

$$a_3 = \frac{b^3(m_3' - 3m_1'm_2' + 2m_1'^3)}{s^3}\qquad(1.108)$$

$$a_3 = \frac{0.5^3 \cdot (5.40 - 3 \cdot 0.45 \cdot 2.75 + 2 \cdot 0.45^3)}{0.5082} = 0.460.$$

4. The kurtosis

$$a_4 = \frac{b^4 \cdot (m_4' - 4 \cdot m_1'm_3' + 6 \cdot m_1'^2 m_2' - 3 \cdot m_1'^4)}{s^4} - 3\qquad(1.109)$$

$$a_4 = \frac{0.5^4 \cdot (22.85 - 4 \cdot 0.45 \cdot 5.40 + 6 \cdot 0.45^2 \cdot 2.75 - 3 \cdot 0.45^4)}{0.4055} - 3$$

$$a_4 = -0.480.$$

The sums $\sum f_i z_i$, $\sum f_i z_i^2$, $\sum f_i z_i^3$, and $\sum f_i z_i^4$ can also be determined with the help of the **summation procedure** introduced in Section 1.3.6.4. In addition to the quantities $\delta_{1,2}$ and $\varepsilon_{1,2}$ we determine, in terms of columns S_4 and S_5, the four sums ζ_1 and ζ_2 (Greek zeta 1 and 2) as well as η_1 and η_2 (Greek eta 1 and 2) (see Table 24) and obtain

Table 24

f_i	S_1	S_2	S_3	S_4	S_5
4	4	4	4	4	4
8	12	16 = δ_1	20 = ε_1	24 = ζ_1	28 = η_1
11					
7	17	34 = δ_2	60 = ε_2	97 = ζ_2	147 = η_2
5	10	17	26	37	50
3	5	7	9	11	13
2	2	2	2	2	2

$\sum f_i z_i = \delta_2 - \delta_1 = 34 - 16 = 18,$

$\sum f_i z_i^2 = 2\varepsilon_2 + 2\varepsilon_1 - \delta_2 - \delta_1 = 2 \cdot 60 + 2 \cdot 20 - 34 - 16 = 110,$

$\sum f_i z_i^3 = 6\zeta_2 - 6\zeta_1 - 6\varepsilon_2 + 6\varepsilon_1 + \delta_2 - \delta_1,$

$\sum f_i z_i^4 = 24\eta_2 + 24\eta_1 - 36\zeta_2 - 36\zeta_1 + 14\varepsilon_2 + 14\varepsilon_1 - \delta_2 - \delta_1,$

$\sum f_i z_i^3 = 6 \cdot 97 - 6 \cdot 24 - 6 \cdot 60 + 6 \cdot 20 + 34 - 16 = 216,$

$\sum f_i z_i^4 = 24 \cdot 147 + 24 \cdot 28 - 36 \cdot 97 - 36 \cdot 24 + 14 \cdot 60 + 14 \cdot 20 - 34 - 16 = 914.$

The statistics can then be found by the formulas (1.101) to (1.109).

When dealing with very extensive samples, and provided the sample distribution exhibits no asymmetry, you should use the moments modified according to Sheppard:

$$\boxed{s^2_{\text{mod}} = s^2 - b^2/12,} \qquad (1.46)$$

$$\boxed{m'_{4,\text{mod}} = m'_4 - (1/2)m'_2 b^2 + (7/240)b^4.} \qquad (1.110)$$

The measures for the skewness and kurtosis arrived at by way of moments have the advantage that the standard errors are known.

Summary

If the data are grouped into classes of class width b, with class means x_i and frequencies f_i, then the mean, variance, and moment coefficients for skewness and kurtosis can be estimated according to

$$\boxed{\begin{aligned}
\bar{x} &= d + b\left(\frac{\sum fz}{n}\right), & (1.111) \\
s^2 &= b^2\left(\frac{\sum fz^2 - (\sum fz)^2/n}{n-1}\right), & (1.112) \\
a_3 &= \frac{b^3}{s^3}\left(\frac{\sum fz^3}{n} - 3\left(\frac{\sum fz^2}{n}\right)\left(\frac{\sum fz}{n}\right) + 2\left(\frac{\sum fz}{n}\right)^3\right), & (1.113) \\
a_4 &= \frac{b^4}{s^4}\left(\frac{\sum fz^4}{n} - 4\left(\frac{\sum fz^3}{n}\right)\left(\frac{\sum fz}{n}\right) + 6\left(\frac{\sum fz^2}{n}\right)\left(\frac{\sum fz}{n}\right)^2 - 3\left(\frac{\sum fz}{n}\right)^4\right) - 3, & (1.114)
\end{aligned}}$$

where d = assumed mean, usually the mean of the most strongly occupied class; b = class width; f = class frequencies, more precisely f_i; and z = deviations $z_i = (x_i - d)/b$: the class containing the mean d gets the number $z = 0$, with classes below being assigned the numbers $z = -1, -2, \ldots$ in descending order, while those in ascending order the numbers $z = 1, 2, \ldots$.

The method of moments was introduced by Karl Pearson (1857–1936). The notions of standard deviation and normal distribution also originated with him.

We are now in a position to discuss at length a one dimensional frequency distribution along with the tabulated and graphical presentation in terms of the four parameter types: means, measures of variance, measures of skewness, and measures of kurtosis.

The statistics $x_{\min}, Q_1, \tilde{x}, Q_3, x_{\max}$, and other measures based on them are sufficient for a survey and appropriate for every distribution type. A good insight into the form of a distribution can generally be gained from quantiles (Section 1.3.8.3). Also measures based on them are often more informative than those based on the mean and standard deviation, the latter being strongly influenced by extreme values. In the case of multimodal distribution, estimates of the modes have to be listed as well.

The more obvious deviations from normal distribution (e.g., left-steepness (see pages 99, 100), right-steepness (seldom!) or/and multimodality) which are already apparent in a counting table, are tabulated or, better yet, graphed—for small sample size, as points on a line, (or, for a two dimensional distribution, as points in the plane; cf., e.g., Section 5.4.4, Fig. 51); for large sample size, as a **histogram** (see E. S. Pearson and N. W. Please 1975 [Biometrika **62**, 223–241, D. W. Scott 1979 [Biometrika **66**, 605–610] and Gawronski and Stadtmüller 1981) or as a two dimensional frequency profile (cf., also Sachs 1977).

Remarks

1. To describe the problem and the data: As soon as a frequency distribution is characterized in the way described above, at least the following **unavoidable questions** arise (cf., also Sachs 1977 [8:2a]): (1) What are the occasion and purpose of the investigation? (2) Can the frequency distribution be interpreted as a representative random sample from the population being studied or from a hypothetical population (cf., Section 1.3.1), or is it merely a nonrepresentative sample? (3) What do we decide are the defining characteristics of the population and the units of analysis and observation?

2. Significant figures (cf., Section 0.4–5) of characteristic values: Mean values and standard deviations are stated with **one or at most two decimal places more precision** than the original data. The last is appropriate in particular when the sample size is large. Dimensionless constants like skewness and kurtosis, correlation and regression coefficient, etc., **should be stated with two or at most four significant figures**. In order to attain this precision it is frequently necessary to compute power moments or other intermediate results correctly to two or three additional decimal places.

1.3.9 The log normal distribution

Many distributions occurring in nature are **left-steep** [flat on the right]. Replacing each measurement by its logarithm will often result in distributions looking more like a normal distribution, especially if the coefficient of variation $V = s/\bar{x} > 0.3$. The logarithm transforms the positive axis $(0, \infty)$ into the real axis $(-\infty, \infty)$ and $(0, 1)$ into $(-\infty, 0)$.

A logarithmic normal distribution, lognormal distribution for short, results (Aitchison and Brown 1957) when many random quantities cooperate multiplicatively so that the effect of a random change is in every case proportional to the previous value of the quantity. In contrast to this, the normal distribution is generated by additive cooperation of many random quantities. Therefore it is not surprising that the **lognormal distribution** predominates, **in particular, for economic and biological attributes**. An example is the sensitivity to drugs of any given species of living beings—from bacteria to large mammals. Such **characteristics in humans** include body height (of children), size of heart, chest girth measurement, pulse frequency, systolic and diastolic blood pressure, sedimentation rate of the red blood corpuscles, and percentages of the individual white blood corpuscle types, in particular the eosinophiles and the stab neutrophiles as well as the proportions of the various components of the serum, as for example glucose, calcium and bilirubin. Other examples are survival times. **Economic statistics** with lognormal distributions include gross monthly earnings of employees, turnovers of businesses, and acreages of cultivation of various types of fruit in a county. The lognormal distribution is often approximated also by attributes that can assume integral values only, as e.g., the number of breeding sows in a census district and the number of fruit trees in a region. In particular, lognormal distributions arise in the study of dimensions of particles under pulverization.

Williams (1940) analyzed a collection of 600 sentences from G. B. Shaw's "An Intelligent Woman's Guide to Socialism" which consisted of the first 15 sentences in each of sections 1 to 40, and found the distribution of the length of the sentences to be

$$y = \frac{1}{0.29 \cdot \sqrt{2\pi}} e^{-(x-1.4)^2/(2 \cdot 0.29^2)}$$

(y = frequency and x = logarithm of the number of words per sentence), a lognormal density. Generally, the number of letters (or of phonemes) per word in English colloquial speech follows a lognormal distribution remarkably well (Herdan 1958, 1966). Lognormal distributions also come up, as mentioned earlier, in studies of precipitation and survival time analyses—in reliability theory—as well as in **analytic chemistry**: in qualitative and quantitative analysis for a very wide range of concentrations (over several powers of ten), when one works in a neighborhood of zero or one hundred percent (e.g., in purity testing) and when the random error of a procedure is comparable to the measured values themselves.

The true lognormal distribution is

$$y = \frac{1}{\sqrt{2\pi\sigma^2}} \cdot \frac{1}{x} e^{-(\ln x - \mu)^2/(2\sigma^2)} \quad \text{for} \quad x > 0. \tag{1.115}$$

To test whether an attribute follows a lognormal distribution, one uses the **logarithmic probability grid**, which has a logarithmically scaled abscissa (cf., Section 1.3.7). The cumulative frequencies are always paired with the upper (lower) class limit, the limit value of the attributes combined in that particular class. The class limit always lies to the right (left) if frequencies are added according to increasing (decreasing) size of the attributes. If the plotted values lie approximately on a straight line, we have at least an approximate lognormal distribution. If the line is bent upward (downward) in the lower region, then the cumulative percentages are taken as the ordinates paired with the abscissae $\log(g + F)$ [or $\log(g - F)$] instead of the originally given limit value $\log g$. The vanishing point F, the lower bound of the distribution, always lies on the steep side of the curve. It is determined by trial and error: if with two F-values one obtains left curvature once and right curvature the other time, the value of F sought is straddled and can be easily found by interpolation. Occasionally it is easy to recognize the physical meaning of F. To determine the parameters graphically, a straight line best fitting the points is drawn; the (median)/(dispersion factor), median, and (median)(dispersion factor) are the abscissae of the points of intersection with the 16%, 50%, and 84% line respectively. Important for a lognormal distribution is the central 68% of its mass, written

$$(\text{median})(\text{dispersion factor})^{\pm 1},$$

which involves an interval, reduced by the extreme values, of "still typical values."

The dispersion factor is presented in greater detail in the formula (1.117). To estimate the parameters mathematically, the data are classified in the usual way with constant class width, the logarithms of the class means ($\log x_j$) are determined, the products $f_j \log x_j$ and $f_j(\log x_j)^2$ (f_j = frequency in class j) are formed, summed, and inserted in the following formulas:

$$\text{median}_L = \text{antilog } \bar{x}_{\log x_j} = \text{antilog}\left(\frac{\sum f_j \log x_j}{n}\right), \quad (1.116)$$

$$\text{dispersion factor} = \text{antilog } \sqrt{s^2_{\log x_j}}$$
$$= \text{antilog } \sqrt{\frac{\sum f_j(\log x_j)^2 - (\sum f_j \log x_j)^2/n}{n-1}}, \quad (1.117)$$

$$\text{mean}_L = \text{antilog}(\bar{x}_{\log x_j} + 1.1513 s^2_{\log x_j}), \quad (1.118)$$

$$\text{mode}_L = \text{antilog}(\bar{x}_{\log x_j} - 2.3026 s^2_{\log x_j}). \quad (1.119)$$

For samples of small size, the logarithms of the individual values are used in place of the logarithms of the class means; the frequencies (f_j) are then all equal to one ($f_j = 1$). The dispersion factor is an estimate of antilog $s_{\log x_j}$. Thus with increasing dispersion factor the arithmetic mean is shifted to the right with respect to the median, and the mode about twice that amount to the left (cf., also the end of Section 7.3.2, Mann et al., 1974 [8:2d], Thöni 1969, King 1971, Hasselblad 1980, and Lee 1980).

EXAMPLE. The following table contains 20 measured values x_j, ordered by magnitude, which are approximately lognormally distributed. Estimate the parameters.

x_j	$\log x_j$	$(\log x_j)^2$
3	0.4771	0.2276
4	0.6021	0.3625
5	0.6990	0.4886
5	0.6990	0.4886
5	0.6990	0.4886
5	0.6990	0.4886
5	0.6990	0.4886
6	0.7782	0.6056
7	0.8451	0.7142
7	0.8451	0.7142
7	0.8451	0.7142
7	0.8451	0.7142
8	0.9031	0.8156
8	0.9031	0.8156
9	0.9542	0.9105
9	0.9542	0.9105
10	1.0000	1.0000
11	1.0414	1.0845
12	1.0792	1.1647
14	1.1461	1.3135
$\sum$	16.7141	14.5104

Mantissas rounded off to two decimal places ($\log 3 = 0.48$) are almost always adequate.

The coefficient of variation of the original data (x_j) is $V = 2.83/7.35 = 38.5\%$, clearly above the 33% bound. We have:

$$\text{median}_L = \text{antilog}\left\{\frac{16.7141}{20}\right\} = \text{antilog } 0.8357 = 6.850,$$

$$\text{dispersion factor} = \text{antilog}\sqrt{\frac{14.5104 - 16.7141^2/20}{20 - 1}} = \text{antilog }\sqrt{0.02854},$$

dispersion factor = antilog $0.1690 = 1.476$.

The central 68% of the mass lies between $6.850/1.476 = 4.641$ and $(6.850)(1.476) = 10.111$ [i.e., $(6.850)(1.476)^{\pm 1}$]. Five values lie outside this region, whereas $(0.32)(20) = 6$ values were to be expected. We have

$$\text{mean}_L = \text{antilog}(0.8357 + 1.1513 \cdot 0.02854) = \text{antilog } 0.8686,$$

$$\text{mean}_L = 7.389,$$

$$\text{mode}_L = \text{antilog}(0.8357 - 2.3026 \cdot 0.02854),$$

$$\text{mode}_L = \text{antilog } 0.7700 = 5.888.$$

Nonsymmetric 95%-confidence interval for μ

Frequently a nonsymmetric confidence interval (e.g., a 95% CI) is given for μ (cf., Sections 141, 151, and 311). It is simply the symmetric confidence interval for $\mu_{\log x}$ of the (approximately) normally distributed random variable log x transformed back into the original scale:

$$95\% \text{ CI: } \text{antilog}[\bar{x}_{(\log x_j)} \pm t_{n-1;0.05}\sqrt{s^2_{(\log x_j)}/n}].$$

pp. 136/137

For the example with the 20 observations and $\bar{x} = 7.35$ there results (cf., Section 7.3.3)

$$[] = 0.8357 \pm 2.093\sqrt{0.02854/20} = 0.7566, 0.9148$$
$$95\% \text{ CI: } 5.71 \leq \mu \leq 8.22.$$

Remarks

1. If you frequently have to compare empirical distributions with normal distributions and/or lognormal distributions, use, e.g., the **evaluation forms** (AFW 172a and 173a), issued by Beuth–Vertrieb (for address see References, Section 7).

2. Moshman (1953) has provided tables for the **comparison of the central tendency** of empirical lognormal distributions (of approximately the same shape).

3. The distribution of extreme values—the high water marks of rivers, annual temperatures, crop yields, etc.,—frequently approximates a lognormal distribution. Since the standard work by Gumbel (1958) would be difficult for the beginner, the readily comprehensible graphical procedures by Botts (1957) and Weiss (1955, 1957) are given as references. Gumbel (1953, 1958; cf., also Weibull 1961) illustrated the use of **extreme-value probability paper** (produced by Technical and Engineering Aids to Management; see References, Section 7) on which a certain distribution function of extreme values is stretched into a straight line (a more detailed discussion of probability grids can be found in King 1971). For an excellent bibliography of extreme-value theory see Harter (1978).

4. Certain socioeconomic quantities such as personal incomes, the assets of businesses, the sizes of cities, and numbers of businesses in branches of industry follow distributions that **are flat to the right**, and which can be approximated over large intervals by the **Pareto distribution** (cf., Quandt 1966)—which exists only for values above a certain threshold (e.g., income > $800)—or other strongly right skewed distributions. If the lognormal distribution is truncated on the left of the mode (restricted to the interval right of the mode), then it is very similar to the Pareto distribution over a wide region.

5. In typing a letter a secretary may make typographical errors each of which prolongs the task by the time necessary to correct it. Highly skewed and roughly L-shaped distributions tend to occur when time scores are recorded for a task subject to infrequent but time-consuming errors (vigilance task conditions). More on bizarre distribution shapes can be found in Bradley (1977).

6. For the three-parameter lognormal distribution see Kübler (1979), Griffiths (1980) and Kane (1982).

7. If some of the observed values we want to transform by $x'_j = \log x_j$ lie between 0 and 1, all the data are multiplied by an appropriate power of 10 so that all the x-values become larger than 1 and all the characteristic numbers are positive (cf., Section 7.3.3).

1.4 THE ROAD TO THE STATISTICAL TEST

1.4.1 The confidence coefficient

Inferences on a parameter from characteristic numbers. The characteristic numbers determined from different samples will in general differ. Hence the characteristic number (e.g., the mean $\bar{x}$) determined from some sample is only an estimate for the mean μ of the population from which the sample originated. In addition to the estimate there is specified an interval which includes neighboring larger and smaller values and which presumably includes also the "unknown" parameter of the population. This interval around the characteristic number is called the **confidence interval**. By changing the size of the confidence interval with the help of an appropriate factor, the reliability of the statement that the confidence interval includes the parameter of the population can be preassigned. If we choose the factor in such a way that the statement is right in 95% and wrong in 5% of all similar cases, then we say: The confidence interval calculated from a sample contains the parameter of the population with the statement probability or confidence probability or **confidence coefficient** S of 95%. Thus the assertion that the parameter lies in the confidence interval is false in 5% of all cases. Hence we choose the factor so that the probability for this does not exceed a given small value α (Greek alpha: $\alpha \leq 5\%$, i.e., $\alpha \leq 0.05$) and call α the level of significance. For the case of a normally distributed population, Table 25 gives a survey of the confidence intervals for the mean μ of the population:

$$\boxed{\bar{X} \pm z\frac{\sigma}{\sqrt{n}}, \quad \text{where} \quad P\left(\bar{X} - z\frac{\sigma}{\sqrt{n}} \leq \mu \bar{X} + z\frac{\sigma}{\sqrt{n}}\right) = S = 1 - \alpha.}$$

(1.120a,b)

The value z is found in a table of the standard normal distribution [cf., Tables 13 and 14 (Section 1.3.4) and Table 44 (Section 2.1.6)]. Sigma (σ) is the standard deviation, which is known or is estimated from a very large number of sample values.

Equation (1.120a,b) implies that with probability α the parameter μ in question fails to lie in the given confidence interval [that with probability α the estimator $\bar{X}$ of μ is off from the true value by more than the (additive) factor $z\sigma/\sqrt{n}$], i.e., if we repeat the experiment m times, we can expect that the resulting confidence intervals do not contain the true value μ in $m\alpha$ of the cases. By looking more closely at Table 25 we recognize that S (or α, the two adding to 100% or to the value 1) determines the confidence in the statistical statement. The larger the confidence coefficient S, the larger will be the confidence interval for given standard deviation and given sample size. This implies that there exists a conflict between the precision of a statement

1.4 The Road to the Statistical Test

Table 25

Confidence interval for the mean μ of a normally distributed population	Confidence probability or confidence coefficient S	Level of significance α
$\bar{X} \pm 2 \dfrac{\sigma}{\sqrt{n}}$	95.44% = 0.9544	4.56% = 0.0456
$\bar{X} \pm 3 \dfrac{\sigma}{\sqrt{n}}$	99.73% = 0.9973	0.27% = 0.0027
$\bar{X} \pm 1.645 \dfrac{\sigma}{\sqrt{n}}$	90% = 0.9	10% = 0.10
$\bar{X} \pm 1.960 \dfrac{\sigma}{\sqrt{n}}$	95% = 0.95	5% = 0.05
$\bar{X} \pm 2.576 \dfrac{\sigma}{\sqrt{n}}$	99% = 0.99	1% = 0.01
$\bar{X} \pm 3.2905 \dfrac{\sigma}{\sqrt{n}}$	99.9% = 0.999	0.1% = 0.001
$\bar{X} \pm 3.8906 \dfrac{\sigma}{\sqrt{n}}$	99.99% = 0.9999	0.01% = 0.0001

and the certainty attached to this statement: **precise statements are uncertain, while statements that are certain are imprecise.** The usual significance levels are $\alpha = 0.05$, $\alpha = 0.01$, and $\alpha = 0.001$, depending on how much weight one wishes to attach to the decision based on the sample. In certain cases, especially when danger to human life is involved in the processes under study, a substantially smaller level of significance must be specified; in other cases a significance level of 5% might be unrealistically small. The concept of the confidence interval will again be considered at some length in Chapter 3 (Section 3.1.1).

Inferences based on the parameters concerning their estimates. The parameters of a population are known from theoretical considerations. What needs to be determined is the region in which the estimators (e.g., the means $\bar{X}_i$) derived from the individual samples lie. Therefore a **tolerance interval** is defined, containing the theoretical value of the parameter, within which the estimators are to be expected with a specified probability. The limits of the interval are called tolerance limits. With a normal distribution (σ known, or else estimated from a very large sample) they are given for the sample mean by

$$\mu \pm z \dfrac{\sigma}{\sqrt{n}}, \quad \text{where} \quad P\left(\mu - z \dfrac{\sigma}{\sqrt{n}} \leq \bar{X} \leq \mu + z \dfrac{\sigma}{\sqrt{n}}\right) = S = 1 - \alpha.$$

(1.121a,b)

If the symbols μ and $\bar{X}$ are interchanged in Table 25, then it is also valid in this context. An arbitrary sample mean $\bar{X}$ is covered by a tolerance interval with confidence coefficient S, i.e., in $(S)(100)\%$ of all cases it is expected that $\bar{X}$ will be within the specified tolerance limits. If the sample mean $\bar{X}$ falls within the tolerance interval, the deviation from the mean μ of the population will be regarded as random. If however, $\bar{X}$ does not lie in the tolerance region, we shall consider the departure of $\bar{X}$ from μ significant and conclude with a confidence of $S\%$ that the given sample was drawn from a different population. Occasionally only one tolerance limit is of interest; it is then tested whether a specific value ("theoretical value," e.g., the mean of an output) is not fallen short of or exceeded.

1.4.2 Null hypotheses and alternative hypotheses

The hypothesis that two populations agree with regard to some parameter is called the **null hypothesis**. It is assumed that the actual difference is zero. Since statistical tests cannot ascertain agreements, but only differences between the populations being compared (where one population might be fully known), the null hypothesis is, as a rule, brought in to be rejected. It is the aim of the experimental or **alternative hypothesis** to prove it "null and void." When can we, using only a statistical test, reject the null hypothesis and accept the alternative hypothesis? Only if an authentic difference exists between the two populations. Often, however, we have only the two samples at our disposal and not the populations from which they came. We must then consider the sampling variation, where we have varying statistics even for samples from a **single** population. This shows that we can practically **always** expect **differences**. To decide whether the difference is intrinsic or only random, we must state, or (better) **agree upon**, what we wish to regard as the limit (critical value) of the manifestation of chance "as a rule," as far as can be foreseen.

We propose the null hypothesis and reject it precisely when a result that arises from a sample is improbable under the proposed null hypothesis. We must define what we will consider improbable. Assume we are dealing with a normal distribution. For the frequently used 5% level, $(\pm)1.96\sigma$ is the critical value ($S = 95\%$). In earlier times the three sigma rule—i.e., with a level of significance $\alpha = 0.0027$ (or confidence coefficient $S = 99.73\%$), corresponding to the 3σ limit—was used almost exclusively.

We can require, e.g., that the probability of the observed outcome (the "worse" one in comparison with the null hypothesis) must be less than 5% under the null hypothesis if this null hypothesis is to be rejected. This probability requirement states that we will consider an outcome random if in four tosses of a fair coin it lands tail (head) up four times. If however the coin lands tail up in five out of five tosses the outcome is viewed as "beyond pure chance," i.e., as in contradiction to the null hypothesis. The probability

1.4 The Road to the Statistical Test

that a fair coin tossed four or five times respectively always lands on the same side is

$$P_{4x} = (1/2)^4 = 1/16 = 0.06250$$
$$P_{5x} = (1/2)^5 = 1/32 = 0.03125,$$

i.e., about 6.3% or about 3.1%. Thus if a factual statement is said to be assured of being beyond pure chance with a confidence coefficient of 95%, this means that its random origin would be as improbable as the event that on tossing a coin five times one always gets tails. The probability that in n tosses of a coin tails come up every time can be found in Table 26 [$2^{-n} = (1/2)^n$].

Table 26 The probability P that a coin tossed n times always falls on the same side, as a prototype for a random event.

n	2^n	2^{-n}	P	level	
1	2	0.50000			
2	4	0.25000			
3	8	0.12500			
4	16	0.06250	<	10 %	
5	32	0.03125	<	5 %	
6	64	0.01562			
7	128	0.00781	<	1 %	
8	256	0.00391	<	0.5 %	
9	512	0.00195			
10	1024	0.00098	≈	0.1 %	$2^{10} \approx 10^3$
11	2048	0.00049	≈	0.05 %	
12	4096	0.00024			
13	8192	0.00012			
14	16384	0.00006	<	0.01 %	
15	32768	0.00003			

If a test with a level of significance of, for example, 5% (**significance level** $\alpha = 0.05$) leads to the detection of a difference, the null hypothesis is rejected and the alternative hypothesis—the populations differ—accepted. The difference is said to be important or **statistically significant at the 5% level**, i.e., a valid null hypothesis is rejected in 5% of all cases of differences as large as those observed in the given samples, and such differences are **so rarely produced by random processes alone** that:

a. we will not be convinced that **random processes alone** give rise to the **data** or, formulated differently,
b. it is assumed that the difference in question is not based solely on a random process but rather on a **difference between the populations**.

Sampling results lead to only two possible statements:

1. The decision on **retaining or rejecting the null hypothesis**.
2. The specification of the confidence intervals.

A comparison of two or more confidence intervals leads to another method of testing whether the differences found are only random or are in fact statistically significant.

Null hypotheses and alternate hypotheses form a net, which we toss out so as to seize "the world"—to rationalize it, to explain it, and to master it prognostically. Science makes the mesh of the net ever finer as it seeks, with all the tools in its logical-mathematical apparatus and in its technical-experimental apparatus, to **reformulate** new, more specific and more general **null hypotheses**—negations of the corresponding alternate hypotheses—of as simple a nature as possible (improved testability) and to **disprove** these null hypotheses. The conclusions drawn are never absolutely certain, but are rather provisional in principle and lead to new and more sharply formulated hypotheses and theories, which will undergo ever stricter testing and will facilitate scientific progress and make possible an improved perception of reality. **It should be the aim of scientific inquiry to explain a maximum number of empirical facts using a minimum number of hypotheses and theories** and then again to question these. What is really creative here is the formulation of the **hypotheses**. Initially simple **assumptions**, they become **empirical generalizations** verified to a greater or lesser degree. If the hypotheses are **ordered** by **rank** and if there are **deductive relations** among them (i.e., if from a general hypothesis particular hypotheses can be deduced), then what we have is a **theory**. To prove theorems within the framework of theories and to synthesize a **scientific model of the world** from individual isolated theories are further goals of scientific research.

Remark: The randomly obtained statistically significant result

It is in the nature of the significance level that in a large number of samples from a common population one or another could have deviated entirely at random. **The probability of randomly obtaining a significant result by a finite number n of inquiries** can be determined from an expansion of the **binomial** $(\alpha + (1 - \alpha))^n$. For a significance level of size $\alpha = 0.01$ for two independent and identical trials, we have, by the known binomial expansion $(a + b)^2 = a^2 + 2ab + b^2$, the relation $(0.01 + 0.99)^2 = (0.01)^2 + 2(0.01)(0.99) + (0.99)^2 = 0.0001 + 0.0198 + 0.9801 = 1.0000$, i.e.,:

1. The probability that under the null hypothesis both inquiries yield significant results, with $P_{H_0} = 0.0001$, is very small.
2. The probability that under H_0 one of the two trials proves significant is $P_{H_0} = 0.0198$, or nearly 2%, about two hundred times as large.
3. Of course with the largest probability under H_0, neither of the two inquiries will yield significant results ($P_{H_0} = 0.9801$).

1.4 The Road to the Statistical Test

Associated probabilities can be determined for other significance levels as well as for 3 or more trials. As an exercise, the probabilities for $\alpha = 0.05$ and 3 trials can be calculated: Recalling that

$$(a + b)^3 = a^3 + 3a^2b + 3ab^2 + b^3,$$

we get

$$(0.05 + 0.95)^3 = 0.05^3 + 3 \cdot 0.05^2 \cdot 0.95 + 3 \cdot 0.05 \cdot 0.95^2 + 0.95^3$$

$$= 0.000125 + 0.007125 + 0.135375 + 0.857375 = 1.$$

The probability that, under the null hypothesis and with $\alpha = 0.05$, out of three trials: (a) one proves to be entirely randomly statistically significant, is 13.5%; (b) at least one proves to be entirely randomly statistically significant is 14.3% (0.142625 = 0.000125 + 0.007125 + 0.135375 = 1 − 0.857375) [see Section 1.2.3, "Examples of the multiplication rule," No. 4]: $P = 1 - (1 - 0.05)^3 = 0.142625$. As an approximation for arbitrary α and n we have the **Bonferroni inequality**: the probability of falsely rejecting at least one of the n null hypotheses is not greater than the sum of the levels of significance, i.e., $0.143 < 0.15 = 0.05 + 0.05 + 0.05$.

1.4.3 Risk I and risk II

In the checking of hypotheses (by means of a test), two erroneous decisions are possible:

1. The unwarranted rejection of the null hypothesis: **error of Type I**.
2. The unwarranted retention of the null hypothesis: **error of Type II**.

Since reality presents two possibilities: (1) the null hypothesis (H_0) is true and (2) the null hypothesis is false, the test can lead to two kinds of erroneous decisions: (1) to retain the null hypothesis or (2) to reject the null hypothesis, i.e., to accept the alternative hypothesis (H_A). The four possibilities correspond to the following decisions:

Decision	State of nature	
	H_0 true	H_0 false
H_0 rejected	ERROR OF TYPE I	Correct decision
H_0 retained	Correct decision	ERROR OF TYPE II

If, e.g., it is found by a comparison that a new medicine is better when in fact the old was just as good, an error of Type I is committed; if it is found by a comparison that the two medicines are of equal value when actually the new one is better, an error of Type II is committed. The two probabilities associated with the erroneous decisions are called risk I and II:

> **The risk I,** the small probability that a valid null hypothesis is rejected, obviously equals the significance level α:
>
> $\alpha = P(\text{decision to reject } H_0 | H_0 \text{ is true}) = P(H_A | H_0)$.
>
> **The risk II,** the probability that an invalid null hypothesis is retained, is noted by β:
>
> $\beta = P(\text{decision not to reject } H_0 | H_0 \text{ is false}) = P(H_0 | H_A)$.

Since α must be greater than zero, if for $\alpha = 0$ the null hypothesis were always retained, a risk of error would always be present. If α and the size n are given, β is determined; the smaller the given α, the larger will be the β. The α and β can be chosen arbitrary small only if n is allowed to grow without bounds, i.e., for very small α and β one can reach a decision only with very large sample sizes. With small sample sizes and small α the conclusion that there exists no significant difference must then be considered with caution. **The nonrejection of a null hypothesis implies nothing about its validity as long as β is unknown.** Wherever in this book we employ the term "significant," it means always and exclusively "statistically significant."

> Depending on which faulty decision has more serious consequences, the α and β are so specified in a particular case that the critical probability is ≤ 0.01 and the other probability is ≤ 0.10. In practice, α is specified in such a way that, if serious consequences result from a
>
> Type I error, $\qquad \alpha = 0.01 \quad \text{or} \quad \alpha = 0.001$;
>
> Type II error, $\qquad \alpha = 0.05 \quad (\text{or } \alpha = 0.10)$.

According to Wald (1950) one must take into account the **gains** and **losses** due to faulty decisions, including the costs of the test procedure, which can depend on the nature and size of the sample. Consider e.g., the production of the new vaccines. The different batches should practically be indistinguishable. Unsound lots must in due time be recognized and eliminated. The unjustified retention of the null hypothesis "vaccine is sound" means a dangerous production error. Thus β is chosen as small as possible, since the rejection of good lots brings on expenses, to be sure, but generally has no serious consequences (then $\alpha = 0.10$ say).

Suppose that on the basis of very many trials with a certain coin, we get to know the probability π of the event "tails"—but tell a friend only that π equals either 0.4 or 0.5. Our friend decides on the following experimental design for testing the null hypothesis $\pi = 0.5$. The coin is to be tossed $n = 1000$ times. If $\pi = 0.05$, tails would presumably appear about 500 times. Under the alternative hypothesis $\pi = 0.04$, about 400 tails would be expected. The friend thus chooses the following decision

1.4 The Road to the Statistical Test

process: If the event tails comes up fewer than 450 times he then rejects the null hypothesis $\pi = 0.05$ and accepts the alternative hypothesis $\pi = 0.4$. If, on the contrary, it comes up 450 or more times, he retains the null hypothesis.

A Type I error—rejecting the valid null hypothesis—is made if π in fact equals 0.5 and in spite of this fewer than 450 tails occur in some particular experiment. A Type II error is committed if in fact $\pi = 0.4$ and during the testing 450 or more tails show up. In this example, we chose risk I and risk II of about the same size (npq equals 250 in one case, 240 in the other). The Type I error can however be reduced, even with the sample size n given, by enlarging the acceptance region for the null hypothesis. It can for example be agreed upon that the null hypothesis $\pi = 0.5$ is rejected only if fewer than 430 tails result. However, with constant sample size n, the Type II error— the retention of the false null hypothesis—then becomes that much more likely.

If $\alpha = \beta$ is chosen, the probabilities for faulty decisions of the first and second type are equal. Frequently only a specific α is chosen and the H_0 is granted a special status, since the H_A is in general not precisely specified. Thus several standard statistical procedures with preassigned α and uncertain β decide in favor of the H_0: they are therefore known as conservative tests.

According to Neyman's rule, α is given a specific value and β should be kept as small as possible. It is assumed that an important property of the test is known, the so-called power function (cf., Section 1.4.7).

Moreover let us point out the difference between **statistical significance and "practical" significance**, which is sometimes overlooked: differences significant in practice must already be discernible in samples of not very large size.

In summing up, let us emphasize: A true H_0 is retained with the probability (confidence coefficient) $S = 1 - \alpha$ and rejected with the probability (level of significance) $\alpha = 1 - S$; thus $\alpha = 5\% = 0.05$ and $S = 95\% = 0.95$ means a true H_0 is rejected in 5% of all cases.

Errors of Type III and IV are discussed by Marascuilo and Levin (1970). Birnbaum (1954), Moses (1956), and Lancaster (1967) consider the combination of independent significance probabilities P_i (we give an approximation to the solutions in Section 4.6.4). Advantages and limitations of nine methods of combining the probabilities of at least two independent studies are discussed in Rosenthal (1978). Here, in contrast with a specific preassigned level of significance α, P is the empirical level of significance under the null hypothesis for a given sample, called the **descriptive or nominal significance level** for short [cf., pages 120 and 266].

Two strategies can in principle be distinguished, that of the "discoverer" and that of the "critic." The discoverer wishes to reject a null hypothesis; thus he prefers a large risk I and a small risk II. The opposite is true of the critic: he prevents the acceptance of a false alternate hypothesis by adopting a small risk I and a large risk II, which allows the null hypothesis to be erroneously retained.

Outside the realm of science one is generally content with a relatively large risk I, thus behaving as discoverer rather than as critic.

1.4.4 The significance level and the hypotheses are, if possible, to be specified before collecting the data

Most people who deal with mathematical statistics emphasize that the **level of significance is to be specified before the data are gathered**. This requirement sometimes leads to a number of puzzles for the practioner (cf. also McNemar 1969).

> If in some exceptional case a significance level cannot be fixed in advance, we can proceed in two ways: (1) Determine the P-value, the nominal level of significance, on the basis of the data. This has the advantage of a full description of the situation. Moreover it permits the experimenter to fix his own level of significance appropriate to the problem and to compare the two. The following procedure is, however, preferred because it prevents the reproach of prejudice: (2) Determine, in terms of the critical 5% (or 10%), 1%, and 0.1% bounds, the limits between which P lies, and mark the result using a three level asterisk scheme: $P > 0.05$; [*] $0.05 \geq P > 0.01$; [**] $0.01 \geq P < 0.001$; [***] $P \leq 0.001$. In general the first category (without asterisk) will be regarded as (statistically) not significant (ns); the last, [***], as unquestionably statistically significant. In other words: the evidence against H_0 is (a) moderate [*], (b) strong [**], and (c) very strong [***].

It is expedient, **before** the statistical analysis of the data, to formulate all hypotheses that according to our knowledge, could be relevant, and choose the appropriate test methods. **During** the analysis the data are to be carefully examined to see whether they suggest still other hypotheses. Such hypotheses as are drawn from the material must be formulated and tested with greater care, since each group of numbers exhibits random extremes. The risk of Type I error is larger, but by an unknown amount, than if the hypotheses are formulated in advance. **The hypotheses drawn from the material** can become important as new hypotheses for subsequent studies.

1.4.5 The statistical test

The following amusing story is due to R. A. Fisher (1960). At a party, Lady X maintained that if a cup of tea, to which milk had been added, were set before her, she could unerringly taste whether the tea or the milk was poured in first. How is such an assertion to be tested? Certainly not by placing before her only two cups, completely alike on the outside, into one of which first milk and then tea (sequence MT) was poured, and into the other first tea and then milk (sequence TM). If the lady were now asked to choose, she would have a 50% chance of giving the right answer even if her assertion

1.4 The Road to the Statistical Test

were false. The following procedure is better: Take eight cups, completely alike on the outside. Four cups of the set of eight are filled in the sequence MT, four in the sequence TM. Then the cups are placed in random order in front of the lady. She is informed that four of the cups are of type TM and four MT, and that she is to find the four TM type cups. The probability of hitting without special talent on the right choice becomes greatly reduced. That is, from among 8 cups, $(8)(7)(6)(5)/(4)(3)(2) = 70$ types of choices of 4 can be made; only one of these choices is correct. The probability of hitting without special talent, randomly, on the right choice, is $\frac{1}{70} = 0.0143$ or about 1.4%, hence very small. If indeed the lady now chooses the 4 correct cups, the null hypothesis—Lady X does not have these special talents—is dropped and her unusual ability recognized. A significance level of at least 1.4% is there assumed. We can of course reduce this significance level still further by increasing the number of cups (e.g., with 12, half of which are filled according to TM and half according to MT, the significance level is $\alpha \simeq 0.1\%$). It is characteristic of our procedure that **we first state the null hypothesis, then reject it if and only if a result occurs that is unlikely under the null hypothesis**. If we state a null hypothesis that we wish to test using statistical methods, it will be interesting to know whether or not some existing sample supports the hypothesis. In the teacup example we would have rejected the null hypothesis if the lady had chosen the 4 correct cups. The null hypothesis is retained in all other cases. We must thus come to a decision with every possible sample. In the example, the decision of rejecting the null hypothesis if the lady chose at least 3 correct cups would also be defensible. More on the "tea test" problem can be found in Neyman (1950), Gridgeman (1959), and Fisher (1960).

In order to avoid the difficulty of having to set down the decision for every possible outcome, we are interested in procedures that always bring about such a decision. One such procedure, which induces a decision on whether or not the sample outcome supports the hypothesis, is called a **statistical test**. Many tests require that the observations be **INDEPENDENT**, as is the case with the so-called random samples. Most statistical tests are carried out with the aid of a **test statistic**. Each such test statistic is a prescription, according to which a number is computed from a given sample. The test now consists of decision based on the value of the test statistic.

For example, let X be a normally distributed random variable. With known standard deviation σ the H_0, $\mu = \mu_0$ or $\mu - \mu_0 = 0$, is proposed, i.e., the mean μ of the population, which is estimated from a random sample, coincides with a desired theoretical value μ_0. The H_A is the negation of the H_0, i.e., $\mu \neq \mu_0$ or $\mu - \mu_0 \neq 0$. As a test statistic for the so-called **one sample Gauss test** we use ($n =$ sample size)

$$\frac{\bar{X} - \mu_0}{\sigma}\sqrt{n} = \hat{Z}. \tag{1.122}$$

Theoretically $\hat{Z}$ is, given H_0, standard normally distributed, i.e., with mean 0, variance 1. The value of the test statistic $\hat{Z}$, which depends on the particular observations, deviates more or less from zero. We take the absolute value $|\hat{Z}|$ as a measure of the deviation. A critical value z depending on the **previously chosen significance level** α can now be specified in such a way that we have under H_0

$$P(|\hat{Z}| \geq z) = \alpha. \qquad (1.123)$$

If our sample yields a value $\hat{z}$ of the test statistic which is smaller in absolute value than the critical value z_α, ($|\hat{z}| < z_\alpha$—e.g., for $\alpha = 0.01$ there results $z = 2.58$), we conclude that this deviation from the value zero of the hypothesis is random. We then say H_0 is not contradicted by the sample. H_0 is retained pending additional tests and, so to speak, for lack of proof, not necessarily because it is true. The $100\alpha\%$ level will in the following stand for the percentage corresponding to the probability α (e.g., for $\alpha = 0.01$, the corresponding $100\alpha\% = (0.01)(100\%) = 1\%$).

A deviation of $|\hat{Z}| > z_\alpha$ (e.g., $|\hat{Z}| > 2.58$ for the 1% level) is, though not impossible under H_0, "improbably" large, the probability of a random occurrence of this situation being less than α. It is more likely in this case that H_0 is not correct, i.e., **for $|\hat{Z}| \geq z_\alpha$ it is decided that the null hypothesis must be rejected at the $100\alpha\%$ level.**

We shall later come to know test statistics other than the one described above in (1.122) (cf., also Section 4.6.3). For all of them, however, the distributions specified for the test statistics are strictly correct only if H_0 is true (cf., also Zahlen 1966 and Calot 1967).

EXAMPLE. Given:

$$\mu_0 = 25.0; \quad \sigma_0 = 6.0 \text{ and } n = 36, \quad \bar{x} = 23.2,$$
$$H_0: \mu = \mu_0 \quad (H_A: \mu \neq \mu_0), \quad \alpha = 0.05 \quad (S = 0.95),$$
$$|\hat{z}| = \frac{|23.2 - 25.0|}{6}\sqrt{36} = 1.80.$$

Since $|\hat{z}| = 1.80 < 1.96 = z_{0.05}$, the H_0 of equal population means cannot be rejected at the 5% significance level, i.e., the H_0 is retained. A nonrejected H_0, since it could be true and since it does not contradict the available data, is retained for the time being. More important however than the possible correctness of H_0 is the fact that we lack sufficient data to reject it. If the amount of data is enlarged, a new verification of H_0 is possible. It is often not easy to decide how many data should be collected to test H_0, for, with sufficiently large sample sizes, almost all H_0 can be rejected. (In Section 3.1 several formulas are given for the choice of appropriate sample sizes.)

1.4 The Road to the Statistical Test

EXAMPLE. Given:

$\mu_0 = 25.0; \quad \sigma_0 = 6.0 \text{ and } n = 49, \quad \bar{x} = 23.2,$

$H_0: \mu = \mu_0 \quad (H_A: \mu \neq \mu_0), \quad \alpha = 0.05 \quad (S = 0.95),$

$|\hat{z}| = \dfrac{|23.2 - 25.0|}{6} \sqrt{49} = 2.10.$

Since $|\hat{z}| = 2.10 > 1.96 = z_{0.05}$, the null hypothesis is rejected at the 5% level (with a confidence coefficient of 95%).

Another simple test is contained in Remark 3 in Section 2.4.2.

> The test theory was developed in the years around 1930 by J. Neyman and E. S. Pearson (1928, 1933; cf., Neyman 1942, 1950 as well as Pearson and Kendall 1970, and Cox 1958, 1977).

Types of statistical tests

If only a single hypothesis, the null hypothesis, is proposed with the "tea test" and the trial carried out serves only to test whether this hypothesis is to be rejected, we speak of a **significance test**. Trials that serve to verify hypotheses on some parameter (e.g., $H_0: \mu = \mu_0$) are called **parameter tests**. A **goodness of fit test** checks whether an observed distribution is compatible with a theoretical one. The question of whether a characteristic is normally distributed plays a special role, since many tests assume this. If a test makes no assumptions about the underlying distribution, it is called distribution-free. Goodness of fit tests are among the distribution-free procedures.

We now also see that optimal tests would be insensitive or robust with respect to deviations from specific assumptions (e.g., normal distribution) but sensitive to the deviations from the null hypothesis. A test is **robust** relative to a certain assumption if it provides sufficiently accurate results even when this assumption is violated, i.e., if the real probability of error corresponds to the preassigned significance level.

> Generally a statistical procedure is described as robust if it is not very sensitive to departure from the assumptions on which it depends.

Mathematical statistics

> **Statistics** can be defined as the method or art of gathering and analyzing data to attain new knowledge, with mathematical treatment of random occurrences (random samples from populations or processes) in the foreground. The branch of science that concerns itself with the mathematical treatment of random occurrences is called **mathematical statistics** and comprises probability theory, statistics and their applications.

On the one hand conclusions about a population are drawn inductively from a **relevant statistic** of a random sample (which can be considered a representative of the population studied); the theory of probability, on the other hand, allows the deduction, based on a theoretical population, the model, of characteristics of a random sample from this theoretical population:

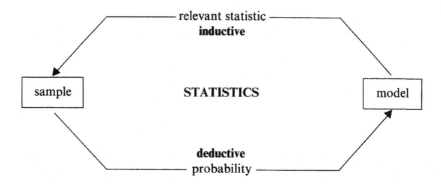

The relevant statistic has two tasks:

1. The estimation of unknown parameters of the population and the confidence limits (**estimation procedure**).
2. The testing of hypotheses concerning the population (**test procedure**).

The more properties of the population are known on the basis of plausible theories or from previous experience, at least in broad outline, the more precise will be the chosen probabilistic model and the more precisely can the results of the test and estimation procedures be grasped. **The connection of inductive and deductive procedures is essential for the scientific method**: **Induction**, which presupposes a more and more refined analysis, has the task of establishing a model based on empirical observations, of testing this model and of improving it. To **deduction** falls the task of pointing out latent consequences of the model chosen on the basis of hitherto existing familiarity with the material, selecting the best procedure for computing the estimates of the parameters of the model from the sample, and deriving the statistical distribution of these estimates for random samples.

1.4.6 One sided and two sided tests

If the objective of some experiment is to establish that a difference exists between two treatments or, better, between two populations created by different treatments, the sign of a presumable difference of the two parameters—say the means of two sets of observations—will in general not be

1.4 The Road to the Statistical Test

known. H_0, which we hope to disprove—the two means originate in the same population ($\mu_1 = \mu_2$), is confronted with H_A: the two means come from different populations ($\mu_1 \neq \mu_2$), because we do not know which parameter has the larger value. Sometimes a **substantiated hypothesis** allows us to make certain predictions about the sign of the expected difference—say, the mean of population I is larger than the mean of population II ($\mu_1 > \mu_2$) or the opposite assertion ($\mu_1 < \mu_2$). In both cases H_0 consists of the antithesis of the alternative hypothesis, i.e., contains the situation that is not included in the alternative hypothesis. If H_A reads $\mu_1 > \mu_2$, then the corresponding H_0 is $\mu_1 \leq \mu_2$. The $H_0: \mu_1 \geq \mu_2$ corresponds to $H_A: \mu_1 < \mu_2$. If H_A reads $\mu_1 \neq \mu_2$, we speak of a **two sided alternative**, because the rejection of $H_0: \mu_1 = \mu_2$ means either $\mu_1 > \mu_2$ or $\mu_1 < \mu_2$. We speak of two sided problems and of **two sided tests**. For the one sided problem—one parameter is larger than the other—$H_A: \mu_1 > \mu_2$ is contrasted with $H_0: \mu_1 \leq \mu_2$ (or $\mu_1 < \mu_2$ with $\mu_1 \geq \mu_2$).

If the sign of a presumable difference of the two parameters—for example means or medians—is known, then a one sided test is decided on **before** statistical analysis. Let $H_0: \pi = \pi_0$ mean, e.g., that two treatments are of equal therapeutic value; $\pi \neq \pi_0$ implies that the remedies are different, the new one being either better or not as good as the standard treatment. Assume that it is known from previous experiences or preliminary tests that the hypothesis of the new remedy being inferior to the standard treatment can in practice be rejected. Then the one sided test $\pi - \pi_0 > 0$ is preferred to the two sided test because it has higher proof: it is more sensitive to (positive) differences.

If it is not clear whether the problem is one or two sided, a two sided test must be used (cf., Section 1.4.7) because the alternative hypothesis must be the antithesis of the null hypothesis.

1.4.7 The power of a test

In decision problems, two types of error are to be taken into consideration: Errors of type I and II. The connection between them is shown in Figure 21. The density functions of a statistic with respect to two different models are plotted as bell-shaped curves; the one on the left represents H_0, T_{S_1}, the one on the right the simple alternate hypothesis, T_{S_2}. We obtain a critical value for the test statistic by prescribing the size of the error of the first kind,

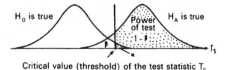

Figure 21 Power as area under a sampling distribution.

Critical value (threshold) of the test statistic T_s

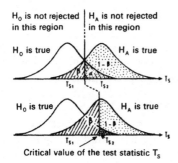

Figure 22 The critical value of the test statistic in its dependence on α (and β).

and compare it with the empirical test statistic, based on the sample. If this value of the test statistic equals or exceeds the critical value, H_0 is rejected. If the critical value is not attained by the test statistic, there is then no cause for rejecting H_0, i.e., it is retained. Figure 22 shows that, depending on the location of the critical value of the test statistic, the value of β (the risk II) increases as the level of significance α becomes smaller.

The risk II, the small probability β of retaining a false H_0, **depends on**:

1. The size n of the sample: the larger the sample, the sooner will a difference between two populations be detected, given a significance level α (risk I).
2. The degree of the difference δ between the hypothetical and the true condition, that is, the amount δ, by which H_0 is false.
3. The property of the test referred to as the **power**.

The power increases: (a) with n, (b) with δ,
(c) with the amount of information in the sample that is incorporated in the test statistic—it increases in the sequence: frequencies, ranks, and measurements (cf., sections 1.4.8 and 3.9);
(d) with the number of assumptions on the distribution of the statistic: a test that requires the normal distribution and homogeneity of variance (homoscedasticity) is in general substantially more powerful than one that makes no assumptions.

The power of a test is the probability of rejecting H_0 under the simple alternate hypothesis H_A. Thus it depends at least on δ, α, n and on the type of the test (simple, two sided, or one sided):

$$\text{Power} = P(\text{reject } H_0 | H_A \text{ is true}) = 1 - \beta. \qquad (1.124)$$

The smaller the probability β, the more sharply does the test separate H_0 and H_A when α is fixed. A test is called powerful if compared to other tests of size smaller or equal to α, it exhibits a relatively high power. If H_0 is true, the maximum power of a test equals α. If a very small α is given, statistical

1.4 The Road to the Statistical Test

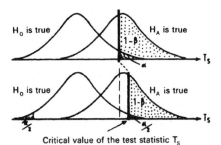

Figure 23 Dependence of the power on the one or two sidedness.

Critical value of the test statistic T_s

significance occurs only for large n or for large difference δ. **Therefore, the 5% level and a power of at least 70% or, better yet, of about 80% are often considered satisfactory.** More on this can be found in Cohen (1977) (cf., also Lehmann 1958 as well as Cleary and Linn 1969). The power can be raised by an arbitrary amount only through an increase in sample size. We recall that random samples with independent observations were assumed (cf., also Section 4.7.1). Powers of tests are compared in terms of the asymptotic relative efficiency (Pitman efficiency; cf., Sections 1.4.8 and 3.9.4). The power is diminished when a one sided problem is replaced by a two sided problem. For Figure 23 this would mean: The "triangle" α is halved; the critical value T_s shifts to the right (increases); β becomes larger and the power smaller. With equal sample size, the one sided test is always more powerful than the two sided. The strongly schematized power curves drawn in Figure 24 show the power as a function of the difference between the two means. The power of a test with given parameter difference increases with n and α. For α, the region of variation at our disposal is of course small, because in most cases we will only reluctantly allow the risk of rejecting a true H_0 to grow beyond 5%:

1. If there is no difference between the means of the populations we will, when working with the significance level α, wrongly reject H_0 in $\alpha\%$ of the cases: rejection probability = risk I.

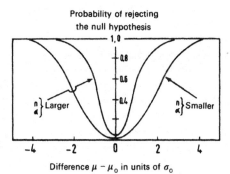

Figure 24 Sketch of power curves under different conditions for the two sided problems, the mean ordinate giving the level of significance for both curves ($\alpha \simeq 0.01$) (resp. $\alpha \simeq 0.03$). The bowl-shaped curves approach their symmetry axis, the ordinate, with increasing α and n.

Difference $\mu - \mu_0$ in units of σ_0

2. If there is a difference between the means of 1.5 units of σ_0, then the more powerful test, the narrower upside down bell-shaped curve in Figure 24, will point out the existence of a difference 80 times in 100 samples (power = 0.80). On the other hand, the weaker test, the wider upside down curve, will pretty well fail; it reveals the difference in only 30% of the cases (power = 0.30).
3. If there exists a very large difference between the means, then both curves have power 1.

Thus we see that, for the two sided test, the probability of rejecting H_0 increases with increasing distance $\mu - \mu_0$, and that a true alternate hypothesis is less likely to be adopted when the significance level becomes smaller as well as when the sample size becomes smaller. From this we see also that to realize a good power, the largest possible sample sizes are to be employed. **If the sample size is small, then the significance level must not be too small**, because a small sample together with a small significance level manifests itself in an undesirable reduction in power. The one sided test is, as we saw, distinguished by a power larger than the two sided. Since the one sided test discloses existing differences sooner than the two sided one, the one sided test is preferred if certain alternatives are of no significance or interest. If, for example, a new therapy is compared with one generally practiced, the only interesting question is whether the new therapy is better. If the new method is less effective or as effective, there is no cause to relinquish the old method. If two new methods are to be compared, only the two sided question makes sense; the one sided test would not treat the therapies symmetrically.

Distribution-free tests, in particular rapid tests, are characterized by an inferior power in comparison with the parametric tests. If data that indeed come from a normally distributed population, or any other homogeneous population with known distribution, are to be analyzed, higher Type II errors have to be put up with when distribution-free tests are used. The statistical decision is then conservative, i.e., H_0 is not as quickly rejected and significant results show up somewhat less frequently—in other words, larger samples are needed to rejct H_0. **If small samples are used ($n < 15$), distribution-free tests are often more efficient than the otherwise optimal parametric tests**, which are most efficient, and also simpler to manage for $n \gtrsim 80$.

If for some analysis there are several tests available, that test is generally preferred which most completely utilizes the information contained in the data. Of course the assumptions of the statistical model on which the test is based have to be satisfied by the data. If the assumptions of a test procedure are not or are only partially fulfilled, this must be taken into consideration in the appropriately cautious interpretation of the result. **It is advisable to list all the assumptions that might have been violated**. For example: "Under the assumption that both samples originated in normally distributed populations, there is ..." (see also Sachs 1984, pp. 100-105).

The following warning helps in avoiding mistakes (cf., Section 1.4.5) and false conclusions (cf., also Section 1.2.1).

> **It is not permitted to work through several tests:** The choice of a test on the basis of the results and the almost exclusive use of one sided tests might in practice lead to effective significance levels which are twice as large as the given significance level (Walter, 1964).

The operating characteristic

Figure 24 gives the power function—i.e., the power as a function of the mean difference in units of the standard deviation $[(\mu - \mu_0)/\sigma_0]$. Its complement, the probability of retaining a false null hypothesis, i.e., making a Type II error, is called **the operating characteristic** OC, the OC-curve, or the acceptance line; formulated somewhat loosely,

$$\text{Operating characteristic} = 1 - \text{power function.} \qquad (1.125)$$

OC curves are, with two sided questions, **bell-shaped** complements of the bowl-shaped power functions.

We can now invoke one of these two functions to characterize a test and e.g., in terms of the OC for given risk I and n, read off the unavoidable risk II in distinguishing between the null and alternative hypothesis, in determining the difference Δ (Greek delta). If for given risk I with small risk II the sample size needed to detect Δ becomes too large, risk I must be increased (Table 52a gives the sample sizes for the comparison of two means from normal distributions with same but unknown variance, for given risk I, risk II, and difference Δ [there termed d]). Indeed, one can sometimes also use a more powerful test. With equal sample size the OC would then vary more steeply and thus provide better detection of a difference. If an experiment is completed, the OC indicates what chance one has of detecting a difference of size Δ. A small sample size, together with small risk I, will lead to a large risk II, and the retention of H_0 is to be considered only with caution because, under these conditions, even a pronounced difference could hardly be detected. The OC is very important in setting up sampling schemes for quality control, in particular in acceptance inspection. Examples of the construction of OC curves are given by Yamane (1964). OC curves for the most important tests are given by Ferris et al., (1946), Owen (1962), Natrella (1963), and Beyer (1968 [cited in Section 2 of the Bibliography: Tables]). (cf., also Guenther 1973, Hodges and Lehmann 1968 as well as Morice 1968). **Comprehensive power tables** are provided by Cohen (1977).

1.4.8 Distribution-free procedures

The classical statistical procedures are usually based on normal distributions. In nature, however, normal distributions do not occur. Therefore, application of normal theory imparts a feeling of uneasiness. For this reason the development of distribution-free or distribution-independent methods met with much interest. No assumptions are made on the underlying distribution. We only need to be assured that the **random samples** we want to compare belong to the same basic population (Walter, 1964), that they can be interpreted (Lubin 1962) as **homomer**. Since parameters hardly play a role (nonparametric hypotheses), the distribution-free methods can also be referred to as **parameter-free or nonparametric methods**. They are, for the most part, very easily dealt with numerically. Their advantage lies in the fact that one need have practically no knowledge whatsoever about the distribution function of the population. Moreover, these quite easily understood procedures can also be applied to **rank data** and qualitative information.

A classical method of comparing means, Student's t-test, can only be applied under the following conditions:

1. The data must be independent (random samples).
2. The characteristic must be measurable in units of a metric scale.
3. The populations must be (at least nearly) normally distributed.
4. The variances must be equal ($\sigma_1^2 = \sigma_2^2$).

The distribution-free procedures for the same problem merely require independent data. Whether the data readings are mutually independent must be deduced from the way they were gathered. Thus we need only assume that **all data or data pairs are drawn randomly and independently of one another from one and the same basic population** of data, and this must be guaranteed by the structure and the realization of the experiment. A distribution-free test, when applied to data from a known family of distributions, is always weaker than the corresponding parametric test (cf., Section 1.4.7). Pitman (1949) defines the index

$$E_n = \frac{n \text{ for the parametric test}}{n \text{ for the nonparametric test}} \qquad (1.126)$$

as the "**efficiency**" of the nonparametric test. Here n denotes the sample size needed to realize a given power. The concept "asymptotic efficiency" is defined as the efficiency of the test in the limiting case of a sample of normally distributed data with size tending to infinity. It becomes apparent, in terms of this index, how effective or how efficient a distribution-free test is, if it is applied, in place of a classical test, to normally distributed data. An asymptotic efficiency of $E = 0.95$—which for example the U-test exhibits—means: If in applying the nonparametric test, a sample of $n = 100$ data readings is, on the average, required for a certain significance level, then with the

1.4 The Road to the Statistical Test

application of the corresponding parametric test, $n = 95$ measured values would suffice. The so-called rank tests (see Section 3.9) assume continuous distributions; the recurrence of some observations has little effect on the validity of the continuity assumption, emphasizing rather the inaccuracy of the method of measurement. Distribution-free procedures are indicated if (a) the parametric procedure is sensitive to certain deviations from the assumptions, or if (b) the forcing of these assumptions by an appropriate transformation (b_1) or by elimination of outliers (b_2) creates difficulties; in general, therefore, such procedures are indicated (1) by nonnormality, (2) by data originating from a rank scale or a nominal scale (see below), (3) as a check of a parametric test, and (4) as a rapid test.

Distribution-free tests, which distinguish themselves by computational brevity, are referred to as rapid tests. The peculiarity of these tests, besides their **computational economy**, is their wide **assumption-free** applicability. Their drawback is their **small power**, since only a part of the information contained in the data is utilized in the statistical decision.

In comparison with the relevant optimal parametric or nonparametric test, the statistical decision of a rapid test is **conservative**; i.e., it retains the null hypothesis longer than necessary: larger samples of data (rank data or binary data) are required in order to reject the null hypothesis. More on this [and also on the so-called randomization test (permutation test), see Biometrics **38** [1982], 864–867)] can be found in the books listed in [8:1b] of the bibliography. We discuss the most important distribution-free tests in Sections 3.9, 4.4, 4.6–8, 5.3, 6.1–2, 7.5, and 7.6. For graphical methods, see Fisher (1983).

Indications for distribution-free rapid tests, according to Lienert (1962), are as follows:

1. The most important area in which rapid tests are employed is the **approximate** assessment of the significance of parametric as well as nonparametric data sequences. The rapid test is used here to investigate whether it is really worthwhile to carry out a time-consuming optimal test. As to the outcome of a rapid test, there are three possibilities:
    a. The result can be clearly significant. Testing with a more powerful test is then unnecessary, because the goal of the testing is already realized by the weak test.
    b. The result can be absolutely insignificant, i.e., there is no chance at all of it being significant; a stronger test is likewise unnecessary in this case.
    c. The result can show borderline significance. A verification using the time-consuming optimal method is reasonable (cf. end of Section 1.4.7).
2. An additional area in which distribution-free rapid tests are indicated is the assessment of significance of data obtained from **preliminary trials**. Results from preliminary surveys must be well founded if the subsequent experiment is to lead to reliable answers.

3. Finally, rapid tests can be used without hesitation to obtain a **definitive** assessment of significance whenever large samples of data are available, i.e., samples of size perhaps $n > 100$.

Of the three possible applications, the first has undoubtedly the greatest practical importance.

Remark: Systems of measurements

The occupations of individuals being surveyed can in no way be used to arrange these individuals in a unique and objective sequence. Classifications of this sort—we are speaking of the **nominal scales**—are present in the listing of groups of races, occupations, languages, and nationalities. Frequently an order relevant to the objective of the study presents itself: If, for example, the objects under study are arranged in an impartial sequence according to age or according to some other property where, however, the distances on the **rank scale or ordinal scale** represents no true distance (only the relative position). Thus, on a rank scale ordered by age, a twenty year old can be followed by a thirty year old, who is then followed by a thirty-two year old.

If consecutive **intervals** are of equal length (here we have the conventional Celsius temperature scale in mind), the interval scale still permits no meaningful comparison: It is incorrect to assert that 10 degree Celsius is twice as warm as 5 degrees Celsius. Only an interval scale with absolute zero makes meaningful comparison possible. Properties for which such a zero can be specified are, for example, temperature measured in degrees Kelvin, length, weight, and time. Scales of this sort are the most useful and are called **ratio scales**. When one ratio scale is transformed into another under multiplication by a positive constant (for example, 1 U.S. mile = 1.609347 kilometers), i.e., $y = ax$, the ratio of two numerical observations remains unchanged, whereas on an interval scale (e.g., conversion from x degrees Celsius to y degrees Fahrenheit: $y = ax + b$ with $a = \frac{9}{5}$ and $b = 32$), the ratio will change.

The admissible scale transformations (ST) are thus: (sequence-altering) permutations (nominal scale); all ST that do not change the order of the elements, e.g., raising a positive number to a power (ordinal scale); addition of a constant (interval scale); multiplication by a constant (ratio scale).

With the four types of scales recognized by Stevens (1946) one can associate the following statistical notions.

1. **Nominal scale**: Licence plate numbers and zip codes (arbitrary numbering); marital status; occupational and color classifications. Ideas: frequency data, χ^2 tests, the binomial and Poisson distributions, and, as a location parameter, the mode.
2. **Rank scale**: School grades and other particulars that set up a ranking; ranking tests such as the sign test, the run test, the U-test, the H-test, the rank analysis of variance, and the rank correlation. Ideas: deciles such as the median.
3. **Interval scale**: (zero point conventionally set, intervals with empirical meaning, direct construction of a ratio not allowed): Calendar date; intelligence quotient; temperature measurement in degrees Celsius or Fahrenheit. Ideas: typical parameters like the arithmetic mean, the

standard deviation, the correlation coefficient, and the regression coefficient, as well as the usual statistical tests like the t-test and the F-test.
4. **Ratio scale**: (with true zero point): Temperature measurement in degrees Kelvin; physical quantities in units such as m, kg, s. Ideas: in addition to the characteristics listed under 3, the geometric and harmonic mean as well as the coefficient of variation.

It is important to realize that to data belonging to a **nominal scale** or a **rank scale** only **distribution-free** tests may be applied, while the values of an interval or ratio scale can be analyzed by parametric as well as by distribution-free tests. More on scaling can be found in Fraser (1980).

1.4.9 Decision principles

Many of our decisions can be interpreted in terms of the so-called minimax philosophy of Abraham Wald (1902–1950). According to the **minimax principle** (cf., von Neumann 1928), that decision is preferred which minimizes the maximum (the worst case) of the expected loss. The decision which causes the smallest possible risk (expected loss) will be adopted. It is optimal in the sense of insisting on the largest possible safeguards against risk; this leads, in many cases, to a scarcely tolerable disregard of important opportunities. Only a chronic pessimist would always act in this way. On the other hand, this principle minimizes the chances of a catastrophic loss. **Thus a minimaxer is someone who decides in such a way as to defend himself as well as possible (maximally) against the worst conceivable situation (minimum)**. According to the minimax criterion, every judge will avoid sending innocent people to jail. Acquittal of not fully convicted criminals is the price of such a course of action. A "minimaxer" has a motive to insure: Let us assume that a workshop valued at $100,000 is insured against loss due to fire by payment of a $5,000 premium. The probability of fire destroying the workshop is 1%. If the loss is to be the smallest possible, one must keep in mind that on taking out insurance a definite loss of $5,000 is experienced, while without insurance one would be faced with an **expected** loss of one percent, which is only $1,000. The **actual** loss is however either zero or $100,000. The minimaxer thus prefers the certain loss of $5,000.

If not one but rather many objects—say 80 ships belonging to a large shipping firm—are to be insured, it can then be expedient to have only particular ships insured or even to take out no insurance. Debt-free objects need not be insured. Nothing is insured by the government.

The full-blown optimist—in our manner of speaking, **a "maximaxer"**— **chooses the decision that yields the best results (maximum) under the most favorable conditions (maximum)** and rejects the notion of taking insurance, since a workshop fire is "improbable." The maximax criterion promises success whenever large gains are possible with relatively small losses. The "maximaxer" buys lottery tickets because the almost certain insignificant loss is more than made up for by the very improbable large gain. This decision principle in which the **largest possible** gain settles things—goes back to Bayes (1702–1761) and Laplace (1749–1827). Barnett (1982) provides a summary.

We cannot here delve into the application of the two decision principles. The interested and to some extent mathematically versed reader is referred, with regard to these as well as other **decision criteria**, to Kramer (1966), who distinguishes a total of twelve different criteria; and to the specialized literature (Bühlmann et al., 1967, Schneeweiss 1967, Bernard 1968, Chernoff and Moses 1959, and the bibliography of Wasserman and Silander 1964). Important particular aspects are treated by Raiffa and Schlaifer (1961), Ackoff (1962), Hall (1962), Fishburn (1964), Theil (1964) and de Groot (1970). An overview is provided by Keeney (1982). For risk and insurance, see Beard et al. (1984, cited in [8:2d]).

Science arrives at conclusions by way of decisions. **Decisions** are of the form "we decide now as if". By the restrictions "deal with as if" and "now" we do "our best" in the present situation without at the same time making a judgment as to the "truth" in the sense of $6 > 4$. On the other hand, **conclusions**—the maxims of science—are drawn while paying particular attention to evidence gathered from specific observations and experiments. Only the "truth" contained in the experiment is relevant. **CONCLUSIONS ARE DEFERRED IF SUFFICIENT EVIDENCE IS NOT AVAILABLE.** A conclusion is a statement that can be taken as applicable to the conditions of the experiment or to some observation, so long as there is not an unusually large amount of evidence to the contrary. This definition sets forth three crucial points: It emphasises "acceptance" in the strict sense of the word, speaks of "unusually strong evidence," and it includes the possibility of subsequent rejection (cf. Tukey, 1960).

1.5 THREE IMPORTANT FAMILIES OF TEST DISTRIBUTIONS

In this section the distribution of **test statistics** is examined. The value of the test statistic, a scalar, is calculated for a given sample. Thus the sample mean, the sample variance or the ratio of the variances of two samples, all of these being estimates or functions of **sample functions**, can be interpreted as test statistics. The test statistic is a random variable. The probability distributions of these test statistics are the foundations of the tests based on them. Because the normal distribution plays a special role, sample functions of normally distributed random variables (cf., end of Section 1.5.3) are called **test distributions**. An important survey is due to Haight (1961). Extensive tables are provided, e.g., Pearson and Hartley (Vol. I, II; 1969, 1972 [cited on page 571]).

1.5.1 Student's *t*-distribution

W. S. Gosset (1876–1937), writing under the pseudonym "Student," proved in 1908 that for given n, the standardized difference (1.127)—the difference between the estimate $\bar{x}$ of the mean and the known mean μ, divided by the standard deviation $\sigma_{\bar{x}}$ of the mean (right side of (1.127)—has a standard normal distribution only when the x's are normally distributed and both parameters (μ, σ) are known. When σ is unknown and replaced by the estimate s (standard deviation of a sample), the quotient (1.128) follows the **"Student" distribution or *t*-distribution** (it is assumed that the individual observations are independent and (approximately) normally distributed):

$$\frac{\text{difference between the estimate and the true mean}}{\text{standard deviation of the mean}}$$
$$= \frac{\bar{x} - \mu}{\sigma}\sqrt{n} = \frac{\bar{x} - \mu}{\sigma/\sqrt{n}} = \frac{\bar{x} - \mu}{\sigma_{\bar{x}}}, \quad (1.127)$$

On page 155 above you find the correct estimator notation.

$$t = \frac{\bar{x} - \mu}{s/\sqrt{n}} = \frac{\bar{x} - \mu}{s_{\bar{x}}}. \quad (1.128)$$

(For definition of t see (1.131) below.)

Remark: (1.127) tends, generally, with increasing n, more or less rapidly toward a normal distribution, in accordance with the type of population form which the samples are drawn; the right side of (1.128) is (a) for small n and for populations with distributions not differing greatly from the normal, distributed approximately as t, (b) for large n and for almost all populations, distributed approximately standard normally.

The *t*-distribution (cf., Figure 25) is very similar to the standard normal distribution [$N(0, 1)$ distribution]. Like the normal distribution, it is continuous, symmetric, and bell-shaped with range from minus infinity to plus infinity. It is, however, **independent of μ and σ**. The shape of the *t*-distribution is determined solely by the so-called degrees of freedom.

Degrees of freedom. The number of degrees of freedom DF or v (Greek nu) of a random variable is defined as the number of "free" available observations—the sample size n minus the number a of a parameters estimated from the sample:

$$DF = v = n - a. \quad (1.129)$$

Recall that $s^2 = [\sum (x_i - \bar{x})^2/(n-1)]$. Since the mean value must be estimated from the sample, $a = 1$, so that the random variable (1.128) is distinguished by $v = n - 1$ degrees of freedom. Instructions on how the

Table 27 Two sided and upper percentage points of the Student distribution excerpted from Fisher and Yates (1963), p. 46, Table III

DF \ α	0.50	0.20	0.10	0.05	0.02	0.01	0.005	0.002	0.001	0.0001
1	1.000	3.078	6.314	12.706	31.821	63.657	127.321	318.309	35.619	6,366.198
2	0.816	1.886	2.920	4.303	6.965	9.925	14.089	22.327	31.598	99.992
3	0.765	1.638	2.353	3.182	4.541	5.841	7.453	10.214	12.924	28.000
4	0.741	1.533	2.132	2.776	3.747	4.604	5.598	7.173	8.610	15.544
5	0.727	1.476	2.015	2.571	3.365	4.032	4.773	5.893	6.869	11.178
6	0.718	1.440	1.943	2.447	3.143	3.707	4.317	5.208	5.959	9.082
7	0.711	1.415	1.895	2.365	2.998	3.499	4.029	4.785	5.408	7.885
8	0.706	1.397	1.860	2.306	2.896	3.355	3.833	4.501	5.041	7.120
9	0.703	1.383	1.833	2.262	2.821	3.250	3.690	4.297	4.781	6.594
10	0.700	1.372	1.812	2.228	2.764	3.169	3.581	4.144	4.587	6.211
11	0.697	1.363	1.796	2.201	2.718	3.106	3.497	4.025	4.437	5.921
12	0.695	1.356	1.782	2.179	2.681	3.055	3.428	3.930	4.318	5.694
13	0.694	1.350	1.771	2.160	2.650	3.012	3.372	3.852	4.221	5.513
14	0.692	1.345	1.761	2.145	2.624	2.977	3.326	3.787	4.140	5.363
15	0.691	1.341	1.753	2.131	2.602	2.947	3.286	3.733	4.073	5.239
16	0.690	1.337	1.746	2.120	2.583	2.921	3.252	3.686	4.015	5.134
17	0.689	1.333	1.740	2.110	2.567	2.898	3.222	3.646	3.965	5.044
18	0.688	1.330	1.734	2.101	2.552	2.878	3.197	3.610	3.922	4.966
19	0.688	1.328	1.729	2.093	2.539	2.861	3.174	3.579	3.883	4.897
20	0.687	1.325	1.725	2.086	2.528	2.845	3.153	3.552	3.850	4.837
21	0.686	1.323	1.721	2.080	2.518	2.831	3.135	3.527	3.819	4.784
22	0.686	1.321	1.717	2.074	2.508	2.819	3.119	3.505	3.792	4.736
23	0.685	1.319	1.714	2.069	2.500	2.807	3.104	3.485	3.767	4.693
24	0.685	1.318	1.711	2.064	2.492	2.797	3.091	3.467	3.745	4.654
25	0.684	1.316	1.708	2.060	2.485	2.787	3.078	3.450	3.725	4.619
26	0.684	1.315	1.706	2.056	2.479	2.779	3.067	3.435	3.707	4.587
27	0.684	1.314	1.703	2.052	2.473	2.771	3.057	3.421	3.690	4.558
28	0.683	1.313	1.701	2.048	2.467	2.763	3.047	3.408	3.674	4.530
29	0.683	1.311	1.699	2.045	2.462	2.756	3.038	3.396	3.659	4.506

Significance level α for the two sided test

1.5 Three Important Families of Test Distributions

DF \ α	0.25	0.10	0.05	0.025	0.01	0.005	0.0025	0.0005	0.00005
30	0.683	1.310	1.697	2.042	2.457	2.750	3.030	3.385	3.646	4.482
32	0.682	1.309	1.694	2.037	2.449	2.738	3.015	3.365	3.622	4.441
34	0.682	1.307	1.691	2.032	2.441	2.728	3.002	3.348	3.601	4.405
35	0.682	1.306	1.690	2.030	2.438	2.724	2.996	3.340	3.591	4.389
36	0.681	1.306	1.688	2.028	2.434	2.719	2.990	3.333	3.582	4.374
38	0.681	1.304	1.686	2.024	2.429	2.712	2.980	3.319	3.566	4.346
40	0.681	1.303	1.684	2.021	2.423	2.704	2.971	3.307	3.551	4.321
42	0.680	1.302	1.682	2.018	2.418	2.698	2.963	3.296	3.538	4.298
45	0.680	1.301	1.679	2.014	2.412	2.690	2.952	3.281	3.520	4.269
47	0.680	1.300	1.678	2.012	2.408	2.685	2.946	3.273	3.510	4.251
50	0.679	1.299	1.676	2.009	2.403	2.678	2.937	3.261	3.496	4.228
55	0.679	1.297	1.673	2.004	2.396	2.668	2.925	3.245	3.476	4.196
60	0.679	1.296	1.671	2.000	2.390	2.660	2.915	3.232	3.460	4.169
70	0.678	1.294	1.667	1.994	2.381	2.648	2.899	3.211	3.435	4.127
80	0.678	1.292	1.664	1.990	2.374	2.639	2.887	3.195	3.416	4.096
90	0.677	1.291	1.662	1.987	2.368	2.632	2.878	3.183	3.402	4.072
100	0.677	1.290	1.660	1.984	2.364	2.626	2.871	3.174	3.390	4.053
120	0.677	1.289	1.658	1.980	2.358	2.617	2.860	3.150	3.373	4.025
200	0.676	1.286	1.653	1.972	2.345	2.601	2.838	3.131	3.340	3.970
500	0.675	1.283	1.648	1.965	2.334	2.586	2.820	3.107	3.310	3.922
1000	0.675	1.282	1.646	1.962	2.330	2.581	2.813	3.098	3.300	3.906
∞	0.675	1.282	1.645	1.960	2.326	2.576	2.807	3.090	3.290	3.891
	0.25	0.10	0.05	0.025	0.01	0.005	0.0025	0.0031	0.0005	0.00005

Significance level α for the one sided test

With $\nu \geq 30$ degrees of freedom one also uses the approximation $t_{\nu;\alpha} = z_\alpha + (z_\alpha^3 + z_\alpha)/4\nu$; the values of z_α can be taken from Table 43 in Section 2.1.6. Example: $t_{30;0.05} = 1.96 + (1.96^3 + 1.96)/(4)(30) = 2.039$ (exact value: 2.0423). For $t_{90;0.05}$ we get 1.9864 (exact value: 1.9867). Better approximations are given by Dudewicz and Dalal (1972), Ling (1978), and Koehler (1983).

Application: Every computed $\hat{t}$ value is based on ν degrees of freedom (DF). On the basis of this quality, the preselected level of significance α and the given one or two sided question, one determines the tabulated value $t_{\nu;\alpha}$: $\hat{t}$ is significant at the 100α% level provided $\hat{t} \geq t_{\nu;\alpha}$. For example, $\hat{t} = 2.00$ with 60 degrees of freedom: the two sided test gives a significant result at the 5% level, the one sided test at the 2.5% level (cf. Sections 3.1.4, 3.6).

number of degrees of freedom is to be determined for particular cases of this random variable (as well as for other test statistics) will be given later for the various cases as they arise.

The smaller the number of degrees of freedom, the greater is the departure from the $N(0, 1)$ distribution, and the flatter are the curves—i.e., in contrast with the $N(0, 1)$ distribution there is more probability concentrated in the tails and less in the central part (cf., Figure 25). **With a large number of degrees of freedom the t-distribution turns into the $N(0, 1)$ distribution.** The primary application of the t-distribution is in the comparison of means.

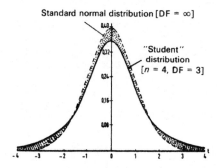

Figure 25 The probability density of the $N(0, 1)$ distribution and the Student distribution with 3 degrees of freedom ($n = 4$ observations). With a decreasing number of degrees of freedom, the maximum of the Student distribution drops and the shaded area grows. In comparison with the $N(0, 1)$ distribution, more probability is concentrated in the tails and less in the central part.

When the number of degrees of freedom is small the Student distribution has, in comparison with the $N(0, 1)$ distribution, with little height a substantially larger spread. Whereas for the normal curve 5% and 1% of the total area lies outside the critical values ± 1.96 and ± 2.58, the corresponding values for 5 degrees of freedom are ± 2.57 and ± 4.03. For 120 degrees of freedom, they are ± 1.98 and ± 2.62, and thus almost coincide with the critical values of the $N(0, 1)$ distribution.

Table 27 gives selected percentage points of the t-distribution. This t-table gives, over a large range of degrees of freedom, the probabilities of exceeding t-values entirely by chance at specific significance levels. One begins with v the number of degrees of freedom; the probabilities that a random variable with a t-distribution assumes an (absolute) value of at least t is indicated at the top of this table. Thus for 5 degrees of freedom (DF = 5 or $v = 5$) the crossing probability P for $t = 2.571$ is found to be 0.05 or 5%. P is that portion of the total area which lies under both tail ends of the t-distribution; it is the probability that the tabulated value t is exceeded by a random variable with a t-distribution ($t_{5;0.05} = 2.57$); $t_{60;0.05} = 2.000$; $t_{\infty;\alpha} = z_\sigma$; (cf., also Sections 3.2, 4.6.1, 4.6.2).

Table 27 lists percentage points for two-sided and one-sided problems. We can, for example, for the one-sided test, read off both of the following t-values: $t_{30;0.05} = 1.697$ and $t_{120;0.01} = 2.358$. The first index indicates the number of degrees of freedom; the second, the selected level of significance. Extensive tables of the Student distribution are given in Federighi (1959), Smirnov (1961), and Hill (1972). For approximations see page 137.

1.5.2 The χ^2 distribution

If s^2 is the variance of a random sample of size n taken from a population with variance σ^2, then the random variable

$$\chi^2 = \frac{(n-1)s^2}{\sigma^2} \qquad (1.130)$$

(n independent observations assumed) follows a χ^2 **distribution (chi square distribution)** with the parameter $v = n - 1$, v degrees of freedom. The χ^2 distribution (cf., Figure 26) is a continuous **nonsymmetric** distribution. Its range extends from zero to infinity. With increasing number of degrees of freedom it ("slowly") approaches the normal distribution. The mean and variance of this asymptotic distribution are, respectively, v and $2v$ (cf., also the end of Section 1.5.3). We see that the shape of the χ^2 distribution depends only on the **number of degrees of freedom**, just as for the Student distribution. For $v \leq 2$ the χ^2 distribution is L-shaped.

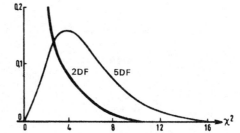

Figure 26 Probability density of the χ_v^2 distribution for $v = 2$ and $v = 5$.

As v increases the skewed, singly peaked ($v > 2$) curve becomes flatter and more symmetric. An essential property of the χ^2 distribution is its **additivity**: If two independent random variables have χ^2 distributions with v_1 and v_2 degrees of freedom, their sum has a χ^2 distribution with $v_1 + v_2$ degrees of freedom. The principal application of this distribution, which was discovered by I. J. Bienayme (1858), F. R. Helmert (1876), and K. Pearson (1900), is (cf. e.g., Section 4.5.5) in testing contingency tables.

χ^2 with v degrees of freedom is defined as the sum of the squares of v independent standard normal variables [cf., also (1.187) in Section 1.6.6.2, as well as Section 1.5.3]:

$$\chi_v^2 = \sum_{i=1}^{v} Z_i^2 \qquad (1.131)$$

(definition of $t: t_v = Z/\sqrt{\chi_v^2/v}$). When more than 30 degrees of freedom are present, the following approximations apply [$v = DF$; $z =$ standard normal variable (see Table 43); one sided test, e.g., $z_{0.05;\text{one sided}} = 1.645$: other (p. 217)

Table 28 Percentage points of the χ^2 distribution [excerpted from Fisher and Yates (1963), p. 47, Table IV]

DF \ α	0.99	0.975	0.95	0.90	0.80	0.70	0.50	0.30	0.20	0.10	0.05	0.025	0.01	0.001
1	0.00016	0.00098	0.0039	0.0158	0.064	0.148	0.455	1.07	1.64	2.71	3.84	5.02	6.63	10.83
2	0.0201	0.0506	0.1026	0.2107	0.446	0.713	1.39	2.41	3.22	4.61	5.99	7.38	9.21	13.82
3	0.115	0.216	0.352	0.584	1.00	1.42	2.37	3.66	4.64	6.25	7.81	9.35	11.34	16.27
4	0.297	0.484	0.711	1.064	1.65	2.20	3.36	4.88	5.99	7.78	9.49	11.14	13.28	18.47
5	0.554	0.831	1.15	1.61	2.34	3.00	4.35	6.06	7.29	9.24	11.07	12.83	15.09	20.52
6	0.872	1.24	1.64	2.20	3.07	3.83	5.35	7.23	8.56	10.64	12.59	14.45	16.81	22.46
7	1.24	1.69	2.17	2.83	3.82	4.67	6.35	8.38	9.80	12.02	14.07	16.01	18.48	24.32
8	1.65	2.18	2.73	3.49	4.59	5.53	7.34	9.52	11.0	13.36	15.51	17.53	20.09	26.13
9	2.09	2.70	3.33	4.17	5.38	6.39	8.34	10.7	12.2	14.68	16.92	19.02	21.67	27.88
10	2.56	3.25	3.94	4.87	6.18	7.27	9.34	11.8	13.4	15.99	18.31	20.48	23.21	29.59
11	3.05	3.82	4.57	5.58	6.99	8.15	10.3	12.9	14.6	17.28	19.68	21.92	24.73	31.26
12	3.57	4.40	5.23	6.30	7.81	9.03	11.3	14.0	15.8	18.55	21.03	23.34	26.22	32.91
13	4.11	5.01	5.89	7.04	8.63	9.93	12.3	15.1	17.0	19.81	22.36	24.74	27.69	34.53
14	4.66	5.63	6.57	7.79	9.47	10.8	13.3	16.2	18.2	21.06	23.68	26.12	29.14	36.12
15	5.23	6.26	7.26	8.55	10.3	11.7	14.3	17.3	19.3	22.31	25.00	27.49	30.58	37.70
16	5.81	6.91	7.96	9.31	11.2	12.6	15.3	18.4	20.5	23.54	26.30	28.85	32.00	39.25
17	6.41	7.56	8.67	10.08	12.0	13.5	16.3	19.5	21.6	24.77	27.59	30.19	33.41	40.79
18	7.01	8.23	9.39	10.86	12.9	14.4	17.3	20.6	22.8	25.99	28.87	31.53	34.81	42.31
19	7.63	8.91	10.12	11.65	13.7	15.4	18.3	21.7	23.9	27.20	30.14	32.85	36.19	43.82
20	8.26	9.59	10.85	12.44	14.6	16.3	19.3	22.8	25.0	28.41	31.41	34.17	37.57	45.31
22	9.54	10.98	12.34	14.04	16.3	18.1	21.3	24.9	27.3	30.81	33.92	36.78	40.29	48.27
24	10.86	12.40	13.85	15.66	18.1	19.9	23.3	27.1	29.6	33.20	36.42	39.36	42.98	51.18
26	12.20	13.84	15.38	17.29	19.8	21.8	25.3	29.2	31.8	35.56	38.89	41.92	45.64	54.05
28	13.56	15.31	16.93	18.94	21.6	23.6	27.3	31.4	34.0	37.92	41.34	44.46	48.28	56.89
30	14.95	16.79	18.49	20.60	23.4	25.5	29.3	33.5	36.2	40.26	43.77	46.98	50.89	59.70
35	18.51	20.57	22.46	24.80	27.8	30.2	34.3	38.9	41.8	46.06	49.80	53.20	57.34	66.62
40	22.16	24.43	26.51	29.05	32.3	34.9	39.3	44.2	47.3	51.81	55.76	59.34	63.69	73.40
50	29.71	32.36	34.76	37.69	41.4	44.3	49.3	54.7	58.2	63.17	67.50	71.42	76.15	86.66
60	37.48	40.48	43.19	46.46	50.6	53.8	59.3	65.2	69.0	74.40	79.08	83.30	88.38	99.61
80	53.54	57.15	60.39	64.28	69.2	72.9	79.3	86.1	90.4	96.58	101.88	106.63	112.33	124.84
100	70.06	74.22	77.93	82.36	87.9	92.1	99.3	106.9	111.7	118.50	124.34	129.56	135.81	149.45
120	86.92	91.57	95.70	100.62	106.8	111.4	119.3	127.6	132.8	140.23	146.57	152.21	158.95	173.62
150	112.67	117.99	122.69	128.28	135.3	140.5	149.3	158.6	164.3	172.58	179.58	185.80	193.21	209.26
200	156.43	162.73	168.28	174.84	183.0	189.0	199.3	210.0	216.6	226.02	233.99	241.06	249.45	267.54

DF \ α	0.10	0.05	0.01	0.001	0.0001
1	2.7055	3.8415	6.6349	10.8276	15.1367
2	4.6052	5.9915	9.2103	13.8155	18.4207
3	6.2514	7.8147	11.3449	16.2662	21.1075
4	7.7794	9.4877	13.2767	18.4668	23.5127
5	9.2364	11.0705	15.0863	20.5150	25.7448
6	10.6446	12.5916	16.8119	22.4577	27.8563

Application: $P(\hat{\chi}^2 \geq$ tabulated value) $= \alpha$; e.g., for 4 degrees of freedom we thus have $P(\hat{\chi}^2 \geq 9.49) = 0.05$; i.e., a $\hat{\chi}^2$ value greater than or equal to 9.49 for DF = 4 is significant at the 5% level. Examples can be found in Sections 33, 34, 43, 46, 61, 62, and 76.

1.5 Three Important Families of Test Distributions

Table 28a Selected percentage points (5%, 1%, and 0.1% levels) of the χ^2 distribution

DF	5 %	1 %	0.1 %	DF	5 %	1 %	0.1 %	DF	5 %	1 %	0.1 %
1	3.84	6.63	10.83	51	68.67	77.39	87.97	101	125.46	136.97	150.67
2	5.99	9.21	13.82	52	69.83	78.61	89.27	102	126.57	138.13	151.88
3	7.81	11.34	16.27	53	70.99	79.84	90.57	103	127.69	139.30	153.10
4	9.49	13.28	18.47	54	72.15	81.07	91.87	104	128.80	140.46	154.31
5	11.07	15.09	20.52	55	73.31	82.29	93.17	105	129.92	141.62	155.53
6	12.59	16.81	22.46	56	74.47	83.51	94.46	106	131.03	142.78	156.74
7	14.07	18.48	24.32	57	75.62	84.73	95.75	107	132.15	143.94	157.95
8	15.51	20.09	26.13	58	76.78	85.95	97.04	108	133.26	145.10	159.16
9	16.92	21.67	27.88	59	77.93	87.16	98.32	109	134.37	146.26	160.37
10	18.31	23.21	29.59	60	79.08	88.38	99.61	110	135.48	147.41	161.58
11	19.68	24.73	31.26	61	80.23	89.59	100.89	111	136.59	148.57	162.79
12	21.03	26.22	32.91	62	81.38	90.80	102.17	112	137.70	149.73	163.99
13	22.36	27.69	34.53	63	82.53	92.01	103.44	113	138.81	150.88	165.20
14	23.68	29.14	36.12	64	83.68	93.22	104.72	114	139.92	152.04	166.41
15	25.00	30.58	37.70	65	84.82	94.42	105.99	115	141.03	153.19	167.61
16	26.30	32.00	39.25	66	85.97	95.62	107.26	116	142.14	154.34	168.81
17	27.59	33.41	40.79	67	87.11	96.83	108.52	117	143.25	155.50	170.01
18	28.87	34.81	42.31	68	88.25	98.03	109.79	118	144.35	156.65	171.22
19	30.14	36.19	43.82	69	89.39	99.23	111.05	119	145.46	157.80	172.42
20	31.41	37.57	45.31	70	90.53	100.42	112.32	120	146.57	158.95	173.62
21	32.67	38.93	46.80	71	91.67	101.62	113.58	121	147.67	160.10	174.82
22	33.92	40.29	48.27	72	92.81	102.82	114.83	122	148.78	161.25	176.01
23	35.17	41.64	49.73	73	93.95	104.01	116.09	123	149.89	162.40	177.21
24	36.42	42.98	51.18	74	95.08	105.20	117.35	124	150.99	163.55	178.41
25	37.65	44.31	52.62	75	96.22	106.39	118.60	125	152.09	164.69	179.60
26	38.89	45.64	54.05	76	97.35	107.58	119.85	126	153.20	165.84	180.80
27	40.11	46.96	55.48	77	98.49	108.77	121.10	127	154.30	166.99	181.99
28	41.34	48.28	56.89	78	99.62	109.96	122.35	128	155.41	168.13	183.19
29	42.56	49.59	58.30	79	100.75	111.14	123.59	129	156.51	169.28	184.38
30	43.77	50.89	59.70	80	101.88	112.33	124.84	130	157.61	170.42	185.57
31	44.99	52.19	61.10	81	103.01	113.51	126.08	131	158.71	171.57	186.76
32	46.19	53.48	62.49	82	104.14	114.69	127.32	132	159.81	172.71	187.95
33	47.40	54.77	63.87	83	105.27	115.88	128.56	133	160.92	173.85	189.14
34	48.60	56.06	65.25	84	106.40	117.06	129.80	134	162.02	175.00	190.33
35	49.80	57.34	66.62	85	107.52	118.23	131.04	135	163.12	176.14	191.52
36	51.00	58.62	67.98	86	108.65	119.41	132.28	136	164.22	177.28	192.71
37	52.19	59.89	69.34	87	109.77	120.59	133.51	137	165.32	178.42	193.89
38	53.38	61.16	70.70	88	110.90	121.77	134.74	138	166.42	179.56	195.08
39	54.57	62.43	72.05	89	112.02	122.94	135.98	139	167.52	180.70	196.27
40	55.76	63.69	73.40	90	113.15	124.12	137.21	140	168.61	181.84	197.45
41	56.94	64.95	74.74	91	114.27	125.29	138.44	141	169.71	182.98	198.63
42	58.12	66.21	76.08	92	115.39	126.46	139.67	142	170.81	184.12	199.82
43	59.30	67.46	77.42	93	116.51	127.63	140.89	143	171.91	185.25	201.00
44	60.48	68.71	78.75	94	117.63	128.80	142.12	144	173.00	186.39	202.18
45	61.66	69.96	80.08	95	118.75	129.97	143.34	145	174.10	187.53	203.36
46	62.83	71.20	81.40	96	119.87	131.14	144.57	146	175.20	188.67	204.55
47	64.00	72.44	82.72	97	120.99	132.31	145.79	147	176.29	189.80	205.73
48	65.17	73.68	84.04	98	122.11	133.47	147.01	148	177.39	190.94	206.91
49	66.34	74.92	85.35	99	123.23	134.64	148.23	149	178.49	192.07	208.09
50	67.50	76.15	86.66	100	124.34	135.81	149.45	150	179.58	193.21	209.26

Examples: $\chi^2_{15;0.05} = 25.00$ and $\chi^2_{47;0.05} = 64.00$.

sometimes necessary one sided bounds are $z_{0.095} = 1.3106$, $z_{0.0975} = 1.2959$, $z_{0.098} = 1.2930$, and $z_{0.099} = 1.2873$].

$$\chi^2 \simeq \tfrac{1}{2}(z + \sqrt{2v-1})^2, \qquad \hat{z} \simeq \sqrt{2\chi_v^2} - \sqrt{2v-1}, \qquad (1.132)$$

$$\chi^2 \simeq v\left(1 - \frac{2}{9v} + z\left[\sqrt{\frac{2}{9v}}\right]\right)^3, \qquad \hat{z} \simeq 3\left[\sqrt{\frac{v}{2}}\right]\left[\frac{2}{9v} + \sqrt{\frac{\chi_v^2}{v}} - 1\right].$$

(1.132a)

(1.132a) is the better of the two [it was improved by Severo and Zelen (1960) through an additional corrective term; for more on χ^2-approximations see Zar (1978) and Ling (1978)].

One more remark on the manner of writing χ^2. Indexing of the critical value at level α is usually in the form $\chi^2_{v;\alpha}$. If no misunderstanding can occur, a single index suffices or even the one can be omitted.

Further discussion of the χ^2 distribution (cf., also Sections 4.3, 4.6.2, 4.6.4) can be found in Lancaster (1969) (Harter 1964 and Vahle and Tews 1969 give tables; Boyd 1965 provides a nomogram). Tables 28 and 28a list only selected values of the χ^2 value (cf., Table 83 in Section 4.6.1), one must carry out a **logarithmic interpolation** between the neighboring P-values. The necessary natural logarithms can be obtained from Table 29.

Table 29 Selected three-place natural logarithms

n	ln n	n	ln n
0.001	- 6.908	0.50	- 0.693
0.01	- 4.605	0.70	- 0.357
0.025	- 3.689	0.80	- 0.223
0.05	- 2.996	0.90	- 0.105
0.10	- 2.303	0.95	- 0.051
0.20	- 1.609	0.975	- 0.025
0.30	- 1.204	0.99	- 0.010

To find ln n for n-values which are $\frac{1}{10} = 10^{-1}$, $\frac{1}{100} = 10^{-2}$, $\frac{1}{1000} = 10^{-3}$, etc., as large as the tabulated n-values, one subtracts from the tabulated ln n the quantity ln 10 = 2.303 (cf., Section 0.2) 2 ln 10 = 4.605, 3 ln 10 = 6.908, etc. For example: ln 0.02 = ln 0.2 − ln 10 = −1.609 −2.303 = −3.912.

1.5 Three Important Families of Test Distributions

EXAMPLE. Let us suppose we get, for $DF = 10$, a value $\hat{\chi}^2 = 13.4$. To this value there corresponds a P-value between 10% and 30%. The corresponding χ^2 bounds are $\chi^2_{0.10} = 16.0$ and $\chi^2_{0.30} = 11.8$. The value P sought is then given by

$$\boxed{\frac{\ln P - \ln 0.3}{\ln 0.1 - \ln 0.3} = \frac{\hat{\chi}^2 - \chi^2_{0.30}}{\chi^2_{0.10} - \chi^2_{0.30}}}, \qquad (1.133)$$

$$\boxed{\ln P = \frac{(\hat{\chi}^2 - \chi^2_{0.30})(\ln 0.1 - \ln 0.3)}{\chi^2_{0.10} - \chi^2_{0.30}} + \ln 0.3,} \qquad (1.133\,a)$$

$$\ln P = \frac{(13.4 - 11.8)(-2.303 + 1.204)}{16.0 - 11.8} - 1.204,$$

$\ln P = -1.623, \quad \log P = 0.4343(\ln P) = 0.4343(-1.623)$

$\log P = -0.7049 = 9.2951 - 10, \quad \text{or } P = 0.197 \simeq 0.20.$

A glance at Table 28 tells us that $\chi^2_{10;\,0.20} = 13.4$; the approximation is good.

1.5.3 The F-distribution

If s_1^2 and s_2^2 are the variances of independent random samples of sizes n_1 and n_2 from two normally distributed populations with the same variance, then the random variable

$$\boxed{F = \frac{s_1^2}{s_2^2}} \qquad (1.134)$$

follows an **F-distribution** with the parameters $v_1 = n_1 - 1$ and $v_2 = n_2 - 1$. The F-distribution (after R. A. Fisher; cf., Figure 27) is also a continuous, **nonsymmetric** distribution with a range from zero to infinity. The F-distribution is L-shaped for $v_1 \leq 2$ and bell-shaped for $v_1 > 2$. Six tables (30a to 30f)

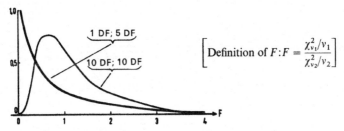

Figure 27 Probability densities of two F distributions: $F(v_1 = 1; v_2 = 5)$ and $F(v_1 = 10; v_2 = 10)$.

Table 30a Upper significance levels of the F-distribution for P = 0.10 (S = 90%); v_1 = degrees of freedom of the numerator; v_2 = degrees of freedom of the denominator. Example: $F_{9;18;0.10} = 2.00$

v_1 \ v_2	1	2	3	4	5	6	7	8	9	10	12	15	20	24	30	40	60	120	∞
1	39.86	49.50	53.59	55.83	57.24	58.20	58.91	59.44	59.86	60.19	60.71	61.22	61.74	62.00	62.26	62.53	62.79	63.06	63.33
2	8.53	9.00	9.16	9.24	9.29	9.33	9.35	9.37	9.38	9.39	9.41	9.42	9.44	9.45	9.46	9.47	9.47	9.48	9.49
3	5.54	5.46	5.39	5.34	5.31	5.28	5.27	5.25	5.24	5.23	5.22	5.20	5.18	5.18	5.17	5.16	5.15	5.14	5.13
4	4.54	4.32	4.19	4.11	4.05	4.01	3.98	3.95	3.94	3.92	3.90	3.87	3.84	3.83	3.82	3.80	3.79	3.78	3.76
5	4.06	3.78	3.62	3.52	3.45	3.40	3.37	3.34	3.32	3.30	3.27	3.24	3.21	3.19	3.17	3.16	3.14	3.12	3.10
6	3.78	3.46	3.29	3.18	3.11	3.05	3.01	2.98	2.96	2.94	2.90	2.87	2.84	2.82	2.80	2.78	2.76	2.74	2.72
7	3.59	3.26	3.07	2.96	2.88	2.83	2.78	2.75	2.72	2.70	2.67	2.63	2.59	2.58	2.56	2.54	2.51	2.49	2.47
8	3.46	3.11	2.92	2.81	2.73	2.67	2.62	2.59	2.56	2.54	2.50	2.46	2.42	2.40	2.38	2.36	2.34	2.32	2.29
9	3.36	3.01	2.81	2.69	2.61	2.55	2.51	2.47	2.44	2.42	2.38	2.34	2.30	2.28	2.25	2.23	2.21	2.18	2.16
10	3.29	2.92	2.73	2.61	2.52	2.46	2.41	2.38	2.35	2.32	2.28	2.24	2.20	2.18	2.16	2.13	2.11	2.08	2.06
11	3.23	2.86	2.66	2.54	2.45	2.39	2.34	2.30	2.27	2.25	2.21	2.17	2.12	2.10	2.08	2.05	2.03	2.00	1.97
12	3.18	2.81	2.61	2.48	2.39	2.33	2.28	2.24	2.21	2.19	2.15	2.10	2.06	2.04	2.01	1.99	1.96	1.93	1.90
13	3.14	2.76	2.56	2.43	2.35	2.28	2.23	2.20	2.16	2.14	2.10	2.05	2.01	1.98	1.96	1.93	1.90	1.88	1.85
14	3.10	2.73	2.52	2.39	2.31	2.24	2.19	2.15	2.12	2.10	2.05	2.01	1.96	1.94	1.91	1.89	1.86	1.83	1.80
15	3.07	2.70	2.49	2.36	2.27	2.21	2.16	2.12	2.09	2.06	2.02	1.97	1.92	1.90	1.87	1.85	1.82	1.79	1.76
16	3.05	2.67	2.46	2.33	2.24	2.18	2.13	2.09	2.06	2.03	1.99	1.94	1.89	1.87	1.84	1.81	1.78	1.75	1.72
17	3.03	2.64	2.44	2.31	2.22	2.15	2.10	2.06	2.03	2.00	1.96	1.91	1.86	1.84	1.81	1.78	1.75	1.72	1.69
18	3.01	2.62	2.42	2.29	2.20	2.13	2.08	2.04	2.00	1.98	1.93	1.89	1.84	1.81	1.78	1.75	1.72	1.69	1.66
19	2.99	2.61	2.40	2.27	2.18	2.11	2.06	2.02	1.98	1.96	1.91	1.86	1.81	1.79	1.76	1.73	1.70	1.67	1.63
20	2.97	2.59	2.38	2.25	2.16	2.09	2.04	2.00	1.96	1.94	1.89	1.84	1.79	1.77	1.74	1.71	1.68	1.64	1.61
21	2.96	2.57	2.36	2.23	2.14	2.08	2.02	1.98	1.95	1.92	1.87	1.83	1.78	1.75	1.72	1.69	1.66	1.62	1.59
22	2.95	2.56	2.35	2.22	2.13	2.06	2.01	1.97	1.93	1.90	1.86	1.81	1.76	1.73	1.70	1.67	1.64	1.60	1.57
23	2.94	2.55	2.34	2.21	2.11	2.05	1.99	1.95	1.92	1.89	1.84	1.80	1.74	1.72	1.69	1.66	1.62	1.59	1.55
24	2.93	2.54	2.33	2.19	2.10	2.04	1.98	1.94	1.91	1.88	1.83	1.78	1.73	1.70	1.67	1.64	1.61	1.57	1.53
25	2.92	2.53	2.32	2.18	2.09	2.02	1.97	1.93	1.89	1.87	1.82	1.77	1.72	1.69	1.66	1.63	1.59	1.56	1.52
26	2.91	2.52	2.31	2.17	2.08	2.01	1.96	1.92	1.88	1.86	1.81	1.76	1.71	1.68	1.65	1.61	1.58	1.54	1.50
27	2.90	2.51	2.30	2.17	2.07	2.00	1.95	1.91	1.87	1.85	1.80	1.75	1.70	1.67	1.64	1.60	1.57	1.53	1.49
28	2.89	2.50	2.29	2.16	2.06	2.00	1.94	1.90	1.87	1.84	1.79	1.74	1.69	1.66	1.63	1.59	1.56	1.52	1.48
29	2.89	2.50	2.28	2.15	2.06	1.99	1.93	1.89	1.86	1.83	1.78	1.73	1.68	1.65	1.62	1.58	1.55	1.51	1.47
30	2.88	2.49	2.28	2.14	2.05	1.98	1.93	1.88	1.85	1.82	1.77	1.72	1.67	1.64	1.61	1.57	1.54	1.50	1.46
40	2.84	2.44	2.23	2.09	2.00	1.93	1.87	1.83	1.79	1.76	1.71	1.66	1.61	1.57	1.54	1.51	1.47	1.42	1.38
60	2.79	2.39	2.18	2.04	1.95	1.87	1.82	1.77	1.74	1.71	1.66	1.60	1.54	1.51	1.48	1.44	1.40	1.35	1.29
120	2.75	2.35	2.13	1.99	1.90	1.82	1.77	1.72	1.68	1.65	1.60	1.55	1.48	1.45	1.41	1.37	1.32	1.26	1.19
∞	2.71	2.30	2.08	1.94	1.85	1.77	1.72	1.67	1.63	1.60	1.55	1.49	1.42	1.38	1.34	1.30	1.24	1.17	1.00

Table 30b Upper significance levels of the F-distribution for P = 0.05 (S = 95%); ν_1 = degrees of freedom of the numerator; ν_2 = degrees of freedom of the denominator. Example: $F_{12;40;0.05} = 2.00$

ν_2 \ ν_1	1	2	3	4	5	6	7	8	9	10	12	15	20	24	30	40	60	120	∞
1	161.4	199.5	215.7	224.6	230.2	234.0	236.8	238.9	240.5	241.9	243.9	245.9	248.0	249.1	250.1	251.1	252.2	253.3	254.3
2	18.51	19.00	19.16	19.25	19.30	19.33	19.35	19.37	19.38	19.40	19.41	19.43	19.45	19.45	19.46	19.47	19.48	19.49	19.50
3	10.13	9.55	9.28	9.12	9.01	8.94	8.89	8.85	8.81	8.79	8.74	8.70	8.66	8.64	8.62	8.59	8.57	8.55	8.53
4	7.71	6.94	6.59	6.39	6.26	6.16	6.09	6.04	6.00	5.96	5.91	5.86	5.80	5.77	5.75	5.72	5.69	5.66	5.63
5	6.61	5.79	5.41	5.19	5.05	4.95	4.88	4.82	4.77	4.74	4.68	4.62	4.56	4.53	4.50	4.46	4.43	4.40	4.36
6	5.99	5.14	4.76	4.53	4.39	4.28	4.21	4.15	4.10	4.06	4.00	3.94	3.87	3.84	3.81	3.77	3.74	3.70	3.67
7	5.59	4.74	4.35	4.12	3.97	3.87	3.79	3.73	3.68	3.64	3.57	3.51	3.44	3.41	3.38	3.34	3.30	3.27	3.23
8	5.32	4.46	4.07	3.84	3.69	3.58	3.50	3.44	3.39	3.35	3.28	3.22	3.15	3.12	3.08	3.04	3.01	2.97	2.93
9	5.12	4.26	3.86	3.63	3.48	3.37	3.29	3.23	3.18	3.14	3.07	3.01	2.94	2.90	2.86	2.83	2.79	2.75	2.71
10	4.96	4.10	3.71	3.48	3.33	3.22	3.14	3.07	3.02	2.98	2.91	2.85	2.77	2.74	2.70	2.66	2.62	2.58	2.54
11	4.84	3.98	3.59	3.36	3.20	3.09	3.01	2.95	2.90	2.85	2.79	2.72	2.65	2.61	2.57	2.53	2.49	2.45	2.40
12	4.75	3.89	3.49	3.26	3.11	3.00	2.91	2.85	2.80	2.75	2.69	2.62	2.54	2.51	2.47	2.43	2.38	2.34	2.30
13	4.67	3.81	3.41	3.18	3.03	2.92	2.83	2.77	2.71	2.67	2.60	2.53	2.46	2.42	2.38	2.34	2.30	2.25	2.21
14	4.60	3.74	3.34	3.11	2.96	2.85	2.76	2.70	2.65	2.60	2.53	2.46	2.39	2.35	2.31	2.27	2.22	2.18	2.13
15	4.54	3.68	3.29	3.06	2.90	2.79	2.71	2.64	2.59	2.54	2.48	2.40	2.33	2.29	2.25	2.20	2.16	2.11	2.07
16	4.49	3.63	3.24	3.01	2.85	2.74	2.66	2.59	2.54	2.49	2.42	2.35	2.28	2.24	2.19	2.15	2.11	2.06	2.01
17	4.45	3.59	3.20	2.96	2.81	2.70	2.61	2.55	2.49	2.45	2.38	2.31	2.23	2.19	2.15	2.10	2.06	2.01	1.96
18	4.41	3.55	3.16	2.93	2.77	2.66	2.58	2.51	2.46	2.41	2.34	2.27	2.19	2.15	2.11	2.06	2.02	1.97	1.92
19	4.38	3.52	3.13	2.90	2.74	1.63	2.54	2.48	2.42	2.38	2.31	2.23	2.16	2.11	2.07	2.03	1.98	1.93	1.88
20	4.35	3.49	3.10	2.87	2.71	2.60	2.51	2.45	2.39	2.35	2.28	2.20	2.12	2.08	2.04	1.99	1.95	1.90	1.84
21	4.32	3.47	3.07	2.84	2.68	2.57	2.49	2.42	2.37	2.32	2.25	2.18	2.10	2.05	2.01	1.96	1.92	1.87	1.81
22	4.30	3.44	3.05	2.82	2.66	2.55	2.46	2.40	2.34	2.30	2.23	2.15	2.07	2.03	1.98	1.94	1.89	1.84	1.78
23	4.28	3.42	3.03	2.80	2.64	2.53	2.44	2.37	2.32	2.27	2.20	2.13	2.05	2.01	1.96	1.91	1.86	1.81	1.76
24	4.26	3.40	3.01	2.78	2.62	2.51	2.42	2.36	2.30	2.25	2.18	2.11	2.03	1.98	1.94	1.89	1.84	1.79	1.73
25	4.24	3.39	2.99	2.76	2.60	2.49	2.40	2.34	2.28	2.24	2.16	2.09	2.01	1.96	1.92	1.87	1.82	1.77	1.71
26	4.23	3.37	2.98	2.74	2.59	2.47	2.39	2.32	2.27	2.22	2.15	2.07	1.99	1.95	1.90	1.85	1.80	1.75	1.69
27	4.21	3.35	2.96	2.73	2.57	2.46	2.37	2.31	2.25	2.20	2.13	2.06	1.97	1.93	1.88	1.84	1.79	1.73	1.67
28	4.20	3.34	2.95	2.71	2.56	2.45	2.36	2.29	2.24	2.19	2.12	2.04	1.96	1.91	1.87	1.82	1.77	1.71	1.65
29	4.18	3.33	2.93	2.70	2.55	2.43	2.35	2.28	2.22	2.18	2.10	2.03	1.94	1.90	1.85	1.81	1.75	1.70	1.64
30	4.17	3.32	2.92	2.69	2.53	2.42	2.33	2.27	2.21	2.16	2.09	2.01	1.93	1.89	1.84	1.79	1.74	1.68	1.62
40	4.08	3.23	2.84	2.61	2.45	2.34	2.25	2.18	2.12	2.08	2.00	1.92	1.84	1.79	1.74	1.69	1.64	1.58	1.51
60	4.00	3.15	2.76	2.53	2.37	2.25	2.17	2.10	2.04	1.99	1.92	1.84	1.75	1.70	1.65	1.59	1.53	1.47	1.39
120	3.92	3.07	2.68	2.45	2.29	2.17	2.09	2.02	1.96	1.91	1.83	1.75	1.66	1.61	1.55	1.50	1.43	1.35	1.25
∞	3.84	3.00	2.60	2.37	2.21	2.10	2.01	1.94	1.88	1.83	1.75	1.67	1.57	1.52	1.46	1.39	1.32	1.22	1.00

$F_{\nu_1;\nu_2;1-\alpha} = 1/F_{\nu_2;\nu_1;\alpha}$ [equation (1.136)].

Table 30c Upper significance levels of the F-distribution for P = 0.025 (S = 97.5%); v_1 = degrees of freedom of the numerator; v_2 = degrees of freedom of the denominator

$v_2 \backslash v_1$	1	2	3	4	5	6	7	8	9	10
1	647.8	799.5	864.2	899.6	921.8	937.1	948.2	956.7	963.3	968.6
2	38.51	39.00	39.17	39.25	39.30	39.33	39.36	39.37	39.39	39.40
3	17.44	16.04	15.44	15.10	14.88	14.73	14.62	14.54	14.47	14.42
4	12.22	10.65	9.98	9.60	9.36	9.20	9.07	8.98	8.90	8.84
5	10.01	8.43	7.76	7.39	7.15	6.98	6.85	6.76	6.68	6.62
6	8.81	7.26	6.60	6.23	5.99	5.82	5.70	5.60	5.52	5.46
7	8.07	6.54	5.89	5.52	5.29	5.12	4.99	4.90	4.82	4.76
8	7.57	6.06	5.42	5.05	4.82	4.65	4.53	4.43	4.36	4.30
9	7.21	5.71	5.08	4.72	4.48	4.32	4.20	4.10	4.03	3.96
10	6.94	5.46	4.83	4.47	4.24	4.07	3.95	3.85	3.78	3.72
11	6.72	5.26	4.63	4.28	4.04	3.88	3.76	3.66	3.59	3.53
12	6.55	5.10	4.47	4.12	3.89	3.73	3.61	3.51	3.44	3.37
13	6.41	4.97	4.35	4.00	3.77	3.60	3.48	3.39	3.31	3.25
14	6.30	4.86	4.24	3.89	3.66	3.50	3.38	3.29	3.21	3.15
15	6.20	4.77	4.15	3.80	3.58	3.41	3.29	3.20	3.12	3.06
16	6.12	4.69	4.08	3.73	3.50	3.34	3.22	3.12	3.05	2.99
17	6.04	4.62	4.01	3.66	3.44	3.28	3.16	3.06	2.98	2.92
18	5.98	4.56	3.95	3.61	3.38	3.22	3.10	3.01	2.93	2.87
19	5.92	4.51	3.90	3.56	3.33	3.17	3.05	2.96	2.88	2.82
20	5.87	4.46	3.86	3.51	3.29	3.13	3.01	2.91	2.84	2.77
21	5.83	4.42	3.82	3.48	3.25	3.09	2.97	2.87	2.80	2.73
22	5.79	4.38	3.78	3.44	3.22	3.05	2.93	2.84	2.76	2.70
23	5.75	4.35	3.75	3.41	3.18	3.02	2.90	2.81	2.73	2.67
24	5.72	4.32	3.72	3.38	3.15	2.99	2.87	2.78	2.70	2.64
25	5.69	4.29	3.69	3.35	3.13	2.97	2.85	2.75	2.68	2.61
26	5.66	4.27	3.67	3.33	3.10	2.94	2.82	2.73	2.65	2.59
27	5.63	4.24	3.65	3.31	3.08	2.92	2.80	2.71	2.63	2.57
28	5.61	4.22	3.63	3.29	3.06	2.90	2.78	2.69	2.61	2.55
29	5.59	4.20	3.61	3.27	3.04	2.88	2.76	2.67	2.59	2.53
30	5.57	4.18	3.59	3.25	3.03	2.87	2.75	2.65	2.57	2.51
40	5.42	4.05	3.46	3.13	2.90	2.74	2.62	2.53	2.45	2.39
60	5.29	3.93	3.34	3.01	2.79	2.63	2.51	2.41	2.33	2.27
120	5.15	3.80	3.23	2.89	2.67	2.52	2.39	2.30	2.22	2.16
∞	5.02	3.69	3.12	2.79	2.57	2.41	2.29	2.19	2.11	2.05

Hald (1952; cf. Cochran 1940) gives for v_1 and v_2 greater than 30 the following approximations, where $g = 1/v_1 - 1/v_2$, $h = 2/(1/v_1 + 1/v_2)$, and $F_\alpha = F_{v_1; v_2; \alpha}$:

$$\log F_{0.5} = -0.290g,$$

$$\log F_{0.3} = \frac{0.4555}{\sqrt{h - 0.55}} - 0.329g,$$

$$\log F_{0.1} = \frac{1.1131}{\sqrt{h - 0.77}} - 0.527g,$$

$$\log F_{0.05} = \frac{1.4287}{\sqrt{h - 0.95}} - 0.681g,$$

$$\log F_{0.025} = \frac{1.7023}{\sqrt{h - 1.14}} - 0.846g,$$

$$\log F_{0.01} = \frac{2.0206}{\sqrt{h - 1.40}} - 1.073g.$$

1.5 Three Important Families of Test Distributions

Table 30c (*continued*)

v_2 \ v_1	12	15	20	24	30	40	60	120	∞
1	976.7	984.9	993.1	997.2	1001	1006	1010	1014	1018
2	39.41	39.43	39.45	39.46	39.46	39.47	39.48	39.49	39.50
3	14.34	14.25	14.17	14.12	14.08	14.04	13.99	13.95	13.90
4	8.75	8.66	8.56	8.51	8.46	8.41	8.36	8.31	8.26
5	6.52	6.43	6.33	6.28	6.23	6.18	6.12	6.07	6.02
6	5.37	5.27	5.17	5.12	5.07	5.01	4.96	4.90	4.85
7	4.67	4.57	4.47	4.42	4.36	4.31	4.25	4.20	4.14
8	4.20	4.10	4.00	3.95	3.89	3.84	3.78	3.73	3.67
9	3.87	3.77	3.67	3.61	3.56	3.51	3.45	3.39	3.33
10	3.62	3.52	3.42	3.37	3.31	3.26	3.20	3.14	3.08
11	3.43	3.33	3.23	3.17	3.12	3.06	3.00	2.94	2.88
12	3.28	3.18	3.07	3.02	2.96	2.91	2.85	2.79	2.72
13	3.15	3.05	2.95	2.89	2.84	2.78	2.72	2.66	2.60
14	3.05	2.95	2.84	2.79	2.73	2.67	2.61	2.55	2.49
15	2.96	2.86	2.76	2.70	2.64	2.59	2.52	2.46	2.40
16	2.89	2.79	2.68	2.63	2.57	2.51	2.45	2.38	2.32
17	2.82	2.72	2.62	2.56	2.50	2.44	2.38	2.32	2.25
18	2.77	2.67	2.56	2.50	2.44	2.38	2.32	2.26	2.19
19	2.72	2.62	2.51	2.45	2.39	2.33	2.27	2.20	2.13
20	2.68	2.57	2.46	2.41	2.35	2.29	2.22	2.16	2.09
21	2.64	2.53	2.42	2.37	2.31	2.25	2.18	2.11	2.04
22	2.60	2.50	2.39	2.33	2.27	2.21	2.14	2.08	2.00
23	2.57	2.47	2.36	2.30	2.24	2.18	2.11	2.04	1.97
24	2.54	2.44	2.33	2.27	2.21	2.15	2.08	2.01	1.94
25	2.51	2.41	2.30	2.24	2.18	2.12	2.05	1.98	1.91
26	2.49	2.39	2.28	2.22	2.16	2.09	2.03	1.95	1.88
27	2.47	2.36	2.25	2.19	2.13	2.07	2.00	1.93	1.85
28	2.45	2.34	2.23	2.17	2.11	2.05	1.98	1.91	1.83
29	2.43	2.32	2.21	2.15	2.09	2.03	1.96	1.89	1.81
30	2.41	2.31	2.20	2.14	2.07	2.01	1.94	1.87	1.79
40	2.29	2.18	2.07	2.01	1.94	1.88	1.80	1.72	1.64
60	2.17	2.06	1.94	1.88	1.82	1.74	1.67	1.58	1.48
120	2.05	1.94	1.82	1.76	1.69	1.61	1.53	1.43	1.31
∞	1.94	1.83	1.71	1.64	1.57	1.48	1.39	1.27	1.00

$$\log F_{0.005} = \frac{2.2373}{\sqrt{h - 1.61}} - 1.250g,$$

$$\log F_{0.001} = \frac{2.6841}{\sqrt{h - 2.09}} - 1.672g,$$

$$\log F_{0.0005} = \frac{2.8580}{\sqrt{h - 2.30}} - 1.857g,$$

Example: $F_{200;100;0.05}$.

$g = 1/200 - 1/100 = -0.005$; $h = 2/(1/200 + 1/100) = 133.333$,

$$\log F_{200;100;0.05} = \frac{1.4284}{\sqrt{133.33 - 0.95}} - 0.681(-0.005) = 0.12755,$$

$F_{200;100;0.05} = 1.34$ (exact value).

Better approximations are described by Johnson (1973) and by Ling (1978).

Table 30d Upper significance levels of the F-distribution for P = 0.01 (S = 99%); v_1 = degrees of freedom of the numerator; v_2 = degrees of freedom of the denominator

v_2 \ v_1	1	2	3	4	5	6	7	8	9	10
1	4052	4999.5	5403	5625	5764	5859	5928	5982	6022	6056
2	98.50	99.00	99.17	99.25	99.30	99.33	99.36	99.37	99.39	99.40
3	34.12	30.82	29.46	28.71	28.24	27.91	27.67	27.49	27.35	27.23
4	21.20	18.00	16.69	15.98	15.52	15.21	14.98	14.80	14.66	14.55
5	16.26	13.27	12.06	11.39	10.97	10.67	10.46	10.29	10.16	10.05
6	13.75	10.92	9.78	9.15	8.75	8.47	8.26	8.10	7.98	7.87
7	12.25	9.55	8.45	7.85	7.46	7.19	6.99	6.84	6.72	6.62
8	11.26	8.65	7.59	7.01	6.63	6.37	6.18	6.03	5.91	5.81
9	10.56	8.02	6.99	6.42	6.06	5.80	5.61	5.47	5.35	5.26
10	10.04	7.56	6.55	5.99	5.64	5.39	5.20	5.06	4.94	4.85
11	9.65	7.21	6.22	5.67	5.32	5.07	4.89	4.74	4.63	4.54
12	9.33	6.93	5.95	5.41	5.06	4.82	4.64	4.50	4.39	4.30
13	9.07	6.70	5.74	5.21	4.86	4.62	4.44	4.30	4.19	4.10
14	8.86	6.51	5.56	5.04	4.69	4.46	4.28	4.14	4.03	3.94
15	8.68	6.36	5.42	4.89	4.56	4.32	4.14	4.00	3.89	3.80
16	8.53	6.23	5.29	4.77	4.44	4.20	4.03	3.89	3.78	3.69
17	8.40	6.11	5.18	4.67	4.34	4.10	3.93	3.79	3.68	3.59
18	8.29	6.01	5.09	4.58	4.25	4.01	3.84	3.71	3.60	3.51
19	8.18	5.93	5.01	4.50	4.17	3.94	3.77	3.63	3.52	3.43
20	8.10	5.85	4.94	4.43	4.10	3.87	3.70	3.56	3.46	3.37
21	8.02	5.78	4.87	4.37	4.04	3.81	3.64	3.51	3.40	3.31
22	7.95	5.72	4.82	4.31	3.99	3.76	3.59	3.45	3.35	3.26
23	7.88	5.66	4.76	4.26	3.94	3.71	3.54	3.41	3.30	3.21
24	7.82	5.61	4.72	4.22	3.90	3.67	3.50	3.36	3.26	3.17
25	7.77	5.57	4.68	4.18	3.85	3.63	3.46	3.32	3.22	3.13
26	7.72	5.53	4.64	4.14	3.82	3.59	3.42	3.29	3.18	3.09
27	7.68	5.49	4.60	4.11	3.78	3.56	3.39	3.26	3.15	3.06
28	7.64	5.45	4.57	4.07	3.75	3.53	3.36	3.23	3.12	3.03
29	7.60	5.42	4.54	4.04	3.73	3.50	3.33	3.20	3.09	3.00
30	7.56	5.39	4.51	4.02	3.70	3.47	3.30	3.17	3.07	2.98
40	7.31	5.18	4.31	3.83	3.51	3.29	3.12	2.99	2.89	2.80
60	7.08	4.98	4.13	3.65	3.34	3.12	2.95	2.82	2.72	2.63
120	6.85	4.79	3.95	3.48	3.17	2.96	2.79	2.66	2.56	2.47
∞	6.63	4.61	3.78	3.32	3.02	2.80	2.64	2.51	2.41	2.32

v_2 \ v_1	12	15	20	24	30	40	60	120	∞
1	6106	6157	6209	6235	6261	6287	6313	6339	6366
2	99.42	99.43	99.45	99.46	99.47	99.47	99.48	99.49	99.50
3	27.05	26.87	26.69	26.60	26.50	26.41	26.32	26.22	26.13
4	14.37	14.20	14.02	13.93	13.84	13.75	13.65	13.56	13.46
5	9.89	9.72	9.55	9.47	9.38	9.29	9.20	9.11	9.02
6	7.72	7.56	7.40	7.31	7.23	7.14	7.06	6.97	6.88
7	6.47	6.31	6.16	6.07	5.99	5.91	5.82	5.74	5.65
8	5.67	5.52	5.36	5.28	5.20	5.12	5.03	4.95	4.86
9	5.11	4.96	4.81	4.73	4.65	4.57	4.48	4.40	4.31
10	4.71	4.56	4.41	4.33	4.25	4.17	4.08	4.00	3.91
11	4.40	4.25	4.10	4.02	3.94	3.86	3.78	3.69	3.60
12	4.16	4.01	3.86	3.78	3.70	3.62	3.54	3.45	3.36
13	3.96	3.82	3.66	3.59	3.51	3.43	3.34	3.25	3.17
14	3.80	3.66	3.51	3.43	3.35	3.27	3.18	3.09	3.00
15	3.67	3.52	3.37	3.29	3.21	3.13	3.05	2.96	2.87
16	3.55	3.41	3.26	3.18	3.10	3.02	2.93	2.84	2.75
17	3.46	3.31	3.16	3.08	3.00	2.92	2.83	2.75	2.65
18	3.37	3.23	3.08	3.00	2.92	2.84	2.75	2.66	2.57
19	3.30	3.15	3.00	2.92	2.84	2.76	2.67	2.58	2.49
20	3.23	3.09	2.94	2.86	2.78	2.69	2.61	2.52	2.42
21	3.17	3.03	2.88	2.80	2.72	2.64	2.55	2.46	2.36
22	3.12	2.98	2.83	2.75	2.67	2.58	2.50	2.40	2.31
23	3.07	2.93	2.78	2.70	2.62	2.54	2.45	2.35	2.26
24	3.03	2.89	2.74	2.66	2.58	2.49	2.40	2.31	2.21
25	2.99	2.85	2.70	2.62	2.54	2.45	2.36	2.27	2.17
26	2.96	2.81	2.66	2.58	2.50	2.42	2.33	2.23	2.13
27	2.93	2.78	2.63	2.55	2.47	2.38	2.29	2.20	2.10
28	2.90	2.75	2.60	2.52	2.44	2.35	2.26	2.17	2.06
29	2.87	2.73	2.57	2.49	2.41	2.33	2.23	2.14	2.03
30	2.84	2.70	2.55	2.47	2.39	2.30	2.21	2.11	2.01
40	2.66	2.52	2.37	2.29	2.20	2.11	2.02	1.92	1.80
60	2.50	2.35	2.20	2.12	2.03	1.94	1.84	1.73	1.60
120	2.34	2.19	2.03	1.95	1.86	1.76	1.66	1.53	1.38
∞	2.18	2.04	1.88	1.79	1.70	1.59	1.47	1.32	1.00

Table 30e Upper significance levels of the F-distribution for $P = 0.005$ (S = 99.5%); v_1 = degrees of freedom of the numerator; v_2 = degrees of freedom of the denominator

v_2 \ v_1	1	2	3	4	5	6	7	8	9	10
1	16211	20000	21615	22500	23056	23437	23715	23925	24091	24224
2	198.5	199.0	199.2	199.2	199.3	199.4	199.4	199.4	199.4	199.4
3	55.55	49.80	47.47	46.19	45.39	44.84	44.43	44.13	43.88	43.69
4	31.33	26.28	24.26	23.15	22.46	21.97	21.62	21.35	21.14	20.97
5	22.78	18.31	16.53	15.56	14.94	14.51	14.20	13.96	13.77	13.62
6	18.63	14.54	12.92	12.03	11.46	11.07	10.79	10.57	10.39	10.25
7	16.24	12.40	10.88	10.05	9.52	9.16	8.89	8.68	8.51	8.38
8	14.69	11.04	9.60	8.81	8.30	7.95	7.69	7.50	7.34	7.21
9	13.61	10.11	8.72	7.96	7.47	7.13	6.88	6.69	6.54	6.42
10	12.83	9.43	8.08	7.34	6.87	6.54	6.30	6.12	5.97	5.85
11	12.23	8.91	7.60	6.88	6.42	6.10	5.86	5.68	5.54	5.42
12	11.75	8.51	7.23	6.52	6.07	5.76	5.52	5.35	5.20	5.09
13	11.37	8.19	6.93	6.23	5.79	5.48	5.25	5.08	4.94	4.82
14	11.06	7.92	6.68	6.00	5.56	5.26	5.03	4.86	4.72	4.60
15	10.80	7.70	6.48	5.80	5.37	5.07	4.85	4.67	4.54	4.42
16	10.58	7.51	6.30	5.64	5.21	4.91	4.69	4.52	4.38	4.27
17	10.38	7.35	6.16	5.50	5.07	4.78	4.56	4.39	4.25	4.14
18	10.22	7.21	6.03	5.37	4.96	4.66	4.44	4.28	4.14	4.03
19	10.07	7.09	5.92	5.27	4.85	4.56	4.34	4.18	4.04	3.93
20	9.94	6.99	5.82	5.17	4.76	4.47	4.26	4.09	3.96	3.85
21	9.83	6.89	5.73	5.09	4.68	4.39	4.18	4.01	3.88	3.77
22	9.73	6.81	5.65	5.02	4.61	4.32	4.11	3.94	3.81	3.70
23	9.63	6.73	5.58	4.95	4.54	4.26	4.05	3.88	3.75	3.64
24	9.55	6.66	5.52	4.89	4.49	4.20	3.99	3.83	3.69	3.59
25	9.48	6.60	5.46	4.84	4.43	4.15	3.94	3.78	3.64	3.54
26	9.41	6.54	5.41	4.79	4.38	4.10	3.89	3.73	3.60	3.49
27	9.34	6.49	5.36	4.74	4.34	4.06	3.85	3.69	3.56	3.45
28	9.28	6.44	5.32	4.70	4.30	4.02	3.81	3.65	3.52	3.41
29	9.23	6.40	5.28	4.66	4.26	3.98	3.77	3.61	3.48	3.38
30	9.18	6.35	5.24	4.62	4.23	3.95	3.74	3.58	3.45	3.34
40	8.83	6.07	4.98	4.37	3.99	3.71	3.51	3.35	3.22	3.12
60	8.49	5.79	4.73	4.14	3.76	3.49	3.29	3.13	3.01	2.90
120	8.18	5.54	4.50	3.92	3.55	3.28	3.09	2.93	2.81	2.71
∞	7.88	5.30	4.28	3.72	3.35	3.09	2.90	2.74	2.62	2.52

v_2 \ v_1	12	15	20	24	30	40	60	120	∞
1	24426	24630	24836	24940	25044	25148	25253	25359	25465
2	199.4	199.4	199.4	199.5	199.5	199.5	199.5	199.5	199.5
3	43.39	43.08	42.78	42.62	42.47	42.31	42.15	41.99	41.83
4	20.70	20.44	20.17	20.03	19.89	19.75	19.61	19.47	19.32
5	13.38	13.15	12.90	12.78	12.66	12.53	12.40	12.27	12.14
6	10.03	9.81	9.59	9.47	9.36	9.24	9.12	9.00	8.88
7	8.18	7.97	7.75	7.65	7.53	7.42	7.31	7.19	7.08
8	7.01	6.81	6.61	6.50	6.40	6.29	6.18	6.06	5.95
9	6.23	6.03	5.83	5.73	5.62	5.52	5.41	5.30	5.19
10	5.66	5.47	5.27	5.17	5.07	4.97	4.86	4.75	4.64
11	5.24	5.05	4.86	4.76	4.65	4.55	4.44	4.34	4.23
12	4.91	4.72	4.53	4.43	4.33	4.23	4.12	4.01	3.90
13	4.64	4.46	4.27	4.17	4.07	3.97	3.87	3.76	3.65
14	4.43	4.25	4.06	3.96	3.86	3.76	3.66	3.55	3.44
15	4.25	4.07	3.88	3.79	3.69	3.58	3.48	3.37	3.26
16	4.10	3.92	3.73	3.64	3.54	3.44	3.33	3.22	3.11
17	3.97	3.79	3.61	3.51	3.41	3.31	3.21	3.10	2.98
18	3.86	3.68	3.50	3.40	3.30	3.20	3.10	2.99	2.87
19	3.76	3.59	3.40	3.31	3.21	3.11	3.00	2.89	2.78
20	3.68	3.50	3.32	3.22	3.12	3.02	2.92	2.81	2.69
21	3.60	3.43	3.24	3.15	3.05	2.95	2.84	2.73	2.61
22	3.54	3.36	3.18	3.08	2.98	2.88	2.77	2.66	2.55
23	3.47	3.30	3.12	3.02	2.92	2.82	2.71	2.60	2.48
24	3.42	3.25	3.06	2.97	2.87	2.77	2.66	2.55	2.43
25	3.37	3.20	3.01	2.92	2.82	2.72	2.61	2.50	2.38
26	3.33	3.15	2.97	2.87	2.77	2.67	2.56	2.45	2.33
27	3.28	3.11	2.93	2.83	2.73	2.63	2.52	2.41	2.29
28	3.25	3.07	2.89	2.79	2.69	2.59	2.48	2.37	2.25
29	3.21	3.04	2.86	2.76	2.66	2.56	2.45	2.33	2.21
30	3.18	3.01	2.82	2.73	2.63	2.52	2.42	2.30	2.18
40	2.95	2.78	2.60	2.50	2.40	2.30	2.18	2.06	1.93
60	2.74	2.57	2.39	2.29	2.19	2.08	1.96	1.83	1.69
120	2.54	2.37	2.19	2.09	1.98	1.87	1.75	1.61	1.43
∞	2.36	2.19	2.00	1.90	1.79	1.67	1.53	1.36	1.00

Table 30f Upper significance levels of the F-distribution for $P = 0.001$ ($S = 99.9\%$); v_1 = degrees of freedom of the numerator; v_2 = degrees of freedom of the denominator. [These tables are excerpted from Table 18 of Pearson and Hartley (1958) and Table V of Fisher and Yates (1963).]

$v_2 \backslash v_1$	1	2	3	4	5	6	7	8	9	10	12	15	20	24	30	40	60	120	∞
1	4053[†]	5000[†]	5404[†]	5625[†]	5764[†]	5859[†]	5929[†]	5981[†]	6023[†]	6056[†]	6107[†]	6158[†]	6209[†]	6235[†]	6261[†]	6287[†]	6313[†]	6340[†]	6366[†]
2	998.5	999.0	999.2	999.2	999.3	999.3	999.4	999.4	999.4	999.4	999.4	999.4	999.4	999.5	999.5	999.5	999.5	999.5	999.5
3	167.0	148.5	141.1	137.1	134.6	132.8	131.6	130.6	129.9	129.2	128.3	127.4	126.4	125.9	125.4	125.0	124.5	124.0	123.5
4	74.14	61.25	56.18	53.44	51.71	50.53	49.66	49.00	48.47	48.05	47.41	46.76	46.10	45.77	45.43	45.09	44.75	44.40	44.05
5	47.18	37.12	33.20	31.09	29.75	28.84	28.16	27.64	27.24	26.92	26.42	25.91	25.39	25.14	24.87	24.60	24.33	24.06	23.79
6	35.51	27.00	23.70	21.92	20.81	20.03	19.46	19.03	18.69	18.41	17.99	17.56	17.12	16.89	16.67	16.44	16.21	15.99	15.75
7	29.25	21.69	18.77	17.19	16.21	15.52	15.02	14.63	14.33	14.08	13.71	13.32	12.93	12.73	12.53	12.33	12.12	11.91	11.70
8	25.42	18.49	15.83	14.39	13.49	12.86	12.40	12.04	11.77	11.54	11.19	10.84	10.48	10.30	10.11	9.92	9.73	9.53	9.33
9	22.86	16.39	13.90	12.56	11.71	11.13	10.70	10.37	10.11	9.89	9.57	9.24	8.90	8.72	8.55	8.37	8.00	8.00	7.81
10	21.04	14.91	12.55	11.28	10.48	9.92	9.52	9.20	8.96	8.75	8.45	8.13	7.80	7.64	7.47	7.30	7.12	6.94	6.76
11	19.69	13.81	11.56	10.35	9.58	9.05	8.66	8.35	8.12	7.92	7.63	7.32	7.01	6.85	6.68	6.52	6.35	6.17	6.00
12	18.64	12.97	10.80	9.63	8.89	8.38	8.00	7.71	7.48	7.29	7.00	6.71	6.40	6.25	6.09	5.93	5.76	5.59	5.42
13	17.81	12.31	10.21	9.07	8.35	7.86	7.49	7.21	6.98	6.80	6.52	6.23	5.93	5.78	5.63	5.47	5.30	5.14	4.97
14	17.14	11.78	9.73	8.62	7.92	7.43	7.08	6.80	6.58	6.40	6.13	5.85	5.56	5.41	5.25	5.10	4.94	4.77	4.60
15	16.59	11.34	9.34	8.25	7.57	7.09	6.74	6.47	6.26	6.08	5.81	5.54	5.25	5.10	4.95	4.80	4.64	4.47	4.31
16	16.12	10.97	9.00	7.94	7.27	6.81	6.46	6.19	5.98	5.81	5.55	5.27	4.99	4.85	4.70	4.54	4.39	4.23	4.06
17	15.72	10.66	8.73	7.68	7.02	6.56	6.22	5.96	5.75	5.58	5.32	5.05	4.78	4.63	4.48	4.33	4.18	4.02	3.85
18	15.38	10.39	8.49	7.46	6.81	6.35	6.02	5.76	5.56	5.39	5.13	4.87	4.59	4.45	4.30	4.15	4.00	3.84	3.67
19	15.08	10.16	8.28	7.26	6.62	6.18	5.85	5.59	5.39	5.22	4.97	4.70	4.43	4.29	4.14	3.99	3.84	3.68	3.51
20	14.82	9.95	8.10	7.10	6.46	6.02	5.69	5.44	5.24	5.08	4.82	4.56	4.29	4.15	4.00	3.86	3.70	3.54	3.38
21	14.59	9.77	7.94	6.95	6.32	5.88	5.56	5.31	5.11	4.95	4.70	4.44	4.17	4.03	3.88	3.74	3.58	3.42	3.26
22	14.38	9.61	7.80	6.81	6.19	5.76	5.44	5.19	4.99	4.83	4.58	4.33	4.06	3.92	3.78	3.63	3.48	3.32	3.15
23	14.19	9.47	7.67	6.69	6.08	5.65	5.33	5.09	4.89	4.73	4.48	4.23	3.96	3.82	3.68	3.53	3.38	3.22	3.05
24	14.03	9.34	7.55	6.59	5.98	5.55	5.23	4.99	4.80	4.64	4.39	4.14	3.87	3.74	3.59	3.45	3.29	3.14	2.97
25	13.88	9.22	7.45	6.49	5.88	5.46	5.15	4.91	4.71	4.56	4.31	4.06	3.79	3.66	3.52	3.37	3.22	3.06	2.89
26	13.74	9.12	7.36	6.41	5.80	5.38	5.07	4.83	4.64	4.48	4.24	3.99	3.72	3.59	3.44	3.30	3.15	2.99	2.82
27	13.61	9.02	7.27	6.33	5.73	5.31	5.00	4.76	4.57	4.41	4.17	3.92	3.66	3.52	3.38	3.23	3.08	2.92	2.75
28	13.50	8.93	7.19	6.25	5.66	5.24	4.93	4.69	4.50	4.35	4.11	3.86	3.60	3.46	3.32	3.18	3.02	2.86	2.69
29	13.39	8.85	7.12	6.19	5.59	5.18	4.87	4.64	4.45	4.29	4.05	3.80	3.54	3.41	3.27	3.12	2.97	2.81	2.64
30	13.29	8.77	7.05	6.12	5.53	5.12	4.82	4.58	4.39	4.24	4.00	3.75	3.49	3.36	3.22	3.07	2.92	2.76	2.59
40	12.61	8.25	6.60	5.70	5.13	4.73	4.44	4.21	4.02	3.87	3.64	3.40	3.15	3.01	2.87	2.73	2.57	2.41	2.23
60	11.97	7.76	6.17	5.31	4.76	4.37	4.09	3.87	3.69	3.54	3.31	3.08	2.83	2.69	2.55	2.41	2.25	2.08	1.89
120	11.38	7.32	5.79	4.95	4.42	4.04	3.77	3.55	3.38	3.24	3.02	2.78	2.53	2.40	2.26	2.11	1.95	1.76	1.54
∞	10.83	6.91	5.42	4.62	4.10	3.74	3.47	3.27	3.10	2.96	2.74	2.51	2.27	2.13	1.99	1.84	1.66	1.45	1.00

[†] These values are to be multiplied by 100

1.5 Three Important Families of Test Distributions

with upper percentage points of the F-distribution for the one-sided test are given here. For example, suppose we wish to find upper percentiles for the ratio of two variances, the variance in the numerator having 12 degrees of freedom and the variance in the denominator having 6 degrees of freedom. For $\alpha = 0.05$ or $P = 0.05$ we enter Table 30b in the column headed $v_1 = 12$, and moving down the left-hand side of the table to $v_2 = 6$, we read $F_{12;6;0.05} = 4.00$. Similarly with Table 30d we find $F_{12;6;0.01} = 7.72$ and with Table 30a $F_{10;10;0.10} = 2.32$. Two F-distribution curves are sketched in Figure 27. Intermediate values are obtained by means of **harmonic interpolation**. Consider, for example, the 1% level for $v_1 = 24$ and $v_2 = 60$. The table (p. 148) specifies the levels for 20 and 60 and also for 30 and 60 degrees of freedom as 2.20 and 2.03. If we denote the value sought for $v_1 = 24$ and $v_2 = 60$ by x, we obtain from (1.135) that $x = 2.115$ (exact value: 2.12):

$$\boxed{\frac{2.20-x}{2.20-2.03} = \frac{1/20-1/24}{1/20-1/30}.} \qquad (1.135)$$

The 1% level for $v_1 = 24$; $v_2 = 200$ is found [with 1.95 for (24; 120) and 1.79 for (24; ∞)] to be $x = 1.79 + (1.95 - 1.79)120/200 = 1.886$ (exact value: 1.89).

F, as a ratio of two squares, can take on only values between zero and plus infinity, and thus, like the χ^2 distribution, can extend only to the right of the origin. In place of the mirror symmetry of the distribution function of e.g., the t-distribution, we have here to a certain extent a "reciprocal symmetry." As t and $-t$ can be interchanged [α replaced by $(1 - \alpha)$], so can F and $1/F$ simultaneously with v_1 and v_2 be interchanged without affecting the corresponding probabilities. We have

$$\boxed{F(v_1, v_2; 1-\alpha) = 1/F(v_2, v_1; \alpha).} \qquad (1.136)$$

With this relation we can, for example, readily determine $F_{0.95}$ from $F_{0.05}$.

EXAMPLE. Given $v_1 = 12$, $v_2 = 8$, $\alpha = 0.05$, so that $F = 3.28$. To find F for $v_1 = 12$, $v_2 = 8$, $\alpha = 0.95$. From $v_1 = 8$, $v_2 = 12$, and $\alpha = 0.05$, whence $F = 2.85$, the F value in question is found to be $F_{12,8;0.95} = 1/2.85 = 0.351$.

When the **number of degrees of freedom is large**, we have the **approximation** (cf., also pages 146–147, below)

$$\boxed{\log F = 0.4343 \cdot z \cdot \left[\sqrt{\frac{2(v_1 + v_2)}{v_1 \cdot v_2}}\right],} \qquad (1.137)$$

where z is the standard normal value for the chosen level of significance of the one-sided question (cf., Table 43, Section 2.1.6). Thus, for example, the value of $F(120, 120; 0.05)$ is seen from

$$\log F = (0.4343)(1.64)\left[\sqrt{\frac{2(120 + 120)}{(120)(120)}}\right] = 0.13004$$

to be $F = 1.35$ (Table 30b).

Interpolation of intermediate values

For the case where neither a particular $v_{\text{numerator}}$ (v_1 or v_n) nor $v_{\text{denominator}}$ (v_2 or v_d) is listed in the table, the neighboring values v'_n, v''_n and v'_d, v''_d ($v'_n < v_n < v''_n$ and $v'_d < v_d < v''_d$) for which the F distribution is tabulated are noted. Interpolation is carried out according to Laubscher (1965) [the formula (1.138) is also valid for nonintegral values of v]:

$$\begin{aligned}
F(v_n, v_d) &= (1 - A) \cdot (1 - B) \cdot F(v'_n, v'_d) \\
&+ A \cdot (1 - B) \cdot F(v'_n, v''_d) \\
&+ (1 - A) \cdot B \cdot F(v''_n, v'_d) \\
&+ A \cdot B \cdot F(v''_n, v''_d) \\
\text{with} \quad A &= \frac{v''_d(v_d - v'_d)}{v_d(v''_d - v'_d)} \quad \text{and} \quad B = \frac{v''_n(v_n - v'_n)}{v_n(v''_n - v'_n)}.
\end{aligned} \quad (1.138)$$

EXAMPLE. Compute

$$F(28, 44; 0.01)$$

given

$$F(20, 40; 0.01) = 2.37$$
$$F(20, 50; 0.01) = 2.27$$
$$F(30, 40; 0.01) = 2.20$$
$$F(30, 50; 0.01) = 2.10$$

with

$$A = \frac{50(44 - 40)}{44(50 - 40)} = \frac{5}{11} \quad \text{and} \quad B = \frac{30(28 - 20)}{28(30 - 20)} = \frac{6}{7}.$$

We get

$$F(28, 44; 0.01) = \frac{6}{11} \cdot \frac{1}{7} \cdot 2.37 + \frac{5}{11} \cdot \frac{1}{7} \cdot 2.27$$

$$+ \frac{6}{11} \cdot \frac{6}{7} \cdot 2.20 + \frac{5}{11} \cdot \frac{6}{7} \cdot 2.10$$

$$= 2.178 \simeq 2.18.$$

1.5 Three Important Families of Test Distributions

The interpolated value equals the tabulated value found in more extensive tables. If the table lists v_n but not v_d, one interpolates according to the formula

$$F(v_n, v_d) = (1 - A) \cdot F(v_n, v'_d) + A \cdot F(v_n, v''_d). \qquad (1.139)$$

For the reverse case (v_n sought, v_d listed), one uses instead

$$F(v_n, v_d) = (1 - B) \cdot F(v'_n, v_d) + B \cdot F(v''_n, v_d). \qquad (1.140)$$

Interpolation of probabilities

We have available the upper significance levels for the 0.1%, 0.5%, 1%, 2.5%, 5%, and 10% level. When it becomes necessary to interpolate the true level of an empirical F-value based on v_1 and v_2 degrees of freedom between the 0.1% and 10% bounds, the procedure suggested by Zinger (1964) is used:

1. Enclose the empirically derived F-value between two tabulated F values (F_1, F_2), with levels of significance α and αm, so that $F_1 < F < F_2$.
2. Determine the quotient k from

$$k = \frac{F_2 - F}{F_2 - F_1}. \qquad (1.141)$$

3. The interpolated probability is then

$$P = \alpha m^k. \qquad (1.142)$$

EXAMPLE. Given: $F = 3.43$, $v_1 = 12$, $v_2 = 12$. Approximate the probability that this F-value will be exceeded.

Solution

1. The observed F value lies between the 1% and 2.5% levels (i.e., $\alpha = 0.01$, $m = 2.5$); $F_1 = 3.28 < F = 3.43 < F_2 = 4.16$.
2. The quotient is $k = (4.16 - 3.43)/(4.16 - 3.28) = 0.8295$.
3. The approximate probability is then found (using logarithms) to be $P = (0.01)(2.5)^{0.8295} = 0.0214$. The exact value is 0.0212.

If the significance of an arbitrary empirical F-value is to be determined, then according to an approximation for $v_2 \geq 3$ proposed by Paulson (1942),

$$\hat{z} = \frac{\left(1 - \dfrac{2}{9v_2}\right)F^{1/3} - \left(1 - \dfrac{2}{9v_1}\right)}{\sqrt{\dfrac{2}{9v_2}F^{2/3} + \dfrac{2}{9v_1}}}. \tag{1.143}$$

If the lower levels of the F-distribution are of interest, we must also have $v_1 \geq 3$.

The cube roots of F and F^2 can be extracted with the help of logarithms.

The relationships of the F-distribution to the other test distributions and to the standard normal distribution are simple and clear.

The F-distribution turns,

$$\begin{aligned}&\text{for } v_1 = 1 \text{ and } v_2 = v, \text{ into the distribution of } t^2;\\&\text{for } v_1 = 1 \text{ and } v_2 = \infty, \text{ into the distribution of } z^2;\\&\text{for } v_1 = v \text{ and } v_2 = \infty, \text{ into the distribution of } \chi^2/v.\end{aligned} \tag{1.144}$$

For example, we get for $F_{10;10;0.05} = 2.98$

$$F_{1;10;0.05} = 4.96, \quad t_{10;0.05} = 2.228, \quad \text{i.e. } t^2_{10;0.05} = 4.96,$$
$$F_{1;\infty;0.05} = 3.84, \quad z_{0.05} = 1.960, \quad \text{i.e. } z^2_{0.05} = 3.84,$$
$$F_{10;\infty;0.05} = 1.83, \quad \chi^2_{10;0.05}/10 = 18.307/10 = 1.83.$$

Thus the Student, standard normal and χ^2 distributions can be traced back to the **F-DISTRIBUTION** and its limiting cases:

$$\begin{array}{c} F \\ \swarrow \quad \downarrow \quad \searrow \\ t_v = \sqrt{F_{1;v}} \quad z = \sqrt{F_{1;\infty}} \quad \chi^2_v = vF_{v;\infty} \end{array} \tag{1.145}$$

or

$$\sqrt{F_{1;\infty}} = t_\infty = z = \sqrt{\chi^2_1}. \tag{1.146}$$

For $v \to \infty$ (or $v_1 \to \infty$ and $v_2 \to \infty$):

1. t_v is asymptotically standard normally distributed
2. χ^2_v is approximately normally distributed
3. F_{v_1, v_2} is asymptotically normally distributed.

Finally, we note that $F_{\infty;\nu;\alpha} = \nu/\chi^2_{\nu;1-\alpha}$ and $F_{\infty;\infty;\alpha} \equiv 1$ just as:

ASSUMED DISTRIBUTION	The TEST STATISTIC computed from $\bar{X}$ and/or S (as well as μ and/or σ) based on n independent observations	
Arbitrary distribution with mean μ and variance σ^2	$\dfrac{\bar{X}-\mu}{\sigma}\sqrt{n}$	is asymptotically N(0;1) distributed, i.e., with n large, the larger the n the more nearly it is standard normally distributed (central limit theorem)
Normal distribution $N(\mu;\sigma^2)$	$\dfrac{\bar{X}-\mu}{\sigma}\sqrt{n}$	is N(0;1) distributed
	$\dfrac{\bar{X}-\mu}{S}\sqrt{n}$	is distributed as t_ν with $\nu = n - 1$
	$\dfrac{S^2}{\sigma^2}(n-1)$	is distributed as χ^2_ν with $\nu = n - 1$
	$\dfrac{S_1^2}{S_2^2}$	is distributed as F_{ν_1,ν_2} with $\nu_1 = n_1 - 1$, $\nu_2 = n_2 - 1$ (two independent samples)

1.6 DISCRETE DISTRIBUTIONS

1.6.1 The binomial coefficient

We denote by $_nC_x$ or $\binom{n}{x}$ (read: *n* over *x*) **the number of combinations of *n* elements in classes of *x* elements each** (or *x* elements at a time). This is the number of *x*-element subsets in a set of *n* elements. The computation proceeds according to (1.147). The numerator and denominator of $\binom{n}{x}$ involve *x* factors each, as we shall see below:

$$_nC_x = \binom{n}{x} = \frac{n!}{x!(n-x)!} \quad \text{with } 1 \leq x \leq n. \tag{1.147}$$

(p. 160)

Here $n!$ (*n* factorial) represents the product of the natural numbers from 1 to *n*, or $n! = (n)(n-1)(n-2) \ldots 1$, e.g., $3! = (3)(2)(1) = 6$ [cf., also (1.152)]. The number of combinations of 5 elements taken 3 at a time is accordingly

$$_5C_3 = \frac{5!}{3!(5-3)!} = \frac{5\cdot 4\cdot 3\cdot 2\cdot 1}{3\cdot 2\cdot 1\cdot 2\cdot 1} = 10,$$

or

$$\binom{5}{3} = \frac{5\cdot 4\cdot 3}{1\cdot 2\cdot 3} = 5\cdot 2 = 10,$$

since

$$\binom{n}{x} = \frac{n\cdots(n-x+1)}{x(x-1)\cdots 1}.$$

For $x > n$ we have obviously $\binom{n}{x} = 0$; for $x < n$,

$$_nC_x = \binom{n}{x} = \frac{n!}{(n-x)!x!} = \binom{n}{n-x} = {_nC_{n-x}}.$$

For example,

$$\binom{5}{3} = \frac{5!}{3!2!} = \frac{5!}{2!3!} = \binom{5}{2}.$$

In particular $_nC_0 = {_nC_n} = 1$, because out of n objects n can be chosen in exactly one way. This also follows from the definition of $0! = 1$. Other ways of writing $_nC_x$ are C_n^x and $C_{n,x}$. A more detailed discussion is to be found in Riordan (1968).

Further examples. How many possibilities are there of selecting a committee consisting of 5 persons from a group of 9?

$\binom{9}{7}$ is computed as $\binom{9}{2} = \frac{9 \cdot 8}{2 \cdot 1} = 36$ [see Table 31].

How many possibilities are there in lottery that involves choosing 6 numbers out of a collection of 49? The number of combinations of 49 elements taken 6 at a time comes to

$$\binom{49}{6} = \frac{49!}{6!43!} \simeq 14 \text{ million.}$$

Pascal's Triangle

The binomial coefficients $\binom{n}{x}$ can be read off from the triangular array of numbers given below, called Pascal's triangle (Pascal, 1623–1662): A number in this array is the sum of the two numbers to its right and its left in the next row above. The first and the last number in any row are ones. The defining law for Pascal's triangle is

$$\boxed{\binom{n}{x} + \binom{n}{x+1} = \binom{n+1}{x+1};} \qquad (1.148)$$

for example,

$$\binom{3}{1} + \binom{3}{2} = 3 + 3 = 6 = \binom{4}{2}.$$

1.6 Discrete Distributions

Binomial coefficients		for	
$\binom{0}{0}$	1	$n=0$	$(a+b)^0 = 1$
$\binom{1}{0} \binom{1}{1}$	1 1	$n=1$	$(a+b)^1 = a+b$
$\binom{2}{0} \binom{2}{1} \binom{2}{2}$	1 2 1	$n=2$	$(a+b)^2 = a^2 + 2ab + b^2$
$\binom{3}{0} \binom{3}{1} \binom{3}{2} \binom{3}{3}$	1 3 3 1	$n=3$	$(a+b)^3 = a^3 + 3a^2b + 3ab^2 + b^3$
$\binom{4}{0} \binom{4}{1} \binom{4}{2} \binom{4}{3} \binom{4}{4}$	1 4 6 4 1	$n=4$	$(a+b)^4 = a^4 + 4a^3b + 6a^2b^2 + 4ab^3 + b^4$
	etc.		

This triangle immediately yields the values of the **probabilities arising in a coin tossing problem**. For example, the sum of the numbers in the fourth line is $1 + 3 + 3 + 1 = 8$. By forming the fractions $\frac{1}{8}$, $\frac{3}{8}$, $\frac{3}{8}$, $\frac{1}{8}$, we get the probabilities for the various possible outcomes in tossing three coins, i.e., three heads ($\frac{1}{8}$), two heads and a tail ($\frac{3}{8}$), one head and two tails ($\frac{3}{8}$), and three tails ($\frac{1}{8}$). Correspondingly, the numbers in the fifth (nth) row, totaling 2^{n-1}, give us the probabilities for head and tail combinations in tossing four $(n - 1)$ coins.

Pascal's triangle thus serves to **identify the probability of combinations**: The probability of a particular boy-girl combination in a family with, say, 4 children, can be quickly determined when independence of the births and equal probabilities are assumed, i.e., $a = b$. First of all, since $n = 4$ is given, the numbers in the bottom row are added; this gives 16. At the ends of the row stand the least likely combinations, i.e., either all boys or all girls, each with the probability of 1 in 16. Going from the outside toward the center, one finds for the next possible combinations, namely 3 boys and 1 girl or vice versa, the probability of 4 in 16 for each. The numbers 6 in the middle corresponds to two boys and two girls; the probability for this is 6 in 16, i.e., nearly 38%.

The coefficients in the expansion of $(a + b)^n$—sums of two terms are called **binomials**, so that this expression is referred to as the nth power of a binomial—can be obtained directly from Pascal's triangle. Note that the first and the last coefficient are always 1; the second and the second to last coefficient always equal the exponent n of the binomial. The coefficient 1 is not written explicitly $[(a + b)^1 = 1a + 1b = a + b]$. The generalization

Table 31 Binomial coefficients $\binom{n}{x} = {}_nC_x = n!/[x!(n-x)!]$. Since $\binom{n}{x} = \binom{n}{n-x}$, we get ${}_6C_4 = \binom{6}{4} = 6!/(4!2!) = (6\cdot5\cdot4\cdot3\cdot2\cdot1)/(4\cdot3\cdot2\cdot1\cdot2\cdot1)$ with $\binom{6}{2} = \binom{6}{6-4}$, the value 15 [note also that $\binom{n}{0} = \binom{n}{n} = 1$ and $\binom{n}{1} = \binom{n}{n-1} = n$]

1	2	3	4	5	6	7	8	9	10	11	12	13	x
1	1	1	1	1	1	1	1	1	1	1	1	1	0
1	2	3	4	5	6	7	8	9	10	11	12	13	1
	1	3	6	10	15	21	28	36	45	55	66	78	2
		1	4	10	20	35	56	84	120	165	220	286	3
			1	5	15	35	70	126	210	330	495	715	4
				1	6	21	56	126	252	462	792	1287	5
					1	7	28	84	210	462	924	1716	6
						1	8	36	120	330	792	1716	7
							1	9	45	165	495	1287	8
								1	10	55	220	715	9
									1	11	66	286	10
										1	12	78	11
											1	13	12
												1	13

x	$\binom{14}{x}$	$\binom{15}{x}$	$\binom{16}{x}$	$\binom{17}{x}$	$\binom{18}{x}$	$\binom{19}{x}$	$\binom{20}{x}$	x
0	1	1	1	1	1	1	1	0
1	14	15	16	17	18	19	20	1
2	91	105	120	136	153	171	190	2
3	364	455	560	680	816	969	1140	3
4	1001	1365	1820	2380	3060	3876	4845	4
5	2002	3003	4368	6188	8568	11628	15504	5
6	3003	5005	8008	12376	18564	27132	38760	6
7	3432	6435	11440	19448	31824	50388	77520	7
8	3003	6435	12870	24310	43758	75582	125970	8
9	2002	5005	11440	24310	48620	92378	167960	9
10	1001	3003	8008	19448	43758	92378	184756	10
11	364	1365	4368	12376	31824	75582	167960	11
12	91	455	1820	6188	18564	50388	125970	12
13	14	105	560	2380	8568	27132	77520	13
14	1	15	120	680	3060	11628	38760	14
15		1	16	136	816	3876	15504	15
16			1	17	153	969	4845	16
17				1	18	171	1140	17
18					1	19	190	18
19						1	20	19
20							1	20

1.6 Discrete Distributions

of the formula to the nth power of a binomial is given by the binomial expansion (Newton, 1643–1727):

$$(a+b)^n = a^n + \binom{n}{1}a^{n-1}b + \binom{n}{2}a^{n-2}b^2 + \ldots + \binom{n}{n-1}ab^{n-1} + b^n$$
$$= \sum_{k=0}^{n} \binom{n}{k} a^{n-k} b^k. \qquad (1.149)$$

For $a \gg b$ we have $(a + b)^n \simeq a^n + na^{n-1}b$. Note that

$$2^n = (1 + 1)^n = \sum_{k=0}^{n} \binom{n}{k}; \quad \text{also} \quad \sum_{k=0}^{n} \binom{n}{k}^2 = \binom{2n}{n}. \qquad (1.150)$$

Table 31 allows us to simply read off the binomial coefficients $_nC_x$. In fact, the results of both examples can be read directly from Table 31. Miller (1954) presented an extensive table of binomial coefficients; their base ten logarithms, which are more manageable, can be found in (e.g.) the Documenta Geigy (1960 and 1968, pp. 70–77). Table 32 gives values of $n!$ and $\log n!$ for $1 \le n \le 100$. When tables of factorials and their logarithms to base ten are unavailable, one can approximate $n!$ according to Stirling's formula

$$n^n e^{-n} \sqrt{2\pi n}. \qquad (1.151)$$

For large values of n the approximation is very good. Besides $\log n$, the following logarithms will be needed:

$$\log \sqrt{2\pi} = 0.39909,$$
$$\log e = 0.4342945.$$

Better than (1.151) is the formula $(n + 0.5)^{n+0.5} e^{-(n+0.5)} \sqrt{2\pi}$, i.e.,

$$\log n! \approx (n + 0.5)\log(n + 0.5) - (n + 0.5)\log e + \log\sqrt{2\pi}. \qquad (1.152)$$

We get, for example, for 100!,

$$\log 100! \approx (100.5)(2.002166) - (100.5)(0.4342945) + 0.39909 = 157.97018$$

i.e.,

$$100! \approx (9.336)(10^{157}).$$

The actual values as tabulated are

$$\log 100! = 157.97000,$$
$$100! = 9.3326 \cdot 10^{157}.$$

Better approximations are given by Abramowitz and Stegun (1968, p. 257 [2]).

Table 32 Factorials and their base ten logarithms

n	n!	log n!	n	n!	log n!
1	1.0000	0.00000	50	3.0414×10^{64}	64.48307
2	2.0000	0.30103	51	1.5511×10^{66}	66.19065
3	6.0000	0.77815	52	8.0658×10^{67}	67.90665
4	2.4000×10	1.38021	53	4.2749×10^{69}	69.63092
			54	2.3084×10^{71}	71.36332
5	1.2000×10^{2}	2.07918	55	1.2696×10^{73}	73.10368
6	7.2000×10^{2}	2.85733	56	7.1100×10^{74}	74.85187
7	5.0400×10^{3}	3.70243	57	4.0527×10^{76}	76.60774
8	4.0320×10^{4}	4.60552	58	2.3506×10^{78}	78.37117
9	3.6288×10^{5}	5.55976	59	1.3868×10^{80}	80.14202
10	3.6288×10^{6}	6.55976	60	8.3210×10^{81}	81.92017
11	3.9917×10^{7}	7.60116	61	5.0758×10^{83}	83.70550
12	4.7900×10^{8}	8.68034	62	3.1470×10^{85}	85.49790
13	6.2270×10^{9}	9.79428	63	1.9826×10^{87}	87.29724
14	8.7178×10^{10}	10.94041	64	1.2689×10^{89}	89.10342
15	1.3077×10^{12}	12.11650	65	8.2477×10^{90}	90.91633
16	2.0923×10^{13}	13.32062	66	5.4435×10^{92}	92.73587
17	3.5569×10^{14}	14.55107	67	3.6471×10^{94}	94.56195
18	6.4024×10^{15}	15.80634	68	2.4800×10^{96}	96.39446
19	1.2165×10^{17}	17.08509	69	1.7112×10^{98}	98.23331
20	2.4329×10^{18}	18.38612	70	1.1979×10^{100}	100.07841
21	5.1091×10^{19}	19.70834	71	8.5048×10^{101}	101.92966
22	1.1240×10^{21}	21.05077	72	6.1234×10^{103}	103.78700
23	2.5852×10^{22}	22.41249	73	4.4701×10^{105}	105.65032
24	6.2045×10^{23}	23.79271	74	3.3079×10^{107}	107.51955
25	1.5511×10^{25}	25.19065	75	2.4809×10^{109}	109.39461
26	4.0329×10^{26}	26.60562	76	1.8855×10^{111}	111.27543
27	1.0889×10^{28}	28.03698	77	1.4518×10^{113}	113.16192
28	3.0489×10^{29}	29.48414	78	1.1324×10^{115}	115.05401
29	8.8418×10^{30}	30.94654	79	8.9462×10^{116}	116.95164
30	2.6525×10^{32}	32.42366	80	7.1569×10^{118}	118.85473
31	8.2228×10^{33}	33.91502	81	5.7971×10^{120}	120.76321
32	2.6313×10^{35}	35.42017	82	4.7536×10^{122}	122.67703
33	8.6833×10^{36}	36.93869	83	3.9455×10^{124}	124.59610
34	2.9523×10^{38}	38.47016	84	3.3142×10^{126}	126.52038
35	1.0333×10^{40}	40.01423	85	2.8171×10^{128}	128.44980
36	3.7199×10^{41}	41.57054	86	2.4227×10^{130}	130.38430
37	1.3764×10^{43}	43.13874	87	2.1078×10^{132}	132.32382
38	5.2302×10^{44}	44.71852	88	1.8548×10^{134}	134.26830
39	2.0398×10^{46}	46.30959	89	1.6508×10^{136}	136.21769
40	8.1592×10^{47}	47.91165	90	1.4857×10^{138}	138.17194
41	3.3453×10^{49}	49.52443	91	1.3520×10^{140}	140.13098
42	1.4050×10^{51}	51.14768	92	1.2438×10^{142}	142.09477
43	6.0415×10^{52}	52.78115	93	1.1568×10^{144}	144.06325
44	2.6583×10^{54}	54.42460	94	1.0874×10^{146}	146.03638
45	1.1962×10^{56}	56.07781	95	1.0330×10^{148}	148.01410
46	5.5026×10^{57}	57.74057	96	9.9168×10^{149}	149.99637
47	2.5862×10^{59}	59.41267	97	9.6193×10^{151}	151.98314
48	1.2414×10^{61}	61.09391	98	9.4269×10^{153}	153.97437
49	6.0828×10^{62}	62.78410	99	9.3326×10^{155}	155.97000
50	3.0414×10^{64}	64.48307	100	9.3326×10^{157}	157.97000

1.6 Discrete Distributions

In applying Stirling's formula it is to be noted that with increasing n, the value of $n!$ grows extraordinarily rapidly and **the absolute error becomes very large, while the relative error (which is about** $1/[12n]$**) tends to zero** and for $n = 9$ it is already less than one percent.

Let us also mention the rough approximation $(n + a)! \approx n! n^a e^r$ with $r = (a^2 + a)/(2n)$.

Further elements of combinatorics

Every listing of n elements in some arbitrary sequence is called a **permutation** of these n elements. With n elements there are $n!$ different permutations (the factorial gives the number of possible sequences). Thus the 3 letters a, b, c, can be ordered in $3! = 6$ ways:

$$\begin{array}{ccc} abc & bac & cab \\ acb & bca & cba. \end{array}$$

If among n elements there are n_1 identical elements of a certain type, n_2 of a second type, and in general n_k of a kth type, then the number of all possible orderings, the number of permutations, equals

$$\boxed{\frac{n!}{n_1! n_2! \cdots n_k!}, \quad \text{where } n_1 + n_2 + n_3 + \cdots + n_k = n.} \quad (1.153)$$

This quotient will be of interest to us later on, in connection with the multinomial distribution.

A selection of k elements from a collection of n elements ($n \geq k$) is called a combination of n elements k at a time, or more simply, a **combination of kth order**. Depending on whether some of the selected elements are allowed to be identical or have all to be different, we speak of combinations **with or without replication**, respectively. If two combinations that in fact consist of exactly the same elements but in different order are treated as distinct, they are called permutations of n elements k at a time; they are also called variations of n elements of kth order, with or without replication. Accordingly, we can distinguish **four models.**

The number of combinations of kth order (k at a time) of n different elements:

1. **without** replication and **without** regard for order is given by the binomial coefficients

$$\boxed{\binom{n}{k},} \quad (1.154)$$

2. **without** replication but **taking** order into account equals

$$\boxed{\frac{n!}{(n-k)!} = \binom{n}{k} k!,} \quad (1.155)$$

3. **with** replication but **without** regard for order equals

$$\boxed{\binom{n+k-1}{k},}\qquad(1.156)$$

4. **with** replication and **taking** order into account equals

$$\boxed{n^k.}\qquad(1.157)$$

EXAMPLE. The number of combinations of second order (in every case consisting of two elements) out of three elements, the letters a, b, c, ($n = 3$, $k = 2$) is as follows:

Model	Replication	Regard to order	Type	Combinations Type	Number
1	without	without	ab ac bc		$\binom{3}{2} = 3$
2	without	with	ab ac bc ba ca cb		$\dfrac{3!}{(3-2)!} = 6$
3		without	aa bb ab ac bc cc		$\binom{3+2-1}{2} = 6$
4	with	with	aa bb ab ac bc cc ba ca cb		$3^2 = 9$

An introduction to combinations is given in Riordan (1958, 1968) and in Wellnitz (1971).

▶ 1.6.2 The binomial distribution

If p represents the probability that a particular trial gives rise to a "success" and $q = 1 - p$ stands for the probability of a "failure" in that trial, then the probability that in n trials there are exactly x successes — x successes and $n - x$ failures occur — is given by the relation

$$\boxed{\begin{aligned}P(X = x \mid p, n) = P_{n,p}(x) &= \binom{n}{x} p^x q^{n-x} \\ &= {}_nC_x p^x q^{n-x} = \frac{n!}{x!(n-x)!} p^x q^{n-x},\end{aligned}}\qquad(1.158)$$

where $x = 0, 1, 2, \ldots, n$.

1.6 Discrete Distributions

The distribution function is given by

$$F(x) = \sum_{k=0}^{x} \binom{n}{k} p^k q^{n-k} \quad \text{and} \quad F(n) = \sum_{k=0}^{n} \binom{n}{k} p^k q^{n-k} = 1.$$

The term **binominal distribution** derives from the **binomial expansion**

$$(p + q)^n = \sum_{x=0}^{n} \binom{n}{x} p^x q^{n-x} = 1 \quad \text{with } p + q = 1. \tag{1.159}$$

Note: We write p (and q) rather than π (and $1 - \pi$) as parameters and $\hat{p}$ (and $\hat{q}$) as estimates of the relative frequencies. The binomial or Bernoulli distribution, which dates back to Jakob Bernoulli (1654–1705), is based on the following underlying assumptions:

1. The trials and the results of these trials are **independent** of one another.
2. The probability p of any particular event remains **constant** for all trials.

This very important discrete distribution is applicable whenever repeated observations on some dichotomies are called for. Since x can take on only certain integral values, probabilities are defined only for positive integral x-values (Figure 28). The binomial distribution is symmetric when $p = 0.5$, is flat to the right when $p < 0.5$ and is flat to the left when $p > 0.5$.

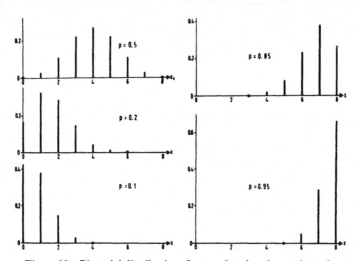

Figure 28 Binomial distributions for $n = 8$ and various values of p.

The parameters of the binomial distribution are n and p, the mean

$$\mu = np, \tag{1.160}$$

and the variance

$$\sigma^2 = np(1-p) = npq. \tag{1.161}$$

Equations (1.160), (1.161) are valid for absolute frequencies; for relative frequencies we have the relations: $\mu = p$ and $\sigma^2 = pq/n$. The coefficient of skewness is

$$\text{Skewness} = \frac{q-p}{\sigma} = \frac{q-p}{\sqrt{npq}}, \qquad (1.162)$$

so that for large n, i.e., for a large standard deviation, the skewness becomes very small and the asymmetry insignificant.

If individual probabilities $P(x)$ are to be calculated, one applies the so-called recursion formula

$$P(x+1) = \frac{n-x}{x+1}\frac{p}{q}P(x). \qquad (1.163)$$

Since $P(0) = q^n$ can, for given q and n, be rapidly computed according to (1.158), it then follows that $P(1) = (n/1)(p/q)P(0)$, $P(2) = \frac{1}{2}(n-1)(p/q)P(1)$, etc.

Tables are provided by the National Bureau of Standards (1950), Romig (1953), Harvard University Computation Laboratory (1955), and Weintraub (1963); Table 33 lists selected binomial probabilities (cf., Examples 1 and 2). Also of importance (cf., Example 2a) is the formula

$$P(X \geq x_0) = P\left(F_{2(n-x_0+1),\, 2x_0} > \frac{q}{p} \cdot \frac{x_0}{n-x_0+1}\right). \qquad (1.164)$$

In the region $0.001 \leq P \leq 0.10$ we interpolate according to (1.141), (1.142).

In Chapters 4 and 6 probabilities are compared in terms of samples from binomial populations; from two binomial distributions with the help of the so-called fourfold test, and from several binomial distributions with the help of the so-called $k \times 2$ χ^2 test.

Approximation of the binomial distribution by the normal distribution

For $npq \geq 9$

$$\hat{z} = \frac{x-np}{\sqrt{npq}} \qquad (1.165)$$

has approximately the standard normal distribution (cf., Examples 4 and 5).

Modification with continuity correction

The exact probability of the binomial variable x, taking integral values, is often approximated by the probability of a normal variable $x - 0.5$ or $x + 0.5$, referred to as corrected for continuity. Probabilities concerning open intervals $P(x_1 < X < x_2)$ or closed intervals $P(x_1 \leq X \leq x_2)$ for any x_1 and x_2 with $0 \leq x_1 < x_2 \leq n$ are thus better approximated by:

$$\text{open} \quad P\left(\frac{x_1 + 0.5 - np}{\sqrt{npq}} < Z < \frac{x_2 - 0.5 - np}{\sqrt{npq}}\right) \quad (1.165a)$$

$$\text{closed} \quad P\left(\frac{x_1 - 0.5 - np}{\sqrt{npq}} \leq Z \leq \frac{x_2 + 0.5 - np}{\sqrt{npq}}\right) \quad (1.165b)$$

Note that (1.165b) is broader than (1.165a).

As an example we evaluate $P(16 < X \leq 26)$ for $n = 100$ and $p = 0.25$ or $np = 25$ and $\sqrt{npq} = 4.330$.

$$P\left(\frac{16 + 0.5 - 25}{4.330} < Z \leq \frac{26 + 0.5 - 25}{4.330}\right) = P(-1.963 < Z \leq 0.346)$$

and with Table 13 and some interpolation we get for $P(16 < X \leq 26) = P(17 \leq X \leq 26)$ the approximated value $(0.5 - 0.0248) + (0.5 - 0.3647) = 0.4752 + 0.1353 = 0.6105$ or 0.61 (exact value 0.62063).

The cumulative binomial probability

$$P(X \leq k|p;n) = \sum_{j=0}^{k} \binom{n}{j} p^j q^{n-j}$$

can be better approximated with the help of the standardized value $\hat{z}$ given by Molenaar (1970):

$$\hat{z} = |\sqrt{q(4k+3.5)} - \sqrt{p(4n-4k-0.5)}|. \quad (1.166)$$

Here (a) for $0.05 \leq P \leq 0.93$, 3.5 is to be replaced by 3 and 0.5 by 1; (b) for extreme P-values, 3.5 is to be replaced by 4 and 0.5 by 0.

EXAMPLE. $P(X \leq 13|0.6; 25) = 0.268$;

$$\hat{z} = |\sqrt{0.4(52 + 3.5)} - \sqrt{0.6(100 - 52 - 0.5)}| = 0.627,$$

i.e., $P = 0.265$; with 3 and 1 the result changes to $\hat{z} = 0.620$, $P = 0.268$.

The confidence limits of the binomial distribution will be examined more thoroughly in Section 4.5. A very useful nomogram of the distribution function on this distribution is given by Larson (1966). Approximations are compared by Gebhardt (1969) and Molenaar (1970).

Remarks

1. With the help of (1.163) a **graphical test** to check whether a sample might come from a binomially distributed population can be carried out: One plots $P(x + 1)/P(x)$ where $P(x)$ is the empirical distribution function, against $1/(x + 1)$, and if the points all lie on roughly a straight line (cf. Chapter 5), then the values follow a binomial distribution (Dubey 1966; cf. also Ord 1967).

2. Mosteller and Tukey (1949), at the suggestion of R. A. Fisher, designed a **binomial probability paper** which, in addition to a graphical assessment of binomial probabilities—(in particular, estimation of the confidence interval of a relative frequency as well as comparison of two relative frequencies), allows also for evaluation of approximate χ^2 probabilities and the variance ratio of F. For suppliers of binomial paper see the References, Section 7. For particulars one must refer to Stange (1965), and also to the pertinent chapters in the book by Wallis and Roberts (1962). Further remarks are given by King (1971).

3. **Functional parameters and explicit parameters.** Parameters that provide information on where the values of the random variables lie on the real line (μ, $\tilde{\mu}$) and how close together they are (σ^2) were called **functional parameters** by Pfanzagl (1966). They can be written as functions of the parameters that appear explicitly in the formula for the density of a distribution. Thus for the binomial distribution

---
n and p are explicit parameters,
$\mu = np$ and $\sigma^2 = np(1 - p)$ are functional parameters,

---

since they can be expressed in terms of the explicit parameters. The density function of the normal distribution also contains two explicit parameters: μ and σ, which are at the same time also functional parameters, as is indicated by the notation.

4. Finally, the winning numbers in roulette are nearly normally distributed even when n is only moderately large. For large n ($n \to \infty$) the percentage of their occurrence is the same. The frequencies of the individual winning numbers are then greatly scattered [they lie, according to (1.161), very far from one another]. Consequently, in cases of completely equal chance (roulette), there is no tendency toward absolute equalization (do equal chances necessarily lead to inequality in society as well?).

5. A more detailed discussion of the binomial distribution can be found in Patil and Joshi (1968 [cited on p. 575]) as well as in Johnson and Kotz (1969 [cited on p. 570]). Two generalizations are given in Altham (1978). Tolerance intervals give Hahn and Chandra (1981).

6. Change-point problem: Methods for testing a change of distribution in a sequence of observations when the initial distribution is unknown are given in Pettitt (1979) and (1980) for zero-one observations, binomial observations and continuous observations.

Examples

1. What is the probability that on tossing an ideal coin ($p = \frac{1}{2}$) three times, (a) three heads, (b) two heads [and a tail] are obtained?

(a) $P = {}_3C_3(\frac{1}{2})^3(\frac{1}{2})^0 = 1 \cdot \frac{1}{8} \cdot 1 = \frac{1}{8} = 0.125$,

(b) $P = {}_3C_2(\frac{1}{2})^2(\frac{1}{2})^1 = 3 \cdot \frac{1}{4} \cdot \frac{1}{2} = \frac{3}{8} = 0.375$.

Note that for $p = \frac{1}{2}$ we have $P(X = x | n; \frac{1}{2}) = \binom{n}{x}(\frac{1}{2})^x(\frac{1}{2})^{n-x} = \binom{n}{x}(\frac{1}{2})^n = \binom{n}{x}/2^n$. See Table 31 in Section 1.61 and Table 26 in Section 1.4.2, as well as the remark below Table 33.

Table 33 Binomial probabilities $\binom{n}{x}p^x(1-p)^{n-x}$ for $n \leq 10$ and for various values of p [taken from Dixon and Massey (1969 [1]]) copyright © April 13, 1965, McGraw-Hill Inc.]

n	x	0.01	0.05	0.10	0.15	0.20	0.25	0.30	1/3	0.35	0.40	0.45	0.50
2	0	0.9801	0.9025	0.8100	0.7225	0.6400	0.5625	0.4900	0.4444	0.4225	0.3600	0.3025	0.2500
	1	0.0198	0.0950	0.1800	0.2550	0.3200	0.3750	0.4200	0.4444	0.4550	0.4800	0.4950	0.5000
	2	0.0001	0.0025	0.0100	0.0225	0.0400	0.0625	0.0900	0.1111	0.1225	0.1600	0.2025	0.2500
3	0	0.9703	0.8574	0.7290	0.6141	0.5120	0.4219	0.3430	0.2963	0.2746	0.2160	0.1664	0.1250
	1	0.0294	0.1354	0.2430	0.3251	0.3840	0.4219	0.4410	0.4444	0.4436	0.4320	0.4084	0.3750
	2	0.0003	0.0071	0.0270	0.0574	0.0960	0.1406	0.1890	0.2222	0.2389	0.2880	0.3341	0.3750
	3	0.0000	0.0001	0.0010	0.0034	0.0080	0.0156	0.0270	0.0370	0.0429	0.0640	0.0911	0.1250
4	0	0.9606	0.8145	0.6561	0.5220	0.4096	0.3164	0.2401	0.1975	0.1785	0.1296	0.0915	0.0625
	1	0.0388	0.1715	0.2916	0.3685	0.4096	0.4219	0.4116	0.3951	0.3845	0.3456	0.2995	0.2500
	2	0.0006	0.0135	0.0486	0.0975	0.1536	0.2109	0.2646	0.2963	0.3105	0.3456	0.3675	0.3750
	3	0.0000	0.0005	0.0036	0.0115	0.0256	0.0469	0.0756	0.0988	0.1115	0.1536	0.2005	0.2500
	4	0.0000	0.0000	0.0001	0.0005	0.0016	0.0039	0.0081	0.0123	0.0150	0.0256	0.0410	0.0625
5	0	0.9510	0.7738	0.5905	0.4437	0.3277	0.2373	0.1681	0.1317	0.1160	0.0778	0.0503	0.0312
	1	0.0480	0.2036	0.3280	0.3915	0.4096	0.3955	0.3602	0.3292	0.3124	0.2592	0.2059	0.1562
	2	0.0010	0.0214	0.0729	0.1382	0.2048	0.2637	0.3087	0.3292	0.3364	0.3456	0.3369	0.3125
	3	0.0000	0.0011	0.0081	0.0244	0.0512	0.0879	0.1323	0.1646	0.1811	0.2304	0.2757	0.3125
	4	0.0000	0.0000	0.0004	0.0022	0.0064	0.0146	0.0284	0.0412	0.0488	0.0768	0.1128	0.1562
	5	0.0000	0.0000	0.0000	0.0001	0.0003	0.0010	0.0024	0.0041	0.0053	0.0102	0.0185	0.0312
6	0	0.9415	0.7351	0.5314	0.3771	0.2621	0.1780	0.1176	0.0878	0.0754	0.0467	0.0277	0.0156
	1	0.0571	0.2321	0.3543	0.3993	0.3932	0.3560	0.3025	0.2634	0.2437	0.1866	0.1359	0.0938
	2	0.0014	0.0305	0.0984	0.1762	0.2458	0.2966	0.3241	0.3292	0.3280	0.3110	0.2780	0.2344
	3	0.0000	0.0021	0.0146	0.0415	0.0819	0.1318	0.1852	0.2195	0.2355	0.2765	0.3032	0.3125
	4	0.0000	0.0001	0.0012	0.0055	0.0154	0.0330	0.0595	0.0823	0.0951	0.1382	0.1861	0.2344
	5	0.0000	0.0000	0.0001	0.0004	0.0015	0.0044	0.0102	0.0165	0.0205	0.0369	0.0609	0.0938
	6	0.0000	0.0000	0.0000	0.0000	0.0001	0.0002	0.0007	0.0014	0.0018	0.0041	0.0083	0.0156
7	0	0.9321	0.6983	0.4783	0.3206	0.2097	0.1335	0.0824	0.0585	0.0490	0.0280	0.0152	0.0078
	1	0.0659	0.2573	0.3720	0.3960	0.3670	0.3115	0.2471	0.2048	0.1848	0.1306	0.0872	0.0547
	2	0.0020	0.0406	0.1240	0.2097	0.2753	0.3115	0.3177	0.3073	0.2985	0.2613	0.2140	0.1641
	3	0.0000	0.0036	0.0230	0.0617	0.1147	0.1730	0.2269	0.2561	0.2679	0.2903	0.2918	0.2734
	4	0.0000	0.0002	0.0026	0.0109	0.0287	0.0577	0.0972	0.1280	0.1442	0.1935	0.2388	0.2734
	5	0.0000	0.0000	0.0002	0.0012	0.0043	0.0115	0.0250	0.0384	0.0466	0.0774	0.1172	0.1641
	6	0.0000	0.0000	0.0000	0.0001	0.0004	0.0013	0.0036	0.0064	0.0084	0.0172	0.0320	0.0547
	7	0.0000	0.0000	0.0000	0.0000	0.0000	0.0001	0.0002	0.0005	0.0006	0.0016	0.0037	0.0078
8	0	0.9227	0.6634	0.4305	0.2725	0.1678	0.1001	0.0576	0.0390	0.0319	0.0168	0.0084	0.0039
	1	0.0746	0.2793	0.3826	0.3847	0.3355	0.2670	0.1977	0.1561	0.1373	0.0896	0.0548	0.0312
	2	0.0026	0.0515	0.1488	0.2376	0.2936	0.3115	0.2965	0.2731	0.2587	0.2090	0.1569	0.1094
	3	0.0001	0.0054	0.0331	0.0839	0.1468	0.2076	0.2541	0.2731	0.2786	0.2787	0.2568	0.2188
	4	0.0000	0.0004	0.0046	0.0185	0.0459	0.0865	0.1361	0.1707	0.1875	0.2322	0.2627	0.2734
	5	0.0000	0.0000	0.0004	0.0026	0.0092	0.0231	0.0467	0.0683	0.0808	0.1239	0.1719	0.2188
	6	0.0000	0.0000	0.0000	0.0002	0.0011	0.0038	0.0100	0.0171	0.0217	0.0413	0.0703	0.1094
	7	0.0000	0.0000	0.0000	0.0000	0.0001	0.0004	0.0012	0.0024	0.0033	0.0079	0.0164	0.0312
	8	0.0000	0.0000	0.0000	0.0000	0.0000	0.0000	0.0001	0.0002	0.0002	0.0007	0.0017	0.0039
9	0	0.9135	0.6302	0.3874	0.2316	0.1342	0.0751	0.0404	0.0260	0.0207	0.0101	0.0046	0.0020
	1	0.0830	0.2985	0.3874	0.3679	0.3020	0.2253	0.1556	0.1171	0.1004	0.0605	0.0339	0.0176
	2	0.0034	0.0629	0.1722	0.2597	0.3020	0.3003	0.2668	0.2341	0.2162	0.1612	0.1110	0.0703
	3	0.0001	0.0077	0.0446	0.1069	0.1762	0.2336	0.2668	0.2731	0.2716	0.2508	0.2119	0.1641
	4	0.0000	0.0006	0.0074	0.0283	0.0661	0.1168	0.1715	0.2048	0.2194	0.2508	0.2600	0.2461
	5	0.0000	0.0000	0.0008	0.0050	0.0165	0.0389	0.0735	0.1024	0.1181	0.1672	0.2128	0.2461
	6	0.0000	0.0000	0.0001	0.0006	0.0028	0.0087	0.0210	0.0341	0.0424	0.0743	0.1160	0.1641
	7	0.0000	0.0000	0.0000	0.0000	0.0003	0.0012	0.0039	0.0073	0.0098	0.0212	0.0407	0.0703
	8	0.0000	0.0000	0.0000	0.0000	0.0000	0.0001	0.0004	0.0009	0.0013	0.0035	0.0083	0.0176
	9	0.0000	0.0000	0.0000	0.0000	0.0000	0.0000	0.0000	0.0001	0.0001	0.0003	0.0008	0.0020
10	0	0.9044	0.5987	0.3487	0.1969	0.1074	0.0563	0.0282	0.0173	0.0135	0.0060	0.0025	0.0010
	1	0.0914	0.3151	0.3874	0.3474	0.2684	0.1877	0.1211	0.0867	0.0725	0.0403	0.0207	0.0098
	2	0.0042	0.0746	0.1937	0.2759	0.3020	0.2816	0.2335	0.1951	0.1757	0.1209	0.0763	0.0439
	3	0.0001	0.0105	0.0574	0.1298	0.2013	0.2503	0.2668	0.2601	0.2522	0.2150	0.1665	0.1172
	4	0.0000	0.0010	0.0112	0.0401	0.0881	0.1460	0.2001	0.2276	0.2377	0.2508	0.2384	0.2051
	5	0.0000	0.0001	0.0015	0.0085	0.0264	0.0584	0.1029	0.1366	0.1536	0.2007	0.2340	0.2461
	6	0.0000	0.0000	0.0001	0.0012	0.0055	0.0162	0.0368	0.0569	0.0689	0.1115	0.1596	0.2051
	7	0.0000	0.0000	0.0000	0.0001	0.0008	0.0031	0.0090	0.0163	0.0212	0.0425	0.0746	0.1172
	8	0.0000	0.0000	0.0000	0.0000	0.0001	0.0004	0.0014	0.0030	0.0043	0.0106	0.0229	0.0439
	9	0.0000	0.0000	0.0000	0.0000	0.0000	0.0000	0.0001	0.0003	0.0005	0.0016	0.0042	0.0098
	10	0.0000	0.0000	0.0000	0.0000	0.0000	0.0000	0.0000	0.0000	0.0000	0.0001	0.0003	0.0010

Table 33 has the three entries (n, x, p). For $n = 3$, $x = 3$, $p = 0.5$ the desired value is found to be 0.1250 and for $n = 3$, $x = 2$, $p = 0.5$ it is 0.3750.

If p is small, there is a preference toward small values of x. For $p = 0.5$ the distribution is symmetric. If p is large, there is a preference toward large values of x: For $p > 0.5$ one therefore replaces (a) p by $1 - p$ and (b) $x = 0, 1, \ldots, n$ by $x = n, n - 1, \ldots, 0$. Example: $n = 7$. $p = 0.85$, $x = 6$: see $n = 7$, $p = 1 - 0.85 = 0.15$, $x = 1$ (previously the second to last value in the column, now the second value from the top); i.e., $P = 0.3960$.

2. Suppose 20% of the pencils produced by a machine are rejects. What is the probability that out of 4 randomly chosen pencils (a) no pencil, (b) one pencil, (c) at most two pencils are rejects? The probability that a reject is produced is $p = 0.2$, while the probability of not producing a reject comes to $q = 1 - p = 0.8$.

(a) $P(\text{no rejects}) = {}_4C_0(0.2)^0(0.8)^4 = 0.4096$,
(b) $P(\text{one reject}) = {}_4C_1(0.2)^1(0.8)^3 = 0.4096$,
(c) $P(\text{two rejects}) = {}_4C_2(0.2)^2(0.8)^2 = 0.1536$.

$P(\text{at most two rejects}) = P(\text{no rejects}) + P(\text{one reject}) + P(\text{two rejects}) = 0.4096 + 0.4096 + 0.1536 = 0.9728$. By Table 33 with $n = 4$, x takes on the values 0, 1, 2 with $p = 0.2$ in every case. The corresponding probabilities can be read off directly. By the recursion formula,

$$p = 0.2 = \frac{1}{5} \quad \text{and} \quad n = 4; \quad \frac{p}{q} = \frac{\frac{1}{5}}{\frac{4}{5}} = \frac{1}{4}; \quad P(x+1) = \frac{4-x}{x+1} \cdot \frac{1}{4} \cdot P_4(x),$$

$$P(0) = 0.8^4 \qquad\qquad\qquad = 0.4096,$$

$$P(1) = \frac{4}{1} \cdot \frac{1}{4} \cdot 0.4096 \qquad = 0.4096,$$

$$P(2) = \frac{3}{2} \cdot \frac{1}{4} \cdot 0.4096 \qquad = 0.1536,$$

$$P(3) = \frac{2}{3} \cdot \frac{1}{4} \cdot 0.1536 \qquad = 0.0256,$$

$$P(4) = \frac{1}{4} \cdot \frac{1}{4} \cdot 0.0256 \qquad = 0.0016,$$

Check: $\qquad\qquad \sum P = 1.0000$.

2a. When the probability of getting at least 3 rejects is sought, we obtain, for $n = 4$ and $p = 0.2$,

$$P(X \geq 3) = P\left(F_{2(4-3+1),\, 2 \cdot 3} > \frac{0.8}{0.2} \cdot \frac{3}{4-3+1}\right) = P(F_{4;\,6} > 6.00).$$

The probability of this F-value (6.00) for $v_1 = 4$ and $v_2 = 6$ degrees of freedom is found by interpolation (cf., Section 1.5.3):

$$\left.\begin{array}{l} F_1 = 4.53 \; (\alpha = 0.05), \\ F_2 = 6.23 \; (\alpha = 0.025), \end{array}\right\} \quad m = 2; \quad k = \frac{6.23 - 6.00}{6.23 - 4.53} = 0.1353,$$

$$P = 0.025 \cdot 2^{0.1353} = 0.0275.$$

The approximation is seen to be good on comparing it with the exact value of 0.0272.

1.6 Discrete Distributions

3. Which is more probable: that tossing (a) 6 ideal dice, at least one six is obtained or (b) 12 ideal dice, at least two sixes turn up?

(a) $\quad P_{\text{no sixes are obtained}} = {}_6C_0(\frac{1}{6})^0(\frac{5}{6})^6 \simeq 0.335$
$\quad P_{\text{one or more sixes are obtained}} = 1 - {}_6C_0(\frac{1}{6})^0(\frac{5}{6})^6 \simeq 0.665$

(b) $\quad P_{\text{two or more sixes are obtained}} = 1 - ({}_{12}C_0(\frac{1}{6})^0(\frac{5}{6})^{12} + {}_{12}C_1(\frac{1}{6})^1(\frac{5}{6})^{11})$
$\quad \simeq 1 - (0.1122 + 0.2692) \simeq 0.619.$

Thus (a) is more probable than (b). To have a coarse estimate of the probability in (a) one can refer to Table 33, using $p' = 0.15$ in place of $p = 0.166 \sim 0.17$.

4. An ideal die is tossed 120 times. We are asked to find the probability that the number 4 appears eighteen times or less. The probability that the 4 comes up from zero to eighteen times ($p = \frac{1}{6}, q = \frac{5}{6}$) equals exactly ${}_{120}C_{18}(\frac{1}{6})^{18}(\frac{5}{6})^{102} + {}_{120}C_{17}(\frac{1}{6})^{17}(\frac{5}{6})^{103} + \cdots + {}_{120}C_0(\frac{1}{6})^0(\frac{5}{6})^{120}$. Since carrying out the computation is rather a waste of time, we resort to the normal distribution as an approximation (cf., $npq = (120) \cdot \frac{1}{6} \cdot \frac{5}{6} = 16.667 > 9$). If we treat the numbers as a continuum, the integers 0 to 18 fours are replaced by the interval -0.5 to 18.5 fours, i.e.,

$$\bar{x} = np = 120(\tfrac{1}{6}) = 20 \quad \text{and} \quad s = \sqrt{npq} = \sqrt{16.667} = 4.08.$$

-0.5 and 18.5 are then transformed into standard units $[z = (x - \bar{x})/s]$; for -0.5 we get $(-0.5 - 20)/4.09 = -5.01$, for 18.5 we get $(18.5 - 20)/4.09 = -0.37$. The probability sought is then given by the area under the normal curve between $z = -5.01$ and $z = -0.37$:

$$P = (\text{area between } z = 0 \text{ and } z = -5.01)$$
$$- (\text{area between } z = 0 \text{ and } z = -0.37),$$
$$P = 0.5000 - 0.1443 = 0.3557.$$

Thus, if we repeatedly take samples of 120 tosses the 4 should appear 18 times or less in about 36% of the cases.

5. It is suspected that a die might no longer be ideal. In 900 tosses a 4 is observed 180 times. Is this consistent with the null hypothesis which says the die is regular? Under the null hypothesis the probability of tossing a 4 is $\frac{1}{6}$. Then $np = 900(\frac{1}{6}) = 150$ and $\sqrt{npq} = \sqrt{900 \cdot \frac{1}{6} \cdot \frac{5}{6}} = 11.18$;

$$\hat{z} = \frac{180 - 150}{11.18} = \frac{30}{11.18} = 2.68; \quad P = 0.0037.$$

Since we have here a two sided question, $P = 0.0074$; hence the result is significant at the 1% level. The die is not unbiased. Problems of this sort can be better analyzed according to Section 4.3.2.

6. We are interested in the number of female offsprings in litters of 4 mice (cf., David 1953, pp. 187 ff.). The results for 200 litters of this type are presented in Table 34.

Table 34 The number of female mice in litters of 4 mice each

Number of female mice/litter	0	1	2	3	4
Number of litters (200 total)	15	63	66	47	9

We now assume that for the strain of mice considered, the probability of being born a female is constant, independent of the number of female animals already born, and in addition, that the litters are independent of each other, thus forming a random process, so that the percentage of female animals in the population can be estimated from the given sample of 200 litters.

The portion of young female animals is

$$\hat{p} = \frac{\text{number of female offsprings}}{\text{total number of offsprings}},$$

$$\hat{p} = \frac{(0)(15) + (1)(63) + (2)(66) + (3)(47) + (4)(49)}{(4)(200)} = 0.465.$$

We know that when the assumptions of the binomial distribution are satisfied, the probabilities of finding 0, 1, 2, 3, 4 females in litters of 4 animals each can be determined with the aid of the binomial expansion of $(0.535 + 0.465)^4$. On the basis of this expansion, the expected numbers for 200 litters of quadruplets are then given by the terms of

$$200 = 200(0.535 + 0.465)^4$$
$$= 200(0.0819 + 0.2848 + 0.3713 + 0.2152 + 0.0468)$$
$$= 16.38 + 56.96 + 74.26 + 43.04 + 9.36.$$

A comparison of the observed and the expected numbers is presented in Table 35.

Table 35 Comparison of the expected numbers with the observed numbers of Table 34

Number of female mice/litter	0	1	2	3	4	Σ
Number of litters:						
observed	15	63	66	47	9	200
expected	16.38	56.96	74.26	43.04	9.36	200

In Section 1.6.7 we will consider a similar example in greater detail and test whether the assumptions of the Poisson distribution are fulfilled, i.e., whether the observations follow a true or compound Poisson distribution.

1.6.3 The hypergeometric distribution

If samples are taken **without replacement** (cf., Section 1.3.1), then the hypergeometric distribution replaces the binomial distribution. This distribution is used frequently in problems relating to quality control. We consider, e.g., drawing 5 balls from an urn with $W = 5$ white and $B = 10$ black balls. We are asked for the probability that exactly $w = 2$ white and $b = 3$ black balls are taken. This probability is given by

$$P(w \text{ out of } W, b \text{ out of } B) = \frac{{}_W C_w \cdot {}_B C_b}{{}_{W+B} C_{w+b}} = \frac{\binom{W}{w}\binom{B}{b}}{\binom{W+B}{w+b}} \quad (1.167)$$

with $0 \le w \le W$ and $0 \le b \le B$.

We get for $P(2$ out of 5 white balls and 3 out of 10 black balls)

$$\frac{{}_5 C_2 \cdot {}_{10} C_3}{{}_{15} C_5} = \frac{(5!/[3! \cdot 2!])(10!/[7! \cdot 3!])}{15!/10! \cdot 5!}$$

$$= \frac{(5 \cdot 4) \cdot (10 \cdot 9 \cdot 8) \cdot (5 \cdot 4 \cdot 3 \cdot 2 \cdot 1)}{(2 \cdot 1) \cdot (3 \cdot 2 \cdot 1) \cdot (15 \cdot 14 \cdot 13 \cdot 12 \cdot 11)} = 0.3996,$$

a probability of around 40%.

With sample sizes $n_1 + n_2 = n$ and corresponding population sizes $N_1 + N_2 = N$, (1.167) can be generalized to

$$P(n_1, n_2 | N_1, N_2) = \frac{\binom{N_1}{n_1}\binom{N_2}{n_2}}{\binom{N}{n}}, \quad (1.167a)$$

$$\text{mean: } \mu = n\frac{N_1}{N} = np, \quad (1.168)$$

$$\text{variance: } \sigma^2 = np(1-p)\frac{N-n}{N-1}. \quad (1.169)$$

If n/N is small, this distribution is practically identical to the binomial distribution. Correspondingly, the variance tends to the variance of the binomial distribution $(N - n)/(N - 1) \simeq 1 - (n/N) \simeq 1$ for $N >> n$).

The generalized hypergeometric distribution (polyhypergeometric distribution)

$$P(n_1, n_2, \ldots, n_k | N_1, N_2, \ldots, N_k) = \frac{\binom{N_1}{n_1}\binom{N_2}{n_2}\cdots\binom{N_k}{n_k}}{\binom{N}{n}} \quad (1.170)$$

gives the probability that in a sample of size n exactly $n_1, n_2, \ldots, n_k$ observations with attributes $A_1, A_2, \ldots, A_k$ are obtained if in the population of size N the frequencies of these attributes are $N_1, N_2, \ldots, N_k$ with $\sum_{i=1}^{k} N_i = N$ and $\sum_{i=1}^{k} n_i = n$. The parameters (for the n_i) are

$$\text{mean:} \quad \mu_i = n \frac{N_i}{N} \tag{1.171}$$

$$\text{variance:} \quad \sigma_i^2 = np_i(1 - p_i) \frac{N - n}{N - 1}. \tag{1.172}$$

The inverse of the hypergeometric distribution, discussed by Guenther (1975), is used, among other things, in quality control and for estimating the unknown size N of a population (e.g., the state of a wild animal population): N_1 individuals are captured, marked, and then released; subsequently some n individuals are captured and the marked individuals counted, yielding n_1; then $\hat{N} \simeq nN_1/n_1$ (cf. also Jolly 1963, Southwood 1966, Roberts 1967, Manly and Parr 1968, as well as Robson 1969).

EXAMPLES

1. Assume we have 10 students, of which 6 study biochemistry and 4 statistics. A sample of 5 students is chosen. What is the probability that among the 5 students 3 are biochemists and 2 are statisticians?

$$P(3 \text{ out of } 6 \text{ B., } 2 \text{ out of } 4 \text{ S.}) = \frac{{}_6C_3 \cdot {}_4C_2}{{}_{6+4}C_{3+2}} = \frac{(6!/[3! \cdot 3!])(4!/[2! \cdot 2!])}{10!/[5! \cdot 5!]}$$

$$= \frac{(6 \cdot 5 \cdot 4) \cdot (4 \cdot 3) \cdot (5 \cdot 4 \cdot 3 \cdot 2 \cdot 1)}{(3 \cdot 2 \cdot 1) \cdot (2 \cdot 1) \cdot (10 \cdot 9 \cdot 8 \cdot 7 \cdot 6)}$$

$$= \frac{20}{42} = 0.4762.$$

The probability thus comes to nearly 50%.

2. The integers from 0 to 49 are given. Six of them are "special". Six are to be drawn. What is the probability of choosing four of the special numbers? In the game Lotto, the integers

$$P(4 \text{ out of } 6, 2 \text{ out of } 43) = \frac{\binom{6}{4}\binom{43}{2}}{\binom{49}{6}} = \frac{15 \cdot 903}{13{,}983{,}816},$$

since

$$\binom{49}{6} = \frac{(49)\cdots(44)}{(1)\cdots(6)} = (49)(47)(46)(3)(44) = 13{,}983{,}816.$$

1.6 Discrete Distributions

For problems of this sort one refers to Tables 31 and 30 (Sections 1.6 and 1.5.3):

$$P \simeq \frac{13.545 \cdot 10^3}{13.984 \cdot 10^6} \simeq 0.967 \cdot 10^{-3},$$

i.e., not quite 0.001. Likewise, the probability that at least four specific numbers are chosen is still below 0.1%. The probability of choosing six specific numbers equals $1/\binom{49}{6} = 1/13{,}983{,}816 \simeq 7 \cdot 10^{-8}$.

3. A population of 100 elements includes 5% rejects. What is the probability that in a sample consisting of 50 elements, (a) no, (b) one reject is found?

Case (a)

$$P(50 \text{ out of } 95, 0 \text{ out of } 5) = \frac{_{95}C_{50} \cdot {_5}C_0}{_{95+5}C_{50+0}} = \frac{95! \cdot 5! \cdot 50! \cdot 50!}{50! \cdot 45! \cdot 5! \cdot 0! \cdot 100!}$$

$$= \frac{95! \cdot 50!}{45! \cdot 100!} \quad \text{(Table 32)}$$

$$= \frac{1.0330 \cdot 10^{148} \cdot 3.0414 \cdot 10^{64}}{1.1962 \cdot 10^{56} \cdot 9.3326 \cdot 10^{157}} = 0.02823.$$

Case (b)

$$P(49 \text{ out of } 95, 1 \text{ out of } 5) = \frac{_{95}C_{49} \cdot {_5}C_1}{_{95+5}C_{49+1}} = \frac{95! \cdot 5! \cdot 50! \cdot 50!}{49! \cdot 46! \cdot 4! \cdot 1! \cdot 100!}$$

$$= 5 \cdot \frac{95! \cdot 50! \cdot 50!}{49! \cdot 46! \cdot 100!} = 0.1529.$$

4. If in the course of a year out of $W = 52$ consecutive issues of a weekly publication $A = 10$ arbitrary issues carry a certain notice, then the probability that someone reading $w = 15$ arbitrary issues does not run across a copy containing the notice ($a = 0$) is

$$P(a \text{ out of } A, w \text{ out of } W) = \frac{\binom{A}{a}\binom{W-A}{w-a}}{\binom{W}{w}}$$

or

$$P(0 \text{ out of } 10, 15 \text{ out of } 42) = \frac{\binom{10}{0}\binom{52-10}{15-0}}{\binom{52}{15}}$$

i.e., since

$$\binom{n}{0} = 1,$$

we have

$$P = \frac{\binom{42}{15}}{\binom{52}{15}} = \frac{42! \cdot 15! \cdot 37!}{15! \cdot 27! \cdot 52!}.$$

This can be calculated as follows:

$$
\begin{aligned}
\log 42! &= 51.14768 \\
\log 15! &= 12.11650 \\
\log 37! &= 43.13874 \\
\hline
&\,106.40292 \\
\log 15! &= 12.11650 \\
\log 27! &= 28.03698 \\
\log 52! &= 67.90665 \\
\hline
&= 108.06013
\end{aligned}
$$

$$
\begin{aligned}
\log P &= 0.34279 - 2 \\
P &= 0.02202 \simeq 2.2\%.
\end{aligned}
$$

Thus the probability of seeing at least one notice comes to nearly 98%. Examples 2 and 3 should be worked out as exercises with the aid of the logarithms to base ten of factorials (Table 32). Problems of this sort can be solved much more quickly by referring to tables (Lieberman and Owen 1961). Nomograms with confidence limits were published by DeLury and Chung (1950).

Approximations (cf., also the end of Section 1.6.5)

1. For N_1 and N_2 large and n small in comparison ($n/N < 0.1$; $N \geq 60$) the hypergeometric distribution is approximated by the **binomial** distribution $p = N_1/(N_1 + N_2)$.

2. For $np \geq 4$

$$\hat{z} = (n_1 - np)/\sqrt{npq(N-n)/(N-1)} \qquad (1.173)$$

can be regarded as having nearly the standard normal distribution. The cumulative probability of the hypergeometric distribution,

$$P(X \leq k | N; N_1; n) = \sum_{n_1=0}^{N_1} \binom{N_1}{n_1}\binom{N_2}{n_2} \Big/ \binom{N}{n},$$

assuming $n \leq N_1 \leq N/2$, can be better approximated (Molenaar 1970) according to

$$\hat{z} = |2[\sqrt{(k + 0.9)(N - N_1 - n + k + 0.9)} - \sqrt{(n - k - 0.1)(N_1 - k - 0.1)}]/\sqrt{N - 0.5}|. \quad (1.173a)$$

In this expression for $0.05 \leq p \leq 0.93$, 0.9 is to be replaced by 0.75, 0.1 by 0.25, and 0.5 by 0; for extreme P values 0.9 is replaced by 1, 0.1 by 0, and 0.5 by 1.

EXAMPLE. $P(X \leq 1|10; 5; 5) = 0.103$; $\hat{z}$ (by 1.173a) $= 1.298$, i.e., $P = 0.0971$; with 0.75, 0.25, and 0 we get $\hat{z} = 1.265$, $P = 0.103$.

3. For p small, n large, and N very large in comparison with n ($n/N \leq 0.05$), the hypergeometric distribution can be approximated by the so-called **Poisson distribution** which is discussed in the next section ($\lambda = np$).
4. The binomial distribution and the Poisson distribution can, for $\sigma^2 = npq \geq 9$ and $\sigma^2 = np = \lambda \geq 9$, be approximated with sufficient accuracy by the normal distribution.

1.6.4 The Poisson distribution

Setting the fairly small value $np = \lambda$ (Greek lambda) in (1.158) and, with $\lambda > 0$ held constant, letting the number n increase to infinity, the binomial distribution with the mean $np = \lambda$ turns into the Poisson distribution with the parameter λ; λ is generally smaller than 10 and is also the mean of this distribution. This distribution was developed by the French mathematician S. D. Poisson (1781–1840). It comes up when the average number of occurrences of an event **is the result of a large collection of situations in which the event could occur and a very small probability for it to occur**. A good example of this is radioactive disintegration: Out of many millions of radium atoms only a very small percentage disintegrates in a small interval of time. It is essential that the disintegration of an individual atom is independent of the number of atoms already disintegrated.

The Poisson distribution is an important distribution. **It is used—as was suggested—to solve problems which arise in the counting of relatively rare and mutually independent events in a unit interval of time, length, area or volume.** One also speaks of isolated events in a continuum. Examples of this discrete distribution are the distribution of the number of raisins in raisin bread, of yeast cells in a suspension, of erythrocytes on the individual fields of a counting chamber, of misprints per page, of the flaws in the insulation on an extension cord, of the surface irregularities on a table top, and of airplane arrivals at an airport; similarly, it can be used for the frequency of sudden storms in a certain region, the contamination of seeds by weed seeds or pebbles, the number of telephone calls occurring in a certain time interval, the number of electrons emitted by a heated cathode in a given time interval, the number of vehicle breakdowns at a large military installation, the number of rejects within a production batch, the number of vehicles per unit distance and unit time, or the number of breakdown points in complex mechanisms. All these quantities are per unit interval. If, however, the probability does not remain constant or the events become dependent, then we are no longer dealing with a proper Poisson distribution. If these possibilities are excluded —and this holds for the given examples—then true Poisson distributions are to be expected. Suicides and industrial accidents per unit of space or time do not follow the Poisson distribution even though they can be conceived of as rare events. In both cases one cannot speak of an "equal chance for each," as there are individual differences with regard to conditions for an accident and suicidal tendencies.

Let us imagine a loaf of raisin bread that has been divided up into small samples of equal size. In view of the random distribution of the raisins it cannot be expected that all the samples contain exactly the same number of raisins. If the mean value λ (lambda) of the number of raisins in these samples is known, the Poisson distribution gives the probability $P(x)$ that a randomly chosen sample contains precisely x ($x = 0, 1, 2, 3, \ldots$) raisins. Another way of putting this: The Poisson distribution indicates the portion, in percent $[100P(x)\%]$, of a long sequence of consecutively chosen samples in which each sample contains exactly $0, 1, 2, \ldots$ raisins. It is given by

$$P(X = x|\lambda) = P(x) = \frac{\lambda^x e^{-\lambda}}{x!}, \qquad (1.174)$$

$$\lambda > 0, \quad x = 0, 1, 2, \ldots.$$

Here $e = 2.718, \ldots$, the base of natural logarithms,

λ = mean,

$x = 0, 1, 2, 3, \ldots$ the precise number of raisins in a single sample; x may be very large,

$x! = (1)(2)(3) \cdots (x-1)(x)$ [e.g., $4! = (1)(2)(3)(4) = 24$].

Remark: $\sum_{x=0}^{\infty} P(X = x|\lambda) = 1$.

The Poisson distribution is defined by the discrete probability function (1.174) This distribution is fully characterized by the parameter λ; it expresses the density of random points in a given time interval or in a unit of length, area, or volume. λ is **simultaneously the mean and variance**, i.e., $\mu = \lambda$, $\sigma^2 = \lambda$ [cf.

1.6 Discrete Distributions

also (1.161) (Section 1.6.2) with $np = \lambda$ and $q = 1 - p = 1 - \lambda/n$: $\sigma^2 = \lambda(1 - \lambda/n)$, for large n σ^2 tends to λ].

This parameter is approximated (for $q \simeq 1$) by

$$\boxed{\hat{\lambda} = np.} \qquad (1.175)$$

If for some discrete distributions the ratio of variance to mean is close to one—say between $\frac{9}{10}$ and $\frac{10}{9}$—then they can be approximated by a Poisson distribution provided the variable X (≥ 0) could assume large values. If $s^2 < \bar{x}$, then the sample could originate from a binomial distribution. In the opposite case, where $s^2 > \bar{x}$, it could originate from a so-called negative binomial distribution (cf., Bliss 1953). It is usually unnecessary to compute the values of $e^{-\lambda}$, since they are tabulated for a whole series of values λ.

Since $e^{-(x+y+z)} = e^{-x}e^{-y}e^{-z}$, with the help of Table 36 we find e.g.,

$$e^{-5.23} = 0.006738 \cdot 0.8187 \cdot 0.9704 = 0.00535.$$

Table 36 is at the same time a table of natural antilogarithms. If for example we set $x = -3$, then $e^{-3} = 1/e^3 = 1/2.718282^3 = 1/20.0855 = 0.049787$, i.e., $\ln 0.049787 = -3.00$.

EXAMPLE. A radioactive preparation gives 10 impulses per minute, on the average. How large is the probability of obtaining 5 impulses in one minute?

$$P = \frac{\lambda^x \cdot e^{-\lambda}}{x!} = \frac{10^5 \cdot e^{-10}}{5!} = \frac{10^5 \cdot 4.54 \cdot 10^{-5}}{5 \cdot 4 \cdot 3 \cdot 2 \cdot 1} = \frac{4.54}{120} = 0.03783 \simeq 0.04.$$

Thus 5 impulses per minute will be counted in about 4% of the cases.

NOTE. Mathijssen and Goldzieher (1965) provide a nomogram for flow scintillation spectrometry that gives the counting duration for a counting rate with preassigned accuracy (cf., also Rigas 1968).

Table 36 Values of $e^{-\lambda}$ for the Poisson distribution

λ	$e^{-\lambda}$	λ	$e^{-\lambda}$	λ	$e^{-\lambda}$	λ	$e^{-\lambda}$	λ	$e^{-\lambda}$
0.01	0.9901	0.1	0.9048	1	0.367879	10	0.0⁴4540	19	0.0⁸5603
0.02	0.9802	0.2	0.8187	2	0.135335	11	0.0⁴1670	20	0.0⁸2061
0.03	0.9704	0.3	0.7408	3	0.049787	12	0.0⁵6144	21	0.0⁹7583
0.04	0.9608	0.4	0.6703	4	0.018316	13	0.0⁵2260	22	0.0⁹2789
0.05	0.9512	0.5	0.6065	5	0.0²6738	14	0.0⁶8315	23	0.0⁹1026
0.06	0.9418	0.6	0.5488	6	0.0²2479	15	0.0⁶3059	24	0.0¹⁰378
0.07	0.9324	0.7	0.4966	7	0.0³9119	16	0.0⁶1125	25	0.0¹⁰139
0.08	0.9231	0.8	0.4493	8	0.0³3355	17	0.0⁷4140	30	0.0¹³936
0.09	0.9139	0.9	0.4066	9	0.0³1234	18	0.0⁷1523	50	0.0²¹193

$e^{-9.85} = e^{-9} \cdot e^{-0.8} \cdot e^{-0.05} = 0.0001234 \cdot 0.4493 \cdot 0.9512 = 0.0000527$

Characteristics of the Poisson distribution

1. It is a discrete **nonsymmetric** distribution. It has the positive skewness $1/\sqrt{\lambda}$ which decreases to zero with increasing λ, i.e., the distribution then becomes nearly symmetric (Figure 29).

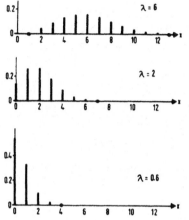

Figure 29 Poisson distributions.

2. For $\lambda < 1$ its **individual probabilities** decrease monotonically with increasing X, while for $\lambda > 1$ they first increase, then decrease.

3. The distribution is **maximum** at the largest integer which is smaller than λ. When λ is a positive integer, the probability is maximum for two neighboring values, namely for $X = \lambda$ and $X = \lambda + 1$.

Table 37 Poisson distributions for small parameters λ and no, one, or more than one event

P(x) \ λ	0.1	0.2	1	2
for x = 0	0.905	0.819	0.368	0.135
for x = 1	0.090	0.164	0.368	0.271
for x > 1	0.005	0.017	0.264	0.594

For example, if the number of misprints per page of a periodical follows a Poisson distribution with $\lambda = 0.2$, then out of 100 pages about 82 pages should exhibit no, 16 one, and about 2 more than one misprint (Table 37). Table 38 shows further that out of 10,000 pages about one can be expected with 4 errors.

1.6 Discrete Distributions

Table 38 Poisson distribution $P(x) = \lambda^x \cdot e^{-\lambda}/x!$ for selected values of λ. As the parameter λ increases the Poisson distribution approaches the normal distribution

x \ λ	0.2	0.5	0.8	1	3	5	8	x
0	0.8187	0.6065	0.4493	0.3679	0.0498	0.0067	0.0003	0
1	0.1637	0.3033	0.3595	0.3679	0.1494	0.0337	0.0027	1
2	0.0164	0.0758	0.1438	0.1839	0.2240	0.0842	0.0107	2
3	0.0011	0.0126	0.0383	0.0613	0.2240	0.1404	0.0286	3
4	0.0001	0.0016	0.0077	0.0153	0.1680	0.1755	0.0573	4
5	0.0000	0.0002	0.0012	0.0031	0.1008	0.1755	0.0916	5
6		0.0000	0.0002	0.0005	0.0504	0.1462	0.1221	6
7			0.0000	0.0001	0.0216	0.1044	0.1396	7
8				0.0000	0.0081	0.0653	0.1396	8
9					0.0027	0.0363	0.1241	9
10					0.0008	0.0181	0.0993	10
11					0.0002	0.0082	0.0722	11
12					0.0001	0.0034	0.0481	12
13					0.0000	0.0013	0.0296	13
14						0.0005	0.0169	14
15						0.0002	0.0090	15
16						0.0000	0.0045	16
17							0.0021	17
18							0.0009	18
19							0.0004	19
20							0.0002	20
21							0.0001	21
22							0.0000	22

For the case where (a) λ is large and (b) $X = \lambda$ we have by Stirling's formula,

$$P(\lambda) = \frac{e^{-\lambda} \cdot \lambda^\lambda}{\lambda!} \simeq \frac{e^{-\lambda} \cdot \lambda^\lambda}{\sqrt{2\pi} \cdot \lambda^{\lambda+1/2} \cdot e^{-\lambda}} = \frac{1}{\sqrt{2\pi\lambda}} \simeq \frac{0.4}{\sqrt{\lambda}},$$

$$\boxed{P(\lambda) \simeq \frac{0.4}{\sqrt{\lambda}},} \qquad (1.176)$$

e.g., $P(X = \lambda = 8) \simeq 0.4/\sqrt{8} = 0.141$; the value listed in Table 38 is 0.1396. A sequence of individual probabilities is obtained by means of the recursion formula

$$\boxed{P(x + 1) = \frac{\lambda}{x + 1} P(x).} \qquad (1.177)$$

A more detailed discussion of this distribution can be found in the monograph by Haight (1967). Extensive tables are given by Molina (1945), Kitagawa (1952), and the Defense Systems Department (1962).

Examples

1. How large is the probability that out of 1,000 persons (a) no one, (b) one person, (c) two, (d) three persons have their birthdays on a particular day? Since $q = \frac{364}{365} \simeq 1$, we can estimate $\hat{\lambda} = np = 1{,}000(\frac{1}{365}) = 2.7397$. We simplify by setting $\hat{\lambda} = 2.74$:

$$P(X = 0) = \frac{\lambda^0 e^{-\lambda}}{0!} = e^{-\lambda} = e^{-2.74} = 0.06457 \simeq 0.065,$$

$$P(X = 1) = \frac{\lambda^1 e^{-\lambda}}{1!} = \lambda e^{-\lambda} \simeq 2.74 \cdot 0.065 = 0.178,$$

$$P(X = 2) = \frac{\lambda^2 e^{-\lambda}}{2!} = \frac{\lambda^2 e^{-\lambda}}{2} \simeq \frac{2.74^2 \cdot 0.065}{2} = 0.244,$$

$$P(X = 3) = \frac{\lambda^3 e^{-\lambda}}{3!} = \frac{\lambda^3 e^{-\lambda}}{6} \simeq \frac{2.74^3 \cdot 0.065}{6} = 0.223.$$

Thus for a given sample of 1,000 people the probability is about 7% that no person has a birthday on a particular day; the probability that one, two, or three persons have their birthdays on a particular day is about 18%, 24%, or 22%, respectively. With the recursion formula (1.177) one obtains the following simplification:

$$P(0) = \text{(cf., above)} \simeq 0.065,$$

$$P(1) \simeq \frac{2.74}{1} 0.065 = 0.178,$$

$$P(2) \simeq \frac{2.74}{2} 0.178 = 0.244,$$

$$P(3) \simeq \frac{2.74}{3} 0.244 = 0.223.$$

Multiplying the probability $P(X = k)$ by n, we get the average number among n samples of 1,000 persons each in which exactly k persons have their birthdays on a particular day.

2. Suppose the probability that a patient does not tolerate the injection of a certain serum is 0.001. We are asked for the probability that out of 2,000 patients (a) exactly three, (b) more than two patients do not tolerate the injection. Since $q = 0.999 \simeq 1$, we get $\hat{\lambda} = np = 2{,}000 \cdot 0.001 = 2$, and

$$P(x \text{ do not tolerate}) = \frac{\lambda^x e^{-\lambda}}{x!} = \frac{2^x e^{-2}}{x!}.$$

Thus

(a) $$P(3 \text{ do not tolerate}) = \frac{2^3 e^{-2}}{3!} = \frac{4}{3e^2} = 0.180;$$

(b)
$$P(0 \text{ do not tolerate}) = \frac{2^0 e^{-2}}{0!} = \frac{1}{e^2},$$

$$P(1 \text{ does not tolerate}) = \frac{2^1 e^{-2}}{1!} = \frac{2}{e^2},$$

$$P(2 \text{ do not tolerate}) = \frac{2^2 e^{-2}}{2!} = \frac{2}{e^2},$$

$P(\text{more than } 2 \text{ do not tolerate}) = 1 - P(0, 1, \text{ or } 2 \text{ do not tolerate}) = 1 - (1/e^2 + 2/e^2 + 2/e^2) = 1 - (5/e^2) = 0.323$. If a large number of samples of 2,000 patients each are available, then with a probability of about 18%, three patients, and with a probability of about 32%, more than two patients will not tolerate the injection. In (a) the computation itself would have been quite formidable if the binomial distribution had been used:

$$P(3 \text{ do not tolerate}) = {}_{2,000}C_3 \cdot 0.001^3 \cdot 0.999^{1,997}.$$

Additional examples are given by G. Bergmann [Metrika **14** (1969), 1–20]. For accidents in which two parties are involved see W. Widdra [Metrika **19** (1972), 68–71].

Note

1. We can find out how large λ must be in order that the event occurs at least once with probability P by observing that

$$P(X = 0) = \frac{e^{-\lambda} \lambda^0}{0!} = e^{-\lambda}$$

so that

$$\boxed{P = 1 - e^{-\lambda}} \quad (1.178)$$

and

$$e^{-\lambda} = 1 - P, \quad \ln e^{-\lambda} = \ln(1 - P)$$

and using the table

P	λ
0.999	6.908
0.99	4.605
0.95	2.996
0.90	2.303
0.80	1.609
0.50	0.693
0.20	0.223
0.05	0.051
0.01	0.010
0.001	0.001

calculated from

$$\lambda = -2.3026 \cdot \log(1 - P). \tag{1.179}$$

For $P = 0.95$, e.g., we find $\lambda = 3$.

2. The following table tells (a) how big a sample should be in order that, with probability $S = 0.95$, at least k rare events (probability of occurrence $p \leq 0.05$) occur, and (b) given p and the sample size n, how many rare events $k(p, n)$ at least can be expected with the same confidence $S = 0.95$ (cf., also Sections 1.2.3 and 2.1.6).

p \ k	0.05	0.04	0.03	0.02	0.01	0.008	0.006	0.004	0.002	0.001
1	60	75	100	150	300	375	499	749	1498	2996
3	126	157	210	315	630	787	1049	1574	3148	6296
5	183	229	305	458	915	1144	1526	2289	4577	9154
10	314	393	524	785	1571	1963	2618	3927	7853	15706
20	558	697	929	1394	2788	3485	4647	6970	13940	27880

If only $k_1 < k(p, n)$ rare events are observed then the null hypothesis $p_1 = p$ will be rejected at the 5% level and the alternate hypothesis $p_1 < p$ accepted. The testing of $\lambda_1 = \lambda_2$ against $\lambda_1 \neq \lambda_2$ is discussed in Section 1.6.6.1.

Confidence intervals for the mean λ

For given values of x there are two kinds of confidence intervals [CIs] for λ:

(1) **Non central (shortest) CIs** following Crow and Gardner, given in Table 80 on pages 344, 345. Examples are given on page 343.
(2) **Central CIs**: calculated according to (1.180), approximated according to (1.181) with the help of Tables 28 and 14 or 43, e.g., the 95% CI, given $x = 10$: $\chi^2_{20;0.975} = 9.59$ and $\chi^2_{22;0.025} = 36.78$ so 95% CI: $4.80 \leq \lambda \leq 18.39$.

Use (1) OR (2) but never both together.

$$90\% \text{ CI}: \tfrac{1}{2}\chi^2_{0.95;2x} \leq \lambda \leq \tfrac{1}{2}\chi^2_{0.05;2(x+1)}, \tag{1.180}$$

$$90\% \text{ CI}: \left(\frac{1.645}{2} - \sqrt{x}\right)^2 \lesssim \lambda \lesssim \left(\frac{1.645}{2} + \sqrt{x+1}\right)^2. \tag{1.181}$$

1.6 Discrete Distributions

On the right of both (1.180) and (1.181) are the (one sided) upper 95% confidence limits: Thus for example for $x = 50$, by (1.180), $2(50 + 1) = 102$, $\chi^2_{0.05;\,102} = 126.57$ (i.e., $\lambda \lesssim 63.3$), and by (1.181), $(1.645/2 + \sqrt{50 + 1})^2 = 63.4$ (i.e., $\lambda \lesssim 63.4$). The upper 90% confidence limits are obtained similarly [(1.180): with $\chi^2_{0.10}$ in place of $\chi^2_{0.05}$; see Tables 28, 28a, Section 1.5.2; (1.181): with 1.282 in place of 1.645; see Table 43, Section 2.1.6].

Table 80 (Section 4.5.4) is also used in testing the null hypothesis: $\lambda = \lambda_x$. The null hypothesis is rejected if the confidence interval for λ_x does not contain the parameter λ.

Tolerance intervals of the Poisson distribution are given in Hahn and Chandra (1981).

▶ 1.6.5 The Thorndike nomogram

This nomogram (Figure 30) provides a means of graphically determining the consecutively added probabilities $e^{-\lambda}\lambda^x/x!$ of the Poisson distribution (Thorndike 1926). Values of λ are marked on the abscissa, and a sequence of curves corresponding to the values $c = 1, 2, 3, \ldots$ runs obliquely across the graph. For various values of λ and c the probability that a variable X

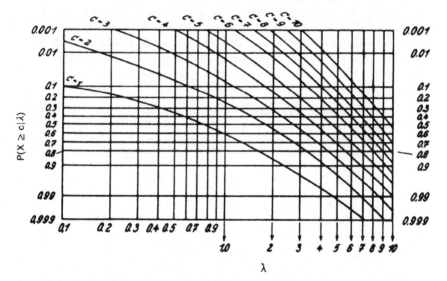

Figure 30 The Thorndike nomogram. **Ordinate:** $P(X \geq c|\lambda)$, the probability that an event occurs c or more times (at least c times). Note that in the nomogram P increases from top to bottom. **Abscissa:** The average frequency λ of occurrence in a large number of trials. The scale is logarithmic. **Curves:** For fixed c the probability $P = \sum_{c+1}^{\infty} e^{-\lambda}\lambda^k/k!$ ($= c! \int_0^{\lambda} e^{-x} x^c dx$) is a (uniquely determined) function of λ; P increases with λ; for given λ, P with increasing c.

is greater than or equal to some c, $P(X \geq c_0 | \lambda_0)$, can be read from the ordinate as follows:

1. Draw the vertical line $\lambda = \lambda_0$ (i.e., through the point $\lambda = \lambda_0$ on the abscissa) to intersect the curve c_0.
2. The ordinate of this point of intersection indicates the probability $P(X \geq c_0 | \lambda_0)$.

Examples

1. A machine produces about 1% rejects. What is the probability that there are at least 6 rejects among 200 items produced?

$p = 0.01$; $n = 200$; $\hat{\lambda} = np = (200)(0.01) = 2$. The ordinate of the point of intersection of the vertical line $\lambda = 2$ and the curve $c = 6$ is $P(X \geq 6) \simeq 0.015$. Thus the probability of finding at least 6 rejects is about 0.015 or 1.5%.

2. An egg wholesaler wants to have not more than 0.5% of all his egg cartons with four or more spoiled eggs. How low must the average percentage of bad eggs be for this quality to be assured? We assume a carton represents a random sample of 250 eggs.

The Thorndike nomogram must be read in a manner "reverse" to that of Example 1. The probability of getting four or more spoiled eggs in a random sample of 250 eggs should not be greater than 0.005. Thus we have $P(X \geq 4) = 0.005$. The average allowed number λ of bad eggs per carton can now be found. The horizontal line extending to the left of 0.005 intersects the curve $c = 4$. The vertical through the point of intersection passes through $\hat{\lambda} \simeq 0.67$. The desired percentage $\hat{p}$ of spoiled eggs which is not to be exceeded is then given by $\hat{\lambda} = n\hat{p}$ or $\hat{p} = \hat{\lambda}/n \simeq 0.67/250 = 0.00268$ or 0.27%, i.e., about 3 per thousand.

3. A hundred light bulbs are delivered together in a carton. The average percentage of defective units is around $p = 1\%$. The probability that a shipment of 100 bulbs contains two or more defective bulbs is to be determined.

Table 39

Light bulbs—number of rejects per 100	Poisson probability
0	0.3679
1	0.3679
2	0.1840
3	0.0613
4	0.0153
5	0.0031
≥ 6	0.0005
	1.0000

1.6 Discrete Distributions

We find the point of intersection of the line $\lambda = 1$ with the curve $c = 2$ and read on the left the ordinate 0.26. Thus out of 100 cartons with 100 bulbs in each, about 26 cartons will contain two or more defective light bulbs. The result by ordinary computation would be $P(X \geq 2; \lambda = 1) = 1 - (P(x = 0|\lambda = 1) + P(x = 1|\lambda = 1)) = 1 - (0.3679 + 0.3679) = 0.2642$. The nomogram can also be used in a similar manner to determine other quantities as e.g., $P(X = 2|\lambda = 1) = P(X \geq 2|\lambda = 1) - P(X \geq 3|\lambda = 1) \simeq 0.26 - 0.08 \simeq 0.18$ (see Table 39).

When extensive calculations are involved tables of the Poisson distribution are usually preferred to the nomogram (cf., Section 1.6.4). The probability **for the occurrence of at least x_0 rare events** is

$$P(X \geq x_0) = 1 - P(\chi^2_{2x_0} \leq 2np). \tag{1.182}$$

We take the last example: $x_0 = 2$, $np = (100)(0.01) = 1$:

$$P(X \geq 2|\hat{\lambda} = 1) = 1 - P(\chi^2_4 \leq 2).$$

Table 28 in Section 1.5.2 gives $P(\chi^2_4 = 2) = 0.73$, i.e.,

$$P(X \geq 2|\hat{\lambda} = 1) \simeq 1 - 0.73 \simeq 0.27.$$

As an exercise, this quick estimate should also be worked out for the other examples.

With the help of (1.177) a graphical test can again be carried out (cf., Section 1.6.2): $P(x)/P(x + 1)$ is plotted against x, and if the points are found to lie on a straight line, then the quantities follow a Poisson distribution (Dubey 1966) (cf., also Ord 1967 and Grimm 1970).

Approximations

An excellent survey is given by Molenaar (1970).

1 Approximating the Binomial Distribution by the Poisson Distribution

Any binomial distribution with large sample size n and small event probability p, so that $q = 1 - p$ practically equals 1 ($p < 0.05$ and $n > 10$, say) can be approximated by the Poisson distribution with $\lambda = np$.

EXAMPLE. In a certain region one house per year out of 2,000 is, on the average, damaged by fire. If there are 4,000 houses in this region, what is the probability that in the course of a year there will be a fire in exactly 5 houses?

$$\hat{\lambda} = np = 4{,}000 \cdot \frac{1}{2{,}000} = 2$$

$$P(X = 5|\hat{\lambda} = 2) = e^{-2} \cdot \frac{2^5}{5!} = 0.036.$$

The probability comes to almost 4%.

2 Approximating the Poisson Distribution by the Normal Distribution

The cumulative Poisson distribution $P(X \leq k|\lambda) = \sum_{j=0}^{k} e^{-\lambda}\lambda^j/j!$ can be approximated according to (1.183) and substantially better according to (1.183a) (Molenaar 1970).

For $\lambda \geq 9$,

$$\hat{z} = |(k - \lambda)/\sqrt{\lambda}|. \qquad (1.183)$$

Examples

1. For $P(X \leq 3|9)$ with $\hat{z} = |(3 - 9)/\sqrt{9}| = 2.000$ we get $P = 0.0228$ (exact value: 0.021226).
2. For $P(X \leq 4|10)$ with $\hat{z} = |(4 - 10)/\sqrt{10}| = 1.897$ we get $P = 0.0289$ (exact value: 0.029253).

For $\lambda \simeq 0.5$,

$$\hat{z} = \left|2\left[\sqrt{k + \frac{t+4}{9}}\right] - 2\left[\sqrt{\lambda + \frac{t-8}{36}}\right]\right| \qquad (1.183a)$$

with $t = (k - \lambda + \tfrac{1}{6})^2/\lambda$.

Example 2, above:

$$t = \frac{(4 - 10 + 1/6)^2}{10} = 3.403$$

$$\hat{z} = \left|2\left[\sqrt{4 + \frac{7.403}{9}}\right] - 2\left[\sqrt{10 - \frac{4.597}{36}}\right]\right| = 1.892, \text{ i.e., } P = 0.0293.$$

1.6.6 Comparison of means of Poisson distributions

1.6.6.1 Comparison of two Poisson distributions

Two Poisson distributions can be compared without any computation with the help of Table 36, pp. (79, 80) 209 in Biometrika Tables by Pearson and Hartley (1966). Two Poisson variables, X_1 and X_2, (with $X_1 > X_2$) can be tested according to

$$\hat{F} = \frac{X_1}{X_2 + 1} \qquad (1.184)$$

($DF = 2(X_2 + 1); 2X_1$), and the null hypothesis ($\lambda_1 = \lambda_2$) can be confronted with the one sided ($\lambda_1 > \lambda_2$) or the two sided ($\lambda_1 \neq \lambda_2$) question. The null

1.6 Discrete Distributions

hypothesis is rejected whenever $\hat{F}$ equals or exceeds the tabulated F-value. We note that the F-values are tabulated for the one sided question.

EXAMPLE. Given $x_1 = 13$ and $x_2 = 4$, test whether the null hypothesis $\lambda_1 = \lambda_2$ can be defended against the alternate hypothesis $\lambda_1 \neq \lambda_2$ ($\alpha = 0.05$). We have

$$\hat{F} = \frac{13}{4+1} = 2.60.$$

Since $2.60 > 2.59 = F_{10;\,26;\,0.025}$, the null hypothesis can still be rejected. [For the one sided question (cf., Section 1.4.6) $\lambda_1 > \lambda_2$ against $\lambda_1 = \lambda_2$ with $F_{10;\,26;\,0.05} = 2.22$, the difference of the λ's can be better guaranteed.]

Comparisons of this sort also go through very well in terms of the standard normal variables for λ not too small ($X_1 + X_2 > 5$):

$$\hat{z} = \frac{X_1 - X_2 - 1}{\sqrt{X_1 + X_2}}. \qquad (1.185)$$

For $X_1 + X_2 > 20$, the following form is preferable:

$$\hat{z} = \frac{X_1 - X_2}{\sqrt{X_1 + X_2}}. \qquad (1.185a)$$

EXAMPLE. We use the last example: $\hat{z} = (13 - 4 - 1)/\sqrt{13 + 4} = 1.940 < 1.960 = z_{0.05;\,\text{twos.}}$. Thus H_0 may not be rejected.

Remark on the Comparison of Two Samples of Relatively Infrequent Events in Time

If x_1 and x_2 are the numbers of occurrences of rare events E_1 and E_2 in time intervals of length t_1 and t_2 respectively, then the null hypothesis (equality of relative frequencies or, better, of probabilities) can be approximately tested by

$$\hat{F} = \frac{t_1(x_2 + 0.5)}{t_2(x_1 + 0.5)} \qquad (1.186)$$

with $(2x_1 + 1, 2x_2 + 1)$ degrees of freedom (Cox 1953).

EXAMPLE. Given:

$$x_1 = 4 \text{ events in } t_1 = 205 \text{ hours},$$
$$x_2 = 12 \text{ events in } t_2 = 180 \text{ hours}.$$

Hypothesis to be tested: Equality of the probabilities (two sided question: $\alpha = 0.05$ [i.e., the upper 2.5% bounds of the F-distribution are to be used]). We find

$$\hat{F} = \frac{205(12 + 0.5)}{180(4 + 0.5)} = 3.16.$$

Since $3.16 > 2.68 = F_{9; 25; 0.025}$, the null hypothesis is rejected.

For the comparison of two relative frequencies $(x_1/n_1 = \hat{p}_1; x_2/n_2 = \hat{p}_2)$ that arise from a binomial $(\hat{p}_1, \hat{p}_2 > 0.05)$, or a Poisson distribution $(\hat{p}_1, \hat{p}_2 \leq 0.05)$ a nomogram given by Johnson (1959) can be used, which allows for an elegant approximate answer to the question of whether $\hat{p}_1$ and $\hat{p}_2$ originate from a common population.

1.6.6.2 Comparison of several Poisson distributions

Comparison of the Expected Number of Events in Several Samples from Poisson Populations. The test of homogeneity on pages 474, 477 is especially useful

If X_i are stochastically independent observations from the same normally distributed population (μ, σ), then the sum of the squared standard deviations,

$$\sum_{i=1}^{v} \left(\frac{X_i - \mu}{\sigma} \right)^2 = \sum_{i=1}^{v} Z_i^2 = \chi_v^2, \qquad (1.187)$$

is χ^2 distributed with v degrees of freedom. For the comparison of k samples $(k \geq 2)$ from arbitrary unit intervals of observation t_i (unit intervals of time, area or volume) in which the event occurs x_i times, one forms $x_i/t_i = \lambda_i^*$ and $(\sum x_i)/(\sum t_i) = \hat{\lambda}$, transforms the x_i according to

$$z_i = 2(\sqrt{x_i + 1} - \sqrt{t_i \hat{\lambda}}) \quad \text{if } \lambda_i^* < \hat{\lambda},$$

$$z_i = 2(\sqrt{x_i} - \sqrt{t_i \hat{\lambda}}) \quad \text{if } \lambda_i^* > \hat{\lambda},$$

and sums the squares of the resulting quantities $\sum z_i^2$. Testing is done in accordance with

$$\hat{\chi}^2 = \sum_{i=1}^{k} z_i^2 \qquad (1.188)$$

(p.141) for $k - 1$ degrees of freedom (one degree of freedom is "lost" in the estimation of the parameter λ; if it is known, there are k degrees of freedom at our disposal).

1.6 Discrete Distributions

EXAMPLE. In order to apply the test to the last example we calculate

$$\lambda_1^* = \frac{4}{205} = 19.51 \cdot 10^{-3},$$

$$\lambda_2^* = \frac{12}{180} = 66.67 \cdot 10^{-3},$$

$$\hat{\lambda} = \frac{4 + 12}{205 + 180} = 41.558 \cdot 10^{-3},$$

$$z_1 = 2(\sqrt{4 + 1} - \sqrt{205 \cdot 41.558 \cdot 10^{-3}}) = -1.366,$$

$$z_2 = 2(\sqrt{12} - \sqrt{180 \cdot 41.558 \cdot 10^{-3}}) = 1.458,$$

$$z_1^2 + z_2^2 = 1.866 + 2.126 = 3.992.$$

Since $3.99 > 3.84 = \chi^2_{1;0.05}$, the null hypothesis is rejected here also.

If the comparison involves only two means, the formula (1.184) is of course used.

1.6.7 The dispersion index

Let us emphasize again: if an empirical distribution is to be described by a Poisson distribution, then the data must satisfy the following conditions:

1. The events under consideration are independent.
2. The average number of events in an interval (of, e.g., time or space) is proportional to the length of the interval (and does not depend on the location of the interval).

If these conditions are satisfied only partially or not at all, then the class zero is often **larger** than can be expected on the basis of the Poisson distribution. If intervals from class zero are shifted to class one, the standard deviation of the distribution becomes smaller. Thus the quotient of the sample **standard deviation and the (estimated) standard deviation of a presumable Poisson distribution**, or more exactly the quotient (one sided question) of the two variances,

$$\frac{\text{sample variance}}{\text{theoretical Poisson variance}} = \frac{\text{sample variance}}{\text{theoretical Poisson mean}} = \frac{s^2}{\hat{\lambda}},$$

(1.189)

is likely to be larger than 1. When sample sizes are large, (1.189) equals the dispersion index. Since however the random samples considered have their own variability, we must answer the following question: How much larger than 1 must this quotient be before we can conclude that the "overdispersed" distribution could not be of the Poisson type? If the quotient is approximately equal to $\frac{9}{10}$ i.e., "underdispersed," a binomial distribution is more probable). approximated by a Poisson distribution (if the quotient is approximately equal to $\frac{9}{10}$ i.e., "underdispersed," a binomial distribution is more probable). The next example will give us an opportunity to apply this rule of thumb.

The **dispersion index** (cf. also Section 3.3) is used in testing whether the data (x_i) originated from a Poisson distribution (with mean λ) (cf., also Rao and Chakravarti 1956 as well as Gbur 1981):

$$\hat{\chi}^2 = \frac{\sum_i (x_i - \bar{x})^2}{\bar{x}} - \frac{n\sum_i x_i^2 - (\sum_i x_i)^2}{\sum_i x_i}, \qquad (1.190)$$

$$\hat{\chi}^2 = \frac{1}{\bar{x}} \sum_i f_i(x_i - \bar{x})^2,$$

with $n - 1$ degrees of freedom. If the empirically estimated value $\hat{\chi}^2$ exceeds the value tabulated (i.e., if the variance is substantially greater than the mean), then we are dealing with a **compound Poisson distribution**: When a rare event occurs at all, it is often immediately followed by several more. One then speaks of **positive probability contagion**. Days on which thunderstorms occur are rare; they occur however in bunches. For this situation the **negative binomial distribution** is better suited. The number of ticks per sheep in a herd is a perfect example. The distributions of other **biological** characteristics are often better approximated by one of the Neyman distributions. Detailed discussions can be found in the works of Neyman (1939), Fisher (1941, 1953), Bliss (1953, 1958), Gurland (1959), Bartko (1966, 1967), and Weber (1972) (cf., also Section 1.6.9). Important tables are given by Grimm (1962, 1964) and also by Williamson and Bretherton (1963).

EXAMPLE. The following is a classic example of a Poisson distribution: Table 40 shows recorded fatalities caused by horses' kicks among the soldiers in 10 army corps during a 20 year period (altogether 200 "army corps years" in the Prussian army 1875–1894). We have

Table 40

Fatalities	0	1	2	3	4	≥5	Σ
Observed	109	65	22	3	1	0	200
Calculated	108.7	66.3	20.2	4.1	0.6	0.1	200

1.6 Discrete Distributions

$$\bar{x} = \frac{\sum x_i f_i}{n} = \frac{0 \cdot 109 + 1 \cdot 65 + 2 \cdot 22 + 3 \cdot 3 + 4 \cdot 1 + 5 \cdot 0}{200} = \frac{122}{200} = 0.61;$$

$$s^2 = \frac{\sum x_i^2 f_i - (\sum x_i f_i)^2/n}{n-1}$$

$$s^2 = \frac{(0^2 \cdot 109 + 1^2 \cdot 65 + 2^2 \cdot 22 + 3^2 \cdot 3 + 4^2 \cdot 1) - 122^2/200}{200 - 1}$$

$$s^2 = \frac{196 - 74.42}{199} = \frac{121.58}{199} = 0.61.$$

We get, by (1.189),

$$\frac{s^2}{\lambda} = \frac{0.61}{0.61} = 1 < \frac{10}{9}$$

and by (1.190),

$$\hat{\chi}^2 = [109(0 - 0.61)^2 + 65(1 - 0.61)^2 + \cdots + 0(5 - 0.61)^2]/0.61$$

$$\hat{\chi}^2 = 199.3 < 233 = \chi^2_{199;\,0.05}.$$

The Poisson distribution, with $\lambda = 0.61$, is thus appropriate in describing the distribution considered. Usually the estimates s^2 and λ^2 will differ (even when the data come from a Poisson population). We obtain

$$P(0) = \frac{0.61^0 \cdot e^{-0.61}}{0!} = 0.5434; \quad 200 \cdot 0.5434 = 108.68 \quad \text{etc.}$$

The completion of Table 40 is recommended as an exercise. The relative frequencies of the probabilities of the Poisson distribution are given by the consecutive terms of the relation

$$\boxed{e^{-\lambda} \sum \frac{\lambda^x}{x!} = e^{-\lambda}\left(1 + \lambda + \frac{\lambda^2}{2!} + \frac{\lambda^3}{3!} + \ldots + \frac{\lambda^x}{x!}\right).} \quad (1.191)$$

The expected frequencies are obtained as products of the individual terms with the total sample size. For example, the expected frequency for the third term is thus found to be

$$ne^{-\lambda}\left(\frac{\lambda^2}{2!}\right) = 200 \cdot 0.54335 \cdot \frac{0.3721}{2} = 20.2 \quad \text{etc.}$$

If given empirical distributions exhibit similarity to Poisson distributions, then λ can be approximately estimated according to

$$\boxed{-\ln\left(\frac{\text{occupation of class zero}}{\text{total of all frequencies}}\right) = \hat{\lambda} = -\ln\left(\frac{n_0}{n}\right)} \quad (1.192)$$

provided class zero (no results) shows the greatest occupation.

Table 41

0	1	2	3	4	5	6	$\sum$
327	340	160	53	16	3	1	900

EXAMPLE. Consider the data in Table 41. A straightforward calculation gives

$$\hat{\lambda} = \frac{1}{900}(0 \cdot 327 + 1 \cdot 340 + \cdots + 6 \cdot 1) = \frac{904}{900} = 1.$$

More concisely,

$$\frac{n_0}{n} = \frac{327}{900} = 0.363 \quad \ln 0.363 = -1.013 \quad \text{so that} \quad \hat{\lambda} = 1.013 \doteq 1.$$

In terms of the base 10 logarithms, this is

$$\log 0.363 = 9.5599 - 10 = -0.4401,$$

$$2.3026 \cdot \log 0.363 = 2.3026(-0.4401) = -1.013.$$

Applying the "quick" method to the example on horse's kicks, we get the estimate

$$\hat{\lambda} = -\ln\left(\frac{109}{200}\right) = -\ln 0.545 = 0.60697,$$

an excellent result.

A homogeneity test that lets one determine the deviations in the occupation of class zero as well as of the other classes is discussed by Rao and Chakravarti (1956). Tables and examples can be found in the original work.

1.6.8 The multinomial coefficient

If n elements are arranged in k groups so that $n_1 + n_2 + \cdots + n_k = n$, where $n_1, n_2, \ldots, n_k$ indicate the number of elements per group, then there are

$$\boxed{\frac{n!}{n_1! \cdot n_2! \cdot \ldots \cdot n_k!}} \tag{1.193}$$

(p.160) different ways of grouping these elements into the k groups (multinomial coefficient).

Examples

1. Ten students are to be separated into two groups, each consisting of five basketball players. How many different teams can be formed?

$$\frac{10!}{5! \cdot 5!} = \frac{3{,}628{,}800}{120 \cdot 120} = 252.$$

2. A deck of 52 playing cards is to be distributed among 4 players so that each gets 13 cards. How many different ways are there of dividing the cards?

$$\frac{52!}{13! \cdot 13! \cdot 13! \cdot 13!} = \frac{8.0658 \cdot 10^{67}}{(6.2270 \cdot 10^9)^4} \simeq 5.36 \cdot 10^{28}.$$

1.6.9 The multinomial distribution

We know that if the probability of choosing a smoker is p while the probability of choosing a nonsmoker is $1 - p$, then the probability of choosing exactly x smokers in n attempts is given by

$$P(x|n, p) = \binom{n}{x} p^x (1-p)^{n-x}. \tag{1.158}$$

The rationale underlying (1.158) can be generalized to situations with more than two events, attributes, items, or classes. Denote by $E_1, E_2, \ldots, E_k$ mutually exclusive and exhaustive events or classes with probabilities $p_1, p_2, \ldots, p_k$, where $0 < p_i < 1$, $\sum_{i=1}^{k} p_i = 1$. The number p_i is the probability of any event being assigned to the ith class, it is the fraction of the total population belonging to the ith class. Then the probability that in a random sample of n independent observations, the event, attribute, or item E_i manifests itself exactly n_i times, $i = 1, 2, \ldots, k$, $\sum_{i=1}^{k} n_i = n$, is given by the multinomial probability

$$P(n_1, n_2, \ldots, n_k | p_1, p_2, \ldots, p_k | n) = \frac{n!}{n_1! \cdot n_2! \cdot \ldots \cdot n_k!} \cdot p_1^{n_1} p_2^{n_2} \cdots p_k^{n_k}. \tag{1.194}$$

Since the terms in the expansion of

$$(p_1 + p_2 + \cdots + p_k)^n = 1$$

are those given by formula (1.194), we call this distribution the multinomial distribution. We have

$$\mu_{E_i} = \mu_{n_i} = np_i, \tag{1.195}$$

$$\sigma_{E_i}^2 = \sigma_{n_i}^2 = np_i(1 - p_i). \tag{1.196}$$

For $k = 2$, formula (1.194) yields (1.158). (1.194) can also be derived from the generalized hypergeometric distribution (1.170) by fixing n and letting N grow.

Parameters of multinomial distributions are compared in Chapter 6 (testing of two way tables for homogeneity or independence).

Examples

1. A box contains 100 pearls, 50 of which are colored red, 30 green, and 20 black. What is the probability that of 6 arbitrarily chosen pearls, 3 are red, 2 green, and 1 black?

Since choice is followed by replacement in every case, the probabilities of choosing 1 red, 1 green, and 1 black pearl are respectively $p_1 = 0.5$, $p_2 = 0.3$, and $p_3 = 0.2$. The probability that a selection of 6 pearls has the aforementioned composition is given by

$$P = \frac{6!}{3! \cdot 2! \cdot 1!}(0.5)^3(0.3)^2(0.2)^1 = 0.135.$$

2. A fair die is tossed twelve times. The probability of the 1, the 2 and the 3 turning up once each and the 4, the 5, and the 6 three times each (note that $1 + 1 + 1 + 3 + 3 + 3 = 12$) is

$$P = \frac{12!}{1! \cdot 1! \cdot 1! \cdot 3! \cdot 3! \cdot 3!}\left(\frac{1}{6}\right)^1\left(\frac{1}{6}\right)^1\left(\frac{1}{6}\right)^1\left(\frac{1}{6}\right)^3\left(\frac{1}{6}\right)^3\left(\frac{1}{6}\right)^3 = 0.001.$$

3. Ten persons vote at random for one of three candidates (A, B, C). What is the probability of the choice: 8A, 1B, and 1C?

$$P = \frac{10!}{8! \cdot 1! \cdot 1!}\left(\frac{1}{3}\right)^8\left(\frac{1}{3}\right)^1\left(\frac{1}{3}\right)^1 = 90 \cdot \frac{1}{6{,}561} \cdot \frac{1}{3} \cdot \frac{1}{3} = 0.00152.$$

The most probable result would be 3A, 3B, 4C, (or 3A, 4B, 3C, or 4A, 3B, 3C) with

$$P = \frac{10!}{3! \cdot 3! \cdot 4!}\left(\frac{1}{3}\right)^3\left(\frac{1}{3}\right)^3\left(\frac{1}{3}\right)^4 = \frac{3{,}628{,}800}{6 \cdot 6 \cdot 24} \cdot \frac{1}{27} \cdot \frac{1}{27} \cdot \frac{1}{81} = \frac{4{,}200}{59{,}049}.$$

Thus $P = 0.07113$, i.e., this result will occur nearly 47 times more frequently than $P_{8A;\,1B;\,1C}$.

A graphical method of determining the sample sizes for confidence intervals of parameters of the multinomial distribution is given by Angers (1974).

> More particulars on discrete distributions can be found in Patil and Joshi (1968 [cited on p. 575]) as well as in Johnson and Kotz (1969 [cited on p. 570]).

2 STATISTICAL METHODS IN MEDICINE AND TECHNOLOGY

> If in the analysis of survival times in medicine or technology some objects are still alive at the end of the study their exact survival times are incomplete. These are called **censored observations** or censored times. More on this and on the **comparison of survival distributions**—see also pages 206, 210 and 235—is provided in the book by Lee (1980) with computer programs for 5 two sample tests and a k sample test [Chapter 5 and Appendix B, with both Peto and Peto's tests: logrank test and generalized Wilcoxon test].

2.1 MEDICAL STATISTICS

The number of hours of sleep gained by means of a soporific (sleep-inducing preparation) will generally vary from person to person. With the help of statistics we would like to make a statement on the average gain in the number of hours of sleep. We must also test whether the gain in the duration of sleep is statistically significant. Analyses of this type require not only knowledge of statistical methods but also **a thorough familiarity with the field of study,** because to determine the unique effects of specified causes we must be able to sort out the more important factors contributing to the phenomenon examined. These factors can be of a psychological or physical nature. In our example confidence in the medication and in the physician, as well as the physician's attitude, are factors in the first category; charges in the diet, and in the daily routine belong to the second. To eliminate two influences of the first type, neither the physician assessing the therapy result nor the patient must know whether a soporific or a placebo is administered. This type of study is called a double blind trial.

Guidelines for medical doctors on the **ethical aspects** of clinical research are the internationally accepted Declarations of Helsinki 1964 and Tokyo 1975 [cf., World Medical Journal **22** (1975), 87–90 and **25** (1978), 58–59].

Another point concerns the following: Suppose the original problem is replaced by the question on the effects particular conditions produce in certain attributes of a given set of objects. The exact state of an attribute is replaced by the observed state; the observations are expressed by symbols. **Errors of substitution** can occur at each point of transition. For many important substitutions the attributes are not closely related to the problem, and are accordingly not very informative. An attribute is informative if it is highly correlated with the parameter under study.

All objects must come from the same, well-defined population by random selection. Measurements or the presence or absence of attributes, as well as complementary data (e.g., physical length if weight is of interest), are recorded (including the value or reading zero). All attributes have to be defined and recorded. Special circumstances like: not checked, doubtful whether checked, checked but not verified, not applicable also have to be recorded.

During the last decades in particular, statistics has been recognized as a valuable tool in the gaining of knowledge in clinical medicine. Statistics provides not only in clinical medicine but in most sciences to a certain extent a **filter** through which new developments must pass before they are recognized and applied on a wider scale.

The statistical and mathematical techniques tailored to problems in the biological sciences, the social sciences, economics, psychology, technology, the scientific literature, and the science of the sciences are called biometrics or biometry, sociometrics, econometrics, psychometrics, technometrics, bibliometry, and scientometry respectively.

2.1.1 Critique of the source material

Sampling errors are due to the fact that only samples are observed and not populations. Errors in sample estimates that cannot be attributed to sampling fluctuations are called **nonsampling errors**. Such errors may arise from different sources. We know the systematic error or bias, an effect that deprives a statistical result of representativeness by systematically distorting it, as distinct from a random error, which may distort on any one occasion but balances out (is self-canceling) on the average. Nonsampling errors are not infrequent in surveys of human populations (e.g., interview bias, a dishonesty effect; cf. Section 2.1.3).

> If a given quantity is measured with an improperly calibrated instrument, the measurement carries a bias, i.e., a systematic error (cf., Section 1.3.5) in addition to a random error. In laboratory settings both errors are monitored by **quality control** [(cf., Section 2.1.2 and 2.4.1; also Clinical Chemistry **22** (1976), 532–540, **24** (1978), 1213–1220, **27** (1981), 798–805, 1536–1545 and **29** (1983), 581] and are reduced by improving the measuring techniques (cf., Section 2.1.2).

Nonsampling errors in survey data may arise through defects in the selection of sample units; double, incomplete, or suppressed recording; contradictory, unqualified, or deliberately false statements; misunderstandings due to ambiguity in the phrasing of questions; informant fatigue, resulting in yea-saying and in using noncommittal midscale ratings; order effects; gaps in memory; and clerical errors. Other pertinent errors may be traced to deficiencies in the formulation of the problem, the guidelines of the survey, the monitoring of the protocol and definitions [e.g., with regard to the population, or the experimental unit (cf., end of Section 1.3.8.7), as well as the identification and influence of target quantities and the possible sources of error in them], the questionnaires, the interviews, the consultant (cf., Section 1.2.4, (Example 3) and Section 2.1.5; also Landis and Koch 1975 [6]), and the processing and tabulating of data. The source material (cf., Section 1.3.3) must in any event be tested for completeness, consistency, and reliability [cf., also Sections 1.2.1, 1.3.2, 1.3.7 (Remark), and 1.3.8.7 (Remark 1), the end of Section 2.2, and the beginning of Chapter 3]. Some further comments are given in Section 2.1.3. This subject crops up continually in medical statistics. Sonquist and Dùnkelberg (1977 [8 : 1]) is indispensable for surveys.

Detailed discussions of the automatic detection and correction of errors are given by Minton (1969, 1970 [8:3]) and by Szameitat and Deininger (1969, [8:3]). For the analysis of surveys, see Yates (1973), Yates (1981 [8:3a]), books on sampling (cited in [8:3a] and 8:1]) and texts in population statistics: Benjamin (1968), Bogue (1969), Cox (1970), Pressat (1972) [8:1], and Keyfitz and Beekman (1983). For other aspects of medical statistics see Cochran (1965, 1968), Burdette and Gehan (1970), Brown (1970/71); also Ryan and Fisher (1974). A survey on nonsampling errors is given by F. Mosteller [in Kruskal and Tanur I (1978) [8:1]]. A bibliography on nonsampling errors in surveys is given by T. Dalenius [International Statistical Review **45** (1977), 71–89, 181–197, 303–317] (see also Strecker 1980 [8 : 3a], cited on page 615). Important warning signals for analytical chemists are provided by Youden (1959/67), and Caulcutt and Boddy (1983).

2.1.2 The reliability of laboratory methods

It is of great importance in medical sciences to know how reliable studies are carried out in the clinical laboratory. The determination of whether or not a result is pathological is based on a thorough knowledge of the reliability of the analytical methods used in the laboratory on the one hand and on a thorough knowledge of the **reference values** on the other [cf., also Clinical Chemistry **24** (1978), 640–651, 772–777 and **28** (1982), 259–265, 422–426, 1432–1433; Castleman et al. 1970, Eilers 1970, Elveback et al. 1970, Williams et al. 1970, Reed et al. 1971, Rümke and Bezemer 1972].

Since the clinically normal values of healthy individuals are usually not normally distributed, the **distribution-free 90% confidence intervals for the quantiles** $\xi_{0.025}$ **and** $\xi_{0.975}$ should be listed (cf., Sections 1.3.8.3 and 3.1.4). Tables are provided by Reed et al. (1971) as well as by Rümke and Bezemer (1972). For instance, the 90% CI for $\xi_{0.025}$ for $n = 120$ (150,300) lies between the values with ranks 1 and 7 (1 and 8, 3 and 13): those for $\xi_{0.975}$ lie between the values with ranks 114 and 120 (143 and 150, 288 and 298).

Thus for the 90% CI, first value $\leq \xi_{0.025} \leq$ 7th value, 114th value $\leq \xi_{0.975} \leq$ 120th value (for $n = 150$ and $n = 300$ respectively).

The reliability of a method of investigation is hard to define, since it is determined by a number of factors the importance of which depends on the medical goal and the one diagnostic value of a particular method. The most important **reliability criteria** are:
1. **Specificity**: characterization of a chemical substance to the exclusion of any other substance (qualitative description).
2. **Accuracy**: determination of the precise amount of the chemical present in the material under study (with due regard for systematic errors). The accuracy can be checked by $(\bar{x} - \mu)/\mu$ with $\mu =$ known true value and $\bar{x} =$ sample mean, and by three simple procedures:
   a. **Comparison tests**: the result of the analysis is compared with the one obtained by another method, possibly one whose reliability is established, or with the results furnished by a series of interlaboratory comparisons or collaborative tests.
   b. **Addition tests**: known quantities of the chemical examined are added to the experimental material.
   c. **Mixture tests**: a serum or urine with a high concentration of the chemical under study and another body fluid with a correspondingly low concentration are mixed in various ratios.
3. **Precision** or reproducibility: The error inherent in the method of analysis due to, e.g., different reagents, different laboratory technicians, different laboratories, different days (e.g., weather, day of the week) [cf., also Clinical Chemistry **24** (1978), 212–222, 1126–1130, 1895–1899, **27** (1981), 202] can be assessed by the standard deviation and the coefficient of variation. If the latter is greater than say 0.05, then double or even triple determinations are necessary. In the case of triple determinations values that seem to be out of line should in general not be discarded, because valuable information—accuracy—might be lost. Large differences between the readings are not at all rare (cf., also Section 3.8). Youden (1962) has described, for **normally distributed** data, how the true mean (μ) and the corresponding confidence intervals can be estimated from double determinations (let x_1 denote the smaller reading, $x_1 \leq x_2$) and triple determinations ($x_1 \leq x_2 \leq x_3$):

(1) μ lies with
   (a) $P = 50\%$ in the interval: $x_1 \leq \mu \leq x_2$, and with
   (b) $P = 75\%$ in the interval: $x_1 \leq \mu \leq x_3$.
(2) The approximate confidence intervals are
   (a) 80% CI: $x_1 - (x_2 - x_1) \leq \mu \leq x_2 + (x_2 - x_1)$ and
   (b) 95% CI: $x_1 - (x_3 - x_1) \leq \mu \leq x_3 + (x_3 - x_1)$.

If the values are at least approximately normally distributed, then, according to McFarren et al. (1970), the overall error G ($=$ random error +

systematic error) of the method of analysis can be given as a percentage of the mean by

$$G = \left[\frac{|\bar{x} - \mu| + 2s}{\mu} \right] 100 \qquad (2.1a)$$

(μ = known true value; $\bar{x}$ and s are computed from not too small a sample). If $G > 50\%$ the procedure is hardly of any value; it is very good if $G < 25\%$.

Example: $\mu = 0.52$, $\bar{x} = 0.50$, $s = 0.05$:

$$G = \left[\frac{|0.50 - 0.52| + 2 \cdot 0.05}{0.52} \right] 100 = 23\%.$$

4. **Sensitivity:** The smallest recognized departure or single result that is statistically significant on the chosen significance level and that can be distinguished from a suitable blank, can be used as a measure for the sensitivity of a method [a better approach might be to use the slope, the regression coefficient, of the standard line].

We assume that blank–sample results and results for samples with discernible values near the blank–sample range are approximately normally distributed with different means and the same variance σ^2, estimated by $s_{corr}^2 = s_D^2 + s_B^2$ with s_D^2 = variance of the discernible values and s_B^2 = variance of the blanks, provided we have sample sizes $n_D \gtrsim 25$ and $n_B \gtrsim 25$. Then the one-sided least significant difference or **detection limit** L of the method is approximated, in the case of $\alpha = 0.05$ and $\beta = 0.05$ (risk I = risk II) and with the arithmetic mean of the blank results $\bar{x}_B$, by

$$L = \bar{x}_B + 2 \cdot 1.645 \sqrt{s_D^2 + s_B^2} = \bar{x}_B + 3.3 \sqrt{s_D^2 + s_B^2}. \qquad (2.1b)$$

Another possibility is to choose the proper one sided statistical test with the appropriate significance level to determine the least significant difference L. More on this can be found as needed in Wilson (1961), Roos (1962), Svoboda and Gerbatsch (1968), Gabriels (1970). To compare two or more methods, the sensitivity ratio of Mandel and Stiehler (1954) (see Mandel 1964) can be employed (cf., also below).

5. **Practical long-range considerations.** Among these are: difficulties in carrying out the experiments, equipment expenditures (e.g., for an autoanalyzer), amount of time required, and other costs. Accuracy and reproducibility are the most important notions in assessing the reliability of measurements. In addition to the standard deviation, which measures the reproducibility, a rough estimate of the systematic error (the bias) should be given.

For this purpose experience is very important. In practice a method that leads to readings with small systematic deviation from the true value and higher precision is preferable to one that yields unbiased values with lower precision; in other words, a result with a small bias and little variability is unquestionably superior to one obtained by a method that furnishes the true value "on the average" but is subject to greater dispersions. We must after all usually content ourselves with few measurements (cf., also Cochran 1968).

More on the **reliability of measurements** can be found in Eisenhart (1963) and in B. A. Barry: Errors in Practical Measurement in Science, Engineering and Technology. Wiley, New York, 1978. Discussing the importance of control serums [cf., Clinical Chemistry **22** (1976), 500–512] or the **comparison of precision and accuracy of a method in various laboratories** is unfortunately beyond the scope of this book. Along with the work of Mandel and Lashof (1959), the publications by Youden (1959–1967) are strongly recommended (cf., also Chun 1966, Kramer 1967 and D. M. Rocke 1983 [Biometrika **70**, 421–431]).

A **comparison of quantitative methods** can be made in accordance with Barnett and Youden (1970) [cf., also Mandel and Stiehler 1954 as well as Clinical Chemistry **20** (1974), 825–833] and with Lawton et al. (1979).

The laboratory control chart

The continuous monitoring of reliability, in particular of the precision of a method of analysis is carried out graphically by using a so-called *control chart*, a graphical chart with control limits and plotted values of some statistical measure, here the mean $\bar{x}$, for a series of subgroups or samples. A central line is commonly shown. A standard sample, one of known content, is analyzed at least 40 times, and the frequency distribution of the resulting data is then drawn. If the graph resembles the Gaussian curve (normal distribution), one may then design a control chart based on the estimated values $\bar{x}$ and s. If there is no maximum or if several maxima occur, then the method is not yet under control.

Following the pattern in Figure 31, limit lines are drawn on a sheet of graph paper (abscissa: days, ordinate: data) at distances $\pm s$ and $\pm 2s$ from the mean $\bar{x}$. Now we know that in the case of a normal distribution at least

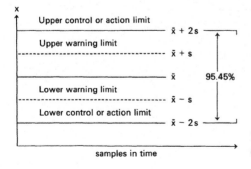

Figure 31 The mean control chart ($\bar{X}$-chart).

2.1 Medical Statistics

68% of all observations will lie between $\bar{x} \pm s$ and at least 95% of all values between $\bar{x} \pm 2s$. Thus we expect that with daily control analyses, out of 100 exact determinations about 32 lie outside $\pm s$ and about 5 outside $\pm 2s$ (cf., Figure 32). If noticeably more than 32% and 5% of the observations fall outside the $\bar{x} \pm s$ and $\bar{x} \pm 2s$ bands, respectively, then every step of the method has to be scrutinized. If the plotted points are not scattered randomly about the mean line but rather form a systematic pattern (e.g., a rising or falling straight line, a sine curve), a time-dependent systematic error might be involved. Should 7 or more consecutive data points lie on the same side of the mean line (cf., also Reynolds 1971) then we may well suspect a systematic error.

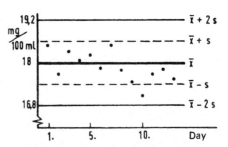

Figure 32 Readings from control experiments carried out daily (or at other regular time intervals) on a potassium standard [schematic; potassium: 18 mg/100 ml (= 18 mg/dl = 180 mg/l) = 0.2557(18) = 4.603 mval/l = 4.603 mmol/l; for the nomenclature associated with medicinal laboratory data see R. Dybkaer, Fed. Proc. **34** (1975), 2116–2122].

The **three sigma limit** ($\bar{x} \pm 3s$) is also considered. As a conservative control limit, it is not prone to cause a false alarm.

In addition to this vital **mean chart** ($\bar{X}$-chart; see Section 2.4.1.1 and Journal of Quality Technology **8** (1976), 183–188, **9** (1977), 166–171, **10** (1978), 20–30 and **12** (1980), 75–87) for controlling accuracy, the **range chart** (R-chart) serves to control the precision by means of double determinations, and the so-called **cumulative sum chart** helps us to recognize systematic deviations early and reliably. The early recognition of a **trend** is exceedingly important for the control of continuous processes: One determines on i consecutive days ($i = 1, 2, 3, \ldots, r$) the difference x_i of the analyzed values of a standard solution from the true concentration, the target value k, and plots the cumulatively summed amounts

$$S_r = \sum_{i=1}^{r}(x_i - k) \tag{2.2}$$

in a diagram like Figure 32: The days are marked on the abscissa and the S_r values on the ordinate—above the abscissa if S_r is positive, below if S_r is negative. In contrast with the usual control charts (e.g., Duncan 1974), the limit lines parallel to the abscissa are missing.

As long as the method of analysis is under control, the cumulative sums will lie on an approximately horizontal line. If the slope of the line becomes markedly different from zero, then the longer the slope does not decrease in absolute value, the more we suspect that the method of analysis is no longer in control. To test whether the slope of the curve exceeds a limiting value, one uses a V-mask whose construction is explained by J. M. Lucas (1976, Journal of Quality Technology **8**, 1–12), Barnard (1959), Kemp (1961), Ewan (1963), and Johnson and Leone (1964) as well as by Woodward and Goldsmith (1964). Further interesting applications of this principle in the context of quality control are given by Page (1963) (cf., also Taylor 1968; Burr 1967, Vessereau 1970) and especially by Dobben de Bruyn (1968) and Bissell (1969), both with important remarks.

Another important control chart for mean value control is described by Reynolds (1971).

2.1.3 How to get unbiased information and how to investigate associations

A sample **survey** is usually an examination of human beings to study human populations with regard to special variables. The objectives are **DESCRIPTION, UNDERSTANDING, EXPLANATION,** and **PREDICTION**. The population, the variables, their level of measurement (cf., end of Section 1.4.8), and the units of measurement must be defined. This holds true for instance for the methods of data collection by interview or questionnaire. **Inaccurate answers** may be attributable to defects of this data collection process: **misunderstanding of a question** (due to its wording, its length, embarrassment, or difficulty) or **dishonesty** (due to the respondent's desire to raise his prestige or to please the interviewer; dishonesty on the part of the investigator is likewise possible). These **response errors** are nonsampling errors. In order to estimate this bias, it is necessary to have information from outside the survey.

In medicine random samples are extremely hard to get. If the investigator is trying to make inferences from his sample he must be very careful to avoid common errors and fallacies (cf., Section 2.1.1 and Sachs 1977). Therefore statisticians are usually unwilling to generalize in medicine.

2.1.3.1 Check points

Results that can be confidently utilized and reliably analyzed are only available after careful consideration and reflection with appropriate attention to all aspects of the design of the research project, taking into account the subtleties and the realities of clinical science. It is useful to check certain features of the observations.

Avoiding nonrepresentativeness and selection bias

Avoiding Nonrepresentativeness

Are all cases with and without the disease representative samples of their respective populations? Did we succeed in avoiding under- and overrepresentation of special parts of the population? Several **different control groups** and/or control groups with different diseases should be used. Are the risks of developing the disease different? Do the volunteers differ in some ways from the nonvolunteers? In prospective studies the percentage of individuals lost to followup—withdrawals for any reason, death included—should not to be greater than 5%. Some further important questions: **Are there other differences between the two groups?** Are there perhaps different frequencies in the characteristics of interest?

Consider a group of people who are being studied; their knowledge of this will change their behavior. Especially disturbing is noncooperation by individuals, called **nonresponse**: either the individuals are unable or unwilling to give the requested information, or they are hard to find (new address unknown). These individuals are sometimes a divergent part of the population. More intensive efforts are necessary to get at least a small sample of the nonrespondents to answer the question: **in which relevant features do the nonrespondents differ from the respondents?** Surveys with high nonresponse rates are particularly prone to systematic errors and are almost useless. Compared with sampling errors and response errors, nonresponse is very hard to overcome.

Avoiding the Selection Bias

Possible nonassociations or associations between diseases or between a characteristic and a disease are difficult to establish. Differences in hospital admission rates or probabilities may artificially build up a spurious association or may conceal an actually existing association in the population. Any characteristic that increases the probability that a diseased individual will be hospitalized may mistakenly be found to be associated with the disease. Elderly people, with a high death rate, by moving to a place renowned as healthy, may raise the local death rate, thus causing the place to appear unhealthy (migration effect). Concerning the sampling process (the selection of the sample unit), it is important that the pertinent actual selection **probabilities**:

1. **should not change** as the survey progresses and
2. **should not be correlated** with measurable characteristics of the units—**with measurements on the units**.

The best way to seek out possible selection biases is to **compare the sample with external data sources** (e.g., true population sex ratio).

Available medical records are often nonrepresentative and afflicted with selection bias. Incomplete, and with time-linked shift of emphasis, they are seldom able to answer special scientific questions raised now, since the documents were not developed to solve the problem at hand. Which findings, facts, medical evidence, though important for the possible solution of the problem, had not been documented?

2.1.3.2 Checking the quality of the data and the size of the sample (cf. Section 2.1.1)

Why are the data wanted? Is there a reconciliation of population specifications? What is known about the data: their source, reliability, methods of measurement, and units? Are the data independent?

The Condition of Independence

Variables that are unrelated in a probabilistic sense are called stochastically independent variables or independent variables. In this sense we may say two characteristics are independent (cf., Sections 1.3.1 [last parts] and 4.1). The tobacco companies claim this for smoking and lung cancer. Different observations of the same kind on the same person are not independent. Observations on parents and on their children may be independent for some characteristics and not independent for others, such as the sex-linked inheritance of X-chromosome-borne hemophilia, or longevity, which is partly determined by heredity. The susceptibility to hypertension and to coronary artery disease is genetically conditioned.

Are the data perhaps from a random sample (cf., Sections 1.3.2 and 7.7)? Is the empirical distribution (histogram) unimodal, left-steep, symmetric, right-steep, or multimodal?

The heterogeneity of a population should be taken into account. The use of only one property, such as the count of white blood cells, size of tumor, or survival time, should be avoided. Other clinically relevant variables may be important, such as pain, the treatment's side effects, and such aspects of the patient's quality of life as relief of symptoms, return of appetite, and ability to work.

2.1.3.3 Surveys to investigate associations

Cohort studies

In cohort studies, the effects of time on individuals are studied, such as weight gain after birth; other examples involve a disturbance such as exposure to a noxious agent or to risk factors, the onset of a disease, or the administration of a treatment, the observed effect or the occurrence of the alleged effect such as the appearance of a disease, the disappearance of symptoms, or the ability to return to work.

People assembled in a cohort should be representative of the group of individuals to whom the results will be extrapolated. This concerns e.g., age, sex, clinical condition, ethnical background, occupation, and other characteristics.

Sometimes two cohorts are assembled, of which one is exposed to a certain risk (e.g., of lung cancer possibly resulting from asbestos dust and/or cigarette smoking, or of thrombophlebitis possibly resulting from oral contraceptive pills), and the other remains unexposed (no asbestos dust and/or no smoking; no pills). **The aim is to determine whether a particular disease develops preferentially in the cohort at risk.** Another technique is to subdivide one cohort consisting of prognostically homogeneous patients by random allocation into groups that receive different treatments, the effects of which are compared. It should be clear that in both comparisons the conditions at the start and afterward must not differ, except that one cohort is subjected to a defined risk, or the two groups of patients to different treatments. That is:

(1) At the onset of the trial the two groups must have equal susceptibility to the target event, (that is, getting the disease, or getting rid of the disease).
(2) One must insist on equal handling and performance of all people during the trial and adherence to the preplanned schedule.
(3) Any change in the detection rate for the target event during the inquiry must be the same for both groups. For instance, pill-takers or smokers should not be pressed harder while under surveillance than non-pill-takers or nonsmokers.

Equal performance is very important. If there are patients with different clinical severity of a given illness, the patients are prognostically heterogeneous. In that case, before dividing the patients into two groups according to the severity of clinical conditions, subgroups or strata are created. The purpose of **stratification** is to achieve similarity of patients. Each patient within a subgroup is then randomly assigned to one of the two treatments. The prognostically disparate strata are thus subdivided into similar (or the same) proportions. Now it is possible to compare the effects of both therapies within the strata (as well as between the strata).

Case-control studies

Case-control studies are also suitable for studying the etiology of a disease. One assembles a group of patients with disease D ("cases" or D-patients) and a group of control persons without disease D. The control persons should be drawn from a wide variety of diseases or admission diagnoses in hospitals and/or from the general population. Then the D group and the non-D group are compared with respect to past and existing features and characteristics judged to be of possible relevance to the etiology of D.

Therefore controls should be similar to the D-patients in all respects except for D and the associated unknown etiological factors. If possible, each D-patient is paired with a control individual who is deliberately chosen to be of the same sex, age, and other possibly relevant features. The procedure of selecting controls such that the control group has the same distribution as the D-group with respect to important characteristics is known as **matching** (cf., Section 1.3.2). It is, however, very difficult to select the appropriate control group and to avoid all sources of bias in case-control studies [see, e.g., J Chronic Diseases **32** (1979), 35–41, 51–63 and 139–144]. More on this is provided by Schlesselman (1982).

Remarks

1. Concerning survival time and **survival probabilities** see R. P. Anderson et al., Journal of Surgical Research **16** (1974), 224–230; D. R. Thomas and G. L. Grunkemeier, Journal of the American Statistical Association **70**, (1975), 865–871: N. E. Breslow, International Statistical Review **43** (1975), 45–57: R. E. Tarone and J. Ware, Biometrika **64** (1977), 156–160. For two **graphical procedures for analyzing distributions of survival time** see D. R. Cox, Biometrika **66** (1979), 188–190. Four tests for **equality of survival curves** in the presence of stratification and censoring are given in L. Lininger et al., Biometrika **66** (1979), 419–428.

2. Measures of **disease incidence** are given in Morgenstern et al. (1980). Important in epidemiological studies (see Lilienfeld and Lilienfeld 1980) are incidence probabilities, the **relative risk** [see Gart [8:4], H. R. Bertell, Experientia **31** (1975), 1–10] and confidence intervals for both [see L. L. Kupper et al., Journal of the American Statistical Association **70** (1975), 524–528 as well as Fleiss 1981, and Hosmer and Hartz 1981].

3. **Matching** is a frequently used technique for controlling variation in medical as well as other investigations involving human populations. Sonja M. McKinlay discusses its advantages and disadvantages [Biometrics **33** (1970), 725–735; see also **38** (1982), 801–812 and American Journal of Epidemiology **116** (1982), 852–866].

2.1.4 Retrospective and prospective comparisons

The most important techniques for etiological studies are retrospective and prospective comparisons (Koller 1963, Cochran 1965) as well as potentially interesting combinations of the two. In **retrospective samples**, using hindsight drawn from medical records with all their shortcomings (e.g. missing and incompatible data), a group of people with the particular illness is compared with a group of people not afflicted by it. We can employ the term "cause" for a limiting factor without which the illness does not occur and whose presence diminishes the effect of other factors. Let us however point out that instead of a causal relationship between factor and illness there can also be a

2.1 Medical Statistics

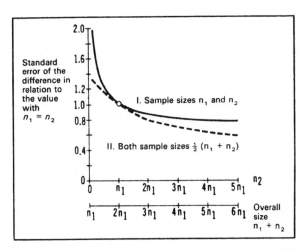

Figure 33 Standard error of the difference of two frequencies with different ratios of sample sizes. From S. Koller, Introduction to the methods of etiological research—statistics and documentation, Method. Inform. Med. 2 (1963), 1–13, Fig. 1, p. 6 (in German).

sequence of other relations (the factor could, e.g., be a symptom or a predisposition).

The frequency of this factor in the two sequences is compared. The control sequence must be at least as large as the test sequence. In Figure 33 the upper curve I indicates the change in the standard error of the difference, if the control sequence n_2 is larger or smaller than the test sequence n_1. If the control sequence is made twice as large as the test sequence, the standard error is reduced by only 13%. Further increases in the size of the control group cause even smaller additional reductions in the standard error. The expenditure is justified only in the case of a rare illness where the size of the test sequence is severely limited. If however the control sequence is smaller than the test sequence the standard error increases sharply, as seen from the left part of the upper curve (I). If sufficient funds are available the test and the control sequences should increase at the same rate. The dashed curve (II), with control and test sequences of equal size, gives the standard error of the difference as a function of the same total sample sizes as curve I. Note that curve II always does better than I, and the better the farther the ratio n_1/n_2 moves away from 1. **Thus, if possible, the sizes of the control and test sequences should be the same.**

Two samples are comparable if they differ only with respect to the attributes we want to compare while being indistinguishable with respect to the other attributes, i.e., the probability distributions of these other attributes must be about the same in the two samples. Three conditions are essential (Koller 1964): **structural homogeneity, uniformity (consistency) in observing** and **representative samples**.

1. **Structural homogeneity:** The frequency distributions of the more important modifying attributes, such as age, sex, and severity of illness, should be the same in the groups we want to compare. For comparisons it is best to pair persons of similar modifying attributes; if there are several choices the pairing is done at random, e.g., with help of random numbers.
2. **Uniformity (or consistency) in observing:** The method of observation and the conditions under which it is carried out must be the same. The factor in question has to be recognized and examined in the same way in all cases. A patient whose physician or who himself knows the hypothesis on the causality will, in general, be questioned differently, more extensively, than the controls; occasionally a patient will be too eager in confirming or concealing the factor. In fact, the results are useful only if both the one who questions and the questioned know neither the exact diagnosis nor the hypothesis on the etiology. The method of observing and measuring is crucial if psychological components are involved, as in retrospective interrogations, e.g., with respect to the set of problems caused by Thalidomide or with respect to the comparison of the success of different therapies. The **interviewer's bias**, well known in relation to questionnaires in the social sciences, belongs to this item, as does the accuracy in diagnosis, which changes (usually increasing) with time, thus distorting studies on the change over time of the causes of death.
3. **Representative samples:** The two groups, control and test group, must be random samples from the same basic population (regional origin, occupation). In the case of prospective etiological studies the two groups are usually drawn from the general population or from a representative portion thereof. It is often very difficult to find a suitable control group for a retrospective study. The control group should be representative for the people who are served by the same hospital and who are not carriers of the factor. Whether the result holds in general is then examined by a separate investigation.

Well-planned studies carried out on **prospective** samples are less subject to error, but the sample sizes have to be much larger. In this scheme two groups of people are observed under the same conditions over equal time intervals. The relevant statistic for the risk caused by the factor studied is given by the ratio of the percentage of the sick persons in the carrier group (having the factor) to the percentage of sick persons in the groups of non-carriers. The risk due to the factor (cf. Remark 2 at the end of Section 2.1.3.3) is recognized and measured directly. The control sample must be representative of all groups without the factor in the population to which the carriers of the factor belong.

In the absence of a particular hypothesis on the etiology we must place special emphasis on **systematic investigation and documentation of the results**. Undirected retrospective analysis of the increased occurrence of deformed limbs in infants led to the discovery of the effect of Thalidomide. A harmful factor such as smoking and the unknown set of illnesses with which it might interact are examined in a prospective study. In both types of investigation it is essential to provide a description of the experimental units (patient, hospital), as well as comprehensive observation and documentation, including a **tabulated breakdown** (cf., e.g., Sachs 1984, p. 3) according to the different combinations of the various attributes. As Lange (1965) in particular has pointed out, it is often rather problematical whether observed associations among illnesses are random phenomena, especially because on the one hand it is difficult to define appropriate control groups and to account for the course of an illness, while on the other hand selection and heterogeneity of the samples may distort the picture (cf., Koller 1963, 1964, 1971; Mainland 1963; Cochran 1965, Rümke 1970, Feinstein 1977, Fleiss 1981, Fienberg and Straf 1982).

Prospective studies are most suitable for investigations on associations of this sort. They are rather time consuming and organizationally demanding; however, the observations are more likely to be consistent, the sample has more the characteristics of a random sample, and it is possible to draw some conclusions about the prevalence of the factor. **Retrospective studies**, which can usually be carried out more quickly and which are also mandatory in the case of rare illnesses, serve quite often as starting points for prospective studies. Since many chronic diseases (e.g., various forms of cancer) are fairly rare and the latent periods are long, a retrospective approach is often unavoidable, as it is with specific high-risk industrial agents like radiation and asbestos.

Remarks on the patients of a clinic

1. The percentages of patients with particular illnesses that are admitted to a clinic are pretty much unknown.

2. Each patient has a different chance of being accepted by a clinic. The patients are *not* a random sample. Due to known and unknown selection factors, at each clinic a definite cluster is assembled (cluster sample; cf., Chapter 3).

(a) In medicine an accessible group of patients is often used as a sample, rather than a random group of patients chosen from a well-defined finite population (the target population).
(b) One needs a sample that is, to some degree, representative of the population. The essential attributes of the individuals of the target population from which a random sample will be drawn have to be listed.
(c) It may be necessary to identify, in a qualified sense, the target population with the sample.

3. The possible selection criteria are: the nature and severity of the affliction; other illnesses; age; sex; occupation; consultation with the physician (as affecting, e.g., the patient's awareness of health problems and of

the accessibility of the physician); diagnosis made by the physician; tendency of the physician to transfer the patient to a hospital; location, condition, and number of beds available in the hospital; diagnostic and therapeutic facilities; and the reputation of the hospital.

4. Therefore a generalization is difficult.

5. Groups of patients at the same hospital cannot be compared if the chances of being admitted to the hospital differ. A comparison is possible if the characteristic considered was itself not a factor in determining admission to the hospital.

6. Relations between illnesses can best be detected by studying cohorts from delivery to death. Longitudinal studies in the population are a useful substitute.

7. Generally it is of no use to collect and combine or pool available medical records from different hospitals, since the data are hardly ever comparable.

2.1.5 The therapeutic comparison

To test the therapeutic value of a medication, it is essential to have a basis of comparison which can be gathered either:

1. from the **outcome** of an illness: good health or death [on morbidity statistics and **mortality** statistics see International Statistical Review **45** (1977), 39–50, Australian Journal of Statistics **20** (1978), 1–42, as well as Armitage 1971 and Hill 1971], [for mortality see page 214 and Watson and Leadbetter 1980 [8:2d]].
2. from its **survival time** (cf. Remark 1 at the end of Section 2.1.3.3; for the comparison of survival distributions see Peto et al. 1976, 1977, Burdette and Gehan 1970, Elandt-Johnson and Johnson 1980, Lee 1980 [8:1], Lawless 1982, Cox and Oakes 1984 [8:2d]) or **duration of recovery**, or
3. from the **course** it takes or the **extent of the recovery** or the **permanent injuries** caused by the illness (cf., also Hinkelmann 1967). In this context the effect of drugs on healthy people (this is an important control group) will gain much importance. For side effects see pages 223, 224, 337.

Criteria that can be measured are of course desirable in each case. One distinguishes **hard and soft data**. Soft data consist of details of the case history, in relation, e.g., to coughing and difficulty in breathing, which greatly depend on the judgement of the patient doing the reporting. Examples of hard data on the other hand are age, weight, height, most of the findings of the medical lab, etc. Evaluation of soft data by counting quantifiable qualitative outcomes does not in general lead to any results worth mentioning.

A critical assessment of therapeutic results, based on comparative observations, includes the task of distinguishing authentic effects (depending on the medication) from spontaneous fluctuations. The most important prerequisites for the statistical methods used are: **homogeneity** of the groups,

random allocation of the individual patients to the various types of treatments, and **reproducibility of the observations**. The requirement that the experimental units, here the patients, be homogeneous in the case of the comparison of two therapies encounters the following difficulties: no two patients suffering from the same illness are entirely alike; no state of a disease repeats itself completely. Only in the course of the **chronic** illness of one particular patient are there time intervals during which the state of the illness is constant. Therefore the so-called **within patient trial**, usually limited to the early stages of drug testing, is preferred for these patients. The patient is treated by the two methods during consecutive time intervals in which the state of the illness does not change markedly. The patient is observed not only during the two intervals in which he receives therapy, but also during the periods preceding the first treatment, between treatments and following the second treatment. In the period preceding therapy the patient is given strictly symptomatic treatment or is only kept under observation. Each period continues until the state of the illness stabilizes under the particular treatment.

Patients with **acute** infectious diseases resemble each other in their clinical picture. It is possible to combine the various patients into two groups with like illnesses. The groups are subjected to the two treatments we want to compare. This is called a **between patients trial** (cf., Martini 1953, 1962). The second requirement, that of **random allocation** of patients to the treatments in the between patients trial or of the order in which the treatments are administered in the within patient trial, is guaranteed by a symmetrical distribution of all secondary causes that interfere with the decision-making process on both comparison groups. The effect error of the secondary causes is thereby neutralized to a great extent. A spontaneous tendency toward recovery is also an important secondary cause.

Formerly an alternating pattern was preferred in which new and standard treatments were assigned alternately to patients and time intervals respectively. The alternating test sequence with equalization consists of a combination of two procedures. In the first, one assigns treatments at random to the patients (e.g., with the help of random digits or by treating the first patient who comes in for observation and treatment with one medication and the next with the other one). In the second, one orders the patients according to the importance for the course of the illness of some characteristic such as sex, age, or state of nutrition. The mixture is useful because in small samples purely random assignments might lead to very unbalanced groups. The characteristic that has the largest influence on the course and prognosis of the illness will be "equidistributed" first (for typhoid fever it is age, for diphtheria the time since infection). In the interest of objectivity the physician who carries out the equalization must, as a precaution, be excluded from any subsequent discussion of the results. This "**equalizing alternation**" is based on the assumption that the samples are essentially random. Differences due to biological (or physiological instead of biological) factors between the two test groups are removed during brief time intervals in order to get similar groups, which can be better compared. If many patients are available for a comparison, it frequently suffices to arrange them in two groups according to the date of birth (even or odd day of the month). **A proper random allocation in a homogeneous group of patients is of course superior to any other scheme.** More on this can be found in the book by Feinstein (1977).

The third requirement, **repeatability** of the observations, encounters difficulties with time and timing: many important aspects of a disease cannot be observed and measured as often and in as quick a succession as one would like, because it would be too much for the patient.

Another requirement that must be met to realize an uncontested therapeutic assessment concerns the use of **representative symptoms** and characteristics of the disease, which permit a quantitative description of the main aspect of the state of the disease. The subjective symptoms can be influenced not only by a patient's self-deception based on his confidence in medical sciences and by an unintentional subconscious suggestive effect of the physician on the patient, but also by autosuggestion of the physician, whose diagnosis, observation and classification of the intensity of the symptoms might be biased because he knows which medications have been administered.

These problems of **unconscious and unintentional error** can only be eliminated by a single or double blind trial (cf., Martini 1957, Schindel 1962). The **single blind trial** simply consists of keeping the patient on whom a medicine is to be tested for effectiveness and usefulness ignorant of the substance and composition of the medicine for the duration of the test; and on top of that he should, if possible, even be kept in the dark about the fact that he will actually be involved in a therapeutic test. The patient is for example, supplied with a disguised medicine to eliminate any bias pro or con. Thus he either gets the medicine or a pseudo medicine called a **placebo** which is composed of pharmacologically inert substances and which looks, smells, and tastes like the active medicine (and, if possible, has the same side effects).

A well-known example is due to Jellinek (1946). Three headache remedies A, B, C and the placebo D were consecutively tested on 199 patients. During a 14 day period each patient was given a certain preparation as soon as he complained of a headache. The ratios of headaches treated successfully to the total numbers treated come to 0.84 for A, 0.80 for B, 0.80 for C, and 0.52 for D. There is thus no significant difference in effectiveness among the three preparations A, B, and C. A more detailed study of the 79 persons whose headaches were not relieved by the placebo reveals success ratios of 0.88 under A, 0.67 under B, and 0.77 under C, for this group of patients. These numbers differ considerably. The success ratios for the remaining 120 patients, those that sometimes found relief from their headaches through preparation D, equal 0.82 for A, 0.87 for B, 0.82 for C and 0.86 for D. All four preparations seem to be equally effective in this group of patients. Thus, before comparing several headache remedies, a placebo is administered to all patients and those responding to the placebo (placebo reactors) are not included in the actual experiment.

About one third, on the average, of every group of patients reacts to a placebo; this reaction comes on quickly but is not long lasting. The dispersion is large. The portion of placebo reactors extends from 0 to 67% for pain in general and from 43 to 73% for headaches. At least 30% of dysmenorrhea cases respond to placebos. Placebos are ineffective with small children, for serious acute illnesses, and for organic diseases with specific causes. Strangely enough, the tests for suggestibility do not agree with the response to placebos (cf., Documenta Geigy 1965), although the medication type (syrup, tablet, colored gelatin capsule) exerts a lot of influence (cf., Schindel 1962). Certain placebo-dependent clinical and, in particular, biochemical results remain a mystery as well (cf., Schindel 1965). Some physicians have used the so-called "**active placebo**," which contains a small amount of effective substance (cf., Lasagna 1962), assuming that small amounts of the active substance cause no effects, either opposite or more or less weakened. For humanitarian and legal reasons the placebo must frequently be replaced by a standard medication.

Going beyond the simple blind study, the **double blind study** makes even more extensive demands; Not only the patients but also the physician (or physicians) who observes and assesses the reactions of the patients must not know which treatments are tested and what specifically is administered to the patients, a medication or a placebo. The physician in charge may neither observe nor give the medication; his not being informed of something involving his patients would not be compatible with his responsibility as a physician. The medications are appropriately administered by nurses, from the same nursing staff that usually dispenses drugs; anything out of the ordinary must be avoided. It is however even more important that these people also not know the medication they give to the patients. It is clear that in this way the patients are also safeguarded to a great extent against unconscious suggestions. The emphasis on such extensive safeguards stems from the belief that not only does prejudice or autosuggestion of the patient add to the effect of a true or pseudo drug, but there are also indirect, conscious or subconscious influences on the patient by the attending physician.

A double blind test is mandatory if the physician is involved in the proper classification, according to subjective criteria, of the reaction to the therapy.

The larger the number of relevant subjective criteria in a research project, the more important it is to apply a double blind test. The simple blind trial is, however, generally adequate if the patients can characterize the symptoms unassisted, without interference of the physician, e.g., in characterizing pain as "better," "unchanged," or "getting worse."

Schindel (1965) commented as follows on a five way blind crossover trial that was once actually carried out: "The authors apparently have the idea that a sufficient amount of blindness generates a kind of occult vision." Further discussion of the therapeutic comparison, see especially Section 2.1.6, can be found in Mainland (1960, 1963), Martini et al. (1968), Burdette and Gehan (1970), Hill (1971), Brown (1972), and Ryan and Fisher (1974) as well as in Lee (1980 [8:1]) and Tygstrup et al. (1982) (cf., also Gehan and Freireich 1974). Mathematical models for clinical trials are given for instance by Canner (1977), Mendoza and Iglewicz (1977) and Glazebrook (1978). A check list for those planning clinical trials is furnished by the British Medical Journal 1977, **I**, pages 1323 and 1324. For sequential clinical trials (pp. 219–223) see Whitehead 1983 [8:2b]. Other helpful hints can be found in The Statistician **31** (1982), 1–142 and in:

(1) Biometrics **35** (1979), 183–197, 503–512; **36** (1980), 69–79, 677–706.
(2) Methods of Information in Medicine **18** (1979), 175–179; **19** (1980), 112–114; **21** (1982), 81–85, 94–95.
(3) New England Journal of Medicine **295** (1976), 74–80; **300** (1979), 73–75, 1242–1245; **301** (1979), 1410–1412.

The large scale **multiclinic trial** (e.g., Peto et al. 1976, 1977) as it has been requested, publicized and carried out for the past two decades in the U.S. by Mainland and in Great Britain by Hill, cannot be discussed here. Let us only mention that "Murphy's Law," as Mainland calls it (a law from the world of the theater: "If something can go wrong it will") applies when several clinics collaborate. How such difficulties can be prevented, especially in the planning, carrying out and evaluation of simple trials [see Controlled Clinical Trials, e.g., 3 (1982), 365–368] and of multiclinic trials has been described repeatedly and extensively by Mainland and Hill.

Helpful hints can be found in Methods of Information in Medicine: e.g., a special number devoted to medical diagnosis, bibliography included: **17**, No. 1 (1978), 1–74. Other important aspects of medical statistics are considered in The Journal of the Royal Statistical Society, Series A, for example **138** (1975): 131–169 Familial Diseases, 239–241 Children's Heights and Weights, 297–337 Ventilatory Function; **139** (1976): 104–107 Diagnosis, 161–182 Multivariate Methods, 218–226 Mortality, 227–245 Epidemiology and Mortality; **140** (1977): 469–491 Cohort Analyses, Asbestosis; **141** (1978): 95–107 Smoking and Lung Cancer, 159–194 Operational Research in the Health Services, 224–235 Mortality Ratios, 323–347 Epidemic Theory, 437–477 Smoking and Lung Cancer; **144** (1981): 94–103 Population Growth, 145–175 Discriminant Analysis, 298–331 Ionizing Radiation and Cancer; **145** (1982): 313–341 and 479–480 Geographic Variation in Cardiovascular Mortality, 395–438 Legal Probability, Evidence, Lawyers and Statisticians.

2.1.6 The choice of appropriate sample sizes for the clinical trial

The answers to the following three questions will essentially determine the sample sizes of the two test groups in a clinical trial, in a comparison of two therapies:

1. How big a risk of ascertaining a difference between two undistinguishable treatments (in other words, of **inventing** a difference) are we willing to put up with? This risk is known as the significance level α.

2.1 Medical Statistics

2. For how big a risk do we allow of missing a substantial difference between two treatments (in other words, concluding that **there is no significant difference** when the two treatments do have different effects)? This risk is called β. We know it as risk II (cf. Section 1.4.3). The **power** of a statistical test is defined as $1 - \beta$. The power of a test for a given alternative hypothesis is the probability of rejecting the null hypothesis when the alternative hypothesis is true. A test has a power of at least 0.95 if it is determined that only one decision out of 20 is wrong insofar as a significant difference is not discovered although it exists.
3. How small a difference should still be recognized as significant? This difference is called δ.

The usual answers to these questions are: (1) zero, (2) zero, (3) any real difference.

The question as to sample size can now be readily answered: Both patient groups should include infinitely many patients. Thus we see that to obtain realistic sample sizes, we must allow for positive risks; moreover, the difference must not be too small. Compare the discussion in Section 1.4.3.

Problems involving the determination of an appropriate sample size are best solved approximately, using a method due to Schneiderman (1964) which assumes the binomial distribution (Sections 1.6.1–2). Figure 34 gives the results for the two sided test ($2\alpha = 0.05$)—whether the therapy under study is better or worse than the standard therapy—and for four levels of risk II (the curves for $\beta = 0.05; 0.10; 0.20; 0.50$) as well as for therapy differences ($p_2 - p_1$) of sizes 5% and 10% (on the left), 15% and 20% (on the right). The recovery percentage p_1 of the standard method is plotted on the abscissa and the required sample sizes on the ordinate; an example can be found in the legend.

In this and in the following section we use the symbols α and 2α (i.e., $\alpha = \alpha_{\text{one S}}$ and $2\alpha = \alpha_{\text{two S}}$.) to distinguish between a one sided and a two sided question.

For an arbitrary risk I, problems of this type are solved according to Table 42 (cf., also the method for the one sided question presented in Section 4.6.1).

To begin with we need some constants, which can be found in Table 43. We denote these constants according to Table 42 by z and z_β. We use again the example in the top portion of Figure 34 to check our estimate with the help of the nomogram. From Table 43 for $2\alpha = 0.05$ we get $z = 1.9600$.

For risk II, $\beta = 0.10$ gives us $z_\beta = 1.2816$. The recovery percentage of the standard therapy is $p_1 = 0.20$. Thus we have the first three items A, B, C. Since we wish to detect an increase of 10% in therapy success, we obtain $p_2 = 0.20 + 0.10 = 0.30$ (D) for the recovery rate of the new method. Following the scheme we arrive at U, the sample size n for each of the two groups. Note that counted values are discrete variables, while z and z_β are based on the continuous normal distribution. The sample size Z is the value adjusted by the continuity correction. Adding a quickly calculated estimate of the continuity correction to the uncorrected estimate of the sample size,

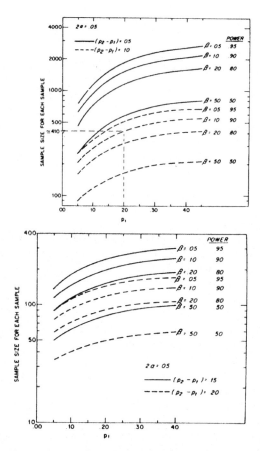

Figure 34 Nomogram for determining the sample size of two differently treated groups of patients for a "success–failure situation." It is meant only to give a general idea; the scheme in Table 42 is preferable. The solid curves in the top figure are for differences of the order of 5%; those in the bottom figure, for differences of the order of 15%. The dashed curves are for differences of the order of 10% (top) and 20% (bottom). The significance level ($2\alpha = 0.05$) holds for the two sided comparison. Four levels of risk II (β) as well as the corresponding "powers" are given in each case. The two dashed straight lines drawn in the lower left corner of the top figure (bottom left) illustrate how the sample sizes of interest are found for $2\alpha = 0.05$, $\beta = 0.10$ (power = 90%); the expected recovery rate p_1 for the standard treatment is 20%; a therapy difference of the order of 10% is called for ($p_2 - p_1 = 0.10$). The ordinate of the point of intersection of the curve $\beta = 0.10$ and the vertical through $p_1 = 0.20$ gives the sample sizes ($n_1 = n_2 \simeq 410$) (Table 42 calls for $n_1 = n_2 = 412$). Taken from Schneiderman, M. A.: The proper size of a clinical trial: "Grandma's strudel" method. J. New Drugs **4** (1964), 3–11.

2.1 Medical Statistics

Table 42 Scheme due to Schneiderman for working out an estimate of the sample size, with example (see legend to Figure 34)

Item	Computation	Example	Item	Computation	Example	
	$\alpha =$	0.025	P:	$\sqrt{M}$	0.6124	
	$2\alpha =$	0.05	Q:	$\sqrt{N}$	0.6083	
	$\beta =$	0.10	R:	$AP + BQ$	1.97990	
A: z		1.9600	S:	$\|C - D\|$	0.10	
B: z_β		1.2816	T:	R/S	19.7990	
C: p_1		0.20		without continuity correction		
D: p_2		0.30	U: n	T^2	392.00	= 392
E: $\bar{p}$	$\frac{C + D}{2}$	0.25		with quick estimate of cont. corr.		
F: q_1	$1 - C$	0.80	V: $n_{k'}$	$U + 2/S$	412.00	= 412
G: q_2	$1 - D$	0.70		with full cont. corr.		
H: $\bar{q}$	$1 - E$	0.75	W:	$R^2 + 4 \cdot S$	4.3200	
J: $\bar{p}\bar{q}$	$E \cdot H$	0.1875	X:	$\sqrt{W}$	2.0785	
K: $p_1 q_1$	$C \cdot F$	0.1600	Y:	$T \cdot X$	411.52	
L: $p_2 q_2$	$D \cdot G$	0.2100	Z: n_k	$\frac{\sqrt{S} + Y}{2}$	411.76	= 412
M: $2\bar{p}\bar{q}$	$2 \cdot J$	0.3750				
N: $\sum p_i q_i$	$K + L$	0.3700				

Table 43 Selected bounds of the standard normal distribution for the two and the one sided test (cf. also Table 13 and Table 14 in Section: 1.3.4; a shorter version of Table 43 can be found on the inside of the book cover)

P	Two-sided	One-sided
0.000001	4.891638	4.753424
0.00001	4.417173	4.264891
0.0001	3.890592	3.719016
0.001	3.290527	3.090232
0.005	2.807034	2.575829
0.01	**2.575829**	**2.326348**
0.02	2.326348	2.053749
0.025	2.241400	1.959964
0.03	2.170090	1.880794
0.04	2.053749	1.750686
0.05	**1.959964**	**1.644854**
0.06	1.880794	1.554774
0.07	1.811911	1.475791
0.08	1.750686	1.405072
0.09	1.695398	1.340755
0.1	1.644854	1.281552
0.2	1.281552	0.841621
0.3	1.036433	0.524401
0.4	0.841621	0.253347
0.5	0.674490	0.000000

we get the estimated sample size **V**. More on sample sizes for clinical trials can be found on pages 350, 351 and in Lee (1980 [8:1]), Tygstrup et al. (1982), in the Journal of Chronic Diseases **21** (1968), 13-24; **25** (1972), 673-681; **26** (1973), 535-560; **27** (1974), 15-24; **34** (1981), 533-544, and in Controlled Clinical Trials **2** (1981), 93-113. Jennie A. Freiman et al., New England Journal of Medicine **299** (1978), 690-694 after having surveyed 71 "negative" trials stress that concern for the probability of missing an important therapeutic improvement because of small sample sizes deserves more attention in the planning of clinical trials.

Remarks

1. **Testing in groups.** During the Second World War a Wassermann test (an indirect test for syphilis) was performed on each American draftee. Positive cases were rare, in the range of 2% of all tests. Since the method is sensitive, to reduce the great expense of the test project it was proposed that combined blood samples of several individuals be tested jointly. If the result were negative, it would mean that all participating individuals were free of syphilis. A positive reaction would mean that all individuals in the group had to be tested again. Now it can be shown (Dorfman 1943) that with a frequency of 2%, the optimal group size is 8; the number of Wassermann tests is thereby reduced by 73%. Dorfman determined the following optimal conditions for other proportions (Table 44). Further discussion (in particular additional tables) can be found in Sobel and Groll (1959, 1966) as well as in Graff and Roeloffs (1972) and especially in Loyer (1983) (cf., Hwang 1976 and C G. Pfeifer and P. Enis, Journal of the American Statistical Association **73** (1978), 588-592). The probability of finding at least one afflicted individual in a random sample of size n is equal to (again, cf., Table 6 in Section 1.2.3)

$$P = 1 - (1 - p)^n,$$

where p = relative frequency of the illness in the population (cf., also Section 1.6.4). Federer (1963) gives a survey and bibliography on **screening** [see also Goldberg and Wittes 1981].

Table 44

Relative frequency p	Optimal group size n	Percentage of tests eliminated
0.01	11	80
0.02	8	73
0.05	5	57
0.10	4	41
0.20	3	18

For $p < 0.11$, $n_{opt} \simeq 0.5 + 1/\sqrt{p}$.

2. **The 37% rule, or the Secretary Problem (or Marriage Problem).** Suppose the personnel director of a business is looking for a new secretary. A hundred applicants show up for the position in question. Suppose further that the personnel director must decide whether to hire a young lady right after she is introduced. Then the probability

that he would thereby choose the best secretary is only 1%. An optimal strategy which increases this probability to almost 37%, consists in having the first 37 young ladies introduce themselves, then hiring the next applicant that surpasses all her predecessors. The number 37 (more precisely 36.7879) is obtained as the quotient of the number of applicants and the constant e, where $e = 2.71828\ldots$ is the base of the natural logarithms. If instead of 100 we say n secretaries apply, the personnel director would accordingly do best to let n/e young ladies pass and offer the position to the next applicant that outshines her predecessors. The probability of having chosen the best from among the n applicants is again 37%. If the personnel director is familiar with the exact "distribution of applicants," then this probability increases to about 58%, as attested to by a study of Gilbert and Mosteller (1966). (See Chow 1964.) Suppose 30 riders and their horses, take part in a tournament. For any particular contest, horses are assigned to riders by lots. The probability that none of the riders gets his own horse, is likewise just under 37%. It is an interesting fact that this probability is around 36.8% for every sample size $n \geq 6$. For large n, it again approaches the value $1/e = 0.367879$. To say it the other way around: if $n \geq 6$ objects are rearranged at random, then with probability $1 - (1/e) = 0.632$ at least one of the objects will occupy its original position.

More on this can be found in Abdel-Hamid et al. (1982); a review is provided by P. R. Freeman (1983, Intern. Statist. Rev. **51**, 189–206).

2.2 SEQUENTIAL TEST PLANS

One branch of statistics—sequential analysis—was developed by A. Wald during the Second World War. Sequential analysis remained a military secret until 1945, since it was immediately recognized as the **most efficient means for continuous quality control in industry**. A very readable elementary but thorough account with many examples was issued by the Statistical Research Group at Columbia University (Sequential Analysis of Statistical Data: Applications, New York: Columbia University Press, 1945). Davies (1956) and Weber (1972) likewise give very good introductions to sequential analysis. Bibliographies (cf., the references in [8:2b]) are found in Jackson (1960), Johnson (1961), Wetherill (1975), and Armitage (1975). Assume that the effects of two treatments differ at least by a given amount. The risks of type I and type II errors are fixed at α and β respectively. The samples are supposed to be random samples from infinitely large populations. **In sequential analysis the sample size is considered a random variable; as such it has a distribution and a mean value**—the expected sample size. Instead of repeating the experiment a given number of times, we check after each additional trial whether we have by now sufficient information to reach a conclusion; i.e., we carry out exactly as many trials as are absolutely necessary to determine with risks α and β which of the two treatments is superior to the other. The advantages of this procedure are obvious when the single experiments are costly and time consuming, but it is also valuable when the number of observations is limited. On the basis of the results of each of the individual outcomes of one particular experiment, it is determined whether the trial or sequence of trials (sequence of experiments) be continued or a decision can be reached. We distinguish between **computational and graphical techniques**, and among these between the so-called **open and closed sequential test plans**: the latter always lead to a decision. The closed sequential test plans will be discussed in greater detail. They permit us to compare two therapies, treatments, or medications, up to now considered interchangeable, **without actual computation**. If a new medicine A is to be compared with another medicine B, then patients are paired off in accordance with the equalizing alternation principle. The two patients are treated either simultaneously or one right after the other. A coin toss determines which patient gets medication A. The result is judged according to the scale:

medicine A is better than B,
medicine B is better than A,
no difference.

A sequential test plan developed by Bross (1952) and adapted to investigations in medicine is shown in Figure 35. If in the first experiment A is better, the field above the black square is marked, if B is better the field to the right of the black square is marked. If there is no difference, no entry is made but the outcome is recorded on a separate sheet.

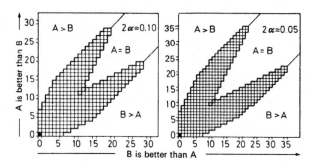

Figure 35 Two sequential test plans due to Bross ($\beta \approx 0.05$); [I. D. J. Bross, Sequential medical plans, *Biometrics* **8**, 188–205 (1952)].

The result of the second trial is introduced in the same way as that of the first trial, the field corresponding to the first result now serving as the reference square, with the third trial, the field marked in the second trial, etc. As soon as a limit is crossed in the course of the test sequence we accept (Fig. 35: left square), with $2\alpha \simeq 10\%$ (two-sided test at the 10% level), one of the following conclusions:

Upper limit: $A > B$, medicine A is better;
Lower limit: $B > A$, medicine B is better;
Middle limit: $A = B$, a significant difference is not apparent.

2.2 Sequential Test Plans

The question of what difference is for us "significant" must still be answered. It is clear that the larger the least significant difference is, the sooner a decision will be reached (given that there is a difference), i.e., the smaller will be the number of trials required; more precisely, the maximum size of the trial sequence depends on this difference. Only our experiment can decide how many trial pairs must be tested in a given case. If we almost always get the result "no difference," it will take us a long time to reach a decision. However, such cases are rather exceptional. Let p_1 and p_2 denote the percentages of patients cured by the standard and the new medication respectively. The outcome of any single trial is one of the possibilities listed in Table 45.

Table 45

No.	Old medicine	New medicine	Probability
1	Cured	Cured	$p_1 p_2$
2	Not cured	Not cured	$(1-p_1)(1-p_2)$
3	Cured	Not cured	$p_1(1-p_2)$
4	Not cured	Cured	$(1-p_1)p_2$

Since we are only interested in cases 3 and 4, the portion of the time that case 4 occurs, written p^+ for short, is found to be

$$p^+ = \frac{p_2(1-p_1)}{p_1(1-p_2)+(1-p_1)p_2}. \tag{2.3}$$

If $p_1 = p_2$, then $p^+ = \frac{1}{2}$ independently of the value assumed by p_1. If the new medicine is better, i.e., $p_2 > p_1$, then p^+ becomes greater than $\frac{1}{2}$. Bross had assumed for the sequential test plan described above that if p_2 is so much larger then p_1 as to yield $p^+ = 0.7$, the difference between the two medicines can be considered "significant". That is, if 10%, 30%, 50%, 70%, or 90% of the treated patients are cured by the old medicine, then the corresponding percentages for the new medicine are 21%, 50%, 70%, 84%, and 95%. We see that the difference between the two methods of treatment is greatest, and thus the maximum sample size is smallest, if 30% to 50% of the patients are cured by the standard mediciation. This is not surprising, for if the treatments are hardly ever or almost always successful, extensive experiments have to be run to get a clear distinction between two therapies. Sequential analysis requires in general only about $\frac{2}{3}$ as many observations as the usual classical procedures.

Let us now return to Figure 35 and investigate the efficiency of this sequential test that was developed for short to medium sized experiments and moderate differences. If there is no difference between the two treatments

($p^+ = 0.5$), a difference is (erroneously) asserted with probability 0.1 at least for either direction ($p_1 > p_2, p_2 > p_1$), i.e., we would conclude correctly that there is no significant difference in not quite 80% of the cases. If there is a significant difference between the two treatments ($p^+ = 0.7$), and if p_2 is "significantly" larger than p_1, then the total probability of reaching the wrong conclusion is only around 10%, so that the superiority of the new method is recognized in 90% of the cases. The chance of coming up with a correct decision thus increases from not quite 80% ($p^+ = 0.5$) to 90% ($p^+ = 0.7$). If the difference between the two medications is slight ($p^+ = 0.6$), then the new treatment is recognized as superior in about 50% of the cases. The probability that we (incorrectly) rate the standard treatment as better is then less than 1%.

If very **small differences** between two therapies have to be discovered, then other sequential test plans with much longer trial sequences must be employed. The symmetric plan for the two sided problem might have to be replaced by one for the one sided problem ($H_0: A > B$, $H_A: A \leq B$), in which the middle region—in Figure 35 the region $A = B$—is combined with the region $B > A$. This is the case if the old treatment has proven itself and is well established and the new treatment will be widely used only if it is shown to be clearly superior. Spicer (1962) has developed a **one sided sequential test plan** (Figure 36) for exactly this purpose. The new method is accepted when $A > B$; it is rejected when $B > A$.

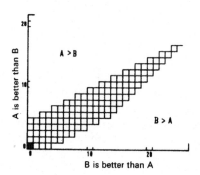

Figure 36 Sequential test plan due to Spicer ($\alpha \simeq 0.05, \beta \simeq 0.05, p^+ = 0.08$); C. C. Spicer: Some new closed sequential designs for clinical trials, *Biometrics* **18** (1962), 203–211.

The one sided test plan of Spicer (1962) (cf., Alling 1966) has the advantage that the maximum sample size is relatively small especially when the new treatment method is in fact not superior to the old one. This plan is therefore particularly well suited for **survey trials**, for example, in tests of several new drug combinations, most of which represent no real improvement. The use of a one sided test can hardly be considered a drawback in clinical experiments of this sort, since we are not interested in finding out whether the new treatment is about the same or worse than the therapy.

A **quick sequential test plan** (Figure 37) devised by Cole (1962) for the purpose of the surveying the ecologically significant differences between groups of organisms, can be used **to detect larger differences**. Overemphasizing minimal differences is deliberately avoided. Here even a fairly large Type II error (accepting a false null hypothesis—the "false negative" in medical diagnosis) is not taken seriously. Thus, should a small difference be discovered, this rapid test, designed for survey trials, is to be replaced by a more sensitive plan.

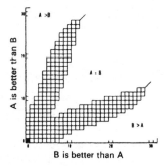

Figure 37 Sequential test plan due to Cole ($2\alpha \simeq 0.10$, $\beta \simeq 0.10$, $p^+ = 0.7$); L. M. C. Cole: A closed sequential test design for toleration experiments, *Ecology* **43** (1962), 749–753.

Assume that one of the three sequential test plans was given or some other one was adopted, that the samples were chosen on the basis of the equalizing alternation principle, and that after the large number of trials we still cannot reach a conclusion. Then it is appropriate and preferable from an ethical point of view to treat each patient with the same therapy as his predecessor if it proved successful; if however the treatment was not successful, then he will undergo the other therapy. The experiment is completed as soon as one of the limits of the sequential test plan is crossed or the number of patients treated by one therapy gets twice as large as the number of patients treated by the other therapy.

Finally, let us emphasize that **natural limits are imposed on sequential analyses employed in medicine**, even when hard data are available. They are, after all, meaningful only if the individual treatment period is short in comparison with the duration of the whole experiment; moreover, a small sample can hardly give information on the secondary and side effects of the new therapy. The decisive advantage of sequential analysis over classical methods, namely that relatively small trial sequences during the experiment, can, without computation, lead to decisions during the experiment, must not lead to indiscriminate use of this method (cf., also Gross and Clark 1975 [8:2d]). J. Whitehead discusses the analysis of sequential clinical trials in Biometrika **66** (1979), 443–452 and design and analysis in his book, cited on page 596.

On clinical testing of drugs for side effects

Probabilistic statements gained in tests on animals cannot be carried over to humans. Harmful side effects (cf., Section 4.5.1) must be taken into account. Their undesirability is based on subjective criteria. The suspicion that a substance produces harmful side effects in humans can be neither confirmed nor denied without a controlled trial involving random allocation; harmlessness cannot be "proved." An important role is played by the problem of distinguishing between random connections, associations by way of a third variable, and possible causal relations. All statements have an inherent uncertainty, which can be narrowed down only through plausibility considerations.

2.3 EVALUATION OF BIOLOGICALLY ACTIVE SUBSTANCES BASED ON DOSAGE–DICHOTOMOUS EFFECT CURVES

Preparations that are destined for pharmaceutical use and that contain a pharmacologically active ingredient are tested on animals, plants, and/or microorganisms. The first step consists in determining the form of the **dosage–effect curve**. This curve represents the observed reactions as a function of the drug dosage. The abscissa carries the dosage scale, the ordinate the intensity or frequency of the reaction. We distinguish between a dosage–dichotomous effect curve and dosage–quantitative effect curves according as the reaction is described by yes–no or by an exact measurement of some quantity, as in the following examples.

Dosage–dichotomous effection relation: In a trial of toxicity, samples of mice are exposed to various concentrations of a toxin. After a certain time the mice that survived and those that died are counted. The test result is "yes" or "no"—that is, an alternative, a dichotomy.

Dosage–quantitative effect relation: Each of several groups of capons receives a certain dose of differently modified testosterone derivatives. The effect is measured in terms of the increase in the length and height of the comb. In pharmacology and toxicology the notion of a mean effective dose (ED_{50}) is important. Precisely defined, it is the dose for which the probability that a test animal indicates an effect is 50%. It is estimated from dosage–dichotomous effect curves.

The percentage of animals affected by a certain dose and higher doses and the percentage of animals not reacting to a certain dose or lower doses can be read off the cumulative percentage curve or the cumulative frequency distribution. Usually the abscissa (dose) carries a logarithmic scale. The symptom may consist in death or survival (for poisons the 50% lethal dose LD_{50} is the dose by which 50% of the animals are killed). Other examples of symptoms are the impairment of driving ability caused by a certain dose of alcohol (per mille content of alcohol in the blood) and the onset of narcosis at a certain dose of a particular narcotic.

2.3 Evaluation of Biologically Active Substances

The value of ED_{50} (LD_{50}) is usually determined by probit analysis, which involves a considerable amount of calculation. Therefore simpler procedures, more suitable for routine tests, have been established that allow us to read the mean and deviation from dosage–effect curves. An adequate estimate of ED_{50} can be obtained provided the three following conditions are met (cf., references in [8:2c], especially Olechnowitz 1958):

1. The doses are symmetrically grouped about the mean. Cumulative percentages of 0 and 100 are included.
2. The spacing between doses, or else the logarithm of the ratio of each consecutive pair of doses, is held constant.
3. Each dose is administered to the same numbers of individuals.

Estimating the mean effective or lethal dose by the Spearman–Kärber method

The Spearman–Kärber method (cf., Bross 1950, Cornfield and Mantel 1950, Brown 1961) is a **rapid distribution-free method** that provides a quick and very good estimate of the mean and standard deviation. If the distribution is symmetric, then the median is estimated. The median effective dose (the median lethal dose) is the dose level at which 50% of the test animals show a reaction (are killed). The conditions stated above, together with the assumption that the distribution is normal rather than lognormal, imply

$$\boxed{LD_{50} \quad \text{or} \quad ED_{50} = m = x_k - d(S_1 - 1/2),} \qquad (2.4)$$

where x_k is the smallest dose such that any equal or larger dose always produces 100% reactions, d is the distance between adjacent doses, and S_1 is the sum of the relative portions of reacting individuals (positive reagents; cf., Table 46) at each dose. The standard deviation $s_m \equiv s_{ED_{50}}$ associated with ED_{50}, is estimated by

$$\boxed{s_{LD_{50}} \quad \text{or} \quad s_{ED_{50}} = s_m = d\sqrt{2S_2 - S_1 - S_1^2 - 1/12},} \qquad (2.5)$$

where S_2 is the sum of the cumulatively added relative portions of reacting individuals.

EXAMPLE. Table 46 indicates the results of a trial for determining the mean lethal dose of an exceptionally efficient anesthetic. Each dose is used on 6 mice. These values can be used to estimate the 90% confidence limits for the true value by $m \pm 1.645 s_m = 30 \pm (1.645)(10.26)$ under an approximately normal distribution,

$$\left.\begin{array}{l} m_{\text{upper}} \\ m_{\text{lower}} \end{array}\right\} = 30 \pm 16.88 = \begin{cases} 46.88 \text{ mg/kg} \\ 13.12 \text{ mg/kg}. \end{cases}$$

Table 46

Dosage mg/kg	Number of mice that died	Relative portions	Cumulative relative portion of mice that died
10	0	0	0
15	0	0	0
20	1	0.17	0.17
25	3	0.50	0.67
30	3	0.50	1.17
35	4	0.67	1.84
40	5	0.83	2.67
45	5	0.83	3.50
50 = x_k	6	1.00	4.50
d = distance from dose to dose = 5		4.50 =S_1	S_2 = 14.52

$m = x_k - d(S_1 - \frac{1}{2})$,
$m = 50 - 5(4.5 - 0.5)$,
$m = 30$,
$s_m = d\sqrt{2S_2 - S_1 - S_1^2 - 1/12}$,
$s_m = 5\sqrt{2 \cdot 14.52 - 4.5 - 4.5^2 - 0.083}$,
$s_m = 10.26$.

We forego **non-bioassay examples**. The tests in question are actually **sensitivity tests** in which an object reacts only above a certain threshold in the way, for instance, that a land mine reacts to a jolt only if it is greater than or equal to a certain intensity (cf., Dixon 1965). Tables for computing LD_{50} estimates for small samples with extreme value response distributions, and an example concerned with the sensitivity of explosives are given by Little (1974). Sometimes these distributions, distinguished by ranges that are small relative to their means, are easily approximated by a normal distribution.

It is characteristic of a bioassay that switching from the linear to the logarithmic dosage scale leads to a "symmetrization" of the distribution of the individual minimum effective doses. Given an approximately lognormal distribution, m and s_m can be determined from

$$m = x_k - d(S - 1/2), \qquad (2.6)$$

$$s_m = \frac{d}{100}\sqrt{\sum \frac{p_i(100-p_i)}{n_i - 1}}, \qquad (2.7)$$

2.3 Evaluation of Biologically Active Substances

where

- m is the estimate of the logarithm of ED_{50} or LD_{50},
- x_k is the logarithm of the smallest dose such that all doses greater than or equal produce 100% reactions (x_0 is the logarithm of the largest dose to which no test animal reacts),
- d is the logarithm of the ratio of each consecutive pair of doses,
- S is the sum of the relative portions of reacting individuals,
- p_i is the frequency, in percent, of reactions with the ith dose ($i = 0, 1, 2, \ldots, k$) (thus $p_0 = 0\%$ and $p_k = 100\%$),
- n_i is the number of test animals tested with the ith dose ($i = 1, 2, \ldots, k$).

Of the three conditions listed near the beginning of this section, only the first two are necessary in this context. Nevertheless it is recommended that samples of approximately equal size n_i be employed. It is sometimes difficult to fulfill requirement 1 in practice, namely to test, under all conditions, at least one dose with 0% reactions and at least one dose with 100% reactions. x_0 and/or x_k is estimated in these cases; the results are then correspondingly less reliable.

EXAMPLE. Table 47 indicates the results of a trial for determining the mean lethal dose of a mildly effective anesthetic. The 95% confidence limits can be estimated by $m \pm 1.96 s_m$ (normal distribution assumed):

$$\left.\begin{array}{r}m_{\text{upper}} \\ m_{\text{lower}}\end{array}\right\} 1.6556 \pm 1.96 \cdot 0.2019 = \begin{cases} 2.0513; & \text{antilog } 2.0513 = 112.54 \text{ mg/kg} \\ 1.2599; & \text{antilog } 1.2599 = 18.19 \text{ mg/kg.} \end{cases}$$

Table 47

Dose mg/kg	Proportion of test animals that died
4	0/8 = 0
16	4/8 = 0.50
64	3/6 = 0.50
256	6/8 = 0.75
1024	8/8 = 1.00
	S = 2.75

$\log \dfrac{16}{4} = \log 4 = 0.6021; \log 1024 = 3.0103,$

$m = \log 1024 - \log 4 (2.75 - 0.5),$

$ = 3.0103 - 0.6021 \cdot 2.25 = 1.6556,$

antilog $1.6556 = 45.25$; $LD_{50} = 45.25$ mg/kg,

$s_m = \dfrac{\log 4}{100} \sqrt{\dfrac{50 \cdot 50}{8-1} + \dfrac{50 \cdot 50}{6-1} + \dfrac{75 \cdot 25}{8-1}},$

$ = 0.2019.$

Let us, for the sake of completeness, also indicate the procedure used for testing the difference between two ED_{50}'s. For two mean effective doses ED'_{50} and ED''_{50} with standard deviations s' and s'', the standard deviation of the difference $ED'_{50} - ED''_{50}$ is

$$\boxed{s_{\text{Diff}} = \sqrt{(s')^2 + (s'')^2}.} \tag{2.8}$$

There is a true difference at the 5% level as soon as we have

$$\boxed{|ED'_{50} - ED''_{50}| > 1.96 s_{\text{Diff}}.} \tag{2.9}$$

To determine the specific biological activity of a preparation, the effect of this preparation on test animals is compared with the effect of a standard preparation. The amount in international units or milligrams, of biologically active substance in the preparation is given by the ratio of the effect of the preparation to that of the standard preparation, the activity of which is known. Confidence limits can then be specified, and the true value can, with a high degree of probability, be expected to lie between them, provided several conditions are fulfilled.

A trimmed Spearman–Kärber method for estimating median lethal concentrations in **toxicity bioassays** is presented by M. A. Hamilton et al., in Environmental Science and Technology **11** (1977), 714–719 and **12** (1978), 417 (cf. Journal of the American Statistical Association **74** (1979), 344–354).

A detailed description of bioassay is given by Finney (1971) and Waud (1972) (cf., also Stammberger 1970, Davis 1971 as well as the special references in [8:2c]). Important tables can be found in Vol. 2 of the Biometrika tables cited (Pearson and Hartley 1972 [2], pp. 306–322 [discussed on pp. 89–97]). More on the **logit transformation** is contained in Ashton (1972) in particular The computation of results from **radioimmunoassays** is given by D. J. Finney in Methods of Information in Medicine **18** (1979), 164–171. For **multivariate bioassay** see Vørlund (1980).

2.4 STATISTICS IN ENGINEERING

Applied statistics as developed over the last 50 years has been instrumental in technological progress. There is by now a collection of statistical methods that is suited or was especially developed for the engineering sciences.

2.4.1 Quality control in industry

The use of statistical methods in the applied sciences is justified by the fact that certain **characteristics** of the output of a production line **follow a probability distribution**. The associated parameters μ and σ are measures of the quality of the output, and σ is a measure of the uniformity of the output. The **distribution** may be viewed **as a calling card of the output**.

2.4.1.1 Control charts

We know that **control charts** (cf., Section 2.1.2) are always necessary **when the output is supposed to be of appropriate quality,** where "quality" in the statistical context means only the "**quality of conformance**" **between the prototype and the manufactured item** (Stange 1965). That the prototype itself can and does admit features that can be deliberately varied in accordance with the changing demand of the buyer, is of no interest to us here.

The standard technique of graphical quality control in industry is based on the **mean**. For continuous quality control of the output one takes small samples at regular intervals, computes the means and records them consecutively in a control chart (Shewhart control chart) in which the warning limits are indicated at $\pm 2\sigma$ and the control or action limits at $\pm 3\sigma$. If a mean value falls outside the 3σ-limits or if two consecutive means cross the 2σ-limits, then it is assumed that the manufacturing process changed. The cause of the strong deviation is traced, the "fault" eliminated, and the process is once again correct.

Instead of a mean value chart ($\bar{X}$-chart), a median chart ($\tilde{X}$-chart) is sometimes used. The standard deviation chart (S-chart) or the range chart (R-chart) can serve to monitor the dispersion of a process. The cumulative sum chart for early detection of a trend has already been referred to (Section 2.1.2).

The range chart

The range chart (R-chart) is used to **localize and remove excessive dispersions**. If the causes of a dispersion are found and eliminated, the R-chart can be replaced by the S-chart. The R-chart is ordinarily used in conjunction with the $\bar{X}$-chart. While the $\bar{X}$-chart controls the variability **between** samples, the R-chart monitors the variability **within** the samples. More on this can be found in Stange (1967), in Hillier (1967, 1969) and also in Yang and Hillier (1970) (see also Sections 7.2.1, 7.3.1).

Preparation and use of the R-chart for the upper limits

Preparation

1. **Repeatedly** take samples of size $n = 4$ (or $n = 10$). A total of 80 to 100 sample values should be made available.
2. Compute the range of each sample and then the **mean range** of all the samples.
3. Multiply the mean range by the constant 1.855 (or 1.518 respectively). The result is the value of the upper 2σ **warning limit**.
4. Multiply this quantity by the constant 2.282 (or 1.777). The result is the value of the upper 3σ **control or action limit**.

Use

Take a random sample of size $n = 4$ (or $n = 10$). Determine the range and record it on the control chart. If it equals or exceeds

(a) the 2σ warning limit, a **new** sample must be taken right after this is found to be the case;
(b) the 3σ action limit, then the process is **out of control.**

	Upper limits	
n	2σ limit	3σ limit
2	2.512	3.267
3	2.049	2.575
4	1.855	2.282
5	1.743	2.115
6	1.670	2.004
7	1.616	1.942
8	1.576	1.864
9	1.544	1.816
10	1.518	1.777
12	1.478	1.716
15	1.435	1.652
20	1.390	1.586

In addition to those control charts for measurable properties, there is also a whole series of special control charts for control of countable properties, i.e. of **error numbers** and of **fractions defective**. In the first case, the quality of the output is rated by the **number of defects per test unit**, e.g. by the number of flaws in the color or in the weave per 100 m length of cloth. Since these flaws are infrequent, the control limits are computed with the help of the Poisson distribution. If each single item of an output is simply rated as flawless or flawed, good or bad, and if the percentage of defective items is chosen as a measure of the quality of the output, then a special chart is used to monitor the **number of defective items (or products)**. The limits are computed with the help of the binomial distribution. Let us call attention to the so-called binomial paper (cf. Section 1.6.2) and the Mosteller-Tukey-Kayser tester (MTK sample tester). A detailed description of the various types of control charts can be found in Rice (1955) and Stange (1975), as well as in the appropriate chapters of books on quality control (e.g., Duncan 1974). **Log-normally distributed data** are controlled as described by Ferrell (1958) and Morrison (1958). An elegant **sequential analytic method** of quality control is presented by Beightler and Shamblin (1965). Knowler et al. (1969) give an outline.

2.4.1.2 Acceptance inspection

Acceptance sampling is the process of evaluating a portion of the product in a shipment or lot for the purpose of accepting or rejecting the entire lot as either conforming or not conforming to a preset quality specification. There are two types of **acceptance sampling plans:** those using measurements of attributes (attributes plans), and those using measurements of variables (variables plans). In both cases it is assumed that the samples drawn are random samples and that the lot consists of a product of homogeneous quality. In **single-sampling plans** the decision to accept or reject a lot is based on the first sample. In **multisampling plans** the results of the first sample may not be decisive. Then a second or perhaps a third sample is necessary to reach a final decision. Increasing the number of possible samples may be accompanied by decreasing the size of each individual sample. In **unit sequential sampling inspection** each item or unit is inspected, and then the decision is made to accept the lot, to reject it, or to inspect another unit. The choice of a particular plan depends upon the amount of protection against sampling errors which both the producer and the consumer require: here α (the rejection of good lots) is termed the **producer's risk** and β (the acceptance of bad lots) is termed the **consumer's risk**. The operating characteristic (OC) curve for a sampling plan quantifies these risks. The OC curve tells the chance of accepting lots that are defective before inspection. For some types of plans, such as **chain sampling plans** and **continuous sampling plans**, it is not the lot quality but the process quality that is concerned.

> In **chain sampling plans** apply the criteria for acceptance and rejection to the cumulative sampling results for the current lot and one or more immediately preceding lots.
>
> In **continuous sampling plans**, applied to a continuous flow of individual units of product, acceptance and rejection are decided on a unit-by-unit basis. Moreover, alternate periods of 100% inspection (all the units in the lot are inspected) and sampling are used. The relative amount of 100% inspection depends on the quality of submitted product. Each period of 100% inspection is continued until a specified number i of consecutively inspected units are found clear of defects.
>
> In **skip-lot sampling plans** some lots in a series are accepted without inspection when the sampling results for a stated number of immediately preceding lots meet stated criteria.

In the simplest form of acceptance sampling a random sample of size n is selected from a lot of size N. The number of defectives in the sample is determined and compared with a predetermined value, termed the **acceptance number** c. If the number of defectives is less than or equal to c the lot is accepted; otherwise it is rejected. Tables exist that enable us to read off n and c for given risks. More on sampling plans can be found in Bowker and Lieberman (1961), Duncan (1974), and other books on quality control.

Recent developments are covered in the Journal of Quality Technology: **8** (1976), 24–33, 37–48, 81–85, 225–231; **9** (1977), 82–88, 188–192; **10** (1978), 47–60, 99–130,

150–154, 159–163, 228; **11** (1979), 36–43, 116–127, 139–148, 169–176, 199–204; **12** (1980), 10–24, 36–46, 53–54, 88–93, 144–149, 187–190, 220–235; **13** (1981), 1–9, 25–41, 131–138, 149–165, 195–200, 221–227; **14** (1982), 34–39, 105–116, 162–171, 211–219. Technical aids given by L. S. Nelson are, e.g., (1) minimum sample sizes for attribute superiority comparisons [**9** (1977), 87–88], (2) a nomograph for samples having zero defectives [**10** (1978), 42–43], (3) a table for testing "too many defectives in too short a time" [**11** (1979), 160–161].

2.4.1.3 Improvement in quality

The improvement of consistent or fluctuating quality is an engineering problem as well as a problem of economics. Before tackling this problem we must determine the factors, sources, or causes to which the excessively large variance σ^2 can be traced. Only then can one decide what has to be improved. The **analysis of variance** (see Chapter 7), with which one answers this question, subdivides the variance of observations into parts, each of which measures variability attributable to some specific factor, source, or cause. The partial variances indicate which factor contributes most to the large variance observed and therefore should be better controlled. Effort spent on improving the control over factors that do not play a big role is wasted. Only the results of analysis of variance carefully carried out provide the necessary tools for a meaningful solution to the technological-economic complex of questions connected with improvement in quality.

Some basics about experimental design that engineers should know are summarized by G. J. Hahn (1977). A particularly interesting and important special case of quality improvement is **guidance toward more favorable working conditions** (cf. Wilde 1964). In technological operations the **target quantity** (for example the yield, the degree of purity, or the production costs) generally depends on numerous influencing factors. The amount of material used, the type and concentration of solvent, the pressure, the temperature, and the reaction time, among other things, all play a role. The influencing factors are chosen (if possible) so that the target quantity is maximized or minimized. To determine the optimal solution experimentally is a difficult, time-consuming, and costly task (cf., Dean and Marks 1965). Methods for which the costs of necessary experimentation are as small as possible are exceptionally valuable in practice. In particular, the **method of steepest ascent**, discussed by Box and Wilson (1951), has proved extremely successful (cf., Brooks 1959). Davies (1956), Box et al. (1969), and Duncan (1974) give a good description, with examples, of the steepest ascent method.

If this not entirely simple method is employed in the **development of new procedures**, one speaks of **"response surface experimentation"** (Hill and Hunter 1966, Burdick and Naylor 1969; cf. Biometrics **31** (1975), 803–851, Technometrics **18** (1976), 411–423 and Math. Scientist **8** (1983), 31–52). Unfortunately, it is difficult, if not impossible, to maintain exact laboratory conditions in a factory; the real conditions always deviate more or less from the ideal ones. If the manufacturing process created in a laboratory is adopted for production and if a succession of small systematic changes of all influencing factors is carried out on methods that are already quite useful, with the result of each change taken into consideration and further adaptations subsequently introduced to gradually optimize the manufacturing process, then we have an **optimal increase in performance** through an **evolutionary operation**. More on this can be found in the publications by Box et al. [pp. 569, 598] (for orthogonality see Box 1952) as well as in the survey article by Hunter and Kittrel (1966). Examples are given by Bingham (1963), Kenworthy (1967), and Peng (1967) (cf., also Ostle 1967, Lowe 1970, and Applied Statistics 23 (1974), 214–226).

2.4.2 Life span and reliability of manufactured products

The **life span** of manufactured products, in many cases measured not in units of time but in units of use (e.g., light bulbs in lighting hours) is an important gauge of quality. If one wishes to compute the annual replacement rate or to properly estimate the amount of warehousing of replacement parts for product types that are no longer manufactured, one must know their mean life span or, better yet, their durability curve or order of depletion. The depletion function [abscissa: time from t_0 to t_{max}; ordinate: relative percentage of elements still available, $F(t) = n(t)100/n_0$ (%), $F(t_0) = 100$, $F(t_{max}) = 0$] is usually ⌣-shaped.

In deciding to what extent new methods of production, other protective measures and means of preservation, new materials, or different economic conditions affect the life span of manufactured articles, a meaningful assertion cannot be made without a knowledge of the depletion function. While the order of dying in a biological population in general changes only gradually with time, the order of depletion of technical and economic populations depends substantially on the state of the art and the economic conditions prevailing at the time. Such depletion functions are thus much less stable. For accurate results they must be watched closely and continuously.

An elegant graphical procedure is here worthy of note. If we let T denote the characteristic life span, t the time and α the rate of depletion, the depletion function $F(t)$ has the simple form

$$F(t) = e^{-(t/T)^\alpha}. \qquad (2.10)$$

On graph paper with appropriately distorted scales, the **Stange life span chart** (1955), this curve is mapped onto a straight line, so that the set of observed points $\{t \mid F(t) = n(t)/n_0\}$—a small number of points suffices—is approximated by a straight line. The associated parameters T and α as well as the life span ratio $\bar{t}/T$ are read off from this graph. The mean life span $\bar{t}$ is then $\bar{t} = (\bar{t}/T)T$. Considerations of accuracy as well as examples of depletion functions for technical commodities and economic conditions as a whole from various fields can be found in the original paper. There also counter-examples are listed so as to avoid giving the impression that all depletion functions can be flattened into straight lines. The life span chart is especially valuable in the analysis of comparison experiments, as it provides the means whereby the question whether a new method prolongs the life span can be answered after a relatively brief period of observation.

In many life span and **failure time** problems the **exponential distribution** is used to get a general idea. Examples of life spans with approximately exponential distribution—the probability density decreases as the variable increases—are the life span of vacuum tubes and the duration of telephone conversations through a certain telephone exchange on any given day. The probability densities and cumulative probability densities

$$\boxed{f(x) = \theta e^{-\theta x}} \quad \boxed{F(x) = 1 - e^{-\theta x}} \quad (2.11, 2.12)$$

$$x \geq 0, \quad \theta > 0$$

of the exponential distribution are structurally simple. The parameter θ yields the mean and variance.

$$\boxed{\mu = \theta^{-1}, \quad \sigma^2 = \theta^{-2}.} \quad (2.13, 2.14)$$

The coefficient of variation equals 1; the median is $(\ln 2)/\theta = 0.69315/\theta$. It can be shown that 63.2% of the distribution lies below the mean, and 36.8% lies above it. For large n, the 95% confidence interval for θ is given approximately by $(1 \pm 1.96/\sqrt{n})/\bar{x}$.

A test for equality of two exponential distributions is given by S. K. Perng in Statistica Neerlandica **32** (1978), 93–102. Other important tests are given by Nelson (1968), Kabe (1970), Kumar and Patel (1971), Mann, Schafer, and Singpurwalla (1974), Gross and Clark (1975), and Lee (1980 [8:1]).

EXAMPLE. It takes 3 hours on the average to repair a car. What is the probability that the repair time is at most two hours?

It is assumed that the time t, measured in hours, needed to repair a car follows the exponential distribution; the parameter is $\theta = 1/$(average repair time) $= 1/3$. We get $P(t \leq 2) = F(2) = 1 - e^{-2/3} = 1 - 0.513 = 0.487$, a probability of barely 50%.

Of considerably greater significance for lifetime and reliability problems is the **Weibull distribution** (Weibull 1951, 1961), which can be viewed as a generalized exponential distribution. It involves 3 parameters, which allow it to approximate the normal distribution and a variety of other, unsymmetric distributions (it also reveals sample heterogeneity and/or mixed distributions). This very interesting distribution [see Mann et al. (1974), Gross and

2.4 Statistics in Engineering

Clark (1975), and Technometrics **18** (1976), 232–235, **19** (1977), 69–75, 323–331] has been tabulated (Plait 1962). For a comparison of two Weibull distributions see Thoman and Bain (1969) and Thoman et al. (1969). The probability density of the Weibull distribution with parameters for location (α), scale (β), and form (γ) reads

$$P(x) = \frac{\gamma}{\beta}\left(\frac{x-\alpha}{\beta}\right)^{\gamma-1} \exp\left[-\left(\frac{x-\alpha}{\beta}\right)^{\gamma}\right] \qquad (2.15)$$

for $x \geq \alpha$, $\beta > 0$, $\gamma > 0$,

where $\exp(t)$ means e^t.

It is generally better to work with the cumulative Weibull distribution:

$$F(x) = 1 - \exp\left[-\left(\frac{x-\alpha}{\beta}\right)^{\gamma}\right]. \qquad (2.16)$$

A nomogram for estimating the three parameters is given by I. Sen and V. Prabhashanker, Journal of Quality Technology **12** (1980), 138–143. A more detailed discussion of the interesting relationships between this distribution and the distributions in the study of life spans and failure times problems (e.g., lognormal in particular)] is presented in Freudenthal and Gumbel (1953) as well as in Lieblein and Zelen (1956). Examples are worked out in both papers. The significance of other distributions in the study of life spans and failure times problems (e.g. lognormal and even normal distributions) can be found in the surveys by Zaludova (1965), Morice (1966), Mann, et al., (1974), and also Gross and Clark (1975). More on survival models with the pertinent distributions (see also pages 107, 111) may be found in Elandt-Johnson and Johnson (1980 [8:2a]), Lee (1980 [8:1]), Sinha and Kale (1980), Lawless (1978, 1982 [8:2a]), Oakes (1983), and Cox and Oakes (1984) (see also Axtell 1963 and Kaufmann 1966, both cited in [8:2c]).

Remarks

1. The mean life spans of several products can be easily computed with the help of the tables provided by Nelson (1963) [8:2d].

2. Since electronic devices (like living beings) are particularly susceptible to breakdown at the beginning and toward the end of their life span, of special interest is the time interval of least susceptibility to breakdown, time of low failure rate, which generally lies between about 100 and 3000 hours. Tables for determining confidence limits of the mean time between failures (MTBF) are provided by Simonds (1963), who also gives examples (cf., also Honeychurch 1965, Goldberg 1981, Durr 1982).

Assume we have n objects with low and constant failure rate λ estimated by $\hat{\lambda} = 1/\bar{x}$; then the 95% CI for λ is given by

$$\frac{\hat{\lambda}}{2n}\chi^2_{2n;\,0.975} \leq \lambda \leq \frac{\hat{\lambda}}{2n}\chi^2_{2n;\,0.025} \qquad (2.17)$$

with χ^2 from Table 28 or from (1.132), (1.132a). Example: Life spans of 50 objects are given. We assume $\lambda \approx$ constant and find from the data a mean survival time of 20 years. Then we have $\hat{\lambda} = 1/20 = 0.05/\text{year}$ and with $\chi^2_{100;0.975} = 74.22$, $\chi^2_{100;0.025} = 129.56$, 95% CI: $(0.05/100)74.22 \leq \lambda \leq (0.05/100)129.56$ or 95% CI: $0.0371 \leq \lambda \leq 0.0648$.

3. It is of interest to compare two failure indices when the probability distribution is unknown [cf., also (1.185) in Section 1.6.6.1]. Let us call the number of failures over a fixed period of time in a piece of equipment the failure index. Then two failure indices x_1 and x_2 (with $x_1 > x_2$ and $x_1 + x_2 \gtrsim 10$) can be compared approximately in terms of

$$\boxed{d = \sqrt{x_1} - \sqrt{x_2}.} \qquad (2.18)$$

If $d > \sqrt{2} = 1.41$, the existence of a genuine difference may be taken as guaranteed at the 5% level.

A more exact test, with $x_1 > x_2$ and $x_1 + x_2 \gtrsim 10$, of whether all observations originated from the same population is based on the relation

$$\hat{z} = \sqrt{2}(\sqrt{x_1 - 0.5} - \sqrt{x_2 + 0.5})$$

underlying (2.18).

EXAMPLE. Two similar machines have $x_1 = 25$ and $x_2 = 16$ breakdowns in a certain month. With regard to the failure indices, are the differences between machine 1 and machine 2 statistically significant at the 5% level? Since $d = \sqrt{25} - \sqrt{16} = 5 - 4 = 1 < 1.41$, the differences are only random; we have $\hat{z} = \sqrt{2}(\sqrt{24.5} - \sqrt{16.5}) = 1.255 < 1.96 = z_{0.05}$.

Since the mean susceptibility to breakdown is seldom constant over a long period of time, as breaking in improves it and aging makes it worse, it should be checked regularly. Naturally these considerations are only completed by fitting Poisson distributions and by an analysis of the duration of the breakdowns by means of a frequency distribution. The mean total loss can be estimated from the product of the mean breakdown frequency and the mean breakdown time.

Reliability

The notion of the **reliability** of a device, in addition to the notion of life span is of great importance. By reliability we mean the probability of breakdown-free operation during a given time interval. Thus a component has a reliability of 0.99 or 99% if on the basis of long experience (or long trial sequences) we know that such a component will work properly over the specified time interval with a probability of 0.99. Straightforward methods

2.4 Statistics in Engineering

and simple auxiliary tools are provided by Eagle (1964), Schmid (1965), Drnas (1966), Prairie (1967), and Brewerton (1970). Surveys are given by Roberts (1964), Barlow and Proschan (1965), Shooman (1968), Amstadter (1970), Störmer (1970), Mann et al. (1974) and in the other books cited on p. 238.

Suppose a device is made up of 300 complicated component parts. If, e.g., 284 of these components could not break down at all, and if 12 had a reliability of 99% and 4 a reliability of 98%, then, given that the reliabilities of the individual components are mutually independent, the reliability of the device would be

$$1.00^{284} 0.99^{12} 0.98^4 = (1)(0.8864)(0.9224) = 0.8176,$$

not quite 82%. No one would buy this device. The manufacturer must therefore see to it that almost all components have a reliability of practically 1. Suppose a device consists of three elements A, B, C, which work perfectly with probabilities p_A, p_B, p_C. The performance of each of these elements is always independent of the state of the other two.

	Model	Reliability	Example $p_A = p_B = p_C = 0.98$
I	—Ⓐ—Ⓑ—Ⓒ—	$P_I = p_A \cdot p_B \cdot p_C$	$P_I^* = 0.94119$
II	Ⓐ—Ⓑ—Ⓒ / Ⓐ—Ⓑ—Ⓒ	$P_{II} = 1 - (1 - P_I)^2$	$P_{II} = 0.99653$
III	Ⓐ Ⓑ Ⓒ / Ⓐ Ⓑ Ⓒ	$P_{III} = (1-(1-p_A)^2) \cdot (1-(1-p_B)^2) \cdot (1-(1-p_C)^2)$	$P_{III} = 0.99930$
IV	Ⓐ Ⓑ Ⓒ / Ⓐ Ⓑ Ⓒ / Ⓐ Ⓑ Ⓒ	$P_{IV} = (1-(1-p_A)^3) \cdot (1-(1-p_B)^3) \cdot (1-(1-p_C)^3)$	$P_{IV} = 0.99999$

* For large survival probabilities p the approximation with the help of the sum of the breakdown probabilities is satisfactory and easier to compute: $P_1 \simeq 1 - (3)(0.02) = 0.94$.

The above reliability table for systems of types I to IV then results. By connecting in parallel a sufficient number of elements of each type—so that the system performs satisfactorily provided at least one of the components functions properly at all times—**the system can be made as reliable as desired**. However, this method of achieving high reliability is limited first by the costs involved, secondly by requirements of space, and thirdly by a **strange phenomenon**: each element has a certain probability of reacting spontaneously when it should not.

It turns out that for very many systems it is optimal to parallel two or, even more frequently, three units of each element (cf., Kapur and Lamberson 1977, Henley and Kumamoto 1980, Tillman et al. 1980, Dhillon and Singh 1981, and Goldberg 1981). For example, the triplex instrument landing system permits fully automatic landing of jet aircraft with zero visibility. Each component of the system is present in triplicate; the failure rate should be less than one failure per ten million landing. Guild and Chips (1977) discuss the reliability improvement in any parallel system by multiplexing. This is important for safety systems (e.g., reactor safety; see Henley and Kumamoto 1980, and Gheorghe 1983).

More on reliability analysis is found in the following books and papers: Amstadter (1970), Mann et al. (1974), Proschan and Serfling (1974), Barlow et al. (1975), Fussell and Burdick (1977), Kapur and Lamberson (1977), Kaufman et al. (1977), Tsokos and Shimi (1977), Henley and Kumamoto (1980), Sinha and Kale (1980), Tillman et al. (1980), Dhillon and Singh (1981), Goldberg (1981), Durr (1982), Nelson (1982), Martz and Waller (1982), and Tillman et al. (1982).

Maintainability

By **maintainability** we mean the property of a device, a plant, or system, that it can be put back into working order in a certain period of time in the field with the help of repair and test equipment according to regulations. Beyond preventive maintenance, costly strategic weapons systems require a complex maintenance policy. Goldman and Slattery (1964) considered five possibilities for submarines: abandoning and scuttling, repair at a friendly or home port, repair at a dockyard, repair involving a repair boat, on-the-spot repair. A mathematical consideration of this decision problem requires the availability of appropriate empirical data (reliability, repair time, type and number of periodic checks, etc.) and profitability studies (e.g., regarding a comparison between automatic control equipment and manual control). Lie et al. (1977) give a survey on availability, which is a combined measure of maintainability and reliability (see also Sherif and Smith 1981, Sherif 1982, Tillman et al. 1982, and Gheorghe 1983).

2.5 OPERATIONS RESEARCH

Operations research or management science, also called industrial planning or methods research, **consists in a systematic study of** contingencies. On the basis of a mathematical-statical model **optimal solutions** are developed **for compound systems, organizations and processes** with the help of an electronic computer. By writing a computer program in accordance with the model and running it with impartial data, the problem is simulated and results are obtained that suit the real system. This could be a traffic network, a chemical manufacturing process, or the flow of blood through the kidneys. As "simulation models" allow for an unrestricted choice of parameter values, complicated problems under various extraneous conditions can be solved without great expense and without the risk of failure. Simulation and linear programming play an important role in operations research (cf. Flagle et al. 1960, Hertz 1964, Sasieni et al. 1965, Stoller 1965, Saaty 1972, and also Müller-Merbach 1973). More on operations research is found in Anderson (1982), Harper and Lim (1982), and Kohlas (1982).

2.5.1 Linear programming

Linear programming (linear optimization) is an interesting method of production planning. It is capable of solving problems involved in the development of an optimal production program on the basis of linear inequalities. Nonlinear relations can sometimes be linearly approximated. By means of linear optimization one can e.g., regulate the manufacture of several products with various profit margins and with given production capacities of the machines so as to maximize the overall profit. Shipments can be so organized that costs or transit times are minimized. This is known as the **problem of the traveling salesman**, who must visit various cities and then return, and who must choose the shortest path for this trip. In the metal industry linear programming is of value in determining workshop loading, in minimizing stumpage and other material losses, and in deciding whether some single component is to be manufactured or purchased. This technique finds very important application in the optimization of the various means of transportation, in particular, the determination of air and sea routes and the arrangement of air and sea shipping plans with fixed as well as uncertain requirements. Models of this sort with **unspecified requirements** or with **variable costs taken into account** are of particular interest for the statistician. Here **uncertainty** appears that is caused by random events (number of tourists, inflationary tendency, employment quota, government policy, weather, accidents, etc.) about whose distribution little or nothing is known. A familiar example is the knapsack problem: The contents may weigh not more than 25 kg but must include all that is "necessary" for a long trip.

Linear programming (cf., Dantzig 1966) is concerned with optimizing (maximizing or minimizing) a certain specific target function of several variables under certain restricting conditions, given as inequalities. The so-called **simplex method**, based on geometrical reasoning, is used to obtain a solution. The auxiliary conditions limit the target function to the interior and surface of a simplex, i.e., a multidimensional convex polyhedron. A certain corner of the polyhedron which a programmed digital computer that follows an interaction method systematically approaches in the course of successive iterations represents the desired optimum (see Kaplan 1982 [8:2f] and Sakarovitch 1983).

2.5.2 Game theory and the war game

While probability theory concerns itself with games of pure chance, game theory considers **strategic games** (von Neumann 1928), games in which the participants have to make **decisions** during the play in accordance with certain rules and can partially influence the result. In some games played with dice, the players decide which pieces are to be moved, but in addition there is the chance associated with a throw of the die that determines how many places the chosen piece must be advanced. Most parlor games involve factors of chance, elements over which the players have no control: in card games, e.g., which cards a player gets; in board games, who has the first move and thus in many cases the advantage of giving the game a certain tack right at the outset.

Games and situations in economics and technology have much in common: chance, incomplete information, conflicts, coalitions, and rational decisions. Game theory thus provides ideas and methods to devise procedures for coping with conflicting business interests. It concerns itself with the question of optimal attitudes for "players" in a wide class of "games," or best "strategies" for resolving conflicting situations. It studies models of economic life as well as problems of military strategy, and determines which behavior of individuals, groups, organizations, managers or military leaders—which comprehensive plan of action, which strategy, applicable in every conceivable situation—is rationally justifiable in terms of a "utility scale." Intrinsic to all of this is the appearance of subjects who have the power to decide and whose objectives differ, whose destinies are closely interwoven, and who, striving for maximal "utility," influence but cannot fully determine the outcome by their modes of behavior. Strategic planning games of an economic or military type—computers permit **"experimentation on the model"**—show the consequences of various decisions and strategies. More on this can be found in Vogelsang (1963) and especially in Williams (1966) (cf., also Charnes and Cooper 1961, David 1963 [8:1], Dresher et al. 1964, Brams 1979, Jones 1979, Packel 1981, Berlekamp et al. 1982, and Kaplan 1982).

At the beginning of the 19th century the Prussian military advisor von Reisswitz devised in Breslau the so-called "sandbox exercise" which, by introduction of rules, was expanded by his son and others into a **war game** and acquired permanent status shortly thereafter; it was, in particular, included in the curriculum of officer's training in Germany. Dice were later introduced to simulate random events; troops were no longer represented by figures but were drawn in with wax crayons on maps coated with plastic. With the help of advanced war games the military campaign of 1941 against the USSR (operation "Barbarossa"), the action "Sea Lion" against Great Britain, and the Ardennes offensives of 1940 and 1944 were "rehearsed from beginning to end" (Young 1959). Pursuit or evasion games, e.g., two "players": one trying to escape, the other trying to shoot him down, were considered by Isaacs (1965). Further discussion of war games is to be found in Wilson (1969) (cf., also Bauknecht 1967, Eckler 1969). After the Second World War, war games were employed in economics, they evolved from stockkeeping and supply games of the U.S. Air Force. Their function is to provide the means whereby management can run an experimental trial of business policies with restricting quantities: output, capacity, prices, capital investments, taxes, profit, ready money, depreciation, share of market, stock prices, etc. on the basis of mathematical models that correspond as closely as possible to reality, models with quick motion and competition effects: the groups of players are competing with each other; the decisions of the groups influence one another. Obviously such simulations can only be carried out with the help of a computer.

2.5.3 The Monte Carlo method and computer simulation

An important task of operations research is to analyze a given complex situation logically and construct an analogous mathematical model, to translate the model into a computer program, and to run it with realistic data: **The original problem is simulated and is guided to an optimal solution.**

If sampling is too costly or not at all feasible, an approximate solution can frequently be obtained from a **simulated sample**, which sometimes yields additional valuable information as well. Simulated sampling ordinarily consists in replacing the actual population, which is characterized by a hypothetical probability distribution, with its theoretical representation, a stochastic "simulation model," and then drawing samples from the theoretical population with the help of random numbers. A digital computer is usually employed, which then also generates pseudo random numbers having the same prescribed statistical distribution as authentic random numbers, e.g., uniform distribution, normal distribution, or Poisson distribution.

Since by a theorem of probability theory every probability density can be transformed into a rectangular distribution between zero and one, a sample whose values follow an arbitrary preselected probability distribution can be obtained by drawing random numbers from the interval between 0 and 1. The so-called **Monte Carlo method** is based on this fact (cf., Hammersley and Handscomb 1964, Buslenko and Schreider 1964, Schreider 1964, Lehmann 1967, Halton 1970, Newman and Odell 1971, Kohlas 1972, Sowey 1972). Examples of applications of this method are simulation and analysis of stochastic processes, computation of critical bounds for test statistics (e.g. t-statistics), estimation of the goodness of a test, and investigation of the influence of different variances on the comparison of two means (Behrens–Fisher problem). This method was quickly extended to the broad field of simulation (cf., Shubik 1960, Guetzkow 1962, Tocher 1963, Teichroew 1965 [8:1], Pritsker and Pegden 1979, Goldberg 1981 [8:2d], Maryanski 1981, Rubinstein 1981, Cellier 1982, Dutter and Ganster 1982, Payne 1982, and Bratley et al. 1983).

Computer simulation is the solution of any mathematical problems by sampling methods. The procedure is to construct an **artificial stochastic model** (a model with random variables) of the mathematical processes and then to perform sampling experiments on it.

Computer simulation examples

1. Test characteristics of the sequential charts designed by Bross (cf., Section 2.2, Figure 35) (Page 1978).
2. Robustness of both one sample and two sample t-tests (cf., Sections 3.1 and 3.6) and the chance of determining departure from normality (cf., Sections 4.3 and 4.4) [Pearson, E. S. and N. W. Please: Biometrika **62** (1975), 223–241].
3. Power of the U test (cf., Section 3.9.4) (Van der Laan and Oosterhoff 1965, 1967).

4. Sensitivity of the distribution of *r* against nonnormality (cf., Sections 5.1 and 5.3; normal correlations analyses should be limited to bivariate normal distributions) (Kowalski 1972).
5. Error rates in multiple comparisons among means (cf., Sections 7.3.2 and 7.4.2) (Thomas 1974).

It is common practice, especially in technology, to study a system by experimenting with a model. An aerodynamic model in a wind tunnel provides information about the properties of an aircraft in the planning stage. In contrast with physical models, abstract models, as simulated by a computer program, are much more flexible. They permit easy, quick, and low cost experimentation. The two principal aims of simulation are assessing the capability of a system before it is realized and ascertaining that the system chosen fulfills the desired criteria. The task of simulation is to provide sufficient data and statistical information on the dynamic operation and capability of a certain system. The system and/or model can be reconsidered in the light of these results, and appropriate modifications can then be introduced. By varying the parameters inherent in the proposed model, the simulated system can be optimally adapted to the desired properties. The simulation of businesses and industries, of traffic flows and nervous systems, of military operations and international crises provides insights into the behavior of a complex system. This is particularly useful when an exact treatment of a system is too costly or not feasible and a relatively quick approximate solution is called for. Analogue computers are also used for such problems.

Examples of digital devices (operated "in final units") are desk calculators, cash registers, bookkeeping machines, and mileage indicators in automobiles. The result is obtained by "counting." In contrast with this, speedometers and other gauges, the needles of which move continuously—**measuring**—function as **analogue** devices. Also to be included here is the **slide rule**, scaled with a continuum of numbers: Each number is assigned an interval, the length of which is proportional to the logarithm of the number. Multiplication of two numbers, for example, is "translated" into adjoining the two corresponding intervals, so that their lengths are thereby added. The **digital computer** (cf., e.g., Richards 1966, and Klerer and Korn 1967) is not based on decimal numbers (0 to 9) but rather on the binary numbers or binary digits zero and one (0, 1) (frequently denoted by the letters O and L to distinguish them more easily) because 1, 0 adapt naturally to any electrical system; thus the construction is simplified and the machine operates more reliably. In writing 365, the following operation is carried out:

$$365 = 300 + 60 + 5 = 3(10^2) + 6(10^1) + 5(10^0).$$

Our notation dispenses with the powers of ten, indicating only the factors, here 3, 6, and 5, in symbolic positions. If 45 is given in powers of 2 (cf., $2^0 = 1, 2^1 = 2, 2^2 = 4, 2^3 = 8, 2^4 = 16, 2^5 = 32$, etc.),

$$45 = 32 + 8 + 4 + 1 = 1(2^5) + 0(2^4) + 1(2^3) + 1(2^2) + 0(2^1) + 1(2^0),$$

and the 2 with its powers from 0 to 5 is dispensed with, then the dual notation for 45 is 101101 or preferably **LOLLOL**. The transformation from decimal to binary representation at the input and the inverse transformation at the output is ordinarily provided by the computer.

The digital computer is indispensable whenever extensive and **very complicated calculations requiring a high degree of accuracy are called for**.

Analogue computers generally work with a continuous electrical signal (cf., Karplus and Soroka 1959, Rogers and Connolly 1960, Fifer 1963, Röpke and Riemann 1969, Wilkins 1970, Adler and Neidhold 1974). A **particular number is represented by a proportional voltage**. We obtain a physical analogue of the given problem in which the varying physical quantities have the same mathematical interdependence as the quantities in the mathematical problem. Hence the name analogue computer. The pressure balance between two gas containers can thus be studied by analogy on two capacitors connected through a resistor. Analogue computers are "living" mathematical models. The immediate display of the solution on a TV screen puts the engineer in a position where he can directly alter the parameters (by turning some knobs) and thereby zero in **very rapidly** on an optimal solution to the problem. The accuracy that can be realized depends on the accuracy of the model, on the noise in the electronic components, on the measurement device and on the tolerances of the electrical and mechanical parts. Although a single computer element (amplifier) can attain an accuracy to at most 4 decimal places or 99.99% (i.e., the computational error is $\geq 0.01\%$ or so) the overall error of about 100 interconnected amplifiers is as large as that of a slide rule. The power of such a computer lies in its ability to handle problems whose solution requires **repeated integration**, i.e., differential equations. **High speed computation, rapid parameter variation, and visual display of results** distinguish analogue computers as "laboratory machines," which are usually less expensive than the hardly comparable digital computers. **Random numbers with preassigned statistical distributions** can be produced by a **random number generator**. Two classified bibliographies on random number generation and testing are given by Sowey (1972, 1978).

Analogue computers can be used for **approximating empirical functions** (i.e., searching for mathematical relations in experimentally determined curves, solving algebraic equations, and integrating ordinary differential equations), for analyzing biological regulating systems, **for designing, controlling and monitoring atomic reactors and particle accelerators**, for monitoring chemical processes and electrical control loops in general, and for **simulations**.

A fusion of the two original principles, digital and analogue, yields the *hybrid computer*. It is characterized by digital-analogue and analogue-digital converters, devices which transform a numerical digit into an analogous potential difference and conversely. A hybrid computer *combines* the advantages of continuous and discrete computational techniques: **the speed of computation and the straightforward methods of altering an analogue computer program with the precision and flexibility of a stored program digital computer**. Hybrid computers are used to solve differential equations and to optimize processes: they regulate hot strip rolling-mill trains, traffic, satellites, and power plants as well as processes in the chemical industry, e.g., crude oil fractionation. We also speak of **process automatization by a "process computer."** Process computer technology represents one of the most radical changes in industrial production. Large hybrid computers with an analogue portion made up of more than 100 amplifiers are used in particular in the aviation and space flight industries, e.g., for the **calculation of rocket and satellite trajectories**. Consult e.g., Anke and Sartorius (1968), Bekey and Karplus (1969), Anke et al. (1970), Barney and Hambury (1970), and Adler and Neidhold (1974) for further discussion on the above.

3 THE COMPARISON OF INDEPENDENT DATA SAMPLES

Special sampling procedures

If we know something about the heterogeneity that is to be expected within the population we wish to study, then there are more effective sampling schemes than total randomization. Of importance is the use of **stratified** samples; here the population is subdivided into relatively homogeneous partial populations (layers or strata), always in accordance with points of view that are meaningful in the study of the variables of interest. If a prediction of election results is called for, then the sample is chosen in such a way as to be a miniature model of the overall population. Thus age stratification, the relative proportion of men and women and the income gradation are taken into account. Also the work force in a modern industrialized nation can be classified according to occupational status as, for example, 50% laborers, 35% white-collar workers, 8% self-employed, and 7% civil servants. Stratification for the most part increases the cost of the sample survey; nevertheless, it is an important device.

In constrast to this, the procedure in a **systematic** sample is such that every qth individual of the population is chosen according to a list of a certain type (quota procedure). Here q is the quotient, rounded off to the nearest integer, which is obtained on dividing the total population by the sample size. Population censuses, candidate lists, or index files of the public health authority can be utilized in choosing a systematic sample. It is of course required that the underlying list be free of periodic variation. Indeed, an **unobjectionable random selection** is possible only if the units—e.g., index cards—are brought into **random order** by **mixing**, whereupon **every qth card is systematically drawn**. Using a systematic sample has the advantage that it is frequently easier to pick out every qth individual than to choose entirely at random. Moreover, the method itself produces indirect stratification in certain cases, for example when the original list is ordered according to residences, occupations, or income groups. Selection procedures not based on the randomness principle, i.e., **most of the quota procedures** and in particular the **choice of typical cases**, do not, however, permit statements as to the reliability of results based on them. They are therefore to be avoided.

Sampling in **clusters** is particularly suited for demographical problems. The population is here subdivided into small, relatively homogeneous groups or clusters which can with economic advantage be jointly studied. A random sample consisting of clusters (families, school grades, houses, villages, blocks of streets, city districts) is then analyzed. Multilevel random selections are feasible (e.g., villages and within them houses, again chosen at random).

Frames for clusters (municipalities, firms, clinics, households) are ordinarily available. Clusters are also more stable in time than the respective units (households, employees, patients, persons). That it is not easy to avoid false conclusions due to the selection used, is illustrated by the following example: Assume two illnesses are independent and the admission probabilities at the clinic differ for the two. The individual groups are differently selected in the process, so that **artificial associations** are created. This **selection correlation**—which, as we said, is not true of the population (cf. also Sections 2.1.4, 5.2)—was recognized by J. Berkson as a source of false conclusions. It results from not taking into account the difference between entrance and exit probabilities.

Some other selection procedures are:

1. **Selection according to final digit** on numbered file cards. If, e.g., a sample with a sampling fraction of 20%, is to be drawn, all cards with final digit 3 or 7 can be chosen. Quota procedures are open to non-random errors.
2. Selection of persons by means of their **birthdays**. In this selection procedure all persons born on certain days of the year are included in the sample. If, e.g., all those born on the 11th of any month are chosen, one gets a sample with a sampling fraction of about $12/365 = 0.033$, i.e., approximately 3%. This procedure can be used only when appropriate frames (e.g., lists, cards) are available for the given class of persons.

Questions connected with the size and accuracy of samples and the expense and economy of **sampling** are considered by Szameitat et al. (1958, 1964). For the class of problems in error control (cf., Sections 2.4 and 2.1.3) and data processing see Szameitat and Deininger (1969) as well as Minton (1969, 1970). More on this can be found in books listed in the bibliography [8:3a]. Ford and Totora (1978) provide a **checklist for designing a survey**. The uncertainties of **opinion polls** are discussed by R. Wilson in New Scientist **82** (1979), 251–253.

3.1 THE CONFIDENCE INTERVAL OF THE MEAN AND OF THE MEDIAN

The notion of **confidence interval** was introduced by J. Neyman and E. S. Pearson (cf., Neyman 1950). It is defined as an interval computed from sample values which includes the true but unknown parameter with a specified probability, the confidence probability. The confidence probability is usually selected to be 0.95 (or 95%); this probability tells us that when the experiment is repeated again and again, the corresponding confidence interval, on the average, includes the parameter in 95% of the cases and fails to include it in only 5% of the cases.

We continue (cf. pp. 46 and 66) to use the estimate notation and not the estimator notation.

▶ 3.1.1 Confidence interval for the mean

Let $x_1, x_2, \ldots, x_n$ be a random sample from a normally distributed population. Assume the mean of the population is unknown. We seek two values, l and u which are to be computed from the sample and which include with a given, not too small probability the unknown parameter μ between them: $l \leq \mu \leq u$. These limits are called **confidence limits**, and they determine the so-called confidence interval. The parameter of interest μ then lies with confidence coefficient S (cf. Section 1.4.2) between the confidence limits

$$\bar{x} \pm \frac{ts}{\sqrt{n}}, \tag{3.1}$$

with $t = t_{n-1;\alpha}$ (the factor of Student's distribution: Table 27, Section 1.5.2), i.e., in $100S\%$ of all samples, on the average, these limits will encompass the true value of the parameter:

$$P\left(\bar{x} - \frac{ts}{\sqrt{n}} \leq \mu \leq \bar{x} + \frac{ts}{\sqrt{n}}\right) = S. \tag{3.1a}$$

In an average of $100(1 - S)\%$ of all samples these limits will not include the parameter, that is, in an average of $100(1 - S)/2 = 100\alpha/2\%$ of all samples it will lie above, and in an average of $100(1 - S)/2 = 100\alpha/2\%$ of all samples below, the confidence interval. Let us recall that for the two sided confidence interval in question we have $\alpha/2 + S + \alpha/2 = 1$. One sided confidence intervals (e.g., upper confidence limits $\mu_{\text{up.}} = \bar{x} + t_{\text{ones.}} s/\sqrt{n}$)

$$P\left(\bar{x} - \frac{ts}{\sqrt{n}} \leq \mu\right) = S \quad \text{or} \quad P\left(\mu \leq \bar{x} + \frac{ts}{\sqrt{n}}\right) = S \tag{3.1b}$$

with $t = t_{n-1, \alpha, \text{ones.}}$ do not include the parameter in an average of $100\alpha\%$ of all cases, but do cover it in an average of $100S\%$ of all cases ($\alpha + S = 1$). If σ is known or if s is computed from a very large n (i.e., $s \simeq \sigma$), then (3.1) is replaced by (z = standard normal variable)

$$\bar{x} \pm z \frac{\sigma}{\sqrt{n}} \quad \text{(sampling with replacement)} \tag{3.2}$$

with $z = 1.96$ ($S = 95\%$), $z = 2.58$ ($S = 99\%$), and $z = 3.29$ ($S = 99.9\%$). These results are all based on the assumption that we sampled from an infinite population or from a finite population with replacement. If the sample originates in a finite population of size N and after drawing and evaluation is not reintroduced into the population, then we have the confidence limits

$$\boxed{\bar{x} \pm z \frac{\sigma}{\sqrt{n}} \sqrt{\frac{N-n}{N-1}}} \quad \text{(sampling without replacement).} \quad (3.2a)$$

The root $\sqrt{(N-n)/(N-1)}$ is referred to as the finite population correction. The quotient $\sigma/\sqrt{n}$ was introduced in Section 1.3.8.4, as the standard error of the mean ($\sigma_{\bar{x}}$). The confidence interval (CI) for μ can thus be written as

(pp. 62, 136–137)

$$\boxed{\bar{x} \pm z\sigma_{\bar{x}} \quad \text{or} \quad \bar{x} \pm ts_{\bar{x}};} \quad (3.2b, 3.1c)$$

if the distribution is not markedly different from a normal distribution, (3.1) through (3.1c) are still approximately valid (cf. also Section 2.1.2).

EXAMPLE. Let a random sample with $n = 200$, $\bar{x} = 320$, $s = 20$ from a large population [$N(\mu, \sigma)$, cf., Section 1.3.4] be given. Determine the 95% confidence interval of the mean.

$$t_{199;0.05} = 1.972,$$
$$t \cdot s_{\bar{x}} = 1.972 \cdot 1.414 = 2.79, \quad s_{\bar{x}} = \frac{s}{\sqrt{n}} = \frac{20}{\sqrt{200}} = 1.414,$$

$$z = 1.96,$$
$$z \cdot s_{\bar{x}} = 1.96 \cdot 1.414 = 2.77, \quad 317 \leq \mu \leq 323.$$

When needed, the seldom used percentage confidence interval is computed according to

$$\frac{t}{\bar{x}} \cdot s_{\bar{x}} = \frac{1.972}{320} \cdot 1.414 = 0.0087 \simeq 0.9\%,$$

or

$$\frac{z}{\bar{x}} \cdot s_{\bar{x}} = \frac{1.96}{320} \cdot 1.414 = 0.0087 \simeq 0.9\%.$$

The 95% CI for μ is stated as "95% CI: $\bar{x} \pm ts_{\bar{x}}$" [cf., (3.1)–(3.1c) with $t = t_{n-1;0.05;\text{twos}}$, or better yet, as "95% CI: $a \leq \mu \leq b$"; e.g. (95% CI: 320 ± 3), 95% CI: $317 \leq \mu \leq 323$. The limits 300 ± 2.78 are called upper and lower 95% confidence limits, and $100(1 - 0.05) = 95\%$ is the confidence coefficient. Statements of the type $317 \leq \mu \leq 322$ are approximately true 95% of the times the method is used, **provided we have random samples of normal populations**. The confidence interval is a random interval: if we draw samples under identical conditions and complete a 95% confidence interval for each sample, then in the long run 95% of these confidence

3.1 The Confidence Interval of the Mean and of the Median

intervals would include the true value of μ. The confidence interval is necessary for **reporting uncertainty** in the value of the parameter.

A useful collection of tables for determining the confidence limits in terms of estimated or known standard deviations is provided by Pierson (1963).

Remark: Inverse and direct inference

If we use the values of a sample and (3.1) to make a statement on the mean of the population, we have an **inverse inference** or, considering the sample as representing the population, a **representative inference**:

$$\bar{x} - t\frac{s}{\sqrt{n}} \leq \mu \leq \bar{x} + t\frac{s}{\sqrt{n}}. \qquad (3.1d)$$

On the other hand, the mean of the sample deduced from the parameters of the population,

$$\mu - z\frac{\sigma}{\sqrt{n}} \leq \bar{x} \leq \mu + z\frac{\sigma}{\sqrt{n}}. \qquad (3.3)$$

is a **direct inference** or, since the population **includes** the sample, an **inclusion inference**. If conclusions about a sample are drawn from the values of another sample originating in the same population, we have a **transposition inference**.

Hahn (1970) gives vital "**prediction intervals**" for transposition inference in normally distributed populations: prediction intervals for future observations as well as for the mean of future observations. A survey with applications is given by G. J. Hahn and W. Nelson, Journal of Quality Technology 5 (1973), 178–188. Tables, and examples for the nonparametric case are given by Hall et al. (1975).

▶ 3.1.2 Estimation of sample sizes

Minimal number of observations for estimating a standard deviation and a mean

The following formulas give the minimal sizes (n_s and $n_{\bar{x}}$) for the estimation of the standard deviation and the mean with specified accuracy d and given confidence coefficient S. The estimates n_s and $n_{\bar{x}}$ are approximations based on the normal distribution; for $d = (s - \sigma)/\sigma$ and $d = \bar{x} - \mu$ respectively,

$$n_s \simeq 1 + 0.5\left(\frac{z_\alpha}{d}\right)^2, \qquad n_{\bar{x}} = \left(\frac{z_\alpha}{d}\right)^2 \cdot \sigma^2 \qquad (3.4, 3.5)$$

values of z_α are given in Table 43 in Section 2.1.6 for a two sided test, $\alpha = 1 - S$. For the examples we use $z_{0.05} = 1.96$ and $z_{0.01} = 2.58$.

Examples

(n_s) To estimate a **standard deviation** with a confidence coefficient of 95% ($\alpha = 0.05$) and an accuracy of $d = 0.2$, about $n_s \simeq 1 + 0.5(1.96/0.2)^2 = 49$ observations are required. For the same confidence coefficient $S = 95\%$ or $S = 0.95$ ($\alpha = 0.05$) but an accuracy of $d = 0.14$, about $n_s \simeq 1 + 0.5 (1.96/0.14)^2 = 99$ observations are called for.

Table 48 gives $n_s = 100$.

($n_{\bar{x}}$) To obtain an estimate of σ^2 one avails oneself of Remarks 5 and 6 in Section 1.3.8.5. Knowing the variance $\sigma^2 = 3$, to estimate a **mean** with a confidence coefficient of 99% ($\alpha = 0.01$) and with an accuracy of $d = 0.5$, about $n_{\bar{x}} = (2.58/0.5)^2(3) = 80$ observations are needed; i.e., with about 80 observations ($2.58\sqrt{3/80} \simeq 0.5$) the 99% CI for μ ($\bar{x} - 0.5 \leq \mu \leq \bar{x} + 0.5$, or equivalently $\mu = \bar{x} \pm 0.5$) of length $2d$ is obtained.

Remark on $n_{\bar{x}}$ (n for short). If n is larger than 10% of the population size N, ($n > 0.1N$), then not n but fewer, namely $n' = n/(1 + n/N)$ observations are sufficient (for the same confidence level and accuracy). For $N = 750$, not 80 but only $80/(1 + (80/750)) = 72$ observations are thus needed.

Other questions relating to the minimal size of samples will be dealt with again later on (Section 3.8; cf., also the remark at the end of the last section and the references to Hahn, Nelson and Hall). More on the choice of appropriate sample sizes can be found in Mace (1964), Odeh and Fox (1975), and Cohen (1977) (cf., also Goldman 1961, McHugh 1961, Guenther 1965 [8:1; see end of Section 1.4.7]), and Winne 1968, as well as Gross and Clark 1975 [8:2d]).

Table 48 Values of d, the half length of the confidence interval for the relative error of the standard deviation [$d = (s - \sigma)/\sigma$] of a normally distributed population for certain confidence coefficients S ($S = 1 - \alpha$) and sample sizes n_s. Compare the second example with (3.4). (From Thompson, W. A., Jr. and Endriss, J.: The required sample size when estimating variances. American Statistician **15** (1961), 22–23, p. 22, Table 1).

n_s \ S	0.99	0.95	0.90	0.80
4	0.96	0.75	0.64	0.50
6	0.77	0.60	0.50	0.40
8	0.66	0.51	0.43	0.34
10	0.59	0.45	0.38	0.30
12	0.54	0.41	0.35	0.27
15	0.48	0.37	0.31	0.24
20	0.41	0.32	0.27	0.21
25	0.37	0.28	0.24	0.18
30	0.34	0.26	0.22	0.17
100	0.18	0.14	0.12	0.09
1000	0.06	0.04	0.04	0.03

Minimal number of observations for the comparison of two means

If a considerable difference is expected between the means of two populations—no overlap of the two data sets—then 3 to 4 ($\alpha = 0.05$) or 4 to 5 ($\alpha = 0.01$) observations should suffice.

To prove there is an actual difference δ (delta) between the means of two normally distributed populations with the same variance about

$$n = 2(z_\alpha + z_\beta)^2 \left[\frac{\sigma^2}{\delta^2}\right], \quad (3.6)$$

independent observations from each population (i.e., $n_1 = n_2 = n$) are required (cf. also Table 52, Section 3.6.2). The values of z_α and z_β—compare what is said at the end of Section 1.43 concerning Type I and Type II errors—are found in Table 43, Section 2.1.6. The value of z_α depends on which type of test, one sided or two sided is planned; z_β is always the value for the one sided test. A sufficiently precise estimate of the common variance σ^2,

$$s^2 = \frac{(n_a - 1)s_a^2 + (n_b - 1)s_b^2}{n_a + n_b - 2},$$

should be available.

EXAMPLE. $\delta = 1.1$, $\alpha = 0.05$ (two sided), i.e., $z_{0.05; \text{two sided}} = 1.960$; $\sigma^2 = 3.0$, $\beta = 0.10$ (one sided), i.e., $z_{0.10; \text{one sided}} = 1.282$.

$$n = 2(1.960 + 1.282)^2 [3.0/1.1^2] = 52.12.$$

About $53 + 53 = 106$ observations have to be taken. Then we can assume that in the case of a two sided problem a true difference of at least 1.1 can be recognized with a probability (power) of at least 90%. Note that for the α and β (or for the Type I and II errors) given in this example we have $n \simeq 21$ (σ^2/δ^2) or $n \simeq 21(3/1.1^2) = 52.1$.

3.1.3 The mean absolute deviation

In distributions with at least one long tail the **mean absolute deviation from the mean** (MD) can also be used as a measure of dispersion. It is defined by

$$MD = \frac{\sum |x_i - \bar{x}|}{n}, \quad (3.7)$$

for grouped observations,

$$MD = \frac{\sum |x_i - \bar{x}| f_i}{\sum f_i}, \quad (3.8)$$

where x_i = class mean, $\sum f_i = n$; but it can more quickly be estimated according to

$$MD = \frac{2}{n} \sum_{x_i > \bar{x}} (x_i - \bar{x}) = 2 \frac{\sum_{x_i > \bar{x}} x_i - n_1 \bar{x}}{n} \quad (n_1 \text{ values } x_i > \bar{x}).$$

(3.8a)

The MD of 1, 2, 3, 4, 5 is thus

$$MD = \frac{2}{5}[(4-3) + (5-3)] = 2[(4+5) - 2 \cdot 3]/5 = 6/5 = 1.2.$$

For small sample sizes (and when the extreme values are suspect) **the MD is superior to the otherwise optimal standard deviation** (cf., Tukey 1960): Values far from the mean are less influential than in the usual estimate, and this is particularly important for distributions that resemble the normal but have heavier tails. Thus the influence of a potential maverick (cf., Section 3.8) is also reduced, and deciding whether to still accept an extreme value or to reject it becomes less critical.

A distribution-free substitute for s and MD is the median deviation (3.11).

Three remarks

(1) MD/σ and kurtosis. The ratio MD/σ has for the uniform distribution the value $\sqrt{3}/2 = 0.86603$, for the triangular distribution $(16/27)\sqrt{2} = 0.83805$, for the normal distribution $\sqrt{2/\pi} = 0.79788$, and for the exponential distribution $2/e = 0.73576$. For samples from approximately normally distributed populations we have $|[MD/s] - 0.7979| < 0.4/\sqrt{n}$. Of course $|[MD/s] - 0.7979|$ measures only the deviation from the kurtosis of a normal distribution. According to D'Agostino (1970), $(a - 0.7979)\sqrt{n}/0.2123$ with $a = 2(\sum_{x_i > \bar{x}} x_i - n_1 \bar{x}\sqrt{n \sum x^2 - (\sum x)^2}$, for n_1 see (3.8a), is approximately standard normally distributed (critical limits are given by Geary 1936) even from small n (kurtosis-related quick test for nonnormality). A test for nonnormality involving kurtosis and skewness is likewise given by D'Agostino (1971, 1972).

(2) 95% confidence interval for μ using the MD. The 95% confidence interval for μ in terms of the MD is found according to

$$\bar{x} \pm (\text{coefficient}) \, MD. \tag{3.9}$$

Coefficients of MD for the sample size n are found in Table 49.

The equality of two or more MD's can be tested by means of tables (Cadwell 1953, 1954). A table for the corresponding one and two sample t-test based on the MD is given by Herrey (1971).

3.1 The Confidence Interval of the Mean and of the Median

Table 49 Coefficients for determining the 95% confidence limits for the mean in terms of the mean absolute deviation. From Herrey E. M. J.: Confidence intervals based on the mean absolute deviation of a normal sample. J. Amer. Statist. Assoc. **60** (1965), p. 267, part of Table 2. Factors for the other usual confidence limits are given by Krutchkoff (1966).

n	Factor	n	Factor
2	12.71	12	0.82
3	3.45	13	0.78
4	2.16	14	0.75
5	1.66	15	0.71
6	1.40	20	0.60
7	1.21	25	0.53
8	1.09	30	0.48
9	1.00	40	0.41
10	0.93	60	0.33
11	0.87	120	0.23

EXAMPLE. Given the eight observations 8, 9, 3, 8, 18, 9, 8, 9 with $\bar{x} = 9$. Determine the 95% confidence interval for μ. First we compute $\sum |x_i - \bar{x}|$:

$$\sum |x_i - \bar{x}| = |8-9| + |9-9| + |3-9| + |8-9| + |18-9| + |9-9| + |8-9| + |9-9|,$$

$$\sum |x_i - \bar{x}| = 1 + 0 + 6 + 1 + 9 + 0 + 1 + 0 = 18,$$

and the mean absolute deviation is, according to (3.7), $MD = 18/8 = 2.25$, or, according to (3.8a), $MD = 2[18 - 1(9)]/8 = 2.25$. For $n = 8$ the factor is found from Table 49 to be 1.09. We then get by (3.9) for the 95% confidence interval the interval $9 \pm (1.09)(2.25) = 9 \pm 2.45$. Thus we have the 95% CI: $6.55 \leq \mu \leq 11.45$.

(3) **50% confidence interval for μ after Peters.** For $n \geq 7$, and for a normal distribution, the approximation (3.10) holds (Peters, 1856)

$$\boxed{\bar{x} \pm 0.84535 \frac{\sum |x_i - \bar{x}|}{n\sqrt{n-1}}.} \tag{3.10}$$

EXAMPLE. We use the data of the last example and find the 50% confidence interval to be $9 \pm 0.84535 \cdot 18/[8\sqrt{8-1}] = 9 \pm 0.72$. 50% CI: $8.28 \leq \mu \leq 9.72$.

The median deviation: An especially robust estimate for dispersion is the median deviation $\tilde{D}$

$$\boxed{\tilde{D} = \text{median}\{|x_i - \tilde{x}|\}} \tag{3.11}$$

with $\tilde{x}$ = sample median [cf., F. R. Hampel, The influence curve and its role in robust estimation, Journal of the American Statistical Association **69** (1974), 383–393].

EXAMPLE. x_i: 3, 9, 16, 25, 60; $\bar{x} = 16$; $|3 - 16| = 13$, $|9 - 16| = 7,\ldots$; from the deviations 7, 9, 13, 44 we have the median $(9 + 13)/2 = 11$, thus $\tilde{D} = 11$.

3.1.4 Confidence interval for the median

The confidence interval for the median replaces (3.1) and (3.2) **when populations are not normally distributed**. If the n observations, ordered by magnitude, are written as $x_{(1)}, x_{(2)}, x_{(3)}, \ldots, x_{(n)}$, then the distribution-free confidence interval for the median, the 95 % CI, and the 99 % CI for $\tilde{\mu}$ are given by (3.12) and Tables 69 and 69a in Section 4.2.4 [see page 319].

$$\boxed{x_{(h)} \leq \tilde{\mu} \leq x_{(n-h+1)}.} \qquad (3.12)$$

For $n > 50$ and the confidence probabilities 90 %, 95 %, and 99 %, h can be approximated by

$$\boxed{h = \frac{n - z\sqrt{n-1}}{2}} \qquad (3.13)$$

with $z = 1.64$, 1.96, and 2.58 respectively. Thus for $n = 300$, the 95% confidence interval lies between the 133rd and the 168th value of the sample ordered by magnitude ($h = [300 - 1.96\sqrt{300 - 1}]/2 \simeq 133$, $n - h + 1 = 300 - 133 + 1 = 168$), e.g., the 95% CI for $\tilde{\mu}$ is $x_{(133)} = 21.3 \leq \tilde{\mu} \leq 95.4 = x_{(168)}$, or 95% CI: $21.3 \leq \tilde{\mu} \leq 95.4$. In giving the result, the last form, omitting $x_{(\text{left})}$ and $x_{(\text{right})}$, is often preferred. Additional tables are found in Mackinnon (1964) and Van der Parren (1970). The procedure of this section also applies to the determination of a 95% confidence interval for a **median difference** $\tilde{\mu}_d$, useful either (I) for differences of paired observations or (II) for all possible differences between two independent (uncorrelated) samples ($n_1 \simeq n_2$). Paired observations with independent pairs may refer to different observations on the same subject (Ia) or to similar observations made on matched subjects (Ib) who have received different treatments at two different times (cf., Sections 2.1.3–2.1.5 and 4.1).

Remark

95 % and 99 % confidence intervals for 18 **other quantiles** (quartiles, deciles, and several percentiles [cf., also Section 2.1.2]) can be found in the Documenta Geigy (1968 [2], p. 104 [cf., p. 162, left side, and p. 188, left side]).

3.2 COMPARISON OF AN EMPIRICAL MEAN WITH THE MEAN OF A NORMALLY DISTRIBUTED POPULATION

The question whether the mean $\bar{x}$ of a sample from a normal distribution differs only randomly or in fact significantly from a specified mean μ_0 can be reformulated: Does the confidence interval for μ computed with $\bar{x}$ include the specified mean μ_0 or not, i.e., is the absolute difference $|\bar{x} - \mu_0|$ greater or less than half the confidence interval $ts/\sqrt{n}$?

Given a sample of size n having standard deviation s, the difference of its mean $\bar{x}$ from the specified mean μ_0 is statistically significant if

$$\boxed{|\bar{x} - \mu_0| > t \frac{s}{\sqrt{n}} \quad \text{or} \quad \frac{|\bar{x} - \mu_0|}{s} \cdot \sqrt{n} > t,} \qquad (3.14)$$

where the quantity t for $n - 1$ degrees of freedom and the required confidence coefficient $S = 1 - \alpha$ is taken from Table 27 in Section 1.5.2. The limit at and above which a difference is significant at the level α and below which it is considered random thus lies at

$$\boxed{t = \frac{|\bar{x} - \mu_0|}{s} \cdot \sqrt{n},} \qquad DF = n - 1. \qquad (3.14a)$$

Thus for testing $H_0: \mu = \mu_0$ against $H_A: \mu \neq \mu_0$ (or $H_A: \mu > \mu_0$) reject H_0 if t, given in (3.14a), surpasses the critical value $t_{n-1;\alpha;\text{two sided}}$ (or $t_{n-1;\alpha;\text{one sided}}$), provided the sample comes from a normally distributed distribution or at least from a distribution with little kurtosis and less skewness, since the latter affects the distribution of t more than kurtosis. With large sample sizes, t can be replaced by a z-value appropriate for the required confidence coefficient. Since parameters are compared—μ_0 with the μ underlying the sample—what we have is a test of the parameter. This test is known as the **one sample t test** for difference in means.

EXAMPLE. A sample of size $n = 25$ yields $\bar{x} = 9$ and $s = 2$. We want to determine whether the null hypothesis $\mu = \mu_0 = 10$—two sided question—can be maintained with $\alpha = 0.05$ or 5% [i.e., with a confidence coefficient of $S = 95\%$]. We have the special value (marked with a caret)

$$\hat{t} = \frac{|9 - 10|}{2}\sqrt{25} = 2.50 > 2.06 = t_{24;0.05}.$$

Since $2.50 > 2.06$ the hypothesis $\mu = \mu_0$ is rejected at the 5% level.

Something should perhaps be said at this point regarding the notion of **function**. A function is an **allocation rule**: In the same way as every seat in a theater is assigned a certain ticket at each performance, a function assigns to every element of a set a certain element of another set. In the simplest case, a certain value of the dependent variable y is assigned to every value of the independent variable x: $y = f(x)$ [read: y equals $f(x)$, short for y is a function of x]; the independent variable x is called the **argument**. For the function $y = x^3$, e.g., the argument $x = 2$ is associated with the **function value** $y = 2^3 = 8$. The symbol t in (3.14a) is defined as a function of μ_0, s, and $\bar{x}$, but recalling that $\bar{x}$ and s are themselves functions of the sample values $x_1, x_2, \ldots, x_n$, we can consider t in (3.14a) as a function of the sample values and the parameter μ_0, $t = f(x_1, x_2, \ldots, x_n; \mu_0)$, while t in (3.14) is a function of the degrees of freedom v and the confidence level $S = 1 - \alpha$, (or α). Since $x_1, x_2, \ldots, x_n$ are the realized values of a random variable, $\hat{t}$ is itself a special realization of a random variable. Under the null hypothesis ($\mu = \mu_0$) this random variable has a t-distribution with $n - 1$ degrees of freedom. If the null hypothesis does not hold ($\mu \neq \mu_0$), it has a noncentral t-distribution and $|\hat{t}|$ is most likely to be **larger** than the corresponding $|t|$-value.

In (3.14a), left side, we might have written $\hat{t}$, since this formula is used for the critical examination of realizations.

Particular function values estimated by means of sample values (or in terms of sample values and one or more parameters) can be marked with a caret to distinguish them from the corresponding **tabulated values** (e.g. of the t, z, χ^2, or F distribution). Some authors do not use these quantities. In their notation, e.g., (3.14a) is stated as: Under the null hypothesis the test statistic

$$\boxed{\frac{|\bar{x} - \mu_0|}{s} \cdot \sqrt{n}} \qquad (3.14b)$$

has a t-distribution with $n - 1$ degrees of freedom (cf., Section 4.6.2).

Another possible way of testing the null hypothesis ($H_0: \mu = \mu_0$ against $H_A: \mu \neq \mu_0$) consists of establishing whether $\bar{x}$ lies within the so-called **acceptance region or NON-REJECTION REGION of H_0**

$$\boxed{\mu_0 - t_{n-1;\alpha} \cdot \frac{s}{\sqrt{n}} \leq \bar{x} \leq \mu_0 + t_{n-1;\alpha} \cdot \frac{s}{\sqrt{n}}.} \qquad (3.15)$$

If this is the case, the null hypothesis cannot be rejected (is retained). Outside the two acceptance limits lies the **critical region**, the upper and lower **rejection region**. If $\bar{x}$ falls in this region, the null hypothesis is rejected. For the one sided question ($H_0: \mu \leq \mu_0$ against $H_A: \mu > \mu_0$) the null hypothesis is retained as long as for the mean $\bar{x}$ of a sample of size n there obtains

$$\boxed{\bar{x} \leq \mu_0 + t_{n-1;\alpha} \cdot \frac{s}{\sqrt{n}},} \qquad (3.15a)$$

3.2 An Empirical Mean with the Mean of a Normally Distributed Population

where the t-value for the one sided test is given by Table 27 in Section 1.5.2. Regions of this sort are important for industrial quality control: they serve to check the stability of "theoretical values" (parameters) such as means or medians, standard deviations or ranges, and relative frequencies (e.g., permissible reject percentages).

This schematic outline of statistics given in Section 1.4.5 can now be provided with further details:

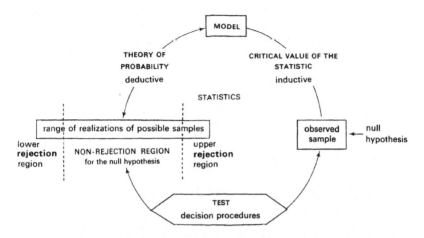

Starting with a null hypothesis and the accompanying **representative** sample —i.e., the sample must, up to random error, fully represent the population— the **stochastic inductive inference** enables us to make a statement about the population underlying the sample, about the stochastic model. A survey of the collection of samples compatible with the model can then be deduced by way of a second stochastic inference, with the help of probability theory in terms of a stochastic variable with a certain distribution (e.g., the t-distribution): the combining of the least expected samples—say the most extreme 5%, 1%, or 0.1% of the cases—to form a rejection region (two sided question) fixes the **NON-REJECTION or acceptance REGION of H_0** (cf., Weiling 1965). The actual test of whether the null hypothesis can be rejected given a sample is carried out by means of a statistical test procedure which establishes the bounds for the **acceptance or the rejection** region. If the observed sample belongs to the acceptance region, then the null hypothesis holds insofar as it is **not refuted** by the sample (acquitted for lack of evidence). **Subject to further investigation, it is decided that the null hypothesis should be retained.** If the sample belongs to the rejection region, it means that whenever the null hypothesis is true, we have the accidental occurrence of a possible but quite improbably large departure. In such a case it is considered more likely that the parameter value of the null hypothesis does not apply to the population under study, hence the deviation. The null hypothesis is then rejected at the preselected level. More on this may be found in Section 3.6.1.

> Confidence intervals and tests for σ, σ^2, and σ_1^2/σ_2^2 are more sensitive (less robust) to deviations from the normal distribution than are procedures which concern the confidence intervals for μ and $\mu_1 - \mu_2$ (t-distribution). Furthermore, one sided procedures are more sensitive than two sided.

▶ 3.3 COMPARISON OF AN EMPIRICAL VARIANCE WITH ITS PARAMETER

For a normally distributed population, the null hypothesis $\sigma = \sigma_0$ or $\sigma^2 = \sigma_0^2$ (as against $\sigma > \sigma_0$ or $\sigma^2 > \sigma_0^2$) is rejected under the following conditions:

Case 1: μ unknown.

$$\hat{\chi}^2 = \frac{\sum(x_i - \bar{x})^2}{\sigma_0^2} = \frac{(n-1)s^2}{\sigma_0^2} > \chi^2_{n-1,\,\alpha}.\qquad(3.16)$$

(p.141)   *Case* 2: μ known.

$$\hat{\chi}^2 = \frac{\sum(x_i - \mu)^2}{\sigma_0^2} = \frac{ns_0^2}{\sigma_0^2} > \chi^2_{n,\,\alpha}.\qquad(3.16a)$$

s_0^2 [cf., (1.33)] can be computed by (3.23) as $s_0^2 = Q/n$. Given extensive samples from a normally distributed population, $H_0: \sigma = \sigma_0$ is rejected and $H_A: \sigma \neq \sigma_0$ accepted at the 5% level when

$$\frac{|s - \sigma_0|}{\sigma_0}\sqrt{2n} > 1.96 \qquad(3.16b)$$

(1% level: replace 1.96 by 2.58).

EXAMPLE. Are the 8 observations 40, 60, 60, 70, 50, 40, 50, 30 ($\bar{x} = 50$) compatible with the null hypothesis $\sigma^2 = \sigma_0^2 = 60$ as against $\sigma^2 > \sigma_0^2 = 60$ ($\alpha = 0.05$)?

$$\hat{\chi}^2 = \frac{(40-50)^2}{60} + \frac{(60-50)^2}{60} + \cdots + \frac{(30-50)^2}{60} = 20.00.$$

Since $\hat{\chi}^2 = 20.00 > 14.07 = \chi^2_{7;\,0.05}$, $H_0: \sigma^2 = \sigma_0^2$ is rejected in favor of $H_A: \sigma^2 < \sigma_0^2$.

A table for testing the (two sided) null hypothesis $\sigma^2 = \sigma_0^2$ is given by Lindley et al. (1960) together with the tables of Rao et al. (1966 [2], p. 67, Table 5.1, middle); a $\hat{\chi}^2$ which lies outside the limits there given is regarded as significant. For our example with $v = n - 1 = 7$ and $\alpha = 0.05$, the limits, which turn out to be 1.90 and 17.39, do not include $\hat{\chi}^2 = 20$ between them, i.e., $\sigma^2 \neq \sigma_0^2$.

3.4 CONFIDENCE INTERVAL FOR THE VARIANCE AND FOR THE COEFFICIENT OF VARIATION

The **confidence interval for** σ^2 can be estimated in terms of the χ^2 distribution according to

$$\frac{s^2(n-1)}{\chi^2_{n-1;\alpha/2}} \leqq \sigma^2 \leqq \frac{s^2(n-1)}{\chi^2_{n-1;1-\alpha/2}}. \qquad (3.17) \quad \text{(p. 140)}$$

For example the 95% confidence interval ($\alpha = 0.05$) for $n = 51$ and $s^2 = 2$, is determined as follows:

$$\chi^2_{50;0.025} = 71.42 \quad \text{and} \quad \chi^2_{50;0.975} = 32.36:$$

$$\frac{2 \cdot 50}{71.42} \leq \sigma^2 \leq \frac{2 \cdot 50}{32.36}$$

$$1.40 \leq \sigma^2 \leq 3.09.$$

> Approximations for $n \geq 150$ as well as tables for the 95% CI and
>
> $$n = 1(1)150(10)200$$
>
> are contained in Sachs (1984).

The estimate for σ^2 is obtained according to

$$\hat{\sigma}^2 = \frac{s^2(n-1)}{\chi^2_{n-1;0.5}} = \frac{2 \cdot 50}{49.335} \simeq 2.03. \qquad (3.17a) \quad \text{(p. 140)}$$

95% confidence interval for σ: $\sqrt{1.40} < \sigma < \sqrt{3.09}$; $1.18 < \sigma < 1.76$. Since the χ^2 distribution is unsymmetric, the estimated parameter (σ) does not lie in the middle of the confidence interval.

The **confidence limits for the coefficient of variation** can be determined by the method described by Johnson and Welch (1940). For $n \gtrsim 25$ and $V < 0.4$ the following approximation is adequate:

$$\frac{V}{1 + z\sqrt{\frac{1 + 2V^2}{2(n-1)}}} \lesssim \gamma \lesssim \frac{V}{1 - z\sqrt{\frac{1 + 2V^2}{2(n-1)}}}. \qquad (3.18)$$

90% CI: $z = 1.64$; 95% CI: $z = 1.96$; 99% CI: $z = 2.58$.

For the (one sided) upper confidence limit (CL$_u$) γ_0 [right side of (3.18)], which is often of interest, the 90% CL$_u$ corresponds to $z = 1.28$; the 95% CL$_u$ to $z = 1.64$; the 99% CL$_u$ to $z = 2.33$.

EXAMPLE. Compute the 90% CI for γ for $n = 25$ and $V = 0.30$.

$$1.64\sqrt{(1 + 2 \cdot 0.3^2)/[2(25 - 1)]} = 0.257$$

$0.3/1.257 = 0.239$, $\quad 0.3/0.743 = 0.404$, $\qquad$ 90%-CI: $\;0.24 \lesssim \gamma \lesssim 0.40$.

0.40 is at the same time the approximate upper 95% CL_u, i.e., 95% CL_u: $\gamma_0 \simeq 0.40$; the coefficient of variation γ lies below 0.40 with a confidence coefficient of $S = 95\%$.

3.5 COMPARISON OF TWO EMPIRICALLY DETERMINED VARIANCES OF NORMALLY DISTRIBUTED POPULATIONS

To investigate whether two independently drawn random samples (cf., also Section 2.7) of sizes n_1 and n_2 originated from a common normally distributed population, one first of all tests their variances (the larger sampling variance is denoted by s_1^2) for equality or homogeneity. The null hypothesis $H_0: \sigma_1^2 = \sigma_2^2$ is rejected as soon as the quantity $\hat{F} = s_1^2/s_2^2$ computed from the two sample variances is larger than the corresponding tabulated quantity $F_{n_1-1;\, n_2-1;\alpha}$ (cf., also Section 4.6.2); the alternative hypothesis $H_A: \sigma_1^2 \neq \sigma_2^2$ is then accepted (two sided problem).

One sided problem: Let (1) denote the population which has under the alternate hypothesis the larger variance (i.e., $H_A: \sigma_1^2 > \sigma^2$). For $\hat{F} > F$ the one sided alternative $H_A: \sigma_1^2 > \sigma_2^2$ is accepted (n_1 should be at least as large as n_2). If a test of this sort is utilized as a preliminary test for a comparison of means (the t-test assumes equality of population variances), then the 10% level is favored because the Type II error (Section 1.4.3) is here the more serious.

In contrast with the two sided t-test, the F-test is very sensitive to deviations from the normal distribution. If normality is not ascertained, then the F-test is replaced by the distribution-free Siegel–Tukey test (Section 3.9.1).

3.5.1 Small to medium sample size

We form the quotient of the two variances s_1^2 and s_2^2, thereby obtaining the test statistic

$$\boxed{\hat{F} = \frac{s_1^2}{s_2^2} \quad \begin{array}{l} \text{with } DF_1 = n_1 - 1 = v_1, \\ \text{with } DF_2 = n_2 - 1 = v_2. \end{array}} \tag{3.19}$$

If the computed $\hat{F}$-value exceeds the tabulated F-value for the pre-selected level of significance and the degrees of freedom $v_1 = n_1 - 1$ and $v_2 = n_2 - 1$, then the hypothesis of homogeneity of population variances ($H_0: \sigma_1^2 = \sigma_2^2$) is abandoned. For $\hat{F} \leq F$ there is no reason to question this hypothesis. If the

3.5 Two Empirically Determined Variances of Normally Distributed Populations

hypothesis is rejected then the confidence interval (CI) for σ_1^2/σ_2^2 is computed by

$$\frac{s_1^2}{s_2^2} \cdot \frac{1}{F_{v_1, v_2}} \leq \frac{\sigma_1^2}{\sigma_2^2} \leq \frac{s_1^2}{s_2^2} \cdot F_{v_2, v_1} \qquad v_1 = n_1 - 1, \quad v_2 = n_2 - 1. \qquad (3.19a)$$

For the 90% CI refer to Table 30b; for the 95% CI to Table 30c (Section 1.5.3). The tables in that section contain the upper significance levels of the F-distribution for the one sided problem usually considered in analysis of variance. In the present case we are interested in departures in both directions, and thus in a two sided test. If we test at the 10% level, the table with the 5% limits is to be used. Analogously the 0.5% limits, Table 30e, apply for the two sided test at the 1% level.

EXAMPLE. Test $H_0: \sigma_1^2 = \sigma_2^2$ against $H_A: \sigma_1^2 \neq \sigma_2^2$ at the 10% level, given

$$n_1 = 21, s_1^2 = 25 \qquad \hat{F} = \frac{25}{16} = 1.56.$$
$$n_2 = 31, s_2^2 = 16$$

Since $\hat{F} = 1.56 < 1.93$ [$= F_{20; 30; 0.10 \text{(two s.)}} = F_{20; 30; 0.05 \text{(one s.)}}$], H_0 cannot be rejected at the 10% level.

For equal sample sizes n, H_0 can also be tested according to

$$\hat{t} = \frac{\sqrt{n-1}(s_1^2 - s_2^2)}{2\sqrt{s_1^2 s_2^2}} \qquad \text{with} \quad v = n - 1 \qquad (3.20)$$

(Cacoullos 1965). A quick test is presented in Section 3.7.1.

EXAMPLE. Test $H_0: \sigma_1^2 = \sigma_2^2$ against $H_A: \sigma_1^2 \neq \sigma_2^2$ at the 10% level, given

$$n_1 = n_2 = 20 = n, \qquad s_1^2 = 8, \qquad s_2^2 = 3,$$

$$\hat{F} = \frac{8}{3} = 2.67 > 2.12, \qquad \hat{t} = \frac{\sqrt{20-1}(8-3)}{2\sqrt{8 \cdot 3}} = 2.22 > 1.729.$$

Since H_0 is rejected at the 10% level, we specify the 90% CI by (3.19a):

$$F_{19; 19; 0.05 \text{(one s.)}} = 2.17 \qquad \frac{2.67}{2.17} = 1.23, \qquad 2.67 \cdot 2.17 = 5.79;$$

$$90\%\text{-CI:} \quad 1.23 \leq \sigma_1^2/\sigma_2^2 \leq 5.79.$$

Distribution-free procedures which replace the F-test

Since the result of the F-test can be strongly influenced even by small deviations from the normal distribution (Cochran 1947, Box, 1953, Box and Anderson 1955), Levene (1960) has proposed an approximate nonparametric procedure: In the individual data sequences that are to be compared, the respective absolute values $|x_i - \bar{x}|$ are formed and subjected to a rank sum test: For two sample sequences, the U-test—see Section 3.9.1—and for more than two sequences, the H-test of Kruskal and Wallis. It is tested whether the absolute deviations $|x_i - \bar{x}|$ for the individual sequences can be regarded as samples from distributions with equal means. The homogeneity of several (k) variances can also be rejected, according to Levene (1960), with the aid of simple analysis of variance, as soon as $\hat{F} > F_{k-1;n-k;\alpha}$ for the n overall absolute deviations of the observations from their k respective means (cf., also Section 7.3.1). More on robust alternatives to the F-test can be found in Shorack (1969).

Minimal sample sizes for the F-test

With every statistical test there are, as we know, two risks to be estimated. An example is given by Table 50. Extensive tables can be found in Davies (1956) (cf., also Tiku 1967).

Table 50 Number of observations needed to compare two variances using the F-test. F-values are tabulated: For $\alpha = 0.05$, $\beta = 0.01$ and $s^2_{numerator}/s^2_{denominator} = F = 4$ the table indicates that in both samples the estimation of the variances is to be based on 30 to 40 degrees of freedom (corresponding to the F-values 4.392 and 3.579)—on at least 35 degrees of freedom, let us say. (Taken from Davies, O. L.: The Design and Analysis of Industrial Experiments. Oliver and Boyd, London, 1956, p. 614, part of Table H.)

DF	$\alpha = 0.05$			
	$\beta = 0.01$	$\beta = 0.05$	$\beta = 0.1$	$\beta = 0.5$
1	654,200	26,070	6,436	161.5
2	1,881	361.0	171.0	19.00
3	273.3	86.06	50.01	9.277
4	102.1	40.81	26.24	6.388
5	55.39	25.51	17.44	5.050
6	36.27	18.35	13.09	4.284
7	26.48	14.34	10.55	3.787
8	20.73	11.82	8.902	3.438
9	17.01	10.11	7.757	3.179
10	14.44	8.870	6.917	2.978
12	11.16	7.218	5.769	2.687
15	8.466	5.777	4.740	2.404
20	6.240	4.512	3.810	2.124
24	5.275	3.935	3.376	1.984
30	4.392	3.389	2.957	1.841
40	3.579	2.866	2.549	1.693
60	2.817	2.354	2.141	1.534
120	2.072	1.828	1.710	1.352
∞	1.000	1.000	1.000	1.000

3.5 Two Empirically Determined Variances of Normally Distributed Populations

Minimal sample sizes for the comparison of two empirical variances for (independent) normally distributed populations can also be determined by means of nomograms by Reiter (1956) or by means of tables by Graybill and Connell (1963).

3.5.2 Medium to large sample size

Nontabulated F-values can be obtained by interpolation when the degrees of freedom are moderately large. When the degrees of freedom are large, the homogeneity of two variances can be tested by

$$\frac{\frac{1}{2}\ln F+\frac{1}{2}\left(\frac{1}{v_1}-\frac{1}{v_2}\right)}{\sqrt{\frac{1}{2}\left(\frac{1}{v_1}+\frac{1}{v_2}\right)}}, \qquad (3.21)$$

which is approximately normally distributed. If tables of natural logarithms are not readily available, replace $\frac{1}{2} \ln F$ with $\frac{1}{2}$ (2.302585) log F to find

$$\hat{z} = \frac{1.1513 \log F + \frac{1}{2}\left(\frac{1}{v_1}-\frac{1}{v_2}\right)}{\sqrt{\frac{1}{2}\left(\frac{1}{v_1}+\frac{1}{v_2}\right)}}, \qquad (3.21a)$$

and evaluate it with the help of a table of the standard normal distribution.

EXAMPLE. We wish to check this formula by means of Table 30. For $v_1 = v_2 = 60$ and $\alpha = 0.05$ we get from the table the value $F = 1.53$. Suppose now we had found this value experimentally for $v_1 = v_2 = 60$ and our table went only to $v_1 = v_2 = 40$. Is the F-value found significant at the 5% level in the one sided problem $\sigma_1^2 = \sigma_2^2$ versus $\sigma_1^2 > \sigma_2^2$? For $F = 1.53$, $v_1 = 60$, and $v_2 = 60$ we obtain

$$\hat{z} = \frac{1.15129 \log 1.53 + \frac{1}{2}\left(\frac{1}{60}-\frac{1}{60}\right)}{\sqrt{\frac{1}{2}\left(\frac{1}{60}+\frac{1}{60}\right)}} = 1.64705,$$

i.e., $\hat{z} = 1.64705 > 1{,}6449$. The value $z = 1.6449$ corresponding to a level of significance of $P = 0.05$ (cf., Table 43, Section 2.1.6) is exceeded, so that the hypothesis of variance homogeneity must be rejected at the 5% level. The approximation by the normal distribution is excellent.

3.5.3 Large to very large sample size (n_1, $n_2 \gtrsim 100$)

$$\hat{z} = \frac{|s_1 - s_2|}{\sqrt{\dfrac{s_1^2}{2n_1} + \dfrac{s_2^2}{2n_2}}}. \tag{3.22}$$

If the statistic (3.22) exceeds the theoretical z-value in the table on the very first page of the book, or in Table 43, for various levels of significance, then the standard deviations σ_1 and σ_2 or the variances σ_1^2 and σ_2^2 are taken to be significantly different or heterogeneous at the level in question; otherwise they are equal or homogeneous.

EXAMPLE. Given $s_1 = 14$, $s_2 = 12$, $n_1 = n_2 = 500$;
Null hypothesis: $\sigma_1^2 = \sigma_2^2$; alternative hypothesis: $\sigma_1^2 \neq \sigma_2^2$; $\alpha = 0.05$; we have

$$\hat{z} = \frac{14 - 12}{\sqrt{\dfrac{14^2}{2 \cdot 500} + \dfrac{12^2}{2 \cdot 500}}} = 3.430 > 1.960 = z_{0.05},$$

i.e., at the 5 % level $H_0: \sigma_1^2 = \sigma_2^2$ is rejected and $H_A: \sigma_1^2 \neq \sigma_2^2$ accepted.

▶ 3.6 COMPARISON OF TWO EMPIRICAL MEANS OF NORMALLY DISTRIBUTED POPULATIONS

3.6.1 Unknown but equal variances

The sum of squares $\sum (x - \bar{x})^2$ is denoted by Q in the following. It is computed according to

$$Q = \sum x^2 - \frac{(\sum x)^2}{n} \quad \text{or} \quad Q = (n - 1)s^2. \tag{3.23, 3.24}$$

For the comparison of the means of two samples of **unequal sample sizes** ($n_1 \neq n_2$) one needs the test statistic (3.25, 3.26) with $n_1 + n_2 - 2$ degrees of freedom for the so-called **two sample t-test for independent random samples** from normally distributed populations with equal variances. Fortunately, in the case of the two sided problem ($H_0: \mu_1 = \mu_2$ vs. $H_A: \mu_1 \neq \mu_2$) and not

3.6 Comparison of Two Empirical Means of Normally Distributed Populations

too small and not too different sample sizes, this test is remarkably robust against departures from the normal distribution (see, e.g., Sachs 1984, p. 51):

$$\hat{t} = \frac{|\bar{x}_1 - \bar{x}_2|}{\sqrt{\left[\frac{n_1 + n_2}{n_1 \cdot n_2}\right] \cdot \left[\frac{Q_1 + Q_2}{n_1 + n_2 - 2}\right]}}$$

$$= \frac{|\bar{x}_1 - \bar{x}_2|}{\sqrt{\left[\frac{n_1 + n_2}{n_1 n_2}\right] \cdot \left[\frac{(n_1 - 1)s_1^2 + (n_2 - 1)s_2^2}{n_1 + n_2 - 2}\right]}}.$$

(3.25, 3.26)

We test the null hypothesis ($\mu_1 = \mu_2$) of equality of the means of the populations with unknown but equal variances underlying the two samples (cf., Sections 1.4.8 and 3.5). In the case of **EQUAL SAMPLE SIZES** ($n_1 = n_2$ is generally preferable, since the Type II error gets minimized), (3.25) and (3.26) reduce to

$$\hat{t} = \frac{|\bar{x}_1 - \bar{x}_2|}{\sqrt{\frac{Q_1 + Q_2}{n(n-1)}}} = \frac{|\bar{x}_1 - \bar{x}_2|}{\sqrt{\frac{s_1^2 + s_2^2}{n}}} \quad (3.27)$$

with $2n - 2$ degrees of freedom, where $n = n_1 = n_2$. If the test quotient exceeds the significance level, then $\mu_1 \ne \mu_2$ applies. If the test quotient is less than this level, then the null hypothesis $\mu_1 = \mu_2$ cannot be rejected.

For $n_1 = n_2 \le 20$ the Lord test (Section 3.7.2) can replace the *t*-test. The comparison of several means is treated in Chapter 7 (cf., also Section 3.7.3). To add variety to this section and make it more understandable, three comments are included after the example: on the test statistic $\hat{t}$ and the decision, on the tabulated value $t_{28;0.05}$ used for the example, and on the comparison of several means.

EXAMPLE. Test $H_0: \mu_1 = \mu_2$ against $H_A: \mu_1 \ne \mu_2$ at the 5% level, given n_1, n_2; $\bar{x}_1, \bar{x}_2$; s_1^2, s_2^2; and (3.24), (3.25):

$n_1 = 16;$ $\bar{x}_1 = 14.5;$ $s_1^2 = 4;$

$n_2 = 14;$ $\bar{x}_2 = 13.0;$ $s_2^2 = 3.$

We have $Q_1 = (16 - 1)(4) = 60$, $Q_2 = (14 - 1)(3) = 39$, which are then substituted together with the other values into (3.25):

$$\hat{t} = \frac{14.5 - 13.0}{\sqrt{\left[\frac{16 + 14}{16 \cdot 14}\right] \cdot \left[\frac{60 + 39}{16 + 14 - 2}\right]}} = 2.180.$$

There are $v = n_1 + n_2 - 2 = 28$ degrees of freedom at our disposal, i.e., $t_{28;0.05} = 2.048$. Since $\hat{t} = 2.180 > 2.048$, the null hypothesis, equality of means, is rejected and the alternative hypothesis $\mu_1 \neq \mu_2$ accepted at the 5% level.

Three comments

I. Comment on the two sample Student's *t*-test

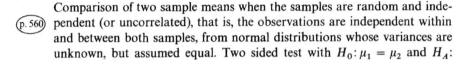

Comparison of two sample means when the samples are random and independent (or uncorrelated), that is, the observations are independent within and between both samples, from normal distributions whose variances are unknown, but assumed equal. Two sided test with $H_0: \mu_1 = \mu_2$ and $H_A: \mu_1 \neq \mu_2$.

1. We assume no difference between population means ($H_0: \mu_1 = \mu_2$).
2. This being the assumption, we calculate for the observations (with n_1, n_2; $\bar{x}_1, \bar{x}_2; s_1^2, s_2^2$) the test statistic $\hat{t}$ (3.26) without the absolute value bars in the numerator, the probability of getting a value greater than $+\hat{t}$, and the probability of getting a value less than $-\hat{t}$. The sum of the two probabilities "more extreme than $\pm t$," P for short, is what emerges in the two sided test.
3. We reject the assumption ($H_0: \mu_1 = \mu_2$) if this probability is low (e.g., <0.05) and decide there is a statistically significant difference ($\mu_1 \neq \mu_2$).

> In other words: With the help of the sampling distribution of the test statistic (3.26) and **assuming that** $H_0: \mu_1 = \mu_2$ is true, the probability P of the test statistic taking a value equal to or more extreme than its numerical value computed from the sample is determined. If P is less than or equal to, say, 0.05, we hold H_0 to be exceptional and hence reject it at the 5% level. A small value of P implies that we have observed something "relatively unlikely", **provided** H_0 is true. Thus statistical significance is only a statement about conditional probability.

Since an assumption or hypothesis is different from a fact, whenever we decide about an assumption we are not proving anything. Moreover we have only two samples and not the population means.

II. Comment on $t_{28;0.05}$

The tabulated value of a two sided Student's t with $v = 28$ degrees of freedom at the 5% level, that is $t_{28;0.05} = 2.048$, is given by

$$P(-2.048 \leq t \leq 2.048) = 0.95,$$
or (a)
$$P(|t| \geq 2.048) = 0.05,$$

3.6 Comparison of Two Empirical Means of Normally Distributed Populations

or, with the density function $f(t)$ and $v = 28$, by

$$\int_{-2.048}^{2.048} f(t)dt = 0.95$$

or

$$\int_{-\infty}^{-2.048} f(t)dt = 0.025 \quad \text{and} \quad \int_{2.048}^{\infty} f(t)dt = 0.025.$$

(b)

The t-distribution is symmetrical about $t = 0$. The integral $\int_a^b f(t)dt$ is a numerical value equal to the proportion of the area under the graph of the function $f(t)$ bounded by the curve, the t-axis, and the lines $t = a$ and $t = b$ to the whole area under the graph of the function. The symbol dt identifies t as a variable.

As v increases the t-distribution tends to the standard normal distribution, for which we have to substitute in (a) and (b) z for t and 1.96 for 2.048.

III. Comment on the comparison of several means for a single set of data with an overall significance level α (cf. Chapter 7 and especially Section 7.4.2)

If we have a large data set and if we plan to do k t-tests from this single set, we should not use the α (for instance 0.05) point of the t-distribution but the α/k (0.05/k) point, e.g., graphically interpolated (Table 27 gives the points 0.02, ..., 0.0001), having then an **overall** 100α% (5%) significance level. For α = 0.05 and k = 50 we use the 0.05/50 = 0.001 point of the t distribution.

Tables of this **Bonferroni t-statistic** are given by Bailey (1977): 100α/k points for α = 0.05, 0.01; v = 2(1)30(5)60(10)120, 250, 500, 1000, ∞; and k = 1(1)20, 21 = $\binom{7}{2}$, 28, 36, 45, 55, 66, 78, 91, 105, 120, 136, 153, 171, 190 = $\binom{20}{2}$. Provided we have four samples of size 15 each and we plan to do all 4(4 − 1)/2 = 6 t-tests with an overall significance level of 5%, then we have by graphical interpolation the critical t-value 2.84, or from the Bailey table, 2.8389 (and not 2.048) for all six tests.

Important remarks (cf. also Sections 2.1.4, 3.1.2, 3.6.2, and 3.9.4)

A The confidence interval for the difference between the means of two samples from normally distributed populations with equal variance is given (e.g., for $S = 0.95$, i.e., α = 0.05, with $t_{v;\,0.05}$) by

$$(\bar{x}_1 - \bar{x}_2) - t_{n_1+n_2-2;\alpha} \cdot s\sqrt{1/n_1 + 1/n_2} \leq \mu_1 - \mu_2$$
$$\leq (\bar{x}_1 - \bar{x}_2) + t_{n_1+n_2-2;\alpha} \cdot s\sqrt{1/n_1 + 1/n_2},$$

(3.28) (pp. 136–137)

$$\text{where } s = \sqrt{\frac{s_1^2(n_1 - 1) + s_2^2(n_2 - 1)}{n_1 + n_2 - 2}} = \sqrt{\frac{Q_1 + Q_2}{n_1 + n_2 - 2}}.$$

If σ is known, t is replaced by the standard normal variable z. If samples of equal size are present, $s\sqrt{1/n_1 + 1/n_2}$ is again replaced by $\sqrt{(s_1^2 + s_2^2)/n}$. A difference between μ_1 and μ_2 is significant at the level employed, provided the confidence interval does not include the value $\mu_1 - \mu_2 = 0$. Statistical test procedures and confidence intervals both lead to decisions: **The confidence interval moreover offers additional information about the parameter or parameters.**

EXAMPLE. We use the last example and obtain the 95% confidence limits for the difference $\mu_1 - \mu_2$ between the two means

$$(\bar{x}_1 - \bar{x}_2) \pm t_{n_1+n_2-2;\alpha} \cdot s\sqrt{1/n_1 + 1/n_2}$$
$$(14.5 - 13.0) \pm 2.048 \cdot 1.880 \cdot \sqrt{1/16 + 1/14}$$
$$1.5 \pm 1.4 \quad \text{i.e.,} \quad 95\%\text{-CI:} \quad 0.1 \leq \mu_1 - \mu_2 \leq 2.9.$$

[cf., $S = 0.95$, or $\alpha = 1 - 0.95 = 0.05$; $t_{28;0.05} = 2.048$.] The null hypothesis ($\mu_1 - \mu_2 = 0$) must, on the basis of the available samples, be rejected at the 5% level.

B A more elegant comparison of the means of two independent samples of different sizes, $n_1 \neq n_2$, with equal variance, is given by

$$\hat{F} = \frac{(n_1+n_2-2)(n_2\sum x_1 - n_1\sum x_2)^2}{(n_1+n_2)[n_1 n_2(\sum x_1^2 + \sum x_2^2) - n_2(\sum x_1)^2 - n_1(\sum x_2)^2]}, \quad (3.29)$$

pp. 144–150

$$DF_1 = 1; \quad DF_2 = n_1 + n_2 - 2,$$

and for the **case $n_1 = n_2 = n$** by

$$\hat{F} = \frac{(n-1)(\sum x_1 - \sum x_2)^2}{n[\sum x_1^2 + \sum x_2^2] - [(\sum x_1)^2 + (\sum x_2)^2]}, \quad (3.30)$$

pp. 144–150

$$DF_1 = 1, \quad DF_2 = 2n - 2.$$

The expressions (3.29) and (3.30) are the rewritten squares of (3.26) and (3.27), and the relation $t_{DF}^2 = F_{DF_1=1, DF_2=DF}$ is introduced. A comparison of the times needed to evaluate (3.26) and (3.29) for the same data shows that up to 30% computing time can be saved by using the somewhat clumsy formulas (3.29) and (3.30). Simple practice problems, which the reader himself can formulate, will confirm this.

3.6 Comparison of Two Empirical Means of Normally Distributed Populations 269

|C| **The comparison of relative frequencies** is dealt with in Sections 4.5.1, 4.6.1, and 6.1.1. The comparison of frequencies is dealt with in Sections 1.6.6, 2.4.2 (Remark 3), 4.5.4 and 4.6 and in Chapter 6.

Mean and variance are independent in samples from a normal population but they are proportional [and not independent] in samples from a binomial population [cf. (1.60), (1.61)]. If we here denote relative frequencies or proportions as $x/n = p$, written without the caret, then the variance of $\sin^{-1}\sqrt{p}$ is independent of the mean of $\sin^{-1}\sqrt{p}$ and for large samples the variance is equal to $820.7/n$ degrees.

Table 51 Angular transformation: the values $x = \sin^{-1}\sqrt{p} =$ arcsin $\sqrt{p}$ with x in degrees; e.g. arcsin $\sqrt{0.25} = 30.0°$; arcsin $\sqrt{1.00} = 90.0°$ [Transformation to arc units (radians): divide the tabulated values by 57.2958.]

p	0.00	0.01	0.02	0.03	0.04	0.05	0.06	0.07	0.08	0.09
0.0	0.000	5.739	8.130	9.974	11.537	12.921	14.179	15.342	16.430	17.457
0.1	18.435	19.370	20.268	21.134	21.973	22.786	23.578	24.350	25.104	25.842
0.2	26.565	27.275	27.972	28.658	29.334	30.000	30.657	31.306	31.948	32.583
0.3	33.211	33.833	34.450	35.062	35.669	36.271	36.870	37.465	38.057	38.646
0.4	39.231	39.815	40.397	40.976	41.554	42.130	42.706	43.280	43.854	44.427
0.5	45.000	45.573	46.146	46.720	47.294	47.870	48.446	49.024	49.603	50.185
0.6	50.769	51.354	51.943	52.535	53.130	53.729	54.331	54.938	55.550	56.167
0.7	56.789	57.417	58.052	58.694	59.343	60.000	60.666	61.342	62.028	62.725
0.8	63.435	64.158	64.896	65.650	66.422	67.214	68.027	68.866	69.732	70.630
0.9	71.565	72.543	73.570	74.658	75.821	77.079	78.463	80.026	81.870	84.261

$\sin^{-1}$ is the real inverse sine; $\sin^{-1}\sqrt{p}$ (written arcus sinus $\sqrt{p}$ or arcsin $\sqrt{p}$), denotes the size (in degrees, as in our brief Table 51, or in radians) of the angle whose sine equals $\sqrt{p}$.

If all sample groups (all binomial proportions x/n) have equal and not too small values n, then the variances of $\sin^{-1}\sqrt{p}$ are equal. [The variance of $x/n = p$ is $p(1-p)/n$ (cf., Section 1.6.2). For 2/100 and 50/100 we get, with $0.02 \cdot 0.98/100 = 2 \cdot 10^{-4}$ as against $0.5 \cdot 0.5/100 = 25 \cdot 10^{-4}$, very different variances.] Therefore we should transform an observed proportion p to $\sin^{-1}\sqrt{p}$, thus **stabilizing the variances**, before computing and comparing means [(3.23 to 3.35)]. In the **angular transformation** $x = \sin^{-1}\sqrt{p}$ the values x will range from 0 degrees to 90 degrees as p ranges from 0 to 1.

Two more transformations used for proportions and with similar effects are the logit transformation and the probit transformation. Extensive tables of all three transformations are given by Fisher and Yates (1963).

Comparison of the means of two independent samples from approximately normally distributed populations

Sample sizes	Variances	
	equal: $\sigma_1^2 = \sigma_2^2$	unequal: $\sigma_1^2 \neq \sigma_2^2$
equal: $n_1 = n_2 = n$	$\hat{t} = \dfrac{|\bar{x}_1 - \bar{x}_2|}{\sqrt{\dfrac{s_1^2 + s_2^2}{n}}}$ $DF = 2n - 2$	$\hat{t} = \dfrac{|\bar{x}_1 - \bar{x}_2|}{\sqrt{\dfrac{s_1^2 + s_2^2}{n}}}$ $DF = n - 1 + \dfrac{2n - 2}{\dfrac{s_1^2}{s_2^2} + \dfrac{s_2^2}{s_1^2}}$
unequal: $n_1 \neq n_2$	$\hat{t} = \dfrac{|\bar{x}_1 - \bar{x}_2|}{\sqrt{\left[\dfrac{n_1 + n_2}{n_1 n_2}\right]\left[\dfrac{(n_1-1)s_1^2 + (n_2-1)s_2^2}{n_1 + n_2 - 2}\right]}}$ $DF = n_1 + n_2 - 2$	$\hat{t} = \dfrac{|\bar{x}_1 - \bar{x}_2|}{\sqrt{\dfrac{s_1^2}{n_1} + \dfrac{s_2^2}{n_2}}}$ $DF = \dfrac{\left(\dfrac{s_1^2}{n_1} + \dfrac{s_2^2}{n_2}\right)^2}{\dfrac{\left(\dfrac{s_1^2}{n_1}\right)^2}{n_1 - 1} + \dfrac{\left(\dfrac{s_2^2}{n_2}\right)^2}{n_2 - 1}}$

The t-distribution is given in Table 27 (Section 1.5.1).

3.6.2 Unknown, possibly unequal variances

We test null hypothesis ($\mu_1 = \mu_2$) of the equality of two means with possibly unequal variances ($\sigma_1^2 \neq \sigma_2^2$). This is the so-called Fisher-Behrens problem (*cf.*, Welch 1937, Breny 1955, Linnik 1966, and Mehta and Srinivasan 1970 as well as Scheffé 1970), for which there is no exact solution. For practical purposes it is appropriate to use

$$\hat{t} = \frac{|\bar{x}_1 - \bar{x}_2|}{\sqrt{\dfrac{s_1^2}{n_1} + \dfrac{s_2^2}{n_2}}} \tag{3.31}$$

with approximately

$$v = \frac{\left(\dfrac{s_1^2}{n_1} + \dfrac{s_2^2}{n_2}\right)^2}{\dfrac{\left(\dfrac{s_1^2}{n_1}\right)^2}{n_1 - 1} + \dfrac{\left(\dfrac{s_2^2}{n_2}\right)^2}{n_2 - 1}} \tag{3.32}$$

degrees of freedom, where v is rounded off to the nearest integer. The value of v always falls between the smaller of $n_1 - 1$ and $n_2 - 1$ and their sum ($n_1 + n_2 - 2$); v is computed only for $\hat{t} > z_\alpha$ (about $1.96 = z_{0.05;\,\text{two s.}}$), since for $\hat{t} < z_\alpha$ the hypothesis $H_0: \mu_1 = \mu_2$ cannot be rejected at the $100\alpha\%$ level. For $n_1 = n_2 = n$ and $\sigma_1^2 = \sigma_2^2$, (3.32) yields $v = 2n - 2$. Other possible ways of solving the two sample problem are indicated by Trickett, Welch, and James (1956) as well as by Banerji (1960). Corresponding to (3.28) the **approximate confidence interval for the difference of two means**, for $\mu_1 - \mu_2$, in the case of possibly unequal variances, is given by (3.28a):

$$(\bar{x}_1 - \bar{x}_2) - t_{v;\alpha} B \leq \mu_1 - \mu_2 \leq (\bar{x}_1 - \bar{x}_2) + t_{v;\alpha} B \tag{3.28a}$$

with v from (3.32) and with B the square root in the denominator of (3.31). When the sample sizes are equal, v is taken from (3.34) and B is the square root in the denominator of (3.33). In the case of equal sample size ($n_1 = n_2 = n$) the formulas (3.31), (3.32) simplify to

$$\hat{t} = \frac{|\bar{x}_1 - \bar{x}_2|}{\sqrt{\dfrac{Q_1 + Q_2}{n(n-1)}}} = \frac{|\bar{x}_1 - \bar{x}_2|}{\sqrt{\dfrac{s_1^2 + s_2^2}{n}}} \tag{3.33}$$

with

$$v = n - 1 + \frac{2n - 2}{\dfrac{Q_1}{Q_2} + \dfrac{Q_2}{Q_1}} = n - 1 + \frac{2n - 2}{\dfrac{s_1^2}{s_2^2} + \dfrac{s_2^2}{s_1^2}} \qquad Q \text{ is computed by (3.23)}$$

$$\tag{3.34}$$

degrees of freedom, where Q is computed according to (3.23). With large sample sizes $\hat{t}$ can again be replaced by $\hat{z}$. Selected values of the standard normal distribution can be taken from Table 14, Section 1.3.4 (or Table 43, Section 2.1.6).

EXAMPLE. A simple numerical example should suffice. Given are the two samples (1) and (2). Test $\mu_1 \leq \mu_2$ against $\mu_1 > \mu_2$ with $\alpha = 0.01$:

$$n_1 = 700, \quad \bar{x}_1 = 18, \quad s_1^2 = 34; \quad n_2 = 1{,}000, \quad \bar{x}_2 = 12, \quad s_2^2 = 73.$$

For the one sided problem we set $\alpha = 0.01$. Because of the large sample sizes we work with the standard normal distribution; therefore we replace the variable t, which follows Student's distribution, by the standard normal variable z

$$\hat{z} = \frac{18 - 12}{\sqrt{\frac{34}{700} + \frac{73}{1{,}000}}} = 17.21 > 2.33 = z_{0.01} \qquad \text{(one sided).}$$

The null hypothesis on homogeneity of the means is rejected at the 1% level, i.e., $\mu_1 > \mu_2$ [we may write "$P \leq 0.01$," that has the meaning of "strong evidence against H_0"].

Small sample sizes ($n_1, n_2 < 9$) with heterogeneous variances can be very elegantly tested for equality of the means by a method derived by McCullough et al. (1960). The tables by Fisher and Yates (1963) offer other possibilities. A number of approximations are available for comparing several means with the variances not necessarily equal (cf., Sachs 1984). A confidence interval for the ratio of two means of independent samples from normally distributed populations (no assumptions are made on the ratio of the two variances) is given by Chakravarti (1971).

Weir (1960) proposed another way for solving the Behrens–Fisher problem. It is of interest to us that a **difference in the means is statistically significant at the 5% level** whenever the following relations hold:

$$\frac{|\bar{x}_1 - \bar{x}_2|}{\sqrt{\frac{Q_1 + Q_2}{n_1 + n_2 - 4}\left[\frac{1}{n_1} + \frac{1}{n_2}\right]}} \geq 2.0,$$

$$\frac{|\bar{x}_1 - \bar{x}_2|}{\sqrt{\frac{(n_1 - 1)s_1^2 + (n_2 - 1)s_2^2}{n_1 + n_2 - 4}\left[\frac{1}{n_1} + \frac{1}{n_2}\right]}} \geq 2.0,$$

(3.35)

where it is required that $n_1 \geq 3$ and $n_2 \geq 3$; if the quotient falls below the value 2, then the null hypothesis $\mu_1 = \mu_2$ cannot be rejected at the 5% level.

EXAMPLE. Comparison of two means at the 5% level:

$$n_1 = 3; \quad 1.0 \quad 5.0 \quad 9.0; \quad \bar{x}_1 = 5.0; \quad Q_1 = 32; \quad s_1^2 = 16;$$
$$n_2 = 3; \quad 10.9 \quad 11.0 \quad 11.0; \quad \bar{x}_2 = 11.0; \quad Q_2 = 0.02; \quad s_2^2 = 0.01.$$

3.6 Comparison of Two Empirical Means of Normally Distributed Populations

Q can be quickly computed by $Q = \sum (x - \bar{x})^2$:

$$\frac{|5.0 - 11.0|}{\sqrt{\frac{32 + 0.02}{3 + 3 - 4}\left[\frac{1}{3} + \frac{1}{3}\right]}} = \frac{6}{3.27} < 2.0.$$

On the basis of the available samples H_0 is not rejected at the 5% level. The standard procedure (3.33), (3.34), i.e.,

$$\hat{t} = \frac{|5.0 - 11.0|}{\sqrt{\frac{32 + 0.02}{3(3-1)}}} = \frac{6}{2.31} < 4.303 = t_{2;0.05} \quad v = 3 - 1 + \frac{2 \cdot 3 - 2}{\frac{32}{0.02} + \frac{0.02}{32}} \simeq 2,$$

leads to the same decision. With (3.28a) we have $5 - 11 = -6$ and $-6 \pm (4.30)(2.31)$ or 95% CI: $-15.9 \leq \mu_1 - \mu_2 \leq +3.9$, thus including zero [according to H_0].

Three further remarks on comparison of means

| 1 | Samples which are not chosen entirely at random are, in comparison with random samples, characterized by greater similarity among the sample elements and less similarity of the sample means. **With nonrandom drawing of samples the standard deviations are thus decreased and the differences among the means increased.** Both effects can therefore contribute to an apparent "significant difference among the means". Consequently great care must be taken in interpreting results which are barely significant if the samples are not properly random.

| 2 | A comparison of two parameters is possible in terms of their confidence intervals: (1) If the confidence intervals intersect, it does not necessarily follow that the parameters do not differ significantly. (2) If the confidence intervals do not intersect, there is at the given significance level a genuine difference between the parameters: For n_1 and $n_2 \lesssim 200$ and $\sigma_1^2 = \sigma_2^2$ there corresponds to two nonintersecting 95% CI (for μ_1 and μ_2) a t-test difference at the 1% level.

| 3 | The number of sample values needed for the comparison of a sample mean with a hypothetical parameter of the population or for the comparison of two sample means is given in Table 52 for controlled errors—Type I error ($\alpha = 0.005$ and 0.025 or $\alpha = 0.01$ and 0.05) and Type II error ($\beta = 0.2; 0.05; 0.01$)—and given standardized deviations.

The use of Table 52 is illustrated by the examples presented in Table 52a (cf., also (3.6) in Section 3.1.2).

Table 52 The approximate sample size n which is necessary in a one sided problem to recognize as significant, at a level α and with power $1 - \beta$, a standardized difference of $d = (\mu - \mu_0)/\sigma$ between the hypothetical mean μ_0 and the actual mean of the population, or of $d = (\mu_1 - \mu_2)/\sigma$ between the means of two populations with the same variance σ^2. For the two sided problem, as an approximation, the significance levels must be doubled. For the two sample test, it is assumed that both samples have the same size, $n_1 = n_2 = n$. (Taken from W. J. Dixon and F. J. Massey: Introduction to Statistical Analysis, New York, 1957, Table A — 12c, p. 425, Copyright McGraw-Hill Book Company, April 21, 1966.)

		One sample test			Two sample test		
α	d \ β / $1-\beta$	0.20 / 0.80	0.05 / 0.95	0.01 / 0.99	0.20 / 0.80	0.05 / 0.95	0.01 / 0.99
0.005	0.1	1173	1785	2403	2337	3567	4806
	0.2	296	450	605	588	894	1206
	0.4	77	115	154	150	226	304
	0.7	28	40	53	50	75	100
	1.0	14	22	28	26	38	49
	2.0	7	8	10	8	11	14
0.025	0.1	788	1302	1840	1574	2603	3680
	0.2	201	327	459	395	650	922
	0.4	52	85	117	100	164	231
	0.7	19	29	40	34	55	76
	1.0	10	16	21	17	28	38
	2.0	-	6	7	6	8	11

Table 52a

Test	Problem	α	β	d	Sample size
One sample test	one sided	0.005	0.20	0.7	n = 28
	two sided	0.01	0.01	1.0	n = 28
Two sample test	one sided	0.025	0.05	1.0	$n_1 = 28, n_2 = 28$
	two sided	0.05	0.05	0.1	$n_1 = 2603, n_2 = 2603$

Remarks

1. Further aids are given by Croarkin (1962), Winne (1963), Owen (1965, 1968), Hodges and Lehmann (1968), Krishnan (1968), Hsiao (1972), and especially Cohen (1977).

2. The nomographic presentation of the t-test (Thöni 1963, Dietze 1967) as well as other test procedures can be found in Wenger (1963), Stammberger (1966, 1967), and Boyd (1969).

3. **The comparison of two coefficients of variation.** The standard error of the coefficient of variation is

$$s_V = \frac{V}{\sqrt{2n}}\sqrt{1 + \frac{2V^2}{10^4}} \simeq \frac{V}{\sqrt{2n}}.$$

The difference between two coefficients of variation with sample sizes not too small ($n_1, n_2 \geq 30$) can thus be tested by

$$\boxed{\frac{|V_1 - V_2|}{\sqrt{V_1^2/2n_1 + V_2^2/2n_2}}} \tag{3.36}$$

and judged in terms of the standard normal distribution. As an example, one gets for (p. 62) $V_1 = 0.10$, $V_2 = 0.13$, and $n_1 = n_2 = 30$

$$\hat{z} = \frac{|0.10 - 0.13|}{\sqrt{0.10^2/60 + 0.13^2/60}} = 1.417.$$

Since $1.42 < 1.96 = z_{0.05}$, there is no reason to doubt the equality of the two coefficients of variation ($\gamma_1 = \gamma_2$). Lohrding (1975) gives an exact test and critical values for small n.

4. One and two sample t-tests in the case of discrete random variables (success percentage) are considered by Weiler (1964).

3.7 QUICK TESTS WHICH ASSUME NEARLY NORMALLY DISTRIBUTED DATA

3.7.1 The comparison of the dispersions of two small samples according to Pillai and Buenaventura

The dispersion of two independent data sets can be compared by means of the **ranges**, R_1 and R_2: In analogy to the F-test the ratio R_1/R_2 is evaluated (assume $R_1 > R_2$) and compared with the corresponding $(n_1, n_2; \alpha)$ bound in Table 53. Homogeneity of the variances is rejected at the α-level if this bound is surpassed. If for example the data set A with $n_1 = 9$ and the data set B with $n_2 = 10$ have the ranges $R_1 = 19$ and $R_2 = 10$, then $R_1/R_2 = 1.9$ is larger than the value 1.82 tabulated for $\alpha = 5\%$. The null hypothesis is thus rejected. The bounds of Table 53 are set up for the one sided problem. If $\sigma_1^2 = \sigma_2^2$ is tested against $\sigma_1^2 \neq \sigma_2^2$, then the 5% and 1% bounds of this table

Table 53 Upper 5% and 1% significance levels of the F'-distribution based on the ranges (from Pillai, K. C. S. and Buenaventura, A. R. Upper percentage points of a substitute F-ratio using ranges, Biometrika **48** (1961), pp. 195 and 196)

n_2 \ n_1	2	3	4	5	6	7	8	9	10
					$\alpha = 5\%$				
2	12.71	19.08	23.2	26.2	28.6	30.5	32.1	33.5	34.7
3	3.19	4.37	5.13	5.72	6.16	6.53	6.85	7.12	7.33
4	2.03	2.66	3.08	3.38	3.62	3.84	4.00	4.14	4.26
5	1.60	2.05	2.35	2.57	2.75	2.89	3.00	3.11	3.19
6	1.38	1.74	1.99	2.17	2.31	2.42	2.52	2.61	2.69
7	1.24	1.57	1.77	1.92	2.04	2.13	2.21	2.28	2.34
8	1.15	1.43	1.61	1.75	1.86	1.94	2.01	2.08	2.13
9	1.09	1.33	1.49	1.62	1.72	1.79	1.86	1.92	1.96
10	1.05	1.26	1.42	1.54	1.63	1.69	1.76	1.82	1.85
					$\alpha = 1\%$				
2	63.66	95.49	116.1	131	143	153	161	168	174
3	7.37	10.00	11.64	12.97	13.96	14.79	15.52	16.13	16.60
4	3.73	4.79	5.50	6.01	6.44	6.80	7.09	7.31	7.51
5	2.66	3.33	3.75	4.09	4.36	4.57	4.73	4.89	5.00
6	2.17	2.66	2.98	3.23	3.42	3.58	3.71	3.81	3.88
7	1.89	2.29	2.57	2.75	2.90	3.03	3.13	3.24	3.33
8	1.70	2.05	2.27	2.44	2.55	2.67	2.76	2.84	2.91
9	1.57	1.89	2.07	2.22	2.32	2.43	2.50	2.56	2.63
10	1.47	1.77	1.92	2.06	2.16	2.26	2.33	2.38	2.44

are interpreted as 10% and 2% levels of the two sided test. The efficiency of the test is adequate for small samples.

3.7.2 The comparison of the means of two small samples according to Lord

To compare the behavior in the central portion of two independent data sets of equal size ($n_1 = n_2 \leq 20$), the difference of the arithmetic means ($\bar{x}_1, \bar{x}_2$) is computed and divided by the arithmetic mean of the ranges (R_1, R_2)

$$\hat{u} = \frac{|\bar{x}_1 - \bar{x}_2|}{(R_1 + R_2)/2}. \tag{3.37}$$

If the test statistic $\hat{u}$ analogous to the t-statistic equals or exceeds the respective bound in Table 54, then the difference of the means is significant at the associated level (Lord 1947). The test assumes normal distribution and **equality of variances**.

Table 54 Bounds for the comparison according to Lord of two means from independent data sets of equal size (from Lord, E.: The use of range in place of the standard deviation in the t-test, Biometrika **34** (1947), 141–67, p. 66, Table 10)

$n_1 = n_2$	One sided test		Two sided test	
	$u_{0.05}$	$u_{0.01}$	$u_{0.05}$	$u_{0.01}$
3	0.974	1.715	1.272	2.093
4	0.644	1.047	0.831	1.237
5	0.493	0.772	0.613	0.896
6	0.405	0.621	0.499	0.714
7	0.347	0.525	0.426	0.600
8	0.306	0.459	0.373	0.521
9	0.275	0.409	0.334	0.464
10	0.250	0.371	0.304	0.419
11	0.233	0.340	0.280	0.384
12	0.214	0.315	0.260	0.355
13	0.201	0.294	0.243	0.331
14	0.189	0.276	0.228	0.311
15	0.179	0.261	0.216	0.293
16	0.170	0.247	0.205	0.278
17	0.162	0.236	0.195	0.264
18	0.155	0.225	0.187	0.252
19	0.149	0.216	0.179	0.242
20	0.143	0.207	0.172	0.232

For the tabulated values n_1 and n_2 the test is just as powerful as the t-test.

EXAMPLE. For the data sets A: 2, 4, 1, 5 and B: 7, 3, 4, 6 ($R_1 = 5 - 1 = 4$, $R_2 = 7 - 3 = 4$), we have

$$\hat{u} = \frac{|3 - 5|}{(4 + 4)/2} = 0.5,$$

a value which, with $n_1 = n_2 = 4$ and the two sided problem, does not permit the rejection of H_0 at the 5% level. Therefore we decide both samples originate in a common population with the mean μ. Moore (1957) also tabulated this test for unequal sample sizes $n_1 + n_2 \leq 39$; an additional table provides estimates of the standard deviation common to both samples.

3.7.3 Comparison of the means of several samples of equal size according to Dixon

If we wish to find out whether the mean ($\bar{x}_1$) of some data set differs substantially from the $k - 1$ mutually different means of other data sets (all data sets of equal size with $3 \leq n \leq 25$), we order them by magnitude—in increasing order if the mean in question is the smallest, in descending order if it is the largest (the problem is not interesting if there are more extreme means). Then we compute (e.g., for $3 \leq n \leq 7$) the test statistic

$$\hat{M} = \left|\frac{\bar{x}_1 - \bar{x}_2}{\bar{x}_1 - \bar{x}_k}\right| \tag{3.38}$$

and decide in terms of the bounds given in Table 55 (Dixon 1950, 1953). For instance, among the four means 157, 326, 177, and 176 the mean $\bar{x}_1 = 326$ stands out; we have $\bar{x}_2 = 177$, $\bar{x}_3 = 176$, and $\bar{x}_4 = 157$ (where $\bar{x}_4 = \bar{x}_k$), and

$$\hat{M} = \left|\frac{\bar{x}_1 - \bar{x}_2}{\bar{x}_1 - \bar{x}_k}\right| = \frac{326 - 177}{326 - 157} = 0.882$$

is a value which exceeds 0.765 (the 5% bound for $n = 4$). The null hypothesis,

Table 55 Significance bounds for the testing of means and of extreme values in the one sided problem. Before the data are gathered it is agreed upon which end of the ordered sequence of means (or observations; cf. Section 3.8) will be tested. For the two sided problem the significance levels must be doubled. (Excerpted from Dixon, W. J.: Processing data for outliers, Biometrics **9** (1953), 74–89, Appendix, p. 89.)

n	α = 0.10	α = 0.05	α = 0.01	Test statistic[a]
3	0.886	0.941	0.988	
4	0.679	0.765	0.889	
5	0.557	0.642	0.780	$\left\|\frac{\bar{x}_1 - \bar{x}_2}{\bar{x}_1 - \bar{x}_k}\right\|$
6	0.482	0.560	0.698	
7	0.434	0.507	0.637	
8	0.479	0.554	0.683	
9	0.441	0.512	0.635	$\left\|\frac{\bar{x}_1 - \bar{x}_2}{\bar{x}_1 - \bar{x}_{k-1}}\right\|$
10	0.409	0.477	0.597	
11	0.517	0.576	0.679	
12	0.490	0.546	0.642	$\left\|\frac{\bar{x}_1 - \bar{x}_3}{\bar{x}_1 - \bar{x}_{k-1}}\right\|$
13	0.467	0.521	0.615	
14	0.492	0.546	0.641	
15	0.472	0.525	0.616	
16	0.454	0.507	0.595	
17	0.438	0.490	0.577	
18	0.424	0.475	0.561	
19	0.412	0.462	0.547	$\left\|\frac{\bar{x}_1 - \bar{x}_3}{\bar{x}_1 - \bar{x}_{k-2}}\right\|$
20	0.401	0.450	0.535	
21	0.391	0.440	0.524	
22	0.382	0.430	0.514	
23	0.374	0.421	0.505	
24	0.367	0.413	0.497	
25	0.360	0.406	0.489	

[a] For the outlier test substitute $x_1, x_2, x_3; x_n, x_{n-1}, x_{n-2}$ for $\bar{x}_1, \bar{x}_2, \bar{x}_3; \bar{x}_k, \bar{x}_{k-1}, \bar{x}_{k-2}$.

according to which the four means originate in a common, at least approximately normally distributed population, must be rejected at the 5% level. (Table 55 also contains test statistics for $8 \leq n \leq 25$.) This test is fortunately rather insensitive to deviations from normality and variance homogeneity, since by the **central limit theorem**, means of nonnormally distributed data sets are themselves approximately normally distributed.

3.8 THE PROBLEM OF OUTLIERS AND SOME TABLES USEFUL IN SETTING TOLERANCE LIMITS

Extremely large or extremely small values, showing perhaps intrinsic variability, within a sequence of the usual moderately varying data may be neglected under certain conditions. Measurement errors, judgement errors, execution faults, computational errors, or a pathological case among sound data can lead to extreme values which, since they originate in populations other than the one from which the sample comes, must be deleted. A general rule says that if there are at least 10 individual values, then a value may be discarded as an **outlier** provided it lies outside the region $\bar{x} \pm 4s$, where the mean and standard deviation are computed without the value suspected of being an outlier. The "**4-sigma region**" ($\mu \pm 4\sigma$) includes 99.99% of the values for a normal distribution and 97% for symmetric unimodal distributions, and even for arbitrary distributions it includes 94% of the values (cf., Section 1.3.4). The presence of outliers may be an indication of **natural variability, weaknesses in the model, the data**, or both.

> Outlier tests are used to (1) routinely inspect the reliability of data, (2) be promptly advised of need to better control the gathering of data, and (3) recognize extreme data which may be important.

The smaller the samples, the less probable are outliers. Table 55 allows the testing of extreme values of a random sample ($n \leq 25$) from a normally distributed population. It is tested whether an extreme value suspected of being an outlier belongs to a population other than the one to which the remaining values of the sample belong (Dixon 1950; cf., also the surveys of Anscombe 1960 and Grubbs 1969 as well as Thompson and Willke 1963).

The individual values of the sample are ordered by magnitude. Let x_1 denote the extreme value, the supposed outlier:

$$x_1 < x_2 < \cdots < x_{n-1} < x_n \quad \text{or} \quad x_1 > x_2 > \cdots > x_{n-1} > x_n.$$

The individual values of the sample are treated like the means in Section 3.7.3. Thus in the numerical sequence 157, 326, 177, 176 the value 326 proves to be an outlier at the 5% level.

Given, for example, the data sequence 1, 2, 3, 4, 5, 9, the value 9 is suspected of being an outlier. On the basis of Table 55 ($n = 6$), $\hat{M} = (9 - 5)/(9 - 1) = 0.5 < 0.560$; thus the null hypothesis of no outliers is not rejected at the 5% level (normal distribution assumed).

For **sample sizes larger than** $n = 25$ the extreme values can be tested with the help of Table 56 by means of the test statistic

$$T_1 = \left|\frac{x_1 - \mu}{\sigma}\right|, \qquad (3.39)$$

where x_1 is the supposed outlier, and where μ and σ are replaced by $\bar{x}$ and s. If M or T_1 equals or exceeds the bound corresponding to the required confidence coefficient S and to the sample size n in the two tables, then we assume that the tested extreme value originated in a population other than the one from which the rest of the sequence came. However, the extreme value, even if it is shown by this test to be an outlier, may be deleted only if it is probable that the values present are **approximately normally distributed** (cf., also Table 72 in Section 4.3.3).

If outliers of this sort are "identified" and excluded from the sample, a remark to this effect must be included in the summary of the analysis of the data; at least their number is not to be concealed. If a sample contains suspected outliers, it is perhaps most expedient to carry out the statistical analysis once with the outliers retained and once with them removed. If the conclusions of the two analyses differ, an exceptionally cautious and guarded interpretation of the data is recommended. Thus it can happen that the outlier carries a lot of information on the typical variability in the population and therefore it can be the cause for some new investigation. More on seven common tests for outlying observations is given by Sheesley (1977). See also Applied Statistics **27** (1978), 10–25, Journal of the Royal Statistical Society **B40** (1978), 85–93, 242–250, and Statistica Neerlandica **22** (1978), 137–148 [and (all three cited in [8:1]) Barnett and Lewis 1978, Hawkins 1980, and Beckman and Cook 1983].

A procedure (Tukey 1962) recommended by Charles P. Winsor is also convenient:
1. Order the sample values by magnitude.
2. Replace the outlier by the adjacent value. This means, e.g., for 26, 18, 21, 78, 23, 17, and the ordered set 17, 18, 21, 23, 26, 78, we get the values 17, 18, 21, 23, 26, 26. The extreme value is here regarded as unreliable; a certain importance is however ascribed to the direction of the deviation.

If this appears inappropriate, the "Winsorization" is abandoned, perhaps in favor of a careful two sided trimming of the rank statistic, i.e., from the upper and lower end of the rank statistic a total of $\leq 3\%$, and in the case of strong inhomogeneity up to 6%, of the sample values are discarded, the same number from each side (cf., Section 1.3.5; see also Dixon and Tukey 1968).

For small samples with values of a high degree of scatter, dispersion, or variation (viewed as inhomogeneous), the mean absolute deviation (or the

3.8 The Problem of Outliers and Some Tables Useful in Setting Tolerance Limits

mean deviation, cf., Section 3.1.3) is a frequently employed measure of dispersion, since it reduces the influence of the extreme values. Analogously to the standard deviation, which is smallest when the deviations are measured from the arithmetic mean, MD is minimal when the deviations are measured from the median. As a rule, in the case of symmetric and weakly skewed distributions the MD amounts to about $\frac{4}{5}$ of the standard deviation (MD/s $\simeq$ 0.8).

For problems related to **quality control** (cf., Section 2.4.1.3), Table 56 is particularly valuable. Assume samples of size $n = 10$ each are drawn from a population with $\bar{x} = 888$ and $s = 44$. On the average, in at most one out of one hundred samples should the smallest sample value fall below $888 - 44 \cdot 3.089 = 752.1$ and the largest exceed $888 + 44 \cdot 3.089 = 1023.9$. If extreme values of this sort come up more often, the production of the population referred to must be examined.

Table 56 Upper significance bounds of the standardized extreme deviation (taken from Pearson and Hartley, E. S. and Hartley, H. O.: Biometrika Tables for Statisticians, Cambridge University Press, 1954, Table 24)

n	S=95%	S=99%	n	S=95%	S=99%
1	1.645	2.326	55	3.111	3.564
2	1.955	2.575	60	3.137	3.587
3	2.121	2.712	65	3.160	3.607
4	2.234	2.806	70	3.182	3.627
5	2.319	2.877	80	3.220	3.661
6	2.386	2.934	90	3.254	3.691
8	2.490	3.022	100	3.283	3.718
10	2.568	3.089	200	3.474	3.889
15	2.705	3.207	300	3.581	3.987
20	2.799	3.289	400	3.656	4.054
25	2.870	3.351	500	3.713	4.106
30	2.928	3.402	600	3.758	4.148
35	2.975	3.444	700	3.797	4.183
40	3.016	3.479	800	3.830	4.214
45	3.051	3.511	900	3.859	4.240
50	3.083	3.539	1000	3.884	4.264

Tolerance limits

Confidence limits relate to a parameter. Limits for a percentage of the population are referred to as **tolerance limits**. Tolerance limits specify the limits within which a certain portion of the population can be expected with preassigned probability $S = 1 - \alpha$. For a normally distributed population, these limits are of the form $\bar{x} \pm ks$, where k is an appropriate constant. For example, to determine a tolerance region—within which the portion $\gamma = 0.90$ of the population lies in 95% of all cases ($S = 0.95$, $\alpha = 0.05$) on the

Table 57 Tolerance factors for the normal distribution. Factors k for the two sided tolerance region for sample means from normally distributed populations: With probability S at least $\gamma \cdot 100\%$ of the elements in the population lie within the tolerance region $\bar{x} \pm ks$; here $\bar{x}$ and s are computed from a sample of size n. Selected, rounded values from Bowker, A. H.: Tolerance Factors for Normal Distributions, p. 102, in (Statistical Research Group, Columbia University), Techniques of Statistical Analysis (edited by Churchill Eisenhart, Millard W. Hastay, and W. Allen Wallis) New York and London 1947, McGraw-Hill Book Company Inc. (copyright March 1, 1966).

n \ γ	S = 0.95				S = 0.99			
	0.90	0.95	0.99	0.999	0.90	0.95	0.99	0.999
3	8.38	9.92	12.86	16.21	18.93	22.40	29.06	36.62
6	3.71	4.41	5.78	7.34	5.34	6.35	8.30	10.55
12	2.66	3.16	4.15	5.29	3.25	3.87	5.08	6.48
24	2.23	2.65	3.48	4.45	2.52	3.00	3.95	5.04
30	2.14	2.55	3.35	4.28	2.39	2.84	3.73	4.77
50	2.00	2.38	3.13	3.99	2.16	2.58	3.39	4.32
100	1.87	2.23	2.93	3.75	1.98	2.36	3.10	3.95
300	1.77	2.11	2.77	3.54	1.82	2.17	2.85	3.64
500	1.74	2.07	2.72	3.48	1.78	2.12	2.78	3.56
1000	1.71	2.04	2.68	3.42	1.74	2.07	2.72	3.47
∞	1.65	1.96	2.58	3.29	1.65	1.96	2.58	3.29

Table 57 can be supplemented, e.g, by pp. 45–46 of the Documenta Geigy (1968 [2]).

average—we read off from Table 57 for a sample size $n = 50$ the factor $k = 2.00$. The tolerance region of interest thus extends from $\bar{x} - 2.00s$ to $\bar{x} + 2.00s$. Here s is the standard deviation estimated from the 50 sample values and $\bar{x}$ is the corresponding mean. Tables for computing k are provided by Weissberg and Beatty (1960) (cf., L. S. Nelson, Journal of Quality Technology 9 (1970), 198–199) as well as by Guttman (1970), who also includes a survey (cf., also Owen and Frawley 1971). Extensive tables of two-sided tolerance factors k for a normal distribution are given by R. E. Odeh in Communications in Statistics—Simulation and Computation B7 (1978), 183–201. See also Odeh and Owen (1980, cited in [8:1]).

Factors for one sided tolerance limits (Lieberman 1958, Bowker and Lieberman 1959, Owen 1963, Burrows 1964) permit, e.g., the assertion that at least the portion γ of the population is expected to be below $\bar{x} + ks$ or above $\bar{x} - ks$ in 95% of all cases, on the average.

For sufficiently large sample size n, $\bar{x} \pm zs$ are approximate tolerance limits. Strictly speaking, this expression holds only for $n = \infty$. For unknown distributions the determination of the value k is irrelevant. In this case the

3.8 The Problem of Outliers and Some Tables Useful in Setting Tolerance Limits

sample size n is chosen so as to ascertain that with confidence probability S the portion γ of the population lies between the smallest and the largest value of the sample (cf., also Weissberg and Beatty 1960, Owen 1968, and Faulkenberry and Daly 1970). R. L. Kirkpatrick gives tables for sample sizes to set tolerance limits, one-sided and two-sided, for a normal distribution and for the distribution-free case (Journal of Quality Technology **9** (1977), 6–12).

Even for distributions which are only slightly different from the normal, the distribution-free procedure is preferred.

G. L. Tietjen and M. E. Johnson (Technometrics **21** (1979), 107–110) derive exact tolerance limits for sample variances and standard deviations arising from a normal distribution.

Distribution-free tolerance limits

If we wish that with a confidence coefficient $S = 1 - \alpha$ the fraction γ of the elements of an arbitrary population lie between the largest and the smallest sample value, the required sample size n can be readily estimated by means of Table 58, which includes sample sizes n for two sided nonparametric tolerance

Table 58 Sample sizes n for two sided non-parametric tolerance limits

S \ γ	0.50	0.90	0.95	0.99	0.999	0.9999
0.50	3	17	34	168	1679	16783
0.80	5	29	59	299	2994	29943
0.90	7	38	77	388	3889	38896
0.95	8	46	93	473	4742	47437
0.99	11	64	130	662	6636	66381
0.999	14	89	181	920	9230	92330
0.9999	18	113	230	1171	11751	117559

limits which satisfy the Wilks equation (Wilks 1941, 1942) $n\gamma^{n-1} - (n-1)\gamma^n = 1 - S = \alpha$. With the confidence coefficient S, on the average at least the portion γ of an arbitrary population lies between the largest and the smallest value of a random sample drawn from it. That is, in about $S \cdot 100\%$ of the cases in which samples of size n are drawn from an arbitrary population, the extreme values of the sample bound at least $\gamma \cdot 100\%$ of the population values. Thus if the values of a sample are ordered by magnitude, then with an average confidence coefficient of $S = 1 - \alpha$ at least $\gamma \cdot 100\%$ of the elements of the population lie within the interval determined by the largest and the smallest value of the sample. Table 59 gives values of γ for various levels of significance α and sample sizes n.

Table 59 Distribution-free tolerance limits (taken from Wetzel, W.: Elementare Statistische Tabellen, Kiel 1965, Berlin, De Gruyter 1966, p. 31)

n \ α	0.200	0.150	0.100	0.090	0.080	0.070	0.060	0.050	0.040	0.030	0.020	0.010	0.005	0.001
3	0.2871	0.2444	0.1958	0.1850	0.1737	0.1617	0.1490	0.1354	0.1204	0.1036	0.0840	0.0589	0.0414	0.0184
4	0.4175	0.3735	0.3205	0.3082	0.2950	0.2809	0.2656	0.2486	0.2294	0.2071	0.1794	0.1409	0.1109	0.0640
5	0.5098	0.4679	0.4161	0.4038	0.3906	0.3762	0.3603	0.3426	0.3222	0.2979	0.2671	0.2221	0.1851	0.1220
6	0.5776	0.5387	0.4897	0.4779	0.4651	0.4512	0.4357	0.4182	0.3979	0.3734	0.3417	0.2943	0.2540	0.1814
7	0.6291	0.5933	0.5474	0.5363	0.5242	0.5109	0.4961	0.4793	0.4596	0.4357	0.4044	0.3566	0.3151	0.2375
8	0.6696	0.6365	0.5938	0.5833	0.5719	0.5594	0.5453	0.5293	0.5105	0.4875	0.4570	0.4101	0.3685	0.2887
9	0.7022	0.6715	0.6316	0.6218	0.6111	0.5993	0.5861	0.5709	0.5530	0.5309	0.5017	0.4560	0.4150	0.3349
10	0.7290	0.7004	0.6632	0.6540	0.6439	0.6328	0.6202	0.6058	0.5888	0.5678	0.5398	0.4956	0.4557	0.3763
11	0.7514	0.7247	0.6898	0.6811	0.6716	0.6611	0.6493	0.6356	0.6195	0.5995	0.5727	0.5302	0.4914	0.4134
12	0.7704	0.7454	0.7125	0.7043	0.6954	0.6855	0.6742	0.6613	0.6460	0.6269	0.6013	0.5605	0.5230	0.4466
13	0.7867	0.7632	0.7322	0.7245	0.7160	0.7066	0.6959	0.6837	0.6691	0.6509	0.6264	0.5872	0.5510	0.4766
14	0.8008	0.7787	0.7493	0.7420	0.7340	0.7250	0.7149	0.7033	0.6894	0.6720	0.6485	0.6109	0.5760	0.5037
15	0.8132	0.7923	0.7644	0.7575	0.7499	0.7414	0.7317	0.7206	0.7073	0.6907	0.6683	0.6321	0.5984	0.5282
16	0.8242	0.8043	0.7778	0.7712	0.7639	0.7558	0.7467	0.7360	0.7234	0.7075	0.6859	0.6512	0.6186	0.5505
17	0.8339	0.8150	0.7898	0.7835	0.7765	0.7688	0.7600	0.7499	0.7377	0.7225	0.7018	0.6684	0.6370	0.5708
18	0.8426	0.8246	0.8005	0.7945	0.7879	0.7805	0.7721	0.7623	0.7507	0.7361	0.7162	0.6840	0.6537	0.5895
19	0.8505	0.8332	0.8102	0.8045	0.7981	0.7910	0.7830	0.7736	0.7624	0.7484	0.7293	0.6982	0.6689	0.6066
20	0.8576	0.8411	0.8190	0.8135	0.8074	0.8006	0.7929	0.7839	0.7731	0.7596	0.7412	0.7112	0.6829	0.6224
21	0.8640	0.8482	0.8271	0.8218	0.8159	0.8093	0.8019	0.7933	0.7829	0.7699	0.7521	0.7232	0.6957	0.6370
22	0.8699	0.8547	0.8344	0.8293	0.8237	0.8174	0.8102	0.8019	0.7919	0.7793	0.7622	0.7342	0.7076	0.6506
23	0.8753	0.8607	0.8412	0.8362	0.8308	0.8247	0.8178	0.8098	0.8002	0.7880	0.7715	0.7443	0.7186	0.6631
24	0.8803	0.8663	0.8474	0.8426	0.8374	0.8315	0.8249	0.8171	0.8078	0.7961	0.7800	0.7538	0.7287	0.6748
25	0.8849	0.8713	0.8531	0.8485	0.8435	0.8378	0.8314	0.8239	0.8149	0.8035	0.7880	0.7625	0.7382	0.6858
26	0.8892	0.8761	0.8585	0.8540	0.8491	0.8437	0.8374	0.8302	0.8215	0.8105	0.7954	0.7707	0.7471	0.6960
27	0.8931	0.8805	0.8634	0.8591	0.8544	0.8491	0.8431	0.8360	0.8276	0.8169	0.8023	0.7783	0.7554	0.7056
28	0.8968	0.8845	0.8681	0.8639	0.8593	0.8542	0.8483	0.8415	0.8333	0.8230	0.8088	0.7854	0.7631	0.7146
29	0.9002	0.8884	0.8724	0.8683	0.8639	0.8589	0.8532	0.8466	0.8387	0.8286	0.8148	0.7921	0.7704	0.7231
30	0.9035	0.8919	0.8764	0.8725	0.8682	0.8633	0.8578	0.8514	0.8437	0.8339	0.8205	0.7984	0.7772	0.7311
31	0.9065	0.8953	0.8802	0.8764	0.8722	0.8675	0.8622	0.8559	0.8484	0.8389	0.8258	0.8043	0.7837	0.7387
32	0.9093	0.8984	0.8838	0.8801	0.8760	0.8714	0.8662	0.8602	0.8528	0.8436	0.8309	0.8099	0.7898	0.7458
33	0.9120	0.9014	0.8872	0.8836	0.8796	0.8751	0.8701	0.8641	0.8570	0.8480	0.8356	0.8152	0.7956	0.7526
34	0.9145	0.9042	0.8903	0.8868	0.8830	0.8786	0.8737	0.8679	0.8610	0.8522	0.8401	0.8202	0.8010	0.7590
35	0.9169	0.9069	0.8934	0.8899	0.8862	0.8819	0.8771	0.8715	0.8647	0.8562	0.8444	0.8249	0.8062	0.7651
36	0.9191	0.9094	0.8962	0.8929	0.8892	0.8851	0.8804	0.8749	0.8683	0.8599	0.8484	0.8290	0.8111	0.7709
37	0.9212	0.9117	0.8989	0.8956	0.8921	0.8880	0.8834	0.8781	0.8716	0.8635	0.8522	0.8337	0.8158	0.7764
38	0.9232	0.9140	0.9015	0.8983	0.8948	0.8909	0.8864	0.8811	0.8748	0.8669	0.8559	0.8377	0.8202	0.7817
39	0.9252	0.9161	0.9039	0.9008	0.8974	0.8935	0.8892	0.8840	0.8779	0.8701	0.8594	0.8416	0.8244	0.7867
40	0.9270	0.9182	0.9062	0.9032	0.8998	0.8961	0.8918	0.8868	0.8808	0.8732	0.8627	0.8453	0.8285	0.7915
41	0.9287	0.9201	0.9084	0.9055	0.9022	0.8985	0.8943	0.8894	0.8836	0.8761	0.8658	0.8488	0.8323	0.7961
42	0.9304	0.9219	0.9106	0.9076	0.9044	0.9008	0.8967	0.8920	0.8862	0.8789	0.8688	0.8521	0.8360	0.8005
43	0.9320	0.9237	0.9125	0.9097	0.9066	0.9031	0.8990	0.8944	0.8887	0.8816	0.8717	0.8554	0.8396	0.8047
44	0.9335	0.9254	0.9145	0.9117	0.9086	0.9052	0.9012	0.8967	0.8911	0.8841	0.8745	0.8584	0.8430	0.8087
45	0.9349	0.9270	0.9163	0.9136	0.9106	0.9072	0.9034	0.8989	0.8934	0.8866	0.8771	0.8614	0.8462	0.8126
46	0.9363	0.9286	0.9181	0.9154	0.9124	0.9091	0.9054	0.9010	0.8957	0.8889	0.8796	0.8642	0.8493	0.8163
47	0.9376	0.9300	0.9197	0.9171	0.9142	0.9110	0.9073	0.9030	0.8978	0.8912	0.8821	0.8669	0.8523	0.8199
48	0.9389	0.9315	0.9214	0.9188	0.9160	0.9128	0.9092	0.9049	0.8998	0.8934	0.8844	0.8695	0.8552	0.8233
49	0.9401	0.9328	0.9229	0.9204	0.9176	0.9145	0.9110	0.9068	0.9018	0.8954	0.8866	0.8721	0.8579	0.8266
50	0.9413	0.9341	0.9244	0.9220	0.9192	0.9162	0.9127	0.9086	0.9037	0.8974	0.8888	0.8745	0.8606	0.8298
60	0.9509	0.9449	0.9367	0.9346	0.9323	0.9298	0.9268	0.9234	0.9192	0.9139	0.9066	0.8944	0.8826	0.8562
70	0.9578	0.9526	0.9456	0.9438	0.9418	0.9396	0.9370	0.9340	0.9304	0.9258	0.9195	0.9089	0.8986	0.8756
80	0.9630	0.9585	0.9522	0.9507	0.9489	0.9470	0.9447	0.9421	0.9389	0.9348	0.9292	0.9199	0.9108	0.8903
90	0.9671	0.9630	0.9575	0.9561	0.9545	0.9527	0.9507	0.9484	0.9455	0.9419	0.9369	0.9285	0.9203	0.9020
100	0.9704	0.9667	0.9617	0.9604	0.9590	0.9574	0.9556	0.9534	0.9509	0.9476	0.9431	0.9355	0.9280	0.9114
200	0.9851	0.9832	0.9807	0.9800	0.9793	0.9785	0.9776	0.9765	0.9752	0.9735	0.9712	0.9673	0.9634	0.9548
300	0.9901	0.9888	0.9871	0.9867	0.9862	0.9856	0.9850	0.9843	0.9834	0.9823	0.9807	0.9781	0.9755	0.9696
400	0.9925	0.9916	0.9903	0.9900	0.9896	0.9892	0.9887	0.9882	0.9875	0.9867	0.9855	0.9835	0.9816	0.9772
500	0.9940	0.9933	0.9922	0.9920	0.9917	0.9914	0.9910	0.9905	0.9900	0.9893	0.9884	0.9868	0.9852	0.9817
600	0.9950	0.9944	0.9935	0.9933	0.9931	0.9928	0.9926	0.9921	0.9917	0.9911	0.9903	0.9890	0.9877	0.9847
700	0.9957	0.9952	0.9945	0.9943	0.9941	0.9938	0.9936	0.9932	0.9929	0.9924	0.9917	0.9906	0.9894	0.9869
800	0.9963	0.9958	0.9951	0.9950	0.9948	0.9946	0.9944	0.9941	0.9937	0.9933	0.9927	0.9917	0.9907	0.9885
900	0.9967	0.9963	0.9957	0.9955	0.9954	0.9952	0.9950	0.9947	0.9944	0.9941	0.9935	0.9926	0.9918	0.9898
1000	0.9970	0.9966	0.9961	0.9960	0.9958	0.9957	0.9955	0.9953	0.9950	0.9947	0.9942	0.9934	0.9926	0.9908
1500	0.9980	0.9978	0.9974	0.9973	0.9972	0.9971	0.9970	0.9968	0.9967	0.9964	0.9961	0.9956	0.9951	0.9939

EXAMPLE 1. For $S = 0.80$ ($\alpha = 0.20$) and $\gamma = 0.90$ a sample of size $n = 29$ is needed. The smallest and largest value of 80% of all random samples of size $n = 29$ enclose at least 90% of their respective populations.

EXAMPLE 2. The smallest and the largest sample value will enclose at least 85% ($\gamma = 0.85$) of the respective population values on the average in 95 out of

100 ($S = 0.95$ or $\alpha = 0.05$) samples of size $n = 30$. If both percentages are set at 70% (90%, 95%, 99%), a random sample of size $n = 8$ (38, 93, 662) is required.

Nelson (1963) (cf., Journal of Quality Technology **6** (1974), 163–164) provides a nomogram for a quick determination of distribution-free tolerance limits. Important tables are given by Danziger and Davis (1964). An extensive table and nomogram for determining one sided distribution-free tolerance limits was presented by Belson and Nakano (1965) (cf., also Harmann 1967 and Guenther 1970). The prediction intervals presented in Section 3.1.1 supplement these methods.

3.9 DISTRIBUTION-FREE PROCEDURES FOR THE COMPARISON OF INDEPENDENT SAMPLES

The simplest distribution-free test for the comparison of two independent samples is due to Mosteller (1948). The two sample sizes are assumed to be equal ($n_1 = n_2 = n$). The null hypothesis that both samples originate in populations with the same distribution is for $n > 5$ rejected at a significance level of 5% if for

$n \leq 25$	the 5 largest or smallest values,
$n > 25$	the 6 largest or smallest values

come from the same sample. Conover (1968) and Neave (1972) give interesting further developments of this test.

The Rosenbaum quick tests

Both tests are distribution-free for independent samples. We assume the sample sizes are equal: $n_1 = n_2 = n$.

Location test. If at least 5 (of $n \geq 16$; $\alpha = 0.05$) [or at least 7 (of $n \geq 20$; $\alpha = 0.01$)] values of **one** sample lie below or above the span of the other sample, (interval determined by the smallest and the largest sample value) then the null hypothesis (equality of medians) is rejected with the specified level of significance. It is assumed that the spans differ only randomly; the significance levels hold for the one sided problems, while for the two sided case they are to be doubled (Rosenbaum 1954).

Variability test. If at least 7 (of $n \geq 25$; $\alpha = 0.05$) [or at least 10 (of $n \geq 51$; $\alpha = 0.01$)] values of one sample (the one with the greater span; one sided problem) lie outside the span of the other sample, then the null hypothesis (equality of variability, equality of dispersion) is rejected at the specified level of significance. The means are assumed to differ only randomly. If it is not known whether both populations have the same location parameter, then this test checks the location **and** variability of both populations. For $7 \leq n \leq 24$ the 7 may be replaced by 6 ($\alpha = 0.05$); for $21 \leq n \leq 50$ (resp. for $11 \leq n \leq 20$), the 10 by 9 (by 8 respectively) (Rosenbaum 1953).

Both papers include critical values for the case of unequal sample sizes.

Rank tests

If n sample values ordered by increasing magnitude are written as $x_{(1)}, x_{(2)}, \ldots, x_{(n)}$, so that

$$x_{(1)} \leq x_{(2)} \leq \ldots \leq x_{(i)} \leq \ldots \leq x_{(n)}$$

holds, then each of the quantities $x_{(i)}$ is called an **order statistic**. The number assigned to each sample value is referred to as the rank. Thus the order statistic $x_{(i)}$ is associated with the rank i. Tests in which ranks are used in place of sample values form a particularly important group of distribution-free tests (cf., Section 1.4.8). Rank tests surprisingly exhibit a fairly high asymptotic efficiency. Moreover, they require no extensive computations. See, e.g., W. J. Conover and R. L. Iman, The American Statistician **35** (1981), 124–133.

3.9.1 The rank dispersion test of Siegel and Tukey

Since the F-test is sensitive to deviations from the normal distribution, Siegel and Tukey (1960) developed a distribution-free procedure based on the Wilcoxon test. It allows to test H_0: both samples belong to the same population against H_A: the two samples come from different populations, where the populations are only characterized by their variability. However, the probability of rejecting H_0 (the null hypothesis) when the variabilities of the two samples are markedly different decreases with increasing difference between the means, i.e., the larger the difference between the means, the larger is also the probability of making a Type II error. This is true in particular when the dispersions are small. If the populations do not overlap, the power is zero. This test, which is thus very sensitive to differences in variability when the localization parameters are almost equal, was generalized to k samples by Meyer–Bahlburg (1970).

To apply this test, the combined samples ($n_1 + n_2$ with $n_1 \leq n_2$) are indexed as follows: the observed extreme values are assigned low indices and the central observations high ones: namely, the smallest value gets rank 1, the largest two values are given the ranks 2 and 3; then 4 and 5 are assigned to the second and the third smallest value, 6 and 7 to the third and fourth largest, etc. If the number of observations is odd, the middle observation is assigned no index, so that the highest rank is always an even number. The

3.9 Distribution-Free Procedures for the Comparison of Independent Samples

sum of the indices (I_1, I_2) is determined for each sample. For $n_1 = n_2$, $I_1 \simeq I_2$ holds under the null hypothesis (H_0); the more strongly the two samples differ in their variability, the more different the index sums should be. (3.40) serves as a control for the rank sums

$$I_1 + I_2 = \frac{(n_1 + n_2)(n_1 + n_2 + 1)}{2}. \tag{3.40}$$

The authors give, for small sample sizes ($n_1 \leq n_2 \leq 20$), exact critical values of I_1 (sums of the ranks of the smaller sample, which enable us to assess the differences); some are shown in the following table:

n_1	4	5	6	7	8	9	10
$n_2 = n_1$	10–26	17–38	26–52	36–69	49– 87	62–109	78–132
$n_2 = n_1 + 1$	11–29	18–42	27–57	38–74	51– 93	65–115	81–139
$n_2 = n_1 + 2$	12–32	20–45	29–61	40–79	53– 99	68–121	84–146
$n_2 = n_1 + 3$	13–35	21–49	31–65	42–84	55–105	71–127	88–152
$n_2 = n_1 + 4$	14–38	22–53	32–70	44–89	58–110	73–134	91–159
$n_2 = n_1 + 5$	14–42	23–57	34–74	46–94	60–116	76–140	94–166

H_0 is rejected ($\alpha = 0.05$ for two sided test, $\alpha = 0.025$ for one sided) if I_1 for $n_1 \leq n_2$ attains or oversteps the bounds.

For sample sizes not to small ($n_1 > 9$, $n_2 > 9$ or $n_1 > 2$, $n_2 > 20$) the dispersion difference can be dealt with with sufficient accuracy in terms of the standard normal variable:

$$\hat{z} = \frac{2I_1 - n_1(n_1 + n_2 + 1) + 1}{\sqrt{n_1(n_1 + n_2 + 1)(n_2/3)}}. \tag{3.41}$$

If $2I_1 > n_1(n_1 + n_2 + 1)$, then the last $+1$ in the numerator of (3.41) above is replaced by -1.

Very different sample sizes. If the sample sizes differ greatly, (3.41) is too inaccurate. Then the following statistic, which is adjusted for sample sizes, is used:

$$\hat{z}_{\text{corr}} = \hat{z} + \left(\frac{1}{10n_1} - \frac{1}{10n_2}\right)(\hat{z}^3 - 3\hat{z}). \tag{3.41a}$$

Many values equal. If more than one-fifth of a sample is involved in **ties** with values of the other sample—ties within a sample do not interfere—then the denominator in the test statistic (3.41) is to be replaced by

$$\sqrt{n_1(n_1 + n_2 + 1)(n_2/3) - 4[n_1 n_2/(n_1 + n_2)(n_1 + n_2 - 1)](S_1 - S_2)}. \tag{3.42}$$

Here S_1 is the sum of the squares of the indices of tied observations, and S_2 is the sum of the squares of the *mean* indices of tied observations. For example, for the sequence 9.7, 9.7, 9.7, 9.7 we obtain as usual the indices 1, 2, 3, 4 or, if we assign mean indices 2.5, 2.5, 2.5, 2.5 (as $1 + 2 + 3 + 4 = 2.5 + 2.5 + 2.5 + 2.5$); correspondingly the sequence 9.7, 9.7, 9.7 supplies the ranks 1, 2, 3 and the mean ranks 2, 2, 2.

EXAMPLE. Given the two samples A and B:

A	10.1	7.3	12.6	2.4	6.1	8.5	8.8	9.4	10.1	9.8
B	15.3	3.6	16.5	2.9	3.3	4.2	4.9	7.3	11.7	13.1

test possible dispersion differences at the 5% level. Since it is unclear whether the samples come from a normally distributed population, we apply the Siegel–Tukey test. We order the values and bring them into a common rank order:

A	2.4	6.1	7.3	8.5	8.8	9.4		9.8	10.1	10.1	12.6
B	2.9	3.3	3.6	4.2	4.9	7.3		11.7	13.1	15.3	16.5
Value	2.4	2.9	3.3	3.6	4.2	4.9	6.1	7.3	7.3	8.5	8.8
Sample	A	B	B	B	B	B	A	A	B	A	A
Index	1	4	5	8	9	12	13	16	17	20	19
Value	9.4	9.8	10.1	10.1	11.7	12.6	13.1	15.3	16.5		
Sample	A	A	A	A	B	A	B	B	B		
Index	18	15	14	11	10	7	6	3	2		

The index sums are found to be

$I_A = 1 + 13 + 16 + 20 + 19 + 18 + 15 + 14 + 11 + 7 = 134,$

$I_B = 4 + 5 + 8 + 9 + 12 + 17 + 10 + 6 + 3 + 2 \quad = 76,$

and their control,

$$134 + 76 = 210 = \frac{(10 + 10)(10 + 10 + 1)}{2};$$

thus we have, since $(2)(134) = 268 > 210 = 10(10 + 10 + 1)$,

$$\hat{z} = \frac{2 \cdot 134 - 10(10 + 10 + 1) - 1}{\sqrt{10(10 + 10 + 1)(10/3)}} = \frac{57}{\sqrt{700}} = 2.154$$

or

$$\hat{z} = \frac{2 \cdot 76 - 10(10 + 10 + 1) + 1}{\sqrt{10(10 + 10 + 1)(10/3)}} = \frac{152 - 210 + 1}{\sqrt{700}} = -2.154.$$

The probability that a random variable with a standard normal distribution assumes a value which is not smaller than 2.154 is, by Table 13, $P = 0.0156$. (p. 62) In short: The probability of a z-value larger than $\hat{z} = 2.154$ is, by Table 13, $P = 0.0156$. For the two sided problem we have with $P \simeq 0.03$, a variability difference statistically significant at the 5% level (cf., also the table below Equation (3.40): $n_1 - n_2 = 10$; $76 < 78$ and $134 > 132$). For the samples in question a dispersion difference of the populations is assured at the 5% level. Although only 10% of the observations are involved in ties between the samples (7.3, 7.3; the tie 10.1, 10.1 disturbs nothing, since it occurs within sample A), we demonstrate the use of the "long root" (3.42): Taking all ties into account, with

$$S_1 = 11^2 + 14^2 + 16^2 + 17^2 \qquad = 862,$$
$$S_2 = 12.5^2 + 12.5^2 + 16.5^2 + 16.5^2 = 857,$$

and

$$\sqrt{10(10 + 10 + 1)(10/3) - 4[10 \cdot 10/(10 + 10)(10 + 10 - 1)](862 - 857)}$$
$$= \sqrt{700 - 100/19} = \sqrt{694.74} = 26.36,$$

we get

$$\hat{z} = -\frac{57}{26.36} = -2.162 \qquad \text{as against} \quad \hat{z} = -2.154,$$

and with $P(Z > 2.162) = 0.0153$ again $P \simeq 0.03$.

3.9.2 The comparison of two independent samples: Tukey's quick and compact test

Two groups of data are the more distinct, the less their values overlap. If one group contains the highest and another the lowest value, then one must count:

1. the a values in one group which **exceed** all the values in the other group,
2. the b values in the other group which **fall below** all values in the first group.

The two frequencies (each must be greater than zero) are added. This leads to the value of the test statistic $T = a + b$. If the two sample sizes are nearly equal, then the critical values of the test statistics are 7, 10, and 13:

> 7 for a two sided test at the 5% level,
> 10 for a two sided test at the 1% level,
> 13 for a two sided test at the 0.1% level (Tukey 1959).

For two equal values, 0.5 is to be taken. If we denote the two sample sizes by n_1 and n_2, where $n_1 \leq n_2$, then the test is valid for sample sizes not too different, in fact precisely for

$$n_1 \leq n_2 \leq 3 + \frac{4n_1}{3}. \tag{3.43}$$

In all other cases a corrective term is subtracted from the computed test statistic T. The adjusted statistic is then compared with 7, 10, or 13. This correction (3.44, 3.45) equals:

$$1, \quad \text{if} \quad 3 + \frac{4n_1}{3} < n_2 < 2n_1, \tag{3.44}$$

$$\text{the largest integer} \leq \frac{n_2 - n_1 + 1}{n_1}, \quad \text{if} \quad 2n_1 \leq n_2. \tag{3.45}$$

For example, the condition in (3.43) is not satisfied by $n_1 = 7$ and $n_2 = 13$, since $3 + (4)(7)/3 = 37/3 < 13$. The inequalities in (3.44) hold; thus the corrective term is 1. The sample sizes $n_1 = 4$ and $n_2 = 14$ satisfy the condition of (3.45); thus $(14 - 4 + 1)/4 = 11/4 = 2.75$ furnishes the corrective term 2. If the difference between the sample sizes is at least 9 ($n_2 - n_1 \geq 9$), then the critical value 14 is to be used in place of the value 13 for the 0.1% level. Critical values for the one sided test (cf., also the beginning of Section 3.9; only one distribution tail is of interest, and hence only a or b) are given by Westlake (1971): 4 for $10 \leq n_1 = n_2 \leq 15$ and 5 for $n_1 = n_2 \geq 16$ ($\alpha = 0.05$), and 7 for $n_1 = n_2 \geq 20$ ($\alpha = 0.01$).

EXAMPLE. The following values are available:

A: 14.7 15.3 16.1 14.9   15.1 14.8   16.7 17.3* 14.6* 15.0
B: 13.9 14.6 14.2 15.0* 14.3 13.8* 14.7 14.4

We distinguish the highest and the lowest value of each row with an asterisk. There are 5 values (underlined) larger than 15.0*, and the value 15.0 in sample A (points underneath) is counted as half a value. There are likewise $5 + \frac{1}{2}$ values not larger than 14.6*. We get $T = 5\frac{1}{2} + 5\frac{1}{2} = 11$. A correction is unnecessary, since $(n_1 \leq n_2 \leq 3 + 4n_1/3)$ $8 < 10 < 41/3$. Since $T = 11 > 10$, the null hypothesis (equality of the distribution functions underlying the two samples) must be rejected at the 1% level.

Exact critical bounds for small sample sizes are given in the original paper. A further development of this test is described by Neave (1966), who likewise provides tables (cf., also Granger and Neave 1968, as well as Neave and Granger 1968). A similar test is due to Haga (1960).

The graphical version of the Tukey test is described by Sandelius (1968).

3.9.3 The comparison of two independent samples according to Kolmogoroff and Smirnoff

If two independent samples from populations with continuous or discrete distributions, but both of the same type, are to be compared as to whether they were drawn from the same population, then the test of Kolmogoroff (1933) and Smirnoff (1939) applies as the sharpest homogeneity test. It covers all sorts of differences in the shape of the distribution, in particular differences in the midrange behavior (mean, median), the dispersion, the skewness, and the excess, i.e., differences in the distribution function (cf., also Darling 1957 and Kim 1969).

The greatest observed ordinate difference between the two empirical cumulative distribution functions serves as a test statistic. Here the cumulative frequencies F_1 and F_2 (with equal class limits for both samples) are divided by the corresponding sample sizes n_1 and n_2. Then the differences $F_1/n_1 - F_2/n_2$ are computed at regular intervals. The maximum of the absolute values of these differences (for the two sided problem of primary interest: see page 702) furnishes the test statistic D:

$$\hat{D} = \max\left|\left(\frac{F_1}{n_1} - \frac{F_2}{n_2}\right)\right|. \tag{3.46}$$

Some percentage points of the distribution of D are tabulated (Smirnoff 1948; also Kim 1969, and in the tables of Harter and Owen (1970 [2] Vol. 1, pp. 77–170). The critical value D can be approximated, for medium to large sample sizes ($n_1 + n_2 > 35$), by

$$D_{(\alpha)} = K_{(\alpha)} \cdot \sqrt{\frac{n_1 + n_2}{n_1 \cdot n_2}}, \tag{3.47}$$

where $K_{(\alpha)}$ represents a constant depending on the level of significance α (cf., the remark in Section 4.4) as shown in Table 60. If a value $\hat{D}$ determined from two samples equals or exceeds the critical value $D_{(\alpha)}$, then a significant difference exists between the distributions of the two populations. Siegel (1956) and Lindgren (1960) give a table with the 5% and 1% limits for small sample sizes. For the case of equal sample sizes ($n_1 = n_2 = n$), a number of critical values $D_{n(\alpha)}$ from a table by Massey (1951) are listed in Table 61. The

Table 60

α	0.20	0.15	0.10	0.05	0.01	0.001
$K_{(\alpha)}$	1.07	1.14	1.22	1.36	1.63	1.95

Table 61 Several values of $D_{n(\alpha)}$

n ($= n_1 = n_2$)	10	15	20	25	30
$\alpha = 0.05$ two sided problem	7/10	8/15	9/20	10/25	11/30
$\alpha = 0.01$	8/10	9/15	11/20	12/25	13/30

denominator gives the sample size. The numerator for nontabulated values of $D_{n(\alpha)}$ is found according to

$$K_{(\alpha)}\sqrt{2n}, \quad \text{(increased) to the next integer;}$$

e.g., for $\alpha = 0.05$ and $n = 10$ with $1.36\sqrt{(2)(10)} = 6.08$ we get 7, i.e., $D_{10(0.05)} = 7/10$. If a value of $\hat{D}$ determined from two samples equals or exceeds this critical value $D_{n(\alpha)}$, then a statistically significant difference is present.

EXAMPLE. Two data sets are to be compared. Nothing is known about any kind of possible differences. We test the null hypothesis (equality of the populations) against the alternative hypothesis that the two populations exhibit different distributions ($\alpha = 0.05$ for the two sided problem):

Data set 1: 2.1 3.0 1.2 2.9 0.6 2.8 1.6 1.7 3.2 1.7
Data set 2: 3.2 3.8 2.1 7.2 2.3 3.5 3.0 3.1 4.6 3.2

The 10 data values of each row are ordered by magnitude:

Data set 1: 0.6 1.2 1.6 1.7 1.7 2.1 2.8 2.9 3.0 3.2
Data set 2: 2.1 2.3 3.0 3.1 3.2 3.2 3.5 3.8 4.6 7.2

From the frequency distributions (f_1 and f_2) of the two samples we get the cumulative frequencies F_1 and F_2 and the quotients F_1/n_1 and F_2/n_2 (cf., Table 62). The largest absolute difference is $\hat{D} = 6/10$, a value which does not attain the critical value $D_{10(0.05)} = 7/10$, so that the homogeneity hypothesis is to be retained: In light of the available samples there is no reason to doubt a common population.

Table 62

Region	0.0 - 0.9	1.0 - 1.9	2.0 - 2.9	3.0 - 3.9	4.0 - 4.9	5.0 - 5.9	6.0 - 6.9	7.0 - 7.9
f_1	1	4	3	2	0	0	0	0
f_2	0	0	2	6	1	0	0	1
F_1/n_1	1/10	5/10	8/10	10/10	10/10	10/10	10/10	10/10
F_2/n_2	0/10	0/10	2/10	8/10	9/10	9/10	9/10	10/10
$F_1/n_1 - F_2/n_2$	1/10	5/10	6/10	2/10	1/10	1/10	1/10	0

Here we shall not delve further into the one sided Kolmogoroff–Smirnoff test [(3.47) with $K_{0.05} = 1.22$ or $K_{0.01} = 1.52$], since with distributions of the same form it is inferior to the one sided U-test of Wilcoxon, Mann, and Whitney. Critical bounds for the three sample test are provided by Birnbaum and Hall (1960), who also tabulated the two sample test for the one sided question. In Section 4.4, the Kolmogoroff–Smirnoff test is used to compare an observed and a theoretical distribution.

▶ 3.9.4 Comparison of two independent samples: The U-test of Wilcoxon, Mann, and Whitney

The rank test of Mann and Whitney (1947), based on the so-called Wilcoxon test for independent samples, is the distribution-free (or better, nearly assumption-free) counterpart to the parametric t-test for the comparison of the means of two **continuous** distributions. This continuity assumption is, strictly speaking, never fulfilled in practice, since all results of measurements are rounded-off numbers. The asymptotic efficiency of the U-test is nearly $100(3/\pi) \simeq 95\%$, i.e., this test based on 1,000 values has about the same power as the t-test based on $0.95(1,000) = 950$ values, when a normal distribution is in fact present. It is therefore to our advantage, even in the case of normal distributions, to apply the U-test [e.g., as a rough computation or as a check of highly significant t-test results in which one does not have real confidence]. It is assumed that samples being compared exhibit **the same form of distribution** (Gibbons 1964, Pratt 1964, Edington 1965). If not, proceed according to Remark 6 below. [The asymptotic efficiency of the U-test, like that of the H-test, cannot fall below 86.4% for any population distribution (Hodges and Lehmann 1956); for the more involved tests of Van der Waerden (X-test, cf., 1965) and of Terry-Hoeffding and Bell-Doksum (see, e.g., Bradley 1968) it is at least 100%. Worked-out examples and remarks concerning important tables are also given by Rytz (1967, 1968) as well as Penfield and McSweeney (1968)].

The U-test of Wilcoxon, Mann, and Whitney tests against the following alternative hypothesis: The probability that an observation from the first population is greater than an arbitrary observation from the second population, does not equal $\frac{1}{2}$. The test is sensitive to differences of the medians—only for $n_1 = n_2$, $H_0: \tilde{\mu}_1 = \tilde{\mu}_2$ is tested and remarkably robust against $\sigma_1^2 \neq \sigma_2^2$—less sensitive to differences in skewness, and insensitive to differences in variance (when needed, these are tested according to Siegel and Tukey; cf., Section 3.9.1).

To compute the test statistic U the $(m + n)$ elements of the combined sample are doubly indexed by their rank and the population to which they belong (cf., Section 3.9). Let R_1 be the sum of the ranks falling to sample 1, and R_2 the sum of the ranks falling to sample 2. The expressions in (3.48) are then worked out, and the computation is checked by (3.49)

$$U_1 = mn + \frac{m(m+1)}{2} - R_1, \qquad U_2 = mn + \frac{n(n+1)}{2} - R_2, \qquad (3.48)$$

$$U_1 + U_2 = mn. \qquad (3.49)$$

The test statistic U is the smaller of the two quantities U_1 and U_2 $[U = \min(U_1, U_2)]$. The null hypothesis is abandoned if the computed value of U is **less than or equal** to the critical value $U(m, n; \alpha)$ from Table 63; extensive tables can be found in Selected Tables cited in [2] (Harter and Owen 1970, Vol. 1, pp. 177–236 [discussed on pp. 171–174]). For larger sizes ($m + n > 60$) the following excellent approximation holds:

$$U(m, n; \alpha) = \frac{nm}{2} - z \cdot \left[\sqrt{\frac{nm(n + m + 1)}{12}} \right] \qquad (3.50)$$

Appropriate values of z for the two and the one sided question are contained in Table 43 in Section 2.1.6. The following approximation is used in place of (3.50) if one cannot or does not wish to specify an α or if no tables of the critical value $U(m, n; \alpha)$ are available, provided the sample sizes are not too small ($m \geq 8, n \geq 8$; Mann and Whitney 1947):

$$\hat{z} = \frac{\left| U - \frac{mn}{2} \right|}{\sqrt{\frac{mn(m+n+1)}{12}}}. \qquad (3.51)$$

We may rewrite (3.51) without the absolute signs in the numerator as

$$\hat{z} = \frac{\dfrac{R_1}{m} - \dfrac{R_2}{n}}{\sqrt{\left[\dfrac{(m+n)^2 + 1}{12}\right]\left[\dfrac{1}{m} + \dfrac{1}{n}\right]\left[\dfrac{m+n}{(m+n) - 1}\right]}}$$

$$= \frac{\bar{R}_1 - \bar{R}_2}{\sqrt{\dfrac{(m+n)^2(m+n+1)}{12mn}}}. \qquad (3.51a)$$

The value $\hat{z}$ is compared with the critical z_α-values of the standard normal distribution (Table 14, Section 1.3.4, or Table 43, Section 2.1.6). A U-test with homogeneous sample subgroups (Wilcoxon: Case III, groups of replicates [randomized blocks]) is discussed in greater detail by Lienert and Schulz (1967) and in particular by Nelson (1970).

3.9 Distribution-Free Procedures for the Comparison of Independent Samples

EXAMPLE. Test the two samples A and B with their values ordered by size for equality of means against $H_A: \mu_A > \mu_B$ (one sided problem with $\alpha = 0.05$):

A: 7 14 22 36 40 48 49 52   ($m = 8$) [sample 1]
B: 3 5 6 10 17 18 20 39   ($n = 8$) [sample 2]

Since normality is not presumed, the t-test is replaced by the U-test, which compares the distribution functions and for $n_1 = n_2$ the medians ($H_0: \tilde{\mu}_A = \tilde{\mu}_B$):

Rank	1	2	3	4	5	6	7	8	9	10	11	12	13	14	15	16
Sample value	3	5	6	7	10	14	17	18	20	22	36	39	40	48	49	52
Sample	B	B	B	A	B	A	B	B	B	A	A	B	A	A	A	A
$R_1 = 89$ $R_2 = 47$	1	+2	+3	4	+5	+6	+7	+8	+9	+10	+11	+12	+13	+14	+15	+16

$$U_1 = 8 \cdot 8 + \frac{8(8+1)}{2} - 89 = 11,$$

$$U_2 = 8 \cdot 8 + \frac{8(8+1)}{2} - 47 = 53,$$

$$U_1 + U_2 = 64 = mn.$$

Since $11 < 15 = U(8; 8; 0.05$; one sided test), the null hypothesis $\tilde{\mu}_A = \tilde{\mu}_B$ is rejected; the alternate hypothesis $\tilde{\mu}_A > \tilde{\mu}_B$ is accepted at the 5% level. (3.51) with

$$\hat{z} = \frac{\left|11 - \frac{8 \cdot 8}{2}\right|}{\sqrt{\frac{8 \cdot 8(8 + 8 + 1)}{12}}} = 2.205$$

and $P = 0.014 < 0.05$ leads to the same decision ($z_{0.05;\text{one sided}} = 1.645$).

$$\hat{z} = \frac{\frac{89}{8} - \frac{47}{8}}{\sqrt{\left[\frac{(8+8)^2 + 1}{12}\right]\left[\frac{1}{8} + \frac{1}{8}\right]\left[\frac{8+8}{(8+8)-1}\right]}}$$

$$= \frac{11.125 - 5.875}{\sqrt{\frac{(8+8)^2(8+8+1)}{(12)(8)(8)}}} = 2.205.$$

Table 63 Critical values of U for the Wilcoxon-Mann-Whitney test for the one sided problem ($\alpha = 0.10$) or the two sided problem ($\alpha = 0.20$). (Taken from Milton, R. C.: An extended table of critical values for the Mann-Whitney (Wilcoxon) two-sample statistic, J. Amer. Statist. Ass. **59** (1964), 925–934.)

m \ n	1	2	3	4	5	6	7	8	9	10	11	12	13	14	15	16	17	18	19	20
1	-																			
2	-	-																		
3	-	0	1																	
4	-	0	1	3																
5	-	1	2	4	5															
6	-	1	3	5	7	9														
7	-	1	4	6	8	11	13													
8	-	2	5	7	10	13	16	19												
9	0	2	5	9	12	15	18	22	25											
10	0	3	6	10	13	17	21	24	28	32										
11	0	3	7	11	15	19	23	27	31	36	40									
12	0	4	8	12	17	21	26	30	35	39	44	49								
13	0	4	9	13	18	23	28	33	38	43	48	53	58							
14	0	5	10	15	20	25	31	36	41	47	52	58	63	69						
15	0	5	10	16	22	27	33	39	45	51	57	63	68	74	80					
16	0	5	11	17	23	29	36	42	48	54	61	67	74	80	86	93				
17	0	6	12	18	25	31	38	45	52	58	65	72	79	85	92	99	106			
18	0	6	13	20	27	34	41	48	55	62	69	77	84	91	98	106	113	120		
19	1	7	14	21	28	36	43	51	58	66	73	81	89	97	104	112	120	128	135	
20	1	7	15	22	30	38	46	54	62	70	78	86	94	102	110	119	127	135	143	151
21	1	8	15	23	32	40	48	56	65	73	82	91	99	108	116	125	134	142	151	160
22	1	8	16	25	33	42	51	59	68	77	86	95	104	113	122	131	141	150	159	168
23	1	9	17	26	35	44	53	62	72	81	90	100	109	119	128	138	147	157	167	176
24	1	9	18	27	36	46	56	65	75	85	95	105	114	124	134	144	154	164	174	184
25	1	9	19	28	38	48	58	68	78	89	99	109	120	130	140	151	161	172	182	193
26	1	10	20	30	40	50	61	71	82	92	103	114	125	136	146	157	168	179	190	201
27	1	10	21	31	41	52	63	74	85	96	107	119	130	141	152	164	175	186	198	209
28	1	11	21	32	43	54	66	77	88	100	112	123	135	147	158	170	182	194	206	217
29	2	11	22	33	45	56	68	80	92	104	116	128	140	152	164	177	189	201	213	226
30	2	12	23	35	46	58	71	83	95	108	120	133	145	158	170	183	196	209	221	234
31	2	12	24	36	48	61	73	86	99	111	124	137	150	163	177	190	203	216	229	242
32	2	13	25	37	50	63	76	89	102	115	129	142	156	169	183	196	210	223	237	251
33	2	13	26	38	51	65	78	92	105	119	133	147	161	175	189	203	217	131	245	259
34	2	13	26	40	53	67	81	95	109	123	137	151	166	180	195	209	224	238	253	267
35	2	14	27	41	55	69	83	98	112	127	141	156	171	186	201	216	230	245	260	275
36	2	14	28	42	56	71	86	100	115	131	146	161	176	191	207	222	237	253	268	284
37	2	15	29	43	58	73	88	103	119	134	150	166	181	197	213	229	244	260	276	292
38	2	15	30	45	60	75	91	106	122	138	154	170	186	203	219	235	251	268	284	301⁺
39	3	16	31	46	61	77	93	109	126	142	158	175	192	208	225	242	258	275	292	309⁺
40	3	16	31	47	63	79	96	112	129	146	163	180	197	214	231	248	265	282	300⁺	317⁺

The *U*-test with tied ranks

If in a sample (or combined sample), the elements of which are ranked by size, two or more values coincide (we speak of a tie), then they are assigned the same averaged rank. For example, for the two sided problem with $\alpha = 0.05$ and the following values:

Sample value	3	3	4	5	5	5	5	8	8	9	10	13	13	13	15	16
Sample	B	B	B	B	B	A	A	A	B	B	A	A	A	A	A	B
Rank	1.5	1.5	3	5.5	5.5	5.5	5.5	8.5	8.5	10	11	13	13	13	15	16

the first two *B*-values get the rank $(1 + 2)/2 = 1.5$; the 4 fives each get the rank $5.5 = (4 + 5 + 6 + 7)/4$; both eights get the rank 8.5; the value 13

Table 63 (*1st continuation*). Critical values of U for the Wilcoxon-Mann-Whitney test for the one sided problem ($\alpha = 0.05$) and the two sided problem ($\alpha = 0.10$)

m \ n	1	2	3	4	5	6	7	8	9	10	11	12	13	14	15	16	17	18	19	20
1	-																			
2	-	-																		
3	-	-	0																	
4	-	-	0	1																
5	-	0	1	2	4															
6	-	0	2	3	5	7														
7	-	0	2	4	6	8	11													
8	-	1	3	5	8	10	13	15												
9	-	1	4	6	9	12	15	18	21											
10	-	1	4	7	11	14	17	20	24	27										
11	-	1	5	8	12	16	19	23	27	31	34									
12	-	2	5	9	13	17	21	26	30	34	38	42								
13	-	2	6	10	15	19	24	28	33	37	42	47	51							
14	-	3	7	11	16	21	26	31	36	41	46	51	56	61						
15	-	3	7	12	18	23	28	33	39	44	50	55	61	66	72					
16	-	3	8	14	19	25	30	36	42	48	54	60	65	71	77	83				
17	-	3	9	15	20	26	33	39	45	51	57	64	70	77	83	89	96			
18	-	4	9	16	22	28	35	41	48	55	61	68	75	82	88	95	102	109		
19	0	4	10	17	23	30	37	44	51	58	65	72	80	87	94	101	109	116	123	
20	0	4	11	18	25	32	39	47	54	62	69	77	84	92	100	107	115	123	130	138
21	0	5	11	19	26	34	41	49	57	65	73	81	89	97	105	113	121	130	138	146
22	0	5	12	20	28	36	44	52	60	68	77	85	94	102	111	119	128	136	145	154
23	0	5	13	21	29	37	46	54	63	72	81	90	98	107	116	125	134	143	152	161
24	0	6	13	22	30	39	48	57	66	75	85	94	103	113	122	131	141	150	160	169
25	0	6	14	23	32	41	50	60	69	79	89	98	108	118	128	137	147	157	167	177
26	0	6	15	24	33	43	53	62	72	82	92	103	113	123	133	143	154	164	174	185
27	0	7	15	25	35	45	55	65	75	86	96	107	117	128	139	149	160	171	182	192
28	0	7	16	26	36	46	57	68	78	89	100	111	122	133	144	156	167	178	189	200
29	0	7	17	27	38	48	59	70	82	93	104	116	127	138	150	162	173	185	196	208
30	0	7	17	28	39	50	61	73	85	96	108	120	132	144	156	168	180	192	204	216
31	0	8	18	29	40	52	64	76	88	100	112	124	136	149	161	174	186	199	211	224
32	0	8	19	30	42	54	66	78	91	103	116	128	141	154	167	180	193	206	218	231
33	0	8	19	31	43	56	68	81	94	107	120	133	146	159	172	186	199	212	226	239
34	0	9	20	32	45	57	70	84	97	110	124	137	151	164	178	192	206	219	233	247
35	0	9	21	33	46	59	73	86	100	114	128	141	156	170	184	198	212	226	241	255
36	0	9	21	34	48	61	75	89	103	117	131	146	160	175	189	204	219	233	248	263
37	0	10	22	35	49	63	77	91	106	121	135	150	165	180	195	210	225	240	255	271
38	0	10	23	36	50	65	79	94	109	124	139	154	170	185	201	216	232	247	263	278
39	1	10	23	38	52	67	82	97	112	128	143	159	175	190	206	222	238	254	270	286[a]
40	1	11	24	39	53	68	84	99	115	131	147	163	179	196	212	228	245	261	278	294[a]

[a] In terms of approximate values based on the normal distribution

occurs three times and is assigned the rank $(12 + 13 + 14)/3 = 13$. Ties influence the value U only when they arise between the two samples, not if they are observed within one or within both samples. If there are ties between the two samples, then the correct formula for the U-test with rank allocation and the sum $S = m + n$ reads

$$\hat{z} = \frac{\left| U - \frac{mn}{2} \right|}{\sqrt{\left[\frac{mn}{S(S-1)} \right] \cdot \left[\frac{S^3 - S}{12} - \sum_{i=1}^{i=r} \frac{t_i^3 - t_i}{12} \right]}}. \quad (3.52)$$

In the corrective term $\sum_{i=1}^{r} (t_i^3 - t_i)/12$ (Walter 1951, following a suggestion by Kendall 1945), r stands for the number of ties, while t_i denotes the multiplicity of the ith tie. Thus for each group $(i = 1, \ldots, r)$ of ties we determine the number t_i of occurrences of that particular tied value and calculate $(t_i^3 - t_i)/12$. The sum of these r quotients forms the corrective term.

Table 63 (*2nd continuation*). Critical values of U for the Wilcoxon-Mann-Whitney test for the one sided problem ($\alpha = 0.025$) and the two sided problem ($\alpha = 0.05$)

m \ n	1	2	3	4	5	6	7	8	9	10	11	12	13	14	15	16	17	18	19	20
1	-																			
2	-	-																		
3	-	-	-																	
4	-	-	-	0																
5	-	-	0	1	2															
6	-	-	1	2	3	5														
7	-	-	1	3	5	6	8													
8	-	0	2	4	6	8	10	13												
9	-	0	2	4	7	10	12	15	17											
10	-	0	3	5	8	11	14	17	20	23										
11	-	0	3	6	9	13	16	19	23	26	30									
12	-	1	4	7	11	14	18	22	26	29	33	37								
13	-	1	4	8	12	16	20	24	28	33	37	41	45							
14	-	1	5	9	13	17	22	26	31	36	40	45	50	55						
15	-	1	5	10	14	19	24	29	34	39	44	49	54	59	64					
16	-	1	6	11	15	21	26	31	37	42	47	53	59	64	70	75				
17	-	2	6	11	17	22	28	34	39	45	51	57	63	69	75	81	87			
18	-	2	7	12	18	24	30	36	42	48	55	61	67	74	80	86	93	99		
19	-	2	7	13	19	25	32	38	45	52	58	65	72	78	85	92	99	106	113	
20	-	2	8	14	20	27	34	41	48	55	62	69	76	83	90	98	105	112	119	127
21	-	3	8	15	22	29	36	43	50	58	65	73	80	88	96	103	111	119	126	134
22	-	3	9	16	23	30	38	45	53	61	69	77	85	93	101	109	117	125	133	141
23	-	3	9	17	24	32	40	48	56	64	73	81	89	98	106	115	123	132	140	149
24	-	3	10	17	25	33	42	50	59	67	76	85	94	102	111	120	129	138	147	156
25	-	3	10	18	27	35	44	53	62	71	80	89	98	107	117	126	135	145	154	163
26	-	4	11	19	28	37	46	55	64	74	83	93	102	112	122	132	141	151	161	171
27	-	4	11	20	29	38	48	57	67	77	87	97	107	117	127	137	147	158	168	178
28	-	4	12	21	30	40	50	60	70	80	90	101	111	122	132	143	154	164	175	186
29	-	4	13	22	32	42	52	62	73	83	94	105	116	127	138	149	160	171	182	193
30	-	5	13	23	33	43	54	65	76	87	98	109	120	131	143	154	166	177	189	200
31	-	5	14	24	34	45	56	67	78	90	101	113	125	136	148	160	172	184	196	208
32	-	5	14	24	35	46	58	69	81	93	105	117	129	141	153	166	178	190	203	215
33	-	5	15	25	37	48	60	72	84	96	108	121	133	146	159	171	184	197	210	222
34	-	5	15	26	38	50	62	74	87	99	112	125	138	151	164	177	190	203	217	230
35	-	6	16	27	39	51	64	77	89	103	116	129	142	156	169	183	196	210	224	237
36	-	6	16	28	40	53	66	79	92	106	119	133	147	161	174	188	202	216	231	245
37	-	6	17	29	41	55	68	81	95	109	123	137	151	165	180	194	209	223	238	252
38	-	6	17	30	43	56	70	84	98	112	127	141	156	170	185	200	215	230	245	259
39	0	7	18	31	44	58	72	86	101	115	130	145	160	175	190	206	221	236	252	267
40	0	7	18	31	45	59	74	89	103	119	134	149	165	180	196	211	227	243	258	274

For the above example, the corrective term results from $r = 4$ groups of ties as follows:

Group 1: $t_1 = 2$: the value 3 twice with the rank 1.5.
Group 2: $t_2 = 4$: the value 5 four times with the rank 5.5.
Group 3: $t_3 = 2$; the value 8 twice with the rank 8.5.
Group 4: $t_4 = 3$: the value 13 three times with the rank 13.

$$\sum_{i=1}^{i=4} \frac{t_i^3 - t_i}{12} = \frac{2^3 - 2}{12} + \frac{4^3 - 4}{12} + \frac{2^3 - 2}{12} + \frac{3^3 - 3}{12}$$

$$= \frac{6}{12} + \frac{60}{12} + \frac{6}{12} + \frac{24}{12} = 8.00$$

A: $m = 8, R_1 = 83.5$ B: $n = 8, R_2 = 52.5$

$$U_1 = 8 \cdot 8 + \frac{8(8+1)}{2} - 83.5 = 16.5 \qquad U_2 = 8 \cdot 8 + \frac{8(8+1)}{2} - 52.5$$

$$U_2 = 47.5$$

Table 63 (*3rd continuation*). Critical values of U for the Wilcoxon-Mann-Whitney test for the one-sided problem ($\alpha = 0.01$) and the two sided problem ($\alpha = 0.02$)

m\n	1	2	3	4	5	6	7	8	9	10	11	12	13	14	15	16	17	18	19	20
1	-																			
2	-	-																		
3	-	-	-																	
4	-	-	-	-																
5	-	-	-	0	1															
6	-	-	-	1	2	3														
7	-	-	,0	1	3	4	6													
8	-	-	0	2	4	6	7	9												
9	-	-	1	3	5	7	9	11	14											
10	-	-	1	3	6	8	11	13	16	19										
11	-	-	1	4	7	9	12	15	18	22	25									
12	-	-	2	5	8	11	14	17	21	24	28	31								
13	-	0	2	5	9	12	16	20	23	27	31	35	39							
14	-	0	2	6	10	13	17	22	26	30	34	38	43	47						
15	-	0	3	7	11	15	19	24	28	33	37	42	47	51	56					
16	-	0	3	7	12	16	21	26	31	36	41	46	51	56	61	66				
17	-	0	4	8	13	18	23	28	33	38	44	49	55	60	66	71	77			
18	-	0	4	9	14	19	24	30	36	41	47	53	59	65	70	76	82	88		
19	-	1	4	9	15	20	26	32	38	44	50	56	63	69	75	82	88	94	101	
20	-	1	5	10	16	22	28	34	40	47	53	60	67	73	80	87	93	100	107	114
21	-	1	5	11	17	23	30	36	43	50	57	64	71	78	85	92	99	106	113	121
22	-	1	6	11	18	24	31	38	45	53	60	67	75	82	90	97	105	112	120	127
23	-	1	6	12	19	26	33	40	48	55	63	71	79	87	94	102	110	118	126	134
24	-	1	6	13	20	27	35	42	50	58	66	75	83	91	99	108	116	124	133	141
25	-	1	7	13	21	29	36	45	53	61	70	78	87	95	104	113	122	130	139	148
26	-	1	7	14	22	30	38	47	55	64	73	82	91	100	109	118	127	136	146	155
27	-	2	7	15	23	31	40	49	58	67	76	85	95	104	114	123	133	142	152	162
28	-	2	8	16	24	33	42	51	60	70	79	89	99	109	119	129	139	149	159	169
29	-	2	8	16	25	34	43	53	63	73	83	93	103	113	123	134	144	155	165	176
30	-	2	9	17	26	35	45	55	65	76	86	96	107	118	128	139	150	161	172	182
31	-	2	9	18	27	37	47	57	68	78	89	100	111	122	133	144	156	167	178	189
32	-	2	9	18	28	38	49	59	70	81	92	104	115	127	138	150	161	173	185	196
33	-	2	10	19	29	40	50	61	73	84	96	107	119	131	143	155	167	179	191	203
34	-	3	10	20	30	41	52	64	75	87	99	111	123	135	148	160	173	185	198	210
35	-	3	11	20	31	42	54	66	78	90	102	115	127	140	153	165	178	191	204	217
36	-	3	11	21	32	44	56	68	80	93	106	118	131	144	158	171	184	197	211	224
37	-	3	11	22	33	45	57	70	83	96	109	122	135	149	162	176	190	203	217	231
38	-	3	12	22	34	46	59	72	85	99	112	126	139	153	167	181	195	209	224	238
39	-	3	12	23	35	48	61	74	88	101	115	129	144	158	172	187	201	216	230	245
40	-	3	13	24	36	49	63	76	90	104	119	133	148	162	177	192	207	222	237	252

$U_1 + U_2 = 64 = mn$ and

$$\hat{z} = \frac{\left| 16.5 - \frac{8 \cdot 8}{2} \right|}{\sqrt{\left[\frac{8 \cdot 8}{16(16-1)} \right] \cdot \left[\frac{16^3 - 16}{12} - 8.00 \right]}} = 1.647 \quad \text{or} \quad 1.65.$$

Since $1.65 < 1.96$, the null hypothesis ($\tilde{\mu}_A = \tilde{\mu}_B$) is retained in the two sided problem ($\alpha = 0.05$).

The U-test is one of the most powerful nonparametric tests. Since the test statistic U is a rather complicated function of the mean, the kurtosis, and the skewness, it must be emphasized that the significance levels (regarding the hypothesis on the difference of two medians or means alone) become more unreliable with increasing difference in the form of the distribution function of the two populations.

Table 63 (*4th continuation*). Critical values of U for the Wilcoxon-Mann-Whitney test for the one sided problem ($\alpha = 0.005$) and the two sided problem ($\alpha = 0.01$)

m	\ n	1	2	3	4	5	6	7	8	9	10	11	12	13	14	15	16	17	18	19	20
1		-																			
2		-	-																		
3		-	-	-																	
4		-	-	-	-																
5		-	-	-	-	0															
6		-	-	-	0	1	2														
7		-	-	-	0	1	3	4													
8		-	-	-	1	2	4	6	7												
9		-	-	0	1	3	5	7	9	11											
10		-	-	0	2	4	6	9	11	13	16										
11		-	-	0	2	5	7	10	13	16	18	21									
12		-	-	1	3	6	9	12	15	18	21	24	27								
13		-	-	1	3	7	10	13	17	20	24	27	31	34							
14		-	-	1	4	7	11	15	18	22	26	30	34	38	42						
15		-	-	2	5	8	12	16	20	24	29	33	37	42	46	51					
16		-	-	2	5	9	13	18	22	27	31	36	41	45	50	55	60				
17		-	-	2	6	10	15	19	24	29	34	39	44	49	54	60	65	70			
18		-	-	2	6	11	16	21	26	31	37	42	47	53	58	64	70	75	81		
19		-	0	3	7	12	17	22	28	33	39	45	51	57	63	69	74	81	87	93	
20		-	0	3	8	13	18	24	30	36	42	48	54	60	67	73	79	86	92	99	105
21		-	0	3	8	14	19	25	32	38	44	51	58	64	71	78	84	91	98	105	112
22		-	0	4	9	14	21	27	34	40	47	54	61	68	75	82	89	96	104	111	118
23		-	0	4	9	15	22	29	35	43	50	57	64	72	79	87	94	102	109	117	125
24		-	0	4	10	16	23	30	37	45	52	60	68	75	83	91	99	107	115	123	131
25		-	0	5	10	17	24	32	39	47	55	63	71	79	87	96	104	112	121	129	138
26		-	0	5	11	18	25	33	41	49	58	66	74	83	92	100	109	118	127	135	144
27		-	1	5	12	19	27	35	43	52	60	69	78	87	96	105	114	123	132	142	151
28		-	1	5	12	20	28	36	45	54	63	72	81	91	100	109	119	128	138	148	157
29		-	1	6	13	21	29	38	47	56	66	75	85	94	104	114	124	134	144	154	164
30		-	1	6	13	22	30	40	49	58	68	78	88	98	108	119	129	139	150	160	170
31		-	1	6	14	22	32	41	51	61	71	81	92	102	113	123	134	145	155	166	177
32		-	1	7	14	23	33	43	53	63	74	84	95	106	117	128	139	150	161	172	184
33		-	1	7	15	24	34	44	55	65	76	87	98	110	121	132	144	155	167	179	190
34		-	1	7	16	25	35	46	57	68	79	90	102	113	125	137	149	161	173	185	197
35		-	1	8	16	26	37	47	59	70	82	93	105	117	129	142	154	166	179	191	203
36		-	1	8	17	27	38	49	60	72	84	96	109	121	134	146	159	172	184	197	210
37		-	1	8	17	28	39	51	62	75	87	99	112	125	138	151	164	177	190	203	217
38		-	1	9	18	29	40	52	64	77	90	102	116	129	142	155	169	182	196	210	223
39		-	2	9	19	30	41	54	66	79	92	106	119	133	146	160	174	188	202	216	230
40		-	2	9	19	31	43	55	68	81	95	109	122	136	150	165	179	193	208	222	237

More than two independent samples may be compared, by comparing the samples pairwise. A **simultaneous** nonparametric comparison of several samples can be carried out with the *H*-test of Kruskal and Wallis (cf., Section 3.9.5). A one sample test corresponding to the *U*-test (cf., also Section 4.2.4) is due to Carnal and Riedwyl (1972) (cf., G. Rey, Biometrical Journal **21** (1979), 259–276). The comparison of two sets of data with clumpings at zero is possible by means of a χ^2 approximation (Lachenbruch 1976).

Remarks

1. The original **two sample test of Wilcoxon** (cf., Jacobson 1963) is now also **completely tabulated** (Wilcoxon et al. 1963; cf., also 1964). Approximations to the Wilcoxon–Mann–Whitney distribution are compared by H. K. Ury in Communications in Statistics–Simulation and Computation B6 (1977), 181–197.

2. Since the assignment of the ranks to large sized samples of grouped data can be very time-consuming Raatz (1966) has proposed a substantially **simpler procedure** which is exact if all the data fall into few classes; if few or no equal data come up, this test offers a good approximation. The procedure can also be applied to the *H*-test of Kruskal and Wallis.

Table 63 (*5th continuation*). Critical values of U for the Wilcoxon-Mann-Whitney test for the one sided problem ($\alpha = 0.001$) and the two sided problem ($\alpha = 0.002$)

m \ n	1	2	3	4	5	6	7	8	9	10	11	12	13	14	15	16	17	18	19	20
1	-																			
2	-	-																		
3	-	-	-																	
4	-	-	-	-																
5	-	-	-	-	-															
6	-	-	-	-	-	-														
7	-	-	-	-	-	-	0	1												
8	-	-	-	-	-	0	1	2	4											
9	-	-	-	-	-	1	2	3	5	7										
10	-	-	-	-	0	1	3	5	6	8	10									
11	-	-	-	-	0	2	4	6	8	10	12	15								
12	-	-	-	-	0	2	4	7	9	12	14	17	20							
13	-	-	-	-	1	3	5	8	11	14	17	20	23	26						
14	-	-	-	-	1	3	6	9	12	15	19	22	25	29	32					
15	-	-	-	-	1	4	7	10	14	17	21	24	28	32	36	40				
16	-	-	-	-	2	5	8	11	15	19	23	27	31	35	39	43	48			
17	-	-	-	0	2	5	9	13	17	21	25	29	34	38	43	47	52	57		
18	-	-	-	0	3	6	10	14	18	23	27	32	37	42	46	51	56	61	66	
19	-	-	-	0	3	7	11	15	20	25	29	34	40	45	50	55	60	66	71	77
20	-	-	-	0	3	7	12	16	21	26	32	37	42	48	54	59	65	70	76	82	88
21	-	-	-	1	4	8	12	18	23	28	34	40	45	51	57	63	69	75	81	87	94
22	-	-	-	1	4	8	13	19	24	30	36	42	48	54	61	67	73	80	86	93	99
23	-	-	-	1	4	9	14	20	26	32	38	45	51	58	64	71	78	85	91	98	105
24	-	-	-	1	5	10	15	21	27	34	40	47	54	61	68	75	82	89	96	104	111
25	-	-	-	1	5	10	16	22	29	36	43	50	57	64	72	79	86	94	102	109	117
26	-	-	-	1	6	11	17	24	31	38	45	52	60	68	75	83	91	99	107	115	123
27	-	-	-	2	6	12	18	25	32	40	47	55	63	71	79	87	95	104	112	120	129
28	-	-	-	2	6	12	19	26	34	41	49	57	66	74	83	91	100	108	117	126	135
29	-	-	-	2	7	13	20	27	35	43	52	60	69	77	86	95	104	113	122	131	140
30	-	-	-	2	7	14	21	29	37	45	54	63	72	81	90	99	108	118	127	137	146
31	-	-	-	2	7	14	22	30	38	47	56	65	75	84	94	103	113	123	132	142	152
32	-	-	-	2	8	15	23	31	40	49	58	68	77	87	97	107	117	127	138	148	158
33	-	-	-	3	8	15	24	32	41	51	61	70	80	91	101	111	122	132	143	153	164
34	-	-	-	3	9	16	25	34	43	53	63	73	83	94	105	115	126	137	148	159	170
35	-	-	-	3	9	17	25	35	45	55	65	76	86	97	108	119	131	142	153	165	176
36	-	-	-	3	9	17	26	36	46	57	67	78	89	101	112	123	135	147	158	170	182
37	-	-	-	3	10	18	27	37	48	58	70	81	92	104	116	127	139	151	164	176	188
38	-	-	-	3	10	19	28	39	49	60	72	83	95	107	119	131	144	156	169	181	194
39	-	-	-	4	11	19	29	40	51	62	74	86	98	110	123	136	148	161	174	187	200
40	-	-	-	4	11	20	30	41	52	64	76	89	101	114	127	140	153	166	179	192	206

3. Further special **modifications** of the U-test are given by Halperin (1960) and Saw (1966). A Wilcoxon two sample "sequential test scheme" for the comparison of two therapies, which reduces the number of necessary observations considerably in certain cases, is described by Alling (1963, cf., also Chun 1965). A modified U-test with improved asymptotic relative efficiency is given by H. Berchtold, Biometrical Journal **21** (1979), 649–655. A modified U-test for samples of possibly different distributions is given by J. R. Green, Biometrika **66** (1979), 645–653.

4. Two interesting two sample rank-sequential tests have been presented (Wilcoxon et al. 1963, Bradley et al. 1965, 1966).

5. **Median tests.** The median test is quite simple: The combined $n_1 + n_2$ values from samples I and II are ordered by increasing size, the median $\tilde{x}$ is determined, and the values in each sample are then arranged according to whether they are larger or smaller than the common median in the following scheme (a, b, c, d are frequencies):

	Number of occurrences of the value	
	$<\tilde{x}$	$>\tilde{x}$
Sample I	a	b
Sample II	c	d

The computation for small sample sizes (compare the more detailed material in Section 4.6.1) are found in Section 4.6.7 (exact test according to Fisher); for large n, in Section 4.6.1 (χ^2 test or G-test, with or without continuity correction respectively). If the result is significant, the null hypothesis $\tilde{\mu}_1 = \tilde{\mu}_2$ is rejected at the level employed. The asymptotic efficiency of the median test is $2/\pi \simeq 64\%$, i.e., this test applied to 1,000 observations has about the same power as the t-test applied to $0.64(1,000) = 640$ observations, if in fact a normal distribution is present. For other distributions the proportion can be entirely different. The median test is therefore used also for rough estimates; moreover it serves to examine highly significant results in which one has little confidence. If it leads to a different result, the computations must be verified.

The main range of application of the median test is the comparison of two medians when the **distributions differ** considerably; then the U-test is of little value.

EXAMPLE. We use the example for the U-test (without rank allocation) and obtain $\tilde{x} = 19$ as well as the following fourfold table:

	$<\tilde{x}$	$>\tilde{x}$
A	2	6
B	6	2

which by Section 4.6.7 with $P = 0.066$ does not permit the rejection of the null hypothesis at the 5% level.

The testing of k rather than 2 independent samples involves the **generalized median test**. The values of the k samples are ordered by magnitude, the common median is determined, and it is seen how many data points in each of the k samples lie above the median and how many lie below the median. The null hypothesis that the samples originated from a common population can be tested under the assumption that the resulting $2 \times k$ table is sufficiently occupied (all expected frequencies must be >1) by the methods given in Sections 6.1.1, 6.1.2, or 6.2.5. The alternate hypothesis then says: Not all k samples originate from a common population (cf., also Sachs 1982). The corresponding optimal distribution-free procedure is the H-test of Kruskal and Wallis.

6. A so-called "**median quartile test**," for which the combined observed values of two independent samples are reduced by its three quartiles: (Q_1, $Q_2 = \tilde{x}$, and Q_3) to the frequencies of a 2×4 table is discussed by Bauer (1962). Provided the table is

Q \ n	$\leq Q_1$	$\leq Q_2$	$\leq Q_3$	$> Q_3$
n_1				
n_2				

sufficiently occupied (all expected frequencies must be >1), the null hypothesis (same underlying population) is tested against the alternate hypothesis (of different

3.9 Distribution-Free Procedures for the Comparison of Independent Samples 303

underlying populations) according to Section 6.1.1, 6.1.2, or 6.2.5. This very useful test examines not only differences in location, but also differences in dispersion and certain differences in the shape of distributions. For an ungrouped ranked sample of size n, Q_1 and Q_3 are the sample values with ranks $n/4$ and $3n/4$ rounded to the next larger integer. If, e.g., $n = 13$, then $Q_1 = 0.25(13) = 3.25$ is the sample value with rank 4. This test may be generalized to three or more samples by methods given in Section 6.2.1 or 6.2.5.

7. **Confidence intervals for differences between medians.** A confidence interval for the difference of two medians can be determined with the help of the U-test ($\tilde{\mu}_1 - \tilde{\mu}_2 = \Delta$, with $\tilde{\mu}_1 > \tilde{\mu}_2$), $k_{min} < \Delta < k_{max}$, as follows: (1) a constant k is added to all values of the second sample, and a U-test is carried out using this and the first sample; (2) the left and right bounds of the confidence interval for Δ are the smallest and largest values of k (k_{min}, k_{max}) which do not permit the rejection of the null hypothesis of the U-test for the two sided problem at the chosen significance level; (3) appropriate extreme values of k which barely lead to an insignificant result are obtained by skillful trials (beginning, say, with $k = 0.1, k = 1, k = 10$). A thorough survey is given by Laan (1970).

▶ ### 3.9.5 The comparison of several independent samples: The H-test of Kruskal and Wallis

The H-test of Kruskal and Wallis (1952) is a generalization of the U-test. It tests against the alternate hypothesis that the k samples do not originate in a common population. Like the U-test, the H-test also has an asymptotic efficiency of $100(3/\pi) \simeq 95\%$ when compared to the analysis of variance procedure, which is optimal for the normal distribution (Chapter 7). The $n = \sum_{i=1}^{k} n_i$ observations, random samples of ordinal data (ranked data: e.g., marks, grades, points) or measured data, of sizes $n_1, n_2, \ldots, n_k$ from large populations, identical in form, with continuous or discrete distribution are ranked (1 to n) as in the U-test. Let R_i be the sum of the ranks in the ith sample: Under the null hypothesis the test statistic

$$\hat{H} = \left[\frac{12}{n(n+1)}\right] \cdot \left[\sum_{i=1}^{k} \frac{R_i^2}{n_i}\right] - 3(n+1) \quad (3.53)$$

($\hat{H}$ is the variance of the sample rank sums R_i) has, for large n (i.e., in practice for $n_i \geq 5$ and $k \geq 4$), a χ^2 distribution with $k - 1$ degrees of freedom; H_0 is rejected whenever $\hat{H} > \chi^2_{k-1;\alpha}$ (cf., Table 28a, Section 1.5.2). For $n_i \leq 5$ and $k = 3$, Table 65 below lists the exact probabilities (H_0 is rejected with P if $\hat{H} \geq H$ where $P \leq \alpha$).

To test the computations of the R_i's the relation

$$\sum_{i=1}^{k} R_i = n(n+1)/2 \quad (3.54)$$

can be used. If the **samples are of equal size**, so that $n_i = n/k$, the following simplified formula is more convenient:

$$\hat{H} = \left[\frac{12k}{n^2(n+1)}\right] \cdot \left[\sum_{i=1}^{k} R_i^2\right] - 3(n+1). \qquad (3.53a)$$

If more than 25% of all values are involved in **ties** (i.e., come in groups of equal ranks), then $\hat{H}$ must be corrected. The formula for $\hat{H}$ adjusted for ties reads

$$\hat{H}_{corr} = \frac{\hat{H}}{1 - \frac{\sum_{i=1}^{i=r}(t_i^3 - t_i)}{n^3 - n}}, \qquad (3.55)$$

where t_i stands for the respective number of **equal ranks** in the tie i. Since the corrected $\hat{H}$ value is larger than the uncorrected value, $\hat{H}_{corr}$ need not be evaluated when $\hat{H}$ is significant.

EXAMPLE. Test the 4 samples in Table 64 with the H-test ($\alpha = 0.05$).

Table 64 Right next to the observations are the ranks

A		B		C		D	
12.1	10	18.3	15	12.7	11	7.3	3
14.8	12	49.6	21	25.1	16	1.9	1
15.3	13	10.1	6 ½	47.0	20	5.8	2
11.4	9	35.6	19	16.3	14	10.1	6 ½
10.8	8	26.2	17	30.4	18	9.4	5
		8.9	4				
R_i	52.0		82.5		79.0		17.5
R_i^2	2704.00		6806.25		6241.00		306.25
n_i	5		6		5		5
R_i^2/n_i	540.800 + 1134.375 + 1248.200 + 61.250 = 2984.625 = $\sum_{i=1}^{k=4} \frac{R_i^2}{n_i}$						

Inspection of the computations:

$$52.0 + 82.5 + 79.0 + 17.5 = 231 = 21(21 + 1)/2,$$

$$\hat{H} = \left[\frac{12}{21(21 + 1)}\right] \cdot [2{,}984.625] - 3(21 + 1) = 11.523.$$

3.9 Distribution-Free Procedures for the Comparison of Independent Samples

Table 65 Significance levels for the H-test after Kruskal and Wallis (from Kruskal, W. H. and W. A. Wallis 1952, 1953; cf. 1975 with H- and P-values for 6,2,2 through 6,6,6*)

$n_1\ n_2\ n_3$	H	P	$n_1\ n_2\ n_3$	H	P	$n_1\ n_2\ n_3$	H	P	$n_1\ n_2\ n_3$	H	P
2 1 1	2.7000	0.500	4 3 2	6.4444	0.008	5 2 2	6.5333	0.008	5 4 4	5.6571	0.049
				6.3000	0.011		6.1333	0.013		5.6176	0.050
2 2 1	3.6000	0.200		5.4444	0.046		5.1600	0.034		4.6187	0.100
				5.4000	0.051		5.0400	0.056		4.5527	0.102
2 2 2	4.5714	0.067		4.5111	0.098		4.3733	0.090			
	3.7143	0.200		4.4444	0.102		4.2933	0.122	5 5 1	7.3091	0.009
3 1 1	3.2000	0.300								6.8364	0.011
			4 3 3	6.7455	0.010	5 3 1	6.4000	0.012		5.1273	0.046
3 2 1	4.2857	0.100		6.7091	0.013		4.9600	0.048		4.9091	0.053
	3.8571	0.133		5.7909	0.046		4.8711	0.052		4.1091	0.086
				5.7273	0.050		4.0178	0.095		4.0364	0.105
3 2 2	5.3572	0.029		4.7091	0.092		0.0400	0.133			
	4.7143	0.048		4.7000	0.101				5 5 2	7.3385	0.010
	4.5000	0.067				5 3 2	6.9091	0.009		7.2692	0.010
	4.4643	0.105	4 4 1	6.6667	0.010		6.8218	0.010		5.3385	0.047
				6.1667	0.022		5.2509	0.049		5.2462	0.051
3 3 1	5.1429	0.043		4.9667	0.048		5.1055	0.052		4.6231	0.097
	4.5714	0.100		4.8667	0.054		4.6509	0.091		4.5077	0.100
	4.0000	0.129		4.1667	0.082		4.4945	0.101			
				4.0667	0.102				5 5 3	7.5780	0.010
3 3 2	6.2500	0.011				5 3 3	7.0788	0.009		7.5429	0.010
	5.3611	0.032					6.9818	0.011		5.7055	0.046
	5.1389	0.061	4 4 2	7.0364	0.006		5.6485	0.049		5.6264	0.051
	4.5556	0.100		6.8727	0.011		5.5152	0.051		4.5451	0.100
	4.2500	0.121		5.4545	0.046		4.5333	0.097		4.5363	0.102
				5.2364	0.052		4.4121	0.109			
3 3 3	7.2000	0.004		4.5545	0.098				5 5 4	7.8229	0.010
	6.4889	0.011		4.4455	0.103	5 4 1	6.9545	0.008		7.7914	0.010
	5.6889	0.029					6.8400	0.011		5.6657	0.049
	5.6000	0.050	4 4 3	7.1439	0.010		4.9855	0.044		5.6429	0.050
	5.0667	0.086		7.1364	0.011		4.8600	0.056		4.5229	0.099
	4.6222	0.100		5.5985	0.049		3.9873	0.098		4.5200	0.101
				5.5758	0.051		3.9600	0.102			
4 1 1	3.5714	0.200		4.5455	0.099				5 5 5	8.0000	0.009
				4.4773	0.102	5 4 2	7.2045	0.009		7.9800	0.010
4 2 1	4.8214	0.057					7.1182	0.010		5.7800	0.049
	4.5000	0.076	4 4 4	7.6538	0.008		5.2727	0.049		5.6600	0.051
	4.0179	0.114		7.5385	0.011		5.2682	0.050		4.5600	0.100
				5.6923	0.049		4.5409	0.098		4.5000	0.102
4 2 2	6.0000	0.014		5.6538	0.054		4.5182	0.101			
	5.3333	0.033		4.6539	0.097				6 6 6*	8.2222	0.010
	5.1250	0.052		4.5001	0.104	5 4 3	7.4449	0.010		8.1871	0.010
	4.4583	0.100					7.3949	0.011		6.8889	0.025
	4.1667	0.105	5 1 1	3.8571	0.143		5.6564	0.049		6.8772	0.026
			5 2 1	5.2500	0.036		5.6308	0.050		5.8011	0.049
4 3 1	5.8333	0.021		5.0000	0.048		4.5487	0.099		5.7193	0.050
	5.2083	0.050		4.4500	0.071		4.5231	0.103			
	5.0000	0.057		4.2000	0.095	5 4 4	7.7604	0.009			
	4.0556	0.093		4.0500	0.119		7.7440	0.011			
	3.8889	0.129									

A table with additional P-levels is included in the book by Kraft and Van Eeden (1968 [8:1b], pp. 241–261); Hollander and Wolfe (1973 [8:1b], pp. 294–310) incorporate these tables and also give tables (pp. 328, 334) for multiple comparisons. More critical values are given by W. V. Gehrlein and E. M. Saniga in Journal of Quality Technology **10** (1978), 73–75.

Since $\hat{H} = 11.523 > 7.815 = \chi^2_{3;0.05}$, it is assumed that the 4 samples do not originate in a common population. In the case of a statistically significant $\hat{H}$-value, **pairwise comparisons** of the mean ranks ($\bar{R}_i = R_i/n_i$) follow. The null hypothesis, equality of both expected mean ranks, is rejected at the 5% level for the difference

$$|\bar{R}_i - \bar{R}_{i'}| > \sqrt{d\chi^2_{k-1;0.05}\left[\frac{n(n+1)}{12}\right]\left[\frac{1}{n_i} + \frac{1}{n_{i'}}\right]}; \quad (3.56)$$

then this difference is statistically different from zero. The value d is usually equal to one. If there are many ties, then d is the denominator of (3.55), the corrected value, and is smaller than one. For our example we get ($n_{A,C,D} = 5$; $n_B = 6$)

$$\bar{R}_D = 3.50, \quad \bar{R}_A = 10.40, \quad \bar{R}_B = 13.75, \quad \bar{R}_C = 15.80,$$

and for $n_i = n_{i'} = 5$,

$$\sqrt{7.815 \left[\frac{21(21+1)}{12}\right]\left[\frac{1}{5} + \frac{1}{5}\right]} = 10.97,$$

whereas for $n_i = 5, n_{i'} = 6$,

$$\sqrt{7.815 \left[\frac{21(21+1)}{12}\right]\left[\frac{1}{5} + \frac{1}{6}\right]} = 10.50.$$

Only D and C are statistically different at the 5% level (15.80 − 3.50 = 12.30 > 10.97).

For Fisher's least significant difference multiple comparisons, computed on the ranks, see Conover (1980, pp. 229–237 [8:1b]).

Remarks (cf., also Remark 2 in Section 2.9.4)

1. More on **pairwise** and **multiple comparisons** is found in Section 7.5.2, Sachs (1984, pp. 95–96) and in J. H. Skillings (1983, Communications in Statistics—Simulation and Computation **12**, 373–387).

2. The power of the H-test can be increased if the null hypothesis, equality of the means (or of the distribution functions), can be confronted with a specific alternate hypothesis: the presence of **a certain ranking**, or the descent (falling off) of the medians (or of the distribution functions), provided the sample sizes are equal. For a generalization of a one sided test, Chacko (1963) gives a test statistic which is a modified version of (3.53a).

3. An H-test for the case where k heterogeneous sample groups can each be subdivided into m homogeneous subgroups corresponding to one another and of n values each is described by Lienert and Schulz (1967).

4. Tests competing with the H-test are analyzed by Bhapkar and Deshpande (1968).

5. For the case where not individual observations, but rather **data pairs** are given, Glasser (1962) gives a modification of the H-test which permits the testing of paired observations for independence.

6. Two correlated samples (paired data, matched pairs) are compared in the first few sections of Chapter 4. The nonparametric comparison of several correlated samples (Friedman rank test) and the parametric comparison of several means (analysis of variance) come later (Chapter 7). Let us emphasize that there is, among other things, an **intimate relation** between the Wilcoxon test for paired data, the Friedman test, and the H-test.

7. For dealing with distributions of different form, the H-test is replaced by the corresponding **4 × k median–quartile test** (Remark 6 in Section 3.9.4 generalized to more than 2 samples; see also the Sections 6.2.1 and 6.2.5).

4 FURTHER TEST PROCEDURES

4.1 REDUCTION OF SAMPLING ERRORS BY PAIRING OBSERVATIONS: PAIRED SAMPLES

When the two different methods of treatment are to be compared for effectiveness, preliminary information is in many cases obtained by experiments on laboratory animals. Suppose we are interested in two ointment preparations. The question arises: Does there or does there not exist a difference in the effectiveness of the two preparations? There are test animals at our disposal on which we can produce the seat of a disease. Let the measure of effectiveness be the amount of time required for recovery:

1. The simplest approach would be to divide a group of test animals randomly into two **subgroups** of equal size, treat one group by method one and the other by method two, and then compare the results of the therapies.
2. The following approach is more effective: Test animals are **paired** in such a way that the individual pairs are as homogeneous as possible with regard to sex, age, weight, activity, etc. The partners are then assigned randomly (e.g., by tossing a coin) to the two treatments. The fact that the experimenter hardly ever has a completely homogeneous collection of animals at his disposal is taken into account in this procedure.
3. The following procedure is considerably more effective: A group of test animals is chosen and a so-called **right-left comparison** carried out. That is, we produce on the right and left flank of each individual (or any such natural homogeneous subgroup of size two, like a pair of twins or the two hands of the same person) two mutually independent seats of a disease, and allot the two treatments to the two flanks, determining by a

random process which is to be treated by the one method and which by the other (cf., also Section 7.7).

Where does the advantage of the pairwise comparison actually lie? The comparison is more precise, since the dispersion existing among the various experimental units is reduced or eliminated. Indeed, in pairwise comparisons—we speak of paired observations and paired samples—the number of degrees of freedom is decreased and the accuracy reduced. For the comparison of means there are in the case of homogeneous variances $n_1 + n_2 - 2$ degrees of freedom at our disposal; in contrast to this, the number of degrees of freedom for the paired samples equals the number of pairs, or differences, minus one, i.e., $(n_1 + n_2)/2 - 1$. If we set $n_1 + n_2 = n$, then the ratio of the number of degrees of freedom for independent samples to that for paired samples is given by $(n - 2)/(n/2 - 1) = 2/1$. The number of degrees of freedom in the paired groups is half as large as in independent groups (with the same number of experimental units). Loss of degrees of freedom means less accuracy, but on the other hand accuracy is gained by a decrease in the within treatment error because the variance between test animals (blocks) is larger than between the flanks (units in a block) of the single animals. In general, the larger the ratio of the variance between the test animals to the variance between the two flanks, the more we gain by using paired samples.

Assume now that the two flanks of each of the n animals were treated differently. Denote the variance between sums and differences of the pairs of animals by s_s^2 and s_d^2 respectively. The experiment with paired samples is superior to the experiment on independent samples if for the same total sample size (n pairs) the following inequality holds:

$$\boxed{\frac{n(2n + 1)[(n - 1)s_s^2 + ns_d^2]}{(n + 2)(2n - 1)(2n - 1)s_d^2} > 1.} \tag{4.0}$$

The values in Table 66 furnish an example: $s_d^2 = [20.04 - (9.2)^2/8]/7 = 1.35$; for s_s^2 the sum of the two treatment effects, $x_i + y_i$, is needed, so

$$s_s^2 = \frac{\sum (x_i + y_i)^2 - (\sum (x_i + y_i))^2/n}{n - 1} = \frac{545.60 - 65.0^2/8}{7} = 2.5,$$

whence the ratio

$$\frac{8 \cdot 17[7 \cdot 2.5 + 8 \cdot 1.35]}{10 \cdot 15 \cdot 15 \cdot 1.35} = 1.27 > 1$$

i.e., paired observations are to be preferred for future tests as well.

Paired samples are obtained according to the two following principles. The setup of experiments with replication on one and the same experimental unit is known. Test persons are e.g., first examined under normal conditions and then under treatment. Note that factors such as exercise or fatigue must

4.2 Observations Arranged in Pairs

be eliminated. The second principle consists in the organization of paired samples with the help of a preliminary test or a measurable or estimable characteristic which is correlated as strongly as possible with the characteristic under study. The individuals are brought into a rank sequence on the basis of the preliminary test. Two individuals with consecutive ranks form pair number i; it is decided by a random process—by a coin toss, say—which partner receives which treatment. We have used

$$s_{\bar{x}_1-\bar{x}_2} = s_{\text{Diff.}} = \sqrt{\frac{s_1^2}{n_1} + \frac{s_2^2}{n_2}} = \sqrt{s_{\bar{x}_1}^2 + s_{\bar{x}_2}^2} \tag{4.1}$$

as an estimate of the standard deviation of the difference between the means of two independent samples [see (3.31), Section 3.6.2]. If the samples are not independent but correlated in pairs as described, then the standard deviation of the difference is reduced (when the correlation is positive) and we get

$$s_{\bar{d}} = \sqrt{s_{\bar{x}_1}^2 + s_{\bar{x}_2}^2 - 2r s_{\bar{x}_1} s_{\bar{x}_2}}. \tag{4.2}$$

The size of the subtracted term depends on the size of the correlation coefficient r, which expresses the degree of connection (Chapter 5). When $r = 0$, i.e., when sequences are completely independent of each other, the subtracted term becomes zero; when $r = 1$, i.e., with maximal correlation or complete dependence, the subtracted term attains its maximum and the standard deviation of the difference its minimum.

4.2 OBSERVATIONS ARRANGED IN PAIRS

If each of two sleep-inducing preparations is tested on the same patients, then for the number of hours the duration of sleep is extended, we have **paired data**, i.e., **paired** samples, also called connected samples.

4.2.1 The *t*-test for data arranged in pairs

4.2.1.1 Testing the mean of pair differences for zero

The data in the connected samples are the pairs (x_i, y_i). We are interested in the difference μ_d of the treatment effects. The null hypothesis is $\mu_d = 0$; the alternative hypothesis can be $\mu_d > 0$ or $\mu_d < 0$ in the one sided test and $\mu_d \neq 0$ in the two sided test. Then the test statistic is given by

$$\hat{t} = \frac{\bar{d}}{s_{\bar{d}}} = \frac{(\sum d_i)/n}{\sqrt{\frac{\sum d_i^2 - (\sum d_i)^2/n}{n(n-1)}}}, \qquad DF = n - 1. \tag{4.3}$$

pp. 136–137

$\hat{t}$ is the quotient of the mean of the n differences and the associated standard error with $n - 1$ degrees of freedom. The differences are assumed to come from random samples from an (at least approximately) normally distributed population. The CI (4.4) is computed always after the test.

Simpler to handle than (4.3) is the test statistic $\hat{A} = \sum d^2/(\sum d)^2$ with tabulated critical values A (see also Runyon and Haber 1967), which is due to Sandler (1955).

EXAMPLE. Table 66 contains data (x_i, y_i) for material that was handled in two ways, i.e., for untreated (x_i) and treated (y_i) material. The material numbers correspond to different origins. Can the null hypothesis of no treatment difference (no treatment effect) be guaranteed at the 5% level?

Table 66

No.	x_i	y_i	d_i $(x_i - y_i)$	d_i^2
1	4.0	3.0	1.0	1.00
2	3.5	3.0	0.5	0.25
3	4.1	3.8	0.3	0.09
4	5.5	2.1	3.4	11.56
5	4.6	4.9	-0.3	0.09
6	6.0	5.3	0.7	0.49
7	5.1	3.1	2.0	4.00
8	4.3	2.7	1.6	2.56
$n = 8$			$\sum d_i = 9.2$	$\sum d_i^2 = 20.04$

We have

$$\hat{t} = \frac{9.2/8}{\sqrt{\frac{20.04 - 9.2^2/8}{8(8-1)}}} = \frac{1.15}{0.4110} = 2.798 \text{ or } 2.80$$

and, since $\hat{t} = 2.798 > 2.365 = t_{7;0.05;\text{two sided}}$, the treatment difference (treatment effect) is statistically significant at the 5% level.

By comparing the completely randomized procedure (3.25), (3.31) with the paired sample procedure, we see that often annoying dispersions within the treatment groups are eliminated by the second method. Moreover the **assumptions are weakened**: the data sets x and y might be far from normally distributed, but the distribution of the differences will be approximately normal.

The confidence interval for the true mean difference of paired observations is given by

$$\boxed{\bar{d} \pm (t_{n-1;\alpha})s_{\bar{d}}} \qquad (4.4)$$

with

$$\bar{d} = \frac{\sum d}{n} \quad \text{and} \quad s_{\bar{d}} = \frac{s_d}{\sqrt{n}} = \sqrt{\frac{\sum d_i^2 - (\sum d_i)^2/n}{n(n-1)}}$$ (pp. 136–137)

and a t for the two sided test. For our example the 95% confidence interval is $1.15 \pm (2.365)(0.411)$ or 1.15 ± 0.97, 95% CI: $0.18 \leq \mu_d \leq 2.12$. Corresponding to the result of the test the value 0 is not included. One sided confidence limits (CL) can of course be stated too. As the upper 95% CL we find with $t_{7;0.05;\text{one sided}} = 1.895$ the value $1.15 + (1.895)(0.411) = 1.93$, thus $\mu_d \leq 1.93$.

Large paired (connected) samples are frequently analyzed by distribution-free tests.

4.2.1.2 Testing the equality of variances of paired observations

If a comparison is to be made of the variability of a characteristic before (x_i) and after (y_i) an aging process or a treatment, then the variances of two paired sets have to be compared. The test statistic is

$$\hat{t} = \frac{|(Q_x - Q_y) \cdot \sqrt{n-2}|}{2\sqrt{Q_x Q_y - (Q_{xy})^2}}$$ (4.5)

with $n - 2$ degrees of freedom. Q_x and Q_y are computed according to (3.23), (3.24). Q_{xy} is correspondingly obtained from (pp. 136–137)

$$Q_{xy} = \sum xy - \frac{\sum x \sum y}{n}.$$ (4.6)

As an example, we have for

$$\begin{array}{c|cccc|l} x_i & 21 & 18 & 20 & 21 & \sum x = 80 \\ \hline y_i & 26 & 33 & 27 & 34 & \sum y = 120 \end{array}$$

with $Q_x = 6$, $Q_y = 50$, and

$$Q_{xy} = [(21)(26) + (18)(33) + (20)(27) + (21)(34)] - (80)(120)/4 = -6;$$

$$\hat{t} = \frac{|(6 - 50) \cdot \sqrt{4 - 2}|}{2 \cdot \sqrt{6 \cdot 50 - (-6)^2}} = 1.91 < 4.30 = t_{2;0.05;\text{two sided}}.$$

For the two sided problem we conclude that the null hypothesis (equality of the two variances) must be retained.

For the one sided problem with $\sigma_x^2 = \sigma_y^2$ versus $\sigma_x^2 < \sigma_y^2$, and this example, the critical bound would be $t_{2;\,0.05;\,\text{one sided}} = 2.92$.

4.2.2 The Wilcoxon matched pair signed-rank test

Optimal tests for the comparison of paired observations (of matched pairs of two sets of matched observations) are the t-test for normally distributed differences (4.3) and the Wilcoxon matched pair signed-rank test with non-normally distributed differences. This test, known as the Wilcoxon test for pair differences, can also be applied to ranked data. It requires, in comparison with the t-test, substantially less computation, and it tests normally distributed differences with just as much power; its efficiency is around 95% for large and small sizes.

The test permits us to check whether the differences for pairwise arranged observations are symmetrically distributed with respect to the median equal to zero, i.e., under the null hypothesis the pair differences d_i originate in a population with symmetric distribution function $F(d)$ or symmetric density $f(d)$:

$$H_0: F(+d) + F(-d) = 1 \quad \text{or} \quad f(+d) = f(-d),$$

respectively. If H_0 is rejected, then either the population is not symmetric with respect to the median — i.e., the median of the differences does not equal zero ($\tilde{\mu}_d \neq 0$) — or different distributions underlie the two differently treated samples. Pairs with equal individual values are discarded (however, cf., Cureton 1967), for the n remaining pairs the differences

$$d_i = x_{i1} - x_{i2} \tag{4.7}$$

are found, and the absolute values $|d_i|$ are ordered in an increasing rank sequence: the smallest is given the rank 1, ..., and the largest the rank n. **Mean ranks** are assigned to equal absolute values. Every rank is associated with the sign of the corresponding difference. Then the sums of the positive and the negative ranks ($\hat{R}_p$ and $\hat{R}_n$) are formed, and the computations are checked by

$$\hat{R}_p + \hat{R}_n = n(n+1)/2. \tag{4.8}$$

The smaller of the two rank sums, $\hat{R} = \min(\hat{R}_1, \hat{R}_2)$, is used as test statistic. The null hypothesis is abandoned if the computed $\hat{R}$-value is **less than or equal to** the critical value $R(n;\alpha)$ in Table 67. For $n > 25$ we have the approximation

$$R(n;\alpha) = \frac{n(n+1)}{4} - z \cdot \sqrt{\frac{1}{24}n(n+1)(2n+1)}. \tag{4.9}$$

Table 67 Critical values for the Wilcoxon matched pair signed-rank test: (taken from McCornack, R. L.: Extended tables of the Wilcoxon matched pair signed rank statistic. J. Amer. Statist. Assoc. **60** (1965), 864–871, pp. 866 + 867)

Test	Two sided			One sided		Test	Two sided			One sided	
n	5%	1%	0.1%	5%	1%	n	5%	1%	0.1%	5%	1%
6	0			2		56	557	484	402	595	514
7	2			3	0	57	579	504	420	618	535
8	3	0		5	1	58	602	525	438	642	556
9	5	1		8	3	59	625	546	457	666	578
10	8	3		10	5	60	648	567	476	690	600
11	10	5	0	13	7	61	672	589	495	715	623
12	13	7	1	17	9	62	697	611	515	741	646
13	17	9	2	21	12	63	721	634	535	767	669
14	21	12	4	25	15	64	747	657	556	793	693
15	25	15	6	30	19	65	772	681	577	820	718
16	29	19	8	35	23	66	798	705	599	847	742
17	34	23	11	41	27	67	825	729	621	875	768
18	40	27	14	47	32	68	852	754	643	903	793
19	46	32	18	53	37	69	879	779	666	931	819
20	52	37	21	60	43	70	907	805	689	960	846
21	58	42	25	67	49	71	936	831	712	990	873
22	65	48	30	75	55	72	964	858	736	1020	901
23	73	54	35	83	62	73	994	884	761	1050	928
24	81	61	40	91	69	74	1023	912	786	1081	957
25	89	68	45	100	76	75	1053	940	811	1112	986
26	98	75	51	110	84	76	1084	968	836	1144	1015
27	107	83	57	119	92	77	1115	997	862	1176	1044
28	116	91	64	130	101	78	1147	1026	889	1209	1075
29	126	100	71	140	110	79	1179	1056	916	1242	1105
30	137	109	78	151	120	80	1211	1086	943	1276	1136
31	147	118	86	163	130	81	1244	1116	971	1310	1168
32	159	128	94	175	140	82	1277	1147	999	1345	1200
33	170	138	102	187	151	83	1311	1178	1028	1380	1232
34	182	148	111	200	162	84	1345	1210	1057	1415	1265
35	195	159	120	213	173	85	1380	1242	1086	1451	1298
36	208	171	130	227	185	86	1415	1275	1116	1487	1332
37	221	182	140	241	198	87	1451	1308	1146	1524	1366
38	235	194	150	256	211	88	1487	1342	1177	1561	1400
39	249	207	161	271	224	89	1523	1376	1208	1599	1435
40	264	220	172	286	238	90	1560	1410	1240	1638	1471
41	279	233	183	302	252	91	1597	1445	1271	1676	1507
42	294	247	195	319	266	92	1635	1480	1304	1715	1543
43	310	261	207	336	281	93	1674	1516	1337	1755	1580
44	327	276	220	353	296	94	1712	1552	1370	1795	1617
45	343	291	233	371	312	95	1752	1589	1404	1836	1655
46	361	307	246	389	328	96	1791	1626	1438	1877	1693
47	378	322	260	407	345	97	1832	1664	1472	1918	1731
48	396	339	274	426	362	98	1872	1702	1507	1960	1770
49	415	355	289	446	379	99	1913	1740	1543	2003	1810
50	434	373	304	466	397	100	1953	1779	1578	2445	1850
51	453	390	319	486	416						
52	473	408	335	507	434						
53	494	427	351	529	454						
54	514	445	368	550	473						
55	536	465	385	573	493						

Table 68 The Wilcoxon matched pair signed-rank test applied to testosterone data

Sample number	1	2	3	4	5	6	7	8	9
A (mg/day)	0.47	1.02	0.33	0.70	0.94	0.85	0.39	0.52	0.47
B (mg/day)	0.41	1.00	0.46	0.61	0.84	0.87	0.36	0.52	0.51
$A - B = d_i$	0.06	0.02	−0.13	0.09	0.10	−0.02	0.03	0	−0.04
Rank for the d_i	5	1.5	8	6	7	1.5	3		4
$\hat{R}_p = 22.5$	(+)5	(+)1.5		(+)6	(+)7		(+)3		
$\hat{R}_n = 13.5$			(−)8			(−)1.5			(−)4
Check	\multicolumn{9}{l}{$22.5 + 13.5 = 36 = 8(8+1)/2$, i.e. $\hat{R} = 13.5$}

4.2 Observations Arranged in Pairs

Appropriate values of z for the one and the two sided test can be found in Table 43 in Section 2.1.6. If one cannot or does not wish to specify an α (and $n > 25$), the following equivalent form is used instead of (4.9):

$$\hat{z} = \frac{\left|\hat{R} - \frac{n(n+1)}{4}\right|}{\sqrt{\frac{n(n+1)(2n+1)}{24}}}. \qquad (4.10)$$

The value $\hat{z}$ is compared with the critical z-value of the standard normal distribution (Table 14, Section 1.3.4). The Friedman test (Section 7.6.1) is a generalization of this test.

EXAMPLE. A biochemist compares two methods A and B employed for the determination of testosterone (male sex hormone) in urine in 9 urine samples in a two sided test at the 5% level. It is not known whether the values are normally distributed. The values in Table 68 are given in milligrams in the urine secreted over 24 hours.

Since $13.5 > 3 = R(8; 0.05)$, the null hypothesis cannot be rejected at the 5% level.

When ties are present (cf., Section 3.9.4), the $\sqrt{A}$ in (4.9; 4.10) is replaced by $\sqrt{A - B/48}$, where $B = \sum_{i=1}^{i=r}(t_i^3 - t_i)/12$ (r = number of ties, t_i = multiplicity of the ith tie). A review of this test and some improvements in the presence of ties is given by W. Buck, Biometrical Journal **21** (1979), 501–526. An extended table ($4 \leq n \leq 100$; 17 significance levels between $\alpha = 0.45$ and $\alpha = 0.00005$) is provided by McCornack (1965; cf., [8:4]).

Examples of quick distribution-free procedures for evaluating the differences of paired observations are the very convenient maximum test and the sign test of Dixon and Mood, which can also be applied to other questions.

4.2.3 The maximum test for pair differences

The **maximum test** is a very simple test for the comparison of two paired data sets. We have only to remember that the effects of two treatments differ at the 10% significance level if the five largest absolute differences come from differences with the same sign. For 6 differences of this sort the distinction is significant at the 5% level, for 8 differences at the 1% level, and for 11 differences at the 0.1% level. These numbers 5, 6, 8, 11 are the critical numbers for the two sided problem and sample size $n \geq 6$. For the one sided problem, of course, the 5%, 2.5%, 0.5%, and 0.05% levels correspond to these numbers. If there should be two differences with opposite signs but the same absolute value, they are ordered in such a way as to break a possibly existing sequence of differences of the same sign (Walter 1951, 1958). The maximum test serves to **verify independently** the result of a t-test but does not replace it (Walter, 1958).

EXAMPLE. The sequence of differences +3.4; +2.0; +1.6; +1.0; +0.7; +0.5; −0.3; +0.3 — note the unfavorable location of −0.3 — leads, with 6 typical differences in a two sided problem, to rejection of $H_0: \tilde{\mu}_d = 0$ at the 5% level.

Remarks

1. Assume the paired observations in Tables 66 and 68 are not (continuously) measured data but rather integers used for grading or scoring; equal spacing (as 1, 2, 3, 4, 5, 6, say) is not necessary. The statistic $\hat{z} = [\sum d_i]/\sqrt{\sum d_i^2}$, with which the null hypothesis $H_0: \tilde{\mu}_d = 0$ can be tested, is for $n \geq 10$ approximately normally distributed; thus reject H_0 at the level 100α if $\hat{z} > z_\alpha$.

2. A special χ^2-test for testing the symmetry of a distribution was introduced by Walter (1954): If one is interested in whether medication M influences, e.g., the LDH (lactatedehydrogenase) content in the blood, then the latter is measured before and after administering a dose of M. If M exerts no influence, the pairwise differences of the measurements (on individuals) are symmetrically distributed with respect to zero.

3. A straightforward nonparametric test for testing the independence of paired observations is described by Glasser (1962). Two examples, fully worked out, and a table of critical bounds illustrate the application of the method.

4.2.4 The sign test of Dixon and Mood

The name "sign test" refers to the fact that only the signs of differences between observations are evaluated. It is assumed the random variables are continuous. The test serves, first of all, as a quick method to recognize the differences in the overall tendency between the two data sets which make up the paired samples (Dixon and Mood 1946). In contrast with the t-test and the Wilcoxon test, **the individual pairs need not originate in a common population**; they could for example belong to different populations with regard to age, sex, etc. It is essential that the outcomes of the individual pairs be independent of each other. The null hypothesis of the sign test is that the differences of paired observations are on the average equal to zero; one expects about half of the differences to be less than zero (negative signs) and the other half to be greater than zero (positive signs). The sign test thus tests the **null hypothesis that the distribution of the differences has median zero**. Bounds or confidence bounds for the median are found in Table 69. The null hypothesis is rejected if the number of differences of one sign is too large or too small, i.e., if this number **falls short of or exceeds** the respective bounds in Table 69. Possible zero differences are ingored. The effective sample size is the number of nonzero differences. The probability that a certain number of plus signs occurs is given by the binomial distribution with $p = q = \frac{1}{2}$. The table of binomial probabilities in Section 1.6.2 (Table 33, last column, $p = 0.5$) shows that at least 6 pairs of observations must be available if in a

Table 69 Bounds for the sign test (from Van der Waerden (1969) [8:4], p. 353, Table 9)

Two sided	5%	2%	1%	Two sided	5%	2%	1%
n = 5	0 5	0 5	0 5	n = 53	19 34	18 35	17 36
6	1 5	0 6	0 6	54	20 34	19 35	18 36
7	1 6	1 6	0 7	55	20 35	19 36	18 37
8	1 7	1 7	1 7	56	21 35	19 37	18 38
9	2 7	1 8	1 8	57	21 36	20 37	19 38
10	2 8	1 9	1 9	58	22 36	20 38	19 39
11	2 9	2 9	1 10	59	22 37	21 38	20 39
12	3 9	2 10	2 10	60	22 38	21 39	20 40
13	3 10	2 11	2 11	61	23 38	21 40	21 40
14	3 11	3 11	2 12	62	23 39	22 40	21 41
15	4 11	3 12	3 12	63	24 39	22 41	21 42
16	4 12	3 13	3 13	64	24 40	23 41	22 42
17	5 12	4 13	3 14	65	25 40	23 42	22 43
18	5 13	4 14	4 14	66	25 41	24 42	23 43
19	5 14	5 14	4 15	67	26 41	24 43	23 44
20	6 14	5 15	4 16	68	26 42	24 44	23 45
21	6 15	5 16	5 16	69	26 43	25 44	24 45
22	6 16	6 16	5 17	70	27 43	25 45	24 46
23	7 16	6 17	5 18	71	27 44	26 45	25 46
24	7 17	6 18	6 18	72	28 44	26 46	25 47
25	8 17	7 18	6 19	73	28 45	27 46	26 47
26	8 18	7 19	7 19	74	29 45	27 47	26 48
27	8 19	8 19	7 20	75	29 46	27 48	26 49
28	9 19	8 20	7 21	76	29 47	28 48	27 49
29	9 20	8 21	8 21	77	30 47	28 49	27 50
30	10 20	9 21	8 22	78	30 48	29 49	28 50
31	10 21	9 22	8 23	79	31 48	29 50	28 51
32	10 22	9 23	9 23	80	31 49	30 50	29 51
33	11 22	10 23	9 24	81	32 49	30 51	29 52
34	11 23	10 24	10 24	82	32 50	31 51	29 53
35	12 23	11 24	10 25	83	33 50	31 52	30 53
36	12 24	11 25	10 26	84	33 51	31 53	30 54
37	13 24	11 26	11 26	85	33 52	32 53	31 54
38	13 25	12 26	11 27	86	34 52	32 54	31 55
39	13 26	12 27	12 27	87	34 53	33 54	32 55
40	14 26	13 27	12 28	88	35 53	33 55	32 56
41	14 27	13 28	12 29	89	35 54	34 55	32 57
42	15 27	14 28	13 29	90	36 54	34 56	33 57
43	15 28	14 29	13 30	91	36 55	34 57	33 58
44	16 28	14 30	14 30	92	37 55	35 57	34 58
45	16 29	15 30	14 31	93	37 56	35 58	34 59
46	16 30	15 31	14 32	94	38 56	36 58	35 59
47	17 30	16 31	15 32	95	38 57	36 59	35 60
48	17 31	16 32	15 33	96	38 58	37 59	35 61
49	18 31	16 33	16 33	97	39 58	37 60	36 61
50	18 32	17 33	16 34	98	39 59	38 60	36 62
51	19 32	17 34	16 35	99	40 59	38 61	37 62
52	19 33	18 34	17 35	100	40 60	38 62	37 63
One sided	2.5%	1%	0.5%	One sided	2.5%	1%	0.5%

If the number of positive (say) differences falls outside the bounds, an effect is guaranteed at the respective level. This table is supplemented by Table 69a.

Table 69a

| \multicolumn{18}{c}{Left bounds for the two sided test} |

n	5%	1%	n	5%	1%	n	5%	1%	n	5%	1%	n	5%	1%	n	5%	1%
101	41	38	121	50	46	141	59	55	161	68	64	181	77	73	210	91	86
102	41	38	122	50	47	142	59	56	162	69	65	182	78	74	220	95	91
103	42	38	123	51	47	143	60	56	163	69	65	183	78	74	230	100	96
104	42	39	124	51	48	144	60	57	164	69	66	184	79	75	240	105	100
105	42	39	125	52	48	145	61	57	165	70	66	185	79	75	250	110	105
106	43	40	126	52	49	146	61	57	166	70	66	186	80	75	260	114	109
107	43	40	127	52	49	147	62	58	167	71	67	187	80	76	270	119	114
108	44	41	128	53	49	148	62	58	168	71	67	188	81	76	280	124	118
109	44	41	129	53	50	149	63	59	169	72	68	189	81	77	290	128	123
110	45	42	130	54	50	150	63	59	170	72	68	190	82	77	300	133	128
111	45	42	131	54	51	151	63	60	171	73	69	191	82	78	350	157	151
112	46	42	132	55	51	152	64	60	172	73	69	192	82	78	400	180	174
113	46	43	133	55	52	153	64	61	173	74	70	193	83	79	450	204	198
114	47	43	134	56	52	154	65	61	174	74	70	194	83	79	500	228	221
115	47	44	135	56	53	155	65	62	175	75	71	195	84	80	550	252	245
116	47	44	136	57	53	156	66	62	176	75	71	196	84	80	600	276	268
117	48	45	137	57	53	157	66	62	177	75	71	197	85	80	700	324	316
118	48	45	138	58	54	158	67	63	178	76	72	198	85	81	800	372	364
119	49	46	139	58	54	159	67	63	179	76	72	199	86	81	900	421	411
120	49	46	140	58	55	160	68	64	180	77	73	200	86	82	1000	469	459

The value of the right bound (RB) is computed from this table in terms of n and the value of the left bound (LB), and equals n − LB + 1.

two sided test a decision has to be reached at the 5% level: $n = 6$, $x = 0$ or 6. The tabulated P-value is to be doubled for the two sided test: $P = 2(0.0156) = 0.0312 < 0.05$. The other bounds in Table 69 are found in a similar manner. The efficiency of the sign test drops with increasing sample size from 95% for $n = 6$ to 64% as $n \to \infty$. We shall return to this test in Section 4.6.3. An extensive table for the sign test ($n = 1(1)1000$) is given by MacKinnon (1964).

EXAMPLE. Suppose we observe 15 matched pairs in a two sided problem at the 5% level, obtaining two zero differences and 13 nonzero differences, of which 11 have the plus and 2 the minus sign. For $n = 13$, Table 69 gives the bounds 3 and 10. Our values lie outside the limits; i.e., $H_0: \tilde{\mu}_d = 0$ is rejected at the 5% level; the two samples originated in different populations, populations with different medians ($\tilde{\mu}_d \neq 0$; $P < 0.05$). If Tables 69 and 69a are not at hand or are insufficient, not too small samples ($n \gtrsim 30$) of differences can be tested by the following statistic $\hat{z}$, which is approximately normally distributed:

$$\hat{z} = \frac{|2x - n| - 1}{\sqrt{n}}, \qquad (4.11)$$

where x is the observed frequency of the less frequent sign and n the number of pairs minus the number of zero differences.

A **modification** suggested by Duckworth and Wyatt (1958) can be used as a quick estimate. The test statistic $\hat{T}$ is the absolute value of the difference of the signs [i.e., |(number of plus signs) − (number of minus signs)|]. The 5% level of this difference corresponds to the bound $2\sqrt{n}$, the 10% level to $1.6\sqrt{n}$, where n is the number of nonzero differences. Then, in a two sided test, the null hypothesis is rejected at the given level if $\hat{T} > 2\sqrt{n}$ or $\hat{T} > 1.6\sqrt{n}$ respectively. For the example just presented the statistic is $T = 11 - 2 = 9$ and $2\sqrt{n} = 2\sqrt{13} = 7.21$; since $9 > 7.21$, the conclusion is the same as under the maximum test.

Confidence interval (CI) for the median ($\tilde{\mu}$): the 95% CI and 99% CI for $\tilde{\mu}$ (see Section 3.1.4) are found for

$n \leq 100$ by means of Table 69 above, 5% and 1% columns, according to LB $\leq \tilde{\mu} \leq 1 +$ RB;

e.g., $n = 60$, 95% CI: (22nd value) $\leq \tilde{\mu} \leq$ (39th value).

$n > 100$ by means of Table 69a, 5% and 1% columns, by LB $\leq \tilde{\mu} \leq n -$ LB $+ 1$;

e.g., $n = 300$, 95% CI: (133rd value) $\leq \tilde{\mu} \leq$ (168th value).

The ( ) are then replaced by the corresponding ordered data values.

REMARK: The null hypothesis of the sign test can be written $H_0: P(Y > X) = \frac{1}{2}$ (see Section 1.2.5 regarding Y, X). The test is also applicable if H_0 concerns a certain difference between or a certain percentage of X and Y. We might perhaps allow Y to be 10% larger than X (both positive) on the average or let Y be 5 units smaller than X on the average; i.e., $H_0: P(Y > 1.10X) = \frac{1}{2}$ or $H_0: P(Y > [X - 5]) = \frac{1}{2}$. The signs of the differences $Y - 1.10X$ or $Y - X + 5$, respectively, are then counted.

Further applications of the sign test for rapid orientation

1. Comparison of two independent samples. Should we only be interested in comparing two populations with respect to their central tendency (location differences) then the computation of the means is not necessary. The values of the two samples are paired at random and then the methods pertinent to paired samples can be applied.

2. Testing membership for a certain population.

EXAMPLE 1. Could the twenty-one values 13, 12, 11, 9, 12, 8, 13, 12, 11, 11, 12, 10, 13, 11, 10, 14, 10, 10, 9, 11, 11 have come from a population with arithmetic mean $\mu_0 = 10$ ($H_0: \mu = \mu_0$; $H_A: \mu \neq \mu_0$; $\alpha = 0.05$)? We count the values that are less than 10 and those greater than 10, form the difference, and test it:

$$\hat{T} = 14 - 3 = 11 > 8.2 = 2\sqrt{17}.$$

It thus cannot be assumed that the above sample originated in a population with $\mu_0 = 10$ (H_0 is rejected, H_A is accepted; $P < 0.05$) (cf., also the single sample test mentioned in Section 3.9.4 just before the remarks in fine print).

EXAMPLE 2. Do the twenty values obtained in the sequence 24, 27, 26, 28, 30, 35, 33, 37, 36, 37, 34, 32, 32, 29, 28, 28, 31, 28, 26, 25 come from a stable or a time-dependent population? To answer this question Taylor (cf., Duckworth and Wyatt 1958) recommended another modification of the sign test, aimed at **assessing the variability of the central tendency within a population**. First the median of the sample is determined; then by counting it is found how many successive data pairs enclose the median. We call this number x^*. If a trend is present, i.e., if the mean (median) of the population considered changes with time, then x^* is small compared to the sample size n. The null hypothesis (the presence of a random sample from some population) is rejected at the 5% level if

$$|n - 2x^* - 1| \geq 2\sqrt{n - 1}. \qquad (4.12)$$

The median of the sample with size $n = 20$ is $\tilde{x} = 29\frac{1}{2}$. The trend changes at the $x^* = 4$ underlined pairs of numbers. We obtain $n - 2x^* - 1 = 20 - 8 - 1 = 11$ and $2\sqrt{n-1} = 2\sqrt{20-1} = 8.7$. Since $11 > 8.7$, we conclude at the 5% level that the observations come from a time-dependent population.

4.3 THE χ^2 GOODNESS OF FIT TEST

> A beginner should read Section 4.3.2 first.

> Reasons for fitting a distribution to a set of data are: the desire for objectivity (the need for automating the data analysis) and interest in the values of the distribution parameters for future prediction in the absence of major changes in the system.

Assume we have a sample from a population with unknown distribution function $F(x)$ on the one hand and a well-defined distribution function $F_0(x)$ on the other hand. A goodness of fit test assesses the null hypothesis $H_0: F(x) = F_0(x)$ against the alternate hypothesis $H_A: F(x) \neq F_0(x)$. Even if the null hypothesis cannot be rejected on the basis of the test, we must be extremely cautious in interpreting the sample in terms of the distribution $F_0(x)$.

4.3 The χ^2 Goodness of Fit Test

The test statistic (4.13), written $\hat{\chi}^2$ for short,

$$\sum_{i=1}^{k} \frac{(O_i - E_i)^2}{E_i} = \sum_{i=1}^{k} \frac{O_i^2}{E_i} - n \quad \text{or} \quad \sum_{i=1}^{k} \frac{(n_i - np_i)^2}{np_i} = \frac{1}{n} \sum_{i=1}^{k} \frac{n_i^2}{p_i} - n,$$

(p. 140)

(4.13)

has under H_0 asymptotically ($n \to \infty$) a χ^2-distribution with v degrees of freedom; thus for not too small n (cf., remark below) $\hat{\chi}^2$ can be compared with the critical values of the χ_v^2-distribution; reject H_0 at the $100\alpha\%$-level when $\hat{\chi}^2 > \chi_{v;\alpha}^2$ (Table 28a, Section 1.5.2). Here

k = number of classes in the sample of size n;
$O_i = n_i$ = observed frequency (occupation number) of the class i,

$$\sum_{i=1}^{k} n_i = n;$$

$E_i = np_i$ = expected frequency under H_0 (in the case of a discrete distribution and a null hypothesis which prescribes hypothetical or otherwise given cell probabilities $p_i > 0$, $i = 1, \ldots, k$, $\sum_{i=1}^{k} p_i = 1$, the observed cell frequencies n_i are compared with the expected cell frequencies np_i);

$v = k - 1$ (if a total of a unknown parameters is estimated from the sample—e.g., the p_i as $\hat{p}_i$—then v is reduced to $v = k - 1 - a$; e.g., $a = 1$ when the sample is compared to a Poisson distribution with single parameter λ; $a = 2$ when it is compared to a normal distribution with parameters μ and σ).

For a **goodness of fit** test of this sort, the total sample size must not be too small, and the average expected frequency under the null hypothesis must not fall below 5. If they do, they are increased to the required level by combining adjacent classes. This however is necessary only if the number of classes is small. For the case $v \gtrsim 8$ and not too small sample size ($n \gtrsim 40$), the expected frequencies in isolated classes may drop below 1. For n large and $\alpha = 0.05$, choose 16 classes.

When computing $\hat{\chi}^2$ note the **signs of the differences** $O - E$: + and − should not exhibit any systematic patterns. We shall take up this subject again in Section 4.3.4. See also Remark 3 on page 493.

4.3.1 Comparing observed frequencies with their expectations

In an experiment in genetics planned as a preliminary experiment 3 phenotypes in the proportion 1:2:1 are expected; the frequencies 14:50:16 are observed (Table 70). Does the proportion found correspond to the 1:2:1

Table 70 Experiment in genetics

O	E	O − E	(O − E)²	(O − E)² / E
14	20	−6	36	1.80
50	40	10	100	2.50
16	20	−4	16	0.80
80	80	$\hat{\chi}^2 = \sum \frac{(O-E)^2}{E} =$		5.10

splitting law H_0: The observed frequencies do not differ significantly from their expectations? No particular significance level is fixed, since the trial should give us the initial information.

Table 28 tells us that $0.05 < P < 0.10$ for $k - 1 = 3 - 1 = 2$ DF and $\hat{\chi}^2 = 5.10$. H_0 is not rejected (cf., Table 70) at the 5% level, but would be at the 10% level.

4.3.2 Comparison of an empirical distribution with the uniform distribution

A die being tested is tossed 60 times. The observed frequencies (O) of the 6 faces are:

Number of spots on face	1	2	3	4	5	6
Frequency	7	16	8	17	3	9

We are dealing with a "fair" die (the probability of each outcome will be very close to 1/6). Thus the null hypothesis predicts for each outcome a theoretical or [under H_0] expected frequency (E) of 10, a so-called uniform distribution. We test at the 5% level and by (4.13) get

$$\hat{\chi}^2 = \sum \frac{(O - E)^2}{E} = \frac{(7 - 10)^2}{10} + \frac{(16 - 10)^2}{10} + \cdots + \frac{(9 - 10)^2}{10}.$$

$\hat{\chi}^2 = 14.8$, a value larger than the tabulated χ^2 value (11.07 from Table 28a) for $k - 1 = 6 - 1 = 5$ degrees of freedom and $\alpha = 0.05$: H_0 is rejected (see also Sections 4.4 and 6.2.5) at the 5% level.

4.3.3 Comparison of an empirical distribution with the normal distribution

Experience indicates that frequency distributions from scientific data, data sequences, or frequencies seldom resemble normal distributions very much. The following procedures are thus particularly useful in practice if the normal probability plot method is too inaccurate. If n independent observa-

4.3 The χ^2 Goodness of Fit Test

tions of a random variable are available we may wish to know whether this sample has been drawn from a normal population. It is impossible to conclude this definitely. All we can hope for is to be in a position to determine when the population is probably not normal. Three tests for departure from normality are (1) the chi-square goodness of fit test, (2) the method based on the standardized 3rd and 4th moments, and especially (3) the Liliefors method with the Kolmogoroff–Smirnoff goodness of fit test (Section 4.4).

We give a simple numerical example for (1): Column 1 of the following Table gives the class means x, the class width b being $b = 1$. The observed frequencies are listed in column 2. The 3rd, 4th, and 5th columns serve for computing $\bar{x}$ and s. Columns 6, 7, and 8 indicate the sequence of computations necessary to determine the probability density of the standard normal variable Z at $Z = z$ (Table 20). The multiplication by the constant K in column 9 adjusts the overall number of expected frequencies. Classes with $E < 1$ are combined with adjacent classes. For the table on page 324 we have then $k = 5$ classes. From the classified data $\bar{x}$ and s we estimate $a = 3$ DF are necessary. [For $\bar{x}$ and s computed from the unclassified values we would need 2 DF; if μ or σ is known, then we only need 1 DF.] So we have $2 = k - 1 - a = 5 - 1 - 3 = 1$ DF. With $2.376 < 2.701 = \chi^2_{1;0.10}$ there is no objection to the hypothesis of normality. This refers to our simple numerical example. We note the loss of sensitivity through grouping together small tail frequencies.

In the practical case this test for nonnormality calls for

1. $n \gtrsim 60$,
2. $k \gtrsim 7$,
3. $\alpha = \mathbf{0.10}$ or 0.05 or 0.01.

A similar procedure for the comparison of an empirical distribution with a **lognormal distribution** is described by Croxton and Cowden (1955, pp. 616–619).

J. S. Ramberg et al. (Technometrics **21** (1973), 201–214) present a four-parameter probability function and a table facilitating parameter estimation using the first four sample moments. A wide variety of curve shapes is possible with this distribution. Moreover it gives good approximations to normal, lognormal, Weibull, and other distributions. An example is given with the moments, calculated, e.g., by (1.106) through (1.109), the four lambda values, the histogram, the probability density curve corresponding to the lambda values, the observed and expected frequencies, and the χ^2 goodness of fit test.

Rule of thumb. When $0.9 < (\tilde{x}/\bar{x}) < 1.1$ and $3s < \bar{x}$, a sample distribution is assumed to be approximately normally distributed.

With the presented data and Equation (1.70) we have $\tilde{x} = 2.5 + 1\{([40/2] - 5)/16\} = 3.4375$ or 3.44 and $\tilde{x}/\bar{x} = 3.44/3.60 = 0.956$ or 0.96; $0.9 < 0.96 < 1.1$ and $3s = 3 \cdot 1.127 = 3.381 < 3.60 = \bar{x}$.

x	O	x²	Ox	Ox²	x − x̄	$\left|\frac{x-\bar{x}}{s}\right| = z$	Ordinate f(z)	f(z)·K	E	O − E	(O − E)²	(O − E)²/E
(1)	(2)	(3)	(4)	(5)	(6)	(7)	(8)	(9)	(10)	(11)	(12)	(13)
1	1	1	1	1	−2.6	2.31	0.0277	0.983	6.15	−1.15	1.322	0.215
2	4	4	8	16	−1.6	1.42	0.1456	5.168	12.30	3.70	13.690	1.113
3	16	9	48	144	−0.6	0.53	0.3467	12.305	13.32	−3.32	11.022	0.827
4	10	16	40	160	0.4	0.35	0.3752	13.317	6.56	0.44	0.194	0.030
5	7	25	35	175	1.4	1.24	0.1849	6.562	1.47	0.53	0.281	0.191
6	2	36	12	72	2.4	2.13	0.0413	1.466				
n = ∑O = 40			144	568					39.80	+0.02		$\hat{\chi}^2 = 2.376$
						[class width b = 1]			≃ 40	≃ 0		$\nu = 5 − 4 = 1$

$$\bar{x} = \frac{\sum Ox}{n} = \frac{144}{40} = 3.60 \qquad K = \frac{nb}{s} = \frac{40 \cdot 1}{1.127} = 35.492$$

$$s = \sqrt{\frac{\sum Ox^2 - (\sum Ox)^2/n}{n-1}} = \sqrt{\frac{568 - 144^2/40}{39}} = 1.127 \qquad \hat{\chi}^2 = 2.376 < 2.706 = \hat{\chi}^2_{0.10}$$

4.3 The χ^2 Goodness of Fit Test

Quantiles of the normal distribution. It is sometimes worthwhile to consider the deciles and other quantiles of a normal distribution (cf. Sections 1.3.4 and 1.3.8.3: $DC_5 = Q_2 = \mu$):

Deciles: $DC_{1;9} = \mu \mp 1.282\sigma$, $DC_{2;8} = \mu \mp 0.842\sigma$,
$DC_{3;7} = \mu \mp 0.525\sigma$, $DC_{4;6} = \mu \mp 0.253\sigma$.
Quartiles: $Q_{1;3} = \mu \mp 0.674\sigma$.

Nonnormality due to skewness and kurtosis

A distribution may depart from a null hypothesis of normality by skewness or kurtosis or both (see Section 1.3.8.7). Table 71 contains percentiles for the tails of the distribution of the standardized third and fourth moments [cf., (1.97), (1.98)]:

$$\sqrt{b_1} = a_3 = \frac{\sqrt{n} \sum_{i=1}^{n}(x_i - \bar{x})^3}{\sqrt{\left[\sum_{i=1}^{n}(x_i - \bar{x})^2\right]^3}}, \quad (4.14)$$

$$b_2 = a_4 + 3 = \frac{n \sum_{i=1}^{n}(x_i - \bar{x})^4}{\left[\sum_{i=1}^{n}(x_i - \bar{x})^2\right]^2} \quad (4.14\text{a})$$

for a normal distribution. The **expected values for a normal population** are

$$\sqrt{\beta_1} = \alpha_3 = 0, \quad (4.15)$$
$$\beta_2 = 3 \quad [\text{or} \quad \alpha_4 = \beta_2 - 3 = 0]. \quad (4.15\text{a})$$

Departure of $\sqrt{b_1}$ from zero is an indication of skewness in the sample population, while departure of b_2 from the value 3 is an indication of kurtosis. Moments lying outside the values of Table 71 give evidence for nonnormality due to skewness and nonnormality due to kurtosis. Table 71a gives 15 examples.

A decision on whether to apply a parametric procedure (preliminary test, cf., Section 3.5.1) should be reached at the 10% significance level.

A very simple method of **rapidly testing a sample for nonnormality** is due to David et al. (1954). These authors have studied the distribution of the ratios

$$\frac{\text{range}}{\text{standard deviation}} = \frac{R}{s} \quad (4.16)$$

Table 71 Lower and upper percentiles of the standardized 3rd and 4th moments, $\sqrt{b_1}$ and b_2, for tests for departure from normality. [From Pearson, E. S. and H. O. Hartley (Eds.): Biometrika Tables for Statisticians. Vol. I 3rd ed., Cambridge Univ. Press 1970, pp. 207-8, Table 34 B and C and from D'Agostino, R. B. and G. L. Tietjen (a): Approaches to the null distribution of $\sqrt{b_1}$. Biometrika **60** (1973), 169-173, p. 172, Table 2. (b) Simulation probability points of b_2 for small samples. Biometrika **58** (1971), 669-672, p. 670, Table 1; and from F. Gebhardt: Verteilung und Signifikanzschranken des 3 und 4. Stichprobenmomentes bei normalverteilten Variablen. Biom. Z. **8** (1966), 219-241, p. 235, Table 4, pp. 238, 239, Table 6.]

Size of sample n	Skewness [$\sqrt{b_1}$] Upper percentiles			Kurtosis [b_2] Lower percentiles			Upper percentiles		
	10%	5%	1%	1%	5%	10%	10%	5%	1%
7	0.787	1.008	1.432	1.25	1.41	1.53	3.20	3.55	4.23
10	0.722	0.950	1.397	1.39	1.56	1.68	3.53	3.95	5.00
15	0.648	0.862	1.275	1.55	1.72	1.84	3.62	4.13	5.30
20	0.593	0.777	1.152	1.65	1.82	1.95	3.68	4.17	5.36
25	0.543	0.714	1.073	1.72	1.91	2.03	3.68	4.16	5.30
30	0.510	0.664	0.985	1.79	1.98	2.10	3.68	4.11	5.21
35	0.474	0.624	0.932	1.84	2.03	2.14	3.68	4.10	5.13
40	0.45	0.587	0.870	1.89	2.07	2.19	3.67	4.06	5.04
45	0.43	0.558	0.825	1.93	2.11	2.22	3.65	4.00	4.94
50	0.41	0.534	0.787	1.95	2.15	2.25	3.62	3.99	4.88
70	0.35	0.459	0.673	2.08	2.25	2.35	3.58	3.88	4.61
75	0.34	—	—	2.08	2.27	—	—	3.87	4.59
100	0.30	0.389	0.567	2.18	2.35	2.44	3.52	3.77	4.39
125	—	0.350	0.508	2.24	2.40	2.50	3.48	3.71	4.24
150	0.249	0.321	0.464	2.29	2.45	2.54	3.45	3.65	4.13
175	—	0.298	0.430	2.33	2.48	2.57	3.42	3.61	4.05
200	0.217	0.280	0.403	2.37	2.51	2.59	3.40	3.57	3.98
250	—	0.251	0.360	2.42	2.55	2.63	3.36	3.52	3.87
300	0.178	0.230	0.329	2.46	2.59	2.66	3.34	3.47	3.79
400	—	0.200	0.285	2.52	2.64	2.70	3.30	3.41	3.67
500	0.139	0.179	0.255	2.57	2.67	2.73	3.27	3.37	3.60
700	—	0.151	0.215	2.62	2.72	2.77	3.23	3.31	3.50
1000	0.099	0.127	0.180	2.68	2.76	2.81	3.19	3.26	3.41
2000	0.070	0.090	0.127	2.77	2.83	2.86	3.14	3.18	3.28

Since the sampling distribution of $\sqrt{b_1}$ is symmetrical about zero, the same values, with negative sign, correspond to the lower percentiles. The dash (—) symbolizes yet unknown percentiles.

4.3 The χ² Goodness of Fit Test

Table 71a Samples (n = 20), their standardized 3rd and 4th moments ($\sqrt{b_1} = a_1$ and $b_2 = a_4 + 3$) and test results for departure from normality, with the code: °)0.1 > P > 0.05, *)0.05 ≥ P > 0.01, **)0.01 ≥ P > 0.001

Sample No.	x_i: 30	40	50	60	n = 20 70	80	90	100	110	$\bar{x}$	s	Skewness $\sqrt{b_1}$	Kurtosis b_2
1	1	0	1	0	16	0	1	0	1	70.0	14.51	0	6.80**
2	1	1	1	1	12	1	1	1	1	70.0	17.77	0	3.93°
3	16	0	1	0	1	0	1	0	1	40.0	22.94	2.15**	6.25**
4	12	1	1	1	1	1	1	1	1	48.0	27.07	1.16**	2.87
5	1	1	12	1	1	1	1	1	1	59.0	20.49	1.27**	3.66
6	1	1	1	1	1	1	1	12	1	86.5	23.46	−1.30**	3.28
7	1	7	6	1	1	1	1	1	1	56.0	22.34	1.22**	3.31
8	1	1	7	6	1	1	1	1	1	61.5	19.81	1.04*	3.54
9	1	1	1	1	2	4	4	4	2	81.0	22.22	−0.81*	2.89
10	1	1	2	3	7	2	2	1	1	69.5	19.32	0.07	2.96
11	1	1	2	2	8	2	2	1	1	70.0	19.19	0	3.03
12	2	2	2	2	2	2	6	1	1	70.0	25.34	0	1.91°
13	1	1	7	1	1	1	1	6	1	69.0	23.37	0.10	1.65**
14	1	7	1	1	1	1	1	6	1	68.5	28.89	0.10	1.28**
15	8	2	1	1	1	1	2	1	3	60.0	32.61	0.48	1.55**

Table 72 Critical values for the quotient R/s. If in a sample the ratio of the range of the standard deviation, R/s, is less than the lower bound or greater than the upper bound, then we conclude at the given significance level that the sample does not come from a normally distributed population. If the upper critical bound is exceeded, mavericks are usually present. The 10% bounds are especially important. (From Pearson, E. S. and Stephens, M. A.: The ratio of range to standard deviation in the same normal sample. Biometrika **51** (1964) 484–487, p. 486, table 3.)

Sample size = n	Lower bounds						Upper bounds					
	Significance level α											
	0.000	0.005	0.01	0.025	0.05	0.10	0.10	0.05	0.025	0.01	0.005	0.000
3	1.732	1.735	1.737	1.745	1.758	1.782	1.997	1.999	2.000	2.000	2.000	2.000
4	1.732	1.83	1.87	1.93	1.98	2.04	2.409	2.429	2.439	2.445	2.447	2.449
5	1.826	1.98	2.02	2.09	2.15	2.22	2.712	2.753	2.782	2.803	2.813	2.828
6	1.826	2.11	2.15	2.22	2.28	2.37	2.949	3.012	3.056	3.095	3.115	3.162
7	1.871	2.22	2.26	2.33	2.40	2.49	3.143	3.222	3.282	3.338	3.369	3.464
8	1.871	2.31	2.35	2.43	2.50	2.59	3.308	3.399	3.471	3.543	3.585	3.742
9	1.897	2.39	2.44	2.51	2.59	2.68	3.449	3.552	3.634	3.720	3.772	4.000
10	1.897	2.46	2.51	2.59	2.67	2.76	3.57	3.685	3.777	3.875	3.935	4.243
11	1.915	2.53	2.58	2.66	2.74	2.84	3.68	3.80	3.903	4.012	4.079	4.472
12	1.915	2.59	2.64	2.72	2.80	2.90	3.78	3.91	4.02	4.134	4.208	4.690
13	1.927	2.64	2.70	2.78	2.86	2.96	3.87	4.00	4.12	4.244	4.325	4.899
14	1.927	2.70	2.75	2.83	2.92	3.02	3.95	4.09	4.21	4.34	4.431	5.099
15	1.936	2.74	2.80	2.88	2.97	3.07	4.02	4.17	4.29	4.44	4.53	5.292
16	1.936	2.79	2.84	2.93	3.01	3.12	4.09	4.24	4.37	4.52	4.62	5.477
17	1.944	2.83	2.88	2.97	3.06	3.17	4.15	4.31	4.44	4.60	4.70	5.657
18	1.944	2.87	2.92	3.01	3.10	3.21	4.21	4.37	4.51	4.67	4.78	5.831
19	1.949	2.90	2.96	3.05	3.14	3.25	4.27	4.43	4.57	4.74	4.85	6.000
20	1.949	2.94	2.99	3.09	3.18	3.29	4.32	4.49	4.63	4.80	4.91	6.164
25	1.961	3.09	3.15	3.24	3.34	3.45	4.53	4.71	4.87	5.06	5.19	6.93
30	1.966	3.21	3.27	3.37	3.47	3.59	4.70	4.89	5.06	5.26	5.40	7.62
35	1.972	3.32	3.38	3.48	3.58	3.70	4.84	5.04	5.21	5.42	5.57	8.25
40	1.975	3.41	3.47	3.57	3.67	3.79	4.96	5.16	5.34	5.56	5.71	8.83
45	1.978	3.49	3.55	3.66	3.75	3.88	5.06	5.26	5.45	5.67	5.83	9.38
50	1.980	3.56	3.62	3.73	3.83	3.95	5.14	5.35	5.54	5.77	5.93	9.90
55	1.982	3.62	3.69	3.80	3.90	4.02	5.22	5.43	5.63	5.86	6.02	10.39
60	1.983	3.68	3.75	3.86	3.96	4.08	5.29	5.51	5.70	5.94	6.10	10.86
65	1.985	3.74	3.80	3.91	4.01	4.14	5.35	5.57	5.77	6.01	6.17	11.31
70	1.986	3.79	3.85	3.96	4.06	4.19	5.41	5.63	5.83	6.07	6.24	11.75
75	1.987	3.83	3.90	4.01	4.11	4.24	5.46	5.68	5.88	6.13	6.30	12.17
80	1.987	3.88	3.94	4.05	4.16	4.28	5.51	5.73	5.93	6.18	6.35	12.57
85	1.988	3.92	3.99	4.09	4.20	4.33	5.56	5.78	5.98	6.23	6.40	12.96
90	1.989	3.96	4.02	4.13	4.24	4.36	5.60	5.82	6.03	6.27	6.45	13.34
95	1.990	3.99	4.06	4.17	4.27	4.40	5.64	5.86	6.07	6.32	6.49	13.71
100	1.990	4.03	4.10	4.21	4.31	4.44	5.68	5.90	6.11	6.36	6.53	14.07
150	1.993	4.32	4.38	4.48	4.59	4.72	5.96	6.18	6.39	6.64	6.82	17.26
200	1.995	4.53	4.59	4.68	4.78	4.90	6.15	6.39	6.60	6.84	7.01	19.95
500	1.998	5.06	5.13	5.25	5.37	5.49	6.72	6.94	7.15	7.42	7.60	31.59
1000	1.999	5.50	5.57	5.68	5.79	5.92	7.11	7.33	7.54	7.80	7.99	44.70

in samples of size n from a normally distributed population with standard deviation σ. They give a table of critical bounds for these ratios. If the quotient does not lie between the tabulated critical values, then the hypothesis of normality is rejected at the respective significance level. Extensive tables for this procedure, which can also be interpreted as a homogeneity test, were presented by Pearson and Stephens (1964).

Applying these methods to the example $n = 40$, $R = 5$, $s = 1.127$, we get the test ratio $R/s = 5/1.127 = 4.44$.

For $n = 40$ Table 72 gives the bounds in Table 73.

4.3 The χ^2 Goodness of Fit Test

Table 73

α	Region
0%	1.98–8.83
1%	3.47–5.56
5%	3.67–5.16
10%	3.79–4.96

Our ratio lies within even the smallest of these regions. The test allows, strictly speaking, only a statement on the range of the sample distribution. The present data are in fact approximately normally distributed.

Let us emphasize that the lower bounds for a significance level $\alpha = 0\%$ for $n \geq 25$ lie above 1.96 and below 2.00 (e.g., 1.990 for $n = 100$); the upper 0% bounds can be readily estimated by $\sqrt{2(n-1)}$ (e.g., 4 for $n = 9$); these bounds ($\alpha = 0.000$) hold for arbitrary populations (Thomson 1955).

In addition to the D'Agostino test for nonnormality referred to in Section 3.1.3, let us in particular mention the W-test of Shapiro and Wilk (1965, 1968, cf., also Wilk and Shapiro 1968); methodology and tables can also be found in Vol. 2 of the Biometrika Tables (Pearson and Hartley 1972 [2], pp. 36–40, 218–221).

4.3.4 Comparison of an empirical distribution with the Poisson distribution

We take the example that deals with getting kicked by a horse (Table 40), combine the three weakly occupied end classes, and obtain Table 74. There are $k = 4$ classes; $a = 1$ parameter was estimated (λ by $\hat{\lambda} = \bar{x}$). Thus we have $v = k - 1 - a = 4 - 1 - 1 = 2$ DF at our disposal. The $\hat{\chi}^2$-value found, $\hat{\chi}^2 = 0.319$, is so low ($\chi^2_{2;0.05} = 5.991$) that the agreement must be regarded as good.

Table 74

O	E	O − E	$(O-E)^2$	$(O-E)^2/E$
109	108.7	0.3	0.09	0.001
65	66.3	−1.3	1.69	0.025
22	20.2	1.8	3.24	0.160
4	4.8	−0.8	0.64	0.133
200	200.0	0	$\hat{\chi}^2 =$	0.319

The last examples are distinguished by the fact that a larger number of classes can arise. The run test allows us to determine whether the signs of the differences $O - E$ can be considered random or due to some nonrandom influences. There is of course a difference between the case where this sign is frequently or almost always positive or negative, and the case where both signs occur about equally often and at random. For given differences $O - E$, the more regular the change in sign, the better is the fit (cf., also the test due to David 1947).

4.4 THE KOLMOGOROFF–SMIRNOFF GOODNESS OF FIT TEST

The test of Kolmogoroff (1941) and Smirnoff (1948) (cf., Section 3.9.3) tests how well an observed distribution fits a theoretically expected one (cf., Massey 1951). This test is distribution-free; it corresponds to the χ^2 goodness of fit test. The Komogoroff–Smirnoff test (K–S test) is more likely to detect deviations from the normal distribution, particularly when sample sizes are small. The χ^2 test is better for detecting irregularities in the distribution, while the K–S test is more sensitive to departures from the shape of the distribution function. This test is, strictly speaking, derived for continuous distributions. It is nevertheless applicable to discrete distributions (cf., e.g. Conover 1972). The null hypothesis that the sample originated in a population with known distribution function $F_0(x)$ is tested against the alternate hypothesis that the population underlying the sample does not have $F_0(x)$ as its distribution function. One determines the absolute frequencies E expected under the null hypothesis, forms the cumulative frequencies of these values, namely F_E, and of the observed absolute frequencies O, namely F_O, and then forms the differences $F_O - F_E$ and divides the difference largest in absolute value by the sample size n. The test ratio

$$\hat{D} = \frac{\max|F_O - F_E|}{n} \qquad (4.17)$$

(for relative frequencies $\hat{D} = \max |F_O - F_E|$) is, for sample sizes $n > 35$, assessed by means of the critical values in Table 75.

Table 75

Bounds for D	Significance level α
$1.073/\sqrt{n}$	0.20
$1.138/\sqrt{n}$	0.15
$1.224/\sqrt{n}$	0.10
$1.358/\sqrt{n}$	0.05
$1.628/\sqrt{n}$	0.01
$1.949/\sqrt{n}$	0.001

4.4 The Kolmogoroff-Smirnoff Goodness of Fit Test

Table 76 Critical values of D for the Kolmogoroff-Smirnoff goodness of fit test (from Miller, L. H.: Table of percentage points of Kolmogorov statistics. J. Amer. Statist. Assoc. **51** (1956), 111–121, 113–115, part of Table 1)

n	$D_{0.10}$	$D_{0.05}$	n	$D_{0.10}$	$D_{0.05}$	n	$D_{0.10}$	$D_{0.05}$	n	$D_{0.10}$	$D_{0.05}$
3	0.636	0.708	23	0.247	0.275	13	0.325	0.361	33	0.208	0.231
4	0.565	0.624	24	0.242	0.269	14	0.314	0.349	34	0.205	0.227
5	0.509	0.563	25	0.238	0.264	15	0.304	0.338	35	0.202	0.224
6	0.468	0.519	26	0.233	0.259	16	0.295	0.327	36	0.199	0.221
7	0.436	0.483	27	0.229	0.254	17	0.286	0.318	37	0.196	0.218
8	0.410	0.454	28	0.225	0.250	18	0.278	0.309	38	0.194	0.215
9	0.387	0.430	29	0.221	0.246	19	0.271	0.301	39	0.191	0.213
10	0.369	0.409	30	0.218	0.242	20	0.265	0.294	40	0.189	0.210
11	0.352	0.391	31	0.214	0.238	21	0.259	0.287	50	0.170	0.188
12	0.338	0.375	32	0.211	0.234	22	0.253	0.281	100	0.121	0.134

Critical bounds for smaller sample sizes can be found in the tables by Massey (1951) and Birnbaum (1952). Miller (1956) gives exact critical values for $n = 1$ to 100 and $\alpha = 0.20, 0.10, 0.05, 0.02$ and 0.01. The particularly important 10% and 5% bounds for small and moderate sample sizes are here reproduced with three decimals (Table 76). An observed $\hat{D}$-value which equals or exceeds the tabulated value is significant at the corresponding level. For other values of α, the numerator of the bound is obtained from $\sqrt{-0.5\ln(\alpha/2)}$ (called $K_{(\alpha)}$ in Section 3.9.3); e.g., $\alpha = 0.10$, $\ln(0.10/2) = \ln 0.05 = -2.996$ (Section 1.5.2, Table 29; or Section 0.2), i.e.,

$$\sqrt{(-0.5)(-2.996)} = 1.224.$$

If the sample distribution is **compared with a normal distribution**, the parameters of which have to be estimated from the sample values, then the results based on Table 75 are very conservative; exact bounds for this K − S test are presented by Lilliefors (1967). Some D-values:

n	10%	5%	1%
5	0.315	0.337	0.405
8	0.261	0.285	0.331
10	0.239	0.258	0.294
12	0.223	0.242	0.275
15	0.201	0.220	0.257
18	0.184	0.200	0.239
20	0.174	0.190	0.231
25	0.158	0.173	0.200
30	0.144	0.161	0.187

For $n > 30$ we have accordingly $0.805/\sqrt{n}$ ($\alpha = 0.10$), $0.866/\sqrt{n}$ ($\alpha = 0.05$), and $1.031/\sqrt{n}$ ($\alpha = 0.01$). The comparison of the sample distribution with an exponential distribution is considered by Finkelstein and Schafer (1971).

EXAMPLE 1. We use the example of Section 4.3.3. The computations are shown in Table 77. The test $2.55/40 = 0.063 < 0.127 = 0.805/\sqrt{40}$ leads to the same result: The null hypothesis cannot be rejected at the 10% level.

Table 77

O E	1 0.98	4 5.17	16 12.30	10 13.32	7 6.56	2 1.47
F_O F_E	1 0.98	5 6.15	21 18.45	31 31.77	38 38.33	40 39.80
$|F_O - F_E|$	0.02	1.15	2.55	0.77	0.33	0.20

EXAMPLE 2. A die is tossed 120 times for control. The frequencies for the 6 faces are 18, 23, 15, 21, 25, 18. Do the proportions found correspond to the null hypothesis of a fair die? In Table 78 we test with $\alpha = 0.01$ the frequencies arranged in order of increasing magnitude: 15, 18, 18, 21, 23, 25. Since $9/120 = 0.075 < 0.1486 = 1.628/\sqrt{120} = D_{120;\,0.01}$, the null hypothesis is not rejected.

Table 78

F_E	20	40	60	80	100	120
F_O	15	33	51	72	95	120
$|F_E - F_O|$	5	7	9	8	5	0

Let us note that—strictly speaking—the χ^2-test requires an infinitely large sample size n, and the K-S goodness of fit test requires infinitely many classes k. Still, both tests can be employed even for small samples with few classes ($n \gtrsim 10, k \gtrsim 5$) as was clearly shown by Slakter (1965); nonetheless the χ^2 goodness of fit test or the corresponding likelihood ratio $2\hat{I}$ test (cf., Section 6.2.5) is preferred in these cases. All three goodness of fit tests assess only the closeness of the fit. The knowledge of the "randomness of the fit" is lost. There is of course a difference between the case where, for example for the χ^2-test, the differences $O - E$ almost without exception have positive resp. negative values and where both signs appear randomly. The more regularly the signs change, the better is the fit with given deviations $O - E$. A simple means of testing the randomness of a fit is provided by the run test (cf., Section 4.7.2).

Other important goodness of fit tests (cf., also Darling 1957) are due to David (1950; cf., also Nicholson 1961, as well as the one and the two sample empty field test with the tables and examples in Csorgo and Guttman 1962) and to Quandt (1964, 1966); cf., also Stephens (1970).

4.5 THE FREQUENCY OF EVENTS

4.5.1 Confidence limits of an observed frequency for a binomially distributed population. The comparison of a relative frequency with the underlying parameter

If x denotes the number of successes in n Bernoulli ("either- or," "success-failure") trials, then $\hat{p} = x/n$ is the relative frequency. The percentage frequency of hits in the sample is

$$\hat{p}\% = \frac{x}{n} 100 \quad \text{with } n \geq 100, \tag{4.18}$$

for $n < 70$ "x out of n" or x/n is given, (for $n \geq 70$ you may write, if needed for a comparison, "($p\%$)," e.g., 29/80 = 0.3625, written as "(36%)"), for percentage with $70 \lesssim n < 150$ places beyond the decimal point are ignored, the first two being included only from about $n = 2,000$ on. Example: $\hat{p} = 20/149 = 13.42\%$ is stated as a relative frequency of 0.13 or as 13%.

Confidence intervals (cf., Sections 1.4.1, 3.1.1, 3.2, 3.6.2) of the binomial distribution are given by Crow (1956), Blyth and Hutchinson (1960), Documenta Geigy (1968, pp. 85–98), Pachares (1960 [8:1]) and especially Blyth and Still (1983). Figure 38 in Section 4.5.2 or the table on page 703 frequently serves as an outline.

Exact two sided limits, the upper and lower limits (π_l, π_u), for the confidence interval (CI) of the parameter π [cf., (4.19)]

$$\text{CI: } \pi_l \leq \pi \leq \pi_u \tag{4.19}$$

can be computed according to

$$\pi_u = \frac{(x+1)F}{n - x + (x+1)F} \quad \text{with} \quad F_{\{DF_1 = 2(x+1), DF_2 = 2(n-x)\}},$$

$$\pi_l = \frac{x}{x + (n - x + 1)F} \quad \text{with} \quad F_{\{DF_1 = 2(n-x+1), DF_2 = 2x\}}. \tag{4.20}$$

EXAMPLE. Compute the 95% confidence interval for π with $\hat{p} = x/n = 7/20 = 0.35$ (F-values are taken from Table 30c in Section 1.5.3).

F-values:

$$2(7 + 1) = 16, \quad 2(20 - 7) = 26, \quad F_{16; 26; 0.025} = 2.36,$$

$$2(20 - 7 + 1) = 28, \quad 2(7) = 14, \quad F_{28; 14; 0.025} = 2.75;$$

CI bounds:

$$\pi_u = \frac{(7+1)2.36}{20-7+(7+1)2.36} = 0.592,$$

$$\pi_l = \frac{7}{7+(20-7+1)2.75} = 0.154;$$

95% CI:

$$0.154 \leq \pi \leq 0.592 \quad (15.4\% \leq \pi \leq 59.2\%).$$

Remarks

1. It is assumed that $\hat{p} = x/n$ was estimated from a random sample.
2. The confidence limits are symmetric with respect to $\hat{p}$ only if $\hat{p} = 0.5$ (cf., above example: $0.592 - 0.350 = 0.242 > 0.196 = 0.350 - 0.154$).

Approximations using the normal distributions: (4.21 to 4.23a)

A good approximation for the 95% confidence interval of not too extreme π-values—$0.3 \leq \pi \leq 0.7$ when $n \geq 10$, $0.05 \leq \pi \leq 0.95$ when $n \geq 60$—is given by [cf., Table 43, Section 2.1.6, with $z_{0.05} = 1.96$; we have $1.95 = (1.96^2 + 2)/3$ and $0.18 = (7 - 1.96^2)/18$]

$$\pi_u = \frac{x + 1.95 + 1.96\sqrt{(x+1-0.18)(n-x-0.18)/(n+11\cdot 0.18-4)}}{n + 2\cdot 1.95 - 1},$$

$$\pi_l = \frac{x - 1 + 1.95 - 1.96\sqrt{(x-0.18)(n+1-x-0.18)/(n+11\cdot 0.18-4)}}{n + 2\cdot 1.95 - 1}$$

(4.21)

(Molenaar 1970).

EXAMPLE. 95% CI for π with $\hat{p} = x/n = 7/20 = 0.35$:

$$\pi_u = \frac{[7 + 1.95 + 1.96\sqrt{(7+1-0.18)(20-7-0.18)/(20+11\cdot 0.18-4)}]}{(20 + 2\cdot 1.95 - 1)},$$

$$\pi_l = \frac{[7 - 1 + 1.95 - 1.96\sqrt{(7-0.18)(20+1-7-0.18)/(20+11\cdot 0.18-4)}]}{(20 + 2\cdot 1.95 - 1)}$$

95% CI:

$$0.151 \leq \pi \leq 0.593 \ (15.1\% \leq \pi \leq 59.3\%).$$

4.5 The Frequency of Events

For sample sizes n not too small and relative frequencies $\hat{p}$ not too extreme, i.e., for $n\hat{p} > 5$ and $n(1 - \hat{p}) > 5$, formula (4.22) can be used for a rough survey [cf., (4.22a below)]:

$$\boxed{\begin{aligned} \pi_u &\approx \left(\hat{p} + \frac{1}{2n}\right) + z \cdot \sqrt{\frac{\hat{p}(1-\hat{p})}{n}}, \\ \pi_l &\approx \left(\hat{p} - \frac{1}{2n}\right) - z \cdot \sqrt{\frac{\hat{p}(1-\hat{p})}{n}}. \end{aligned}} \quad (4.22)$$

This approximation (drawing samples with replacement; cf., Remark 2 below) serves for general orientation; if the conditions for Table 79 are fulfilled, (4.22) is still good though inferior to (4.21).

The corresponding 95% CI is

$$\boxed{95\% \, \text{CI}: \; \left(\hat{p} - \frac{1}{2n}\right) - 1.96 \cdot \sqrt{\frac{\hat{p}(1-\hat{p})}{n}} \lesssim \pi \lesssim \left(\hat{p} + \frac{1}{2n}\right) + 1.96 \cdot \sqrt{\frac{\hat{p}(1-\hat{p})}{n}}.}$$

(4.22a)

(The value $z = 1.96$ comes from Table 43, Section 2.1.6; for the 90% CI 1.96 is replaced by 1.645; for the 99% CI, by 2.576).

Examples

1. 95% CI for π with $\hat{p} = x/n = 7/20 = 0.35$ [check: $(20)(0.35) = 7 > 5$]; $0.35 - 1/[2(20)] = 0.325$; $1.96\sqrt{(0.35)(0.65)/20} = 0.209$;

$$95\% \, \text{CI}: \quad 0.325 \pm 0.209 \qquad (0.116 \lesssim \pi \lesssim 0.534).$$

(Compare the exact limits above).

2. 99% CI for π with $\hat{p} = x/n = 70/200 = 0.35$ or 35% (check: conditions of Table 79 fulfilled): $0.35 - 1/[2(200)] = 0.3475$; $2.576\sqrt{(0.35)(0.65)/200} = 0.0869$;

$$99\% \, \text{CI}: \quad 0.3475 \pm 0.0869 \qquad (0.261 \lesssim \pi \lesssim 0.434, \; 26.1\% \lesssim \pi \lesssim 43.4\%);$$

(the exact limits are 26.51% and 44.21%). The corresponding 95% CI, $28.44\% \leq \pi \leq 42.06\%$, can be found in the table on page 703.

Remarks

1. The quantity $1/2n$ is referred to as the *continuity correction*. It *widens* the confidence interval. The initial values are frequencies and thus discrete variables; for the confidence interval we use the standard normal variable, a continuous distribution. The error we make in going over from the discrete to the normal distribution is diminished by the continuity correction.

2. For **finite populations** of size N, (4.23) can be used for general orientation; $\sqrt{(N-n)/(N-1)}$ is the **finite population correction**, which tends to one as $N \to \infty$ (since $\sqrt{} = \sqrt{(1-n/N)/(1-1/N)} \to \sqrt{1} = 1$) and may then be neglected [cf., e.g., (4.22), (4.22a)]. This is also true for the case when N is sufficiently large in comparison with n, i.e., when, e.g., n is less than 5% of N. The approximation (4.23) can be employed only if the requirements given in Table 79 are met.

For finite populations (cf. Remark 2)

Equations (4.23), (4.23a) and Table 79 describe sampling without replacement.

Table 79 (from Cochran 1963, p. 57, Table 3.3)

For $\hat{p}$ equal to	and $n\hat{p}$ as well as $n(1-\hat{p})$ equal to at least	with n greater than or equal to
0.5	15	30
0.4 or 0.6	20	50
0.3 or 0.7	24	80
0.2 or 0.8	40	200
0.1 or 0.9	60	600
0.05 or 0.95	70	1400

(4.23) may be applied

$$\pi_u \approx \left(\hat{p}+\frac{1}{2n}\right)+z \cdot \sqrt{\left\{\frac{\hat{p}(1-\hat{p})}{n}\right\}\left\{\frac{N-n}{N-1}\right\}}$$
$$\pi_l \approx \left(\hat{p}-\frac{1}{2n}\right)-z \cdot \sqrt{\left\{\frac{\hat{p}(1-\hat{p})}{n}\right\}\left\{\frac{N-n}{N-1}\right\}}$$
(4.23)

$$\left(\hat{p}-\frac{1}{2n}\right)-z \cdot \sqrt{\left\{\frac{\hat{p}(1-\hat{p})}{n}\right\}\left\{\frac{N-n}{N-1}\right\}} \lessapprox \pi \lessapprox \left(\hat{p}+\frac{1}{2n}\right)+z \cdot \sqrt{\left\{\frac{\hat{p}(1-\hat{p})}{n}\right\}\left\{\frac{N-n}{N-1}\right\}}.$$
(4.23a)

Special cases: $\hat{p} = 0$ resp. $\hat{p} = 1$ (with 4 examples)

The one sided upper confidence limit (CL) for $\hat{p} = 0$ (complete failure; cf., table below) is given by

$$\pi_u = \frac{F}{n+F} \quad \text{with} \quad F_{(DF_1=2;\, DF_2=2n)}.$$
(4.24)

4.5 The Frequency of Events

Compute the one sided upper 95% confidence limit π_u with $\hat{p} = 0$ for $n = 60$.

$$F_{2;120;0.05} = 3.07 \quad \text{(Table 30b, Section 1.5.3)}$$

95% CL: $\pi_u = \dfrac{3.07}{60 + 3.07} = 0.0487 \quad$ [i.e., $\pi \leq 0.049$].

The one sided lower confidence limit for $\hat{p} = 1$ (complete success, cf., the table below) is given by

$$\boxed{\pi_l = \frac{n}{n + F} \quad \text{with} \quad F_{(DF_1 = 2;\, DF_2 = 2n)}.} \tag{4.25}$$

Compute the one sided lower 99% confidence limit π_l with $\hat{p} = 1$ for $n = 60$.

$$F_{2;120;0.01} = 4.79 \quad \text{(Table 30d, Section 1.5.3)}$$

99% CL: $\pi_l = \dfrac{60}{60 + 4.79} = 0.9261 \quad$ [i.e., $\pi \geq 0.93$].

For the one sided 95% confidence limits (CL) with $n > 50$ and

$$\boxed{\begin{aligned} \hat{p} &= 0 \text{ we have approximately } \pi_u \simeq \frac{3}{n}, \\ \hat{p} &= 1 \text{ we have approximately } \pi_l \simeq 1 - \frac{3}{n}. \end{aligned}} \tag{4.26}$$

$\hat{p} = 0,\; n = 100;\; 95\%\,\text{CL}:\; \pi_u \approx 3/100 = 0.03,$

$\hat{p} = 1,\; n = 100;\; 95\%\,\text{CL}:\; \pi_l \approx 1 - (3/100) = 0.97.$

In comparison: $F_{2;200;0.05} = 3.04$ and hence by (4.24, 4.25)

$\hat{p} = 0;\; 95\%\,\text{CL}:\; \pi_u = 3.04/(100 + 3.04) = 0.0295,$

$\hat{p} = 1;\; 95\%\,\text{CL}:\; \pi_l = 100/(100 + 3.04) = 0.9705.$

Thus, if no undesirable side effects (cf., end of Section 2.2) occur on 100 patients treated with a certain medicine, then we may ascertain at the 5% level that at most 3% of the patients who will be treated with this medication will suffer undesirable side effects.

One sided upper and lower 95% and 99% confidence limits for the special cases $\hat{p} = 0$ respectively $\hat{p} = 1$ ($\alpha = 0.05$; $\alpha = 0.01$), in percent, for certain sample sizes n are as follows:

α	n	10	30	50	80	100	150	200	300	500	1000
5%	π_u	26	9.5	5.8	3.7	3.0	2.0	1.5	0.99	0.60	0.30
	π_l	74	90.5	94.2	96.3	97.0	98.0	98.5	99.01	99.40	99.70
1%	π_u	37	14	8.8	5.6	4.5	3.0	2.3	1.5	0.92	0.46
	π_l	63	86	91.2	94.4	95.5	97.0	97.7	98.5	99.08	99.54

Comparison of two relative frequencies

The comparison of two relative frequencies is a **comparison of the probabilities of two binomial distributions**. Exact methods (cf., Section 4.6.7) and good approximating procedures for such comparisons (cf., Section 4.6.1) are known. For not too small sample sizes [with $n\hat{p}$ as well as $n(1-\hat{p}) > 5$] an approximation with the help of the standard normal distribution is also possible:

1. **Comparison of a relative frequency $\hat{p}_1$ with the underlying parameter π** without or with a finite population correction (4.27 resp. 4.27a) (cf., the examples below):

$$\hat{z} = \frac{|\hat{p}_1 - \pi| - \frac{1}{2n}}{\sqrt{\frac{\pi(1-\pi)}{n}}}, \tag{4.27}$$

$$\hat{z} = \frac{|\hat{p}_1 - \pi| - \frac{1}{2n}}{\sqrt{\left\{\frac{\pi(1-\pi)}{n}\right\}\cdot\left\{\frac{N-n}{N-1}\right\}}}, \tag{4.27a}$$

where z has (approximately) a standard normal distribution. Null hypothesis: $\pi_1 = \pi$. The alternative hypothesis is $\pi_1 \neq \pi$ (or in a one sided problem $\pi_1 > \pi$ or $\pi_1 < \pi$) (cf., also Section 4.5.5).

2. **Comparing two relative frequencies $\hat{p}_1$ and $\hat{p}_2$** (comparing two percentages). It is assumed that (a) $n_1 \geq 50$, $n_2 \geq 50$; (b) $n\hat{p} > 5$, $n(1-\hat{p}) > 5$ (see also below) we have

$$\hat{z} = \frac{|\hat{p}_1 - \hat{p}_2|}{\sqrt{\hat{p}(1-\hat{p})[(1/n_1)+(1/n_2)]}} \tag{4.28}$$

where $\hat{p}_1 = x_1/n_1$, $\hat{p}_2 = x_2/n_2$, $\hat{p} = (x_1 + x_2)/(n_1 + n_2)$. Null hypothesis: $\pi_1 = \pi_2$; alternative hypothesis: $\pi_1 \neq \pi_2$ (for one sided question $\pi_1 > \pi_2$ or $\pi_1 < \pi_2$). Thus for $n_1 = n_2 = 300$, we have $\hat{p}_1 = 54/300 = 0.18$, $\hat{p}_2 = 30/300 = 0.10$ [note that $n\hat{p}_2 = (300)(0.10) = 30 > 5$], $\hat{p} = (54+30)/(300+300) = 0.14$, $\hat{z} = (0.18 - 0.10)/\sqrt{0.14(0.86)(2/300)} = 2.82$, i.e., $P \approx 0.005$.

Note that computations can also be carried out in terms of percentages [$\hat{z} = (18 - 10)/\sqrt{14(86)(2/300)} = 2.82$], and that (for $n_1 = n_2$) differences greater than or equal to D (in %) are significant at the 5% level. (Tables for $n_1 = n_2 \geqq 50$ and $n_1 > n_2 \geq 100$ are included in my booklet, Sachs (1976), Appendix, Table C):

n_1	50	100	150	200	300	500	1000	5000
D	20	14	11.5	10	8	6.3	4.5	2

4.5 The Frequency of Events

If both percentages to be compared lie below 40% or above 60%, the corresponding P-values are substantially smaller than 5% (for our example above $18\% - 10\% = 8\%$ with $P \simeq 0.005$).

Somewhat more precise than (4.28) and not subject to requirements as stringent [$n\hat{p}$ and $n(1 - \hat{p}) \geq 1$ for n_1 and $n_2 \geq 25$] is an approximation based on the arcsine transformation (Table 51, Section 3.6.1):

$$\hat{z} = (|\arcsin\sqrt{\hat{p}_1} - \arcsin\sqrt{\hat{p}_2}|)/28.648\sqrt{1/n_1 + 1/n_2};$$

for the example we have $\hat{z} = (25.104 - 18.435)/28.648\sqrt{2/300} = 2.85$ (cf., also the Remarks in Section 4.6.1).

To test $H_0: \pi_1 - \pi_2 = d_0$ against the alternative hypothesis $\pi_1 - \pi_2 \neq d_0$ ($\pi_1 - \pi_2 < d_0$ or $> d_0$), use ($\hat{p}_1 = x_1/n_1$, $\hat{p}_2 = x_2/n_2$, $\hat{q}_1 = 1 - \hat{p}_1$, $\hat{q}_2 = 1 - \hat{p}_2$)

$$\hat{z} = \frac{|(\hat{p}_1 - \hat{p}_2) - d_0|}{\sqrt{(\hat{p}_1\hat{q}_1/n_1) + (\hat{p}_2\hat{q}_2/n_2)}}. \qquad (4.28a) \quad \text{(p. 62)}$$

Examples

1. In a certain large city $\pi = 20\%$ of the families received a certain periodical. There are reasons for assuming that the number of subscribers is now below 20%. To check this hypothesis, a random sample consisting of 100 families is chosen and evaluated, and $\hat{p}_1 = 0.16$ (16%) is found. The null hypothesis $\pi_1 = 20\%$ is tested against the alternative hypothesis $\pi_1 < 20\%$ (significance level $\alpha = 0.05$). We can omit the finite population correction, since the population is very large in comparison with the sample. Since $n\hat{p}_1 > 5$ and $n(1 - \hat{p}_1) > 5$, we use the approximation involving the (p. 62) normal distribution (4.27):

$$\hat{z} = \frac{|\hat{p}_1 - \pi| - \frac{1}{2n}}{\sqrt{\frac{\pi(1 - \pi)}{n}}} = \frac{|0.16 - 0.20| - \frac{1}{2 \cdot 100}}{\sqrt{\frac{0.20 \cdot 0.80}{100}}} = 0.875.$$

The value $z = 0.875$ corresponds to a level of significance $P\{\hat{p}_1 \leq 0.16 | \pi = 0.20\} = 0.19 > 0.05$. Thus 19 out of 100 random samples from a population with $\pi = 0.20$ exhibit a subscriber portion $\hat{p}_1 \leq 0.16$. We therefore retain the null hypothesis.

2. Out of 2,000 dealers $\pi = 40\%$ decide to increase their orders. A short time later there are indications that the percentage of dealers who increase their orders has risen again.

A random sample of 400 dealers indicates that with $\hat{p}_1 = 46\%$ the percentage is in fact higher. It is asked whether this increase can be deemed significant. The null hypothesis $\pi_1 = 0.40$ is tested against the alternative hypothesis $\pi_1 > 0.40$ with $\hat{p}_1 = 0.46$ (significance level $\alpha = 0.05$). Since the sample includes 20% of the population, the finite population correction and thus (4.27a) must be employed:

$$\hat{z} = \frac{|\hat{p}_1 - \pi| - \frac{1}{2n}}{\sqrt{\left[\frac{\pi(1-\pi)}{n}\right] \cdot \left[\frac{N-n}{N-1}\right]}} = \frac{|0.46 - 0.40| - \frac{1}{2 \cdot 400}}{\sqrt{\left[\frac{0.40 \cdot 0.60}{400}\right] \cdot \left[\frac{2000 - 400}{2000 - 1}\right]}} = 2.68,$$

$$P\{\hat{p}_1 \leq 0.46 | \pi = 0.40\} = 0.0037 < 0.05.$$

The null hypothesis is rejected at the 5% level: There is an actual increase.

4.5.2 Clopper and Pearson's quick estimation of the confidence intervals of a relative frequency

A rapid method for drawing inferences on the population parameter from the portion or percentage in the sample (indirect inference) is offered by Figure 38, due to Clopper and Pearson. This diagram gives the confidence

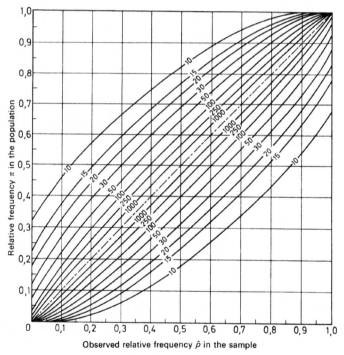

Figure 38 95% confidence interval for relative frequencies. The numbers on the curves indicate the sample size n. (From Clopper, C. J. and Pearson, E. S.: The use of confidence or fiducial limits illustrated in the case of the binomial. Biometrika **26** (1934) 404–413, p. 410.)

4.5 The Frequency of Events

limits for a relative frequency $\hat{p} = x/n$ with a confidence coefficient of 95%, i.e., the 95% confidence interval for π. The numbers on the curves indicate the sample size. **The confidence limits become tighter and more symmetric with increasing sample size n**, since the binomial distribution goes over into a normal distribution; for $\hat{p} = 0.5$ the confidence interval is symmetric even for small values of n. The graph also lets us read off the n required to attain a certain accuracy.

For practical work, the table on page 703 or Table 41 (2 charts) of the Biometrika Tables, Vol. I (Pearson and Hartley 1966, 1970) is preferred.

Examples

1. In a sample of $n = 10$ values the event x was observed 7 times, i.e., $\hat{p} = x/n = 7/10 = 0.70$. Fig. 38: The points of intersection of the vertical above 0.7 with the upper and the lower $n = 10$ curve then determine the limits of the 95% confidence interval for the parameter π of the population: $0.34 \lesssim \pi \lesssim 0.93$.

2. A percentage lying in the vicinity of 40% is to be estimated in such a way that the resulting 95% CI forms a 20% region. By Figure 38, this condition is fulfilled about when $n \approx 100$.

4.5.3 Estimation of the minimum size of a sample with counted data

The expression $\hat{p} \pm z\sqrt{\hat{p}(1 - \hat{p})/n}$ (based on the normal distribution; cf., (4.22)) for the confidence limits implies

$$\hat{p} + z\sqrt{\hat{p}(1 - \hat{p})/n} - (\hat{p} - z\sqrt{\hat{p}(1 - \hat{p})/n}) = 2a,$$

i.e., $z\sqrt{\hat{p}(1 - \hat{p})/n} = a$, whence $n = z^2\hat{p}(1 - \hat{p})/a^2$.

For $S = 95\%$ we have $z = 1.96 \simeq 2$ and therefore n must be at least

$$\boxed{\hat{n} = \frac{4 \cdot \hat{p} \cdot (1 - \hat{p})}{a^2}}. \qquad (4.29)$$

Since n attains its maximum when $\hat{p}(1 - \hat{p})$ is largest (which is the case for $\hat{p} = 50\%$), if we set $\hat{p} = 50\%$, then the sample size becomes larger than is generally necessary and

$$\hat{n} = \frac{4 \cdot 0.5^2}{a^2} \qquad \boxed{\hat{n} = \frac{1}{a^2}}. \qquad (4.30)$$

If we write (4.23) with the simplified population correction

$$\sqrt{\frac{N-n}{N}} \quad \text{instead of} \quad \sqrt{\frac{N-n}{N-1}}$$

and if we drop $1/2n$, then we have

$$\hat{n} = \frac{N}{1+a^2 N} \tag{4.31}$$

for the estimated minimum size.

Examples

1. Suppose we are interested in the percentage of families in a carefully delimited rural district that watches a certain television program. About 1,000 families live there. Polling (cf. last sentence before Section 3.1) all the families appears too tedious. The investigators decide to draw a sample and estimate with a deviation a of $\pm 10\%$ and a confidence coefficient of 95%. How large must the sample be? By (4.31) we have

$$\hat{n} = \frac{1{,}000}{1 + (0.10)^2(1{,}000)} \simeq 91.$$

Thus only 91 families need be polled. An estimate of π with an error of $a = 0.10$ and a confidence coefficient of 95% is obtained. By (4.30) we would have very roughly obtained $n = 1/0.10^2 = 1/0.01 = 100$. If we know that $\pi = 0.30$, our estimated sample size is of course too large, and we then need only about $n' = 4n\pi(1 - \pi) = 4(91)(0.3)(0.7) = 76$ individual values:

$$\hat{n}' = 4n\hat{p}(1 - \hat{p}). \tag{4.32}$$

For $\hat{n} > 0.05N$, (4.29) is replaced by (4.29a)

$$\hat{n}_{\text{corr.}} = \frac{N(a^2/4) + Np - Np^2}{N(a^2/4) + p - p^2}, \tag{4.29a}$$

i.e.,

$$\hat{n}_{\text{corr}} = \frac{1{,}000(0.10^2/4) + 1{,}000 \cdot 0.30 - 1{,}000 \cdot 0.30^2}{1{,}000(0.10^2/4) + 0.30 - 0.30^2} \simeq 74.$$

If required, the "4" in each of the formula is replaced by the appropriate value of z^2:

$2.6896 \ (S = 90\%), \quad 3.8416 \ (S = 95\%), \quad 6.6564 \ (S = 99\%).$

4.5 The Frequency of Events

2. We are asked for the percentage of families in a certain small town of 3,000 residents that watched a certain television program. A confidence coefficient of 95% with a deviation of ±3% is called for. Thus with (4.31)

$$\hat{n} = \frac{N}{1 + a^2 N} = \frac{3,000}{1 + 0.0009 \cdot 3,000} \simeq 811.$$

After taking a sample of 811 families it turns out that 243 families had watched the television program, i.e., $\hat{p} = 243/811 \simeq 0.30$. The 95% confidence interval is thus found to be

$$0.30 - 0.03 \leq \pi \leq 0.30 + 0.03,$$

$$95\% \text{ CI: } 0.27 \leq \pi \leq 0.33.$$

4.5.4 The confidence interval for rare events

Here we tie in with the discussion in Section 1.6.4 on the confidence limits of the Poisson distribution and illustrate the application of Table 80: In an 8 hour **observation unit**, 26 events were registered. The 95% limits ($x = 26$) for (a) the observation unit are $16.77 \simeq 17$ and $37.67 \simeq 38$ events and for (b) one hour are $16.77/8 \simeq 2$ and $37.67/8 \simeq 5$ events.

Examples

1. In a certain district four floods were observed in the course of a century. If it is assumed that the number of floods in various centuries follows a Poisson distribution, it can be counted on that the number of floods lies outside the limits $1.366 \simeq 1$ and $9.598 \simeq 10$ in only one century out of 20 on the average; i.e., 95% CI: $1 \lesssim \lambda \lesssim 10$.

2. A telephone exchange handles 23 calls in one minute. We wish to find the 95% confidence limits for the expected number of calls in 1 minute and in 1 hour. If we assume the number of calls in the time interval considered is fairly constant and (since, let us say, 1,000 calls/minute can be dealt with) follow a Poisson distribution, then the 95% confidence limits for 1 minute (according to Table 80) are $14.921 \simeq 15$ and $34.048 \simeq 34$. In one hour $60(14.921) \simeq 895$ to $60(34.048) \simeq 2,043$ calls are to be expected ($S = 0.95$). Thus we have 95% CI: $15 \lesssim \lambda_{1\,\text{min}} \lesssim 34$ and $895 \lesssim \lambda_{1\,\text{h}} \lesssim 2,043$.

Table 80 can also be used to test the null hypothesis: $\lambda = \lambda_x$ (λ is given; x is the number of observed results, λ_x is the associated parameter). If the CI for λ_x does not include the parameter λ, the null hypothesis is abandoned in favor of $\lambda \neq \lambda_x$.

Table 80 Confidence intervals for the mean of a Poisson distribution (taken from Crow E. L. and Gardner, R. S.: Confidence intervals for the expectation of a Poisson variable, Biometrika **46** (1959), 441–453). This table does not permit the assignment of one sided confidence limits.

x	95		99		x	95		99		x	95		99	
0	0	3.285	0	4.771	100	80.25	120.36	76.61	127.31	200	172.38⁵	227.73	164.31	238.01
1	0.051	5.323	0.010	6.914	101	81.61	121.06	76.61	128.70	201	173.79	228.99	165.33	239.46
2	0.355	6.686	0.149	8.727	102	83.14	122.37	77.15	130.27⁵	202	175.48⁵	230.28	166.71	241.32
3	0.818	8.102	0.436	10.473	103	84.57	123.77	78.71	131.50	203	176.23	231.65	168.29	241.32
4	1.366	9.598	0.823	12.347	104	84.57	125.46	80.06	131.82	204	176.23	233.19	169.49	242.01
5	1.970	11.177	1.279	13.793	105	84.67	126.26	80.06	133.21	205	176.23	234.53	169.49	243.31⁵
6	2.613	12.817	1.785	15.277	106	86.01	126.48	80.65	134.79	206	177.48	234.53	169.64	244.69
7	3.285	13.765	2.330	16.801	107	87.48	127.78	82.21	135.99	207	178.77	235.14⁵	170.98	246.24
8	3.285	14.921	2.906	18.362	108	89.23	129.14	83.56	136.30	208	180.14	236.39	172.41	247.54⁵
9	4.460	16.768	3.507	19.462	109	89.23	130.68	83.56	137.68	209	181.67	237.67	174.36	247.54⁵
10	5.323	17.633	4.130	20.676	110	89.23	132.03	84.12	139.24	210	183.05	239.00	174.36	248.62
11	5.323	19.050	4.771	22.042	111	90.37	132.03	85.65	140.54	211	183.05	240.45	174.36	249.94
12	6.686	20.335	4.771	23.765	112	91.78	133.14⁵	87.12	140.76	212	183.05	242.27	175.25	251.35
13	6.686	21.364	5.829	24.925	113	93.48	134.48	87.12	142.12	213	183.86	242.53	176.61	253.14
14	8.102	22.945	6.668	25.992	114	94.23	135.92	87.55	143.64	214	185.13	242.53	178.11	253.65
15	8.102	23.762	6.914	27.718	115	94.23	137.79	89.05	145.13	215	186.46	243.76	179.67	253.92
16	9.598	25.400	7.756	28.852	116	94.70⁵	137.79	90.72	145.19	216	187.89	245.02	179.67	255.20
17	9.598	26.306	8.727	29.900	117	96.06	138.49	90.72	146.54	217	189.83	246.32⁵	179.67	256.54
18	11.177	27.735	8.727	31.839	118	97.54⁵	139.79	90.96	148.01	218	189.83	247.70	180.84	258.00
19	11.177	28.966	10.009	32.547	119	99.17	141.16	92.42	149.76	219	189.83	249.28	182.22	259.78
20	12.817	30.017	10.473	34.183	120	99.17	142.70	94.34⁵	149.76	220	190.21	250.43	183.81	259.78
21	12.817	31.675	11.242	35.204	121	99.17	144.01	94.34⁵	150.93	221	191.46	250.43	184.97⁵	260.47
22	13.765	32.277	12.347	36.544	122	100.32	144.01	94.35	152.35⁵	222	192.76	251.11	184.97⁵	261.77
23	14.921	34.048	12.347	37.819	123	101.71	145.08	95.76	154.18	223	194.11⁵	252.35	185.08	263.12⁵
24	14.921	34.665	13.793	38.939	124	103.31⁵	146.39	97.42	154.60	224	195.63	253.63	186.40	264.63
25	16.768	36.030	13.793	40.373	125	104.40	147.80	98.36	155.31	225	197.09	254.95	187.81	266.15
26	16.77	37.67	15.28	41.39	126	104.40	149.53	98.36	156.69	226	197.09	256.37	189.50	266.15
27	17.63	38.16⁵	15.28	42.85	127	104.58	150.19	99.09	158.25	227	197.09	258.34	190.28	267.01
28	19.05	39.76	16.80	43.91	128	105.90⁵	150.36	100.61	159.53	228	197.78	258.34	190.28	268.31
29	19.05	40.94	16.80	45.26	129	107.32	151.63	102.16⁵	159.67	229	199.04	258.45	190.61⁵	269.68
30	20.33⁵	41.75	18.36	46.50	130	109.11	152.96	102.16⁵	161.01	230	200.35	259.67	191.94	271.22
31	21.36	43.45	18.36	47.62	131	109.61	154.39	102.42	162.46	231	201.73	260.92	193.36	272.56
32	21.36	44.26	19.46	49.13	132	109.61	156.32	103.84	164.31	232	203.35⁵	262.20	195.19	272.56
33	22.94⁵	45.28	20.28⁵	49.96	133	110.11	156.32	105.66	164.31	233	204.36	263.54	195.59	273.53
34	23.76	34.665	20.68	51.78	134	111.44	156.87	106.12	165.33	234	204.36	265.00	195.59	274.83
35	23.76	47.69	22.04	52.28	135	112.87	158.15	106.12	166.71	235	204.36	266.71	196.13	276.20⁵
36	25.40	48.74	22.04	54.03	136	114.84	159.48	107.10	168.29	236	205.31⁵	266.71	197.46	277.77
37	26.31	50.42	23.76⁵	54.74	137	114.84	160.92⁵	108.61⁵	169.49	237	206.58	266.97	198.88	279.01⁵
38	26.31	51.29	23.76⁵	56.14	138	114.84	162.79	110.16	169.64	238	207.90	268.19	200.84	279.01⁵
39	27.73⁵	52.15	24.92⁵	57.61⁵	139	115.60⁵	162.79	110.16	170.98	239	209.30	269.44	200.94	280.02
40	28.97	53.72	25.83	58.35	140	116.93	163.35	110.37	172.41	240	211.03	270.73	200.94	281.22
41	28.97	54.99	25.99	60.39	141	118.35	164.63	111.78	174.36	241	211.69	272.08	201.62	282.70
42	30.02	55.51	27.72	60.59	142	120.36	165.96	113.45	174.36	242	211.69	273.57	202.94	284.25
43	31.67⁵	56.99	27.72	62.13	143	120.36	167.39	114.33	175.25	243	211.69	275.15	204.36	285.53
44	31.67⁵	58.72	28.85	63.63⁵	144	120.36	169.33	114.33	176.61	244	212.82	275.15	206.19	285.53
45	32.28	58.84	29.90	64.26	145	121.06	169.33	114.99	178.11	245	214.09	275.46	206.60	286.50
46	34.05	60.24	29.90	65.96	146	123.77	169.80	116.44	179.67	246	215.40	276.69	206.60	287.79
47	34.66⁵	61.90	31.84	66.81⁵	147	123.77	171.07	118.33	179.67	247	216.81	277.94	207.08	289.16
48	34.66⁵	62.81	31.84	67.92	148	125.46	172.38⁵	118.33	180.84	248	216.81	279.22	208.40	290.68
49	36 03	63.49	32.55	69.83	149	126.26	173.79	118.33	182.22	249	219.16	280.57	209.81	292.10
50	37.67	64.95	35.48⁵	70.05	150	126.26	175.48⁵	119.59	183.81	250	219.16	282.05	211.50	292.10
51	37.67	66.76	34.18	71.56	151	126.48	176.23	121.09	184.97⁵	251	219.16	283.67	212.29	292.95
52	38.16⁵	66.76	35.20	73.20	152	127.78	176.23	122.69	185.08	252	220.29	283.67	212.29	294.24
53	39.76	68.10	36.54	73.62	153	129.14	177.48	122.69	186.40	253	221.56	283.93	212.53	295.59
54	40.94	69.62	36.54	75.16	154	130.68	178.77	122.78	187.81	254	222.86⁵	285.15	213.84	297.07
55	40.94	71.09	37.82	76.61	155	132.03	180.14	124.16	189.50	255	224.26	286.40	215.22	298.71
56	41.75	71.28	38.94	77.15	156	132.03	181.67	125.70	190.28	256	225.90⁵	287.68	216.80	298.71
57	43.45	72.66	38.94	78.71	157	132.03	183.05	127.07	190.61⁵	257	226.81	289.01	217.98	299.39
58	44.26	74.22	40.37	80.06	158	133.14⁵	183.05	127.07	191.94	258	226.81	290.46	217.98	300.67
59	44.26	75.49	41.39	80.65	159	134.48	183.86	127.31	193.36	259	226.81	292.26	217.98	302.00
60	45.28	75.78⁵	41.39	82.21	160	135.92	185.13	128.70	195.19	260	227.73	292.26	219.25	303.43
61	47.02⁵	77.16	42.85	83.56	161	137.79	186.46	130.27⁵	195.59	261	228.99	292.37	220.61	305.35
62	47.69	78.73	43.91	84.12	162	137.79	187.89	131.50	196.13	262	230.28	293.59	222.10⁵	305.35
63	47.69	79.98	43.91	85.65	163	137.79	189.83	131.50	197.46	263	231.65	294.82⁵	223.67⁵	305.81
64	48.74	80.25	45.26	87.12	164	138.49	189 83	131.82	198.88	264	233.19	296.09	223.67⁵	307.07
65	50.42	81.61	46.50	87.55	165	139.79	190.21	133.21	200.84	265	234.53	297.41	223.67⁵	308.38
66	51.29	84.04	46.50	89.05	166	141.16	191.46	134.79	200.94	266	234.53	298.81	224.65	309.77⁵
67	51.29	84.57	47.62	90.72	167	142.70	192.76	135.99	201.62	267	234.53	300.56	225.98	311.41
68	52.15	84.67	49.13	90.96	168	144.01	194.11⁵	135.99	202.94	268	235.14⁵	301.16	227.41	312.38

4.5 The Frequency of Events

Table 80 (*continued*)

x	95		99		x	95		99		x	95		99	
69	53.72	86.01	49.13	92.42	169	144.01	195.63	136.30	204.26	269	236.39	301.16	229.37	312.38
70	54.99	87.48	49.96	94.34⁵	170	144.01	197.09	137.68	206.19	270	237.67	302.00	229.37	313.46
71	54.99	89.23	51.78	94.35	171	145.08	197.09	139.24	206.60	271	239.00	303.22	229.37	314.75⁵
72	55.51	89.23	51.78	95.76	172	146.39	197.78	140.54	207.08	272	240.45	304.48	230.03	316.11
73	56.99	90.37	52.28	97.42	173	147.80	199.04	140.54	208.40	273	242.27	305.77	231.33	317.60
74	58.72	91.78	54.03	98.36	174	149.53	200.35	140.76	209.81	274	242.27	307.13	232.71	319.19
75	58.72	93.48	54.74	99.09	175	150.19	201.73	142.12	211.50	275	242.27	308.64⁵	234.28	319.19
76	58.84	94.23	54.74	100.61	176	150.19	203.35⁵	143.64	212.29	276	242.53	310.07	235.50	319.84
77	60.24	94.70⁵	56.14	102.16⁵	177	150.36	204.36	145.13	212.53	277	243.76	310.07	235.50	321.11
78	61.90	96.06	57.61⁵	102.42	178	151.63	204.36	145.13	213.84	278	245.02	310.38	235.50	322.43
79	62.81	97.54⁵	57.61⁵	103.84	179	152.96	205.31⁵	145.19	215.22	279	246.32⁵	311.60	236.68	323.84
80	61.81	99.17	58.35	105.66	180	154.39	206.58	146.54	216.80	280	247.70	312.83⁵	238.01	325.58
81	63.49	99.17	60.39	106.12	181	156.32	207.90	148.01	217.98	281	249.28	314.10	239.46	326.21
82	64.95	100.32	60.39	107.10	182	156.32	209.30	149.70	217.90	282	250.40	315.42	241.02	326.21
83	66.76	101.71	60.59	108.61⁵	183	156.32	211.03	149.76	219.25	283	250.43	316.83	241.32	327.46
84	66.76	103.31⁵	62.13	110.16	184	156.87	211.69	149.76	220.61	284	250.43	318.63	241.32	328.75
85	66.76	104.40	63.63⁵	110.37	185	158.15	211.69	150.93	222.10⁵	285	251.11	319.09	242.01	330.10
86	68.10	104.58	63.63⁵	111.78	186	159.48	212.82	152.35⁵	223.67⁵	286	252.35	319.09	243.31⁵	331.59
87	69.62	105.90⁵	64.26	113.45	187	160.92⁵	214.09	154.18	223.67⁵	287	253.63	319.95	244.69	333.20
88	71.09	107.32	65.96	114.33	188	162.79	215.40	154.60	224.65	288	254.95	321.17	246.24	333.20
89	71.09	109.11	66.81⁵	114.99	189	162.79	216.81	154.60	225.98	289	256.37	322.42	247.54⁵	333.80
90	71.28	109.61	66.81⁵	116.44	190	162.79	218.56	155.31	227.41	290	258.34	323.70	247.54⁵	335.06⁵
91	72.66	110.11	67.92	118.33	191	163.35	219.16	156.69	229.37	291	258.34	325.04	247.54⁵	336.37
92	74.22	111.44	69.83	118.33	192	164.63	219.16	158.25	229.37	292	258.34	326.50	248.62	337.76
93	75.49	112.87	69.83	119.59	193	165.96	220.29	159.53	230.03	293	258.45	328.21	249.94	339.38
94	75.49	114.84	70.05	121.09	194	167.39	221.56	159.53	231.33	294	259.67	328.21	251.35	340.41
95	75.78⁵	114.84	71.56	122.69	195	169.33	222.86⁵	159.67	232.71	295	260.92	328.28⁵	253.14	340.41
96	77.16	115.60⁵	73.20	122.78	196	169.33	224.26	161.01	234.28	296	262.20	329.49	253.65	341.38
97	78.73	116.93	73.20	124.16	197	169.33	225.90⁵	162.46	235.50	297	263.54	330.72	253.65	342.65
98	79.98	118.35	73.62	125.70	198	169.80	226.81	164.31	235.50	298	265.00	331.97	253.92	343.98
99	79.98	120.36	75.16	127.07	199	171.07	226.81	164.31	236.68	299	266.71	333.26	255.20	345.41
100	80.25	120.36	76.61	127.31	200	172.38⁵	227.73	164.31	238.01	300	266.71	334.62	256.54	347.37⁵

The special case x = 0

For $x = 0$ the one sided lower confidence limit is $\lambda_l = 0$, while the upper (one sided) confidence limit λ_u can be found in the little table computed by (1.179) in Section 1.6.4 (e.g. for $S = 95\%$, $\lambda_u = 2.996 \simeq 3.00$) or computed by the formula $\lambda_u = \frac{1}{2}\chi^2_{2;\alpha}$ [$\chi^2_{2;0.05} = 5.99$; $\lambda_u = 0.5 (5.99) \simeq 3.00$].

4.5.5 Comparison of two frequencies; testing whether they stand in a certain ratio

The question sometimes asked, whether two observed frequencies (a and b, where $a \leq b$ [for a comparison of the two see e.g., (2.17)]) form a certain ratio $\beta/\alpha = \xi$ (Greek xi), is settled approximately by the statistic $\hat{\chi}^2$:

$$\hat{\chi}^2 = \frac{\{|\xi a - b| - (\xi + 1)/2\}^2}{\xi \cdot (a+b)}; \quad (4.33a)$$

for large values of a and b, without continuity correction, (p.141)

$$\hat{\chi}^2 = \frac{(\xi a - b)^2}{\xi(a+b)}. \quad (4.33)$$

$\hat{\chi}^2$ has a χ^2 distribution with one degree of freedom. If the computed $\hat{\chi}^2$ is less than or equal to $\chi^2 = 3.841$, then the null hypothesis (the observed frequencies form the ratio ξ) cannot be rejected at the 5% level.

EXAMPLE. Do the two frequencies $a = 6$ and $b = 25$ form the ratio $\xi = \beta/\alpha = 5/1$ ($\alpha = 0.05$)?

$$\hat{\chi}^2 = \frac{\{|5 \cdot 6 - 25| - (5 + 1)/2\}^2}{5(6 + 25)} = \frac{4}{155} < 3.841.$$

The departure ($25/6 = 4.17$ as against 5.00) is of a random nature ($P < 0.05$): The ratio of the observed frequencies is compatible with the theoretical ratio $5:1$.

4.6 THE EVALUATION OF FOURFOLD TABLES

4.6.1 The comparison of two percentages—the analysis of fourfold tables

The comparison of two relative frequencies determined from frequencies is important particularly in medicine. A new medicine or a new surgical procedure is developed: 15 out of 100 patients died previously, but only 4 out of 81 died under the new treatment. Does the new treatment promise greater success, or are we dealing with a spurious result? The classification of n objects according to two pairs of characteristics generally leads to four classes—the observed frequencies a, b, c, d—and thus to a so-called fourfold table (Table 81). Borderline cases, half of each being assigned to the two possible classes, can lead to half-integer values.

Table 81 Fourfold table for the comparison of two samples or for testing the statistical or stochastical independence between two attributes. Table 89 is a simplified version; Table 82 gives an example.

Characteristic (pair) II / Characteristic (pair) I	Events (+)	Complementary events (−)	Total
First sample	a	b	$a + b = n_1$
Second sample	c	d	$c + d = n_2$
Total	a + c	b + d	$n_1 + n_2 = n$

The two samples of data are then studied to determine whether they can be viewed as random samples from a population represented by the marginal sums, i.e., whether the 4 occupancy numbers (e.g., from Table 82) are distributed proportionally to the marginal sums, and whether the deviations

4.6 The Evaluation of Fourfold Tables

of the ratios a/n_1 and c/n_2 from the ratio $(a + c)/n$ [null hypothesis of equality or homogeneity: roughly $a/n_1 = c/n_2 = (a + c)/n$] can be regarded as random deviations.

Table 82 Fourfold table

Treatment	Patients		Total
	Died	Recovered	
Usual therapy	16	85	100
New therapy	4	77	81
Total	19	162	181

The example above leads to the fourfold scheme (Table 82) and the question: Is the low relative frequency of deaths under the new treatment due to chance?

The null hypothesis reads: **The percentage of cured patients is independent of the therapy employed**, or: Both samples, the group of conventionally treated patients and the group of patients treated with the new therapy, originate in a common population with regard to the therapy effect, i.e., **the therapy effect is the same with both treatments** (cf., also Section 6.2.1).

The two treatment groups are in fact samples from two binomial distributions. **Thus, the probabilities of binomial distributions are compared**, i.e.,

Null hypothesis:	Both samples originate in a common population with success probability π
Alternate hypothesis:	The two samples originate in two different populations with success probabilities π_1 and π_2

The null hypothesis on equality or homogeneity of the two parameters (π_1, π_2) or independence between two attributes (cf., also Section 6.2.1) is rejected or not rejected on the basis of the **APPROXIMATE CHI-SQUARE TEST** (we discussed exact χ^2-tests in Sections 3.3 and 3.4).

Turning from frequencies to relative frequencies and to probabilities we have:

$$\begin{array}{|cc|} a & b \\ c & d \\ \hline & n \end{array} \rightarrow \begin{array}{|cc|} a/n & b/n \\ c/n & d/n \\ \hline & 1 \end{array} \quad n \rightarrow N \quad \begin{array}{|cc|} P_a & P_b \\ P_c & P_d \\ \hline & 1 \end{array}$$

For the null hypothesis of independence we make use of the cross product of the probabilities $[P_a P_d]/[P_b P_c]$:

Fourfold table	Null hypothesis	
a b c d ――― n	Independence of two attributes:	$[P_a P_d]/[P_b P_c] = 1$ with the proportions P_a, P_b, P_c, P_d of the total population
a b a + b = n_1 c d c + d = n_2 ―――	Equality or homogeneity of two expectations:	$E\left(\dfrac{a}{a+b}\right) = E\left(\dfrac{c}{c+d}\right)$ with the specified expected proportions of both populations

Returning to Table 82: the following questions must be answered: Are the field frequencies distributed in proportion to the marginal sums? To resolve this question, we determine the frequencies (called expected frequencies—E for short) expected under the assumption that they are proportional. We multiply the row sum by the column sum of the field a $[(100)(19) = 1{,}900]$ and divide the product by the size n of the combined sample $(1{,}900/181 = 10.497$; $E_a = 10.50)$. We proceed analogously with the remaining fields, obtaining $E_b = 89.50$, $E = 8.50$, $E_d = 72.50$. To assess whether the observed values a, b, c, d agree with the expected values E_a, E_b, E_c, E_d in the sense of the null hypothesis, we form the test statistic $\hat{\chi}^2$:

$$\hat{\chi}^2 = \frac{(a - E_a)^2}{E_a} + \frac{(b - E_b)^2}{E_b} + \frac{(c - E_c)^2}{E_c} + \frac{(d - E_d)^2}{E_d},$$

which, after several transformations, becomes

$$\hat{\chi}^2 = \Delta^2\left(\frac{1}{E_a} + \frac{1}{E_b} + \frac{1}{E_c} + \frac{1}{E_d}\right), \tag{4.34}$$

where $|\Delta| = |a - E_a| = |b - E_b| = |c - E_c| = |d - E_d|$ or

$$\hat{\chi}^2 = \frac{n(ad - bc)^2}{(a+b)(c+d)(a+c)(b+d)} \tag{4.35}$$

where $n = a + b + c + d$ [Note the remarks under (4.35ab) and those following the example]. The fourfold χ^2 has only **one degree of freedom**, since with marginal sums given only one of the four frequencies can be freely chosen. **For n small, n in (4.35) has to be replaced by $(n - 1)$**; the resulting formula is generally applicable, provided $n_1 \geq 6$, $n_2 \geq 6$; it is better if for small n we also have $n_1 \simeq n_2$ or $n_2 \gtrsim \sqrt{n_1}$ for $n_1 > n_2$ (Van der Waerden 1965, Berchtold 1969, Sachs 1974 [see also Rhoades and Overall 1982 and Upton 1982 [8:6]]). With n still smaller, we have to use Fisher's exact test or Gart's F-test.

4.6 The Evaluation of Fourfold Tables

Instead of the χ^2 test we may use the version of the G-test presented in Section 4.6.7, in which the effective level of significance corresponds better to what is given, even for small n. However, the G-test and the χ^2 test are both approximations.

For $n_1 = n_2$ (4.35) becomes (4.35ab)

$$\hat{\chi}^2 = \frac{n(a-c)^2}{(a+c)(b+d)}, \quad \text{or for small } n, \quad \hat{\chi}^2_* = \frac{(n-1)(a-c)^2}{(a+c)(b+d)}.$$

(4.35ab)

The null hypothesis on independence or homogeneity is rejected as soon as the $\hat{\chi}^2$ computed according to (4.34), (4.35), or (4.35ab) [if $n-1$ used in place of n in (4.35), one calls it $\hat{\chi}^2_*$] is greater than the critical value in the following table:

Level of significance α	0.05	0.01	0.001
Two sided test ($H_0: \pi_1 = \pi_2$, $H_A: \pi_1 \neq \pi_2$)	3.841	6.635	10.828
One sided test ($H_0: \pi_1 = \pi_2$, $H_A: \pi_1 > \pi_2$ or $\pi_2 > \pi_1$)	2.706	5.412	9.550

(p.140)

Testing is generally two sided (cf., the remarks in Section 1.4.7, at the beginning and in the box at the end). Table 84 shows that with small sample sizes and $\alpha = 0.05$ in nearly all cases the power of the test is extremely low and

Table 83 χ^2 **table for one degree of freedom** (taken from Kendall and Stuart (1973) [1], Vol. II, pp. 651, 652): two sided probabilities

χ^2	P	χ^2	P	χ^2	P	χ^2	P	χ^2	P
0	1.00000	2.1	0.14730	4.0	0.04550	6.0	0.01431	8.0	0.00468
0.1	0.75183	2.2	0.13801	4.1	0.04288	6.1	0.01352	8.1	0.00443
0.2	0.65472	2.3	0.12937	4.2	0.04042	6.2	0.01278	8.2	0.00419
0.3	0.58388	2.4	0.12134	4.3	0.03811	6.2	0.01207	8.3	0.00396
0.4	0.52709	2.5	0.11385	4.4	0.03594	6.4	0.01141	8.4	0.00375
0.5	0.47950	2.6	0.10686	4.5	0.03389	6.5	0.01079	8.5	0.00355
0.6	0.43858	2.7	0.10035	4.6	0.03197	6.6	0.01020	8.6	0.00336
0.7	0.40278	2.8	0.09426	4.7	0.03016	6.7	0.00964	8.7	0.00318
0.8	0.37109	2.9	0.08858	4.8	0.02846	6.8	0.00912	8.8	0.00301
0.9	0.34278	3.0	0.08326	4.9	0.02686	6.9	0.00862	8.9	0.00285
1.0	0.31731	3.1	0.07829	5.0	0.02535	7.0	0.00815	9.0	0.00270
1.1	0.29427	3.2	0.07364	5.1	0.02393	7.1	0.00771	9.1	0.00256
1.2	0.27332	3.3	0.06928	5.2	0.02259	7.2	0.00729	9.2	0.00242
1.3	0.25421	3.3	0.06928	5.3	0.02133	7.3	0.00690	9.3	0.00229
1.4	0.23672	3.4	0.06520	5.4	0.02014	7.4	0.00652	9.4	0.00217
1.5	0.22067	3.5	0.06137	5.5	0.01902	7.5	0.00617	9.5	0.00205
1.6	0.20590	3.6	0.05778	5.6	0.01796	7.6	0.00584	9.6	0.00195
1.7	0.19229	3.7	0.05441	5.7	0.01697	7.7	0.00552	9.7	0.00184
1.8	0.17971	3.8	0.05125	5.8	0.01603	7.8	0.00522	9.8	0.00174
1.9	0.16808	3.9	0.04829	5.9	0.01514	7.9	0.00494	9.9	0.00165
2.0	0.15730	4.0	0.04550	6.0	0.01431	8.0	0.00468	10.0	0.00157

Table 84 Sample size per group required to obtain a specific power when $\alpha = 0.05$ (one-tailed). Some values from J. K. Haseman: Exact sample sizes for use with the Fisher-Irwin test for 2 × 2 tables, Biometrics **34** (1978), 106–109, part of Table 1, p. 107. Example: see remark 6.

π_2 \ π_1	0.9	0.7	0.5	0.3	0.1
0.8	232				
	173	upper figure: Power = 0.9			
0.6	39	408	lower figure: Power = 0.8		
	30	302		$\alpha = 0.05$	
0.4	17	53	445	(one sided test)	
	13	41	321		
0.2	10	18	47	338	
	8	15	36	249	
0.05	6	10	18	42	503
	5	9	14	34	371

thus the test is of no use. Table 83 gives exact probabilities for $\chi^2 = 0.0$ through $\chi^2 = 10.0$ in increments of 0.1.

EXAMPLE. We test Table 82 at the 5% significance level (one sided test; alternative: the new therapy is superior):

$$\hat{\chi}^2 = \frac{181(15 \cdot 77 - 4 \cdot 85)^2}{100 \cdot 81 \cdot 19 \cdot 162} = 4.822.$$

Since $\hat{\chi}^2 = 4.822 > 2.706 = \chi^2_{0.05}$, the hypothesis of independence or homogeneity is rejected at the 5% level on the basis of the available data. There is a dependence between the new treatment and the reduction in the mortality.

For a generalization of the fourfold χ^2 test see Section 6.2.1.

Remarks

1. In **preliminary trials** where significance levels α are not specified beforehand, the $\hat{\chi}^2$-value found is compared with that given in Table 83 (two sided question).

2. We note that the numerical value of the quotient (4.35) does not change if the four inner field frequencies (a, b, c, d) and the four marginal frequencies $(a + b, c + d, a + c, b + d)$ are all divided by a constant k (the sample size n is not to be divided by k), so that **the amount of computation can be significantly reduced**. In an approximate calculation of $\hat{\chi}^2$ one can, moreover, round off the frequencies that are divided by k. For large n the computation in (4.34 or 4.35) is however too tedious, and the formula (4.28) or, even more (4.36a) is preferred.

3. Since the fourfold χ^2 test represents an approximation, the corrected formulas (4.34a), (4.35c) (the quantities $\frac{1}{2}$ and $n/2$ are called continuity corrections) were

4.6 The Evaluation of Fourfold Tables

recommended by Yates (1934, p. 30) if at least one of the four expected frequencies is smaller than 500:

$$\hat{\chi}^2 = \left(|\Delta| - \frac{1}{2}\right)^2 \left(\frac{1}{E_a} + \frac{1}{E_b} + \frac{1}{E_c} + \frac{1}{E_d}\right) \quad (4.34a)$$

$$\hat{\chi}^2 = \frac{n\left(|ad - bc| - \frac{n}{2}\right)^2}{(a+b)(c+d)(a+c)(b+d)} \quad (4.35c)$$

Grizzle (1968) has shown that one can do without (4.34a), (4.35c). They are appropriate only if the probabilities of the exact test of Fisher (cf., Sections 4.6.6.7), a conservative procedure, must be approximated (cf. Adler 1951, Cochran 1952, Vessereau 1958, Plackett 1964, 1974 [8:6]). Then, however, the F-test due to Gart [formulas (4.37) and (4.38) given in Section 4.6.2] is more convenient.

4. The **standardization of fourfold tables** (overall sum equals 1 and all 4 marginal sums equal 0.5) is obtained by way of $a_{\text{standardized}} = (v - \sqrt{v})/[2(v - 1)]$ with $v = ad/bc$. For Table 82 we find with $v = 3.397$ and $a_{\text{standardized}} = 0.324$ the values $d_{\text{st.}} = a_{\text{st.}} = 0.324$; $b_{\text{st.}} = c_{\text{st.}} = 0.176$.

To standardize square tables (all marginal sums equal 100) each row is multiplied by the associated value (100/row sum), the columns are dealt with accordingly, and the procedure is then iterated until, e.g., all marginal sums are equal to 100.00.

5. **Additional remarks** are found in Sections 4.6.2, 4.6.7, 5.4.4, and 6.2.1; in [8:6] we cite a work by Mantel and Haenszel (1959) which is very informative in particular for medical students [see also Fleiss 1981 and Schlesselman 1982, both in [8:2a]].

6. **Sample sizes**: According to Table 84 at least $n_1 = n_2 = 53$ observations are needed for the test $H_0: \pi_1 = \pi_2$ vs. $H_A: \pi_1 > \pi_2$ with $\pi_1 = 0.7$, $\pi_2 = 0.4$, $\alpha = 0.05$, and a power of 0.9 or 90%, i.e., if there are for the test two random samples of such sizes from populations with $\pi_1 = 0.7$ and $\pi_2 = 0.4$ at our disposal, then the chance of detecting a difference $\delta = \pi_1 - \pi_2 = 0.7 - 0.4 = 0.3$ is 90% in a one sided test with a significance level of 5%. More exact values for $\alpha = 0.05$ and $\alpha = 0.01$ are given by Haseman (1978) and by Casagrande et al. (1978).

The Woolf G-test

Our modified version of the Woolf G-test (1957) is superior to the fourfold χ^2 test (with "n" or "$n - 1$" in the numerator of (4.35)). $\hat{G}$ is defined by (4.36).

$$\hat{G} = 2 \sum \text{observed} \, (\ln \text{observed} - \ln \text{expected}). \quad (4.36)$$

This ought not to be studied in greater detail. It is essential that the values $2n \ln n$ needed in this test, called g-values for short, were made available in tabulated form by Woolf. Fourfold tables for $n_1 \geq 4$, $n_2 \geq 4$ can then be tested for independence or homogeneity as follows:

1. For the frequencies a, b, c and d and for the frequencies a', b', c', d' corrected according to Yates [cf., Table 82a, that which is enclosed in ( )] the eight g-values in Tables 85 and 86 are written down and their sum, divided by 2, is called S_1 (cf., item 6).

pp. 353-357

2. The tabulated value corresponding to the overall sample size n can be found in Table 85; we denote it by S_2.
3. The tabulated values corresponding to the four marginal sum frequencies are likewise obtained from Table 85, and their sum is S_3.
4. The test statistic $\hat{G}$ is then defined by

$$\hat{G} = S_1 + S_2 - S_3. \qquad (4.36a)$$

p.364

5. The test statistic $\hat{G}$ is distributed like χ^2, with one degree of freedom, for not too weakly occupied fourfold tables.
6. If all 4 expected frequencies E are larger than 30, then computations are carried out using the observed frequencies a, b, c, d; the corresponding g-values are taken from Table 85, and their sum is S_1.

Table 82a Fourfold table. The values adjusted according to Yates are in brackets. Values which are smaller than the corresponding expected frequencies (cf. Table 82) are increased by $\frac{1}{2}$; values which are larger are decreased by $\frac{1}{2}$

Treatment\Patients	Died	Recovered	Total
Usual therapy	**15** (14 1/2)	**85** (85 1/2)	100
New therapy	**4** (4 1/2)	**77** (76 1/2)	81
Total	19	162	181

EXAMPLE. We take our last example (Table 82), as shown in Table 82a. We have

from Table 85:
$$\begin{cases} 15 \to & 81.2415 \\ 85 \to & 755.2507 \\ 4 \to & 11.0904 \\ 77 \to & 668.9460 \end{cases}$$

p.358

from Table 86:
$$\begin{cases} 14\tfrac{1}{2} \to & 77.5503 \\ 85\tfrac{1}{2} \to & 760.6963 \\ 4\tfrac{1}{2} \to & 13.5367 \\ 76\tfrac{1}{2} \to & 663.6055 \end{cases}$$

$$2S_1 = 3{,}031.9174$$
$$S_1 = 1{,}515.9587$$

4.6 The Evaluation of Fourfold Tables

Table 85 $2n \ln n$ for $n = 0$ to $n = 399$ (from Woolf, B.: The log likelihood ratio test (the G-test). Methods and tables for tests of heterogeneity in contingency tables, Ann. Human Genetics **21**, 397–409 (1957), Table 1, p. 400–404)

	0	1	2	3	4	5	6	7	8	9
0	0·0000	0·0000	2·7726	6·5917	11·0904	16·0944	21·5011	27·2427	33·2711	39·5500
10	46·0517	52·7537	59·6378	66·6887	73·8936	81·2415	88·7228	96·3293	104·0534	111·8887
20	119·8293	127·8699	136·0059	144·2327	152·5466	160·9438	169·4210	177·9752	186·6035	195·3032
30	204·0718	212·9072	221·8071	230·7695	239·7925	248·8744	258·0134	267·2079	276·4565	285·7578
40	295·1104	304·5129	313·9642	323·4632	333·0087	342·5996	352·2350	361·9139	371·6353	381·3984
50	391·2023	401·0462	410·9293	420·8509	430·8103	440·8067	450·8394	460·9078	471·0114	481·1494
60	491·3213	501·5266	511·7647	522·0350	532·3370	542·6703	553·0344	563·4288	573·8530	584·3067
70	594·7893	605·3005	615·8399	626·4071	637·0016	647·6232	658·2715	668·9460	679·6466	690·3728
80	701·1243	711·9008	722·7020	733·5275	744·3772	755·2507	766·1477	777·0680	788·0113	798·9773
90	809·9657	820·9764	832·0091	843·0635	854·1394	865·2366	876·3549	887·4939	898·6536	909·8337
100	921·0340	932·2543	943·4945	954·7542	966·0333	977·3317	988·6491	999·9854	1011·3403	1022·7138
110	1034·1057	1045·5157	1056·9437	1068·3896	1079·8532	1091·3344	1102·8329	1114·3487	1125·8816	1137·4314
120	1148·9980	1160·5813	1172·1811	1183·7974	1195·4298	1207·0784	1218·7430	1230·4235	1242·1197	1253·8316
130	1265·5590	1277·3017	1289·0597	1300·8329	1312·6211	1324·4242	1336·2421	1348·0748	1359·9220	1371·7838
140	1383·6599	1395·5503	1407·4549	1419·3736	1431·3062	1443·2528	1455·2131	1467·1872	1479·1748	1491·1760
150	1503·1906	1515·2185	1527·2597	1539·3140	1551·3814	1563·4618	1575·5551	1587·6612	1599·7800	1611·9115
160	1624·0556	1636·2122	1648·3812	1660·5626	1672·7562	1684·9620	1697·1799	1709·4099	1721·6519	1733·9058
170	1746·1715	1758·4489	1770·7381	1783·0389	1795·3512	1807·6751	1820·0104	1832·3570	1844·7149	1857·0841
180	1869·4645	1881·8559	1894·2584	1906·6719	1919·0964	1931·5317	1943·9778	1956·4346	1968·9022	1981·3804
190	1993·8691	2006·3684	2018·8782	2031·3984	2043·9290	2056·4698	2069·0209	2081·5823	2094·1537	2106·7353
200	2119·3269	2131·9286	2144·5402	2157·1616	2169·7930	2182·4341	2195·0850	2207·7456	2220·4158	2233·0957
210	2245·7852	2258·4841	2271·1926	2283·9105	2296·6377	2309·3744	2322·1203	2334·8755	2347·6398	2360·4134
220	2373·1961	2385·9879	2398·7888	2411·5986	2424·4174	2437·2452	2450·0818	2462·9273	2475·7816	2488·6447
230	2501·5165	2514·3970	2527·2861	2540·1839	2553·0903	2566·0052	2578·9286	2591·8605	2604·8008	2617·7496
240	2630·7067	2643·6721	2656·6458	2669·6279	2682·6181	2695·6165	2708·6231	2721·6378	2734·6607	2747·6915
250	2760·7305	2773·7774	2786·8323	2799·8951	2812·9658	2826·0444	2839·1308	2852·2251	2865·3271	2878·4369
260	2891·5544	2904·6797	2917·8125	2930·9530	2944·1011	2957·2568	2970·4200	2983·5908	2996·7690	3009·9547
270	3023·1479	3036·3484	3049·5563	3062·7716	3075·9942	3089·2241	3102·4613	3115·7057	3128·9573	3142·2162
280	3155·4822	3168·7553	3182·0356	3195·3229	3208·6174	3221·9188	3235·2273	3248·5428	3261·8652	3275·1946
290	3288·5309	3301·8741	3315·2242	3328·5811	3341·9449	3355·3155	3368·6928	3382·0769	3395·4677	3408·8653
300	3422·2695	3435·6804	3449·0979	3462·5221	3475·9528	3489·3902	3502·8341	3516·2845	3529·7415	3543·2049
310	3556·6748	3570·1512	3583·6340	3597·1233	3610·6188	3624·1208	3637·6291	3651·1437	3664·6647	3678·1919
320	3691·7254	3705·2652	3718·8112	3732·3634	3745·9218	3759·4864	3773·0571	3786·6340	3800·2169	3813·8060
330	3827·4012	3841·0024	3854·6096	3868·2229	3881·8422	3895·4675	3909·0987	3922·7359	3936·3790	3950·0281
340	3963·6830	3977·3438	3991·0105	4004·6831	4018·3615	4032·0456	4045·7356	4059·4314	4073·1329	4086·8402
350	4100·5532	4114·2719	4127·9963	4141·7264	4155·4622	4169·2036	4182·9507	4196·7033	4210·4616	4224·2255
360	4237·9949	4251·7699	4265·5504	4279·3365	4293·1280	4306·9251	4320·7276	4334·5356	4348·3490	4362·1679
370	4375·9922	4389·8219	4403·6570	4417·4975	4431·3433	4445·1945	4459·0510	4472·9129	4486·7800	4500·6524
380	4514·5302	4528·4131	4542·3013	4556·1948	4570·0935	4583·9974	4597·9064	4611·8207	4625·7401	4639·6647
390	4653·5945	4667·5293	4681·4693	4695·4144	4709·3645	4723·3198	4737·2801	4751·2454	4765·2158	4779·1912

Table 85 (continuation 1) 2n ln n for n = 400 to 799

n	0	1	2	3	4	5	6	7	8	9
400	4793.1716	4807.1570	4821.1475	4835.1429	4849.1432	4863.1485	4877.1588	4891.1739	4905.1940	4919.2190
410	4933.2489	4947.2836	4961.3232	4975.3677	4989.4170	5003.4712	5017.5301	5031.5939	5045.6625	5059.7358
420	5073.8140	5087.8968	5101.9845	5116.0769	5130.1740	5144.2758	5158.3823	5172.4935	5186.6095	5200.7300
430	5214.8553	5228.9852	5243.1197	5257.2589	5271.4027	5285.5510	5299.7040	5313.8616	5328.9238	5342.1905
440	5356.3618	5370.5376	5384.7179	5398.9028	5413.0922	5427.2861	5441.4845	5455.6874	5469.8947	5484.1066
450	5498.3228	5512.5435	5526.7687	5540.9983	5555.2323	5569.4707	5583.7134	5597.9606	5612.2122	5626.4681
460	5640.7284	5654.9930	5669.2620	5683.5353	5697.8129	5712.0948	5726.3810	5740.6715	5754.9663	5769.2654
470	5783.5687	5797.8763	5812.1882	5826.5042	5840.8245	5855.1491	5869.4778	5883.8107	5898.1479	5912.4892
480	5926.8347	5941.1843	5955.5381	5969.8961	5984.2582	5998.6244	6012.9948	6027.3693	6041.7478	6056.1305
490	6070.5173	6084.9081	6099.3031	6113.7020	6128.1051	6142.5122	6156.9233	6171.3385	6185.7577	6200.1809
500	6214.6081	6229.0393	6243.4745	6257.9137	6272.3569	6286.8040	6301.2551	6315.7102	6330.1692	6344.6321
510	6359.0989	6373.5697	6388.0444	6402.5230	6417.0055	6431.4919	6445.9822	6460.4763	6474.9744	6489.4762
520	6503.9820	6518.4915	6533.0050	6547.5222	6562.0433	6576.5682	6591.0969	6605.6294	6620.1657	6634.7058
530	6649.2496	6663.7973	6678.3487	6692.9038	6707.4628	6722.0254	6736.5918	6751.1620	6765.7358	6780.3134
540	6794.8947	6809.4797	6824.0683	6838.6607	6853.2568	6867.8565	6882.4599	6897.0670	6911.6777	6926.2921
550	6940.9101	6955.5318	6970.1570	6984.7860	6999.4185	7014.0546	7028.6943	7043.3377	7057.9846	7072.6351
560	7087.2892	7101.9469	7116.6081	7131.2729	7145.9412	7160.6131	7175.2885	7189.9675	7204.6499	7219.3359
570	7234.0255	7248.7185	7263.4150	7278.1150	7292.8185	7307.5255	7322.2360	7336.9500	7351.6674	7366.3883
580	7381.1126	7395.8404	7410.5716	7425.3063	7440.0443	7454.7859	7469.5308	7484.2791	7499.0309	7513.7860
590	7528.5446	7543.3065	7558.0719	7572.8406	7587.6126	7602.3881	7617.1669	7631.9490	7646.7345	7661.5234
600	7676.3156	7691.1111	7705.9100	7720.7121	7735.5176	7750.3264	7765.1385	7779.9540	7794.7727	7809.5947
610	7824.4199	7839.2485	7854.0803	7868.9154	7883.7538	7898.5954	7913.4403	7928.2884	7943.1397	7957.9943
620	7972.8522	7987.7132	8002.5775	8017.4450	8032.3157	8047.1896	8062.0667	8076.9470	8091.8304	8106.7171
630	8121.6070	8136.5000	8151.3962	8166.2956	8181.1981	8196.1037	8211.0126	8225.9245	8240.8396	8255.7579
640	8270.6793	8285.6038	8300.5314	8315.4621	8330.3960	8345.3329	8360.2730	8375.2161	8390.1623	8404.1117
650	8420.0641	8435.0196	8449.9781	8464.9397	8479.9044	8494.8722	8509.8430	8524.8168	8539.7937	8554.7736
660	8569.7566	8584.7426	8599.7316	8614.7236	8629.7187	8644.7168	8659.7178	8674.7219	8689.7290	8704.7391
670	8719.7521	8734.7682	8749.7872	8764.8092	8779.8342	8794.8621	8809.8930	8824.9269	8839.9637	8855.0035
680	8870.0462	8885.0919	8900.1405	8915.1920	8930.2464	8945.3038	8960.3641	8975.4273	8990.4934	9005.5625
690	9020.6344	9035.7092	9050.7870	9065.8676	9080.9511	9096.0375	9111.1267	9126.2188	9141.3139	9156.4117
700	9171.5125	9186.6161	9201.7225	9216.8318	9231.9439	9247.0589	9262.1767	9277.2974	9292.4208	9307.5471
710	9322.6763	9337.8082	9352.9429	9368.0805	9383.2209	9398.3640	9413.5100	9428.6588	9443.8103	9458.9646
720	9474.1217	9489.2816	9504.4443	9519.6097	9534.7779	9549.9489	9565.1226	9580.2991	9595.4783	9610.6603
730	9625.8450	9641.0325	9656.2227	9671.4156	9686.6113	9701.8096	9717.0107	9732.2146	9747.4211	9762.6303
740	9777.8423	9793.0569	9808.2743	9823.4943	9838.7171	9853.9425	9869.1706	9884.4014	9899.6348	9914.8710
750	9930.1098	9945.3513	9960.5954	9975.8422	9991.0917	10006.3438	10021.5986	10036.8560	10052.1160	10067.3787
760	10082.6440	10097.9120	10113.1826	10128.4558	10143.7316	10159.0100	10174.2911	10189.5748	10204.8610	10220.1499
770	10235.4414	10250.7355	10266.0321	10281.3314	10296.6333	10311.9377	10327.2447	10342.5543	10357.8665	10373.1812
780	10388.4985	10403.8184	10419.1408	10434.4658	10449.7933	10465.1234	10480.4561	10495.7913	10511.1290	10526.4693
790	10541.8121	10557.1574	10572.5052	10587.8556	10603.2085	10618.5640	10633.9219	10649.2824	10664.6453	10680.0108

4.6 The Evaluation of Fourfold Tables

Table 85 (continuation 2) $2n \ln n$ for $n = 800$ to 1199

	0	1	2	3	4	5	6	7	8	9
800	10695.3788	10710.7492	10726.1222	10741.4977	10756.8756	10772.2561	10787.6390	10803.0244	10818.4123	10833.8026
810	10849.1955	10864.5908	10879.9886	10895.3888	10910.7915	10926.1966	10941.6042	10957.0143	10972.4268	10987.8417
820	11003.2591	11018.6789	11034.1012	11049.5259	11064.9530	11080.3826	11095.8146	11111.2490	11126.6858	11142.1250
830	11157.5667	11173.0107	11188.4572	11203.9060	11219.3573	11234.8110	11250.2670	11265.7255	11281.1863	11296.6496
840	11312.1152	11327.5832	11343.0535	11358.5263	11374.0014	11389.4789	11404.9588	11420.4410	11435.9256	11451.4125
850	11466.9018	11482.3934	11497.8874	11513.3838	11528.8825	11544.3835	11559.8869	11575.3926	11590.9006	11606.4110
860	11621.9237	11637.4387	11652.9561	11668.4758	11683.9977	11699.5220	11715.0487	11730.5776	11746.1088	11761.6424
870	11777.1782	11792.7163	11808.2568	11823.7995	11839.3445	11854.8918	11870.4414	11885.9933	11901.5474	11917.1039
880	11932.6626	11948.2235	11963.7868	11979.3523	11994.9201	12010.4901	12026.0624	12041.6370	12057.2138	12072.7929
890	12088.3742	12103.9578	12119.5436	12135.1316	12150.7219	12166.3145	12181.9092	12197.5062	12213.1054	12228.7069
900	12244.3106	12259.9165	12275.5246	12291.1349	12306.7475	12322.3622	12337.9792	12353.5984	12369.2198	12384.8434
910	12400.4692	12416.0972	12431.7273	12447.3597	12462.9943	12478.6310	12494.2700	12509.9111	12525.5544	12541.1999
920	12556.8476	12572.4974	12588.1494	12603.8036	12619.4599	12635.1184	12650.7791	12666.4419	12682.1069	12697.7740
930	12713.4433	12729.1148	12744.7884	12760.4641	12776.1420	12791.8220	12807.5042	12823.1885	12838.8749	12854.5635
940	12870.2542	12885.9470	12901.6419	12917.3390	12933.0382	12948.7395	12964.4429	12980.1485	12995.8561	13011.5659
950	13027.2778	13042.9917	13058.7078	13074.4260	13090.1463	13105.8687	13121.5931	13137.3197	13153.0483	13168.7791
960	13184.5119	13200.2468	13215.9838	13231.7229	13247.4640	13263.2072	13278.9525	13294.6999	13310.4493	13326.2008
970	13341.9544	13357.7100	13373.4677	13389.2274	13404.9892	13420.7531	13436.5190	13452.2869	13468.0569	13483.8290
980	13499.6030	13515.3792	13531.1573	13546.9375	13562.7198	13578.5040	13594.2903	13610.0787	13625.8690	13641.6614
990	13657.4558	13673.2522	13689.0506	13704.8511	13720.6536	13736.4580	13752.2645	13768.0730	13783.8835	13799.6960
1000	13815.5106	13831.3271	13847.1456	13862.9661	13878.7886	13894.6131	13910.4396	13926.2680	13942.0985	13957.9309
1010	13973.7653	13989.6017	14005.4401	14021.2805	14037.1228	14052.9671	14068.8134	14084.6616	14100.5118	14116.3640
1020	14132.2181	14148.0742	14163.9322	14179.7923	14195.6543	14211.5182	14227.3840	14243.2519	14259.1216	14274.9934
1030	14290.8670	14306.7426	14322.6201	14338.4996	14354.3810	14370.2644	14386.1497	14402.0369	14417.9260	14433.8171
1040	14449.7101	14465.6050	14481.5018	14497.4006	14513.3012	14529.2038	14545.1083	14561.0147	14576.9231	14592.8333
1050	14608.7454	14624.6595	14640.5754	14656.4933	14672.4130	14688.3347	14704.2582	14720.1836	14736.1110	14752.0402
1060	14767.9713	14783.9043	14799.8391	14815.7759	14831.7145	14847.6551	14863.5975	14879.5417	14895.4879	14911.4359
1070	14927.3858	14943.3376	14959.2912	14975.2467	14991.2040	15007.1633	15023.1244	15039.0873	15055.0521	15071.0187
1080	15086.9873	15102.9576	15118.9298	15134.9039	15150.8798	15166.8575	15182.8371	15198.8186	15214.8018	15230.7869
1090	15246.7739	15262.7626	15278.7533	15294.7457	15310.7400	15325.7361	15342.7340	15358.7338	15374.7354	15390.7388
1100	15406.7440	15422.7510	15438.7599	15454.7706	15470.7831	15486.7974	15502.8135	15518.8314	15534.8511	15550.8726
1110	15566.8960	15582.9211	15598.9480	15614.9767	15631.0073	15647.0396	15663.0737	15679.1096	15695.1473	15711.1868
1120	15727.2281	15743.2711	15759.3160	15775.3626	15791.4110	15807.4612	15823.5132	15839.5669	15855.6224	15871.6797
1130	15887.7388	15903.7996	15919.8622	15935.9266	15951.9927	15968.0606	15984.1303	16000.2017	16016.2749	16032.3498
1140	16048.4265	16064.5049	16080.5851	16096.6671	16112.7508	16128.8362	16144.9234	16161.0123	16177.1030	16193.1954
1150	16209.2896	16225.3855	16241.4832	16257.5825	16273.6836	16289.7865	16305.8911	16321.9974	16338.1054	16354.2152
1160	16370.3267	16386.4399	16402.5548	16418.6715	16434.7898	16450.9099	16467.0317	16483.1553	16499.2805	16515.4075
1170	16531.5361	16547.6665	16563.7986	16579.9324	16596.0679	16612.2051	16628.3439	16644.4845	16660.6268	16676.7708
1180	16692.9165	16709.0639	16725.2130	16741.3638	16757.5162	16773.6704	16789.8262	16805.9838	16822.1430	16838.3039
1190	16854.4664	16870.6307	16886.7966	16902.9642	16919.1335	16935.3045	16951.4771	16967.6515	16983.8274	17000.0051

Table 85 (continuation 3) 2n ln n for n = 1200 to 1599

	0	1	2	3	4	5	6	7	8	9
1200	17016.1844	17032.3654	17048.5480	17064.7324	17080.9183	17097.1060	17113.2953	17129.4862	17145.6788	17161.8731
1210	17178.0690	17194.2666	17210.4659	17226.6667	17242.8693	17259.0734	17275.2793	17291.4867	17307.6958	17323.9066
1220	17340.1190	17356.3330	17372.5487	17388.7660	17404.9849	17421.2055	17437.4277	17453.6516	17469.8770	17486.1041
1230	17502.3328	17518.5632	17534.7952	17551.0288	17567.2640	17583.5008	17599.7393	17615.9794	17632.2211	17648.4644
1240	17664.7093	17680.9559	17697.2040	17713.4538	17729.7051	17745.9581	17762.2127	17778.4689	17794.7267	17810.9861
1250	17827.2471	17843.5097	17859.7739	17876.0397	17892.3071	17908.5760	17924.8466	17941.1188	17957.3925	17973.6679
1260	17989.9448	18006.2234	18022.5035	18038.7852	18055.0685	18071.3533	18087.6398	18103.9278	18120.2174	18136.5086
1270	18152.8013	18169.0957	18185.3916	18201.6890	18217.9881	18234.2887	18250.5909	18266.8947	18283.2000	18299.5069
1280	18315.8153	18332.1253	18348.4369	18364.7500	18381.0647	18397.3810	18413.6988	18430.0181	18446.3391	18462.6615
1290	18478.9855	18495.3111	18511.6382	18527.9669	18544.2971	18560.6289	18576.9622	18593.2970	18609.6334	18625.9713
1300	18642.3108	18658.6518	18674.9944	18691.3384	18707.6841	18724.0312	18740.3799	18756.7301	18773.0819	18789.4351
1310	18805.7899	18822.1463	18838.5041	18854.8635	18871.2244	18887.5868	18903.9508	18920.3162	18936.6832	18953.0517
1320	18969.4217	18985.7933	19002.1663	19018.5409	19034.9169	19051.2945	19067.6736	19084.0542	19100.4363	19116.8199
1330	19133.2050	19149.5916	19165.9798	19182.3694	19198.7605	19215.1531	19231.5473	19247.9429	19264.3400	19280.7386
1340	19297.1387	19313.5403	19329.9434	19346.3480	19362.7540	19379.1616	19395.5706	19411.9812	19428.3932	19444.8067
1350	19461.2217	19477.6381	19494.0561	19510.4755	19526.8964	19543.3187	19559.7426	19576.1679	19592.5947	19609.0230
1360	19625.4527	19641.8840	19658.3166	19674.7508	19691.1864	19707.6235	19724.0621	19740.5021	19756.9435	19773.3865
1370	19789.8309	19806.2768	19822.7241	19839.1729	19855.6231	19872.0748	19888.5279	19904.9825	19921.4386	19937.8961
1380	19954.3550	19970.8154	19987.2773	20003.7406	20020.2053	20036.6715	20053.1391	20069.6082	20086.0787	20102.5507
1390	20119.0241	20135.4989	20151.9752	20168.4529	20184.9321	20201.4126	20217.8947	20234.3781	20250.8630	20267.3493
1400	20283.8370	20300.3262	20316.8168	20333.3088	20349.8023	20366.2972	20382.7935	20399.2912	20415.7903	20432.2909
1410	20448.7929	20465.2963	20481.8011	20498.3073	20514.8149	20531.3240	20547.8345	20564.3464	20580.8597	20597.3744
1420	20613.8905	20630.4080	20646.9270	20663.4473	20679.9691	20696.4922	20713.0168	20729.5427	20746.0700	20762.5988
1430	20779.1290	20795.6606	20812.1935	20828.7279	20845.2636	20861.8008	20878.3393	20894.8792	20911.4206	20927.9633
1440	20944.5074	20961.0529	20977.5997	20994.1480	21010.6977	21027.2487	21043.8011	21060.3549	21076.9101	21093.4667
1450	21110.0246	21126.5840	21143.1447	21159.7067	21175.2702	21192.8350	21209.4012	21225.9688	21242.5378	21259.1081
1460	21275.6798	21292.2529	21308.8273	21325.4031	21342.9803	21358.5588	21375.1387	21391.7200	21408.3026	21424.8866
1470	21441.4720	21458.0587	21474.6468	21491.2362	21507.8270	21524.4191	21541.0126	21557.6075	21574.2037	21590.8013
1480	21607.4002	21624.0005	21640.6021	21657.2051	21673.8094	21690.4151	21707.0221	21723.6304	21740.2401	21756.8512
1490	21773.4636	21790.0773	21806.6924	21823.3088	21839.9265	21856.5456	21873.1661	21889.7878	21906.4109	21923.0354
1500	21939.6612	21956.2883	21972.9167	21989.5465	22006.1776	22022.8100	22039.4438	22056.0789	22072.7153	22089.3530
1510	22105.9921	22122.6325	22139.2742	22155.9172	22172.5616	22189.2073	22205.8543	22222.5026	22239.1522	22255.8032
1520	22272.4555	22289.1091	22305.7640	22322.4202	22339.0777	22355.7366	22372.3967	22389.0582	22405.7209	22422.3850
1530	22439.0504	22455.7171	22472.3851	22489.0544	22505.7251	22522.3970	22539.0702	22555.7447	22572.4205	22589.0977
1540	22605.7761	22622.4558	22639.1368	22655.8192	22672.5028	22689.1877	22705.8739	22722.5614	22739.2502	22755.9403
1550	22772.6317	22789.3243	22806.0183	22822.7135	22839.4100	22856.1079	22872.8070	22889.5074	22906.2090	22922.9120
1560	22939.6162	22956.3218	22973.0286	22989.7366	23006.4460	23023.1567	23039.8686	23056.5818	23073.2962	23090.0120
1570	23106.7290	23123.4473	23140.1669	23156.8877	23173.6099	23190.3332	23207.0579	23223.7838	23240.5110	23257.2395
1580	23273.9692	23290.7002	23307.4324	23324.1660	23340.9008	23357.6368	23374.3741	23391.1127	23407.8525	23424.5936
1590	23441.3360	23458.0796	23474.8244	23491.5706	23508.3179	23525.0666	23541.8164	23558.5676	23575.3200	23592.0736

4.6 The Evaluation of Fourfold Tables

Table 85 *(continuation 4)* $2n \ln n$ values for $n = 1600$ to 2009

n	0	1	2	3	4	5	6	7	8	9
1600	23608.8285	23625.5846	23642.3420	23659.1007	23675.8606	23692.6217	23709.1841	23726.1477	23742.9125	23759.6787
1610	23776.4461	23793.2147	23809.9845	23826.7556	23843.5279	23860.3016	23877.0763	23893.8523	23910.6296	23927.4081
1620	23944.1878	23960.9688	23977.7510	23994.5345	24011.3191	24028.1051	24044.8922	24061.6806	24078.4702	24095.2610
1630	24112.0531	24128.8463	24145.6409	24162.4366	24179.2335	24196.0317	24212.8311	24229.6318	24246.4336	24263.2367
1640	24280.0410	24296.8465	24313.6532	24330.4612	24347.2703	24364.0807	24380.8923	24397.7051	24414.5192	24431.3344
1650	24448.1509	24464.9685	24481.7874	24498.6075	24515.4288	24532.2513	24549.0750	24565.9000	24582.7261	24599.5534
1660	24616.3820	24633.2117	24650.0427	24666.8748	24683.7082	24700.5427	24717.3785	24734.2155	24751.0536	24767.8930
1670	24784.7335	24801.5753	24818.4183	24835.2624	24852.1077	24868.9543	24885.8020	24902.6509	24919.5011	24936.3524
1680	24953.2049	24970.0586	24986.9135	25003.7695	25020.6268	25037.4852	25054.3449	25071.2057	25088.0677	25104.9309
1690	25121.7953	25138.6608	25155.5276	25172.3955	25189.2646	25206.1349	25223.0064	25239.8790	25256.7528	25273.6278
1700	25290.5040	25307.3814	25324.2599	25341.1396	25358.0205	25374.9025	25391.7858	25408.6702	25425.5558	25442.4425
1710	25459.3304	25476.2195	25493.1097	25510.0011	25526.8937	25543.7875	25560.6824	25577.5785	25594.4757	25611.3741
1720	25628.2737	25645.1745	25662.0764	25678.9794	25695.8837	25712.7890	25729.6956	25746.6033	25763.5121	25780.4222
1730	25797.3333	25814.2457	25831.1592	25848.0738	25864.9896	25881.9065	25898.8246	25915.7439	25932.6643	25949.5859
1740	25966.5086	25983.4324	26000.3574	26017.2836	26034.2109	26051.1393	26068.0689	26084.9997	26101.9315	26118.8646
1750	26135.7987	26152.7340	26169.6705	26186.6081	26203.5468	26220.4867	26237.4277	26254.3699	26271.3132	26288.2576
1760	26305.2032	26322.1499	26339.0977	26356.0467	26372.9968	26389.9481	26406.9004	26423.8540	26440.8086	26457.7644
1770	26474.7213	26491.6793	26508.6385	26525.5988	26542.5602	26559.5227	26576.4864	26593.4512	26610.4171	26627.3842
1780	26644.3524	26661.3217	26678.2921	26695.2636	26712.2363	26729.2101	26746.1850	26763.1610	26780.1382	26797.1164
1790	26814.0958	26831.0763	26848.0579	26865.0407	26882.0245	26899.0095	26915.9956	26932.9827	26949.9711	26966.9605
1800	26983.9510	27000.9426	27017.9354	27034.9292	27051.9242	27068.9203	27085.9175	27102.9158	27119.9152	27136.9157
1810	27153.9173	27170.9200	27187.9238	27204.9288	27221.9348	27238.9419	27255.9501	27272.9595	27289.9699	27306.9814
1820	27323.9941	27341.0078	27358.0226	27375.0386	27392.0556	27409.0737	27426.0929	27443.1133	27460.1347	27477.1572
1830	27494.4767	27511.2054	27528.2312	27545.2581	27562.2861	27579.3151	27596.3453	27613.3765	27630.4088	27647.4422
1840	27664.4767	27681.5123	27698.5490	27715.5867	27732.6256	27749.6655	27766.7065	27783.7486	27800.7918	27817.8361
1850	27834.8814	27851.9278	27868.9753	27886.0239	27903.0736	27920.1243	27937.1761	27954.2290	27971.2830	27988.3380
1860	28005.3942	28022.4514	28039.5096	28056.5690	28073.6294	28090.6909	28107.7535	28124.8171	28141.8818	28158.9476
1870	28176.0145	28193.0824	28210.1514	28227.2214	28244.2926	28261.3648	28278.4380	28295.5124	28312.5878	28329.6642
1880	28346.7417	28363.8203	28380.9000	28397.9807	28415.0625	28432.1453	28449.2292	28466.3141	28483.4002	28500.4872
1890	28517.5754	28534.6646	28551.7548	28568.8461	28585.9385	28603.0319	28620.1264	28637.2219	28654.3185	28671.4161
1900	28688.5148	28705.6146	28722.7154	28739.8172	28756.9201	28774.0241	28791.1291	28808.2351	28825.3422	28842.4504
1910	28859.5596	28876.6698	28893.7811	28910.8934	28928.0068	28945.1212	28962.2367	28979.3532	28996.4707	29013.5893
1920	29030.7090	29047.8297	29064.9514	29082.0742	29099.1980	29116.3228	29133.4487	29150.5756	29167.7036	29184.8326
1930	29201.9626	29219.0936	29236.2258	29253.3589	29270.4931	29287.6283	29304.7645	29321.9018	29339.0401	29356.1794
1940	29373.3198	29390.4612	29407.6037	29424.7471	29441.8916	29459.0371	29476.1837	29493.3312	29510.4799	29527.6295
1950	29544.7801	29561.9318	29579.0845	29596.2383	29613.3930	29630.5488	29647.7056	29664.8634	29682.0223	29699.1821
1960	29716.3430	29733.5049	29750.6679	29767.8318	29784.9968	29802.1628	29819.3298	29836.4978	29853.6668	29870.8369
1970	29888.0080	29905.1800	29922.3531	29939.5273	29956.7024	29973.8785	29991.0557	30008.2338	30025.4130	30042.5932
1980	30059.7744	30076.9566	30094.1398	30111.3241	30128.5093	30145.6955	30162.8828	30180.0711	30197.2603	30214.4506
1990	30231.6419	30248.8342	30266.0274	30283.2217	30300.4170	30317.6133	30334.8106	30352.0089	30369.2082	30386.4085
2000	30403.6098	30420.8121	30438.0154	30455.2197	30472.4251	30489.6314	30506.8386	30524.0469	30541.2562	30558.4665

Table 86 $2n \ln n$ for $n = \tfrac{1}{2}$ to $n = 299\tfrac{1}{2}$ (from Woolf, B.: The log likelihood ratio test (the G-Test). Methods and tables for tests of heterogeneity in contingency tables, Ann. Human Genetics **21**, 397–409 (1957), Table 2, p. 405)

	1/2	1 1/2	2 1/2	3 1/2	4 1/2	5 1/2	6 1/2	7 1/2	8 1/2	9 1/2
0	−0.6931	1.2164	4.5815	8.7693	13.5367	18.7522	24.3334	30.2235	36.3811	42.7745
10	49.3789	56.1740	63.1432	70.2226	77.5503	84.9660	92.5109	100.1770	107.9575	115.8462
20	123.3374	131.9263	140.1082	148.3790	156.7350	165.1726	173.6887	182.2802	190.9445	199.6790
30	208.4813	217.3492	226.2806	235.2735	244.3262	253.4368	262.6038	271.8256	281.1007	290.4278
40	299.8055	309.2326	318.7078	328.2202	337.7985	347.4118	357.0690	366.7693	376.5117	386.2953
50	396.1193	405.9829	415.8854	425.8259	435.8039	445.8185	455.8692	465.9553	476.0761	486.2312
60	496.4198	506.6416	516.8958	527.1821	537.4998	547.8486	558.2279	568.6372	579.0762	589.5444
70	600.0414	610.5667	621.1201	631.7010	642.3091	652.9441	663.6055	674.2931	685.0065	695.7454
80	706.5094	717.2983	728.1117	738.9494	749.8110	760.6963	771.6050	782.5368	793.4915	804.4687
90	815.4683	826.4900	837.5336	848.5988	859.6854	870.7931	881.9218	893.0712	904.2411	915.4314
100	926.6417	937.8719	949.1219	960.3913	971.6801	982.9880	994.3149	1005.6605	1017.0248	1028.4075
110	1039.8084	1051.2275	1062.6645	1074.1193	1085.5916	1097.0815	1108.5887	1120.1130	1131.6544	1143.2126
120	1154.7876	1166.3792	1177.9872	1189.6116	1201.2521	1212.9087	1224.5813	1236.2697	1247.9737	1259.6933
130	1271.4284	1283.1788	1294.9444	1306.7251	1318.5208	1330.3313	1342.1566	1353.9966	1365.8511	1377.7200
140	1389.6033	1401.5008	1413.4125	1425.3382	1437.2778	1449.2312	1461.1985	1473.1793	1485.1737	1497.1816
150	1509.2029	1521.2374	1533.2852	1545.3461	1557.4200	1569.5068	1581.6065	1593.7190	1605.8442	1617.9820
160	1630.1324	1642.2952	1654.4704	1666.6578	1678.8576	1691.0695	1703.2934	1715.5294	1727.7773	1740.0371
170	1752.3087	1764.5921	1776.8870	1789.1936	1801.5117	1813.8413	1826.1823	1838.5346	1850.8981	1863.2729
180	1875.6588	1888.0558	1900.4638	1912.8828	1925.3127	1937.7534	1950.2048	1962.6671	1975.1399	1987.6234
190	2000.1175	2012.6220	2025.1370	2037.6624	2050.1981	2062.7441	2075.3003	2087.8667	2100.4433	2113.0299
200	2125.6265	2138.2331	2150.8497	2163.4761	2176.1123	2188.7583	2201.4141	2214.0795	2226.7546	2239.4392
210	2252.1335	2264.8372	2277.5503	2290.2729	2303.0049	2315.7462	2328.4967	2341.2565	2354.0255	2366.8036
220	2379.5909	2392.3872	2405.1926	2418.0069	2430.8302	2443.6624	2456.5035	2469.3534	2482.2120	2495.0795
230	2507.9556	2520.8405	2533.7340	2546.6360	2559.5467	2572.4658	2585.3935	2598.3296	2611.2742	2624.2271
240	2637.1884	2650.1580	2663.1358	2676.1220	2689.1163	2702.1188	2715.1295	2728.1482	2741.1751	2754.2100
250	2767.2529	2780.3038	2793.3627	2806.4295	2819.5041	2832.5867	2845.6770	2858.7752	2871.8811	2884.9947
260	2898.1161	2911.2451	2924.3818	2937.5261	2950.5780	2963.8375	2977.0045	2990.1790	3003.3609	3016.5504
270	3029.7472	3042.9511	3056.1630	3069.3820	3082.6082	3095.8418	3109.0826	3122.3306	3135.5859	3148.8483
280	3162.1179	3175.3946	3188.5784	3201.9693	3215.2672	3228.5722	3241.8842	3255.2031	3268.5291	3281.8619
290	3295.2017	3308.5483	3321.9018	3335.2622	3348.6293	3362.0033	3375.3840	3388.7715	3402.1657	3415.5665

4.6 The Evaluation of Fourfold Tables

from Table 85: $\quad \dfrac{181 \to S_2 = 1{,}881.8559}{S_1 + S_2 = 3{,}397.8146}$ (pp. 353–357)

from Table 85: $\quad \dfrac{\begin{cases} 100 \to 921.0340 \\ 81 \to 711.9008 \\ 19 \to 111.8887 \\ 162 \to 1648.3812 \end{cases} \quad \left. \begin{array}{l} S_1 + S_2 = 3{,}397.8146 \\ S_3 = 3{,}393.2047 \end{array} \right]}{S_3 = 3{,}393.2047 \qquad \hat{G} = \quad 4.6099}$

Then $\hat{G} = S_1 + S_2 - S_3 = 4.610 > 2.706$.

Woolf (1957) gives g-values for $n = 1$ to $n = 2{,}009$ (Table 85) and for $n = \tfrac{1}{2}$ to $n = 299\tfrac{1}{2}$ (Table 86). Kullback et al., (1962) give tables for $n = 1$ to $n = 10{,}000$. The tables provided by Woolf are generally adequate; moreover, Woolf gives auxiliary tables which, for $n > 2{,}009$, permit us to find

Table 87 Auxiliary table for computing large values of 2 ln p (from Woolf, B.: The log likelihood ratio test (the G-Test). Methods and tables for tests of heterogeneity in contingency tables, Ann. Human Genetics **21** 397–409 (1957), Table 5, p. 408)

p	2 ln p
2	1.386294361
3	2.197224577
4	2.772588722
5	3.218875825
6	3.583518938
7	3.891820306
8	4.158883083
9	4.394449155
10	4.605170186
11	4.795790556
13	5.129898725
17	5.666426699
19	4.888877971
20	5.991464547
40	7.377758908
50	7.824046011
100	9.210340372

without a great deal of computation any needed g-values up to $n \simeq 20{,}000$ accurate to 3 decimal places, and up to $n \simeq 200{,}000$ accurate to 2 decimal places: n is divided by a number p so that $n/p = q$ falls within the range of Table 85. The desired function g of n is

$$g(n) = 2n \ln n = p(2q) \ln q + n(2) \ln p = p \cdot g(q) + n(2) \ln p.$$

To minimize the rounding error, the integer p is chosen as small as possible. Table 87 gives, for integral values of p, the corresponding values of $2 \ln p$.

EXAMPLE. Determine the value of $2n \ln n$ for $n = 10{,}000$ accurately to 3 decimal places. We choose $p = 10$ and obtain $q = n/p = 10{,}000/10 = 1{,}000$:

$$\begin{aligned} g(q) &= 13{,}815.5106 \\ p \cdot g(q) &= 138{,}155.106 \\ 2 \ln p &= 4.605170187 \\ n \cdot 2 \ln p &= 46{,}051.70187 \\ \hline g(n) &\simeq 184{,}206.808 \end{aligned}$$

The Kullback tables indicate that $g(n) = 184{,}206.807$. For the case where p is not a factor of n, there are two other auxiliary tables given by Woolf (1957) which can be found in the original work.

4.6.2 Repeated application of the fourfold χ^2 test

> In this section we point out a frequently made error and show how it can easily be avoided. Then the Gart approximation to the exact Fisher test follows a remark on the most important test statistics as well as three remarks on the fourfold χ^2 test.

The small table printed below (4.35ab) in Section 4.6.1 is appropriate, as are many others in this book (e.g., Tables 27, 28, 28a, 30a–f, 83) for the single "isolated" application of the test in question and not for a sequence of tests. **The tabulated significance levels of the selected distribution refer to a single test carried out in isolation.**

Suppose we are given data which lead to τ (Greek tau) $= 30$ fourfold χ^2 tests (two sided). When we consider the 30 tests with the bound 3.841 simultaneously, the actual significance level is considerably higher. Therefore it would not be correct to use 3.841 as critical value for each of the 30 tests. By Bonferroni's inequality the proper bound (for $\alpha = 0.05$, $\tau = 30$) is $S_{0.05} = 9.885$ (cf., Table 88). More on this can be found in Section 6.2.1. 10%, 5%, and 1% bounds for $v = 1$ and $\tau \leq 12$ are contained in Table 141 there, and its source is also the source of the bounds S for $\tau > 12$ given in Table 88.

4.6 The Evaluation of Fourfold Tables

Table 88 Supplements and is supplemented by Table 141 (Section 6.2.1)

τ	13	14	15	16	18	20	22	24	26	28	30
$S_{0.10}$	7.104	7.237	7.360	7.477	7.689	7.879	8.052	8.210	8.355	8.490	8.615
$S_{0.05}$	8.355	8.490	8.615	8.733	8.948	9.141	9.315	9.475	9.622	9.758	9.885
$S_{0.01}$	11.314	11.452	11.580	11.700	11.919	12.116	12.293	12.456	12.605	12.744	12.873

Some important remarks: (1) F-test due to Gart and (2) exact and approximate test statistics

A fourfold table such as Table 95a in Section 4.6.7 is arranged in such a way—through transposition if necessary—that the following inequalities hold (see Table 89):

$$\boxed{a + c \leq b + d \quad \text{and} \quad a/n_1 \leq c/n_2.} \tag{4.37}$$

Table 89

		Dichotomy		
		1	2	Σ
Sample or dichotomy	1	a	b	n_1
	2	c	d	n_2
	Σ	a + c	b + d	n

If $ac/(n_1 + n_2) < 1$, then the following F-test due to Gart (1962, formula (11)) is a good approximation: H_0 is rejected when

$$\boxed{\hat{F} = \frac{c(2n_1 - a)}{(a + 1)(2n_2 - c + 1)} > F_{v_1; v_2; \alpha} \quad \text{with } v_1 = 2(a + 1), v_2 = 2c.} \quad \text{(p.145)}$$

(4.38)

EXAMPLE (Section 4.6.7, Table 95a)

$$\begin{array}{c|c|c} 2 & 8 & 10 \\ 10 & 4 & 14 \\ \hline 12 & 12 & 24 \end{array} \qquad \frac{2 \cdot 10}{10 + 14} < 1;$$

$$\hat{F} = \frac{10(2 \cdot 10 - 2)}{(2 + 1)(2 \cdot 14 - 10 + 1)} = 3.158 \begin{array}{l} > 2.60 = F_{6; 20; 0.05} \\ < 3.87 = F_{6; 20; 0.01}. \end{array}$$

For $n_1 = n_2 \leq 20$ there is an exact quick test due to Ott and Free (1969), which I have included in my booklet (Sachs 1984, p. 66).

Estimates of parameters are sometimes (cf., e.g., Sections 1.3.1 and 1.6.2) distinguished from the true value of the parameter by a caret, e.g., $\hat{p}$ for the estimate of p. We use this notation (from Section 1.4.5 on) also to differentiate test statistics such as the $\hat{F} = 3.158$ above (or e.g., $\hat{z}$, $\hat{t}$, $\hat{\chi}^2$), estimated (computed) in terms of concrete sampled values, from the tabulated critical limits such as $z_{0.05;\,\text{one sided}} = 1.645$ (or $t_{v;\,\alpha}$, $F_{v_1;\,v_2;\,\alpha}$, $\chi^2_{v;\,\alpha}$). **Note that under H_0 the following are distributed exactly like the corresponding theoretical distributions: (1) $\hat{z}$, only for $n \to \infty$, (2) $\hat{\chi}^2$, only for $n \to \infty$** [i.e., large expected frequencies under H_0; exceptions: (3.16) and (3.17)], **(3) $\hat{t}$ and $\hat{F}$, for arbitrary n.**

More Remarks

1. Le Roy (1962) has proposed a simple χ^2 test for **comparing two fourfold tables**. Null hypothesis: two analogous samples which give rise to two fourfold tables originate in one and the same population (alternate hypothesis: they stem from different populations).

Denote the two tables by I, II;

a_1	b_1	I
c_1	d_1	
n_1		

a_2	b_2	II
c_2	d_2	
n_2		

The **equivalence** of the two fourfold tables can be tested by means of

$$\hat{\chi}^2 = \frac{n_1 + n_2}{n_1 n_2} \cdot (n_1 Q + n_2 q) - (n_1 + n_2), \tag{4.39}$$

DF = 3.

Table 90 shows the computation of the product sums q and Q in terms of the quotients a, b, c, d (column 4) and the differences A, B, C, D (column 5) [(4.39) is identical to (6.1), (6.1a) in Sections 6.1.1 and 6.1.3 for $k = 4$].

Table 90

1	2	3	4	5	6	7
a_1	a_2	$a_1 + a_2$	$a_1/(a_1 + a_2) = a$	$A = 1 - a$	$a_1 a$	$a_2 A$
b_1	b_2	$b_1 + b_2$	$b_1/(b_1 + b_2) = b$	$B = 1 - b$	$b_1 b$	$b_2 B$
c_1	c_2	$c_1 + c_2$	$c_1/(c_1 + c_2) = c$	$C = 1 - c$	$c_1 c$	$c_2 C$
d_1	d_2	$d_1 + d_2$	$d_1/(d_1 + d_2) = d$	$D = 1 - d$	$d_1 d$	$d_2 D$
n_1	n_2	$n_1 + n_2$	-	-	q	Q

4.6 The Evaluation of Fourfold Tables

If none of the eight frequencies or occupation numbers is <3, then this test may be carried out; it must however be regarded as only an approximate test statistic for weakly occupied tables.

2. If the frequencies of a fourfold table can be subdivided by taking another variable into consideration, then a **generalized sign test** described by Bross (1964) (cf. also Ury 1966) is recommended. An instructive example is contained in Bross's article.

3. Fourfold tables with specified probabilities were partitioned by Rao (1973) into three χ^2 components.

4. More on fourfold tables may be found in the reviews cited on page 462.

4.6.3 The sign test modified by McNemar

Two trials on the same individuals: Significance of a change in the frequency ratio of two dependent distributions of binary data; McNemar's test for correlated proportions in a 2 × 2 table

If a sample is studied **twice**—separated by a certain interval of time or under different conditions, say—with respect to the strength of some binary characteristic, then we are no longer dealing with independent but rather with **dependent samples**. Every experimental unit provides a pair of data. The frequency ratio of the two alternatives will change more or less from the first to the second study. The **intensity of this change** is tested by the sign test known as the McNemar χ^2 test (1947) which, more precisely, exhausts the information as to how many individuals are transferred into another category between the first and the second study. We have a fourfold table with one entry for the first study and with a second entry for the second study, as shown in Table 91.

Table 91

Study I \ Study II	+	−
+	a	b
−	c	d

The null hypothesis is that the frequencies in the population do not differ for the two studies, i.e., the frequencies b and c indicate only random variations in the sample. Since these two frequencies represent the only possible frequencies which change from study I to study II, where b changes from + to − and c from − to +, McNemar, together with Bennett and Underwood

(1970) (cf., also Gart 1969, Maxwell 1970, and Bennett 1971), was able to show that changes of this sort can be tested for $(b + c) \geq 30$ by

$$\hat{\chi}^2 = \frac{(b-c)^2}{b+c+1}, \quad DF = 1 \qquad (4.40)$$

and for $8 \leq b + c < 30$ with continuity correction by

$$\hat{\chi}^2 = \frac{(|b-c|-1)^2}{b+c+1}, \quad DF = 1. \qquad (4.40a)$$

Thus the frequencies b and c are compared and their ratio tested with respect to 1:1 (cf., also Section 4.5.5). Under the null hypothesis both frequencies b and c have the same expected value $(b + c)/2$. The more b and c deviate from this expected value, the less confidence one has in the null hypothesis. If a sound assumption as to the direction of the change to be expected can be made even prior to the experiment, a one sided test may be made. A computation of the associated confidence interval can be found in Sachs (1984, pp. 74, 75).

EXAMPLE. A medication and a placebo are compared on a sample of 40 patients. Half the patients begin with one preparation and half with the other. Between the two phases of the therapy a sufficiently long therapy-free phase is inserted. The physician grades the effect as "[at best] weak" or "strong" based on the statements of the patients.

The null hypothesis (the two preparations have equal effect) is set ($\alpha = 0.05$) against a one sided alternate hypothesis (the preparation is more effective than the neutral preparation). The fourfold scheme in Table 92 is obtained, and

Table 92

		Placebo effect (effect of neutral prep.)	
		strong	weak
Effect of the preparation	strong	8 (a)	16 (b)
	weak	5 (c)	11 (d)

$$\hat{\chi}^2 = \frac{(|16 - 5| - 1)^2}{16 + 5 + 1} = 4.545.$$

With help of the table in the middle of page 349:

$$\hat{\chi}^2 = 4.545 > 2.706 = \chi^2_{1;\, 0.05;\, \text{one sided}}.$$

The value $\hat{\chi}^2 = 4.545$ corresponds, by Table 83 for the one sided test, to a probability $P \simeq 0.0165$.

4.6 The Evaluation of Fourfold Tables

Let us consider the example in somewhat greater detail: In Table 92 the 11 patients that reacted weakly to both preparations, and the 8 patients that experienced a strong effect in both cases told us nothing about the possible difference between the preparation and the placebo. The essential information is taken from fields b and c:

Weak placebo effect and strong preparation effect:	16 patients
Weak preparation effect and strong placebo effect:	5 patients
Altogether	21 patients

If there was no real difference between the two preparations, then we should expect the frequencies b and c to be related as 1:1. Deviations from this ratio can also be tested with the aid of the binomial distribution. For the one sided question we obtain

$$P(X \leq 5 \mid n = 21, p = 0.5) = \sum_{x=0}^{x=5} \binom{21}{x}\left(\frac{1}{2}\right)^x\left(\frac{1}{2}\right)^{21-x} = 0.0133$$

or by means of the approximation involving the normal distribution,

$$\hat{z} = \frac{|5 + 0.5 - 21 \cdot 0.5|}{\sqrt{21 \cdot 0.5 \cdot 0.5}} = 2.182, \quad \text{i.e.,} \quad P(X \leq 5) = 0.0146.$$

This **sign test**, known in psychology as the McNemar test, is based on the signs of the differences of paired observations. It is a frequently used form of the test introduced above. The plus signs and the minus signs are counted. The null hypothesis (the two signs are equally likely) is tested against the alternative hypothesis (the two signs occur with different probability) with the help of a χ^2 test adjusted for continuity:

$$\boxed{\hat{\chi}^2 = \frac{(|n_{\text{Plus}} - n_{\text{Minus}}| - 1)^2}{n_{\text{Plus}} + n_{\text{Minus}} + 1}.} \tag{4.40b}$$

The null hypothesis is then

$$\frac{n_{\text{Plus}}}{n_{\text{Plus}} + n_{\text{Minus}}} = \frac{1}{2} \quad \text{or} \quad \frac{n_{\text{Minus}}}{n_{\text{Plus}} + n_{\text{Minus}}} = \frac{1}{2}.$$

The alternate hypothesis is the negation of this statement. Thus we have to test Prob(+ sign) = $\frac{1}{2}$ (cf., Table 69).

A generalization of this test for the comparison of several percentages in matched samples is the Q-test of Cochran (1950) which we give in Section 6.2.4 (cf., also Seeger 1966, Bennett 1967, Marascuilo and McSweeney 1967, Seeger and Gabrielsson 1968, and Tate and Brown 1970).

4.6.4 The additive property of χ^2

A sequence of experiments carried out on heterogeneous material which cannot be jointly analyzed could yield the $\hat{\chi}^2$ values $\hat{\chi}_1^2, \hat{\chi}_2^2, \hat{\chi}_3^2, \ldots$ with $v_1, v_2, v_3, \ldots$ degrees of freedom. If there is a systematic tendency toward deviations in the same direction the overall result may then be combined into a single $\hat{\chi}^2$ statistic $\hat{\chi}_1^2 + \hat{\chi}_2^2 + \cdots$ with $v_1 + v_2 + \cdots$ degrees of freedom. When combining the $\hat{\chi}^2$-values from fourfold tables, make sure they are not adjusted by Yates's correction, as that leads to overcorrection. When combining $\hat{\chi}^2$ values from sixfold or larger tables (cf., Chapter 6) other methods are better (cf., Koziol and Perlman (1978)).

EXAMPLE. To test a null hypothesis ($\alpha = 0.05$) an experiment is carried out four times—in a different locale and on different material, let us say. The corresponding $\hat{\chi}^2$-values are 2.30, 1.94, 3.60, and 2.92, with one degree of freedom in each case. The null hypothesis cannot be rejected. On the basis of the additive property of χ^2 the results can be combined:

$$\hat{\chi}^2 = 2.30 + 1.94 + 3.60 + 2.92 = 10.76 \quad \text{with } 1 + 1 + 1 + 1 = 4 \text{ DF.}$$

$\hat{\chi}^2 > \chi^2_{4;0.05}$ does not imply that the null hypothesis has to be rejected in all 4 experiments, but H_0 does not hold simultaneously for all 4 experiments. Since for 4 DF we have $\chi^2_{4;0.05} = 9.488$ (Table 28, lower part), the null hypothesis must be rejected at the 5% level for at least one experiment.

Remark: Combining comparable test results, that is, combining exact probabilities from independent tests of significance with the same H_A.

Occasionally several studies of certain connections (e.g., smoking and lung cancer) which have been evaluated by means of different statistical tests (e.g., U-test and t-test) are available. The **comparable** statistical statements with equal tendencies can be combined into a single statement.

Small independent values of the attained significance level P_i with all deviations in the same direction may be combined to test the combined null hypothesis. Fisher's combination procedure rejects this combined null hypothesis if the product of the P_i is small, if

$$\boxed{-2 \sum_{i=1}^{n} \ln P_i \geq \chi^2_{2n;\alpha}.}$$

This is an approximation.

EXAMPLE. Combine $P_1 = 0.06$, $P_2 = 0.07$, $P_3 = 0.08$; $n = 3$, $2n = 6$; $\alpha = 0.05$

$$\ln 0.06 = -2.8134$$
$$\ln 0.07 = -2.6593$$
$$\ln 0.08 = -2.5257$$
$$\overline{-7.9984}$$

4.6 The Evaluation of Fourfold Tables   367

Since $(-2)(-7.9984) = 15.997 > 12.59 = \chi^2_{6;0.05}$ the combined null hypothesis is rejected at the 5% level.

Further methods for combining independent χ^2 tests are given in Koziol and Perlman (1978) [see also the end of Section 1.4.3, Good (1958), and Kincaid (1962)].

4.6.5 The combination of fourfold tables

If several fourfold tables are available that cannot be regarded as replications because the conditions on samples 1 and 2 (with $n_1 + n_2 = n$), which make up one experiment, vary from table to table, then Cochran (1954) recommends both of the following procedures as sufficiently accurate approximate solutions (cf., also Radhakrishna 1965, Fleiss 1981 [8:2a], and Sachs 1984):

I. The **sample sizes** n_k of the i fourfold tables ($k = 1, \ldots, i$) **do not differ very greatly from one another** (at most by a factor of 2); the ratios $a/(a + b)$ and $c/(c + d)$ (Table 81) lie between about 20% and 80% in all the tables. Then the result can be tested on the basis of i combined fourfold tables in terms of the normal distribution according to

$$\boxed{\hat{z} = \frac{\sum \hat{\chi}}{\sqrt{i}} \quad \text{or} \quad \hat{z} = \frac{\sum \sqrt{\hat{G}}}{\sqrt{i}}.} \qquad (4.41\text{ab})$$

The test in detail:

1. Take the square root of the $\hat{\chi}^2$ or $\hat{G}$ values, determined for the i fourfold tables (cf., Section 4.6.1, second half) without the Yates correction.
2. The signs of these values are given by the signs of the differences

$$a/(a + b) - c/(c + d).$$

3. Sum the $\hat{\chi}$ or $\sqrt{\hat{G}}$ values (keeping track of signs).
4. Take the square root of the number of combined fourfold tables.
5. Construct the quotients $\hat{z}$ by the above formula.
6. Test the significance of $\hat{z}$ by means of tables of standard normal variables (Table 14 or Table 43).

II. **No assumptions are made on the sample sizes** n_{i1} and n_{i2} of the 2 × 2 tables or the respective proportions $a/(a + b)$ and $c/(c + d)$. Here the question regarding the significance of a result can be tested in terms of the normal distribution by

$$\boxed{\hat{z} = \frac{\sum W_i \cdot D_i}{\sqrt{\sum W_i \cdot p_i(1 - p_i)}},} \qquad (4.42)$$

where W_i is the weight of the ith sample [for the i-th 2×2 table] with frequencies $a_i, b_i, c_i,$ and d_i (Table 81), defined as

$$W_i = \frac{(n_{i1})(n_{i2})}{n_i}$$

with $n_{i1} = a_i + b_i$; $n_{i2} = c_i + d_i$ and $n_i = n_{i1} + n_{i2}$; p_i is the average ratio

$$p_i = \frac{a_i + c_i}{n_i},$$

and D_i is the difference between the ratios:

$$D_i = \frac{a_i}{n_{i1}} - \frac{c_i}{n_{i2}}.$$

We give the example cited by Cochran as an illustration.

EXAMPLE. Erythroblastosis of the newborn is due to the incompatibility between the Rh-negative blood of the mother and the Rh-positive blood of the embryo that, among other things, leads to the destruction of embryonic

Table 93 Mortality by sex of donor and severity of disease. The sample sizes vary only from 33 to 60; the portion of deaths however varies between 3% and 46%, so that the 4 tables are combined in accordance with the second procedure.

Symptoms	Sex of the donor	Number		Total	% Deaths
		Deaths	Survivals		
none	male	2	21	$23 = n_{11}$	$8.7 = p_{11}$
	female	0	10	$10 = n_{12}$	$0.0 = p_{12}$
	total	2	31	$33 = n_1$	$6.1 = p_1$
slight	male	2	40	$42 = n_{21}$	$4.8 = p_{21}$
	female	0	18	$18 = n_{22}$	$0.0 = p_{22}$
	total	2	58	$60 = n_2$	$3.3 = p_2$
moderate	male	6	33	$39 = n_{31}$	$15.4 = p_{31}$
	female	0	10	$10 = n_{32}$	$0.0 = p_{32}$
	total	6	43	$49 = n_3$	$12.2 = p_3$
very pronounced	male	17	16	$33 = n_{41}$	$51.5 = p_{41}$
	female	0	4	$4 = n_{42}$	$0.0 = p_{42}$
	total	17	20	$37 = n_4$	$45.9 = p_4$

erythrocytes, a process which, after birth, is treated by exchange transfusion: The blood of the infant is replaced with blood of Rh-negative donors of the same blood type.

As was observed on 179 newborns at a Boston clinic (Allen, Diamond, and Watrous: The New Engl. J. Med. **241** [1949] pp. 799-806), the blood of female donors is more compatible with the infants than that of male donors. The question arises: Is there an association between the sex of the donor and the alternatives "survival" or "death"? The 179 cases cannot be considered as a unit, because of the difference in severity of the condition. They are therefore divided, according to the severity of the symptoms as a possible intervening variable, into 4 internally more homogeneous groups. The results are summarized in Table 93.

Table 94

Symptoms	D_i	p_i	$p_i(H - p_i)$	$W_i = \dfrac{n_{i1} \cdot n_{i2}}{n_i}$	$W_i D_i$	$W_i p_i(H - p_i)$
none	8.7 - 0.0 = 8.7	6.1	573	7.0	60.90	4011.0
slight	4.8 - 0.0 = 4.8	3.3	319	12.6	60.48	4019.4
moderate	15.4 - 0.0 = 15.4	12.2	1071	8.0	123.20	8568.0
very pronounced	51.5 - 0.0 = 51.5	45.9	2483	3.6	185.40	8938.8
					429.98 $\sum W_i D_i$	25537.2 $\sum W_i p_i(H - p_i)$

By means of an auxiliary table (Table 94) with p_i in % and $H = 100$, we obtain $\hat{z} = 429.98/\sqrt{25{,}537.2} = 2.69$. For the two sided question under consideration, there corresponds to this $\hat{z}$-value a significance level of 0.0072. We may thus confirm that the blood of female donors constitutes a better replacement in the case of fetal erythroblastosis than that of males. The difference in compatibility is particularly pronounced in severe cases.

Incidentally, this result is not confirmed by other authors: In fact the gender of the donor in no way influences the prognosis of fetal erythroblastosis.

Turning once again to the original table, we note that the relatively high proportion of male blood donors is conspicuous ($>76\%$) and increases with increasing severity of symptoms implying that conditions are more favorable with female donors. Nevertheless, these findings are difficult to interpret.

4.6.6 The Pearson contingency coefficient

What characterizes the fourfold table viewed as a contingency table is that both entries are occupied by characteristic alternatives. In frequency comparison one entry is occupied by a characteristic alternative, the other by a sample dichotomy. χ^2 tests and the G test can show the existence of a connection. They will say nothing about the strength of the connection or association. A statistic for the **degree of association**—if a relation (a contingency) is ascertained between the two characteristics—is the Pearson contingency coefficient. It is a measure of the consistency of the association between the two characteristics of fourfold and manifold tables and is obtained from the $\hat{\chi}^2$ value by the formula

$$CC = \sqrt{\frac{\hat{\chi}^2}{n + \hat{\chi}^2}}. \qquad (4.43)$$

(For fourfold tables (4.35) with "n" in the numerator is used to compute $\hat{\chi}^2$). The maximal contingency coefficient of the fourfold table is 0.7071; it always occurs with perfect contingency, i.e., when the fields b and c remain unoccupied. Square manifold tables with unoccupied diagonal fields from lower left to upper right exhibit a maximal contingency coefficient, given by

$$CC_{max} = \sqrt{(r-1)/r}, \qquad (4.44)$$

where r is the number of rows or columns, i.e., for the fourfold table

$$CC_{max} = \sqrt{(2-1)/2} = \sqrt{1/2} = 0.7071. \qquad (4.45)$$

(p.483) Section 6.2.2 supplements these considerations.

Remarks

1. The exact computation of the correlation coefficients developed by Pearson (cf., Chapter 5) for fourfold tables is exceedingly involved; a straightforward and sufficiently accurate method for estimating fourfold correlations with the help of two diagrams is presented by Klemm (1964). More on this can be found, e.g., in the book by McNemar (1969 [8:1]).

2. A test for comparing the associations in two independent fourfold tables is given in G. A. Lienert et al., Biometrical Journal **21** (1979), 473–491.

4.6.7 The exact Fisher test of independence, as well as an approximation for the comparison of two binomially distributed populations (based on very small samples)

For fourfold tables with very small n (cf., Section 4.6.1), begin with the field with (1) the smallest diagonal product and (2) the smallest frequency [Table 95: (2)(4) < (8)(10), thus 2] and, while holding constant the four marginal sums $(a+b, c+d, a+c, b+d)$, construct all fourfold tables with an even smaller frequency in the field considered. Among all such fourfold tables,

4.6 The Evaluation of Fourfold Tables

those in which the field with the smallest observed frequency is even less occupied have probability P. In other words, if one takes the marginal sums of the fourfold table as given and looks for the probability that the observed occupancy of the table or one even less likely comes about entirely at random (one sided question), then this probability P turns out to be the **sum of some terms of the hypergeometric distribution**:

$$P = \frac{(a+b)!(c+d)!(a+c)!(b+d)!}{n!} \sum_i \frac{1}{a_i!b_i!c_i!d_i!}. \qquad (4.46)$$

The index i indicates that the expression under the summation sign is computed for each of the above described tables, and then included in the sum. Significance tables obtained in this way or with help of recursion formulas are contained in a number of collections of tables (e.g., Documenta Geigy 1968). Particularly extensive tables to $n = 100$ are given by Finney, Latscha, Bennett, and Hsu (1963, with supplement by Bennett and Horst 1966). The probabilities can be read off directly. Unfortunately, the tables allow no two sided tests at the 5% and the 1% level for sample sizes $31 \leq n \leq 100$. More on the two sided test can be found in Johnson (1972) (cf., also below).

EXAMPLE

Table 95

2	8	10
10	4	14
12	12	24

a

1	9	10
11	3	14
12	12	24

b

0	10	10
12	2	14
12	12	24

c

From Table 95 we obtain two tables with more extreme distributions. The probability that the observed table occurs is

$$P = \frac{10! \cdot 14! \cdot 12! \cdot 12!}{24!} \cdot \frac{1}{2! \cdot 8! \cdot 10! \cdot 4!}.$$

The total probability for both the observed and more extreme distributions is given by

$$P = \frac{10!14!12!12!}{24!} \left(\frac{1}{2!8!10!4!} + \frac{1}{1!9!11!3!} + \frac{1}{0!10!12!2!} \right),$$

$$P = 0.0018 \qquad \text{(one sided test; use, e.g. Table 32, Section 1.6.1).}$$

For a symmetric hypergeometric distribution (i.e., here, row or column sums are equal $[a + b = c + d$ or $a + c = b + d]$) we can easily treat the two sided problem; the significance level of the two sided test is twice that of the one sided, e.g., $P = 0.036$ in the example. The null hypothesis ($\pi_1 = \pi_2$ or independence; see Section 4.6.1) is rejected at the 5% level in both cases (because $P < 0.05$). More on the two sided Fisher test may be found in H. P. Krüger, EDV in Medizin und Biologie **10** (1979), 19–21 and in T. W. Taub, Journal of Quality Technology **11** (1979), 44–47.

Computation tools [see Section 4.6.2, (4.37) and (4.38)] for *n* small or large

1 Recursion formula

The computations are carried out **more rapidly** with the help of a **recursion formula** (Feldman and Klinger 1963)

$$P_{i+1} = \frac{a_i \cdot d_i}{b_{i+1} \cdot c_{i+1}} P_i. \tag{4.47}$$

Identifying a, b, c above with 1, 2, 3, we compute the probabilities successively, starting with the observed table (Table 95a) and the above given expression for P_1:

$$P_1 = \frac{10! \cdot 14! \cdot 12! \cdot 12! \cdot 1}{24! \cdot 2! \cdot 8! \cdot 10! \cdot 4!} = 0.016659;$$

for Table 95b, by (4.47),

$$P_{1+1} = P_2 = \frac{2 \cdot 4}{9 \cdot 11} \cdot P_1 = 0.0808 \cdot 0.016659 = 0.001346;$$

and for Table 95c, by (4.47),

$$P_{2+1} = P_3 = \frac{1 \cdot 3}{10 \cdot 12} \cdot P_2 = 0.0250 \cdot 0.001346 = 0.000034.$$

Altogether: $P = P_1 + P_2 + P_3 = 0.0167 + 0.0013 + 0.0000 = 0.018$.

2 Collections of tables

The Finney tables for the one sided test use the fourfold scheme in Table 96,

Table 96

a	A - a	A
b	B - b	B
r	N - r	N

with $A \geq B$ and $a \geq b$ or $A - a \geq B - b$, in the last case writing $A - a$ as a, $B - b$ as b, and the two remaining fields of the table as differences. After performing the required change in notation of the 4 frequencies on page 14 of the tables, our example, as shown in Table 97, supplies the exact probability that $b \leq 2$ for a significance level of 5% with $P = 0.018$. An important aid is also provided by the hypergeometric distribution tables of Lieberman and Owen, mentioned in Section 1.6.3.

Table 97

10	4	14
2	8	10
12	12	24

3 Binomial coefficients

Problems of this sort for sizes up to $n = 20$ can, with the help of Table 31 (Section 1.6.1), be easily solved by

$$P = \frac{\binom{10}{2}\binom{14}{10} + \binom{10}{1}\binom{14}{11} + \binom{10}{0}\binom{14}{12}}{\binom{24}{12}} = 0.01804.$$

Binomial coefficients for larger values of n ($20 < n \leq 100$), such as $\binom{24}{12} = 2{,}704{,}156$, are computed sufficiently accurately by means of Table 32 [Section 1.6.1; cf., also (1.152)].

More on this test (cf., I. Clarke, Applied Statistics **28** (1979), p. 302) can be found in the books by Lancaster (1969, [8:1], pp. 219–225, 348) as well as Kendall and Stuart (Vol. 2, 1973[1], pp. 567–575).

A quick test is presented by Ott and Free (1969) (see, e.g., Sachs 1984, p. 66). Also of particular importance are the tables and nomograms given by Patnaik (1948) as well as Bennett and Hsu (1960), which supplement Table 84 in Section 4.6.1.

4.7 TESTING THE RANDOMNESS OF A SEQUENCE OF DICHOTOMOUS DATA OR OF MEASURED DATA

4.7.1 The mean square successive difference

A straightforward trend test (von Neumann et al. 1941; cf., also Moore 1955) in terms of the dispersion of sample values $x_1, x_2, \ldots, x_i, \ldots, x_n$, consecutive in time, which originate in a normally distributed population, is based on the variance, determined as usual, and the mean square of the $n-1$ differences of consecutive values, called the **mean square successive difference** Δ^2 (delta-square):

$$\Delta^2 = [(x_1 - x_2)^2 + (x_2 - x_3)^2 + (x_3 - x_4)^2 + \cdots + (x_i - x_{i+1})^2 + \cdots + (x_{n-1} - x_n)^2]/(n-1),$$

i.e.,

$$\Delta^2 = \sum (x_i - x_{i+1})^2/(n-1). \qquad (4.48)$$

If the consecutive values are independent, then $\Delta^2 \simeq 2s^2$ or $\Delta^2/s^2 \simeq 2$. Whenever a **trend** is present $\Delta^2 < 2s^2$, i.e., $\Delta^2/s^2 < 2$, since adjacent values are then more similar than distant ones. The null hypothesis (consecutive values are independent) must be abandoned in favor of the alternative hypothesis (there exists a trend) if the quotient

$$\Delta^2/s^2 = \sum (x_i - x_{i+1})^2 / \sum (x_i - \bar{x})^2 \qquad (4.49)$$

drops to or below the critical bounds of Table 98.

For example, for the sequence 2, 3, 5, 6 we find $\sum (x_i - \bar{x})^2 = 10$ and $\sum (x_i - x_{i+1})^2 = (2-3)^2 + (3-5)^2 + (5-6)^2 = 6$; hence

$$\Delta^2/s^2 = 6/10 = 0.60 < 0.626,$$

and the null hypothesis can be rejected at the 1% level. For large sample sizes, approximate bounds can be computed with the help of the normal distribution by

$$2 - 2z \cdot \sqrt{\frac{n-2}{(n-1)(n+1)}} \qquad (4.50)$$

or by

$$2 - 2z \cdot \frac{1}{\sqrt{n+1}}. \qquad (4.50a)$$

where the standard normal variable z equals 1.645 for the 5% bound, 2.326 for the 1% bound, and 3.090 for the 0.1% bound. For example, we get as an approximate 5% bound for $n = 200$ from (4.50, 4.50a)

$$2 - 2 \cdot 1.645 \cdot \sqrt{\frac{200 - 2}{(200-1)(200+1)}} = 1.77,$$

$$2 - 2 \cdot 1.645 \cdot \frac{1}{\sqrt{200 + 1}} = 1.77$$

Table 98 Critical bounds for the ratio of the mean square successive difference and the variance, extracted, and modified by the factor $(n - 1)/n$, from the tables by Hart, B. I.: Significance levels for the ratio of the mean square successive difference to the variance. Ann. Math. Statist. **13** (1942), 445–447

n	0.1%	1%	5%	n	0.1%	1%	5%
4	0.5898	0.6256	0.7805	33	1.0055	1.2283	1.4434
5	0.4161	0.5379	0.8204	34	1.0180	1.2386	1.4511
6	0.3634	0.5615	0.8902	35	1.0300	1.2485	1.4585
7	0.3695	0.6140	0.9359	36	1.0416	1.2581	1.4656
8	0.4036	0.6628	0.9825	37	1.0529	1.2673	1.4726
9	0.4420	0.7088	1.0244	38	1.0639	1.2763	1.4793
10	0.4816	0.7518	1.0623	39	1.0746	1.2850	1.4858
11	0.5197	0.7915	1.0965	40	1.0850	1.2934	1.4921
12	0.5557	0.8280	1.1276	41	1.0950	1.3017	1.4982
13	0.5898	0.8618	1.1558	42	1.1048	1.3096	1.5041
14	0.6223	0.8931	1.1816	43	1.1142	1.3172	1.5098
15	0.6532	0.9221	1.2053	44	1.1233	1.3246	1.5154
16	0.6826	0.9491	1.2272	45	1.1320	1.3317	1.5206
17	0.7104	0.9743	1.2473	46	1.1404	1.3387	1.5257
18	0.7368	0.9979	1.2660	47	1.1484	1.3453	1.5305
19	0.7617	1.0199	1.2834	48	1.1561	1.3515	1.5351
20	0.7852	1.0406	1.2996	49	1.1635	1.3573	1.5395
21	0.8073	1.0601	1.3148	50	1.1705	1.3629	1.5437
22	0.8283	1.0785	1.3290	51	1.1774	1.3683	1.5477
23	0.8481	1.0958	1.3425	52	1.1843	1.3738	1.5518
24	0.8668	1.1122	1.3552	53	1.1910	1.3792	1.5557
25	0.8846	1.1278	1.3671	54	1.1976	1.3846	1.5596
26	0.9017	1.1426	1.3785	55	1.2041	1.3899	1.5634
27	0.9182	1.1567	1.3892	56	1.2104	1.3949	1.5670
28	0.9341	1.1702	1.3994	57	1.2166	1.3999	1.5707
29	0.9496	1.1830	1.4091	58	1.2227	1.4048	1.5743
30	0.9645	1.1951	1.4183	59	1.2288	1.4096	1.5779
31	0.9789	1.2067	1.4270	60	1.2349	1.4144	1.5814
32	0.9925	1.2177	1.4354	∞	2.0000	2.0000	2.0000

4.7.2 The run test for testing whether a sequence of dichotomous data or of measured data is random

The run test, like the two subsequent tests (Sections 4.7.3 and 4.8), is distribution-free. It serves to test the **independence** (the random order) of sampled values. A run is a sequence of identical symbols preceded or followed by other symbols. Thus the sequence (coin tossing with head [H] or tail [T])

$$\underbrace{T, T, T;}_{1} \quad \underbrace{H;}_{2} \quad \underbrace{T, T;}_{3} \quad \underbrace{H, H}_{4}$$

consists of $\hat{r} = 4$ runs ($n = 8$). Runs are obtained not only for dichotomous data but also for measured data that are divided into two groups by the median. For given n, a small $\hat{r}$ indicates clustering of similar observations, and a large $\hat{r}$ indicates regular change. The null hypothesis H_0 that the sequence is random (even though a random sequence can be ordered, it can still be random as far as values go, though not in random sequence) is in a two sided problem opposed by the alternate hypothesis H_A that the given sequence is not random. In the one sided question the H_0 is opposed either by H_{A1} ("cluster effect") or H_{A2} ("regular change"). The critical bounds $r_{\text{lower}} = r_l$ and $r_{\text{upper}} = r_u$ for $n_1, n_2 \leq 20$ and for $20 \leq n_1 = n_2 \leq 100$ are found in Table 99 (where n_1 and n_2 are the numbers of times the two symbols

Table 99 Critical values for the run test (from Swed, F. S. and Eisenhart, C.: Tables for testing randomness of grouping in a sequence of alternatives, Ann. Math. Statist., **14**, (1943), 66–87)

P = 0.01

lower 0.5%-bounds $r_{l;\,0.5\%}$

n_1/n_2	6	7	8	9	10	11	12	13	14	15	16	17	18	19	20
5	2														
6	2	2	3												
7	2	3	3	3											
8	3	3	3	3											
9	3	3	3	4	4										
10	3	3	4	4	5	5									
11	3	4	4	5	5	5	6								
12	3	4	4	5	6	6	6	6							
13	3	4	5	5	5	6	6	7							
14	4	4	5	5	6	6	7	7	7						
15	4	4	5	6	6	7	7	7	8	8					
16	4	5	5	6	6	7	7	8	8	9	9				
17	4	5	5	6	7	7	8	8	8	9	9	10			
18	4	5	6	6	7	7	8	8	9	9	10	10	11		
19	4	5	6	6	7	8	8	9	9	10	10	10	11	11	
20	4	5	6	7	7	8	8	9	9	10	10	11	11	12	12

upper 0.5%-bounds $r_{u;\,0.5\%}$

n_1/n_2	6	7	8	9	10	11	12	13	14	15	16	17	18	19	20
5	11														
6	12														
7	13	13													
8	13	14	15												
9		15	15	16											
10		15	16	17	17										
11		15	16	17	18	19									
12			17	18	19	19	20								
13			17	18	19	20	21	21							
14			17	18	19	20	21	22	23						
15				19	20	21	22	22	23	24					
16				19	20	21	22	23	23	24	25				
17				19	20	22	22	23	24	25	26	26			
18					21	22	23	24	25	25	26	27	27		
19					21	22	23	24	25	26	27	27	28	29	
20					21	22	23	24	25	26	27	28	29	29	30

P = 0.05

lower 2.5%-bounds $r_{l;\,2.5\%}$

n_1/n_2	2	3	4	5	6	7	8	9	10	11	12	13	14	15	16	17	18	19	20
5				2	2														
6				2	2	3	3												
7				2	2	3	3	3											
8				2	3	3	3	3	4	4									
9				2	3	3	4	4	5	5									
10				2	3	3	4	5	5	5	6								
11				2	3	4	4	5	5	6	6	7							
12		2	2	3	4	4	5	6	6	7	7	7							
13		2	2	3	4	5	5	6	6	7	7	8	8						
14		2	2	3	4	5	5	6	7	7	8	8	9	9					
15		2	3	3	4	5	6	6	7	8	8	9	9	10					
16		2	3	4	4	5	6	6	7	8	8	9	10	10	11				
17		2	3	4	4	5	6	7	7	8	9	9	10	10	11	11			
18		2	3	4	5	5	6	7	8	8	9	9	10	11	11	12	12		
19		2	3	4	5	6	6	7	8	8	9	10	10	11	12	12	13	13	
20		2	3	4	5	6	6	7	8	9	9	10	11	12	12	13	13	13	14

upper 2.5%-bounds $r_{u;\,2.5\%}$

n_1/n_2	4	5	6	7	8	9	10	11	12	13	14	15	16	17	18	19	20
5		9	10														
6		9	10	11													
7			11	12	13												
8			11	12	13	14											
9				13	14	14	15										
10				13	14	15	16	16									
11				13	14	15	16	17	17								
12					13	14	16	16	17	18	19						
13						15	16	17	18	19	19	20					
14						15	16	17	18	19	20	20	21				
15						15	16	18	18	19	20	21	22	22			
16							17	18	19	20	21	22	22	23	23		
17							17	18	19	20	21	22	23	23	24	25	
18							17	18	19	20	21	22	23	24	25	26	
19							17	18	20	21	22	23	23	24	25	26	27
20							17	18	20	21	22	23	24	25	25	26	27	27	28

P = 0.10

lower 5%-bounds $r_{l;\,5\%}$

n_1/n_2	4	5	6	7	8	9	10	11	12	13	14	15	16	17	18	19	20
4	2																
5	2	3															
6	3	3	3														
7	3	3	4	4													
8	3	3	4	4	5												
9	3	4	4	5	5	6											
10	3	4	5	5	6	6	6										
11	3	4	5	5	6	6	7	7									
12	4	4	5	6	6	7	7	8	8								
13	4	4	5	6	6	7	8	8	9	9							
14	4	5	5	6	7	7	8	8	9	9	10						
15	4	5	6	6	7	8	8	9	9	10	10	11					
16	4	5	6	6	7	8	8	9	10	10	11	11	11				
17	4	5	6	7	7	8	9	9	10	10	11	11	12	12			
18	4	5	6	7	8	8	9	10	10	11	11	12	12	13	13		
19	4	5	6	7	8	8	9	10	10	11	12	12	13	13	14	14	
20	4	5	6	7	8	9	9	10	11	11	12	12	13	13	14	14	15

upper 5%-bounds $r_{u;\,5\%}$

n_1/n_2	4	5	6	7	8	9	10	11	12	13	14	15	16	17	18	19	20
4	8																
5	9	9															
6	9	10	11														
7		9	10	11	12												
8			11	12	13	13											
9			11	12	13	14	14										
10			11	12	13	14	15	16									
11				13	14	15	15	16	17								
12				13	14	15	16	17	17	18							
13				13	14	15	16	17	18	18	19						
14				13	14	16	17	17	18	19	20						
15					15	16	17	18	19	20	21	21					
16					15	16	17	18	19	20	21	21	22	23			
17					15	16	17	18	19	20	21	22	23	23	24		
18					15	16	18	19	20	21	22	22	23	24	24	25	
19					15	16	18	19	20	21	22	23	23	24	25	25	26
20					15	17	18	19	20	21	22	23	24	25	25	26	27	27

4.7 Testing the Randomness of a Sequence of Dichotomous Data

Table 99 (*continuation*)

$$n_1 = n_2 = n$$

n	P=0,10	P=0,05	P=0,02	P=0,01	n	P=0,10	P=0,05	P=0,02	P=0,01
20	15–27	14–28	13–29	12–30	60	51– 71	49– 73	47– 75	46– 76
21	16–28	15–29	14–30	13–31	61	52– 72	50– 74	48– 76	47– 77
22	17–29	16–30	14–32	14–32	62	53– 73	51– 75	49– 77	48– 78
23	17–31	16–32	15–33	14–34	63	54– 74	52– 76	50– 78	49– 79
24	18–32	17–33	16–34	15–35	64	55– 75	53– 77	51– 79	49– 81
25	19–33	18–34	17–35	16–36	65	56– 76	54– 78	52– 80	50– 82
26	20–34	19–35	18–36	17–37	66	57– 77	55– 79	53– 81	51– 83
27	21–35	20–36	19–37	18–38	67	58– 78	56– 80	54– 82	52– 84
28	22–36	21–37	19–39	18–40	68	58– 80	57– 81	54– 84	53– 85
29	23–37	22–38	20–40	19–41	69	59– 81	58– 82	55– 85	54– 86
30	24–38	22–40	21–41	20–42	70	60– 82	58– 84	56– 86	55– 87
31	25–39	23–41	22–42	21–43	71	61– 83	59– 85	57– 87	56– 88
32	25–41	24–42	23–43	22–44	72	62– 84	60– 86	58– 88	57– 89
33	26–42	25–43	24–44	23–45	73	63– 85	61– 87	59– 89	57– 91
34	27–43	26–44	24–46	23–47	74	64– 86	62– 88	60– 90	58– 92
35	28–44	27–45	25–47	24–48	75	65– 87	63– 89	61– 91	59– 93
36	29–45	28–46	26–48	25–49	76	66– 88	64– 90	62– 92	60– 94
37	30–46	29–47	27–49	26–50	77	67– 89	65– 91	63– 93	61– 95
38	31–47	30–48	28–50	27–51	78	68– 90	66– 92	64– 94	62– 96
39	32–48	30–50	29–51	28–52	79	69– 91	67– 93	64– 96	63– 97
40	33–49	31–51	30–52	29–53	80	70– 92	68– 94	65– 97	64– 98
41	34–50	32–52	31–53	29–55	81	71– 93	69– 95	66– 98	65– 99
42	35–51	33–53	31–54	30–56	82	71– 95	69– 97	67– 99	66–100
43	35–53	34–54	32–56	31–57	83	72– 96	70– 98	68–100	66–102
44	36–54	35–55	33–57	32–58	84	73– 97	71– 99	69–101	67–103
45	37–55	36–56	34–58	33–59	85	74– 98	72–100	70–102	68–104
46	38–56	37–57	35–59	34–60	86	75– 99	73–101	71–103	69–105
47	39–57	38–58	36–60	35–61	87	76–100	74–102	72–104	70–106
48	40–58	38–60	37–61	36–63	88	77–101	75–103	73–105	71–107
49	41–59	39–61	38–62	36–64	89	78–102	76–104	74–106	72–108
50	42–60	40–62	38–64	37–65	90	79–103	77–105	74–108	73–109
51	43–61	41–63	39–65	38–66	91	80–104	78–106	75–109	74–110
52	44–62	42–64	40–66	39–67	92	81–105	79–107	76–110	75–111
53	45–63	43–65	41–67	40–68	93	82–106	80–108	77–111	75–113
54	45–65	44–66	42–68	41–69	94	83–107	81–109	78–112	76–114
55	46–66	45–67	43–69	42–70	95	84–108	82–110	79–113	77–115
56	47–67	46–68	44–70	42–72	96	85–109	82–112	80–114	78–116
57	48–68	47–69	45–71	43–73	97	86–110	83–113	81–115	79–117
58	49–69	47–71	46–72	44–74	98	87–111	84–114	82–116	80–118
59	50–70	48–72	46–74	45–75	99	87–113	85–115	83–117	81–119
60	51–71	49–73	47–75	46–76	100	88–114	86–116	84–118	82–120

appear); for n_1 or $n_2 > 20$ one may use the approximation (4.51) or (4.51a) (cf., Table 14, Section 1.3.4, or Table 43, Section 2.1.6):

$$\hat{z} = \frac{|\hat{r} - \mu_r|}{\sigma_r} = \frac{\left|\hat{r} - \left(\frac{2n_1 n_2}{n_1 + n_2} + 1\right)\right|}{\sqrt{\frac{2n_1 n_2 (2n_1 n_2 - n_1 - n_2)}{(n_1 + n_2)^2 (n_1 + n_2 - 1)}}} = \frac{|n(\hat{r} - 1) - 2n_1 n_2|}{\sqrt{\frac{2n_1 n_2 (2n_1 n_2 - n)}{n - 1}}}; \quad (n = n_1 + n_2)$$

(4.51)

for $n_1 = n_2 = n/2$ (i.e., $n = 2n_1 = 2n_2$),

$$\hat{z} = \frac{\left|\hat{r} - \left(\frac{n}{2} + 1\right)\right|}{\sqrt{\frac{n(n-2)}{4(n-1)}}}.$$

(4.51a)

Two sided test: For $r_l < \hat{r} < r_u$, H_0 is retained; H_0 is rejected if either

$$\hat{r} \leq r_l \quad \text{or} \quad \hat{r} \geq r_u \quad \text{or} \quad \hat{z} \geq z_{\text{two sided}}.$$

One sided test: H_0 is rejected against H_{A1} (respectively H_{A2}) as soon as $\hat{r} \leq r_l$ (respectively $\hat{r} \geq r_u$) or $\hat{z} \geq z_{\text{one sided}}$.

More on this can be found in the work of Stevens (1939), Bateman (1948), Kruskal (1952), Levene (1952), Wallis (1952), Ludwig (1956), Olmstead (1958), and Dunn (1969).

The run test can also serve to test the null hypothesis that two samples of about the same size originate in the same population ($n_1 + n_2$ observations ordered by magnitude; then the values are dropped and only the population labels retained, to which the run test is applied; H_0 is abandoned for small $\hat{r}$).

Examples

1. Testing data for nonrandomness ($\alpha = 0.10$). The 11 observations 18, 17, 18, 19, 20, 19, 19, 21, 18, 21, 22, are obtained in sequence; let L denote a value larger or equal, S a value smaller than the median $\tilde{x} = 19$. For $n_1 = 4(S)$, $n_2 = 7(L)$ with $\hat{r} = 4$ the sequence SSSLLLLLSLL is compatible with the randomness hypothesis at the 10% level (Table 99; $P = 0.10$; $r_{l;5\%} = 3$ is not attained: $3 = r_{l;5\%} < \hat{r} < r_{u;5\%} = 9$).

2. Testing observations for noncluster effect ($\alpha = 0.05$) (i.e., testing H_0 against H_{A1} at the 5% level in terms of the lower 5% bounds of Table 99 or the standard normal distribution). Suppose two combined random samples of sizes $n_1 = 20$, $n_2 = 20$ form $\hat{r} = 15$ runs. Since by Table 99 $r_{l;5\%} = 15$ and H_0 is rejected for $\hat{r} \leq r_{l;5\%}$, the cluster effect hypothesis ($P = 0.05$) is accepted. This result is also obtained from (4.51a) and (4.51):

$$\hat{z} = \frac{|15 - (20 + 1)|}{\sqrt{40(40-2)/[4(40-1)]}} = 1.922,$$

$$\hat{z} = \frac{|40(15-1) - 2 \cdot 20 \cdot 20|}{\sqrt{[2 \cdot 20 \cdot 20(2 \cdot 20 \cdot 20 - 40)]/(40-1)}} = 1.922,$$

since by Table 43 (Section 2.1.6) $z_{5\%;\text{one sided}} = 1.645$ and H_0 is rejected for $\hat{z} \geq z_{5\%;\text{one sided}}$.

4.7.3 The phase frequency test of Wallis and Moore

This test evaluates the deviation of a sequence of measured data $x_1, x_2, \ldots, x_i, \ldots, x_n$ ($n > 10$) from **randomness**. The indices $1, 2, \ldots, i, \ldots, n$ denote a time sequence. If the sample is independent of time, then the signs of the differences $(x_{i+1} - x_i)$ are random (null hypothesis). The alternative hypothesis would then be: The sequence of plus and minus signs deviates significantly from random. The present test is thus to be regarded as a **difference–sign run test**.

A sequence of like signs is referred to as a **phase** according to Wallis and Moore (1941); the test is based on the frequency of the plus and minus phases. If the overall number of phases is denoted by h (small h is an indication of trend persistence), where the initial and the final phase are omitted, then under the assumption of randomness of a data sequence the test statistic (4.52a) for $n > 10$ is approximately standard normally distributed;

$$\hat{z} = \frac{\left|h - \frac{2n-7}{3}\right| - 0.5}{\sqrt{\frac{16n-29}{90}}}. \tag{4.52a}$$

For $n > 30$ the continuity correction can be omitted:

$$\hat{z} = \frac{\left|h - \frac{2n-7}{3}\right|}{\sqrt{\frac{16n-29}{90}}}. \tag{4.52}$$

EXAMPLE. Given the following sequence consisting of 22 values in Table 100.

Table 100

Data	5 6 2 3 5 6 4 3 7 8 9 7 5 3 4 7 3 5 6 7 8 9
Signs	+ - + + + - - + + + - - - - + + - + + + + +
Phase number	1 2 3 4 5 6 7

For $h = 7$,

$$\hat{z} = \frac{\left|7 - \frac{2 \cdot 22 - 7}{3}\right| - 0.5}{\sqrt{\frac{16 \cdot 22 - 29}{90}}} = \frac{4.83}{1.89} = 2.56 > 1.96 = z_{0.05}.$$

The result is significant at the 5% level; the null hypothesis is rejected.

4.8 THE S_3 SIGN TEST OF COX AND STUART FOR DETECTION OF A MONOTONE TREND

A time series is a chronological sequence of (historical) data, a set of observations ordered according to time, for example the monthly unemployment figures in this country. To test a time series (cf., Bihn 1967, Harris 1967,

Jenkins 1968, Jenkins and Watts 1968, and the appropriate chapter in Suits 1963 or Yamane 1967) for monotonic (cf., Section 5.3.1) trend (H_0: no trend [randomness]; H_A: monotonic trend) the n values of the sequence are partitioned into three groups so that the first and last, with $n' = n/3$, have an equal number of data values. The middle third is, for sample sizes n not divisible by 3, reduced by one or two values. Every kth observation in the first third of the data sequence is compared with the corresponding ($\frac{2}{3}n + k$)th observation in the last third of the data sequence, and a "plus" is marked next to an ascending trend, a "minus" next to a descending trend, the mark thus depending on whether a positive or a negative difference appears (Cox and Stuart 1955). The sum S of the plus or the minus signs is approximately normally distributed with an expected value of $n/6$ and a standard deviation of $\sqrt{n/12}$, so that

$$\hat{z} = \frac{|S - n/6|}{\sqrt{n/12}}. \qquad (4.53)$$

For small samples ($n < 30$) this is corrected according to Yates:

$$\hat{z} = \frac{|S - n/6| - 0.5}{\sqrt{n/12}}. \qquad (4.53a)$$

EXAMPLE. We use the values of the last example. Since 22 is not divisible by 3, we measure off both thirds as if n were equal to 24 (see Table 101). We find that 7 of 8 signs are positive. The test for ascending trend yields

$$\hat{z} = \frac{\left|7 - \frac{22}{6}\right| - 0.5}{\sqrt{22/12}} = \frac{2.83}{1.35} = 2.10.$$

To $\hat{z} = 2.10$ there corresponds, by Table 13 for a two sided question, a random probability $P \simeq 0.0357$. The increasing trend is ascertained at the 5% level.

Table 101

Data values of the last third	4	7	3	5	6	7	8	9
Data values of the first third	5	6	2	3	5	6	4	3
Sign of the difference	−	+	+	+	+	+	+	+

Remarks

1. If the mean of a data sequence changes abruptly at a certain instant of time, after n_1 observation say, then the difference of the two means, $\bar{x}_1 - \bar{x}_2$, where $\bar{x}_2$ is the mean of the subsequent n_2 observations, can be tested (with one degree of freedom) (Cochran 1954) by

$$\hat{\chi}^2 = \frac{n_1 n_2}{n} \cdot \frac{(\bar{x}_1 - \bar{x}_2)^2}{\bar{x}}, \qquad (4.54)$$

with $n = n_1 + n_2$ and $\bar{x}$ the common mean of all data values. The difference between the two portions of the time sequence can be assessed by a one sided test, provided a substantiated assumption on the direction of the change is available; otherwise the two sided question is chosen (cf., also Section 5.6 and the end of Section 7.4.3).

2. Important partial aspects of trend analysis (cf., also the end of Section 5.2) are considered by Weichselberger (1964), Parzen (1967), Bredenkamp (1968), Nullau (1968), Sarris (1968), Bogartz (1968), Jesdinsky (1969), Rehse (1970), and Box and Jenkins (1976).

3. A survey on time series is given by Makridakis (1976). An update and evaluation of time series analysis and forecasting is given by S. Makridakis in International Statistical Review **46** (1978), 255–278 [for time series analysis, see also **49** (1981), 235–264 and **51** (1983), 111–163]. An empirical investigation on the accuracy of forecasting is given by S. Makridakis and M. Hibon, Journal of the Royal Statistical Society A **142** (1979), 97–145. D. R. Cox provides a selective review of the statistical analysis of time series, Scandinavian Journal of Statistics **8** (1981), 93–115.

4. Analytical procedures for cross-sectional time series in which the sample size is large and the number of observations per case is relatively small are given by D. K. Simonton, Psychological Bulletin **84** (1977), 489–502.

5 MEASURES OF ASSOCIATION: CORRELATION AND REGRESSION

5.1 PRELIMINARY REMARKS AND SURVEY

In many situations it is desirable to learn something about the association between two attributes of an individual, a material, a product, or a process. In some cases it can be ascertained by theoretical considerations that two attributes are related to each other. The problem then consists of determining the nature and degree of the relation. First the pairs of values (x_i, y_i) are plotted in a coordinate system in a two dimensional space. The resulting scatter diagram gives us an idea about the dispersion, the form and the direction of the point "cloud".

1. The length and weight of each piece of wire in a collection of such pieces (of uniform material, constant diameter) is measured. The points lie on a straight line. With increasing length the weight increases proportionally: equal lengths always give the same weights and conversely. The weight y of the piece of wire is a function of its length x. There exists a **functional relation** between x and y. Here it does not matter which variable is assigned values and which is measured. Likewise the area A of a circle is a function of the radius r and conversely ($A = \pi r^2$, $r = \sqrt{A/\pi}$, with $\pi \simeq 3.1416$): To every radius there corresponds a uniquely determined area and conversely.
2. If errors in measurement occur, then to a given length there do not always correspond equal weights. The plot reveals a point cloud with a clear trend (cf., e.g., Figure 39): In general, the weight gets larger with increasing length. The so-called equalizing line, drawn through the point cloud for visual estimation, allows one to read off: (1) what y-value can be expected for a specified x-value and (2) what x-value can be expected for

5.1 Preliminary Remarks and Survey

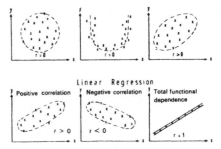

Figure 39 The sample correlation coefficient r indicates the degree of association between sample values of the random variables X and Y [x and y are their realizations]; it is a measure of the linear relationship between X and Y. The plot in the middle of the top row implies a U-shaped relation.

a specified y-value. In place of the functional relation there is here a more or less loose connection, which we refer to as a **stochastic relation**.

3. In fields such as biology and sociology the considerable natural variation in the objects under study often contributes more to the error than inaccuracies in measurements and/or observations (in the example of the wire, this would be due to nonuniform material with varying diameter). The point cloud becomes larger and perhaps loses its clearly recognizable trend. In stochastic relations (cf. also Sections 4.6.1, 4.6.6, 4.7) one distinguishes correlation (does there exist a stochastic relation between x and y? how strong is it?) and regression (what kind of relation exists between x and y? can y be estimated from x?). Let us first give a survey.

I. Correlation analysis

Correlation analysis investigates stochastic relations between random variables of equal importance on the basis of a sample. A statistic for the strength of the **LINEAR** relationship between two variables is the product moment correlation coefficient of Bravais and Pearson, called correlation coefficient for short. It equals zero if there is no linear relation (cf., Figure 39).

For the correlation coefficient ρ (the parameter is denoted by the Greek letter rho) of the two random variables (cf. Section 1.2.5) X and Y we have:

(1) $-1 \leq \rho \leq 1$ (ρ is a dimensionless quantity).
(2) For $\rho = \pm 1$ there exists a functional relationship between X and Y, all [empirical] points (x, y) lie on a straight line (cf., II, 7).
(3) If $\rho = 0$ then we say X and Y are uncorrelated (independent random variables are uncorrelated; two random variables are the more strongly correlated the closer $|\rho|$ is to 1).
(4) For a **bivariate (two-dimensional) normal distribution** ρ is a MEASURE of LINEAR INTERDEPENDENCE; and $\rho = 0$ implies the **stochastic independence** of X and Y.

The two-dimensional normal distribution is a bell shaped surface in space (cf., Figure 47 below; there $\rho \simeq 0$) which is characterized by ρ (and four additional parameters: $\mu_x, \mu_y, \sigma_x, \sigma_y$). The cross section parallel to the X, Y plane is a circle for $\rho = 0$ and $\sigma_x = \sigma_y$, and an ellipse for $\sigma_x \neq \sigma_y$ or $\rho \neq 0$, which becomes narrower as $\rho \to 1$.

The parameter ρ is estimated by the sample correlation coefficient **r** (Section 5.4.1); r is, for nonnormally distributed random variables with approximately linear regression (cf., II, 2 below), a measure of the strength of the stochastic relation.

We consider:

1. The correlation coefficient (Section 5.4.1).
2. The partial correlation coefficient (Section 5.8).
3. The multiple correlation coefficient (Section 5.8).
4. The rank correlation coefficient of Spearman (Sections 5.3 and 5.3.1).
5. The quadrant correlation (Section 5.3.2) and the corner test (Section 5.3.3). Both allow one to test the presence of correlation without computation but through the analysis of the point cloud alone. In the corner test points lying "furthest out" are decisive. The exact values need not be known for either procedure.

Several remarks on avoiding incorrect interpretations in correlation analysis are made in Section 5.2. Contingency coefficients (cf., Section 4.6.6) will be discussed in Section 6.2.2 of the next chapter. The last remark in Section 5.5.9 concerns bivariate normality.

II. Regression analysis

1. In regression analysis a regression equation is fitted to an observed point cloud.
2. A straight line

$$Y = \alpha + \beta X \tag{5.1}$$

describes a linear relationship—a linear regression—between the dependent (random) variable Y (predictand, regressand or response variable) and the independent (not random) variable X (regressor, predictor or explanatory variable). If Y and X are the components of a bivariate normally distributed vector, then the regression line (5.1) can be written as $(Y - \mu_y)/\sigma_y = \rho(X - \mu_x)/\sigma_x$, or $Y = \mu_y + \rho(\sigma_y/\sigma_x)(X - \mu_x)$.

5.1 Preliminary Remarks and Survey

3. The parameters [e.g., α and β in (5.1)] are estimated from the sample values, usually by the method of least squares with the help of the so-called normal equations (Section 5.4.2), or else by the maximum likelihood method.
4. Estimations and tests of parameters are discussed in Sections 5.4 and 5.5. Often only the pairs (x_i, y_i) of data are known (but not the causal relationship nor the underlying bivariate distribution). Since by II, 2 the variable X is fully specified and Y is a random variable, it is to our advantage to get several measurements of y at the same x_i and average these values to $\bar{y}_i$. Then we investigate the changes in the target variable $\bar{y}_i$ as a function of the changes in the influence variable x (regression as "dependence in the mean"; cf., also Section 5.5.3).
5. Frequently it is impossible to specify x without error (observation error, measurement error). The influence quantities and target quantities are then afflicted with error. Special methods (see below) deal with this situation.
6. In addition to the straightforward linear regression, one distinguishes nonlinear (curvilinear) regression (Section 5.6) and multiple regression, characterized by several influence quantities (Section 5.8).
7. Correlation and regression: If the two variables are the components of a two dimensional normally distributed random variable, then there exist two regression lines (see Figures 43 through 46 below and Section 5.4.2). The first infers Y (target quantity $\hat{Y}$) from X, the second X (target quantity $\hat{X}$) from Y (see below, and the example at the end of Section 5.4.4). The two regression lines intersect at the center of gravity $(\bar{X}, \bar{Y})$ and form a "pair of scissors" (cf., Figure 46 below): the narrower they are, the tighter is the stochastic relation. For $|\rho| = 1$ they close up, we have $\hat{Y} = \hat{X}$, and the two regression lines coincide: there exists a linear relation. Thus ρ is a measure of the **linear relation** between $\hat{X}$ and $\hat{Y}$. For $\rho = 0$, the two regression lines are perpendicular to each other and run parallel to the coordinate axes (stochastic independence) (cf., Figure 45, below).

It is the aim of regression analysis to find a **functional relationship** between Y and X by means of an empirical function $\bar{y}_i(x_i)$, the graphical presentation of the conditional mean $\bar{y}_i(x_i)$ as a function of x_i, which allows us to estimate for preassigned values (arbitrary values) of the independent variable x the respective dependent variable y.

If data pairs (x_i, y_i) are given, then the simplest relationship $y_i(x_i)$ between y_i and x_i, i.e., y_i as a function of x_i, is described by the **equation of a straight line** (Figure 40).

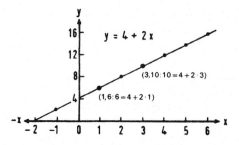

Figure 40 The line $y = 4 + 2x$.

The general equation of a straight line (Figure 41) can be written as $y = a + bx$; a and b are the parameters: a stands for the segment of the y-axis extending from the origin 0 to the intersection of the line with the y-axis, and is referred to as the **intercept** (on the ordinate); b specifies how much y grows when x increases by one unit, and is called the **slope**. In the case of a regression line the slope is called the **coefficient of regression**. A negative

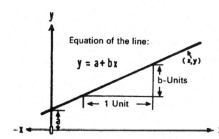

Figure 41 The equation of the straight line.

value of the regression coefficient means that the predictand y decreases when the regressor x increases (Figure 42 with $b < 0$).

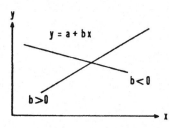

Figure 42 The sign of the regression coefficient b determines whether with increasing x values the associated y values increase (b positive) or decrease (b negative).

To estimate the parameters of the regression line, more precisely, the regression line of "y on x" (indicated by the double index yx: $y = a_{yx} + b_{yx}x$), we adopt the principle that the straight line should fit the empirical y-values as well as possible. The sum of the squares of the vertical deviations (d) (Figure 43) of the empirical y-values from the estimated straight line is to be smaller than from any other straight line. By this "**method of least squares**" (see Harter 1974, 1975 [8:1]) one can determine both coefficients a_{yx} and b_{yx} for the prediction of y from x.

5.1 Preliminary Remarks and Survey

If for a point cloud, as for example the one given in Fig. 39 (lower left), from specified or arbitrary values of the independent variable y, the various values of the dependent variable x are to be estimated (for example the dependence of the duration of the gestation period on the bodily length of the newborn)—i.e., if the parameters a_{xy} (here denoted by a') and b_{xy} of the regression line of x on y are to be estimated (Figure 44):

$$\hat{x} = a' + b_{xy}y,$$

—then the sum of the squares of the horizontal deviations (d') is made a minimum.

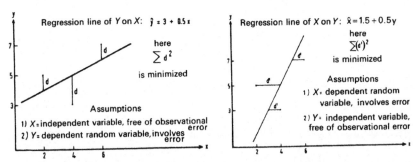

Figures 43, 44 The two regression lines: interchange of the dependent and independent variables. The estimation of $\hat{y}$ from given x-values is not the inverse of the estimation of $\hat{x}$ from y-values: If we estimate $\hat{y}$ from x with the help of the regression line of Y on X then we make the sum of the vertical squares d^2 a minimum; if we estimate $\hat{x}$ from y with the help of the regression of X on Y, then we make the sum of the horizontal squares $(d')^2$ a minimum.

It is sometimes difficult to decide which regression equation is appropriate. Of course it depends on whether x or y is to be predicted. **In the natural sciences** every equation connects only precise quantities, and **the question of which variable is independent is often irrelevant**. The measurement errors are usually small, the correlation is pronounced, and the difference between the regression lines is negligible. If the point cloud in Figure 39 (lower left) is made to condense into a straight line—perfect functional dependence (cf., Figure 39, lower right)—then the two regression lines coincide (Figure 45). We thus obtain a correlation coefficient of $r = 1$. As r increases, the angle between the regression lines becomes smaller (Figure 46).

$$\boxed{\hat{x} = a_{xy} + b_{xy}y = a' + b_{xy}y} \qquad (5.2)$$

$$\boxed{\hat{y} = a_{yx} + b_{yx}x} \qquad (5.3)$$

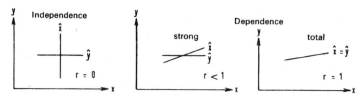

Figure 45 With increasing dependence or correlation the two regression lines $\hat{y} = a_{yx} + b_{yx}x$ and $\hat{x} = a_{xy} + b_{xy}y$ approach each other. When r is near zero they are approximately at right angles to each other. As r increases, the regression lines come closer together until they coincide for $r = 1$.

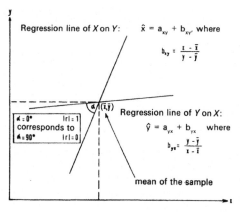

Figure 46 The connection between correlation and regression: The absolute values of the correlation coefficient can be taken as a measure of the angle α between the two regression lines. The smaller α is, the larger r is. Moreover, $\tan \alpha = (1 - r^2)/2r$, or $r = \sqrt{1 + \tan^2 \alpha} - \tan \alpha$. For $r = 0$ with $\alpha = 90°$ both straight lines are orthogonal.

It can further be shown that the correlation coefficient is the geometric mean of the two regression coefficients b_{yx} and b_{xy}:

$$\boxed{r = \sqrt{b_{yx}b_{xy}}.} \tag{5.4}$$

Since $b_{yx}b_{xy} = r^2 \leq 1$, one of the two regression coefficients must be less than unity and the other greater than unity or both will be unity (cf., examples in Section 5.4.2).

The following formula emphasizes once again the close connection between the correlation and regression coefficients:

$$\boxed{r = b_{yx}\frac{s_x}{s_y} \quad \text{or} \quad b_{yx} = r\frac{s_y}{s_x}.} \tag{5.5}$$

Since standard deviations are positive, this relation implies that r and b must have the same sign. If the two variables have the same dispersion, i.e., $s_x = s_y$, then the correlation coefficient is identical to the regression coefficient b_{yx}.

5.1 Preliminary Remarks and Survey

The ratio

$$r^2 = \frac{\sum(\hat{y}_i - \bar{y})^2}{\sum(y_i - \bar{y})^2} = 1 - \frac{\sum(y_i - \hat{y}_i)^2}{\sum(y_i - \bar{y})^2} \tag{5.6}$$

is called the **coefficient of determination**. The equivalent expressions emphasize two aspects: in the first the dispersion of the predicted values is compared with the dispersion of the observations. The second term in the expression on the right is a measure of how well the predicted values fit. The less the observed values depart from the fitted line, the smaller this ratio is and the closer r^2 is to 1. Thus r^2 can be considered **a measure for how well the regression line explains the observed values**. If $r^2 = 0.9^2 = 0.81$, then 81% of the variance of the target quantity y can be accounted for by the linear regression between y and the influence quantity x [cf., (5.6), middle]. For interesting remarks concerning r^2, see D. Griffiths, The Statistician **31** (1982), 268–270.

Comments regarding the coefficient of determination

If the random variables X and Y (cf., Section 1.2.5) have a bivariate normal distribution with the variances σ_x^2 and σ_y^2, and if we denote the variance of Y with X given by $\sigma_{y.x}^2$ and the variance of X with Y given by $\sigma_{x.y}^2$, then in terms of the coefficient of determination ρ^2 we have

$$\begin{aligned}\sigma_{y.x}^2 &= \sigma_y^2(1 - \rho^2), \\ \sigma_{x.y}^2 &= \sigma_x^2(1 - \rho^2),\end{aligned} \tag{5.6a}$$

and for:

1. $\rho = 0$ (points not on a straight line but widely dispersed), we have

$$\sigma_{y.x}^2 = \sigma_y^2 \quad \text{and} \quad \sigma_{x.y}^2 = \sigma_x^2;$$

2. $\rho = 1$ (points on a straight line), we have

$$\sigma_{y.x}^2 = 0 \quad \text{and} \quad \sigma_{x.y}^2 = 0.$$

We see that $\sigma_{y.x}^2$ and $\sigma_{x.y}^2$ are the variances of the estimation of Y and X, respectively, by linear regression (cf., Section 5.4.3).

For (5.6a) we may write

$$\rho^2 = \frac{\sigma_y^2 - \sigma_{y.x}^2}{\sigma_y^2} = \frac{\sigma_x^2 - \sigma_{x.y}^2}{\sigma_x^2}. \tag{5.6b}$$

Thus ρ^2 **is a measure of the linearity of the points** that shows the relative reduction of the total error when a regression line is fitted. For example, if $\rho^2 = 0.8$, it means that 80% of the variations in Y are "explained" by the variation in X.

If the quantities x and y are measured for every element of a random sample, then the errors in measurement are disregarded in view of the much higher variability between the individual x- and y-values. The classic example is the relationship between body size and body weight in men. Both quantities are random variables. Figure 47 gives an idealized frequency surface for distributions of this sort. Only in this situation are there **two regression lines**: One for estimating $\hat{y}$ when x is given, and another for estimating $\hat{x}$ from y.

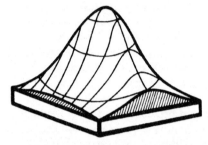

Figure 47 Ideal symmetric ("normal") frequency surface with extreme regions cut off: Truncated two dimensional normal distribution.

Only in this case does the **correlation coefficient** r of the sample have a meaning as a measure of association between X and Y in the population. If the samples are not fully random in both variables, but one, x say, is deterministic (e.g., all men with heights x exactly 169.5 cm to 170.5 cm, 179.5 to 180.5 cm, etc., are chosen and their body weights analyzed), then:

1. no correlation coefficient can be computed,
2. nor can a regression line for estimating $\hat{x}$ from y be determined;
3. only the regression line for estimating $\hat{y}$ from x can be worked out:

$$\hat{y} = a_{yx} + b_{yx} x.$$

We repeat: this is the case if the values of the attribute y of sample elements with particular x-values are examined, in other words after a preselection from the sample on the basis of the value of the variable x.

Estimates of the coefficient of correlation and of regression lines according to standard methods are given in Section 5.4. Bartlett and Kerrich's quick estimates of the regression line in the case where both x and y are subjected to error are described below. (Cf., Tukey 1951, Acton 1959, Madansky 1959, Carlson et al., 1966). For collinearity and robust regression, see Hocking and Pendleton (1983).

5.1.1 The Bartlett procedure

Partition the n points into 3 groups according to the magnitude of x, of the same size if possible, where the first and the third group contain exactly k points (with the k smallest and largest x-components, respectively). The regression coefficient is then estimated by

$$\boxed{\hat{b} = \frac{\bar{y}_3 - \bar{y}_1}{\bar{x}_3 - \bar{x}_1},} \tag{5.7}$$

5.1 Preliminary Remarks and Survey

where $\bar{y}_3$ = the mean of y in the third group, $\bar{y}_1$ = the mean of y in the first group; $\bar{x}_3$ = the mean of x in the third group, and $\bar{x}_1$ = the mean of x in the first group. The y-intercept is determined by

$$\hat{a} = \bar{y} - \hat{b}\bar{x}, \tag{5.8}$$

where $\bar{x}$ and $\bar{y}$ stand for the overall means.

This method is surprisingly effective if the distance between consecutive values of x is held constant. Wendy M. Gibson and G. H. Jowett (1957) mention in an interesting study that the ratio of the sizes of the three groups should be roughly $1:2:1$. However, the result based on the ratio $1:1:1$ does not differ critically: This ratio is optimal for U-shaped and rectangular distributions, while the $1:2:1$ ratio is preferable for J-shaped and skewed distributions as well as for a normal distribution.

For verification, the rapid estimate $\hat{b} \simeq \sum y / \sum x$ can be used. If the line does not pass through the origin, then the parameters a and b are estimated from the upper 30% and the lower 30% of the values (Cureton 1966):

$$\hat{b} \simeq \frac{\sum y_{u.} - \sum y_{1.}}{\sum x_{u.} - \sum x_{1.}}, \tag{5.9}$$

$$\hat{a} \simeq \sum y_{1.} - \hat{b} \sum x_{1.}. \tag{5.10}$$

EXAMPLE. Estimating the regression line if both variables (x, y) are subjected to measurement errors: The comparison of two methods of measuring between which a linear relation is assumed. For the data in Table 102, the

Table 102

Sample (No.)	Method I (x)	Method II (y)
1	38.2	54.1
2	43.3	62.0
3	47.1	64.5
4	47.9	66.6
5	55.6	75.7
6	64.0	83.3
7	72.8	91.8
8	78.9	100.6
9	100.7	123.4
10	116.3	138.3

fitted line goes through the point $(\bar{x}, \bar{y})$ with the values $\bar{x} = 66.48$ and $\bar{y} = 86.03$. We estimate the regression coefficients in terms of the means of the first and last thirds of the two sequences according to (5.7):

$$\hat{b} = \frac{\bar{y}_3 - \bar{y}_1}{\bar{x}_3 - \bar{x}_1} = \frac{120.767 - 60.200}{98.633 - 42.867} = 1.0861.$$

The y-intercept is found by (5.8) in terms of the overall means: $\hat{a} = \bar{y} - \hat{b}\bar{x} = 86.03 - (1.0861)(66.48) = 13.826$. The fitted regression line is thus given by $\hat{y} = 13.833 + 1.0861x$. The graphical presentation of this problem and the computation according to Cureton of (5.9), (5.10) are recommended as exercises.

The calculation of the confidence ellipses for the estimated parameters (cf., Mandel and Linning 1957) can be found in Bartlett (1949).

5.1.2 The Kerrich procedure

If both variables are subject to error, only positive values of x_i and y_i come up, and the point cloud hugs a line $(y = bx)$ passing through the origin, then one can use the following elegant procedure (Kerrich 1966) for estimating b: For n independent data pairs (x_i, y_i) one forms the differences $d_i = \log y_i - \log x_i$, their mean $\bar{d}$, and the standard deviation

$$s_d = \sqrt{\sum (d_i - \bar{d})^2/(n-1)}. \tag{5.11}$$

Since each quotient y_i/x_i represents an estimate of b, each d_i is an estimate of $\log b$. A useful estimate of $\log b$ is $\bar{d}$, particularly if the quantities x_i and y_i exhibit small coefficients of variation. It is assumed that $\log y_i$ and $\log x_i$ are at least approximately normally distributed.

The 95% confidence interval for β is given by

$$\log b \pm s_d t_{n-1;\,0.05}/\sqrt{n}. \tag{5.12}$$

EXAMPLE. Given: $n = 16$ data pairs (the fitted line passes through the origin) with $\bar{d} = 9.55911 - 10 = \log b$ and $s_d = 0.00555$—i.e., $t_{15;\,0.05} = 2.131$ and hence $s_d t_{n-1;\,0.05}/\sqrt{n} = 0.00555 \cdot 2.131/\sqrt{16} = 0.00296$. The 95% confidence interval for $\log \beta$ is $9.55911 - 10 \pm 0.00296$; i.e., $\hat{b} = 0.362$, $0.359 \leq \beta \leq 0.365$.

Special considerations in fitting the no intercept model are discussed by G. J. Hahn, Journal of Quality Technology **9** (1977), 56–61 and G. Casella, The American Statistician **37** (1983), 147–152.

5.2 HYPOTHESES ON CAUSATION MUST COME FROM OUTSIDE, NOT FROM STATISTICS

One speaks of stochastic dependence, or of a stochastic relation, if the null hypothesis that there is stochastic independence is disproved.

The factual interpretation of any statistical relations, and their testing for possible causal relations, lies outside the scope of statistical methodology. Even when stochastic dependence appears certain, one must bear in mind that the existence of a functional relation—for example the increase in the number of storks and newborns during a certain period of time in Sweden—says nothing about a causal relation. There can exist a pronounced positive correlation between the dosage of some medicine and the lethality of a disease although the lethality increases not because of the larger dosage but in spite of it. A correlation can be conditioned by direct **causal relations** between x and y, by a **joint dependence on a third quantity**, or by **heterogeneity of the material**—or it can be **purely formal**. **Causal correlations** exist, e.g., between ability and achievement, between dosage and effect of a remedy, between working time involved and the price of a product. Examples of **simultaneous correlation** are the relationships among bodily dimensions: e.g. between the length of the left and the right arm, or between height and body weight, as well as correlations between time series, such as the decrease in the number of stork nests in East Prussia and the decrease in the number of births, caused by the growing industrialization.

If we combine three groups, say, each with $r \simeq 0$, and if the 3 data point clouds happen to lie nearly on a straight line, then the resulting large value of r for all groups is meaningless. Thus **inhomogeneity correlation** can occur when the data space dissociates into several regions. The overall correlation might be completely different from the correlation within the single regions. Examples are given by J. N. Morgan and J. A. Sonquist, [Journal of the American Statistical Association **58**, 1963, pp. 415–434]. The following example is particularly impressive: The hemoglobin content of the blood and the surface areas of the blood corpuscles show no correlation in the newborn or in men or in women. The coefficient values are -0.06, -0.03, and -0.07 respectively. If however one were to combine the data, one would find a correlation coefficient of -0.75.

If for example x and y are percentages which add up to 100%, then there must necessarily be a negative correlation between them, as for the protein and fat content of foodstuff, etc. The expression "spurious correlation" is normally used for these relationships; it should however be avoided, since in fact a "spurious correlation" between two such percentages is not an illusion but is quite real. Besides this formal correlation there is, as indicated above, a whole collection of additional noncausal correlations.

For the interpretation of correlations in real life data Koller (1955, 1963) provides guidelines which enable us to recognize proper, or rather causal, correlations by excluding other possibilities (cf., selection correlation, at the beginning of Chapter 3). In order to interpret a correlation we can ask first whether the correlation might be only formal. If this is not the case we proceed according to the scheme below:

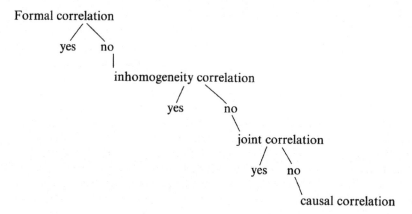

The recognition of a causal correlation thus follows from the exclusion of other possibilities. Because of the possible overlapping of types, the scheme cannot always be applied as strictly, in such well-defined steps, as presented in the model. Frequently one cannot really advance to the causal correlation type but must stop at a preceding type, not being able to disprove this type for the particular case. The size of the correlation coefficient will rarely make any difference in this context.

Causal statements. A remark on statistical investigation of causes. A cause is any condition which cannot be conceptually traced further back—the fundamental component through which the effect enters in. One must test whether the available statistical information is sufficient for adopting a causation adequate to the effect. A more detailed discussion of the possible statements in statistical investigations of causes can be found in Koller (1971 [8 : 2a]). A survey on causal interpretations of statistical relationships with comments on spurious correlation, a typology of causal relations, and a typology of three-variable analyses is given by Nowak (1975).

Remark. Correlation among time series. Time series (for bibliography see Section 4.8 as well as Brown 1962, Pfanzagl 1963, Ferguson 1965, and Kendall 1973) almost always exhibit a general trend, a rise or a fall. If one correlates two increasing series (e.g., the population, the energy production, the price index, and the number of traffic accidents) or two decreasing series (e.g., infant mortality and the proportion of agricultural workers in the population), one finds positive correlation, which may even be quite large (joint correlation). Misinterpretations are common here. One can guard against overrating time correlations by considering additional control variables with the same trend. If the correlation determined originally (e.g., the growth of a disease and increase in the consumption of a certain luxury food) does not differ very substantially from the control correlations (e.g., production of television sets), or if the partial correlation (cf., Section 5.8) with the control variables held constant is noticeably smaller than the original correlation, then joint correlation can be ruled out.

5.3 DISTRIBUTION-FREE MEASURES OF ASSOCIATION

A test on the correlation between the two components of a set of data that is based on the product moment correlation r, as an estimate of the parameter ρ, presupposes a population with an approximately bivariate normal distribution. But this assumption is often not satisfied or only partially satisfied.

If the condition does not hold, the rank correlation coefficient of Spearman (r_S) is used in general, whereby transformations—otherwise perhaps necessary to achieve approximate normality—can be avoided and a substantial amount of time can be saved. The test is exact also for small sample sizes and nonnormal data; moreover the effects of outliers, which greatly influence the size of the product moment correlation, are weakened. A further advantage lies in the invariance of r_S under monotone transformations of x; the product moment correlation is not invariant when x is replaced by $f(x)$. For large sized samples from bivariate normal populations with sufficiently small product moment correlation coefficients ($|\rho| < 0.25$), the test based on r_S has the same power as a test based on r from a sample of size $0.91n$. The rank correlation thus uses 91% of the information on the correlation in the sample. Because the slight loss of accuracy is connected with a significant saving in time, r_S frequency serves for rapid orientation and possibly as an estimate of the usual correlation coefficients in the population. In the case of data from a normal distribution $|\rho|$ will be somewhat overestimated. With increasing sample size r_S does not approach ρ as r does, but rather approaches ρ_S. **The difference between ρ and ρ_S is however always less than** 0.018 (cf., Walter 1963).

There are considerable advantages in applying r_S in nonlinear monotone regression: e.g., for attributes between which there exists a logarithmic or exponential relationship, so that when one variable increases the other either almost always increases or almost always decreases. If we want to use r as a correlation measure, the data have to be transformed so as to render the relationship linear. The use of r_S leads to a significant saving in time.

Also very handy is the medial or **quadrant correlation** of Quenouille, which is appropriate for survey purposes and which evolved from the **corner test**. If a normal distribution is present, then the quadrant correlation coefficient (r_Q) can also be used for estimating the usual correlation coefficient ρ, although in this case the test is not particularly powerful, since it utilizes only 41% of the information in the sample. Like the rank correlation coefficient, however, the quadrant correlation coefficient has the advantage of providing a **valid test** regardless of the underlying distribution function, **decreasing the effects of outliers** and being **invariant under transformations**.

▶ 5.3.1 The Spearman rank correlation coefficient

If the bivariate sample (x, y), the relationship between whose components we wish to examine, originates in a population with continuous nonnormal distribution, then the mutual dependence of x and y can be assessed through the Spearman rank correlation coefficient r_S:

$$r_S = 1 - \frac{6 \sum D^2}{n(n^2 - 1)} \qquad (5.13)$$

where D is the difference of the pair of rankings, the rank difference for short. Note that $-1 \leq r_S \leq 1$. [(5.13) is identical to (5.18) in Section 5.4.1 if in (5.18) the measured values are replaces by rank numbers]. To compute the rank correlation coefficient both the x-set and the y-set are ranked, and then the difference D_i between the ranks of the components of each sample point is formed, squared, and summed as indicated by the above formula. Mean ranks are assigned to equal values within a set ("ties"); in either of the two sequences, at most about 1/5 of the observations are allowed to be of equal rank. With ties present it is best to use (5.16).

If two rank orders are equal, the differences are zero, i.e., $r_S = 1$. If one rank order is the reverse of the other, so that there is total discrepancy, we get $r_S = -1$. This test thus allows one to answer the question whether a positive or a negative correlation is present.

It is supposed that the following assumptions hold (concerning X and Y, see Section 1.2.5):

1. X and Y are continuous random variables. They are at least ranked (ordinal) data.
2. The data are independent paired observations.

Then we may test:

$$H_0: X \text{ and } Y \text{ are independent (or } \rho_S = 0).$$

Two sided case

H_A: X and Y are correlated (or there is a linear or at least monotonic relationship, or $\rho_S \neq 0$). [Monotonic: The sequence $x_1 \leq x_2 \leq x_3 \leq \ldots$ is called monotonic increasing; it is never decreasing. The sequence $x_1 \geq x_2 \geq x_3 \geq \ldots$ is called monotonic decreasing; it is never increasing].

One sided case

H_{A1}: Small X and small Y tend to occur together, as do large X and large Y. In short: X and Y are positively correlated (or there is a positive linear or at least a positive monotonic relationship, or $\rho_S > 0$).

H_{A2}: Small X and large Y tend to occur together, as do large X and small Y. In short: X and Y are negatively correlated (or there is a negative linear or at least a negative monotonic relationship, or $\rho_S < 0$).

Decision

For the two sided and for the one sided test H_0 is rejected at the $100\alpha\%$ level for

$$|r_S| \geq \text{critical value } r^*_{S;n;\alpha},$$

from Table 103.

For $n > 100$, H_0 is rejected at the $100\alpha\%$ level with the formula (5.15) and $\hat{t} \geq t_{n-2;\alpha}$.

For $n > 100$ the significance of r_S can be tested with sufficient accuracy according to (5.14) [cf., also (5.15)] on the basis of the standard normal (p.217) distribution

$$\boxed{\hat{z} = |r_S| \cdot \sqrt{n-1}.} \qquad (5.14)$$

If, for example, for $n = 30$ and a one sided test a value of $r_S = 0.307$ obtains, then $|0.307| \cdot \sqrt{30-1} = 1.653 > 1.645 = z_{0.05;\text{one-sided}}$ implies that one has a positive correlation, significant at the 5% level ($r_S = 0.307 > 0.306 = r^*_S$ from Table 103). For the 7 observations x, y in Table 107 (Section 5.4.2) and the two sided test, $H_0: \rho_S = 0$ must be retained at the 5% level (cf., Table 107a): (p.411)

$$r_S = 1 - \frac{6(15.5)}{7(49-1)} = 0.7232 < 0.786 = r^*_S$$

($n = 7$, $\alpha_{0.025;\text{one sided}} = \alpha_{0.05;\text{two sided}}$); there is no true correlation. Since $\frac{2}{7}$ of the x-values here are involved in ties, formula (5.16) should have been applied.

Remarks concerning ρ_S and ρ

1. In comparison with r, r_S estimates the parameter ρ with an asymptotic efficiency of $9/\pi^2$ or 91.2% for very large n and a bivariate normal population with $\rho = 0$.
2. For increasing n and binormally distributed random variables, $2\sin(\frac{1}{6}\pi r_S)$ is asymptotically like r. For $n \geq 100$ one may thus specify r in addition to r_S. Hence one obtains for $r_S = 0.840$, with $\frac{1}{6}\pi = 0.5236$,

$$r = 2\sin[(0.5236)(0.840)] = 2\sin 0.4398 = 2(0.426) = 0.852.$$

Table 103 Critical values of the Spearman rank correlation coefficient for sample sizes n and for one sided ($\alpha_{one\,s.}$; above) and two sided ($\alpha_{two\,s.}$; below) tests. From Jerrold H. Zar, Biostatistical Analysis, © 1974, pp. 498–499. Reprinted by permission of Prentice-Hall, Inc., Englewood Cliffs, New Jersey.

$\alpha_{one\,s.}$	0.25	0.10	0.05	0.025	0.01	0.005	0.0025	0.001	0.0005
n: 4	0.600	1.000	1.000						
5	0.500	0.800	0.900	1.000	1.000				
6	0.371	0.657	0.829	0.886	0.943	1.000	1.000		
7	0.321	0.571	0.714	0.786	0.893	0.929	0.964	1.000	1.000
8	0.310	0.524	0.643	0.738	0.833	0.881	0.905	0.952	0.976
9	0.267	0.483	0.600	0.700	0.783	0.833	0.867	0.917	0.933
10	0.248	0.455	0.564	0.648	0.745	0.794	0.830	0.879	0.903
11	0.236	0.427	0.536	0.618	0.709	0.755	0.800	0.845	0.873
12	0.217	0.406	0.503	0.587	0.678	0.727	0.769	0.818	0.846
13	0.209	0.385	0.484	0.560	0.648	0.703	0.747	0.791	0.824
14	0.200	0.367	0.464	0.538	0.626	0.679	0.723	0.771	0.802
15	0.189	0.354	0.446	0.521	0.604	0.654	0.700	0.750	0.779
16	0.182	0.341	0.429	0.503	0.582	0.635	0.679	0.729	0.762
17	0.176	0.328	0.414	0.485	0.566	0.615	0.662	0.713	0.748
18	0.170	0.317	0.401	0.472	0.550	0.600	0.643	0.695	0.728
19	0.165	0.309	0.391	0.460	0.535	0.584	0.628	0.677	0.712
20	0.161	0.299	0.380	0.447	0.520	0.570	0.612	0.662	0.696
21	0.156	0.292	0.370	0.435	0.508	0.556	0.599	0.648	0.681
22	0.152	0.284	0.361	0.425	0.496	0.544	0.586	0.634	0.667
23	0.148	0.278	0.353	0.415	0.486	0.532	0.573	0.622	0.654
24	0.144	0.271	0.344	0.406	0.476	0.521	0.562	0.610	0.642
25	0.142	0.265	0.337	0.398	0.466	0.511	0.551	0.598	0.630
26	0.138	0.259	0.331	0.390	0.457	0.501	0.541	0.587	0.619
27	0.136	0.255	0.324	0.382	0.448	0.491	0.531	0.577	0.608
28	0.133	0.250	0.317	0.375	0.440	0.483	0.522	0.567	0.598
29	0.130	0.245	0.312	0.368	0.433	0.475	0.513	0.558	0.589
30	0.128	0.240	0.306	0.362	0.425	0.467	0.504	0.549	0.580
31	0.126	0.236	0.301	0.356	0.418	0.459	0.496	0.541	0.571
32	0.124	0.232	0.296	0.350	0.412	0.452	0.489	0.533	0.563
33	0.121	0.229	0.291	0.345	0.405	0.446	0.482	0.525	0.554
34	0.120	0.225	0.287	0.340	0.399	0.439	0.475	0.517	0.547
35	0.118	0.222	0.283	0.335	0.394	0.433	0.468	0.510	0.539
36	0.116	0.219	0.279	0.330	0.388	0.427	0.462	0.504	0.533
37	0.114	0.216	0.275	0.325	0.383	0.421	0.456	0.497	0.526
38	0.113	0.212	0.271	0.321	0.378	0.415	0.450	0.491	0.519
39	0.111	0.210	0.267	0.317	0.373	0.410	0.444	0.485	0.513
40	0.110	0.207	0.264	0.313	0.368	0.405	0.439	0.479	0.507
41	0.108	0.204	0.261	0.309	0.364	0.400	0.433	0.473	0.501
42	0.107	0.202	0.257	0.305	0.359	0.395	0.428	0.468	0.495
43	0.105	0.199	0.254	0.301	0.355	0.391	0.423	0.463	0.490
44	0.104	0.197	0.251	0.298	0.351	0.386	0.419	0.458	0.484
45	0.103	0.194	0.248	0.294	0.347	0.382	0.414	0.453	0.479
46	0.102	0.192	0.246	0.291	0.343	0.378	0.410	0.448	0.474
47	0.101	0.190	0.243	0.288	0.340	0.374	0.405	0.443	0.469
48	0.100	0.188	0.240	0.285	0.336	0.370	0.401	0.439	0.465
49	0.098	0.185	0.238	0.282	0.333	0.366	0.397	0.434	0.460
50	0.097	0.184	0.235	0.279	0.329	0.363	0.393	0.430	0.456
$\alpha_{two\,s.}$	0.50	0.20	0.10	0.05	0.02	0.01	0.005	0.002	0.001

5.3 Distribution-Free Measures of Association

Table 103 (*continued*) Critical values of the Spearman rank correlation coefficient for sample sizes n and for one sided ($\alpha_{one\,s.}$; above) and two sided ($\alpha_{two\,s.}$; below) tests

$\alpha_{one\,s}$	0.25	0.10	0.05	0.025	0.01	0.005	0.0025	0.001	0.0005
n: 51	0.096	0.182	0.233	0.276	0.326	0.359	0.390	0.426	0.451
52	0.095	0.180	0.231	0.274	0.323	0.356	0.386	0.422	0.447
53	0.095	0.179	0.228	0.271	0.320	0.352	0.382	0.418	0.443
54	0.094	0.177	0.226	0.268	0.317	0.349	0.379	0.414	0.439
55	0.093	0.175	0.224	0.266	0.314	0.346	0.375	0.411	0.435
56	0.092	0.174	0.222	0.264	0.311	0.343	0.372	0.407	0.432
57	0.091	0.172	0.220	0.261	0.308	0.340	0.369	0.404	0.428
58	0.090	0.171	0.218	0.259	0.306	0.337	0.366	0.400	0.424
59	0.089	0.169	0.216	0.257	0.303	0.334	0.363	0.397	0.421
60	0.089	0.168	0.214	0.255	0.300	0.331	0.360	0.394	0.418
61	0.088	0.166	0.213	0.252	0.298	0.329	0.357	0.391	0.414
62	0.087	0.165	0.211	0.250	0.296	0.326	0.354	0.388	0.411
63	0.086	0.163	0.209	0.248	0.293	0.323	0.351	0.385	0.408
64	0.086	0.162	0.207	0.246	0.291	0.321	0.348	0.382	0.405
65	0.085	0.161	0.206	0.244	0.289	0.318	0.346	0.379	0.402
66	0.084	0.160	0.204	0.243	0.287	0.316	0.343	0.376	0.399
67	0.084	0.158	0.203	0.241	0.284	0.314	0.341	0.373	0.396
68	0.083	0.157	0.201	0.239	0.282	0.311	0.338	0.370	0.393
69	0.082	0.156	0.200	0.237	0.280	0.309	0.336	0.368	0.390
70	0.082	0.155	0.198	0.235	0.278	0.307	0.333	0.365	0.388
71	0.081	0.154	0.197	0.234	0.276	0.305	0.331	0.363	0.385
72	0.081	0.153	0.195	0.232	0.274	0.303	0.329	0.360	0.382
73	0.080	0.152	0.194	0.230	0.272	0.301	0.327	0.358	0.380
74	0.080	0.151	0.193	0.229	0.271	0.299	0.324	0.355	0.377
75	0.079	0.150	0.191	0.227	0.269	0.297	0.322	0.353	0.375
76	0.078	0.149	0.190	0.226	0.267	0.295	0.320	0.351	0.372
77	0.078	0.148	0.189	0.224	0.265	0.293	0.318	0.349	0.370
78	0.077	0.147	0.188	0.223	0.264	0.291	0.316	0.346	0.368
79	0.077	0.146	0.186	0.221	0.262	0.289	0.314	0.344	0.365
80	0.076	0.145	0.185	0.220	0.260	0.287	0.312	0.342	0.363
81	0.076	0.144	0.184	0.219	0.259	0.285	0.310	0.340	0.361
82	0.075	0.143	0.183	0.217	0.257	0.284	0.308	0.338	0.359
83	0.075	0.142	0.182	0.216	0.255	0.282	0.306	0.336	0.357
84	0.074	0.141	0.181	0.215	0.254	0.280	0.305	0.334	0.355
85	0.074	0.140	0.180	0.213	0.252	0.279	0.303	0.332	0.353
86	0.074	0.139	0.179	0.212	0.251	0.277	0.301	0.330	0.351
87	0.073	0.139	0.177	0.211	0.250	0.276	0.299	0.328	0.349
88	0.073	0.138	0.176	0.210	0.248	0.274	0.298	0.327	0.347
89	0.072	0.137	0.175	0.209	0.247	0.272	0.296	0.325	0.345
90	0.072	0.136	0.174	0.207	0.245	0.271	0.294	0.323	0.343
91	0.072	0.135	0.173	0.205	0.244	0.269	0.293	0.321	0.341
92	0.071	0.135	0.173	0.205	0.243	0.268	0.291	0.319	0.339
93	0.071	0.134	0.172	0.204	0.241	0.267	0.290	0.318	0.338
94	0.070	0.133	0.171	0.203	0.240	0.265	0.288	0.316	0.336
95	0.070	0.133	0.170	0.202	0.239	0.264	0.287	0.314	0.334
96	0.070	0.132	0.169	0.201	0.238	0.262	0.285	0.313	0.332
97	0.069	0.131	0.168	0.200	0.236	0.261	0.284	0.311	0.331
98	0.069	0.130	0.167	0.199	0.235	0.260	0.282	0.310	0.329
99	0.068	0.130	0.166	0.198	0.234	0.258	0.281	0.308	0.327
100	0.068	0.129	0.165	0.197	0.233	0.257	0.279	0.307	0.326
$\alpha_{two\,s.}$	0.50	0.20	0.10	0.05	0.02	0.01	0.005	0.002	0.001

EXAMPLE. Table 104 indicates how ten alphabetically aranged students are ordered according to rank on the basis of performances in a practical course and in a seminar. Can a positive correlation be ascertained at the 1% level?

Table 104

| Practical course | 7 | 6 | 3 | 8 | 2 | 10 | 4 | 1 | 5 | 9 |
| Seminar | 8 | 4 | 5 | 9 | 1 | 7 | 3 | 2 | 6 | 10 |

Null hypothesis: Between the two performances there is no positive correlation, but rather independence. We determine the differences in rank, their squares, and the sum thereof in Table 104a.

Table 104a

											Σ
Rank differences D	-1	2	-2	-1	1	3	1	-1	-1	-1	0
D²	1	4	4	1	1	9	1	1	1	1	24

Verification of the computations: The sum of the D-values must equal zero. We get

$$r_S = 1 - \frac{6 \sum D^2}{n(n^2 - 1)} = 1 - \frac{6 \cdot 24}{10(10^2 - 1)} = 0.8545.$$

A rank correlation coefficient of this size, computed from a sample of $n = 10$, is, according to Table 103, statistically significant at the 1% level ($0.8545 > 0.745 = r^*_{S;10;0.01;\text{one sided}}$). There is an authentic correlation $P < 0.01$) between the two performances.

Given at least 30 pairs of values ($n \geq 30$), the randomness of occurrence of a certain r_S value can also be judged on the basis of Student's distribution, by

$$\hat{t} = |r_S| \cdot \sqrt{\frac{n-2}{1-r_S^2}} \qquad (5.15)$$

with $(n - 2)$DF. For the example, with $n = 10$ (note $10 < 30$ and (5.15) is, strictly speaking, not applicable)

$$\hat{t} = 0.8545 \cdot \sqrt{\frac{10 - 2}{1 - 0.8545^2}} = 4.653,$$

5.3 Distribution-Free Measures of Association

and $4.653 > 2.896 = t_{8;\,0.01;\,\text{one sided}}$, we obtain a confirmation of our results. It is emphasized that (5.14) and (5.15) represent only approximations; (5.15) is the better one.

Spearman's rank correlation with ties

Only if ties (equal values) occur in aggregates is it worth while to employ the test statistic (cf., Kendall 1962, Yule and Kendall 1965)

$$
\boxed{\begin{aligned}
\text{TIES} \quad & r_{S,\,\text{ties}} = \frac{M - (\sum D^2 + T_{x'} + T_{y'})}{\sqrt{(M - 2T_{x'})(M - 2T_{y'})}} \\
& \text{with } M = \tfrac{1}{6}(n^3 - n), \\
& T_{x'} = \tfrac{1}{12}\sum(t_{x'}^3 - t_{x'}), \\
& T_{y'} = \tfrac{1}{12}\sum(t_{y'}^3 - t_{y'}),
\end{aligned}} \qquad (5.16)
$$

where $t_{x'}$ (the prime on the x indicates we are dealing with rank quantities) equals the number of ties in consecutive groups (equal rank quantities) of the x' series, and $t_{y'}$ equals the number of ties in consecutive groups (equal rank quantities) of the y' series. Thus one counts how often the same value appears in the first group, cubes this frequency, and then subtracts the frequency. One proceeds analogously with all the groups, and then forms the sums $T_{x'}$ and $T_{y'}$.

EXAMPLE. Testing the independence of mathematical and linguistic aptitude of 8 students (S) on the basis of grades in Latin (L, $[x]$) and in mathematics (M, $[y]$) (two sided test with $\alpha = 0.05$; R are the rank quantities):

S	D	B	G	A	F	E	H	C	$n = 8$
L	1	2	2	2	3	3	4	4	
M	2	4	1	3	4	3	4	3	
R_L	1	3	3	3	5.5	5.5	7.5	7.5	
R_M	2	7	1	4	7	4	7	4	
D	-1	-4	2	-1	-1.5	1.5	0.5	3.5	$\sum D = 0$
D^2	1	16	4	1	2.25	2.25	0.25	12.25	

$\sum D^2 = 39$, $M = \tfrac{1}{6}(8^3 - 8) = 84$,

$$T_L = \tfrac{1}{12}[(3^3 - 3) + (2^3 - 2) + (2^3 - 2)] = 3,$$

$$T_M = \tfrac{1}{12}[(3^3 - 3) + (3^3 - 3)] = 4; \quad r_{S,\,\text{ties}} = \frac{84 - (39 + 3 + 4)}{\sqrt{(84 - 6)(84 - 8)}} = 0.4935.$$

Without regard for the ties,

$$r_S = 1 - \frac{(6)(39)}{8^3 - 8} = 0.536 \quad (0.536 > 0.494);$$

the correlation is overestimated. Since $0.494 < 0.738$, the independence hypothesis cannot be disproved at the 5% level by means of the grades. [For the one sided test ($0.494 < 0.643$) the same decision would be reached].

Instead of (5.16) we can, with $R(X_i)$ and $R(Y_i)$ [$i = 1, 2, \ldots, n$] representing the ranks assigned to the ith value of X_i and Y_i respectively, compute the usual product moment correlation coefficient of Bravais and Pearson on ranks and average ranks by (R):

$$r_S = \frac{\sum R(X_i)R(Y_i) - \frac{n(n+1)^2}{4}}{\sqrt{\left[\sum R^2(X_i) - \frac{n(n+1)^2}{4}\right]\left[\sum R^2(Y_i) - \frac{n(n+1)^2}{4}\right]}} \quad (R)$$

Our example: $8(8+1)^2/4 = 162$

									Σ
$R^2(X_i)$	1	9	9	9	30.25	30.25	56.25	56.25	201
$R(X_i)$	1	3	3	3	5.5	5.5	7.5	7.5	—
$R(Y_i)$	2	7	1	4	7	4	7	4	—
$R^2(Y_i)$	4	49	1	16	49	16	49	16	201
$R(X_i)R(Y_i)$	2	21	3	12	38.5	22	52.5	30	181

$$r_S = \frac{181 - 162}{\sqrt{[201 - 162][200 - 162]}} = 0.4935.$$

The rank correlation coefficient (Spearman 1904) can also be used:
1. If a **quick approximate estimate of the correlation coefficient is desired** and the exact computation is very costly.
2. If the **agreement between two judges as to the chosen rank order of objects** is to be examined, for example in a beauty contest. It can also be used to test the reasoning faculty by ordering a collection of objects and comparing this rank order with a standardized rank order. The arrangements by children of building blocks of various sizes serves as an example.
3. If a **monotone trend is suspected**: The n measured values, transformed to their ranks, are correlated with the natural number sequence from 1 to n, and the coefficient is tested for significance.
4. If two independent samples of equal size are given, then $H_0: \rho_{S_1} = \rho_{S_2}$ can be rejected with the help of the U-test **applied to absolute rank differences**.

5. From a **bivariate frequency table** one determines r_S in accordance with Raatz (1971) [cf., also Stuart (1963)].
6. For the following situation T. P. Hutchinson [Applied Statistics **25** (1976), 21–25] proposes a test: Judges are presented with stimuli which are ordered along some dimension, such as large to small or pleasant to unpleasant, and are asked to rank them. There is reason to believe that some judges will tend to rank the stimuli in one order: 1, 2, ..., n, while others will order them oppositely: n, $n-1$, ..., 1. Hutchinson tests whether the judges can detect the ordered nature of the stimuli. Critical values, a normal approximation, and two examples of the **combined two tailed Spearman rank-correlation statistics** are given. In example 2 eight models of cars are ordered in terms of their accident rates. They are also ordered in terms of certain of their design and handling parameters, such as weight, ratio of height of center of gravity to track, understeer and braking instability. The question to be answered by the test: is there evidence that these parameters affect the accident rate?

The rank correlation coefficient τ (Kendall's tau) proposed by Kendall (1938) is more difficult to calculate than r_S. Griffin (1957) describes a graphical procedure for estimating τ. A simplified computation of τ is given by Lieberson (1961) as well as by Stilson and Campbell (1962).

A discussion of certain advantages of τ over ρ and ρ_S can be found in Schaeffer and Levitt (1956); however, the power of the test (testing for the condition non-null), for the same level of significance, is smaller for τ than for ρ_S.

For partial and multiple rank correlation coefficients see R. Lehmann, Biometrical Journal **19** (1977), 229–236.

5.3.2 Quadrant correlation

This quick test (Blomqvist 1950, 1951) checks whether two attributes x and y, known through data, are independent. First plot the pairs of values (x_i, y_i) as a point cloud in a coordinate system which is partitioned by the two medians $\tilde{x}$ and $\tilde{y}$ into four quadrants, i.e., twice into halves, in such a way that each half contains the same number of pairs of values. If the number of pairs of observations is odd, then the horizontal median line passes through a point, which is subsequently ignored. A significant relationship between the attributes is ascertained as soon as the number of points in the single quadrants does not lie within the bounds given in Table 105. If we are dealing with samples from a two-dimensional normal distribution, then this test has an asymptotic efficiency of $(2/\pi)^2 = 0.405$ or 41% in comparison with the t-test of the product-moment correlation coefficient. More on this can be found in Konijn (1956) and Elandt (1962).

Table 105 Upper and lower critical bounds for a quadrant for the assessment of quadrant correlation (taken from Quenouille, M. H.: Rapid Statistical Calculations, Griffin, London 1959, Table 6)

	Critical bound							
n	lower		upper		n	lower		upper	
	5%	1%	5%	1%		5%	1%	5%	1%
8-9	0	-	4	-	74-75	13	12	24	25
10-11	0	0	5	5	76-77	14	12	24	26
12-13	0	0	6	6	78-79	14	13	25	26
14-15	1	0	6	7	80-81	15	13	25	27
16-17	1	0	7	8	82-83	15	14	26	27
18-19	1	1	8	8	84-85	16	14	26	28
20-21	2	1	8	9	86-87	16	15	27	28
22-23	2	2	9	9	88-89	16	15	28	29
24-25	3	2	9	10	90-91	17	15	28	30
26-27	3	2	10	11	92-93	17	16	29	30
28-29	3	3	11	11	94-95	18	16	29	31
30-31	4	3	11	12	96-97	18	17	30	31
32-33	4	3	12	13	98-99	19	17	30	32
34-35	5	4	12	13	100-101	19	18	31	32
36-37	5	4	13	14	110-111	21	20	34	35
38-39	6	5	13	14	120-121	24	22	36	38
40-41	6	5	14	15	130-131	26	24	39	41
42-43	6	5	15	16	140-141	28	26	42	44
44-45	7	6	15	16	150-151	31	29	44	46
46-47	7	6	16	17	160-161	33	31	47	49
48-49	8	7	16	17	170-171	35	33	50	52
50-51	8	7	17	18	180-181	37	35	53	55
52-53	8	7	18	19	200-201	42	40	58	60
54-55	9	8	18	19	220-221	47	44	63	66
56-57	9	8	19	20	240-241	51	49	69	71
58-59	10	9	19	20	260-261	56	54	74	76
60-61	10	9	20	21	280-281	61	58	79	82
62-63	11	9	20	22	300-301	66	63	84	87
64-65	11	10	21	22	320-321	70	67	90	93
66-67	12	10	21	23	340-341	75	72	95	98
68-69	12	11	22	23	360-361	80	77	100	103
70-71	12	11	23	24	380-381	84	81	106	109
72-73	13	12	23	24	400-401	89	86	111	114

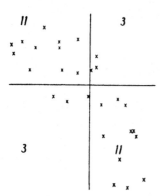

Figure 48 Quadrant correlation (taken from Quenouille, M. H.: Rapid Statistical Calculations, Griffin, London 1959, p. 28).

5.3 Distribution-Free Measures of Association

EXAMPLE. The 28 pairs of observations in Fig. 48 are so distributed among the quadrants that the bounds of Table 105 are attained. The negative correlation is acertained at the 1% level.

This test is essentially the median test of independence, in which the pairs are classified according as the components of a pair are larger or smaller than the respective medians.

		Number of x-values	
		$< \tilde{x}$	$> \tilde{x}$
Number of y-values	$< \tilde{y}$	a	b
	$> \tilde{y}$	c	d

The analysis of the fourfold table is carried out according to Section 4.6.1 (cf., also the comments at the end of Section 3.9.4).

5.3.3 The corner test of Olmstead and Tukey

This test generally utilizes more information than the quadrant correlation. It is particularly suitable for proof of a correlation which is largely caused by pairs of extreme values (Olmstead and Tukey 1947). A test statistic of this important rapid test for independence (asymptotic efficiency: about 25%) is the sum S of 4 "quadrant sums" (see below). For $|S| \geq S_\alpha$, depending on the sign of S, a positive or a negative association is assumed.

1. The n pairs of observations (x_i, y_i) are plotted in a scatter diagram as in the quadrant correlation discussed above, and are then successively split up by the horizontal and by the vertical median line into two groups of equal size.
2. The points in the upper right and in the lower left quadrants are regarded as positive; those in the other two quadrants, as negative.
3. Move along the abscissa until the first point on the other side of the y (horizontal) median line is reached, count the points encountered, and affix the sign appropriate for the particular quadrant to this number. Repeat this counting procedure from below, from the left, and from above:

α	0.10	0.05	0.02	0.01	0.005	0.002	0.001
S_α	9	11	13	14–15	15–17	17–19	18–21

1. For $\alpha \leq 0.01$, the larger value of S_α applies for smaller n, the smaller value for larger n.
2. For $|S| \geq 2n - 6$ one should forgo the test.

EXAMPLE. The 28 pair of observations Fig. 48 are so distributed among the $(-10) + (-11) + (-6) = -35$; the negative correlation is clearly ascertained.

If one denotes the absolute value of the sum of the four countings by k, then for large sample size the probability P can be estimated by

$$P \simeq \frac{9k^3 + 9k^2 + 168k + 208}{216 \cdot 2^k}, \qquad k = |S| > 0. \qquad (5.17)$$

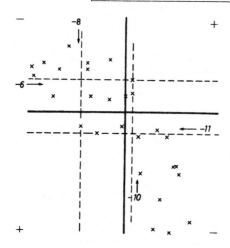

Figure 49 Corner test of Olmstead and Tukey (taken from Quenouille, M. H.: Rapid Statistical Calculations, Griffin, London 1959, p. 29). Move along a median line toward the intersection of the two median lines, and draw a dotted line through the first point which finds itself on the other side of the median line along which you are going. The number of points preceding this dotted line forms a term in the "quadrant sum" (see text).

5.4 ESTIMATION PROCEDURES

5.4.1 Estimation of the correlation coefficient

The correlation coefficient measures the strength of the linear relationship between two variables, say X and Y. We make the following assumptions on r, in addition to the one that X and Y are random variables from a bivariate frequency distribution with random selection of individuals:

1. Equidistant units of measurement for both variables.
2. Linearity of regression.
3. Normality for both variables.

(1) is very important [by the way: it holds also (cf., end of Section 1.4.8) for $\bar{x}$ and s]; if (2) is not true, then the value of r is an underestimate and the trouble is not great; some statisticians omit (3).

5.4 Estimation Procedures

The correlation coefficient is estimated by the right side of (5.18) (for small n one sometimes prefers the first expression):

$$r = \frac{\sum(x-\bar{x})(y-\bar{y})}{\sqrt{\sum(x-\bar{x})^2 \sum(y-\bar{y})^2}} = \frac{\sum xy - \frac{1}{n}(\sum x)(\sum y)}{\sqrt{[\sum x^2 - \frac{1}{n}(\sum x)^2][\sum y^2 - \frac{1}{n}(\sum y)^2]}}. \quad (5.18)$$

Other formulas:

$$r = \frac{n\sum xy - (\sum x)(\sum y)}{\sqrt{[n\sum x^2 - (\sum x)^2][n\sum y^2 - (\sum y)^2]}},$$

$$\left[r = \frac{1}{n-1}\sum\left(\frac{x_i - \bar{x}}{s_x}\right)\left(\frac{y_i - \bar{y}}{s_y}\right) = \frac{\sum(x_i - \bar{x})(y_i - \bar{y})}{ns_x s_y} = \frac{s_{xy}}{s_x s_y}. \right]$$

For small sample size n, r underestimates the parameter ρ. An improved estimate for ρ is obtained by (5.18a) (Olkin and Pratt 1958):

$$r^* = r\left[1 + \frac{1 - r^2}{2(n-3)}\right] \quad \text{for } n \geq 8. \quad (5.18a)$$

Thus, e.g., the following r^* values result:

for $n = 10$ and $r = 0.5$, $r^* = 0.527$,
for $n = 10$ and $r = 0.9$, $r^* = 0.912$,
for $n = 30$ and $r = 0.5$, $r^* = 0.507$,
for $n = 30$ and $r = 0.9$, $r^* = 0.903$.

Tables for finding r^* from r when $8 \leq n \leq 40$ are given by R. Jäger in Biometrische Zeitschrift **16** (1974), 115–124.

Generally one will choose the sample size not too small and do without the correction (5.18a).

Remark on point biserial correlation. If one of the two attributes is dichotomous, then (5.18) is replaced by (5.18b). The relationship between a continuously distributed variable and a dichotomy is estimated by means of the point biserial correlation coefficient (the sample is subdivided according to the presence or absence of the attribute y, with resulting group sizes n_1 and n_2 [$n_1 + n_2 = n$]; then the corresponding means $\bar{x}_1$ and $\bar{x}_2$ and the common standard deviation s of the x-attributes are determined):

$$r_{pb} = \frac{\bar{x}_1 - \bar{x}_2}{ns}\sqrt{n_1 n_2}. \quad (5.18b)$$

The relationship is then tested for significance on the basis of Table 113 or (5.38), (5.38a, b) (Section 5.5.1). The r_{pb} can serve as an estimate of ρ, in particular if $|r_{pb}| < 1$; for $r_{pb} > 1$, ρ is estimated by 1; for $r_{pb} < -1$, correspondingly $\rho = -1$. A more detailed discussion can be found in Tate (1954, 1955), Prince and Tate (1966), and Abbas (1967) (cf., also Meyer-Bahlburg 1969).

▶ 5.4.2 Estimation of the regression line

> The following two models of regression analysis are always to be distinguished:
> Model I: The target quantity Y is a random variable; the values of the influence variable X are always given or X_{fixed} [see (5.3)].
> Model II: Both the variable Y and the variable X are random variables. Two regressions are possible in this case: one of Y on X and one of X on Y [(5.3) and (5.2)].

Axis intercepts and regression coefficients (cf., also Sections 5.4.4, 5.5.3, 5.5.9) are estimated by the following expressions:

$$\hat{y} = a_{yx} + b_{yx}x, \tag{5.3}$$

$$b_{yx} = \frac{n\sum xy - \sum x \sum y}{n\sum x^2 - (\sum x)^2}, \tag{5.19}$$

$$a_{yx} = \frac{\sum y - b_{yx}\sum x}{n}, \tag{5.20}$$

$$\hat{x} = a_{xy} + b_{xy}y, \tag{5.2}$$

$$b_{xy} = \frac{n\sum xy - \sum x \sum y}{n\sum y^2 - (\sum y)^2}, \tag{5.21}$$

$$a_{xy} = \frac{\sum x - b_{xy}\sum y}{n}. \tag{5.22}$$

a_{yx} and a_{xy} can be found directly from the sums

$$a_{yx} = \frac{(\sum y)(\sum x^2) - (\sum x)(\sum xy)}{n\sum x^2 - (\sum x)^2}, \tag{5.20a}$$

$$a_{xy} = \frac{(\sum x)(\sum y^2) - (\sum y)(\sum xy)}{n\sum y^2 - (\sum y)^2}. \tag{5.22a}$$

5.4 Estimation Procedures

The computations can however be carried out more quickly according to (5.20), (5.22). Whenever n is large or multidigit x_i and y_i are involved, (5.19) and (5.21) are replaced by (5.19a) and (5.21a):

$$b_{yx} = \frac{\sum xy - \frac{(\sum x)(\sum y)}{n}}{\sum x^2 - \frac{(\sum x)^2}{n}}, \qquad (5.19\,\text{a})$$

$$b_{xy} = \frac{\sum xy - \frac{(\sum x)(\sum y)}{n}}{\sum y^2 - \frac{(\sum y)^2}{n}}. \qquad (5.21\,\text{a})$$

EXAMPLE 1

Table 106

x	y	xy	x^2	y^2	$\hat{y}$
2	4	8	4	16	4.2
4	5	20	16	25	5.4
6	8	48	36	64	6.6
8	7	56	64	49	7.8
10	9	90	100	81	9.0
30	33	222	220	235	33
$\sum x$	$\sum y$	$\sum xy$	$\sum x^2$	$\sum y^2$	$\sum \hat{y}$

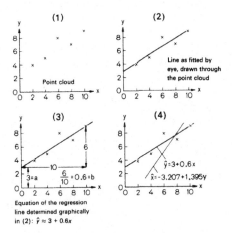

Equation of the regression line determined graphically in (2): $\hat{y} \approx 3 + 0.6x$

Computing the regression lines and the correlation coefficients

(1) $\boxed{\hat{y} = a_{yx} + b_{yx}x}$

$$b_{yx} = \frac{\sum xy - \frac{1}{n}(\sum x)(\sum y)}{\sum x^2 - \frac{1}{n}(\sum x)^2} = \frac{222 - \frac{1}{5}30 \cdot 33}{220 - \frac{1}{5}30^2} = 0.600,$$

$$a_{yx} = \frac{\sum y - b_{yx}\sum x}{n} = \frac{33 - 0.6 \cdot 30}{5} = 3.000$$

$\hat{y} = 3 + 0.6x$, estimated regression line for predicting $\hat{y}$ from x (Table 106, last column); also called regression of y on x (cf. a_{yx}, b_{yx}).

(2) $\boxed{\hat{x} = a_{xy} + b_{xy}y}$

$$b_{xy} = \frac{\sum xy - \frac{1}{n}(\sum x)(\sum y)}{\sum y^2 - \frac{1}{n}(\sum y)^2} = \frac{222 - \frac{1}{5}30 \cdot 33}{235 - \frac{1}{5}33^2} = 1.395,$$

$$a_{xy} = \frac{\sum x - b_{xy}\sum y}{n} = \frac{30 - 1.395 \cdot 33}{5} = -3.207,$$

$\hat{x} = -3.207 + 1.395y$ estimated regression line for predicting $\hat{x}$ from y.

(3)
$$\boxed{r = \frac{\sum xy - \frac{1}{n}(\sum x)(\sum y)}{\sqrt{\left[\sum x^2 - \frac{1}{n}(\sum x)^2\right]\left[\sum y^2 - \frac{1}{n}(\sum y)^2\right]}}},$$

$$r = \frac{222 - \frac{1}{5}30 \cdot 33}{\sqrt{\left[220 - \frac{1}{5}30^2\right]\left[235 - \frac{1}{5}33^2\right]}} = 0.915,$$

$r = 0.915$ estimated correlation coefficient, a measure of *linear* dependence between the two attributes.
$[r^* = 0.952$ cf. (5.18a)]

Checking r, b_{yx} and b_{xy}: $r = \sqrt{b_{yx} \cdot b_{xy}}$, $\sqrt{0.6 \cdot 1.395} = 0.915$.

EXAMPLE 2

We now compute the axis intercept by (5.20a):

$$a_{yx} = \frac{(\sum y)(\sum x^2) - (\sum x)(\sum xy)}{n \sum x^2 - (\sum x)^2} = \frac{98 \cdot 1{,}593 - 103 \cdot 1{,}475}{7 \cdot 1{,}593 - 103^2} = 7.729,$$

and the regression coefficient by (5.19):

$$b_{yx} = \frac{n \sum xy - (\sum x)(\sum y)}{n \sum x^2 - (\sum x)^2} = \frac{7 \cdot 1{,}475 - 103 \cdot 98}{7 \cdot 1{,}593 - 103^2} = 0.426.$$

5.4 Estimation Procedures

Table 107a Belongs to Section 5.3.1, below formula (5.14): given are ranks for x, y of Table 107 and values D and D^2.

Table 107

x	y	x^2	y^2	xy
13	12	169	144	156
17	17	289	289	289
10	11	100	121	110
17	13	289	169	221
20	16	400	256	320
11	14	121	196	154
15	15	225	225	225
103	98	1593	1400	1475

For the example in Section 5.3.1			
Ranks		D	D^2
x	y		
3	2	1	1
5.5	7	−1.5	2.25
1	1	0	0
5.5	3	2.5	6.25
7	6	1	1
2	4	−2	4
4	5	−1	1
		0	15.50

The regression line of y on x then reads

$$\hat{y} = a_{yx} + b_{yx}x \quad \text{or} \quad \underline{\hat{y} = 7.73 + 0.426x}$$

(see Figure 50).

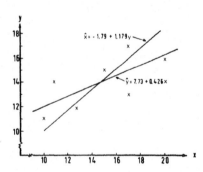

Figure 50 The two regression lines of Example 2.

This can also be done more quickly and in a more elegant manner: First find b_{yx} according to the given relation, then determine the means $\bar{x}, \bar{y}$, and finally use these values in the relation

$$\boxed{a_{yx} = \bar{y} - b_{yx}\bar{x},} \quad (5.23)$$

$$\bar{x} = \frac{103}{7} = 14.714, \quad \bar{y} = \frac{98}{7} = 14;$$

$$a_{yx} = 14 - 0.426 \cdot 14.714 = 7.729.$$

For the regression line of x on y we get, according to (5.22a) and (5.21),

$$a_{xy} = \frac{(\sum x)(\sum y^2) - (\sum y)(\sum xy)}{n\sum y^2 - (\sum y)^2} = \frac{103 \cdot 1{,}400 - 98 \cdot 1{,}475}{7 \cdot 1{,}400 - 98^2} = -1.786$$

$$b_{xy} = \frac{n\sum xy - (\sum x)(\sum y)}{n\sum y^2 - (\sum y)^2} = \frac{7 \cdot 1{,}475 - 103 \cdot 98}{7 \cdot 1{,}400 - 98^2} = 1.179$$

$$\hat{x} = a_{xy} + b_{xy}y \quad \text{or} \quad \hat{x} = -1.79 + 1.179y.$$

Without an electronic pocket computer transformed values may be used (β_{yx} and β_{xy} are unaffected by this) as shown in Table 108, with $x = k_1 x^* + k_2$ [cf., (1.36) to (1.38)], $y = k_3 y^* + k_4$ (x^* and y^* are small integers),

Table 108

x˙ (= x - 15)	y˙ (= y - 14)	x˙²	y˙²	x˙·y˙
-2	-2	4	4	4
2	3	4	9	6
-5	-3	25	9	15
2	-1	4	1	-2
5	2	25	4	10
-4	0	16	0	0
0	1	0	1	0
-2	0	78	28	33

$\bar{y} = k_3 \bar{y}^* + k_4$, $s_y^2 = k_3 s_y^{*2}$, and $s_{xy} = k_1 k_3 s_{xy}^*$ and also $r = s_{xy}^*/(s_x^* s_y^*)$. By these transformations the computations may be simplified:

$$b_{yx} = \frac{n\sum \dot{x}\dot{y} - (\sum \dot{x})(\sum \dot{y})}{n\sum \dot{x}^2 - (\sum \dot{x})^2} = \frac{7 \cdot 33 - (-2)(0)}{7 \cdot 78 - (-2)^2} = 0.426,$$

$$b_{xy} = \frac{n\sum \dot{x}\dot{y} - (\sum \dot{x})(\sum \dot{y})}{n\sum \dot{y}^2 - (\sum \dot{y})^2} = \frac{7 \cdot 33 - (-2) \cdot 0}{7 \cdot 28 - 0^2} = 1.179.$$

Since $\bar{x} = 103/7 = 14.714$ and $\bar{y} = 98/7 = 14$, the regression equations are

$$\boxed{y - \bar{y} = b_{yx}(x - \bar{x}), \quad \text{i.e.,} \quad y = \bar{y} - b_{yx}\bar{x} + b_{yx}x} \qquad (5.2\text{a})$$

or
$$y = 14 - 0.426 \cdot 14.714 + b_{yx}x,$$
$$\hat{y} = 7.73 + 0.426x,$$

and

$$x - \bar{x} = b_{xy}(y - \bar{y}), \quad \text{i.e.,} \quad x = \bar{x} - b_{xy}\bar{y} + b_{xy}y$$

or
$$x = 14.71 - 1.179 \cdot 14 + b_{xy}y,$$
$$\underline{\hat{x} = -1.79 + 1.179y.}$$

5.4 Estimation Procedures

The location of the regression lines in the given system of coordinates is thus determined. We estimate the correlation coefficients from the regression coefficients by (5.4) and by (5.18a):

$$r = \sqrt{b_{yx} \cdot b_{xy}} = \sqrt{0.426 \cdot 1.179} = 0.709 \quad \text{and} \quad r^* = 0.753.$$

5.4.3 The estimation of some standard deviations

The standard deviations s_x and s_y are evaluated from the sums of squares of the deviations of x and y. We recall (cf., Chapter 3)

$$Q_x = \sum(x - \bar{x})^2 = \sum x^2 - (\sum x)^2/n,$$

$$s_x = \sqrt{\frac{Q_x}{n-1}},$$

$$Q_y = \sum(y - \bar{y})^2 = \sum y^2 - (\sum y)^2/n,$$

$$s_y = \sqrt{\frac{Q_y}{n-1}}.$$

Every observation of a bivariate or two-dimensional frequency distribution consists of a pair of observed values (x, y). The product of the two deviations from the respective means is thus an appropriate measure of the degree of common variation of the observations. The sum of products of deviations may be called the "codeviance":

$$Q_{xy} = \sum(x - \bar{x})(y - \bar{y}).$$

On dividing by $n - 1$ one gets a sort of an average codeviance:

$$\boxed{\frac{\sum(x - \bar{x})(y - \bar{y})}{n-1} = \frac{Q_{xy}}{n-1} = s_{xy}.} \quad (5.24)$$

(5.24) is an estimate of the so-called **covariance** σ_{xy}. The computation of the codeviance, Q_{xy} for short, can be facilitated by use of the following identities:

$$\boxed{Q_{xy} = \sum xy - \bar{x}\sum y,} \quad (5.25\,\text{a})$$

$$\boxed{Q_{xy} = \sum xy - \bar{y}\sum x,} \quad (5.25\,\text{b})$$

$$\boxed{Q_{xy} = \sum xy - \frac{\sum x \sum y}{n}.} \quad (5.25)$$

Equation (5.25) is usually the easiest for computations. In terms of Q_{xy}, one obtains the correlation coefficient r as well as both regression coefficients b_{yx} and b_{xy} according to

$$r = \frac{Q_{xy}}{\sqrt{Q_x \cdot Q_y}}, \qquad (5.26)$$

(cf., formulas (5.19a) and (5.21a))

$$b_{yx} = \frac{Q_{xy}}{Q_x}, \qquad (5.27)$$

$$b_{xy} = \frac{Q_{xy}}{Q_y}. \qquad (5.28)$$

The standard deviation of y, assuming x is deterministic, is

$$s_{y.x} = \sqrt{\frac{\sum (y - \hat{y})^2}{n - 2}} = \sqrt{\frac{\sum (y - a_{yx} - b_{yx}x)^2}{n - 2}}$$
$$= \sqrt{\frac{\sum y^2 - a \sum y - b \sum xy}{n - 2}}. \qquad (5.29)$$

The symbol $s_{y.x}$ for the standard deviation of the y-values for given x is to be read "s y dot x". The numerator under the square root sign consists of the sum of the squares of the deviations of observed y-values for the corresponding values on the regression line. This sum is divided by $n - 2$ and not by $n - 1$, since we had estimated the two parameters a_{yx} and b_{yx}. The value $s_{y.x}$ could be obtained by determining for every x-value the corresponding y-value by means of the regression line, summing the squares of the individual differences, and dividing by the sample size reduced by two. The square root of this would then be $s_{y.x}$. The residual sum of squares (RSS), or error sum of squares, may be computed by

$$\sum (y - \hat{y})^2 = \sum (y - \bar{y})^2 - \frac{[\sum (x - \bar{x})(y - \bar{y})]^2}{\sum (x - \bar{x})^2} = Q_x - \frac{Q_{xy}^2}{Q_x}.$$

(RSS)

The standard error for given values x is thus obtained more quickly according to

$$s_{y.x} = \sqrt{\frac{Q_y - (Q_{xy})^2/Q_x}{n - 2}}. \qquad (5.29a)$$

5.4 Estimation Procedures

Since $s_{y.x}$ is a measure of the inadequacy of fit for the fitted equation $\hat{y} = a + bx$, or of the error which is made in the estimation or prediction of y from given values of x, this standard deviation will also be referred to as the **standard error of estimate** or as the **standard error of prediction**. If we now denote the **standard deviation of the axis intercept a** (on the ordinate) by s_a and the **standard deviation of the regression coefficient** $b_{yx} = b$ by s_b, then we have

$$s_{a_{yx}} = s_{y.x}\sqrt{\frac{1}{n} + \frac{\bar{x}^2}{Q_x}}, \tag{5.30}$$

$$s_{b_{yx}} = \frac{s_{y.x}}{\sqrt{Q_x}}, \tag{5.31}$$

$$s_{a_{yx}} = s_{b_{yx}}\sqrt{\frac{\sum x^2}{n}}. \tag{5.30a}$$

Thus a verification of the computations for s_a and s_b is possible:

$$\frac{s_a}{s_b} = \sqrt{\frac{\sum x^2}{n}}. \tag{5.30b}$$

The square of the standard error of estimation—the dispersion about the regression line—is called the residual variance $s_{y.x}^2$ [cf., (5.6a,b)], often called the residual mean square or error mean square, and is the variance of y when the linear influence of x is accounted for. There is an interesting relation between the two measures:

$$s_{y.x}^2 = (s_y^2 - b_{yx}^2 s_x^2)\frac{n-1}{n-2} = s_y^2(1 - r^2)\frac{n-1}{n-2}. \tag{5.29b}$$

For large sample size, there obtains

$$s_{y.x} \simeq s_y\sqrt{1 - r^2}, \tag{5.32}$$

$$s_{x.y} \simeq s_x\sqrt{1 - r^2}. \tag{5.33}$$

Notice the following connection:

$$s_{y.x} = \sqrt{\frac{\sum(y - \hat{y})^2}{n - 2}}, \tag{5.29}$$

$$s_{b_{yx}} = \sqrt{\frac{\sum(y - \hat{y})^2}{n - 2}}\sqrt{\frac{1}{\sum(x - \bar{x})^2}}, \tag{5.31a}$$

$$s_{a_{yx}} = \sqrt{\frac{\sum(y - \hat{y})^2}{n - 2}}\sqrt{\frac{1}{\sum(x - \bar{x})^2}}\sqrt{\frac{\sum x^2}{n}}. \tag{5.30a}$$

EXAMPLE. Reexamine our last example with $n = 7$ and with the sums

$$\sum x = 103, \quad \sum y = 98,$$
$$\sum x^2 = 1{,}593, \quad \sum y^2 = 1{,}400,$$
$$\sum xy = 1{,}475.$$

We first compute

$$Q_x = 1{,}593 - (103)^2/7 = 77.429,$$
$$Q_y = 1{,}400 - (98)^2/7 = 28,$$
$$Q_{xy} = 1{,}475 - (103)(98)/7 = 33,$$

and, if the correlation coefficients are needed, use this in Equations (5.26) and (5.18a):

$$r = \frac{Q_{xy}}{\sqrt{Q_x Q_y}} = \frac{33}{\sqrt{77.429 \cdot 28}} = 0.709 \quad \text{and} \quad r^* = 0.753.$$

From Q_x and Q_y, the standard deviations of the variables x and y are readily obtained:

$$s_x = \sqrt{\frac{77.429}{6}} = 3.592,$$

$$s_y = \sqrt{\frac{28}{6}} = 2.160,$$

and

$$s_{y.x} = \sqrt{\frac{28 - 33^2/77.429}{5}} = 1.670,$$

and using this, the standard deviation of the axis intercept $(s_{a_{yx}})$ and the standard deviation of the regression coefficient $(s_{b_{xy}})$ are found:

$$s_{a_{yx}} = 1.670 \cdot \sqrt{\frac{1}{7} + \frac{14.714^2}{77.429}} = 2.862,$$

$$s_{b_{yx}} = \frac{1.670}{\sqrt{77.429}} = 0.190.$$

Verification

$$\frac{s_{a_{yx}}}{s_{b_{yx}}} = \frac{2.862}{0.190} \simeq 15 \simeq \sqrt{\frac{1{,}593}{7}} = \sqrt{\frac{\sum x^2}{n}}.$$

Verification

The following relations are used to verify the computations:

(1) $$\boxed{\sum (x+y)^2 = \sum x^2 + \sum y^2 + 2\sum xy,} \quad (5.34)$$

5.4 Estimation Procedures

(2) $$\sum(x+y)^2 - \frac{1}{n}\left[\sum(x+y)\right]^2 = Q_x + Q_y + 2Q_{xy},$$ (5.35)

(3) $$s_{y.x}^2 = \frac{\sum(y-\hat{y})^2}{n-2}.$$ (5.36)

Computational scheme for regression and correlation

Step 1: Computation of $\bar{x}, \bar{y}, Q_x, Q_y, Q_{xy}$ in terms of n and
$$\sum x, \sum y$$
$$\sum x^2, \sum y^2, \sum xy$$

Check of the computations:
$$\sum(x+y)^2 = \sum x^2 + \sum y^2 + 2\sum xy$$
$$\sum(x+y)^2 - \frac{1}{n}\{\sum(x+y)\}^2 = Q_x + Q_y + 2Q_{xy}$$

$$\bar{x} = \frac{1}{n}\sum x \qquad \bar{y} = \frac{1}{n}\sum y$$

$$Q_x = \sum x^2 - \frac{1}{n}(\sum x)^2$$

$$Q_y = \sum y^2 - \frac{1}{n}(\sum y)^2$$

$$Q_{xy} = \sum xy - \frac{1}{n}(\sum x)(\sum y)$$

Step 2: Computation of $Q_{y.x}, b_{yx}, a_{yx}, r, s_x, s_y, s_{xy}, s_{y.x},$
$s_{b_{yx}}$ and $s_{a_{yx}}$

$$Q_{y.x} = Q_y - b_{yx}Q_{xy}$$

$$b_{yx} = \frac{Q_{xy}}{Q_x} \qquad s_x = \sqrt{\frac{Q_x}{n-1}} \qquad s_{y.x} = \sqrt{\frac{Q_{y.x}}{n-2}}$$

$$a_{yx} = \bar{y} - b_{yx}\bar{x} \qquad s_y = \sqrt{\frac{Q_y}{n-1}} \qquad s_{b_{yx}} = \frac{s_{y.x}}{\sqrt{Q_x}}$$

$$r = \frac{Q_{xy}}{\sqrt{Q_x Q_y}} \qquad s_{xy} = \frac{Q_{xy}}{n-1} \qquad s_{a_{yx}} = s_{y.x}\sqrt{\frac{1}{n} + \frac{\bar{x}^2}{Q_x}}$$

Check of the computations:

$$r = \frac{s_{xy}}{s_x s_y} = \sqrt{b_{yx} b_{xy}} \qquad \frac{s_{a_{yx}}}{s_{b_{yx}}} = \sqrt{\frac{\sum x^2}{n}}$$

$$s_{y.x}^2 = \frac{\sum(y-\hat{y})^2}{n-2} \qquad s_{y.x} = s_y\sqrt{(1-r^2)\frac{n-1}{n-2}}$$

Scheme for variance analytic testing of regression

Source	SSD	DF	MS (SSD/DF)	MSR ($MS_{Regr.}/MS_{Resid.}$)	$F_{(1,n-2;\alpha)}$
Regression	$(Q_{xy})^2/Q_x$	1		$\hat{F}$	
Residual	$Q_y - (Q_{xy})^2/Q_x$	n − 2		—	—
Total	Q_y	n − 1	—	—	—

If $MS_{Regr.}/MS_{Resid.} = \hat{F} > F_{(1,n-2;\alpha)}$ then H_0 ($\beta = 0$) is rejected. More on variance analysis can be found in Chapter 7.

EXAMPLE. We check the results of example 2 (Section 5.4.2) and, with the help of Table 109, evaluate $\sum (x + y)$ and $\sum (x + y)^2$. The values $\sum x^2 = 1{,}593$, $\sum y^2 = 1{,}400$, and $\sum xy = 1{,}475$ are known. If we had computed correctly, then, according to the first test equation (5.34), we must have $5{,}943 = 1{,}593 + 1{,}400 + (2)(1{,}475) = 5{,}943$. Now we check the

Table 109

x	y	x + y	$(x + y)^2$
13	12	25	625
17	17	34	1156
10	11	21	441
17	13	30	900
20	16	36	1296
11	14	25	625
15	15	30	900
103	98	201	5943

sums of the squares of the deviations $Q_x = 77.429$, $Q_y = 28$, $Q_{xy} = 33$ according to the second control equation (5.35):

$$5{,}943 - (1/7)(201)^2 = 171.429 = 77.429 + 28 + (2)(33).$$

For the last check we need the values predicted by the regression line $\hat{y} = 7.729 + 0.426x$ for the 7 given x-values (Table 110: note the Remark

Table 110

x	y	$\hat{y}$	$y - \hat{y}$	$(y - \hat{y})^2$
13	12	13.267	− 1.267	1.6053
17	17	14.971	2.029	4.1168
10	11	11.989	− 0.989	0.9781
17	13	14.971	− 1.971	3.8848
20	16	16.249	− 0.249	0.0620
11	14	12.415	1.585	2.5122
15	15	14.119	0.881	0.7762
			+ 0.019 ≈ 0	13.9354

(1) concerning residuals in Section 5.6). For $s_{y \cdot x}$ we had obtained the value 1.67, which we now substitute into the third test equation (5.36):

$$1.67^2 = 2.79 = \frac{13.9354}{5}.$$

In tables **summarizing the results** one should specify both variables, and perhaps a third variable (say age in years) in $k \geq 2$ classes, r, a, b, $s_{y \cdot x}^2$, and confidence intervals, at least:

First variable	Second variable	Third variable	n	r	a	b	$s_{y \cdot x}^2$
(1)	(2)	(3)	(4)	(5)	(6)	(7)	(8)
		≤ 50 yr > 50 yr					

5.4.4 Estimation of the correlation coefficients and the regression lines from a correlation table

Candy boxes can be classified according to the length and width of the base, or human beings according to height and weight. In each case we are dealing with two random variables and the question as to a possible correlation between the two attributes is obvious. Correlation coefficients $\rho = \sigma_{xy}/\sigma_x \sigma_y$ always exist when the variances exist and are different from zero. A clear presentation of a two-dimensional frequency distribution with certain attribute combinations can generally be made in the form of a comprehensive **correlation table** of l rows and k columns. For each of the two attributes, a **constant class width** b must here be chosen. Moreover, b should not be taken too large, since a subdivision into classes of larger size will in general lead to an underestimation of r. The class means are, as usual, denoted by x_i and y_j. From the primary list, a tally chart or a **counting table** (Figure 51) with numbered classes (rows and columns) is constructed. Every field of the table

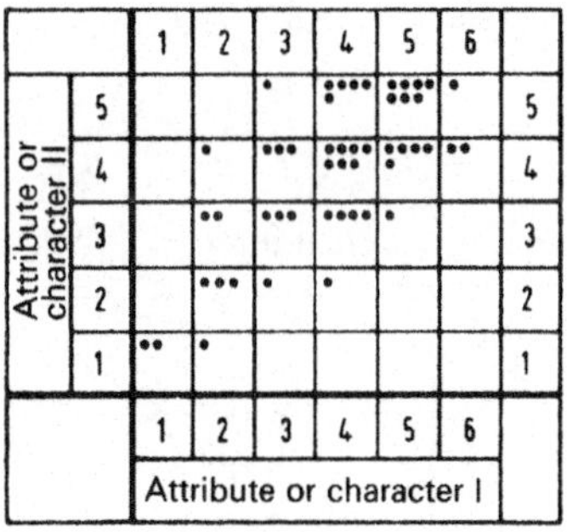

Figure 51

exhibits a certain occupation number; the two regions, lying at the opposite corners of the table are ordinarily unoccupied or sparsely occupied. The occupation number of a field of the ith column (character or attribute I) and the jth row (character or attribute II) is denoted by n_{ij}. Then the

$$\text{row sums} = \sum_{i=1}^{k} n_{ij} = \sum_{i} n_{ij} = n_{.j},$$

$$\text{column sums} = \sum_{j=1}^{l} n_{ij} = \sum_{j} n_{ij} = n_{i.},$$

$$\text{and of course } n = \sum_{i=1}^{k}\sum_{j=1}^{l} n_{ij} = \sum_{i} n_{i.} = \sum_{j} n_{.j}.$$

Table 111 Correlation table

		Attribute or character I				
	Cl. No.	1	...	i	...	k	Row
Cl. No.	$\bar{x}$ / $y\uparrow$	x_1	...	x_i	...	x_k	sum
1	y_1	n_{11}	...	n_{i1}	...	n_{k1}	$n_{.1}$
.	.	.		.		.	.
.	.	.		.		.	.
j	y_j	n_{1j}	...	n_{ij}	...	n_{kj}	$n_{.j}$
.	.	.		.		.	.
.	.	.		.		.	.
l	y_l	n_{11}	...	n_{i1}	...	n_{k1}	$n_{.l}$
Column sum		$n_{1.}$	...	$n_{i.}$	...	$n_{k.}$	n

(Attribute or character II — row label)

With the class widths, b_x and b_y, x_a the column and y_b the row belonging to the largest occupation number (or one of the largest occupation numbers), x_i the columns and y_j the rows, and the definitions

$$v_i = \frac{x_i - x_a}{b_x} \quad \text{and} \quad w_j = \frac{y_j - x_b}{b_y}$$

(v_i and w_j are then integers), the correlation coefficient is given by

$$r = \frac{n\sum_{i}\sum_{j} n_{ij} v_i w_j - (\sum_{i} n_{i.} v_i)(\sum_{j} n_{.j} w_j)}{\sqrt{\left[n\sum_{i} n_{i.} v_i^2 - (\sum_{i} n_{i.} v_i)^2\right]\left[n\sum_{j} n_{.j} w_j^2 - (\sum_{j} n_{.j} w_j)^2\right]}}. \tag{5.37}$$

5.4 Estimation Procedures

Table 112 Areas of the bases of 50 candy boxes with edge lengths x_i and widths y_j measured in cm

	x_i	12	16	20	24	28	32	Row (j) sum $n_{.j}$	$n_{.j}w_j$	$n_{.j}w_j^2$
y_j \ w_j	v_i	-3	-2	-1	0	1	2			
21	1			1	5	7	1	14	14	14
18	0		1	3	7	5	2	18	0	0
15	-1		2	3	4	1		10	-10	10
12	-2		3	1	1			5	-10	20
9	-3	2	1					3	-9	27
Column (i) sum $n_{i.}$		2	7	8	17	13	3	50 n	-15 $\sum n_{.j}w_j$	71 $\sum n_{.j}w_j^2$
$n_{i.}v_i$		-6	-14	-8	0	13	6	-9 $\sum n_{i.}v_i$		
$n_{i.}v_i^2$		18	28	8	0	13	12	79 $\sum n_{i.}v_i^2$		

EXAMPLE. Compute r for the length and width of the base of 50 candy boxes (Table 112; x_i and y_j are class means).

The v_i and the w_j are computed first; we choose $x_a = 24$ and $y_b = 18$:

$$v_i: \quad \frac{12-24}{4} = -3, \quad \frac{16-24}{4} = -2, \quad \text{etc.}$$

$$w_j: \quad \frac{21-18}{3} = 1, \quad \frac{18-18}{3} = 0, \quad \text{etc.}$$

Then the sums (cf., Table 112) of the rows and columns and the four sums of the products are worked out. To compute the sum $\sum_i \sum_j n_{ij} v_i w_j$, we set up a small auxiliary table. For every occupancy number we compute the product $v_i w_j$ and multiply this product by the associated occupancy number n_{ij}:

		-1	0	7	2	8
		0	0	0	0	0
		4	3	0	-1	6
		12	2	0		14
	18	6				24
	18	+22	+4	+0	+6	+2	52

$$\sum_{ij} n_{ij} v_i w_j = 52$$

By (5.37), we then have

$$r = \frac{50 \cdot 52 - (-9)(-15)}{\sqrt{[50 \cdot 79 - (-9)^2][50 \cdot 71 - (-15)^2]}} = 0.6872.$$

One could of course have carried out the computations directly by using the sums

$$\sum_i n_{i.} x_i = 2 \cdot 12 + 7 \cdot 16 + \cdots + 3 \cdot 32 = 1{,}164,$$

$$\sum_i n_{i.} x_i^2 = 2 \cdot 12^2 + 7 \cdot 16^2 + \cdots + 3 \cdot 32^2 = 28{,}336,$$

$$\sum_j n_{.j} y_j = 3 \cdot 9 + 5 \cdot 12 + \cdots + 14 \cdot 21 = 855,$$

$$\sum_j n_{.j} y_j^2 = 3 \cdot 9^2 + 5 \cdot 12^2 + \cdots + 14 \cdot 21^2 = 15{,}219,$$

$$\sum_{ij} x_i(n_{ij} y_j) = 12(2 \cdot 9) + 16(9 + 3 \cdot 12 + 2 \cdot 15 + 18) + \cdots$$

$$+ 32(2 \cdot 18 + 21) = 20{,}496.$$

According to (5.18),

$$r = \frac{\sum x_i y_j - \frac{1}{n} \sum x_i \sum y_j}{\sqrt{\left[\sum x_i^2 - \frac{1}{n}(\sum x_i)^2\right]\left[\sum y_j^2 - \frac{1}{n}(\sum y_j)^2\right]}}$$

$$= \frac{20{,}496 - \frac{1}{50} 1{,}164 \cdot 855}{\sqrt{\left[28{,}336 - \frac{1}{50} 1{,}164^2\right]\left[15{,}219 - \frac{1}{50} 855^2\right]}} = 0.6872.$$

If one of the two quantities under study can be interpreted as being dependent on the other, the computation of the correlation should be supplemented by an analysis of the regression. Letting b_x and b_y be the class widths, one obtains both of the means, the standard deviations, the residual variances, and the regression lines as well as other interesting quantities (cf., also the scheme at the beginning of this section as well as Section 5.5.3) according to

$$\bar{x} = b_x \frac{\sum_i n_{i.} v_i}{n} + x_a = 4 \frac{(-9)}{50} + 24 = 23.28,$$

$$\bar{y} = b_y \frac{\sum_j n_{.j} w_j}{n} + y_b = 3 \frac{(-15)}{50} + 18 = 17.10,$$

$$s_x = b_x \sqrt{\frac{\sum_i n_i \cdot v_i^2}{n} - \left[\frac{\sum_i n_i \cdot v_i}{n}\right]^2} = 4 \cdot \sqrt{\frac{79}{50} - \left[\frac{(-9)}{50}\right]^2} = 4.976,$$

$$s_y = b_y \sqrt{\frac{\sum_j n_{\cdot j} w_j^2}{n} - \left[\frac{\sum_j n_{\cdot j} w_j}{n}\right]^2} = 3 \cdot \sqrt{\frac{71}{50} - \left[\frac{(-15)}{50}\right]^2} = 3.460,$$

$$(s_{y \cdot x})^2 = s_y^2 (1 - r^2) \frac{n-1}{n-2} = 3.46^2 (1 - 0.6872^2) \frac{49}{48} = 6.4497,$$

$$(s_{x \cdot y})^2 = s_x^2 (1 - r^2) \frac{n-1}{n-2} = 4.976^2 (1 - 0.6872^2) \frac{49}{48} = 13.3398,$$

$$b_{yx} = r \frac{s_y}{s_x} = 0.6872 \frac{3.460}{4.976} = 0.4778,$$

$$b_{xy} = r \frac{s_x}{s_y} = 0.6872 \frac{4.976}{3.460} = 0.9883,$$

$$a_{yx} = \bar{y} - b_{yx} \bar{x} = 17.10 - 0.4778 \cdot 23.28 = 5.977,$$

$$a_{xy} = \bar{x} - b_{xy} \bar{y} = 23.28 - 0.9883 \cdot 17.10 = 6.380,$$

i.e.,

$$\hat{y} = 5.977 + 0.478 x \qquad \hat{x} = 6.380 + 0.988 y.$$

5.4.5 Confidence limits of correlation coefficients

The 95% confidence interval for ρ is given in Figure 52 as the interval on the vertical draw above r, between the two curves, corresponding to the n in question. Only when the confidence interval does not include the value $\rho = 0$ are we dealing with a proper ($\rho \neq 0$) correlation. The confidence limits for large n can be found by means of (5.41).

Examples

1. This may be illustrated by an extreme example with $r = 0.5$ and $n = 3$. We carry out the construction in the nomogram at $r = +0.5$ (the middle of the right half of the abscissa) and read from the ordinate the heights of the two $n = 3$ curves at $r = 0.5$: $\rho_1 \simeq -0.91$ and $\rho_2 \simeq +0.98$. The confidence interval is huge (95% CI: $-0.91 \lesssim \rho \lesssim +0.98$) and practically does not allow any conclusion.

2. We obtain the 95% CI for $r = 0.68$ and $n = 50$ (cf., Figure 52): $0.50 \lesssim \rho \lesssim 0.80$, and thus the confirmation of a proper formal correlation ($P = 0.05$).

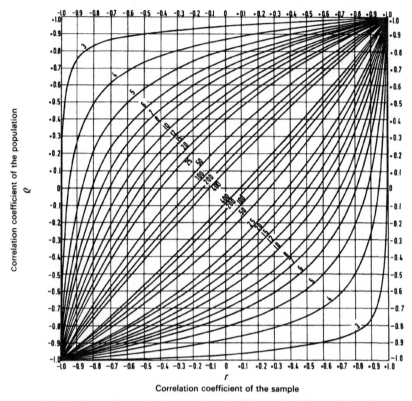

Figure 52 Confidence limits of the correlation coefficients: the 95% confidence interval for ρ. The numbers on the curves indicate the sample size (from F. N. David: Tables of the Ordinates and Probability Integral of the Distribution of the Correlation Coefficient in Small Samples, The Biometrika Office, London 1938).

5.5 TEST PROCEDURES

5.5.1 Testing for the presence of correlation and some comparisons

The null hypothesis that the correlation coefficient determined for a sample is a random deviation from the zero correlation in the population ($\rho = 0$) is tested according to R. A. Fisher by means of the t-distribution with $n - 2$ degrees of freedom:

$$\hat{t} = |r|\left[\sqrt{\frac{n-2}{1-r^2}}\right]. \tag{5.38}$$

For $\hat{t} \geq t_{n-2;\alpha}$, H_0: $\rho = 0$ is rejected [cf., I(4), Section 5.1]. It is simpler to use Table 113.

Table 113 Testing the correlation coefficient r for significance against zero. The null hypothesis ($\rho = 0$) is rejected in favor of the alternative hypothesis (two sided problem: $\rho \neq 0$; one sided problem: $\rho > 0$ or $\rho < 0$) if |r| attains or exceeds the value tabulated for the appropriate problem, the chosen level of significance, and the number of degrees of freedom present (DF = n − 2) (then both regression coefficients β_{yx} and β_{xy} are also different from zero). The one sided test can be carried out only if the sign of the correlation coefficient is known prior to sampling. This table is based on (5.38), solved for r in terms of t^2. Thus r = 0.25 with DF = 60 or n = 62 is statistically significant ($\rho \neq 0$) at the 5% level.

DF	Two sided test			One sided test		
	5 %	1 %	0.1 %	5 %	1 %	0.1 %
1	0.9969	A*	B*	0.9877	0.9995	C*
2	0.9500	0.9900	0.9990	0.9000	0.9800	0.9980
3	0.8783	0.9587	0.9911	0.805	0.934	0.986
4	0.811	0.917	0.974	0.729	0.882	0.963
5	0.754	0.875	0.951	0.669	0.833	0.935
6	0.707	0.834	0.925	0.621	0.789	0.905
7	0.666	0.798	0.898	0.582	0.750	0.875
8	0.632	0.765	0.872	0.549	0.715	0.847
9	0.602	0.735	0.847	0.521	0.685	0.820
10	0.576	0.708	0.823	0.497	0.658	0.795
11	0.553	0.684	0.801	0.476	0.634	0.772
12	0.532	0.661	0.780	0.457	0.612	0.750
13	0.514	0.641	0.760	0.441	0.592	0.730
14	0.497	0.623	0.742	0.426	0.574	0.711
15	0.482	0.606	0.725	0.412	0.558	0.694
16	0.468	0.590	0.708	0.400	0.543	0.678
17	0.456	0.575	0.693	0.389	0.529	0.662
18	0.444	0.561	0.679	0.378	0.516	0.648
19	0.433	0.549	0.665	0.369	0.503	0.635
20	0.423	0.537	0.652	0.360	0.492	0.622
21	0.413	0.526	0.640	0.352	0.482	0.610
22	0.404	0.515	0.629	0.344	0.472	0.599
23	0.396	0.505	0.618	0.337	0.462	0.588
24	0.388	0.496	0.607	0.330	0.453	0.578
25	0.381	0.487	0.597	0.323	0.445	0.568
26	0.374	0.478	0.588	0.317	0.437	0.559
27	0.367	0.470	0.579	0.311	0.430	0.550
28	0.361	0.463	0.570	0.306	0.423	0.541
29	0.355	0.456	0.562	0.301	0.416	0.533
30	0.349	0.449	0.554	0.296	0.409	0.526
35	0.325	0.418	0.519	0.275	0.381	0.492
40	0.304	0.393	0.490	0.257	0.358	0.463
50	0.273	0.354	0.443	0.231	0.322	0.419
60	0.250	0.325	0.408	0.211	0.295	0.385
70	0.232	0.302	0.380	0.195	0.274	0.358
80	0.217	0.283	0.357	0.183	0.257	0.336
90	0.205	0.267	0.338	0.173	0.242	0.318
100	0.195	0.254	0.321	0.164	0.230	0.302
120	0.178	0.232	0.294	0.150	0.210	0.277
150	0.159	0.208	0.263	0.134	0.189	0.249
200	0.138	0.181	0.230	0.116	0.164	0.216
250	0.124	0.162	0.206	0.104	0.146	0.194
300	0.113	0.148	0.188	0.095	0.134	0.177
350	0.105	0.137	0.175	0.0878	0.124	0.164
400	0.0978	0.128	0.164	0.0822	0.116	0.154
500	0.0875	0.115	0.146	0.0735	0.104	0.138
700	0.0740	0.0972	0.124	0.0621	0.0878	0.116
1000	0.0619	0.0813	0.104	0.0520	0.0735	0.0975
1500	0.0505	0.0664	0.0847	0.0424	0.0600	0.0795
2000	0.0438	0.0575	0.0734	0.0368	0.0519	0.0689

A* = 0.999877   B* = 0.99999877   C* = 0.9999951

More critical values may be computed by

$$t_{n-2;\alpha}/\sqrt{(n-2) + t_{n-2;\alpha}^2}.$$

For instance: n = 30, α = 0.05; two sided test: $t_{28;0.05}$ = 2.048, $2.048/\sqrt{(30-2) + 2.048^2} = 0.3609$; one sided test:

$t_{28;0.05 \text{ one sided}} = 1.701$, $1.701/\sqrt{(30-2) + 1.701^2} = 0.3060$.

Examples

1. Suppose $\alpha = 0.01$, $r = 0.47$. Then according to Table 113 there must be at least 29 ($= DF + 2$) observations available to allow the conclusion that the variables are mutually dependent.

2. If from 27 observations an $r = 0.50$ is computed and $\alpha = 0.01$ agreed upon, then the null hypothesis ($\rho = 0$) must be rejected, since 0.50 is larger than the tabulated value (0.487).

REMARKS

1. The test for the null hypothesis can also be written in terms of the F-distribution (5.38a, 5.38b):

$$\hat{F} = \frac{r^2(n-2)}{1-r^2}$$
$$DF_1 = 1, \quad DF_2 = n-2$$
(5.38a)

[Note: (5.38) and (5.38a) are of equal value; cf., (1.145), leftmost part];

$$\hat{F} = \frac{1+r}{1-r}$$
$$DF_1 = DF_2 = n-2$$
(5.38b)

(Kymn 1968).

2. The hypothesis $H_0: \rho = \rho_0$ can be tested according to Samiuddin (1970) by

$$\hat{t} = \frac{(r-\varrho)\sqrt{n-2}}{\sqrt{(1-r^2)(1-\varrho^2)}},$$
$$DF = n-2.$$
(5.39)

3. Two estimated correlation coefficients r_1 and r_2 ($r_1 = r_{AB}$, $r_2 = r_{BC}$, $r_{12} = r_{AC}$) from the same sample (with the three characteristics A, B, and C), can be tested for equality according to Hotelling (1940):

$$\hat{F} = \frac{(r_1-r_2)^2(n-3)(1+r_{12})}{2(1-r_{12}^2-r_1^2-r_2^2+2r_{12}r_1r_2)},$$
$$DF_1 = 1, \quad DF_2 = n-3.$$
(5.39a)

Other tests for the **equality of dependent correlation coefficients** (e.g., $H_0: \rho_{12} = \rho_{13}$) are given by J. J. Neill and O. J. Dunn, Biometrics **31** (1975), 531–543, S. C. Choi, Biometrika **64** (1977), 645–647 and by B. M. Bennett, Statistische Hefte **19** (1978), 71–76 (cf., Psychological Bulletin **87** (1980), 245–251).

4. **Multiple tests of correlations** as well as one-stage and multistage Bonferroni procedures are compared by R. E. Larzelere and S. A. Mulaik, Psychological Bulletin **84** (1977), 557–569.

5. Pairwise comparisons among k independent correlations from samples with unequal sample sizes are given by K. J. Levy, British Journal of Mathematical and Statistical Psychology **30** (1977), 137–139 (cf., Psychological Bulletin **82** (1975), 174–176 and 177–179). More on this is given by P. A. Games, Psychological Bulletin **85** (1978), 661–672.

6. Nomograms for computing and assessing correlation and regression coefficients are given by Friedrich (1970) (cf., also Ludwig (1965)).

The r to $\dot{z}$ transformation

If the correlation coefficient differs significantly from zero, then the smaller the number of observations and the larger the absolute value of the correlation coefficient, the more the distribution of r deviates from the normal. The distribution of the correlation coefficient is approximately normalized by the r to $\dot{z}$ transformation of R. A. Fisher, given by

$$\dot{z} = \frac{1}{2} \ln \frac{1+r}{1-r} = 1.1513 \log \frac{1+r}{1-r} \qquad (F.1)$$

with the standard deviation

$$s_{\dot{z}} = \frac{1}{\sqrt{n-3}}. \qquad (F.2)$$

The goodness of this approximation increases with decreasing absolute value of ρ and with increasing sample size. The interval $-1 < r < 1$ is mapped onto $-\infty < \dot{z} < \infty$.

From

$$\dot{z} = r + \tfrac{1}{3}r^3 + \tfrac{1}{5}r^5 + \tfrac{1}{7}r^7 + \ldots, \qquad (F.3)$$

we see that for

1. $r = \pm 1$ we get $\dot{z} = \pm\infty$,
2. $r < 0.3$ we get $\dot{z} \simeq r$.

This r to $\dot{z}$ transformation requires that x and y have bivariate normal distribution in the population. The larger the sample size, the less stringent is this assumption. The $\dot{z}$ of this transformation (r is the hyperbolic tangent of $\dot{z}$: $r = \tanh \dot{z}$ and $\dot{z} = \tanh^{-1} r$) must not be confused with the standard normal variable z. One uses this transformation only for samples with $n > 10$ from a bivariate normal population. For $n < 50$, Hotelling (1953) suggests replacing $\dot{z}$ by $\dot{z}_H$ and $s_{\dot{z}}$ by $s_{\dot{z}_H}$:

$$\dot{z}_H = \dot{z} - \frac{3\dot{z} + r}{4n}; \qquad s_{\dot{z}_H} = \frac{1}{\sqrt{n-1}}. \qquad (F.4)$$

We do without this correction in the examples. The conversion from r to $\hat{z}$, and vice versa, is carried out with the help of Table 114: The first column of the Table lists the $\hat{z}$-values with one place beyond the decimal, while the second place beyond the decimal can be found in the uppermost row.

The significance of the correlation coefficients (cf., Table 113) can then be tested according to

(p.217)

$$\hat{z} = \frac{\hat{z}}{s_{\hat{z}}} = \hat{z}\sqrt{n-3}. \qquad (5.40)$$

The 95% confidence interval for ρ is given by

$$\hat{z} \pm 1.960 s_{\hat{z}}. \qquad (5.41)$$

With the help of Table 114, we can transform the upper and lower $\hat{z}$-values obtained back into r-values. The unknown correlation coefficient ρ of the population then lies with the required probability in the interval given by the two r-values.

Two better approximations of confidence intervals for ρ are discussed by A. Boomsma, Statistica Neerlandica **31** (1977), 179–185.

Table 114 Transformation of the correlation coefficient $z = \frac{1}{2} \ln[(1+r)/(1-r)]$ (taken from Fisher, R. A. and Yates, F.: Statistical Tables for Biological, Agricultural, and Medical Research, Oliver and Boyd Ltd., Edinburgh 1963, p. 63)

$\hat{z}$	0.00	0.01	0.02	0.03	0.04	0.05	0.06	0.07	0.08	0.09
0.0	0.0000	0.0100	0.0200	0.0300	0.0400	0.0500	0.0599	0.0699	0.0798	0.0898
0.1	0.0997	0.1096	0.1194	0.1293	0.1391	0.1489	0.1586	0.1684	0.1781	0.1877
0.2	0.1974	0.2070	0.2165	0.2260	0.2355	0.2449	0.2543	0.2636	0.2729	0.2821
0.3	0.2913	0.3004	0.3095	0.3185	0.3275	0.3364	0.3452	0.3540	0.3627	0.3714
0.4	0.3800	0.3885	0.3969	0.4053	0.4136	0.4219	0.4301	0.4382	0.4462	0.4542
0.5	0.4621	0.4699	0.4777	0.4854	0.4930	0.5005	0.5080	0.5154	0.5227	0.5299
0.6	0.5370	0.5441	0.5511	0.5580	0.5649	0.5717	0.5784	0.5850	0.5915	0.5980
0.7	0.6044	0.6107	0.6169	0.6231	0.6291	0.6351	0.6411	0.6469	0.6527	0.6584
0.8	0.6640	0.6696	0.6751	0.6805	0.6858	0.6911	0.6963	0.7014	0.7064	0.7114
0.9	0.7163	0.7211	0.7259	0.7306	0.7352	0.7398	0.7443	0.7487	0.7531	0.7574
1.0	0.7616	0.7658	0.7699	0.7739	0.7779	0.7818	0.7857	0.7895	0.7932	0.7969
1.1	0.8005	0.8041	0.8076	0.8110	0.8144	0.8178	0.8210	0.8243	0.8275	0.8306
1.2	0.8337	0.8367	0.8397	0.8426	0.8455	0.8483	0.8511	0.8538	0.8565	0.8591
1.3	0.8617	0.8643	0.8668	0.8692	0.8717	0.8741	0.8764	0.8787	0.8810	0.8832
1.4	0.8854	0.8875	0.8896	0.8917	0.8937	0.8957	0.8977	0.8996	0.9015	0.9033
1.5	0.9051	0.9069	0.9087	0.9104	0.9121	0.9138	0.9154	0.9170	0.9186	0.9201
1.6	0.9217	0.9232	0.9246	0.9261	0.9275	0.9289	0.9302	0.9316	0.9329	0.9341
1.7	0.9354	0.9366	0.9379	0.9391	0.9402	0.9414	0.9425	0.9436	0.9447	0.9458
1.8	0.94681	0.94783	0.94884	0.94983	0.95080	0.95175	0.95268	0.95359	0.95449	0.95537
1.9	0.95624	0.95709	0.95792	0.95873	0.95953	0.96032	0.96109	0.96185	0.96259	0.96331
2.0	0.96403	0.96473	0.96541	0.96609	0.96675	0.96739	0.96803	0.96865	0.96926	0.96986
2.1	0.97045	0.97103	0.97159	0.97215	0.97269	0.97323	0.97375	0.97426	0.97477	0.97526
2.2	0.97574	0.97622	0.97668	0.97714	0.97759	0.97803	0.97846	0.97888	0.97929	0.97970
2.3	0.98010	0.98049	0.98087	0.98124	0.98161	0.98197	0.98233	0.98267	0.98301	0.98335
2.4	0.98367	0.98399	0.98431	0.98462	0.98492	0.98522	0.98551	0.98579	0.98607	0.98635
2.5	0.98661	0.98688	0.98714	0.98739	0.98764	0.98788	0.98812	0.98835	0.98858	0.98881
2.6	0.98903	0.98924	0.98945	0.98966	0.98987	0.99007	0.99026	0.99045	0.99064	0.99083
2.7	0.99101	0.99118	0.99136	0.99153	0.99170	0.99186	0.99202	0.99218	0.99233	0.99248
2.8	0.99263	0.99278	0.99292	0.99306	0.99320	0.99333	0.99346	0.99359	0.99372	0.99384
2.9	0.99396	0.99408	0.99420	0.99431	0.99443	0.99454	0.99464	0.99475	0.99485	0.99495
	0.0	0.1	0.2	0.3	0.4	0.5	0.6	0.7	0.8	0.9
3	0.99505	0.99595	0.99668	0.99728	0.99777	0.99818	0.99851	0.99878	0.99900	0.99918
4	0.99933	0.99945	0.99955	0.99963	0.99970	0.99975	0.99980	0.99983	0.99986	0.99989

5.5 Test Procedures

EXAMPLE. In the example of Section 5.4.4 we obtained a correlation coefficient of $r = 0.6872 \simeq 0.687$ for 50 data points. Does this value differ significantly from zero?

For 48 DF, a correlation coefficient of this size is, according to Table 113, clearly different from zero. Thus the question is answered. We nevertheless wish to determine the 95% confidence interval. From Table 114, $\dot{z} = 0.842$; hence $\hat{z} = \dot{z}\sqrt{n-3} = 0.842\sqrt{47} = 5.772$. To this $\dot{z}$-value there corresponds a $P \ll 0.001$. The 95% confidence interval is obtained from

$$s_{\dot{z}} = \frac{1}{\sqrt{n-3}} = \frac{1}{\sqrt{50-3}} = 0.146$$

and

$$\dot{z} \pm 1.96 \cdot 0.146 = \dot{z} \pm 0.286,$$

$$0.556 \leq \dot{z} \leq 1.128$$

so that we have

$$95\%\text{-CI}: 0.505 \leq \varrho \leq 0.810.$$

The transformation of small values of r ($0 < r < 0.20$) into $\dot{z} = \tanh^{-1} r$ can be carried out with sufficient accuracy according to $\dot{z} = r + (r^3/3)$ (e.g., $\dot{z} = 0.100$ for $r = 0.10$); Values of $\dot{z}$ for $r = 0.00(0.01)0.99$ can be found in the following table (for $r = 1$, $\dot{z} = \infty$):

r	0.00	0.01	0.02	0.03	0.04	0.05	0.06	0.07	0.08	0.09
0.0	0.00000	0.01000	0.02000	0.03001	0.04002	0.05004	0.06007	0.07011	0.08017	0.09024
0.1	0.10034	0.11045	0.12058	0.13074	0.14093	0.15114	0.16139	0.17167	0.18198	0.19234
0.2	0.20273	0.21317	0.22366	0.23419	0.24477	0.25541	0.26611	0.27686	0.28768	0.29857
0.3	0.30952	0.32055	0.33165	0.34283	0.35409	0.36544	0.37689	0.38842	0.40060	0.41180
0.4	0.42365	0.43561	0.44769	0.45990	0.47223	0.48470	0.49731	0.51007	0.52298	0.53606
0.5	0.54931	0.56273	0.57634	0.59015	0.60416	0.61838	0.63283	0.64752	0.66246	0.67767
0.6	0.69315	0.70892	0.72501	0.74142	0.75817	0.77530	0.79281	0.81074	0.82911	0.84796
0.7	0.86730	0.88718	0.90764	0.92873	0.95048	0.97296	0.99622	1.02033	1.04537	1.07143
0.8	1.09861	1.12703	1.15682	1.18814	1.22117	1.25615	1.29334	1.33308	1.37577	1.42193
0.9	1.47222	1.52752	1.58903	1.65839	1.73805	1.83178	1.94591	2.09230	2.29756	2.64665

5.5.2 Further applications of the $\dot{z}$-transformation

1. **The two sided test of the hypothesis that ρ_1 (estimated by r_1) is equal to any value ρ** proceeds on the basis of the standard normal variable z according to

$$\hat{z} = \frac{|\dot{z}_1 - \dot{z}|}{\sqrt{1/(n_1 - 3)}} = |\dot{z}_1 - \dot{z}|\sqrt{n_1 - 3}. \tag{5.42}$$

If the test quantity is below the significance bound (Table 14, Section 1.3.4), then it can be assumed that $\rho_1 = \rho$ (cf., also Section 5.5.1, Formula (5.3.9)).

2. **The two sided comparison of two estimated correlation coefficients** ρ_1 and ρ_2 proceeds according to

$$\hat{z} = \frac{|\dot{z}_1 - \dot{z}_2|}{\sqrt{\dfrac{1}{n_1-3} + \dfrac{1}{n_2-3}}}. \tag{5.43}$$

The sizes of the two samples have to be greater than 20. If the test quotient falls below the significance bound, then it can be assumed that the underlying parameters are equal ($\rho_1 = \rho_2$). Estimation of the **common correlation coefficient** $\bar{r}$ then proceeds by way of $\hat{\dot{z}}$:

$$\hat{\dot{z}} = \frac{\dot{z}_1(n_1-3) + \dot{z}_2(n_2-3)}{n_1 + n_2 - 6} \tag{5.44}$$

with

$$s_{\dot{z}} = \frac{1}{\sqrt{n_1 + n_2 - 6}}. \tag{5.45}$$

The significance of ρ [parameter of $\bar{r}$] can be tested according to

$$\hat{z} = \hat{\dot{z}} \cdot \sqrt{n_1 + n_2 - 6}. \tag{5.46}$$

Examples

1. Given $r_1 = 0.3$, $n_1 = 40$, $\rho = 0.4$. Can $\rho_1 = \rho$ be assumed (two sided test with $\alpha = 0.05$)? By (5.42) (Table 114),

$$\hat{z} = (|0.30952 - 0.42365|)\sqrt{40 - 3} = 0.694 < 1.96.$$

Since the test quantity is smaller than the significance bound, the null hypothesis $\rho_1 = \rho$ cannot be rejected at the 5% level.

2. Given $r_1 = 0.6$, $n_1 = 28$, and $r_2 = 0.8$; $n_2 = 23$. Can it be assumed that $\rho_1 = \rho_2$ (two sided test with $\alpha = 0.05$)? By (5.43),

$$\hat{z} = \frac{|0.6932 - 1.0986|}{\sqrt{\dfrac{1}{28-3} + \dfrac{1}{23-3}}} = 1.35 < 1.96.$$

5.5 Test Procedures

Since $\hat{z} = 1.35 < 1.96$, the null hypothesis $\rho_1 = \rho_2$ cannot be rejected at the 5% level. The 95% confidence interval for ρ is found in terms of $\hat{z}$ (5.44) to be

$$\hat{z} = \frac{17.330 + 21.972}{28 + 23 - 6} = 0.8734,$$

$$s_{\hat{z}} = \frac{1}{\sqrt{28 + 23 - 6}} = 0.1491,$$

$$\hat{z} = 0.8734 + 1.96 \cdot 0.1491,$$

$$\hat{z} = 0.8734 \pm 0.2922,$$

$$0.5812 \le \hat{z} \le 1.1656,$$

95% CI: $0.5235 \le \varrho \le 0.8223$ or $0.52 \le \rho \le 0.82$.

We can test simultaneously whether the k samples come from populations with given hypothetical correlation coefficients. The case where the hypothetical coefficients are all the same is of particular interest (null hypothesis: $\rho_1 = \rho_2 = \cdots = \rho_i = \cdots = \rho_k = \rho_0$, ρ_0 arbitrary but fixed theoretical value); the corresponding test statistic is given by

$$\hat{\chi}^2 = \sum_{i=1}^{k} (n_i - 3)(\dot{z}_i - \dot{z})^2, \qquad (5.47)$$

where $\dot{z} = \dot{z}$-transform of the common correlation coefficient ρ_0; $\hat{\chi}^2$ has an approximate χ^2-distribution with k degrees of freedom. E.g., we have for $\alpha = 0.05$ and $k = 4$ the significance bound $\chi^2_{4;\,0.05} = 9.49$. If the test statistic turns out to be smaller than or equal to the significance bound, then the null hypothesis that the k samples come from bivariate populations with the same correlation coefficient ρ_0 cannot be rejected.

For a **test for homogeneity among the coefficients of correlation**—null hypothesis: $\rho_1 = \cdots = \rho_k = \rho$ [the value of ρ is not known]—we estimate the z-transform of the common coefficient of correlation by

$$\hat{z} = \frac{\sum_{i=1}^{k} \dot{z}_i(n_i - 3)}{\sum_{i=1}^{k} (n_i - 3)}. \qquad (5.48)$$

The associated standard deviation is

$$s_{\hat{z}} = \frac{1}{\sqrt{\sum_{i=1}^{k} (n_i - 3)}}. \qquad (5.49)$$

Then the test statistic for homogeneity is given by

$$\hat{\chi}^2 = \sum_{i=1}^{k} (n_i - 3)(\hat{z}_i - \hat{\bar{z}})^2 \qquad (5.50)$$

but with $DF = k - 1$. If the test quantity is smaller or equal to the significance bound, the null hypothesis may be retained and the **common correlation coefficient** $\bar{r}$ estimated. The confidence limits for the common parameter ρ are obtained in a well-known manner in terms of the associated $\hat{\bar{z}}$-value and standard deviation $s_{\hat{\bar{z}}}$:

[For the 95% CI] $\qquad \hat{\bar{z}} \pm 1.960 s_{\hat{\bar{z}}}, \qquad (5.51)$

[For the 99% CI] $\qquad \hat{\bar{z}} \pm 2.576 s_{\hat{\bar{z}}}, \qquad (5.52)$

by transforming the upper and lower limits into the corresponding r-values.

EXAMPLE

Table 115

r_i	$\hat{z}_i$	n_i	$n_i - 3$	$\hat{z}_i(n_i - 3)$	$\hat{z}_i - \hat{\bar{z}}$	$(\hat{z}_i - \hat{\bar{z}})^2$	$(n_i - 3)(\hat{z}_i - \hat{\bar{z}})^2$
0.60	0.6932	28	25	17.330	0.1777	0.03158	0.7895
0.70	0.8673	33	30	26.019	0.0036	0.00001	0.0003
0.80	1.0986	23	20	21.972	0.2277	0.05185	1.0369
$\sum (n_i - 3) =$			75	65.321		$\hat{\chi}^2 =$	1.8268

Since $\hat{\chi}^2$ is substantially less than $\chi^2_{2;0.05} = 5.99$, a common correlation coefficient may be estimated:

$$\hat{\bar{z}} = \frac{65.321}{75} = 0.8709; \qquad \bar{r} = 0.702$$

$$s_{\hat{\bar{z}}} = 1/\sqrt{75} = 0.115; \qquad \hat{\bar{z}} \pm 1.96 \cdot 0.115 = \hat{\bar{z}} \pm 0.2254$$

$$0.6455 \leq \hat{\bar{z}} \leq 1.0963$$

95% CI: $0.5686 \leq \rho \leq 0.7992 \quad$ or $\quad 0.57 \leq \rho \leq 0.80$.

The estimates of common correlation coefficients can in their turn be used for comparisons between two estimates $r_{(1)}$ and $r_{(2)}$, or for comparisons between an estimate $r_{(1)}$ and a hypothetical correlation coefficient ρ.

▶ 5.5.3 Testing the linearity of a regression

It is possible to test the null hypothesis that a given regression is linear, if the total number n of y-values is larger than the number k of x-values: For every value x_i of the k x-values there are thus n_i y-values present. [If the linearity or nonlinearity is clear from the aggregate of points, the linearity test can be dispensed with]. If we are dealing with a linear regression, then the group means $\bar{y}_i$ must lie on an approximately straight line, i.e., their deviation from the regression line (lack of fit) may not be too large in comparison with the deviations among multiple observations (pure error). Hence if the ratio

$$\frac{\text{Deviation of the means from the regression line}}{\text{Deviation of the } y\text{-values from their group mean}} = \frac{\text{lack of fit}}{\text{pure error}}$$

—in other words, the test quantity

$$\boxed{\hat{F} = \frac{\dfrac{1}{k-2}\sum_{i=1}^{k} n_i(\bar{y}_i - \hat{y}_i)^2}{\dfrac{1}{n-k}\sum_{i=1}^{k}\sum_{j=1}^{n_i}(y_{ij}-\bar{y}_i)^2}} \quad \begin{array}{l} v_1 = k-2, \\ v_2 = n-k \end{array} \quad (5.53)$$

with $(k-2, n-k)$ degrees of freedom—attains or exceeds the significance (p.145) bound, then the linearity hypothesis must be rejected. The numerator and denominator are each unbiased estimates of $\sigma_{y.x}^2$ if the regression function is linear. The denominator is so even if the regression function is not linear, it being a weighted average of independent variance estimates at the individual x values.

A closer look: we denote the individual values found by y_{ij} and the values found with the help of the empirical regression function by $\hat{y}_i$ and write

$$y_{ij} - \hat{y}_i = (y_{ij} - \bar{y}_i) + (\bar{y}_i - \hat{y}_i).$$

Squaring and summing this over i and j gives

$$\sum_{i=1}^{k}\sum_{j=1}^{n_i}(y_{ij}-\hat{y}_i)^2 = \sum_{i=1}^{k}\sum_{j=1}^{n_i}[(y_{ij}-\bar{y}_i)+(\bar{y}_i-\hat{y}_i)]^2,$$

and hence (5.54)

$$\boxed{\sum_{i=1}^{k}\sum_{j=1}^{n_i}(y_{ij}-\hat{y}_i) = \sum_{i=1}^{k}\sum_{j=1}^{n_i}(y_{ij}-\bar{y}_i)^2 + \sum_{i=1}^{k} n_i(\bar{y}_i - \hat{y}_i)^2} \quad (5.54)$$

where the crossproduct term vanishes because $\sum(y_{ij}-\bar{y}_i) = 0$. The first term on the right is a contribution to the variability of the observations about the empirical regression line caused by variability of the observations about the means at the individual x-values. The second term is a contribution caused by variation of the means about the empirical regression line.

EXAMPLE. Given Table 116: $n = 8$ observations were made at $k = 4$ different x's. To test the linearity at the 5% level we first estimate the regression line and then compute for the four x_i-values the corresponding $\hat{y}_i$-values. The sums required for (5.53) can be read off from Tables 117 and 117a.

Table 116   n = 8 observations were made at k = 4 different x's; the x's carry multiple observations

x_i	y_{ij}			n_i
	j = 1	j = 2	j = 3	
1	1	2		2
5	2	3	3	3
9	4			1
13	5	6		2

$$\bar{x} = \frac{\sum_{i=1}^{k} n_i x_i}{n} = \frac{52}{8} = 6.5, \qquad \bar{y} = \frac{\sum_{i=1}^{k} \sum_{j=1}^{n_i} y_{ij}}{n} = \frac{26}{8} = 3.25,$$

$$Q_x = \sum_{i=1}^{k} n_i x_i^2 - \frac{1}{n}\left(\sum_{i=1}^{k} n_i x_i\right)^2 = 496 - \frac{52^2}{8} = 158,$$

$$Q_y = \sum_{i=1}^{k} \sum_{j=1}^{n_i} y_{ij}^2 - \frac{1}{n}\left(\sum_{i=1}^{k} \sum_{j=1}^{n_i} y_{ij}\right)^2 = 104 - \frac{26^2}{8} = 19.5,$$

$$Q_{xy} = \sum_{i=1}^{k} \sum_{j=1}^{n_i} x_i y_{ij} - \frac{1}{n}\left(\sum_{i=1}^{k} n_i x_i\right)\left(\sum_{i=1}^{k} \sum_{j=1}^{n_i} y_{ij}\right) = 222 - \frac{52 \cdot 26}{8} = 53,$$

$$b_{yx} = \frac{Q_{xy}}{Q_x} = \frac{53}{158} = 0.335,$$

$$a_{yx} = \bar{y} - b_{yx}\bar{x} = 3.25 - 0.335 \cdot 6.5 = 1.07,$$

$$\hat{y} = 1.07 + 0.335x.$$

5.5 Test Procedures

The test quantity becomes

$$\hat{F} = \frac{\frac{1}{4-2} 0.0533}{\frac{1}{8-4} 1.67} = 0.064.$$

Since $\hat{F} = 0.064 < 6.94 = F(2; 4; 0.05)$, the linearity hypothesis is retained.

Table 117

x_i	y_{ij}	n_i	$\bar{y}_i$	$\hat{y}_i$	$\|\bar{y}_i - \hat{y}_i\|$	$(\bar{y}_i - \hat{y}_i)^2$	$n_i(\bar{y}_i - \hat{y}_i)^2$
1	1;2	2	1.50	1.41	0.09	0.0081	0.0162
5	2;3;3	3	2.67	2.75	0.08	0.0064	0.0192
9	4	1	4.00	4.09	0.09	0.0081	0.0081
13	5;6	2	5.50	5.43	0.07	0.0049	0.0098
						$\sum_i n_i(\bar{y}_i - \hat{y}_i)^2$	= 0.0533

Table 117a

x_i	y_{ij}	$\bar{y}_i$	$\|y_{ij} - \bar{y}_i\|$	$(y_{ij} - \bar{y}_i)^2$	$\sum_j (y_{ij} - \bar{y}_i)^2$
1	1;2	1.50	0.5;0.5	0.25;0.25	0.50
5	2;3;3	2.67	0.67;0.33;0.33	0.45;0.11;0.11	0.67
9	4	4.00	0	0	0
13	5;6	5.50	0.5;0.5	0.25;0.25	0.50
				$\sum_{ij}(y_{ij} - \bar{y}_i)^2$	= 1.67

Testing the linearity of a regression estimated from a correlation table

If the data are based on a correlation table, then a different modification of the linearity test is common. The starting point is the so-called correlation ratio of y on x, written E_{yx}, which records the degree of deviation of the column frequencies from the column means:

$$\boxed{1 \geq E_{yx}^2 \geq r^2.} \tag{5.55}$$

If a regression in question is linear, then the correlation ratio and the correlation coefficient are approximately equal. The more strongly the column means deviate from a straight line, the more marked is the difference between E_{yx} and r. This difference between the two index numbers can be used in testing the linearity of regression:

$$\boxed{\hat{F} = \frac{\frac{1}{k-2}(E_{yx}^2 - r^2)}{\frac{1}{n-k}(1 - E_{yx}^2)}}, \qquad \begin{array}{l} v_1 = k - 2, \\ v_2 = n - k, \end{array} \tag{5.56}$$

where k is the number of columns. On the basis of the test quantity (5.56), the null hypothesis $\eta_{xy}^2 - \rho^2 = 0$ (i.e., there is a linear relation between x and y) is rejected for $\hat{F} > F_{k-2; n-k; \alpha}$ at the $100\alpha\%$ level; there is then a significant deviation from linearity.

The square of the correlation ratio is estimated by

$$\boxed{E_{yx}^2 = \frac{S_1 - R}{S_2 - R}}, \tag{5.57}$$

where the computation of S_1, S_2, and R can be gathered from the following example. We form with the data from Table 112 for each x_i the sum $\sum_j n_{ij} w_j$, i.e., $\{2(-3)\}$, $\{1(-3) + 3(-2) + 2(-1) + 1(0)\}$, $\{1(-2) + 3(-1) + 3(0) + 1(1)\}$, $\{1(-2) + 4(-1) + 7(0) + 5(1)\}$, $\{1(-1) + 5(0) + 7(1)\}$, $\{2(0) + 1(1)\}$, divide the squares of these sums by the associated $n_{i.}$ and sum the quotients over all i, thus obtaining S_1:

$$S_1 = \frac{(-6)^2}{2} + \frac{(-11)^2}{7} + \frac{(-4)^2}{8} + \frac{(-1)^2}{17} + \frac{6^2}{13} + \frac{1^2}{3} = 40.447.$$

S_2 is presented in Table 112 as $\sum_j n_{.j} w_j^2 = 71$; R can be computed from $\sum_j n_{.j} w_j$ and n, E_{yx}^2 by (5.57), and $\hat{F}$ by (5.56):

$$R = \frac{(\sum n_{.j} w_j)^2}{n} = \frac{(-15)^2}{50} = 4.5,$$

$$E_{yx}^2 = \frac{S_1 - R}{S_2 - R} = \frac{40.447 - 4.5}{71 - 4.5} = 0.541,$$

$$\hat{F} = \frac{\frac{1}{k-2}(E_{yx}^2 - r^2)}{\frac{1}{n-k}(1 - E_{yx}^2)} = \frac{\frac{1}{6-2}(0.541 - 0.472)}{\frac{1}{50-6}(1 - 0.541)} = 1.653.$$

Since $\hat{F} = 1.65 < 2.55 = F_{4; 54; 0.05}$, there is no cause to doubt the linearity hypothesis.

5.5 Test Procedures

If the linearity test reveals significant deviations from linearity, it might be possible to achieve linearity by suitable **transformation** of the variables. We shall delve more deeply into the transformation problem during our discussion of the analysis of variance. If transformations are unsuccessful, a second order model might be tried (cf., Section 5.6).

Assumptions of regression analysis

We have discussed the testing of an important assumption of regression analysis—namely linearity. Other assumptions or suppositions are only briefly indicated, since we assumed them as approximately given in the discussion of the test procedures. Besides the existence of a linear regression in the population for the original or transformed data (correctness of the model), the values of the dependent random variables y_i for given controlled and/or observation-error-free values of the independent variables x must be **mutually independent and normally distributed with the same residual variance** $\sigma_{y.x}^2$ (usually not known), whatever be the value of x. This homogeneity of the residual variance is called **homoscedasticity**. Slight deviations from homoscedasticity or normality can be neglected. More particulars can be found in the specialized literature. For **practical work**, there are also the following essential points: The data **really originate in the population** about which information is desired, and there are **no extraneous variables which degrade the importance of the relationship between** x **and** y.

Remark concerning Sections 5.5.4 through 5.5.8. Formulas (5.58) to (5.60) as well as (5.64) and (5.66) are given for the two-sided test. Hints concerning one-sided tests are given at the end of Sections 5.5.4 and 5.5.5.

▶ 5.5.4 Testing the regression coefficient against zero

If the hypothesis of linearity is not rejected by the test described above, then one tests whether the estimate of the regression coefficient differs statistically from zero ($H_0: \beta_{yx} = 0$, $H_A: \beta_{yx} \neq 0$). For given significance level, Student's t-distribution provides the significance bound:

(p. 136)

$$\hat{t} = \frac{|b_{yx}|}{s_{b_{yx}}} \qquad (5.58)$$

with DF = $n - 2$. If the value of the test statistic equals or exceeds the bound then β_{yx} differs significantly from zero (cf., Section 5.4.4: scheme for variance analytic testing of regression and the caption of Table 113).

EXAMPLE. Given $b_{yx} = 0.426$; $s_{b_{yx}} = 0.190$; $n = 80$, $S = 95\%$ (i.e., $\alpha = 5\% = 0.05$)

$$\hat{t} = \frac{0.426}{0.190} = 2.24 > 1.99 = t_{78;0.05}.$$

$H_0: \beta_{yx} = 0$ is rejected at the 5% level, i.e., the basic parameter β_{yx} differs significantly from zero.

If the sample correlation coefficient r was computed and tested ($H_0: \rho = 0$ against $H_A: \rho \neq 0$), and H_0 could not be rejected, then also β_{yx} (and β_{xy}) = 0.

One-sided tests concerning (5.58):

H_0	H_A	H_0 is rejected for
$\beta_{yx} \leq 0$	$\beta_{yx} > 0$	$b_{yx}/s_{b_{yx}} \geq t_{n-2;\alpha;\text{one s.}}$
$\beta_{yx} \geq 0$	$\beta_{yx} < 0$	$b_{yx}/s_{b_{yx}} \leq -t_{n-2;\alpha;\text{one s.}}$

5.5.5 Testing the difference between an estimated and a hypothetical regression coefficient

To test whether an estimated regression coefficient is compatible with a theoretical parameter value $\beta_{0,xy}$ (null hypothesis $H_0: \beta_{0,yx} = \beta_{yx}$, alternative hypothesis $H_A: \beta_{0,yx} \neq \beta_{yx}$),

> Compatibility here and in the following means that, provided the null hypothesis is correct, the parameter belonging to the estimate (e.g., b_{yx}) is identical to the theoretical parameter (i.e., here $\beta_{0;yx}$); i.e., for example, $H_0: \beta_{0;yx} = \beta_{yx}$ [as well as $H_A: \beta_{0;yx} \neq \beta_{yx}$ (incompatibility)].

we use the fact that, H_0 given, the test quantity

$$\frac{b_{yx} - \beta_{yx}}{s_{b_{yx}}}$$

(p.136) exhibits a t-distribution with DF = $n - 2$:

$$\hat{t} = \frac{|b_{yx} - \beta_{yx}|}{s_{y.x}/s_x} \cdot \sqrt{n-1} = \frac{|b_{yx} - \beta_{yx}|}{\sqrt{1-r^2}} \cdot \frac{s_x}{s_y} \cdot \sqrt{n-2} = \frac{|b_{yx} - \beta_{yx}|}{s_{b_{yx}}}. \quad (5.59)$$

EXAMPLE. Given $b_{yx} = 0.426$, $\beta_{yx} = 0.5$, $s_{b_{yx}} = 0.190$, $n = 80$, we have
$S = 95\%$, i.e., $t_{78;0.05} = 1.99$,
$$\hat{t} = \frac{|0.426 - 0.500|}{0.190} = 0.39 < 1.99.$$
The null hypothesis is not rejected on the 5% level.

One-sided tests concerning (5.59):

H_0	H_A	H_0 is rejected for
$\beta_{0;yx} \leq \beta_{yx}$	...>...	$(b_{yx} - \beta_{yx})/s_{b_{yx}} \geq t_{n-2;\alpha;\text{one s.}}$
$\beta_{0;yx} \geq \beta_{yx}$	...<...	$(b_{yx} - \beta_{yx})/s_{b_{yx}} \leq -t_{n-2;\alpha;\text{one s.}}$

5.5.6 Testing the difference between an estimated and a hypothetical axis intercept

To test the null hypothesis: a_{yx} is compatible with α_{yx} (H_A: a_{yx} is not compatible with α_{yx}), one uses for the two sided test

$$\hat{t} = \frac{|a_{yx} - \alpha_{yx}|}{s_{a_{yx}}} \qquad (5.60)$$

with DF $= n - 2$.

EXAMPLE. Given: $a_{yx} = 7.729$; $\alpha_{yx} = 15.292$; $s_{a_{yx}} = 2.862$; $n = 80$
$$S = 95\%;$$
thus the significance bound is
$t_{78;0.05} = 1.99$,
$$\hat{t} = \frac{|7.729 - 15.292|}{2.862} = 2.64 > 1.99.$$

The hypothetical and the actual axis intercepts are not compatible at the 5% level.

5.5.7 Confidence limits for the regression coefficient, for the axis intercept, and for the residual variance

The confidence intervals for regression coefficients and axis intercepts are given by (5.61) and (5.62); for both t, DF $= n - 2$:

$$b_{yx} \pm t \cdot s_{b_{yx}} \quad \text{and} \quad a_{yx} \pm t \cdot s_{a_{yx}}. \qquad (5.61, 5.62)$$

EXAMPLES FOR 95% CONFIDENCE INTERVALS ($S = 0.95$, $\alpha = 1 - 0.95 = 0.05$)

Given: $b_{yx} = 0.426$, $s_{b_{yx}} = 0.190$, $n = 80$, $S = 95\%$, we have

$$t_{78; 0.05} = 1.99,$$
$$1.99 \cdot 0.19 = 0.378,$$
$$b_{yx} \pm t s_{b_{yx}} = 0.426 \pm 0.378,$$
$$95\% \text{ CI}: \quad 0.048 \leq \beta_{yx} \leq 0.804.$$

Given: $a_{yx} = 7.729$, $s_{a_{yx}} = 2.862$, $n = 80$, $S = 95\%$, we have

$$t_{78; 0.05} = 1.99,$$
$$1.99 \cdot 2.862 = 5.695,$$
$$a_{yx} \pm t s_{a_{yx}} = 7.729 \pm 5.695,$$
$$95\% \text{ CI}: \quad 2.034 \leq \alpha_{yx} \leq 13.424.$$

The **confidence interval for the residual variance** $\sigma^2_{y \cdot x}$ is obtained from

(p.140)

$$\boxed{\frac{s^2_{y \cdot x}(n-2)}{\chi^2_{(n-2; \alpha/2)}} \leq \sigma^2_{y \cdot x} \leq \frac{s^2_{y \cdot x}(n-2)}{\chi^2_{(n-2; 1-\alpha/2)}}.} \tag{5.63}$$

EXAMPLE. Given: $s^2_{y \cdot x} = 0.138$; $n = 80$; $S = 95\%$ (i.e., $\alpha = 5\% = 0.05$; $\alpha/2 = 0.025$; $1 - 0.025 = 0.975$), we have

$$\chi^2_{78; 0.025} = 104.31, \qquad \chi^2_{78; 0.975} = 55.47.$$

The 95% confidence interval thus reads

$$\frac{0.138 \cdot 78}{104.31} \leq \sigma^2_{y \cdot x} \leq \frac{0.138 \cdot 78}{55.47},$$

$$95\% \text{ CI}: \quad 0.103 \leq \sigma^2_{y \cdot x} \leq 0.194.$$

▶ 5.5.8 Comparing two regression coefficients and testing the equality of more than two regression lines

Two regression coefficients, b_1 and b_2, can be compared by means of

(p.417)

$$\boxed{\hat{t} = \frac{|b_1 - b_2|}{\sqrt{\dfrac{s^2_{y_1 \cdot x_1}(n_1 - 2) + s^2_{y_2 \cdot x_2}(n_2 - 2)}{n_1 + n_2 - 4} \left[\dfrac{1}{Q_{x_1}} + \dfrac{1}{Q_{x_2}}\right]}}} \tag{5.64}$$

5.5 Test Procedures

with $n_1 + n_2 - 4$ degrees of freedom (null hypothesis: $\beta_1 = \beta_2$). The samples (n_1, n_2) from populations with the same residual variances ($\sigma^2_{y_1 \cdot x_1} = \sigma^2_{y_2 \cdot x_2}$) are assumed to be independent.

Example

Given

$$n_1 = 40, \quad s^2_{y_1 \cdot x_1} = 0.14, \quad Q_{x_1} = 163, \quad b_1 = 0.40,$$
$$n_2 = 50, \quad s^2_{y_2 \cdot x_2} = 0.16, \quad Q_{x_2} = 104, \quad b_2 = 0.31.$$

Null hypothesis. (a) $\beta_1 \leq \beta_2$; (b) $\beta_1 = \beta_2$.
  (a) One sided problem ($\alpha = 0.05$): Alternative hypothesis: $\beta_1 > \beta_2$.
  (b) Two sided problem ($\alpha = 0.05$): Alternative hypothesis: $\beta_1 \neq \beta_2$.

We have

$$\hat{t} = \frac{|0.40 - 0.31|}{\sqrt{\dfrac{0.14(40-2) + 0.16(50-2)}{40+50-4}\left(\dfrac{1}{163} + \dfrac{1}{104}\right)}} = 1.85.$$

For a: Since $\hat{t} = 1.85 > 1.66 = t_{86; 0.05; \text{one sided}}$, the null hypothesis is rejected at the 5% level.

For b: Since $\hat{t} = 1.85 < 1.99 = t_{86; 0.05; \text{two sided}}$, the null hypothesis is not rejected.

For the case of unequal residual variances, i.e., if

$$\boxed{\frac{s^2_{y_1 \cdot x_1}}{s^2_{y_2 \cdot x_2}} > F_{(n_1 - 2; n_2 - 2; 0.10)},} \qquad (5.65) \quad \text{(p.145)}$$

the comparison can be carried out approximately according to

$$\boxed{\hat{z} = \frac{|b_1 - b_2|}{\sqrt{\dfrac{s^2_{y_1 \cdot x_1}}{Q_{x_1}} + \dfrac{s^2_{y_2 \cdot x_2}}{Q_{x_2}}}}} \qquad (5.66) \quad \text{(p.62)}$$

provided both sample sizes are > 20. If a sample size is smaller, then the distribution of the test quantity can be approximated by the t-distribution (p.136) with ν degrees of freedom, where

$$\boxed{\nu = \frac{1}{\dfrac{c^2}{n_1 - 2} + \dfrac{(1-c)^2}{n_2 - 2}} \quad \text{with } c = \frac{\dfrac{s^2_{y_1 \cdot x_1}}{Q_{x_1}}}{\dfrac{s^2_{y_1 \cdot x_1}}{Q_{x_1}} + \dfrac{s^2_{y_2 \cdot x_2}}{Q_{x_2}}}, \quad n_1 \leq n_2,}$$

$$(5.67)$$

always lies between $\min(n_1 - 2, n_2 - 2)$ [the smaller of the two] and $n_1 + n_2 - 4$ (cf., also Potthoff 1965).

Testing the equality of more than two regression lines

The null hypothesis H_0 = equality of k regression lines is rejected at the 5% level for

$$\hat{F} = \frac{\dfrac{1}{2k-2}\left[Q_{y\cdot x; T} - \sum_{i=1}^{k} Q_{y\cdot x; i}\right]}{\dfrac{1}{n-2k}\sum_{i=1}^{k} Q_{y\cdot x; i}} > F_{2k-2;\, n-2k;\, 0.05} \quad \text{(R1)}$$

where

k = number of linear regression functions $\hat{y} = a_i + b_i x$ with $i = 1, 2, \ldots, k$.
n = total number of all pairs (x, y) necessary for the calculation of the total regression line T; this is done by summing the individual sums $\sum x$, $\sum y$, $\sum x^2$, $\sum y^2$, $\sum xy$ of the k regression lines, and computing $\hat{y}_T = a + bx$ and $Q_{y\cdot x; T} = \sum_{i=1}^{n}(y_i - \hat{y}_T)^2 = s^2_{y\cdot x; T}(n-2)$.
$Q_{y\cdot x; i}$ = the value $Q_{y\cdot x} = \sum(y - \hat{y})^2 = Q_y(1 - r^2) = Q_y - (Q^2_{xy}/Q_x)$ for the ith regression line.

If it is of interest whether or not the regression lines are parallel, the equality of the β_i is tested. If it is not possible to reject $H_0: \beta_i = \beta$ then (especially in the case of unequal regression lines) $H_0: \alpha_i = \alpha$ is tested.

$H_0: \beta_i = \beta$ is rejected at the 5% level for

$$\hat{F} = \frac{\dfrac{1}{k-1}\left[A - \sum_{i=1}^{k} Q_{y\cdot x; i}\right]}{\dfrac{1}{n-2k}\sum_{i=1}^{k} Q_{y\cdot x; i}} > F_{k-1;\, n-2k;\, 0.05} \quad \text{(R2)}$$

with

$$A = \sum_{i=1}^{k} Q_{y; i} - \frac{\sum_{i=1}^{k} Q_{xy; i}}{\sum_{i=1}^{k} Q_{x; i}}.$$

If $\beta_i = \beta$ holds, then it is possible to reject $H_0: \alpha_i = \alpha$ at the 5% level for

$$\hat{F} = \frac{\dfrac{1}{k-1}[Q_{y\cdot x; T} - A]}{\dfrac{1}{n-2k}\sum_{i=1}^{k} Q_{y\cdot x; i}} > F_{k-1;\, n-2k;\, 0.05}. \quad \text{(R3)}$$

For a comparison of three [two] regression lines with ordered alternative see Biometrics **38** (1982), 837–841 [827–836].

▶ 5.5.9 Confidence interval for the regression line

We recall that a confidence interval is a random interval containing the unknown parameter. A **prediction interval** is a statement about the value to be taken by a random variable. Prediction intervals are wider than the corresponding confidence intervals.

A change in $\bar{y}$ causes a parallel translation, upwards or downwards, of the regression line; a change in the regression coefficient effects a rotation of the regression line about the center of gravity $(\bar{x}, \bar{y})$ (cf., Figure 53).

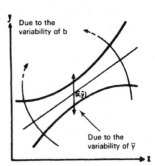

Figure 53 Confidence region for linear regression

First we need two standard deviations:

1. The standard deviation for the predicted mean $\hat{y}$ at the point x:

$$\boxed{s_{\hat{y}} = s_{y \cdot x} \sqrt{\frac{1}{n} + \frac{(x - \bar{x})^2}{Q_x}}.} \qquad (5.68)$$

2. The standard deviation for a predicted observation $\hat{y}$ at the point x:

$$\boxed{s_{\hat{y}} = s_{y \cdot x} \sqrt{1 + \frac{1}{n} + \frac{(x - \bar{x})^2}{Q_x}}.} \qquad (5.69)$$

The following **confidence limits and confidence intervals** (CI) hold for $(x_{\min} < x < x_{\max})$:

1. **the whole regression line:**

$$\boxed{\hat{y} \pm \sqrt{2 F_{(2, n-2)}} \, s_{\hat{y}}.} \qquad (5.70) \;\; \text{(p.145)}$$

2. **the expected value of y at a point $x = x_0$ (say):**

$$\boxed{\hat{y} \pm t_{(n-2)} s_{\hat{y}}.} \qquad (5.71) \;\; \text{(p.136)}$$

A prediction interval for a future observation of y at a point $x = x_0$ (cf., also Hahn 1972) is

$$\hat{y} \pm t_{(n-2)} s_{\hat{y}}. \qquad (5.72)$$

For a prediction interval for a future sample mean at the point x, based on a sample of m further observations, with our estimated mean $\bar{y} = \hat{y}_m$, we use (5.72) with $\hat{y}_m$ instead of $\hat{y}$ and (5.69) with $1/m$ instead of 1 under the square root. This makes the prediction interval shorter.

These regions hold only for the data space. They are bounded by the branches of a hyperbola which depends on x. Figure 54 indicates the growing

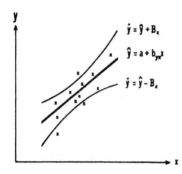

Figure 54 Confidence interval scheme for linear regression with the boundary value B_x depending on x.

uncertainty in any prediction made as x recedes from the center of gravity $(\bar{x}, \bar{y})$ of the regression line. The confidence interval (5.70) is the widest of the three regions, and (5.71) is the narrowest; as $n \to \infty$, (5.70) and (5.71) shrink to zero, and (5.72) shrinks to a strip of width $z\sigma_{y \cdot x}$. Thus limits within which, at the point x, 95% of the values of y lie, are estimated by $a + bx \pm 1.96 s_{y \cdot x}$. A tolerance interval, covering a portion of the population, is approximated by a formula given in Dixon and Massey (1969, [1], pp. 199–200) (cf., end of this section). When there is not one y-value for each value of x, but n_i y-values, then we have not $n - 2$ but $\sum n_i - 2$ degrees of freedom, and in (5.68) and (5.69) instead of $1/n$ we have to write $1/\sum n_i$.

For practical work it is important to note that, e.g., for the 95% CI of (5.71) more than 1 in 20 points representing observations falls outside the CI, because the limits are not for individual values of y but for the expected value of y. Some users of (5.70) may find it disturbing that this CI is considerably wider than (5.71). This is the price that must be paid to be able to include conditional expected values of y for **any** value of x whatsoever with $x_{\min} < x < x_{\max}$.

> Keeping in mind the distinction between X and x (cf., Section 1.2.5), the formula (5.71) gives the CI for the expected value of Y at $x = x_0$ [the mean of the Y-values for all individuals having a particular X-value]; (5.70) may be regarded as a set of CIs for the conditional expected values of Y given $X = x_0$ for all x_0 included in $x_{\min} <$

5.5 Test Procedures

$x_0 < x_{max}$. For modifications of (5.70) see Dunn (1968) and also Hahn and Hendrickson (1971) (cf., Section 7.3.2). Equation (5.72) holds for a new observed value of Y with $X = x_0$.

EXAMPLE. We again take the simple model of Example 2, Section 5.4.2, and pick four x-values at which the associated points of the confidence interval are to be determined (95% CI: i.e., $F_{(2;5;0.025)} = 8.43$). The x-values should lie within the data range and be equally spaced. In Table 118 these four

Table 118

x	$x - \bar{x}$ ($\bar{x} = 14.714$)	$\hat{y}$	$\dfrac{1}{n} + \dfrac{(x-\bar{x})^2}{Q_x}$	$\sqrt{\dfrac{1}{n} + \dfrac{(x-\bar{x})^2}{Q_x}}$	B_x
12	−2.714	12.84	0.2380	0.488	3.35
14	−0.714	13.69	0.1494	0.387	2.65
16	1.286	14.54	0.1642	0.405	2.78
18	3.286	15.40	0.2823	0.531	3.64

Table 119

$\hat{y} - B_x$	$\hat{y} + B_x$
9.49	16.19
11.04	16.34
11.76	17.32
11.76	19.07

x-values form column 1, while their deviations from the mean ($\bar{x} = 14.714$) are listed in the following column. Column 3 contains the $\hat{y}$-values estimated on the basis of the regression line $\hat{y} = 7.729 + 0.426x$ for the chosen x-values. The deviations of the x-values from their mean are squared, divided by $Q_x = 77.429$ and augmented by $1/n = 1/7$. The square root of this intermediate result, when multiplied by $\sqrt{2F}\, s_{y\cdot x} = \sqrt{2 \cdot 8.43} \cdot 1.67 = 6.857$, yields the corresponding B_x-values (compare $\hat{y} \pm B_x$ with

$$B_x = \sqrt{2F_{(2, n-2)}}\, s_{\hat{y}}).$$

Connecting the upper ($\hat{y} + B_x$) and the lower ($\hat{y} - B_x$) bounds of the confidence region, we find the 95% confidence region for the entire line of regression. Note that by symmetry the four B_x-values in our example represent in fact eight B_x-values, and only the four additional $\hat{y}$-values remain to be computed. For example B_x, depending on $(x - \bar{x})$, has the same value for $x = 14$ [i.e., $(\bar{x} - 0.714)$] as for $x = 15.428$ [i.e., $(\bar{x} + 0.714)$].

We determine below both of the other confidence regions ($t_{5;0.05} = 2.57$) for the point $x = 16$, and start by computing $B_{x=16}$ by (5.71) and subsequently $B'_{x=16}$ by (5.71):

$$B_{x=\text{const.}} = ts_{y \cdot x} \sqrt{\frac{1}{n} + \frac{(x - \bar{x})^2}{Q_x}},$$

$$B_{16} = 2.57 \cdot 1.67 \sqrt{\frac{1}{7} + \frac{(16 - 14.714)^2}{77.429}} = 1.74.$$

The 95% confidence region for an estimate of the mean at the point $x = 16$ is given by the interval 14.54 ± 1.74. The bounds of the interval are 12.80 and 16.28.

$$B'_{x=\text{const.}} = ts_{y \cdot x} \sqrt{1 + \frac{1}{n} + \frac{(x - \bar{x})^2}{Q_x}},$$

$$B'_{16} = 2.57 \cdot 1.67 \cdot \sqrt{1 + \frac{1}{7} + \frac{(16 - 14.714)^2}{77.429}} = 4.63.$$

The 95% confidence region for a predicted observation $\hat{y}$ at the value $x = 16$ is given by the interval 14.54 ± 4.63. The bounds of the interval are 9.91 and 19.17. The confidence interval for individual predictands is substantially larger than the one computed above for the predicted mean.

> Details for the construction of confidence and tolerance ellipses can be found in the Geigy tables (Documenta Geigy 1968 [2], 183–184 (cf., also p. 145)). Plotting methods for probability ellipses of a bivariate normal distribution are described by M. J. R. Healy and D. G. Altman, Applied Statistics **21** (1972), 202–204 and **27** (1978), 347–349. Tolerance regions can be given according to Weissberg and Beatty (1960) (cf., Sections 3.8 and 5.4.3). More particulars on linear regression (cf., also Anscombe (1967) [8:5b]) can be found in Williams (1959), Draper and Smith (1981), Stange (1971) [1], Part II, 121–178, and Neter and Wasserman (1974). An excellent **review on the estimation of various linear regressions** appropriate to fifteen different situations especially in biology is given by W. E. Ricker, Journal of the Fisheries Research Board of Canada **30** (1973), 409–434.

Remarks

1. For an **empirical regression curve**, realizations of a continuous two-dimensional random variable, S. Schmerling and J. Peil, Biometrical Journal **21** (1979), 71–78 give a **belt**, the local width of which varies depending on local frequency and variance of the measured points.

2. **Tests for bivariate normality** are compared by C. J. Kowalski, Technometrics **12** (1970), 517–544. A **goodness of fit test for the singly truncated bivariate normal distribution** and the corresponding truncated 95% probability ellipse is given by M. N. Brunden, Communications in Statistics—Theory and Methods **A7** (1978), 557–572.

For **robust regression** see pages 390–392 and Hocking and Pendleton (1983) [cf., also Hogg 1979 and Huber 1981, both cited on p. 66]; for detecting a single **outlier** in linear regression see Barnett and Lewis (1978, Chap. 7, [8:1]) and R. Doornbos (1981, Biometrics **37**, 705–711 [cf., also Technometrics **15** (1973), 717–721, **17**, (1975), 129–132, 473–476, **23** (1981), 21–26, 59–63]).

5.6 NONLINEAR REGRESSION

It is in many cases evident from the graphical representation that the relation of interest cannot be described by a regression line. Very frequently there is a sufficiently accurate correspondence between a **second degree equation** and the actual relation. We shall in the following again avail ourselves of the method of least squares.

The general second degree equation reads

$$y = a + bx + cx^2.\qquad(5.73)$$

The constants a, b, and c for the second degree function sought can be determined from the following **normal equations**:

$$\begin{aligned}
\text{I} & \quad an +b\sum x+c\sum x^2=\sum y, \\
\text{II} & \quad a\sum x+b\sum x^2+c\sum x^3=\sum xy, \\
\text{III} & \quad a\sum x^2+b\sum x^3+c\sum x^4=\sum x^2 y.
\end{aligned} \qquad(5.74\text{a,b,c})$$

Table 120

x	y	xy	x^2	$x^2 y$	x^3	x^4
1	4	4	1	4	1	1
2	1	2	4	4	8	16
3	3	9	9	27	27	81
4	5	20	16	80	64	256
5	6	30	25	150	125	625
15	19	65	55	265	225	979

This is illustrated (cf., e.g., Sachs 1984, pages 92–94) by a simple example (see Table 120): These values are substituted in the normal equations

$$\begin{aligned}
\text{I} & \quad 5a + 15b + 55c = 19, \\
\text{II} & \quad 15a + 55b + 225c = 65, \\
\text{III} & \quad 55a + 225b + 979c = 265.
\end{aligned}$$

The unknown a is first eliminated from I and II as well as from II and III:

$$\begin{aligned}5a + 15b + 55c &= 19 \quad \cdot 3 \\ 15a + 55b + 225c &= 65\end{aligned} \qquad \begin{aligned}15a + 55b + 225c &= 65 \quad \cdot 11 \\ 55a + 225b + 979c &= 265 \quad \cdot 3\end{aligned}$$

$$\begin{aligned}15a + 45b + 165c &= 57 \\ 15a + 55b + 225c &= 65\end{aligned} \qquad \begin{aligned}165a + 605b + 2475c &= 715 \\ 165a + 675b + 2937c &= 795\end{aligned}$$

IV $\qquad 10b + 60c = 8 \qquad$ V $\qquad 70b + 462c = 80$

From IV and V we eliminate b and determine c:

$$\begin{aligned}70b + 462c &= 80 \\ 10b + 60c &= 8 \quad \cdot 7\end{aligned}$$

$$\begin{aligned}70b + 462c &= 80 \\ 70b + 420c &= 56\end{aligned}$$

$$42c = 24$$

$$c = \frac{24}{42} = \frac{12}{21} = \frac{4}{7} (= 0.571).$$

By substituting c in IV we obtain b:

$$10b + 60c = 8,$$

$$10b + \frac{60 \cdot 4}{7} = 8,$$

$$70b + 240 = 56 \quad \text{and} \quad b = \frac{56 - 240}{70} = -\frac{184}{70} = -\frac{92}{35} (= -2.629).$$

By substituting b and c in I we get a:

$$5a + 15 \cdot \left(-\frac{92}{35}\right) + 55\left(\frac{4}{7}\right) = 19,$$

$$5a - \frac{15 \cdot 92}{35} + \frac{55 \cdot 4 \cdot 5}{7 \cdot 5} = 19,$$

$$35 \cdot 5a - 15 \cdot 92 + 55 \cdot 20 = 19 \cdot 35,$$
$$175a - 1380 + 1100 = 665,$$

$$175a - 280 = 665 \quad \text{and} \quad a = \frac{945}{175} = \frac{189}{35} (= 5.400).$$

5.6 Nonlinear Regression

Check (of the computations): Substituting the values in the normal equation I:

$$5 \cdot 5.400 - 15 \cdot 2.629 + 55 \cdot 0.571 = 27.000 - 39.435 + 31.405$$
$$= 18.970 \simeq 19.0.$$

The second order regression reads

$$\hat{y} = \frac{189}{35} - \frac{92}{35}x + \frac{4}{7}x^2 \simeq 5.400 - 2.629x + 0.5714x^2.$$

Table 121 indicates the goodness of fit. The deviations $y - \hat{y}$, called residuals, are considerable. It is sometimes more advantageous to fit $y = a + bx + c\sqrt{x}$ (cf., Table 124).

THREE REMARKS ON NONLINEAR REGRESSION:

1. If the model applies, then for every regression model the **residuals** $y - \hat{y}$ are interpreted as observed random errors. Information in this regard is presented in graphical form: (a) as a histogram, (b) $y_i - \hat{y}_i$ (ordinate) against i, (c) against $\hat{y}_i$, (d) against x_i [(b) and (d) should give "horizontal bands"], and (e) against a possibly important variable which has not as yet been taken into consideration (cf., Draper and Smith 1981, Chap. 3 as well as Cox and Snell 1968). A review of current literature on techniques for the **examination of residuals** in linear and nonlinear models, time serials included, with concentration on graphical techniques and statistical test procedures is given by S. R. Wilson, Australian Journal of Statistics **21** (1979), 18–29 (cf., P. Hackl and W. Katzenbeisser, Statistische Hefte **19** (1978), 83–98).

2. The **nonlinear** [nl] **coefficient of determination** ($B_{nl} = r_{nl}^2$) is generally given by $B_{nl} = 1 - (A/Q_y)$ [cf., Section 5.1, (5.6), and Section 5.4.3] with $A = \sum (y - \hat{y})^2$; for (5.73) there is the elegant formula [cf., (5.74a, b, c), right hand side].

$$A = \sum y^2 - a \sum y - b \sum xy - c \sum x^2 y$$

—i.e., for our example: $A = 87 - (189/35)19 + (92/35)65 - (4/7)265 = 87 - 102.6000 + 170.8571 - 151.4286 = 3.8285$ (cf., Table 121: $A = 3.83$); $Q_y = 87 - (19)^2/5 = 14.8000$ (cf., Section 5.4.4); $B_{nl} = 1 - (3.8285/14.8000) = 0.7413$; and the nonlinear correlation coefficient $r_{nl} = \sqrt{0.7413} = 0.8610$.

3. One can, in summary, give for (5.73), as the **average rate of change**, the gradient $b + 2cx$ of the curve at the point $(x_1 + x_n)/2$.

Table 121

x	y	$\hat{y} = \frac{189}{35} - \frac{92}{35}x + \frac{4}{7}x^2$					$y - \hat{y}$	$(y - \hat{y})^2$
1	4	$\frac{189}{35} - \frac{92}{35} \cdot 1 + $	" $\cdot 1$	$= \frac{117}{35}$	$= 3.343$		0.657	0.432
2	1	" $-$ " $\cdot 2 + $	" $\cdot 4$	$= \frac{85}{35}$	$= 2.429$		-1.429	2.042
3	3	" $-$ " $\cdot 3 + $	" $\cdot 9$	$= \frac{93}{35}$	$= 2.657$		0.343	0.118
4	5	" $-$ " $\cdot 4 + $	" $\cdot 16$	$= \frac{141}{35}$	$= 4.029$		0.971	0.943
5	6	" $-$ " $\cdot 5 + $	" $\cdot 25$	$= \frac{229}{35}$	$= 6.543$		-0.543	0.295
	19					19.00	-0.001	3.830

If the relation between y and x seems to be of exponential type,

$$y = ab^x, \tag{5.75}$$

then taking the logarithm of both sides leads to

$$\log y = \log a + x \log b. \tag{5.75a}$$

The associated **normal equations** are

$$\begin{aligned} \text{I} \quad & n \cdot \log a + (\textstyle\sum x) \cdot \log b = \textstyle\sum \log y, \\ \text{II} \quad & (\textstyle\sum x) \cdot \log a + (\textstyle\sum x^2) \cdot \log b = \textstyle\sum (x \cdot \log y). \end{aligned} \tag{5.76ab}$$

Since the exponential function fitted in this way ordinarily yields somewhat distorted estimates of a and b, it is generally advantageous to replace (5.75) by $y = ab^x + d$ and estimate a, b, d according to Hiorns (1965).

EXAMPLE

Table 122

x	y	log y	x log y	x^2
1	3	0.4771	0.4771	1
2	7	0.8451	1.6902	4
3	12	1.0792	3.2376	9
4	26	1.4150	5.6600	16
5	51	1.7076	8.5380	25
15	99	5.5240	19.6029	55

5.6 Nonlinear Regression

The sums from Table 122 are substituted in the equations

$$\text{I} \quad 5 \log a + 15 \log b = 5.5240 \quad \cdot 3$$
$$\text{II} \quad 15 \log a + 55 \log b = 19.6029$$

$$15 \log a + 45 \log b = 16.5720$$
$$15 \log a + 55 \log b = 19.6029$$

$$10 \log b = 3.0309$$
$$\log b = 0.30309.$$

Substituting in I:

$$5 \log a + 15 \cdot 0.30309 = 5.5240,$$
$$5 \log a + 4.54635 = 5.5240,$$
$$5 \log a = 0.9776,$$
$$\log a = 0.19554.$$

The corresponding antilogarithms are $a = 1.569$ and $b = 2.009$.

The exponential regression based on the above values which estimates y given x thus reads $\hat{y} = (1.569)(2.009)^x$ (see Table 123).

Table 123

x	y	log $\hat{y}$	$\hat{y}$
1	3	0.1955 + 1·0.3031 = 0.4986	3.15
2	7	0.1955 + 2·0.3031 = 0.8017	6.33
3	12	0.1955 + 3·0.3031 = 1.1048	12.73
4	26	0.1955 + 4·0.3031 = 1.4079	25.58
5	51	0.1955 + 5·0.3031 = 1.7110	51.40
	99		99.19

Table 124 gives the normal equations for the functional equations discussed, as well as for other functional relations.

REMARK

In the natural sciences one is quite frequently confronted with the task of comparing an empirical curve with another one, obtained by subjecting half of experimental material to a specified treatment. The means $\bar{y}_{1i}$ and $\bar{y}_{2i}$ for given x_i, for example consecutive days, are available. First the deviation for n days of the sum of squares can be tested according to an approximation given by Gebelein and Ruhenstroth-Bauer (1952):

$$\hat{\chi}^2 = \frac{\sum_{i=1}^{n} (\bar{y}_{1i} - \bar{y}_{2i})^2}{s_1^2 + s_2^2}, \quad DF = n. \tag{5.77}$$

Table 124 Exact and approximate normal equations for the more important functional relations

Functional relation	Normal equations
$y = a + bx$	$a \cdot n + b\sum x = \sum y$ $a\sum x + b\sum x^2 = \sum(xy)$
$\log y = a + bx$	$a \cdot n + b\sum x = \sum \log y$ $a\sum x + b\sum x^2 = \sum(x \log y)$
$y = a + b \log x$	$a \cdot n + b\sum \log x = \sum y$ $a\sum \log x + b\sum(\log x)^2 = \sum(y \log x)$
$\log y = a + b \log x$	$a \cdot n + b\sum \log x = \sum \log y$ $a\sum \log x + b\sum(\log x)^2 = \sum(\log x \log y)$
$y = a \cdot b^x$, or $\log y = \log a + x \log b$	$n \log a + \log b \sum x = \sum \log y$ $\log a \sum x + \log b \sum x^2 = \sum(x \log y)$
$y = a + bx + cx^2$	$a \cdot n + b\sum x + c\sum x^2 = \sum y$ $a\sum x + b\sum x^2 + c\sum x^3 = \sum xy$ $a\sum x^2 + b\sum x^3 + c\sum x^4 = \sum(x^2 y)$
$y = a + bx + c\sqrt{x}$	$a \cdot n + b\sum x + c\sum \sqrt{x} = \sum y$ $a\sum x + b\sum x^2 + c\sum \sqrt{x^3} = \sum xy$ $a\sum \sqrt{x} + b\sum \sqrt{x^3} + c\sum x = \sum(y\sqrt{x})$
$y = a \cdot b^x \cdot c^{x^2}$, or $\log y = \log a + x \log b + x^2 \log c$	$n \log a + \log b \sum x + \log c \sum x^2 = \sum \log y$ $\log a \sum x + \log b \sum x^2 + \log c \sum x^3 = \sum(x \log y)$ $\log a \sum x^2 + \log b \sum x^3 + \log c \sum x^4 = \sum(x^2 \log y)$
$y = b_0 + b_1 x + b_2 x^2 + b_3 x^3$ $\log y = b_0 + b_1 x + b_2 x^2 + b_3 x^3$	$b_0 n + b_1 \sum x + b_2 \sum x^2 + b_3 \sum x^3 = \sum y$ $b_0 \sum x + b_1 \sum x^2 + b_2 \sum x^3 + b_3 \sum x^4 = \sum xy$ $b_0 \sum x^2 + b_1 \sum x^3 + b_2 \sum x^4 + b_3 \sum x^5 = \sum x^2 y$ $b_0 \sum x^3 + b_1 \sum x^4 + b_2 \sum x^5 + b_3 \sum x^6 = \sum x^3 y$ as above with $y = \log y$

The observations of the first two days are initially considered together, then those of the first three, those of the first four, etc. Of course one can also carry out a test on the sum of the squares for an arbitrary interval, say from the 5th to the 12th day, if that seems justified by its relevance.

This procedure permits us to test whether the deviations could be due to random variations. How the development had changed can be determined by testing the arithmetic mean of the deviations of several days. The arithmetic mean of the differences of the means for the first n days, n not too small, can be assessed in terms of the standard normal variable z (two sided problem):

$$\hat{z} = \frac{\sum_{i=1}^{n}(\bar{y}_{1i} - \bar{y}_{2i})}{\sqrt{n(s_1^2 + s_2^2)}}. \tag{5.78}$$

5.7 Some Linearizing Transformations

Both procedures assume independent normally distributed populations with standard deviations σ_1 and σ_2 (cf., also Hoel 1964 as well as Section 7.4.3, Remark 5).

More particulars on **nonlinear regression** can be found in Box et al., (1969 [8 :2d]), Chambers (1973), Gallant (1975), Ostle and Mensing (1976), the books by Snedecor and Cochran (1967 [1], pp. 447–471) and by Draper and Smith (1981, pp. 458–517), and the literature mentioned under "Simplest Multiple Linear Regression" near the end of Section 5.8.

Several nonlinear functions are presented in Figures 55–58.

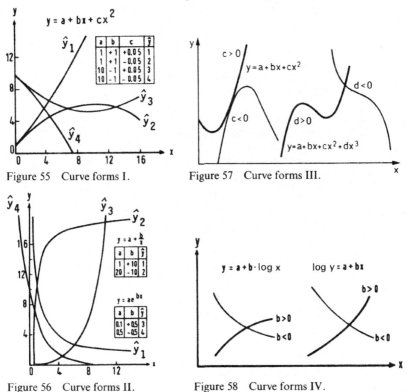

Figure 55 Curve forms I.

Figure 57 Curve forms III.

Figure 56 Curve forms II.

Figure 58 Curve forms IV.

5.7 SOME LINEARIZING TRANSFORMATIONS

If the form of a nonlinear dependence between two variables is known, then it is sometimes possible by transforming one or both of the variables to obtain a linear relation, a straight line. In the equation $y = ab^x$ just discussed, we took the logarithm of both sides to get $\log y = \log a + x \log b$; this is the equation of a straight line with $\log a$ = ordinate intercept and $\log b$-regression coefficient.

If the normal equations are not used in the computation, then the separate steps are:

1. Transform all y-values into log y values and carry out necessary computations on the logarithms of the observed y-values ($y' = \log y$).
2. Estimate the regression line $\hat{y}' = a' + b'x$ as usual.
3. By taking the antilogarithms of $a' = \log a$, $b' = \log b$, obtain estimates of the constants a and b of the original equation $y = ab^x$.

It is recommended that the student carry out these computations using the numerical values of the last example. Table 125 exhibits a number of relations between x and y which can be easily linearized, points out the necessary transformations, and gives the formulas for going over from the

Table 125 Modified and extended table of linearizing transformations according to Natrella, M. G.: Experimental Statistics, National Bureau of Standards Handbook 91, US Government Printing Office, Washington 1963, 5-31

If there is a relation of the form	Introduce the transformed variables into the coordinate system		Determine from a' and b' the constants a and b	
	$y' =$	$x' =$	$a =$	$b =$
$y = a + \dfrac{b}{x}$	y	$\dfrac{1}{x}$	a	b
$y = \dfrac{a}{b + x}$	$\dfrac{1}{y}$	x	$\dfrac{b}{a}$	$\dfrac{1}{a}$
$y = \dfrac{ax}{b + x}$	$\dfrac{1}{y}$	$\dfrac{1}{x}$	$\dfrac{1}{a}$	$\dfrac{b}{a}$
$y = \dfrac{x}{a + bx}$	$\dfrac{x}{y}$	x	a	b
$y = ab^x$	$\log y$	x	$\log a$	$\log b$
$y = ax^b$	$\log y$	$\log x$	$\log a$	b
$y = ae^{bx}$	$\ln y$	x	$\ln a$	b
$y = ae^{b/x}$	$\ln y$	$\dfrac{1}{x}$	$\ln a$	b
$y = a + bx^n$, where n is known	y	x^n	a	b
	and estimate $\hat{y}' = a' + b'x'$			

5.7 Some Linearizing Transformations

parameters of the straight line to the constants of the original relation. A fine comprehensive summary is provided by Hoerl (1954).

These linearizing transformations could also be used to determine the form of a relation by completely empirical means. We now read the table, going from the transformed values to the type of relation:

1. Plot y against $1/x$ in a rectangular coordinate system. If the points lie on a straight line, then the relation $y = a + b/x$ holds.
2. Plot $1/y$ against x in a rectangular coordinate system. If the points lie on a straight line, then the relation $y = a/(b + x)$ holds.
3. Plot y (logarithmic scale) against x (arithmetic scale) on semi-logarithmic paper (exponential paper). If the points lie on a straight line, then the relation $y = ab^x$ or $y = ae^{bx}$ or $a10^{bx}$ holds.
4. Plot y (logarithmic scale) against x (logarithmic scale) [double logarithmic or log-log paper]. If the points lie on a straight line, then the relation $y = ax^b$ [power function] holds.

Graph paper with a coordinate lattice unlike the usual linear grid, that is, with coordinate axes scaled according to arbitrary functions, is referred to as **function paper** (for sources see the references in Section 8.7). Besides the exponential and power function graph paper, there are other important types of graph paper which linearize complicated nonlinear functions. We mention in particular sine paper, which has one axis linear and the other scaled according to a sine function, and on which functions of the form

$$ax + b \sin y + c = 0$$

can be represented by the straight line

$$ax' + by' + c = 0$$
$$[x' = xe_x, y' = (\sin y)e_y \text{ with } e_x = e_y = 1].$$

Exponential paper is important for the study of radioactive and chemical disintegration processes as well as for the analysis of the growth in length of many living beings. In theoretical biology and in physics, exponential laws, and thus also various types of exponential graph paper, can be very useful (cf., also Batschelet 1975 [8:1]). A more detailed discussion of **GROWTH CURVES** can be found in Hiorns (1965), Scharf (1974) and Batschelet (1975 [8:1]), [cf., also Biometrika **30** (1938), 16–28; **51** (1964), 313–326; **52** (1965), 447–458, as well as Biometrics **18** (1962), 148–159; **25** (1969), 357–381; **29** (1973), 361–371, **33** (1977), 653–657; **35** (1979), 255–271, 835–848; Applied Statistics **26** (1977), 143–148; and Biometrical Journal **22** (1980), 23–39].

▶ 5.8 PARTIAL AND MULTIPLE CORRELATIONS AND REGRESSIONS

We must in general allow for the possibility that the correlation between two particular variables might be influenced by additional (recognized or unknown) variables. The methods used in the examination of the dependence between more than two random variables are based on samples from multivariate normal populations. In this case the partial correlation coefficients can be considered as measures of the linear dependence between pairs of variables. They specify the degree of dependence between two variables while the remaining variables are held constant.

If x, y, and z are linearly correlated and if r_{xy}, r_{yz}, and r_{xz} are the three pairwise computed correlation coefficients, then $r_{xy \cdot z}$ is the partial correlation coefficient between x and y when z is held constant:

$$r_{xy \cdot z} = \frac{r_{xy} - r_{xz} \cdot r_{yz}}{\sqrt{(1 - r_{xz}^2)(1 - r_{yz}^2)}}. \tag{5.79}$$

If in place of the letters x, y, z the numbers 1, 2, 3 are used then the partial correlation coefficient between x_1 and x_2, with x_3 remaining constant, is

$$r_{12 \cdot 3} = \frac{r_{12} - r_{13} \cdot r_{23}}{\sqrt{(1 - r_{13}^2)(1 - r_{23}^2)}}, \tag{5.79a}$$

and by cyclic permutation

$$r_{13 \cdot 2} = \frac{r_{13} - r_{12} \cdot r_{23}}{\sqrt{(1 - r_{12}^2)(1 - r_{23}^2)}}, \tag{5.79b}$$

$$r_{23 \cdot 1} = \frac{r_{23} - r_{12} \cdot r_{13}}{\sqrt{(1 - r_{12}^2)(1 - r_{13}^2)}}. \tag{5.79c}$$

A nomogram for determining the partial correlation coefficients is given by Koller (1953, 1969) as well as by Lees and Lord (1962). The computation of partial correlations can clarify the mutual significance of the variables in involved interdependence relations. If for example the correlation between x_1 and x_2 is based only on a common influence due to x_3, then $r_{12 \cdot 3} \simeq 0$. It can also happen that a correlation manifests itself only after the elimination of an interfering variable.

If not just three but four variables are known, then the partial correlation between x_1 and x_2, if the influences of x_3 and x_4 are to be excluded, is computed according to

$$r_{12 \cdot 34} = \frac{r_{12 \cdot 4} - r_{13 \cdot 4} \cdot r_{23 \cdot 4}}{\sqrt{(1 - r_{13 \cdot 4}^2)(1 - r_{23 \cdot 4}^2)}} = \frac{r_{12 \cdot 3} - r_{14 \cdot 3} \cdot r_{24 \cdot 3}}{\sqrt{(1 - r_{14 \cdot 3}^2)(1 - r_{24 \cdot 3}^2)}}. \tag{5.80}$$

5.8 Partial and Multiple Correlations and Regressions

The partial correlation coefficient is tested like the normal correlation coefficient. It is however to be noted that the number of degrees of freedom must be reduced by 1 for each excluded variable. If only one variable is excluded, then the number of degrees is $n - 2 - 1 = n - 3$. The computation of partial correlation coefficients gives in general one way of eliminating the interference due to the factors which can be controlled very little or not at all during the trial.

Methods for testing hypotheses concerning partial correlation are reviewed by K. L. Levy and S. C. Narula [International Statistical Review **46** (1978), 215–218].

Before giving an example, let us first call attention to a procedure that enables us to reduce the number of dependent attributes (or characteristics) observed (on the experimental material) to a smaller number of independent true influence quantities ("factors") by combining attributes which are highly correlated. A more detailed discussion of **factor analysis** can be found e.g., in the books by Lawley and Maxwell (1971 [8 :5a]) and Rummel (1970 [8 :5a]).

EXAMPLE. A detailed analysis was carried out in Iowa and Nebraska on a random sample of 142 elderly women (Swanson et al., 1955 [cf., Snedecor and Cochran 1967 [8:1], p. 401]). Three of the variables were age A, blood pressure B, and cholesterol concentration C in the blood, with correlation coefficients

$$r_{AB} = 0.3332, \quad r_{AC} = 0.5029, \quad r_{BC} = 0.2495.$$

Since a rise in blood pressure could be related to an increase of cholesterol deposits in the walls of the blood vessels, this appears to be an interesting question worthy of further investigation. Since B and C grow with age, the question arises whether the weak connection is entirely traceable to age or whether at every stage of life a real connection is present. The age effect is eliminated in the partial correlation $r_{BC.A}$ [cf., (5.79c)]:

$$r_{BC.A} = \frac{r_{BC} - r_{AB} \cdot r_{AC}}{\sqrt{(1 - r_{AB}^2)(1 - r_{AC}^2)}},$$

$$r_{BC.A} = \frac{0.2495 - 0.3332 \cdot 0.5029}{\sqrt{(1 - 0.3332^2)(1 - 0.5029^2)}} = 0.1005.$$

For $142 - 3 = 139$ DF a true correlation cannot be ascertained at the 5% level.

If we are interested in how the random variable x_1 depends simultaneously on the variables x_2 and x_3, (i.e., if we consider one variable as target variable and at least two other variables as influence variables), then the **multiple correlation coefficient** $R_{1.23}$ comes into play. It measures the dependence of the target variable on the influence variable. This multiple correlation is given by

$$R_{1.23} = \sqrt{\frac{r_{12}^2 + r_{13}^2 - 2r_{12}r_{13}r_{23}}{1 - r_{23}^2}}. \qquad (5.81)$$

The multiple correlation describes the target quantity (the so called regressand) in terms of at least two influence quantities (the so called regressors). The dot in $R_{1.23}$ separates the target quantity, indicated first, from the two influence quantities. There are analogous formulas for $R_{2.13}$ and $R_{3.12}$. The multiple correlation coefficients always lie between 0 and 1. Lord (1955) gives a nomogram for determining $R_{1.23}$. The square of the multiple correlation coefficient is written as a multiple determination measure: $B = R^2$ (Model II, cf., Section 5.4.2). $B = 1$ means that the values of the target quantity are calculable exactly from the values of the influence quantities by a multiple linear regression function (e.g., $\hat{y} = a + b_1 x_1 + b_2 x_2$). In addition to

$$R_{1.23}^2 = r_{12}^2 + r_{13.2}^2 (1 - r_{12}^2) \qquad (5.82)$$

let us also mention the more revealing relations

$$1 - R_{1.23}^2 = (1 - r_{12}^2)(1 - r_{13.2}^2), \qquad (5.83)$$
$$1 - R_{1.234}^2 = (1 - r_{12}^2)(1 - r_{13.2}^2)(1 - r_{14.23}^2). \qquad (5.84)$$

Partial correlation coefficients of second order, e.g., $r_{14.23}$, are obtained from

$$r_{14.23} = \frac{r_{14.2} - r_{13.2} r_{34.2}}{\sqrt{(1 - r_{13.2}^2)(1 - r_{34.2}^2)}} = \frac{r_{14.3} - r_{12.3} r_{24.3}}{\sqrt{(1 - r_{12.3}^2)(1 - r_{24.3}^2)}}, \qquad (5.85)$$

$$r_{12.34} = \frac{r_{12.3} - r_{14.3} r_{24.3}}{\sqrt{(1 - r_{14.3}^2)(1 - r_{24.3}^2)}} = \frac{r_{12.4} - r_{13.4} r_{23.4}}{\sqrt{(1 - r_{13.4}^2)(1 - r_{23.4}^2)}}. \qquad (5.86)$$

Compare also

$$r_{14.23}^2 = \frac{R_{1.234}^2 - R_{1.23}^2}{1 - R_{1.23}^2}. \qquad (5.87)$$

5.8 Partial and Multiple Correlations and Regressions

The null hypothesis, according to which the parameter corresponding to R equals zero (as against > 0), is tested by means of the F-test

$$\hat{F} = \frac{R^2}{1-R^2} \cdot \frac{n-k-1}{k},$$
$$\nu_1 = k, \quad \nu_2 = n-k-1$$
(5.88) (p. 145)

(k = number of independent variables). Testing $H_0: \beta_1 = \beta_2 = 0$ in terms of R may be done by (5.88), since the population multiple correlation between $y(x_1, x_2)$ [generally $y(x_1, x_2, \ldots, x_k)$] is zero if only every $\beta_i = 0$.

Frequently one would like to know whether an R_1 with several influence quantities u_1 is significantly larger than an R_2 with a smaller number u_2. The corresponding F-test is

$$\hat{F} = \frac{(R_1^2 - R_2^2)(n - u_1 - 1)}{(1 - R_1^2)(u_1 - u_2)},$$
$$\nu_1 = u_1 - u_2, \quad \nu_2 = n - u_1 - 1.$$
(5.89) (p. 145)

In particular, if n is small and the number of variables k relatively large, then R^2 must be replaced by the exact (unbiased) estimate ${}_u R^2$:

$$_u R^2 = 1 - (1-R^2)\frac{n-1}{n-k}.$$
(5.90)

Simplest multiple linear regression

Suppose we have three random variables: two influence quantities $[x_1, x_2]$, and one target quantity $[y]$. Then

$$\hat{y} = a + b_1 x_1 + b_2 x_2,$$

$$b_1 = \frac{Q_{yx_1} Q_{x_2} - Q_{yx_2} Q_{x_1 x_2}}{C}, \quad b_2 = \frac{Q_{yx_2} Q_{x_1} - Q_{yx_1} Q_{x_1 x_2}}{C},$$

(p.417)

$$C = Q_{x_1} Q_{x_2} - (Q_{x_1 x_2})^2$$

(for the Q-symbols: c.f., (4.6) and

$$Q_{x_1 x_2} = \sum x_1 x_2 - \frac{1}{n}(\sum x_1)(\sum x_2),$$

as well as Section 5.4.4).

Check (of the computations):

$$b_1 Q_{x_1} + b_2 Q_{x_1 x_2} = Q_{yx_1},$$

$$a = \bar{y} - b_1 \bar{x}_1 - b_2 \bar{x}_2,$$

and

$$b_1 Q_{x_1 x_2} + b_2 Q_{x_2} = Q_{yx_2},$$

$$R^2_{y.x_1 x_2} = B_{y.12} = D/Q_y,$$

$$D = b_1 Q_{yx_1} + b_2 Q_{yx_2}.$$

Test of the regression ($H_0: \beta_1 = \beta_2 = 0$) and hence also whether the parameter corresponding to B differs significantly from zero:

$$\hat{F} = \frac{D(n - 2 - 1)}{(Q_y - D)2}, \quad v_1 = 2, \quad v_2 = n - 2 - 1.$$

One can also test whether the estimate of y from x_1 is improved substantially by adding x_2 to the regression:

$$\hat{F} = \frac{D - E}{Q_y - D} \quad E = \frac{(Q_{yx_1})^2}{Q_{x_1}}$$

with

$$v_1 = 1, \quad v_2 = n - 3.$$

More detailed discussions of multiple regression analysis and related topics can be found in Neter and Wasserman (1974), Draper and Smith (1981), Daniel and Wood (1971), and Searle (1971), as well as in the books by Stange (1971 [1], Part II) and by Dunn and Clark (1974 [1]), (cf., also Hocking 1976, Hahn and Shapiro 1966, and Enderlein et al., 1967, as well as Väliaho (1969), Bliss 1970, Cramer 1972, and the other authors' works mentioned in [8 : 5b]). Recommendations for the selection of variables in multiple regression with examples are given by Thompson (1978). Cole (1959) provides a method of computation which summarizes the elementary approach to multiple correlation and regression for the benefit of workers inexperienced in statistics.

Other techniques related to regression analysis have unfortunately to be foregone in this text, as e.g., **orthogonal polynomials** (Bancroft 1968, Emerson 1968) for the elegant fitting of higher order polynomials (Robson 1959), particularly in the case where the x-values are equally spaced [by means of the tables of Anderson and Houseman (1942), Pearson and Hartley (1966), or Fisher and Yates (1963 [2])]; and **discriminant analysis**, whose task it is to allocate given data belonging to various populations, by means of a discriminant function of the observed characteristics, to the correct populations, at a specified confidence level (cf., Radhakrishna 1964, Cornfield 1967, and P. A. Lachenbruch and M. Goldstein, Biometrics **35** (1979),

69–85). For **trend analysis** (cf., Sections 4.8 and 5.3) the monograph by Gregg, Hossel, and Richardson (1964) is useful, as are the tables by Cowden and Rucker (1965) (cf., also Roos 1955, Salzer et al. 1958, Brown 1962, Ferguson 1965, and also Hiorns 1965). For **calibration** see Biometrics **34** (1978), 39–45, **36** (1980), 729–734, Technometrics **24** (1982), 235–242, and J. Roy. Statist. Soc. B **44** (1982), 287–321.

Remarks

1. **Curve fitting by orthogonal polynomial regression** when the independent variable occurs at unequal intervals and is observed with unequal frequency is discussed and described with a simple example by S. C. Narula, International Statistical Review **47** (1979), 31–36.
2. **Multicollinearity** arises when the independent variables are correlated among themselves. When the independent variables are highly correlated with each other, then least squares estimation of the regression coefficients is biased. In this case a method called **ridge regression** may be useful. More on this may be found in a paper of B. Price, Psychological Bulletin **84** (1977), 759–766 [cf., also **86** (1979), 242–249 and The American Statistician **29** (1975), 3–20].
3. **Canonical correlation analysis** is the general procedure for investigating the relationships between two sets of variables; an interesting review is given by T. R. Knapp, Psychological Bulletin **85** (1978), 410–416 [cf., also M. Krzysko, Biometrical Journal **24** (1982), 211–228].

Multivariate statistical procedures

Figure 51 and Table 112 give two dimensional sampling distributions. The height and weight of each student in a dormitory represent such a distribution. If we add also the age of each student, we have a three dimensional data vector in each case, and thus a three dimensional sampling distribution. The analysis of this and of other, more complicated, n-variate distributions—on a set of persons or objects, several variables are measured and jointly evaluated—forms the domain of multidimensional or multivariate analysis. In other words: multivariate analysis is concerned with the **development of general mathematical models for analyzing a collection of dependent variables**. Parameters are estimated, and relations among the variables are determined. These procedures have gained decisive importance in the analysis of complex questions. One should first master matrix algebra [e.g., start with Chapter 6 in Neter and Wasserman (1974), cited at the end of Section 5.5] and then study, e.g., Kramer (1972) and Kramer and Jensen (1969–1972) as well as Roy (1957), Anderson (1958), Seal (1964), Miller (1966), Saxena and Surendran (1967), Dempster (1968), Krishnaiah (1966, 1969, 1973) Cooley and Lohnes (1971), Puri and Sen (1971), Searle (1971), Press (1972), Bishop et al. (1975 [8:6]), Kendall (1975), and Morrison (1979); Rao (1960, 1972) provides review articles. Important tables can be found in Volume 2 of the Biometrika Tables (Pearson and Hartley 1972 [2], pp. 333–358 [discussed on pp. 98–117]) and in the Kres-Tables (1983 [2]), both cited on page 571. A fine bibliography (articles in periodicals until 1966, books until 1970) is provided by Anderson, Gupta, and Styan (1973). Saxena (1978) gives an annotated bibliography.

6 THE ANALYSIS OF $k \times 2$ AND OTHER TWO WAY TABLES

(p.346) The information content of frequencies is small. Nevertheless, analysis of **fourfold tables** (cf., Section 4.6), the simplest **two way** tables, offers a number of possibilities. We can test these 2 by 2 tables for independence, homogeneity, correlation, and symmetry. These and other tests are discussed in this chapter for tables of size 3 by 2 or greater. Especially important is Section (p.474) 6.2.1.

The testing of a two way table for **trend** offers the possibility of estimating the linear regression portion of the total variation. Comparison of two way tables with respect to their **regression coefficients** supplements the comparison with respect to the amount of correlation by means of the corrected contingency coefficients. Further on, the introduction of the information statistic for the testing of two way tables for independence or homogeneity is presented, and the importance of information analysis of three way and many way tables indicated. A bibliography is provided by Killion and Zahn (1976). The proper use of chi-square for the analysis of contingency tables is reviewed by K. L. Delucchi (1983, Psychological Bulletin **94**, 166–176).

6.1 COMPARISON OF SEVERAL SAMPLES OF DICHOTOMOUS DATA AND THE ANALYSIS OF A $k \times 2$ TWO WAY TABLE

6.1.1 $k \times 2$ tables: The binomial homogeneity test

The fourfold test allows us to investigate whether or not two samples of dichotomous data can be considered random samples from the same population. If we now compare several—let us say k—samples of dichotomous

6.1 Comparison of Several Samples of Dichotomous Data

data, obviously only the two sided question makes sense, and we get for our **initial scheme** a k by 2 table of the following sort (see the Tables 126, 130 below; cf., also Table 129).

Table 126 If n elements of a random sample can be classified according to two characteristics A and B with at least two levels each (see e.g. Section 4.6.1, Table 82: 181 patients, A = treatment, B = course of the illness), then a table of this kind obtains. In this context (A with a dichotomous attribute [+, −]; B with k levels), we can reject the null hypothesis that A and B are independent whenever $\hat{\chi}^2$, computed according to (6.1), is larger than $\chi^2_{k-1,\alpha}$ (cf. also Sections 6.2.1, 6.2.2).

Sample or 2nd attribute	1st attribute level +	−	Σ
1	x_1	$n_1 - x_1$	n_1
2	x_2	$n_2 - x_2$	n_2
.	.	.	.
.	.	.	.
.	.	.	.
j	x_j	$n_j - x_j$	n_j
.	.	.	.
.	.	.	.
.	.	.	.
k	x_k	$n_k - x_k$	n_k
Σ	x	$n - x$	n

For convenience's sake assume that x is less than $n - x$ (Table 126, column 1, "sample"). The null hypothesis reads: The relative share of the character " + " is the same in all k populations. This is estimated in the k independent samples by x/n. Under the null hypothesis, we expect that the k by 2 cells of the table show a frequency distribution which is proportional to the marginal sums. By means of the χ^2 test for k by 2 tables it is thus checked whether the relative frequencies in the k classes deviate more than randomly from the average relative frequency computed over all k classes. We will assume n independent observations as well as mutually exclusive alternatives which exhaust the manifold under observation.

We consider

(p.140)

$$\hat{\chi}^2 = \frac{n^2}{x(n-x)} \left[\sum_{j=1}^{k} \frac{x_j^2}{n_j} - \frac{x^2}{n} \right] \tag{6.1}$$

with $k - 1$ degrees of freedom (DF), and where (cf., Table 126)

$n = $ ("corner-n" or "corner-sum") size of the combined samples,
$n_j = $ size of sample j,
$x = $ total number of sample elements at the $+$ level of the first attribute,
$x_j = $ frequency of the $+$ level of the first attribute [cf., also page 465, below:(6.1*)].

At this point we once again call attention to the difference between $\hat{\chi}^2$ and χ^2. The test statistic $\hat{\chi}^2$ has an approximate χ^2-distribution only for (p.362) large n and not too small expected cell frequencies. The tabulated χ^2-values are critical values for random variables with a χ^2-distribution. The expectation of the cell frequencies is calculated with respect to the null hypothesis of homogeneity of k independent samples from a common binomial population: Under homogeneity (independence) the **expected cell frequencies** of a $k \times 2$ (or, more generally, an $r \times c$) table is computed as the product of the corresponding marginal sums divided by the total sample size [cf., Table 126: The expected frequency E for the field x_j equals $E(x_j) = n_j x/n$]. For small $k \times 2$ tables ($k < 5$) all expected frequencies must be at least equal to 2; if there are at least 4 degrees of freedom at our disposal ($k \geq 5$), then all expected frequencies must be $\gtrsim 1$ (Lewontin and Felsenstein 1965). If these requirements are not met, then the table must be simplified by combining "underoccupied" cells. Only then does the test statistic $\hat{\chi}^2$ computed by the above or by some other formula have an approximate χ^2 distribution.

Remarks

1. Ryan (1960) introduced a simple analysis of variance procedure for the multiple **comparison of k relative frequencies** [see M. Horn, Biometrical Journal **23** (1981), 343–355, 350, 351].

2. Assume that in a $k \times 2$ table for the comparison of relative frequencies or of several means, the null hypothesis that the parameters are all equal is contrasted by the alternate hypothesis that the parameters follow a certain **rank order**. Bartholomew (1959) established a very efficient one sided test for this case. The alternative hypothesis corresponding to the two sided problem reads: the rank order of the parameters is given [see Remark 6 on page 541].

3. If weakly occupied contingency tables of type 3×2 are to be analyzed, one can use the tables prepared by Bennett and Nakamura (1963, cf., also 1964) (for

6.1 Comparison of Several Samples of Dichotomous Data

$n_1 = n_2 = n_3 \leq 20$ and $0.05 \geq \alpha \geq 0.001$). [For $n_1 \geq n_2 \geq n_3$ with $n_1 + n_2 + n_3 = 6(1)15$ and $P \leq 0.2$ see *EDV* in Medizin und Biologie **5** (1974), pp. 73–82].

4. The combination of $k \times 2$ contingency tables with $k =$ constant is considered by Kincaid (1962).

5. The power of tests of homogeneity of k independent samples from a common binomial population is examined by Wisniewski (1972) and by Bennett and Kaneshiro (1978).

6. The Poisson homogeneity test is given in Section 1.6.7 by (1.190), there called the dispersion test (of Poisson frequencies).

EXAMPLE (An extended version is given in Section 6.2.1, Example 2.) (p.477)

Problem: Comparison of two types of therapy.

Design: During an epidemic, a total of 80 persons were treated. Forty patients were given a standard dose of a specific new drug. The other 40 afflicted were treated only symptomatically (treatment only of symptoms but not of cause). (Source: Martini 1953, p. 83, Table 14.) The result of the treatment is presented in terms of cell entries for three classes: quickly recovered, slowly recovered, and not recovered (Table 127).

Table 127

Treatment effect	Treatment		Total
	Symptomatic	Specific (standard dose)	
Recovery within *a* weeks	14	22	36
Recovery between *a*th and (*a* + *b*)th week	18	16	34
No recovery	8	2	10
Total	40	40	80

Null hypothesis: The therapeutic results are the same for both types of therapy.

Alternative hypothesis: The therapeutic results are not the same for the two types of therapy.

Significance level: $\alpha = 0.05$ (two sided).

Choice of test: Only the $k \times 2$ χ^2 test is suitable (cf., expectation frequencies for the patients that didn't recover, Table 127: $x_k = 8, n_k - x_k = 2$; $E(8) = (10)(40)/80 = 5$ and $E(2) = (10)(40)/80 = 5 > 2$).

Results and evaluation: By formula (6.1)

$$\hat{\chi}^2 = \frac{80^2}{40 \cdot 40}\left[\left(\frac{14^2}{36} + \frac{18^2}{34} + \frac{8^2}{10}\right) - \frac{40^2}{80}\right] = 5.495.$$

Remark: For $x = n - x$ or $x = n/2$ (6.1) simplifies to (6.4) [p. 478]:

$$\hat{\chi}^2 = \sum_{j=1}^{k}\{(x_j - [n_j - x_j])^2/n_j\} = (14 - 22)^2/36 + (18 - 16)^2/34 + (8 - 2)^2/10 = 64/36 + 4/34 + 36/10 = 5.495.$$

Decision: Since $\hat{\chi}^2 = 5.495 < 5.99 = \chi^2_{2;0.05}$, we cannot reject the null hypothesis.

Interpretation: On the basis of the given data, a difference between the two types of therapy cannot be guaranteed at the 5% level.

Remark: If a comparison of the mean therapeutic results of the two therapies is of interest, then testing should be carried out according to Cochran (1966, pp. 7–10).

Table 128 (cf. Table 127)

TREATMENT EFFECT Computation of $\hat{\chi}^2$	TREATMENT Symptomatic	Specific (standard dose)	Total
RECOVERED IN *a* WEEKS: Observed O Expected E Difference O − E (Difference)² (O − E)² Chi-square $\frac{(O-E)^2}{E}$	14 18.00 −4.00 16.00 0.8889	22 18.00 4.00 16.00 0.8889	36 36 0.0 1.7778
RECOVERED BETWEEN *a*th AND (*a* + *b*)th WEEK: Observed O Expected E Difference O − E (Difference)² (O − E)² Chi-square $\frac{(O-E)^2}{E}$	18 17.00 1.00 1.00 0.0588	16 17.00 −1.00 1.00 0.0588	34 34 0.0 0.1176
NOT RECOVERED Observed O Expected E Difference O − E (Difference)² (O − E)² Chi-square $\frac{(O-E)^2}{E}$	8 5.00 3.00 9.00 1.8000	2 5.00 −3.00 9.00 1.8000	10 10 0.0 3.6000
Total: O = E $\hat{\chi}^2$-column sums:	40 2.7477	40 2.7477	80 $\hat{\chi}^2$ = 5.4954

In particular, it should be mentioned that every contribution to the $\hat{\chi}^2$-value is relative to the expected frequency E: A large difference $O - E$ with large E may contribute approximately the same amount to $\hat{\chi}^2$ as a small frequency with small E, e.g.,

$$\frac{(15-25)^2}{25} = 4 = \frac{(3-1)^2}{1}.$$

This result could naturally have been obtained also by means of the **general χ^2-formula** (4.13) (Section 4.3). Under null hypothesis for homogeneity or independence, the expected frequencies E will—as remarked previously—be determined as quotients of the products of the corresponding marginal sums of the table and the total sample size. Thus for example we have in the upper left hand corner of Tables 127 and 128 the observed frequency $O = 14$ and the associated expected frequency $E = (36)(40)/80 = 36/2 = 18$. Computing

$$\frac{(\text{observed frequency} - \text{expected frequency})^2}{\text{expected frequency}} = \frac{(O-E)^2}{E}$$

for every cell of the $k \times 2$ table and adding these values, we again find $\hat{\chi}^2$.

$\hat{\chi}^2$ tests the compatibility of observation and theory, of observed and expected frequencies. If the divergence is great and the observed value of $\hat{\chi}^2$ exceeds the tabulated value $\chi^2_{0.05}$ for $k - 1 = 3 - 1 = 2$ degrees of freedom, $\chi^2_{2;0.05} = 5.99$ we say that the difference between observation and theory is statistically significant at the 5% level. The procedure set out in Table 128 displays the contribution to $\hat{\chi}^2$ of each single cell and shows that the difference "not recovered" dominates. Since both groups of patients consisted of 40 persons, the contributions to $\hat{\chi}^2$ are pairwise equal.

6.1.2 Comparison of two independent empirical distributions of frequency data

Given two frequency tables, we are faced with, among other things, the question of whether they originated in different populations. A test for nonequivalence of the underlying populations of the two samples rests on the formula (6.1) (cf., also Section 4.3). The levels of a classifying attribute are again assumed to be mutually exclusive and exhaustive.

EXAMPLE. Do the distributions B_1 and B_2 in Table 129 come from the same population ($\alpha = 0.01$)?

$$\hat{\chi}^2 = \frac{387^2}{200 \cdot 187} \left[\left(\frac{60^2}{108} + \frac{52^2}{102} + \cdots + \frac{5^2}{13} \right) - \frac{200^2}{387} \right] = 5.734,$$

$$\text{or } \hat{\chi}^2 = \frac{387^2}{200 \cdot 187} \left[\left(\frac{48^2}{108} + \frac{50^2}{102} + \cdots + \frac{8^2}{13} \right) - \frac{187^2}{387} \right] = 5.734.$$

Since this $\hat{\chi}^2$-value is substantially smaller than $\chi^2_{6;0.01} = 16.81$, the null hypothesis that both samples were drawn from the same population cannot be rejected at the 1% level.

Table 129

Category	Frequencies B_I	B_{II}	Σ
1	60	48	108
2	52	50	102
3	30	36	66
4	31	20	51
5	10	15	25
6	12	10	22
7	4 }5	8 }8	13
8	1	0	
Σ	$n_1 = 200$	$n_2 = 187$	$n = 387$

6.1.3 Partitioning the degrees of freedom of a $k \times 2$ table

For the $k \times 2$ table we label the frequencies according to the following scheme (Table 130), an extension of Table 126. It allows the direct comparison

Table 130

Sample	Attribute +	−	Total	Level of p_+
1	x_1	$n_1 - x_1$	n_1	$p_1 = x_1/n_1$
2	x_2	$n_2 - x_2$	n_2	$p_2 = x_2/n_2$
.	.	.	.	.
.	.	.	.	.
.	.	.	.	.
j	x_j	$n_j - x_j$	n_j	$p_j = x_j/n_j$
.	.	.	.	.
.	.	.	.	.
k	x_k	$n_k - x_k$	n_k	$p_k = x_k/n_k$
Total	x	$n - x$	n	
		$\hat{p} = x/n$		

of "success" percentages—the relative frequency of the + level—for all samples. The formula for the χ^2-test can then be written

$$\hat{\chi}^2 = \frac{\sum_{j=1}^{k} x_j p_j - x\hat{p}}{\hat{p}(1-\hat{p})} \qquad (6.1\text{a})$$

with $k - 1$ DF.

Here we have:

x = the total number of sample elements at the + level,
x_j = the number of elements of sample j at the + level,
$\hat{p} = x/n$: the relative frequency of the + level in the total population.

Under the null hypothesis that all samples originate in populations with π = const, estimated by $\hat{p} = x/n$, we expect again in all samples a frequency distribution corresponding to this ratio.

Formula (6.1a) is used not only for testing the homogeneity of the sample consisting of k sub-samples, but also of each set of two or more samples—let us say j (with DF = $j - 1$)—which are chosen as a group from the k samples. We thus succeed in partitioning the $k - 1$ degrees of freedom into components $[1 + (j - 1) + (k - j - 1) = k - 1]$ (Table 131). In other words, the total $\hat{\chi}^2$ is partitioned. This provides a

Table 131

Components of $\hat{\chi}^2$	Degrees of freedom
Dispersion of the p's within the first j groups	1
Dispersion of the p's within the last k − j groups	j − 1
Dispersion of the p's between the two groups of samples	k − j − 1
Total $\hat{\chi}^2$	k − 1

test which reflects the change of the p-level in a sequence of samples of dichotomous data. Let's consider a simple example (Table 132):

EXAMPLE

Table 132

No.	x_j	$n_j - x_j$	n_j	$p_j = x_j/n_j$	$x_j p_j$
1	10	10	20	0.500	5.000
2	8	12	20	0.400	3.200
3	9	11	20	0.450	4.050
4	5	15	20	0.250	1.250
5	6	14	20	0.300	1.800
Σ	38	62	100		15.300
			$\hat{p} = 38/100 = 0.380$		

$\hat{\chi}^2$ (overall deviation of the p's from $\hat{p}$) = $\dfrac{15.300 - (38)(0.380)}{(0.380)(0.620)}$ = 3.650,

Table 133

No.	Group	x_i	n_i	$\bar{p}_i$	$x_i p_i$
1+2+3	n_1	27	60	0.450	12.150
4 + 5	n_2	11	40	0.275	3.025
Σ	n	38	100		15.175

$\hat{\chi}^2$ (differences between mean p's of the sample groups n_1 (= numbers 1, 2, 3) and n_2 (= numbers 4, 5)).

REMARK: $\bar{p}_1$ for G_1 (or n_1) is the arithmetic mean of the three p_+-levels $[n_j = 20]$, $(0.500 + 0.400 + 0.450)/3 = 0.450 = 27/60$; the analogous statement holds for p_2 of G_2 (or n_2).

$\hat{\chi}^2$ (difference between the $\bar{p}$'s of groups 1 and 2)

$$= \frac{15.175 - (38)(0.380)}{(0.380)(0.620)} = 3.120,$$

$\hat{\chi}^2$ (variation of the p's within n_1 [group 1])

$$= \frac{12.250 - (27)(0.450)}{(0.380)(0.620)} = 0.424,$$

$\hat{\chi}^2$ (variation of the p's within n_2 [group 2])

$$= \frac{3.050 - (11)(0.275)}{(0.380)(0.620)} = 0.106.$$

The components are collected in Table 134.

Table 134

Source	$\hat{\chi}^2$	DF	Significance level
Variation between G_1 and G_2	3.120	1	$0.05 < P < 0.10$
Variation within G_1	0.424	2	$0.80 < P < 0.90$
Variation within G_2	0.106	1	$P \simeq 0.30$
Total variation of the p's with respect to $\hat{p}$	3.650	4	$0.40 < P < 0.50$

As is indicated by the example, one sometimes succeeds in isolating homogeneous elements from heterogeneous samples. The decisive $\hat{\chi}^2$-component is furnished by the difference between the mean success percentages (Table 133: 0.450 as against 0.275) of the sample groups G_1 and G_2.

6.1 Comparison of Several Samples of Dichotomous Data

With the given significance level of $\alpha = 0.05$, the null hypothesis $\pi_{G_1} = \pi_{G_2}$ is retained. If the direction of a possible difference can be established **before** the experiment is carried out, a one sided test for this component is justified. The value of $\hat{\chi}^2 = 3.120$ would then be significant at the 5% level; the null hypothesis would have to be abandoned in favor of the alternative hypothesis $\pi_{G_1} > \pi_{G_2}$.

Let us demonstrate **another partitioning** with the help of the same example. We will dispense with the general formulation, since the decomposition principle which splits $\hat{\chi}^2$ into independent components with one degree of freedom each is quite simple. Table 132a has been written somewhat differently than Table 132.

Table 132a

Type	A	B	C	D	E	Σ
I	10	8	9	5	6	38
II	10	12	11	15	14	62
Σ	20	20	20	20	20	100

Let us now consider fourfold tables; except for the first one considered they arise through successive combinations. The corresponding $\hat{\chi}^2$ are computed by a formula which resembles (4.35) in Section 4.6.1. First we examine the homogeneity of the samples A and B (with regard to I and II) within the total sample and denote it by $A \times B$. We form the difference of the "diagonal products," square it, and then multiply it by the square of the total sample size ($= 100$ in Table 132a). The divisor consists of the product of 5 factors: the sum of row I, of row II, of column A, of column B, and of the sums A and B which we have enclosed in parentheses:

$$A \times B: \quad \hat{\chi}^2 = \frac{100^2(10 \cdot 12 - 8 \cdot 10)^2}{38 \cdot 62 \cdot 20 \cdot 20(20 + 20)} = 0.4244.$$

The homogeneity of $A + B$, the sum of the columns A and B, in comparison with C, for which the symbol $(A + B) \times C$ is used, can correspondingly be determined by

$$(A + B) \times C: \quad \hat{\chi}^2 = \frac{100^2\{(10 + 8)11 - 9(10 + 12)\}^2}{38 \cdot 62(20 + 20)20(40 + 20)} = 0,$$

and by analogy for

$$(A + B + C) \times D: \quad \hat{\chi}^2 = \frac{100^2\{(10 + 8 + 9)15 - 5(10 + 12 + 11)\}^2}{38 \cdot 62(20 + 20 + 20)20(60 + 20)}$$

$$= 2.5467$$

and for

$(A + B + C + D) \times E$:

$$\hat{\chi}^2 = \frac{100^2\{(10 + 8 + 9 + 5)14 - 6(10 + 12 + 11 + 15)\}^2}{38 \cdot 62(20 + 20 + 20 + 20)20(80 + 20)} = 0.6791.$$

We collect our results in Table 135.

Table 135 $\hat{\chi}^2$-partition for the 5 × 2 table

Source	DF	$\hat{\chi}^2$	P
(1) A×B	1	0.4244	n.s.
(2) (A+B)×C	1	0.0000	n.s.
(3) (A+B+C)×D	1	2.5467	<0.15
(4) (A+B+C+D)×E	1	0.6791	n.s.
Total	4	3.6502	n.s.

The sum of the four $\hat{\chi}^2$-values is 3.650 (cf., Table 134). There are no apparent characteristic differences among the "sample pairs" (1), (2), (3), and (4). An exceptional status of D in the frequency ratio I/II is suggested by (3). To test the homogeneity of few "sample pairs" that are in some sense chosen in advance, the table can be rewritten with the columns interchanged.

6.1.4 Testing a $k \times 2$ table for trend: The share of linear regression in the overall variation

(p.465) We examine Table 127 (Section 6.1.1) again. The results of the therapy can be ordered in a natural way by the categories "no recovery," "slow recovery," and "quick recovery" (see Table 136). We notice immediately that the p-value of the specifically treated group increases as we move from the "no recovery" to the "quick recovery" class: $2/10 < 16/34 < 22/36$.

Table 136

z_j (Score)	x_j	$n_j - x_j$	n_j	$p_j = x_j/n_j$	$x_j z_j$	$n_j z_j$	$n_j z_j^2$
+1	22	14	36	0.611	22	36	36
0	16	18	34	0.471	0	0	0
-1	2	8	10	0.200	-2	-10	10
	x = 40	40	n = 80		20	26	46
		$\hat{p} = x/n = 40/80 = 0.50$					

6.1 Comparison of Several Samples of Dichotomous Data

If the relative frequency increases **with the order of the classes**, then a test for **linear regression** is appropriate. The $\hat{\chi}^2$ can then be split up into two parts: one part decomposes into frequencies assumed linearly increasing, the remainder corresponds to the difference between the observed frequencies and the linearly increasing frequency component. One can thus dissociate the linear regression portion, with one degree of freedom, from the portion determined by the deviation from the regression line. This portion will be regarded as the difference between $\hat{\chi}^2$ and $\hat{\chi}^2$ linear regression.

For the case of a $k \times 2$ table, Cochran (1954) has provided a straightforward method for computing the linear regression component. (For the remaining component $DF = k - 2$ holds.) First the "natural" order of the k levels (categories), in our case the therapeutic results, must be replaced by a sequence of numbers, by "**scores**". Most often sequences which are symmetric with respect to zero, e.g., $-2, -1, 0, 1, 2$ or $-4, -2, 0, 1, 2, 3$, are used, because they simplify the computation; this "scoring" should be made before data are gathered. The points need not be equally spaced. The sequence $-2, -1, 0, 3, 6$ emphasizes the last two categories on the basis of their exceptional properties. For example, we can employ in Table 136 the sequences $-2, 0, 1$ or $-3, 0, 1$ to distinguish the fundamental difference between no and slow recovery from the difference in degree between slower and more rapid recovery. The $\hat{\chi}^2_{\text{linear regression}}$, according to Cochran (1954) [cf., also Armitage (1955), Bartholomew (1959), as well as Bennett and Hsu (1962)], is given by

$$\hat{\chi}^2_{\text{lin. regr.}} = \frac{\left(\sum x_j z_j - \frac{x \sum n_j z_j}{n}\right)^2}{\hat{p}(1-\hat{p})\left(\sum n_j z_j^2 - \frac{(\sum n_j z_j)^2}{n}\right)} \tag{6.2}$$

with $DF = 1$.

One can also estimate $\hat{b} = S_1/S_2$ by $S_1 = \sum n_j(p_j - \hat{p})(z_j - \bar{z})$ and $S_2 = \sum n_j z_j^2 - (\sum n_j z_j)^2/n$, and test $H_0: \beta = 0$ on the basis of the standard normal variable z (Table 43, Section 2.1.6) by $\hat{z} = \hat{b}/s_b$ with

$$s_b = \sqrt{\hat{p}(1-\hat{p})/S_2};$$

note that here the sum of the scores should be nonzero.

EXAMPLE. Applying the formula (6.2) to the values in Table 136, we get for the linear regression portion

$$\hat{\chi}^2_{\text{lin. regr.}} = \frac{\left(20 - \frac{40 \cdot 26}{80}\right)^2}{0.50 \cdot 0.50\left(46 - \frac{26^2}{80}\right)} = 5.220 > 3.84 = \chi^2_{1;\,0.05}.$$

This value is statistically significant at the 5% level. In the example in Section 6.1.1, the homogeneity hypothesis with a significance level of $\alpha = 0.05$ was not rejected as against the general heterogeneity hypothesis for $\hat{\chi}^2 = 5.495$ and DF = 2.

Table 137 shows the decisive part played by the linear regression in the total variation, which is already apparent in the column of p_j-values in Table 136 and which points out the superiority of the specific therapy.

Table 137

Source	$\hat{\chi}^2$	DF	Significance level
Linear regression	5.220	1	$0.01 < P < 0.05$
Departure from linear regression	0.275	1	$P = 0.60$
Total	5.495	2	$0.05 < P < 0.10$

6.2 THE ANALYSIS OF $r \times c$ CONTINGENCY AND HOMOGENEITY TABLES

▶ 6.2.1 Testing for independence or homogeneity

An extension of the fourfold table, as the simplest two way table, to the general case, leads to the $r \times c$ table having r rows and c columns (Table 138). A sample of size n is randomly drawn from some population. Every

Table 138 The levels (classes) of one of the two attributes can also represent different samples. The totals along the edge of the table are called marginal frequencies.

1st attribute r rows \ 2nd attribute c columns	1	2	-	j	-	c	Row totals
1	n_{11}	n_{12}	-	n_{1j}	-	n_{1c}	$n_{1.}$
2	n_{21}	n_{22}	-	n_{2j}	-	n_{2c}	$n_{2.}$
-	-	-	-	-	-	-	-
i	n_{i1}	n_{i2}	-	n_{ij}	-	n_{ic}	$n_{i.}$
-	-	-	-	-	-	-	-
r	n_{r1}	n_{r2}	-	n_{rj}	-	n_{rc}	$n_{r.}$
Column totals	$n_{.1}$	$n_{.2}$	-	$n_{.j}$	-	$n_{.c}$	$n_{..} = n$ Corner-n

6.2 The Analysis of $r \times c$ Contingency and Homogeneity Tables

element of this sample is then classified according to the two different attributes, each of which is subdivided into different categories, classes, or levels. The hypothesis of **independence** (characteristics I and II do not influence each other) is to be tested. In other words, one tests whether the distribution of the classes of an attribute is independent of the distribution of the classes of the other attribute (cf., Section 6.1.1), i.e., whether we are dealing with a frequency distribution which is to a great extent proportional to the marginal sums (cf., the example in Section 6.1.1). We observe that the **comparison of difference samples** of sizes $n_{1.}, n_{2.}, \ldots, n_{i.}, \ldots, n_{r.}$, from r different discrete distributions as to similarity or homogeneity, leads to the same test procedure. Thus the test statistic is the same whether we examine a contingency table for independence or check a set of samples for a common underlying population (comparison of population probabilities of multinomial distributions). This is gratifying, since in many problems it is in no way clear which interpretation is more appropriate. The test statistic is

$$\hat{\chi}^2 = \sum_{i=1}^{r} \sum_{j=1}^{c} \left[\frac{\left(n_{ij} - \frac{n_{i.}n_{.j}}{n}\right)^2}{\frac{n_{i.}n_{.j}}{n}} \right] = n \left[\sum_{i=1}^{r} \sum_{j=1}^{c} \frac{n_{ij}^2}{n_{i.}n_{.j}} - 1 \right] \quad (6.3)$$

[as follows from (4.13), Section 4.3, with $k = rc$, $\sum B = \sum E = n$, and $E_{ij} = (n_{i.}n_{.j})/n$] with $(r-1)(c-1)$ degrees of freedom. Here

n = corner-n, overall sample size,
n_{ij} = occupation number of the cell in the ith row and jth column,
$n_{i.}$ = sum of cell entries of the ith row (row sum),
$n_{.j}$ = sum of cell entries of the jth column (column sum),
$n_{i.}n_{.j}$ = product of marginal sums corresponding to cell (i, j).

The expected frequencies are computed (under the null hypothesis) according to $n_{i.}n_{.j}/n$. The test may be applied when all expected frequencies are ≥ 1. If some expected frequencies are smaller, then the table is to be simplified by **grouping the under-occupied cells**. We note that one should apply the most **objective scheme** possible so as not to influence the result by a more or less deliberate arbitrariness in this grouping. A method of analyzing **unusually weakly occupied** contingency tables which are mostly independent or homogeneous was proposed by Nass (1959). R. Heller presents a FORTRAN program which computes for these cases exact P-values by way of the Freeman-Halton test for $r \times c$ tables; this test coincides with the exact Fisher test for $r = c = 2$ (EDV in Medizin und Biologie **10** (1979), 62–63). See also J. Amer. Statist. Assoc. **76** (1981), 931–934 and **78** (1983), 427–434.

Remarks pertaining to the analysis of **square tables** ($r = c$) can be found in Sections 4.6.1 (Remark 4), 4.6.6, 6.2.1 (Formula 6.5)), 6.2.2, 6.2.4 as well as in Bishop, Fienberg, and Holland (1975).

> ### Models for an $r \times c$ table
>
> Starting with the 2×2 table, we discern three models:
>
> I. **Both** sets of marginal frequencies, the totals for rows and columns, are specified in advance (are fixed): test of **independence**.
> II. **One** set of marginal frequencies, the totals for rows or for columns, is specified in advance: test of **homogeneity**.
> III. **No** set of marginal frequencies is specified in advance: neither totals for rows nor for columns are fixed: only n (corner-n) is specified in advance: test of **bivariate independence**.
>
> It should be stressed that for an $r \times c$ table, $\hat{\chi}^2$ defined in (6.3) is approximately [better: asymptotically] distributed with $(r-1)(c-1)$ degrees of freedom **UNDER ALL THREE MODELS**. A statistically significant large value of $\hat{\chi}^2$ is regarded as evidence of lack of independence or of homogeneity. The result must be approximate, since we apply the continuous χ^2 distribution, in the derivation of which normality is assumed, to discrete variables and assume the expected frequencies E_i are not too small, thus avoiding a greater degree of discontinuity. Moreover, we assume independent observations from random samples.
>
> If one set of marginal frequencies is specified in advance, a test of homogeneity is made by testing whether the various rows or columns of the **homogeneity table** have the same proportions of individuals in the various categories. As mentioned, the procedure, is exactly the same as in the test of independence of a **contingency table**, whether or not both sets of marginal frequencies are fixed.

EXAMPLE 1 ($\alpha = 0.01$)

Table 139

24	7	7	38
76	38	70	184
69	32	82	183
27	9	55	91
196	86	214	496

$(4-1)(3-1) = 6 \text{ DF}$

According to (6.3):

$$\hat{\chi}^2 = 496\left[\frac{24^2}{(38)(196)} + \frac{7^2}{(38)(86)} + \frac{7^2}{(38)(214)} + \frac{76^2}{(184)(196)} + \cdots\right.$$

$$\left. + \frac{27^2}{(91)(196)} + \frac{9^2}{(91)(86)} + \frac{55^2}{(91)(214)} - 1\right]$$

$$\hat{\chi}^2 = 24.939.$$

Since $24.94 > 16.81 = \chi^2_{6;0.01}$, the null hypothesis of independence or homogeneity at the 1% level must be rejected for the two way table in question.

EXAMPLE 2 (See Sections 1.3.2, 2.1.3 through 2.1.5, and 6.1.1.)

Table 140

	Treatment			
		\multicolumn{2}{c}{Specific}	
		Standard	2x Standard	
Treatment effect	Symptomatic	dose	dose	Total
Recovery within a weeks	14	22	32	68
Recovery between ath and (a + b)th week	18	16	8	42
No recovery	8	2	0	10
Total	40	40	40	120

Problem: Comparing three types of therapy.

Design of experiment: Three groups of forty patients were treated. Two groups have been compared in Section 6.1.1. The third group is subjected to the specific therapy with double the normal dose. (Source: Martini 1953, p. 79, Table 13.)

Null hypothesis: no difference among the three treatments; alternative hypothesis: the treatment effects are not all the same.

Significance level: $\alpha = 0.05$.

Test choice: $\hat{\chi}^2$-test. Note: The Dunn test (1964) can be used to advantage, or one can compute r_S according to Raatz (see example in Section 5.3.1).

Results and evaluation: $\hat{\chi}^2 = 120\left[\frac{14^2}{(68)(40)} + \cdots + \frac{0^2}{(10)(40)} - 1\right]$

$= 21.576.$

Degrees of freedom: $(3-1)(3-1) = 4$.

Decision: Since $21.576 > 9.49 = \chi^2_{4;0.05}$, the null hypothesis is rejected.

Interpretation: The association between treatment and effect is ascertained at the 5% level. In view of the previous result concerning symptomatic and specific treatments, we are fairly sure that the specific treatment with double the standard dose has a different effect, and, checking the respective cell entries, we conclude that it is superior to the other two treatments. However, a test of the double dose versus the standard dose should be carried out to substantiate the heuristic reasoning.

There are **special formulas for the cases** $n_1 = n_2$ and $n_1 = n_2 = n_3$:

$$\hat{\chi}^2 = \sum_{j=1}^{k} \frac{(n_{1j} - n_{2j})^2}{n_{1j} + n_{2j}} \quad (6.4)$$

[DF = $k - 1$],

$$\hat{\chi}^2 = \sum_{j=1}^{k} \frac{(n_{1j} - n_{2j})^2 + (n_{1j} - n_{3j})^2 + (n_{2j} - n_{3j})^2}{n_{1j} + n_{2j} + n_{3j}} \quad (6.4a)$$

[DF = $2(k - 1)$].

Repeated application of tests to the same body of data

1. If a total of τ tests at the respective significance levels α_i is run, the overall significance of the τ tests is less than or equal to $\sum_{i=1}^{\tau} \alpha_i$ [cf., Sections 1.4.2 (example), 4.6.2]. The value $\alpha_i = \alpha/\tau$ is usually chosen for each test, and α is then the nominal significance level for this sequence of tests (the Bonferroni procedure; cf., e.g., Dunn 1961, 1974 [8 : 7a]).
2. As part of a survey, τ χ^2-tests are designed (type $k \times 1$, $k \times 2$ and $k \geq 2$, or $r \times c$ with $r, c > 2$) with v_i degrees of freedom respectively. The critical bounds of the Bonferroni χ^2-table (Table 141 as well as Table 88 [Section 4.6.2]) are then applied. The probability of incorrectly rejecting at least one of the null hypotheses is then not greater than the nominal significance level α.

The following table gives an example for $\tau = 12$ tests ($\alpha = 0.05$):

No.	Page	Table	$\hat{\chi}^2$	v	$\chi^2(0.05/12)$	Decision
1	468	129	5.734	6	18.998	H_0
2	476	139	24.939	6	18.998	$\not{H_0}$
3	477	140	21.576	4	15.273	$\not{H_0}$
(9 more tables)						

6.2 The Analysis of $r \times c$ Contingency and Homogeneity Tables

Table 141 Upper bounds of the Bonferroni χ^2-statistics $\chi^2(\alpha/\tau, v)$. From Kramer, C. Y.: A First Course in Methods of Multivariate Analysis, Virginia Polytechnic Institute and State University, Blacksburg 1972, Appendix D: Beus, G. B. and Jensen, D. R., Sept. 1967, pp. 327–351 [$\tau \leq 120$ (42 entries), $\tau \leq 30$ (25 entries) and $\alpha = 0.005; 0.01; 0.025; 0.05; 0.10$]; with the permission of the author.

$\alpha = 0.10, 0.05, 0.10$

v	$\tau:1$	2	3	4	5	6	7	8	9	10	11	12
1	2.706	3.841	4.529	5.024	5.412	5.731	6.002	6.239	6.447	6.635	6.805	6.960
2	4.605	5.991	6.802	7.378	7.824	8.189	8.497	8.764	9.000	9.210	9.401	9.575
3	6.251	7.815	8.715	9.348	9.837	10.236	10.571	10.861	11.117	11.345	11.551	11.739
4	7.779	9.488	10.461	11.143	11.668	12.094	12.452	12.762	13.034	13.277	13.496	13.695
5	9.236	11.070	12.108	12.833	13.388	13.839	14.217	14.544	14.831	15.086	15.317	15.527
6	10.645	12.592	13.687	14.449	15.033	15.506	15.903	16.245	16.545	16.812	17.053	17.272
8	13.362	15.507	16.705	17.535	18.168	18.680	19.109	19.478	19.802	20.090	20.350	20.586
9	14.684	16.919	18.163	19.023	19.679	20.209	20.653	21.034	21.368	21.666	21.934	22.177
10	15.987	18.307	19.594	20.483	21.161	21.707	22.165	22.558	22.903	23.209	23.485	23.736
12	18.549	21.026	22.394	23.337	24.054	24.632	25.115	25.530	25.894	26.217	26.508	26.772
14	21.064	23.685	25.127	26.119	26.873	27.480	27.987	28.422	28.803	29.141	29.446	29.722
15	22.307	24.996	26.473	27.488	28.259	28.880	29.398	29.843	30.232	30.578	30.889	31.171
16	23.542	26.296	27.808	28.845	29.633	30.267	30.796	31.250	31.647	32.000	32.317	32.605
1	3.841	5.024	5.731	6.239	6.635	6.960	7.237	7.477	7.689	7.879	8.052	8.210
2	5.991	7.378	8.189	8.764	9.210	9.575	9.883	10.150	10.386	10.597	10.787	10.961
3	7.815	9.348	10.236	10.861	11.345	11.739	12.071	12.359	12.612	12.838	13.043	13.229
4	9.488	11.143	12.094	12.762	13.277	13.695	14.048	14.358	14.621	14.860	15.076	15.273
5	11.070	12.833	13.839	14.544	15.086	15.527	15.898	16.217	16.499	16.750	16.976	17.182
6	12.592	14.449	15.506	16.245	16.812	17.272	17.659	17.993	18.286	18.548	18.783	18.998
8	15.507	17.535	18.680	19.478	20.090	20.586	21.002	21.360	21.675	21.955	22.208	22.438
9	16.919	19.023	20.209	21.034	21.666	22.177	22.607	22.976	23.301	23.589	23.850	24.086
10	18.307	20.483	21.707	22.558	23.209	23.736	24.178	24.558	24.891	25.188	25.456	25.699
12	21.026	23.337	24.632	25.530	26.217	26.772	27.237	27.637	27.987	28.300	28.581	28.836
14	23.685	26.119	27.480	28.422	29.141	29.722	30.209	30.627	30.993	31.319	31.613	31.880
15	24.996	27.488	28.880	29.843	30.578	31.171	31.668	32.095	32.469	32.801	33.101	33.373
16	26.296	28.845	30.267	31.250	32.000	32.605	33.111	33.547	33.928	34.267	34.572	34.850
1	6.635	7.879	8.615	9.141	9.550	9.885	10.169	10.415	10.633	10.828	11.004	11.165
2	9.210	10.597	11.408	11.983	12.429	12.794	13.102	13.369	13.605	13.816	14.006	14.180
3	11.345	12.838	13.706	14.320	14.796	15.183	15.510	15.794	16.043	16.266	16.468	16.652
4	13.277	14.860	15.777	16.424	16.924	17.331	17.675	17.972	18.233	18.467	18.678	18.871
5	15.086	16.750	17.710	18.386	18.907	19.332	19.690	20.000	20.272	20.515	20.735	20.935
6	16.812	18.548	19.547	20.249	20.791	21.232	21.603	21.924	22.206	22.458	22.685	22.892
8	20.090	21.955	23.024	23.774	24.352	24.821	25.216	25.557	25.857	26.124	26.366	26.586
9	21.666	23.589	24.690	25.462	26.056	26.539	26.945	27.295	27.603	27.877	28.125	28.351
10	23.209	25.188	26.320	27.112	27.122	28.216	28.633	28.991	29.307	29.588	29.842	30.073
12	26.217	28.300	29.487	30.318	30.957	31.475	31.910	32.286	32.615	32.909	33.175	33.416
14	29.141	31.319	32.559	33.426	34.091	34.631	35.084	35.475	35.818	36.123	36.399	36.650
15	30.978	32.801	34.066	34.950	35.628	36.177	36.639	37.037	37.386	37.697	37.978	38.233
16	32.000	34.267	35.556	36.456	37.146	37.706	38.175	38.580	38.936	39.252	39.538	39.798

The test statistic $\hat{\chi}^2$ for independence in a **rectangular table** is bounded by

$$\hat{\chi}^2_{max} = n[\min(c - 1, r - 1)], \quad \text{i.e.,} \quad \hat{\chi}^2_{max} = \begin{cases} n(c - 1) & \text{if } c \leq r, \\ n(r - 1) & \text{if } r \leq c. \end{cases}$$

(6.5)

The maximum is assumed in the case of complete dependence, e.g.,

20	0	20
0	20	20
20	20	40

$\hat{\chi}^2 = 4 \cdot 10 = 40$

$\hat{\chi}^2_{max} = 40(2 - 1) = 40.$

A few additional remarks

1. Several methods have been presented for testing homogeneity or independence or, generally **proportionality** in two way tables. Section 6.2.5 presents yet another economical method of computation. Besides, experience suggests that it is a good (for the beginner an indispensable) habit to check the computations by one of the $\hat{\chi}^2$-formulas unless too much work is involved. If tables with many cells are to be evaluated, then the computational method would be verified on simple—or simplified—tables.

2. If in the course of the analysis of rectangular tables the null hypothesis is rejected in favor of the alternative hypothesis for dependence or heterogeneity, then sometimes there is interest in **localizing the cause of the significance**. This is done by repeating the test for a table from which all suspicious-looking rows or columns are removed one at a time (cf., also the text following (6.1) in Section 6.1.1). Other possibilities for testing interesting partial hypotheses (cf., also Gabriel 1966) are offered by the selection of 4 symmetrically arranged cells (each cell shares 2 rows, and a column, with one of the other three cells); then the resulting 2 × 2 table is analyzed. This should be regarded as "**experimentation**" (cf., Section 1.4.4); the results can only serve as clues for future investigations. A proper (statistical) statement can be made only if associated partial hypotheses have been formulated **before** the data were collected.

   Let us here add another caution. When the dependence seems assured, one must bear in mind that the existence of a **formal relation** says nothing about the **causal relation**. It is entirely possible that indirect relations introduce part (or all) of the dependence (cf., also Sections 5.1 and 5.2).

3. The test statistic $\hat{\chi}^2$ for independence in an $r \times c$ table can always be decomposed into $(r-1)(c-1)$ independent components of one degree of freedom each (cf., Kastenbaum 1960, Castellan 1965 as well as Bresnahan and Shapio 1966 [see also Shaffer 1973 and C. B. Read, Communications in Statistics—Theory and Methods A **6** (1977), 553–562]). With the notation of Table 138, we have, e.g., for a 3 × 3 table, with $2 \cdot 2 = 4$ degrees of freedom, the following four components:

$$(1)\ \hat{\chi}^2 = \frac{n\{n_{2.}(n_{.2}n_{11}-n_{.1}n_{12})-n_{1.}(n_{.2}n_{21}-n_{.1}n_{22})\}^2}{n_{1.}n_{2.}n_{.1}n_{.2}(n_{1.}+n_{2.})(n_{.1}+n_{.2})},$$ (6.6a)

$$(2)\ \hat{\chi}^2 = \frac{n^2\{n_{23}(n_{11}+n_{12})-n_{13}(n_{21}+n_{22})\}^2}{n_{1.}n_{2.}n_{.3}(n_{1.}+n_{2.})(n_{.1}+n_{.2})},$$ (6.6b)

$$(3)\ \hat{\chi}^2 = \frac{n^2\{n_{32}(n_{11}+n_{21})-n_{31}(n_{12}+n_{22})\}^2}{n_{3.}n_{.1}n_{.2}(n_{1.}+n_{2.})(n_{.1}+n_{.2})},$$ (6.6c)

$$(4)\ \hat{\chi}^2 = \frac{n\{n_{33}(n_{11}+n_{12}+n_{21}+n_{22})-(n_{13}+n_{23})(n_{31}+n_{32})\}^2}{n_{3.}n_{.3}(n_{1.}+n_{2.})(n_{.1}+n_{.2})}.$$ (6.6d)

6.2 The Analysis of $r \times c$ Contingency and Homogeneity Tables

We consider Table 140 with the simplified categories (A, B, C versus I, II, III; cf., Table 140a). The following 4 comparisons are possible:

Table 140a

Type	A	B	C	Σ
I	14	22	32	68
II	18	16	8	42
III	8	2	0	10
Σ	40	40	40	120

1. The comparison of I against II with respect to A against B (in symbols I × II ÷ $A \times B$).
2. The comparison of I against II with respect to $(A + B)$ against C (I × II ÷ $\{A + B\} \times C$).
3. The comparison of $\{I + II\}$ against III with respect to A against B ($\{I + II\} \times III \div A \times B$).
4. The comparison of $\{I + II\}$ against III with respect to $(A + B)$ against C ($\{I + II\} \times III \div \{A + B\} \times C$).

See Table 142.

Table 142 $\hat{\chi}^2$-table: decomposition of the $\hat{\chi}^2$-value of a 3 × 3 table (Table 140a) into specific components with one degree of freedom each

Mutually independent components	DF	$\hat{\chi}^2$	P
(1) I × II ÷ A × B	1	1.0637	n.s.
(2) I × II ÷ {A + B} × C	1	9.1673	<0.01
(3) {I + II} × III ÷ A × B	1	5.8909	<0.05
(4) {I + II} × III ÷ {A + B} × C	1	5.4545	<0.05
Total	4	21.5764	<0.001

$$(1) \quad \hat{\chi}^2 = \frac{120\{42(40 \cdot 14 - 40 \cdot 22) - 68(40 \cdot 18 - 40 \cdot 16)\}^2}{68 \cdot 42 \cdot 40 \cdot 40 \cdot (68 + 42)(40 + 40)} = 1.0637,$$

$$(2) \quad \hat{\chi}^2 = \frac{120^2\{8(14 + 22) - 32(18 + 16)\}^2}{68 \cdot 42 \cdot 40 \cdot (68 + 42)(40 + 40)} = 9.1673,$$

$$(3) \quad \hat{\chi}^2 = \frac{120^2\{2(14 + 18) - 8(22 + 16)\}^2}{10 \cdot 40 \cdot 40 \cdot (68 + 42)(40 + 40)} = 5.8909,$$

$$(4) \quad \hat{\chi}^2 = \frac{120\{0(14 + 22 + 18 + 16) - (32 + 8)(8 + 2)\}^2}{10 \cdot 40 \cdot (68 + 42)(40 + 40)} = 5.4545.$$

If other specific comparisons are to be tested, the associated rows or columns (or both) must be interchanged.

> Further remarks on the analysis of contingency tables (cf., also the end of this chapter) can be found especially in the following works: Goodman (1963–1971), Caussinus (1965), Gart (1966), Meng and Chapman (1966), Bhapkar (1968), Bhapkar and Koch (1968), Hamdan (1968), Ku and Kullback (1968), Lancaster (1969), Altham (1970), Odoroff (1970), Goodman and Kruskal (1972), Shaffer (1973), Kastenbaum (1974), Nelder (1974), Bishop et al., (1975), Everitt (1977), Fienberg (1978), Gokhale and Kullback (1978), Upton (1978) and Plackett (1981).

6.2.2 Testing the strength of the relation between two categorically itemized characteristics. The comparison of several contingency tables with respect to the strength of the relation by means of the corrected contingency coefficient of Pawlik

The $\hat{\chi}^2$-value of a contingency table says nothing about the strength of the relation between two characteristics. This is easily seen because for given relative frequencies $\hat{\chi}^2$ is proportional to the total number of observations. If the null hypothesis of independence between the two attributes of a rectangular table is rejected, Pearson's **contingency coefficient**

$$\boxed{CC = \sqrt{\frac{\hat{\chi}^2}{n+\hat{\chi}^2}}} \qquad (6.7)$$

(p. 369) furnishes a measure of the strength of the relation (cf., also Section 4.6.6). This measure of correlation has the value zero when there is total independence. In the case of total dependence of the two qualitative variables, CC does not, however, equal 1 but rather a value less than 1 which varies with the number of cells of the contingency table. Thus different CC-values are comparable as to size only if they were computed on contingency tables of the same values of r and c. This drawback of the CC is compensated for by the fact that for any rectangular table the **largest possible contingency coefficient** CC_{max} is known; thus the observed relative contingency coefficient, CC/CC_{max}, can be given. CC_{max} is defined as that value which CC attains for a table with total dependence among the attributes. For square contingency tables (number of rows = number of columns, i.e., $r = c$), Kendall

6.2 The Analysis of $r \times c$ Contingency and Homogeneity Tables

has shown that the value of CC_{max} depends only on the number of levels, in fact

$$CC_{max} = \sqrt{\frac{r-1}{r}}. \qquad (6.8)$$

The **maximal contingency coefficient of nonsquare contingency tables is**, according to Pawlik (1959), also given by (6.8), where the designation is to be so chosen that $r < c$.

In order to compare CC-values which were computed for contingency tables of various sizes, it is recommended that the CC-value found be expressed as percentage of the corresponding CC_{max}; this **corrected contingency coefficient CC_{corr}** reads

$$CC_{corr} = \frac{CC}{CC_{max}} 100 \quad \text{or} \quad CC_{corr} = \frac{CC}{CC_{max}}. \qquad (6.9)$$

It lies between 0 and 100%, or between 0 and 1, and is independent of the table size. To facilitate computation of CC_{corr} the values of CC_{max} for $r = 2$ to $r = 10$, and based on (6.8), are provided in Table 143 together with each corrective factor $1/CC_{max}$ by which the uncorrected CC-value is to be multiplied.

Table 143

$r = c$	CC_{max}	$\frac{1}{CC_{max}}$
2	0.7071	1.4142
3	0.8165	1.2247
4	0.8660	1.1547
5	0.8944	1.1181
6	0.9129	1.0954
7	0.9258	1.0801
8	0.9354	1.0691
9	0.9428	1.0607
10	0.9487	1.0541

The relation $r \leq c$ can be used to define, according to H. Cramér, a contingency coefficient $K = \sqrt{\hat{\chi}^2/(n[r-1])}$ with $0 \leq K \leq 1$; in a fourfold table $K = \sqrt{\hat{\chi}^2/n} = \sqrt{\hat{\chi}^2/(n-1)}$ (see Section 4.6.1). Examples are given in (p. 349) Table 144.

Table 144

Table No.	Table type	n	$\hat{\chi}^2$	$CC = \sqrt{\frac{\hat{\chi}^2}{n+\hat{\chi}^2}}$	$CC_{corr} = \frac{CC}{CC_{max}}$	$K = \sqrt{\hat{\chi}^2/(n[r-1])}$
139	3 × 4	496	24.932	0.21877	0.26793	0.12944
140	3 × 3	120	20.844	0.38470	0.47114	0.24062

More on measures of association can be found in Mosteller (1968), and Goodman and Kruskal (1972; for the 3 lambda coefficients see Hartwig 1973) as well as in Bishop, Fienberg, and Holland (1975).

The **equivalence of two $r \cdot c$ tables**, each with r degrees of freedom, may be tested by R. B. D'Agostino and B. Rosman (1971, Psychometrika **36**, 251–252).

6.2.3 Testing for trend: The component due to linear regression in the overall variation. The comparison of regression coefficients of corresponding two way tables

Once the dependence between the distribution of the classes of the first attribute and the classes of the second attribute is established by a sufficiently large $\hat{\chi}^2$, the question arises whether **the increase of the frequencies follows a (linear) pattern**; in other words whether the frequencies of the levels of one attribute increase (decrease) linearly with the levels of the other attribute or whether they are related in a more complicated way. The $\hat{\chi}^2$-value can be split into two parts just as in the case of a $k \times 2$ table, one part with a single DF due to the linear trend—the so-called regression line component—and the remaining part due to the difference between the observed frequencies and the estimated linear trend component of the frequencies. It will be computed as the difference between $\hat{\chi}^2$ and $\hat{\chi}^2_{\text{regression}}$.

The different levels of each attribute are assigned scores (x- and y-values) whereby both attributes of an $r \times c$ table are transformed into the simplest possible coordinate system. After this "quantification" of the data the bivariate frequency table will be examined for correlation of two variables. In practice, one proceeds according to Yates (1948) by testing the regression of one of these variables against the other: one determines the regression coefficients b_{yx} ($[b_{xy}]$) and the associated variance $V(b_{yx})[V(b_{xy})]$), and tests the significance of the linear regression by means of

$$\hat{\chi}^2_{\text{lin. regr.}} = \frac{(b_{yx})^2}{V(b_{yx})} = \frac{(b_{xy})^2}{V(b_{xy})} \qquad (6.10)$$

with 1 DF. The regression coefficient of y on x is defined by [see p. 485!]

$$b_{yx} = \sum xy / \sum x^2 \qquad (6.11a)$$

the one of x on y by

$$b_{xy} = \sum xy / \sum y^2. \qquad (6.11b)$$

6.2 The Analysis of $r \times c$ Contingency and Homogeneity Tables

[Note the discussion following (6.12b)]. The variances of the two regression coefficients are under the null hypothesis

$$V(b_{yx}) = \frac{s_y^2}{\sum x^2} = \frac{\sum y^2}{n \sum x^2} \qquad (6.12a)$$

$$V(b_{xy}) = \frac{s_x^2}{\sum y^2} = \frac{\sum x^2}{n \sum y^2}. \qquad (6.12b)$$

In these equations the quantities x and y represent the departure from the mean of the respective variables, s_y^2 is an estimate of the variance of the variable y, and s_x^2 is an estimate of the variance of the variable x. Three frequency distributions, those of the variables x, y, and $x - y$, will be required for the computation of (6.10)–(6.12b): one then obtains $\sum x^2$, $\sum y^2$, and $\sum (x - y)^2$.

EXAMPLE. Consider Table 140. After assigning scores to the categories of both attributes (Table 145), we form the products of the marginal sums and

Table 145

y \ x Score	-1	0	1	$n_{i.}$	$n_{i.}y$	$n_{i.}y^2$
1	14	22	32	68	68	68
0	18	16	8	42	0	0
-1	8	2	0	10	-10	10
$n_{.j}$	40	40	40	120	58	78
$n_{.j}x$	-40	0	40	0		
$n_{.j}x^2$	40	0	40	80		

the associated scores as well as of the marginal sums and the squares of the scores. The sums of these products are (cf., the symbols of Table 138) (p.474)

$$\sum n_{i.}y = 58, \qquad \sum n_{i.}y^2 = 78,$$
$$\sum n_{.j}x = 0, \qquad \sum n_{.j}x^2 = 80.$$

These product sums yield $\sum x^2$ and $\sum y^2$ according to

$$\sum y^2 = \sum n_{i.}y^2 - \frac{(\sum n_{i.}y)^2}{\sum n_{i.}} = 78 - \frac{58^2}{120} = 49.967,$$

$$\sum x^2 = \sum n_{.j}x^2 - \frac{(\sum n_{.j}x)^2}{\sum n_{.j}} = 80 - \frac{0^2}{120} = 80.$$

To calculate $\sum(x - y)^2$, the associated frequency distribution (Table 146) is used. Column 2 of this table lists the "diagonal sums" of Table 145. The "diagonal sums" are taken from lower left to upper right. One thus obtains 14, 18 + 22 = 40, 8 + 16 + 32 = 56, 2 + 8 = 10, and 0.

Table 146

x − y		$n_{\text{diag.}}$	$n_{\text{diag.}}(x - y)$	$n_{\text{diag.}}(x - y)^2$
−1 − (+1)	= −2	14	−28	56
0 − 1 = −1 − 0	= −1	40	−40	40
1 − 1 = 0 − 0 = −1 − (−1)	= 0	56	0	0
1 − 0 = 0 − (−1)	= +1	10	10	10
1 − (−1)	= +2	0	0	0
Total		120	−58	106

Column 1 lists the differences $x - y$ for all the cells of Table 145; each time the "diagonal elements" are combined, because the $(x - y)$-values are constant along each diagonal. For example, one obtains the value zero for the difference $x - y$ for all fields of the main diagonal from lower left to upper right, i.e., for the cells with the cell entries 8, 16, 32:

for cell 8 (lower left), $\quad x = -1, \quad y = -1,$
$$x - y = -1 - (-1) = -1 + 1 = 0;$$
for cell 16 (center of table), $\quad x = 0, \quad y = 0,$
$$x - y = 0 - 0 = 0;$$
for cell 32 (upper right), $\quad x = 1, \quad y = 1,$
$$x - y = 1 - 1 = 0,$$

i.e., $x - y = 0$ holds for $8 + 16 + 32 = 56$, etc. The sums of the products lead to

$$\sum(x - y)^2 = \sum n_{\text{diag.}}(x - y)^2 - \frac{(\sum n_{\text{diag.}}(x - y))^2}{\sum n_{\text{diag.}}}$$

$$= 106 - \frac{(-58)^2}{120}$$

$$= 77.967.$$

Then we obtain by (6.10), (6.11a), (6.12a)

$$\hat{\chi}^2_{\text{lin. regr.}} = \frac{(b_{yx})^2}{V(b_{yx})} = \frac{((80 + 49.967 - 77.967)/2 \cdot 80)^2}{49.967/(120 \cdot 80)} = 20.293,$$

or by (6.10), (6.11b), (6.12b)

$$\hat{\chi}^2_{\text{lin. regr.}} = \frac{(b_{xy})^2}{V(b_{xy})} = \frac{((80 + 49.967 - 77,967)/2 \cdot 49.967)^2}{80/(120 \cdot 49.967)} = 20.293.$$

6.2 The Analysis of $r \times c$ Contingency and Homogeneity Tables

The significance of both regression coefficients ($\chi^2_{1;0.001} = 10.828$) can also be determined in terms of the standard normal distribution:

$$\hat{z} = b/\sqrt{V(b)} \tag{6.13}$$

$$\hat{z} = \frac{b_{yx}}{\sqrt{V(b_{yx})}} = \frac{0.325000}{\sqrt{0.005205}} = 4.505,$$

$$\hat{z} = \frac{b_{xy}}{\sqrt{V(b_{xy})}} = \frac{0.520343}{\sqrt{0.013342}} = 4.505.$$

Naturally the significance level is the same ($z_{0.001} = 3.290$) (Since $z_u^2 = \chi^2_{1,u}$ we have $3.290^2 = 10.828$.) Summarizing the results in Table 147, we notice

Table 147

Source	$\hat{\chi}^2$	DF	Significance level
Linear regression	20.293	1	$P \ll 0.001$
Departure from linear regression	0.551	3	$0.90 < P < 0.95$
Total	20.844	4	$P < 0.001$

that the departure of the frequencies in Table 145 from proportionality is almost fully due to a **linear regression**; the treatment by a double standard dose increases the success (recovery) rate markedly. If this observation sounds trite, one must not overlook the fact that it is (statistically) substantiated only by the results listed in Table 147 (for "P much smaller than 0.001" one writes $P \ll 0.001$).

If the regression lines of two corresponding or matching tables have to be compared, one tests by means of (6.14) whether the regression coefficients differ (Fairfield Smith 1957). The significance of the difference is determined by means of the standard normal distribution.

$$\hat{z} = \frac{|b_1 - b_2|}{\sqrt{V(b_1) + V(b_2)}}. \tag{6.14}$$

EXAMPLE. Assuming that the cell entries listed in Tables 140 and 145 were gathered from a sample of persons of the same race, the same age group, etc., and that we have at our disposal the result of a corresponding trial on people of a different age group:

$$b_1 = 0.325, \qquad b_2 = 0.079,$$
$$V(b_1) = 0.00521, \qquad V(b_2) = 0.00250.$$

Then the null hypothesis of equality of regression coefficients is rejected at the 1% level, with

$$\hat{z} = \frac{0.325 - 0.079}{\sqrt{0.00521 + 0.00250}} = 2.80 \qquad (P = 0.0051).$$

6.2.4 Testing square tables for symmetry

(p.363) The McNemar test gave us the means to test whether a 2 × 2 table is symmetric with respect to its diagonal. An analogous tool to test the **symmetry with respect to the diagonal in an** $r \times r$ **table** is provided by Bowker (1948). This test probes the alternate hypothesis that the pairs of cells located symmetrically with respect to the main diagonal show different entries. The main diagonal is the one which displays the largest frequencies. Under the null hypothesis (symmetry) we expect that

$B_{ij} = B_{ji}$, where
B_{ij} = observed frequency in the cell in the ith row and the jth column,
B_{ji} = observed frequency in the cell in the jth row and ith column.

To resolve the question of whether the null hypothesis can be maintained, one computes

$$\hat{\chi}^2_{\text{sym}} = \sum_{j=1}^{r-1} \sum_{i>j} \frac{(B_{ij} - B_{ji})^2}{B_{ij} + B_{ji}} \tag{6.15}$$

with DF = $r(r-1)/2$. All $r(r-1)/2$ differences of symmetrically located cell entries for which $i > j$ are formed, squared, divided by the sum of the cell entries, and added. If not more than 1/5 of the $r \times r$ cells has expected frequencies $E < 3$, then $\hat{\chi}^2_{\text{sym}}$ is approximately χ^2-distributed and thus can be tested accordingly (cf. also Ireland, Ku, and Kullback 1969, Bennett 1972, Hettmansperger and McKean 1973). Some very interesting extensions are given by Rebecca Zwick et al., Psychological Bulletin **92** (1982), 258–271.

EXAMPLE

Table 148 Since (0 + 2 + 3 + 1) is less than (8 + 4 + 10 + 15), the main diagonal runs from lower left to upper right

0	10	16	15	41
4	2	10	4	20
12	4	3	6	25
8	4	1	1	14
24	20	30	26	100

$$\hat{\chi}^2_{\text{sym}} = \frac{(12-4)^2}{12+4} + \frac{(4-1)^2}{4+1} + \frac{(0-1)^2}{0+1} + \frac{(2-3)^2}{2+3} + \frac{(10-6)^2}{10+6} + \frac{(16-4)^2}{16+4} = 15.2$$

6.2 The Analysis of $r \times c$ Contingency and Homogeneity Tables

Table 148 contains 4 rows and 4 columns; hence there are $4(4-1)/2 = 6$ degrees of freedom at our disposal. The corresponding $\chi^2_{6;0.05}$ equals 12.59; the null hypothesis as to symmetry is thus rejected at the 5% level. In a relatively large group of people the comparison of the perspiration intensity of hands and feet leads to typical symmetry problems in the same way as would a comparison of the visual acuity of the left and right eye or a comparison of the education or hobbies of spouses. Beyond that, almost every square table which is tested for symmetry presents interesting aspects; thus Table 140 exhibits definite asymmetry: (p.477)

$$\hat{\chi}^2_{\text{sym}} = \frac{(18-2)^2}{18+2} + \frac{(14-0)^2}{14+0} + \frac{(22-8)^2}{22+8}$$
$$= 33.333 > 16.266 = \chi^2_{3;0.001}.$$

It is caused by the small number of not recuperating and slowly recuperating patients due to the standard and in particular to the double normal dose.

For generalizations of the Bowker test see (three-way tables) C. B. Read, Psychometrika **43** (1978), 409–420 and (multi-way tables) K.-D. Wall, EDV in Medizin und Biologie **2** (1976), 57–64.

Another test from the class of symmetry tests is the **Q-test due to Cochran** (see end of Section 4.6.3), which is a homogeneity test for s correlated samples (C.S.; (p.365) e.g., methods of treatment or instants of time) of dichotomous data $(+, -)$.
H_A (at least two of the C.S. originated in different populations);
H_0 (all of the C.S. originated in a common population)
will, for large n ($n \cdot s \geq 30$), be rejected at the $100\alpha\%$ level whenever

$$Q = \frac{(s-1)\left[s\sum_{j=1}^{s}T_j^2 - \left(\sum_{j=1}^{s}T_j\right)^2\right]}{s\sum_{i=1}^{n}L_i - \sum_{i=1}^{n}L_i^2} > \chi^2_{s-1;\alpha},$$

where
L_i = sum of the numbers of plus signs of the individual i over all C.S.,
T_j = sum of the numbers of plus signs of the n individuals for the treatment j.

I. \ C.S.	Correlated samples 1 2 . j . s	Σ
individuals 1 2 . i . n		L_i
Σ	T_j	

This test extends the McNemar test to a multivariate distribution of dichotomous variables with H_0: The proportions of success $(+)$ are the same for all treatments [or, there are no treatment effects].

▶ 6.2.5 Application of the minimum discrimination information statistic in testing two way tables for independence or homogeneity

Procedures based on information statistics are quite manageable when the necessary auxiliary tables (Table 85) are available. **Extensive contingency tables** as well as three- or fourfold tables can be analyzed with the help of the **minimum discrimination information statistic** $2I$ ($2I$ is identical to the G-value described in Section 4.6.1); it is based on the measure of information which Kullback and Leibler (1951) introduced to measure the divergence between populations (cf., Gabriel 1966). It is derived and applied to several statistical problems in the book by Kullback (1959). For the two way table (cf., Table 138 and the symbols employed there) it amounts to

$$2\hat{I} = \left(\sum_{i=1}^{r} \sum_{j=1}^{c} 2n_{ij} \ln n_{ij} + 2n \ln n \right) \\ - \left(\sum_{i=1}^{r} 2n_{i.} \ln n_{i.} + \sum_{j=1}^{c} 2n_{.j} \ln n_{.j} \right)$$

(6.16)

or simplified,

$$2\hat{I} = (\text{sum I}) - (\text{sum II}),$$

where the sums are defined as follows.

Sum I: For every n_{ij}-value, i.e., for every cell entry (for each cell of a $k \times 2$ or $r \times c$ table) the associated value is read off from Table 85. The values taken from the table are summed and then the value associated with the overall sample size is added.

Sum II: For every field of the marginal sums (row and column sums) the corresponding tabulated values are determined. These values are summed.

The difference between the two sums yields the value $2\hat{I}$; the caret over the I indicates that one is dealing with a value "estimated" in terms of the observed occupation numbers. Under the null hypothesis of independence or homogeneity, $2\hat{I}$ is asymptotically distributed as χ^2 with $(r-1)(c-1)$ degrees of freedom. For two way tables that are not too weakly occupied ($k \times 2$ or $r \times c$), the approximation of the χ^2-statistic by the minimum discrimination information statistic is excellent. If one or more cells of a table remain unoccupied, one should then apply the correction proposed by Ku (1963): For each zero, a 1 is to be subtracted from the computed minimum discrimination information statistic $2\hat{I}$. For the computation of $2\hat{I}$ there are $(r+1)(c+1)$ individual tabulated values read off; this fact can provide a certain check in the case of large tables.

6.2 The Analysis of $r \times c$ Contingency and Homogeneity Tables

EXAMPLE. We use the cell entries of Table 140 (Section 6.2.1), and obtain the result shown in Table 149. This value is somewhat larger than the

Table 149

14	22	32	68
18	16	8	42
8	2	0	10
40	40	40	120

$\left.\begin{array}{r}73.894\\136.006\\221.807\end{array}\right\}$ 1st row $\qquad$ $\left.\begin{array}{r}573.853\\313.964\\46.052\end{array}\right\}$ row sums

$\left.\begin{array}{r}104.053\\88.723\\33.271\end{array}\right\}$ 2nd row $\qquad$ $\left.\begin{array}{r}295.110\\295.110\\295.110\end{array}\right\}$ column sums

$\left.\begin{array}{r}33.271\\2.773\\0.000\end{array}\right\}$ 3rd row

1,819.199 = Sum II
check (verification): we have read
$(3 + 1)(3 + 1) = 16$ tabulated values

1,148.998 $n = 120$

1,842.796 = Sum I

$\left.\begin{array}{r}1,842.796\\1,819.199\end{array}\right] -$

$\left.\begin{array}{r}23.597\\1.000\end{array}\right] -$ (one zero taken into account)

$2\hat{I} = 22.597$

corresponding $\hat{\chi}^2$-value (21.576), which however in no way influences the decision, since $\chi^2_{4;0.001} = 18.467$ is obviously exceeded by both.

Other problems which lend themselves to equally elegant solutions with the help of the minimum discrimination information statistic are the testing of two distributions of frequency data for homogeneity (cf., Sections 6.1.2 and 4.3.1) and the testing of an empirical distribution for uniform distribution (cf., Section 4.3.2). To compare two frequency distributions we apply the homogeneity test for a $k \times 2$ table. For the example in Section 6.1.2 we get $2\hat{I} = 5.7635$ as against $\hat{\chi}^2 = 5.734$. For tables of this size $2\hat{I}$ is almost always somewhat larger than $\hat{\chi}^2$.

Testing for nonuniform distribution

EXAMPLE. Time is read off a watchmaker's 1000 watches. Time class 1 includes all watches which indicate between 1 : 00 and 1 : 59; the limits for the other k classes are chosen analogously. The frequency distribution is given in Table 150 with $k = 12$ and $n = 1,000$.

Table 150

Time class	1	2	3	4	5	6	7	8	9	10	11	12	n =
Frequency	81	95	86	98	90	73	70	77	82	84	87	77	1,000

The null hypothesis (uniform distribution) is tested at the 5% level:

$$2\hat{I} = \sum_{i=1}^{k} 2f_i \ln f_i - 2n \ln n + 2n \ln k \qquad DF = k-1 \qquad (6.16a)$$

$2\hat{I} = [2 \cdot 81 \ln 81 + \cdots] - 2 \cdot 1{,}000 \ln 1{,}000 + 2(1{,}000 \ln 12).$

REMARK: The last summand $2(1{,}000 \ln 12)$, is not tabulated but must be computed. We require this value to be correct to one place beyond the decimal point, and accordingly round off the other values of $2n \ln n$ read off the table. If no table of natural logarithms is available, ln 12 is determined by the conversion to base 10 logarithms:

$$\ln a = 2.302585 \log a;$$

thus ln 12 = (2.30258)(1.07918) = 2.484898 ≃ 2.48490, whence the last summand becomes (2)(1,000)(2.48490) = 4,969.80, and

$$2\hat{I} = [711.9 + \cdots + 668.9] - 13{,}815.5 + 4{,}969.8 = 9.4,$$
$$2\hat{I} = 9.4 < 19.68 = \chi^2_{11;0.05}.$$

There is thus no reason to reject the null hypothesis of a uniform distribution.

The particular importance of the minimum discrimination information statistic of a three way or multiway table rests on the fact, demonstrated by Kullback (1959) (cf., also Kullback et al., 1962; Ku, Varner, and Kullback 1968, 1971), that it can be relatively easily decomposed into additive components (i.e., components with specific degrees of freedom) which can be individually tested and added, yielding $2\hat{I}$ or partial sums of $2\hat{I}$. These components refer to partial independence, conditional independence, and interaction (cf., also however the methods proposed by Bishop 1969, Grizzle et al., 1969, Goodman 1969, 1970, 1971, Shaffer 1973, Nelder 1974, and Fienberg 1978).

Even for a simple $3 \times 3 \times 3$ table—a **contingency die**—there are already a total of 16 hypotheses to be tested. Analyses of this sort are referred to as analyses of information—they can be regarded as distribution-free analyses of variance. For specifics consult Bishop, Fienberg, and Holland (1975), and Gokhale and Kullback (1978).

Some remarks

1. The analysis of incomplete two and three way contingency tables is discussed by Enke (1977, 1978).

2. Benedetti and Brown (1978) examine and assess methods of model building for multiway contingency table analyses with respect to the final choice of model and with respect to intermediate information available to the data analyst.

3. For testing the equality of two independent χ_2^2 variables see D'Agostino and Rosman (1971, cited on page 484).

4. For graphical analysis and the identification of sources of significance in two-way contingency tables see M. B. Brown, Applied Statistics **23** (1974), 405–413 and R. D. Snee, The American Statistician **28** (1974), 9–12 [cf., also Communications in Statistics—Theory and Methods A **6** (1977), 1437–1451 and A **9** (1980), 1025–1041].

5. Exact tests for trends in ordered contingency tables are given in W. M. Patefield, Applied Statistics **31** (1982), 32–43; a survey of strategies for modeling cross classifications having ordinal variables is given by A. Agresti, Journal of the American Statistical Association **78** (1983), 184–198.

7 ANALYSIS OF VARIANCE TECHNIQUES

▶ 7.1 PRELIMINARY DISCUSSION AND SURVEY

In Chapter 2 we mentioned, under the heading of Response Surface Experimentation, an experimental strategy for quality improvement in the widest sense. An essential part of this special **theory of optimal design** is based on regression analysis and on the so-called analysis of variance, introduced by R. A. Fisher for the planning and evaluation of experiments, in particular of field trials, which allows the detection of factors contributing to or controlling the variation found. The comparison of means plays a particular role. Since analysis of variance, like the t-test, presupposes **normal distribution** and **equality of variances**, we wish to familiarize ourselves first with the procedures which are used for testing the equality or the homogeneity of a number of population variances. If they are equal, then the corresponding means may be compared by analysis of variance. This is the simplest form of variance. If the influence of each of **several** independent factors, at different levels, has to be sorted out properly, it is necessary that the observed values be obtained from special **designs** (cf., Section 7.7).

> The analysis of variance is a tool for the quantitative evaluation of the influence of the independent variables (factors: cf. Section 7.4.1, Model I) on the dependent variable: **The total variation** displayed by a set of observations, as measured by the sums of squares of deviations from the mean, **may** in certain circumstances **be separated into components associated with defined sources of variation** used as criteria of classification for the observations. Such an analysis is called an analysis of variance, although in the strict sense it is an **analysis of sums of squares**. Many standard situations can be reduced to the variance analysis form.

One can gather information on the required sample sizes from the literature cited at the end of Section 7.4.3 (Remark 3). The rapid tests of the analysis of variance are presented in Section 7.5. Ott (1967) gives a simple graphic method. Graphical analyses are often sufficient, for example multicomparative plotting of means which demonstrates trends, curvilinearities, and configurations of interactions (see Enrick 1976).

Independent sample groups with **not necessarily equal variances** (cf., Section 3.6.2) but nearly the same distribution type can be compared by means of the H-test (Section 3.9.5). For correlated groups of samples of nearly the same distribution type, the Friedman test with its associated multiple comparisons is indicated.

The assumption of equal variances may be dropped: an exact analysis of variance with unequal variances is presented by Bishop and Dudewicz (1978); the 10%, 5%, and 1% critical points of the null distribution and an example are given. Multiple comparison procedures of means with unequal variances are compared by A. C. Tamhane, Journal of the American Statistical Association **74** (1979), 471–480. For a survey on robust multiple comparisons see Games et al. (1983) and C. W. Dunnett, Communications in Statistics—Theory and methods **11** (1982), 2611–2629, for more tables see R. R. Wilcox, Technometrics **25** (1983), 201–204.

7.2 TESTING THE EQUALITY OF SEVERAL VARIANCES

In the sequel independent random samples from normally distributed populations will be assumed.

7.2.1 Testing the equality of several variances of equally large groups of samples

A relatively simple test for the rejection of the null hypothesis as to equality or homogeneity of the variances $\sigma_1^2 = \sigma_2^2 = \ldots = \sigma_i^2 = \ldots = \sigma_k^2 = \sigma^2$ has been proposed by Hartley. Under the condition of **equal group sizes** (n_0), this hypothesis can be tested by

$$\hat{F}_{max} = \frac{\text{greatest sample variance}}{\text{smallest sample variance}} = \frac{s_{max}^2}{s_{min}^2}. \qquad (7.1)$$

The distribution of the test statistic F_{max} can be found in Table 151. The parameters of this distribution are the number k of groups and the number of degrees of freedom $v = n_0 - 1$ for every group variance. If, for a given significance level α, $\hat{F}_{max}$ exceeds the tabulated value, then the equality or

Table 151 Critical values for Hartley's test for testing several variances for homogeneity at the 5% and 1% level of significance (from Pearson, E. S. and H. O. Hartley, Biometrika Tables for Statisticians, Vol. 1 (3rd ed.), Cambridge, 1966, Table 31). Values given are for the test statistic $F_{max} = s^2_{max}/s^2_{min}$, where s^2_{max} is the largest and s^2_{min} the smallest in a set of k independent values of s^2, each based on v degrees of freedom.

$\alpha = 0.05$

v \ k	2	3	4	5	6	7	8	9	10	11	12
2	39.0	87.5	142	202	266	333	403	475	550	626	704
3	15.4	27.8	39.2	50.7	62.0	72.9	83.5	93.9	104	114	124
4	9.60	15.5	20.6	25.2	29.5	33.6	37.5	41.1	44.6	48.0	51.4
5	7.15	10.8	13.7	16.3	18.7	20.8	22.9	24.7	26.5	28.2	29.9
6	5.82	8.38	10.4	12.1	13.7	15.0	16.3	17.5	18.6	19.7	20.7
7	4.99	6.94	8.44	9.70	10.8	11.8	12.7	13.5	14.3	15.1	15.8
8	4.43	6.00	7.18	8.12	9.03	9.78	10.5	11.1	11.7	12.2	12.7
9	4.03	5.34	6.31	7.11	7.80	8.41	8.95	9.45	9.91	10.3	10.7
10	3.72	4.85	5.67	6.34	6.92	7.42	7.87	8.28	8.66	9.01	9.34
12	3.28	4.16	4.79	5.30	5.72	6.09	6.42	6.72	7.00	7.25	7.48
15	2.86	3.54	4.01	4.37	4.68	4.95	5.19	5.40	5.59	5.77	5.93
20	2.46	2.95	3.29	3.54	3.76	3.94	4.10	4.24	4.37	4.49	4.59
30	2.07	2.40	2.61	2.78	2.91	3.02	3.12	3.21	3.29	3.36	3.39
60	1.67	1.85	1.96	2.04	2.11	2.17	2.22	2.26	2.30	2.33	2.36
∞	1.00	1.00	1.00	1.00	1.00	1.00	1.00	1.00	1.00	1.00	1.00

$\alpha = 0.01$

v \ k	2	3	4	5	6	7	8	9	10	11	12
2	199	448	729	1036	1362	1705	2063	2432	2813	3204	3605
3	47.5	85	120	151	184	21(6)	24(9)	28(1)	31(0)	33(7)	36(1)
4	23.2	37	49	59	69	79	89	97	106	113	120
5	14.9	22	28	33	38	42	46	50	54	57	60
6	11.1	15.5	19.1	22	25	27	30	32	34	36	37
7	8.89	12.1	14.5	16.5	18.4	20	22	23	24	26	27
8	7.50	9.9	11.7	13.2	14.5	15.8	16.9	17.9	18.9	19.8	21
9	6.54	8.5	9.9	11.1	12.1	13.1	13.9	14.7	15.3	16.0	16.6
10	5.85	7.4	8.6	9.6	10.4	11.1	11.8	12.4	12.9	13.4	13.9
12	4.91	6.1	6.9	7.6	8.2	8.7	9.1	9.5	9.9	10.2	10.6
15	4.07	4.9	5.5	6.0	6.4	6.7	7.1	7.3	7.5	7.8	8.0
20	3.32	3.8	4.3	4.6	4.9	5.1	5.3	5.5	5.6	5.8	5.9
30	2.63	3.0	3.3	3.4	3.6	3.7	3.8	3.9	4.0	4.1	4.2
60	1.96	2.2	2.3	2.4	2.4	2.5	2.5	2.6	2.6	2.7	2.7
∞	1.00	1.0	1.0	1.0	1.0	1.0	1.0	1.0	1.0	1.0	1.0

The numbers in brackets ($\alpha = 0.01$ for $v = 3$, $7 \leq k \leq 12$) are unreliable, e.g., F_{max} for $v = 3$, $k = 7$ is about 216. Bounds for F_{max} for $\alpha = 0.10$ and $\alpha = 0.25$ can be found in R. J. Beckman and G. L. Tietjen (1973, Biometrika 60, 213–214).

homogeneity hypothesis is rejected and the alternative hypothesis $\sigma_i^2 \neq \sigma^2$ for fixed i is accepted (Hartley 1950).

EXAMPLE. Test the homogeneity of three sample groups of size $n_0 = 8$ with $s_1^2 = 6.21$, $s_2^2 = 1.12$, $s_3^2 = 4.34$ ($\alpha = 0.05$).

We have $\hat{F}_{max} = 6.21/1.12 = 5.54 < 6.94 = F_{max}$ (for $k = 3$, $v = n_0 - 1 = 8 - 1 = 7$ and $\alpha = 0.05$). On the basis of the samples in question it is not possible, at the 5% level, to reject the null hypothesis of homogeneity of the variances.

A rapid test based on the quotient of the largest and the smallest ranges was introduced by Leslie and Brown (1966). The upper critical limits for 4 significance levels can be found in the original work.

7.2.2 Testing the equality of several variances according to Cochran

If the variance s_{max}^2 of one group is substantially larger than that of the others, this test (Cochran 1941) is preferred. The test statistic is

$$\hat{G}_{max} = \frac{s_{max}^2}{s_1^2 + s_2^2 + \ldots + s_k^2}. \quad (7.2)$$

The assessment of $\hat{G}_{max}$ follows on the basis of Table 152: if $\hat{G}_{max}$ is greater than the tabulated value for k, the chosen significance level, and $v = n_0 - 1$

Table 152 Critical values for Cochran's test for testing several variances for homogeneity at the 5% and 1% level of significance (from Eisenhart, C., Hastay, M. W., and Wallis, W. A.: Techniques of Statistical Analysis, McGraw-Hill, New York 1947): values given are for the test statistic $G_{max} = s_{max}^2 / \sum s_i^2$, where each of the k independent values of s^2 has v degrees of freedom

$\alpha = 0.05$

k \ v	1	2	3	4	5	6	7	8	9	10	16	36	144	∞
2	0.9985	0.9750	0.9392	0.9057	0.8772	0.8534	0.8332	0.8159	0.8010	0.7880	0.7341	0.6602	0.5813	0.5000
3	0.9669	0.8709	0.7977	0.7457	0.7071	0.6771	0.6530	0.6333	0.6167	0.6025	0.5466	0.4748	0.4031	0.3333
4	0.9065	0.7679	0.6841	0.6287	0.5895	0.5598	0.5365	0.5175	0.5017	0.4884	0.4366	0.3720	0.3093	0.2500
5	0.8412	0.6838	0.5981	0.5441	0.5065	0.4783	0.4564	0.4387	0.4241	0.4118	0.3645	0.3066	0.2513	0.2000
6	0.7808	0.6161	0.5321	0.4803	0.4447	0.4184	0.3980	0.3817	0.3682	0.3568	0.3135	0.2612	0.2119	0.1667
7	0.7271	0.5612	0.4800	0.4307	0.3974	0.3726	0.3535	0.3384	0.3259	0.3154	0.2756	0.2278	0.1833	0.1429
8	0.6798	0.5157	0.4377	0.3910	0.3595	0.3362	0.3185	0.3043	0.2926	0.2829	0.2462	0.2022	0.1616	0.1250
9	0.6385	0.4775	0.4027	0.3584	0.3286	0.3067	0.2901	0.2768	0.2659	0.2568	0.2226	0.1820	0.1446	0.1111
10	0.6020	0.4450	0.3733	0.3311	0.3029	0.2823	0.2666	0.2541	0.2439	0.2353	0.2032	0.1655	0.1308	0.1000
12	0.5410	0.3924	0.3264	0.2880	0.2624	0.2439	0.2299	0.2187	0.2098	0.2020	0.1737	0.1403	0.1100	0.0833
15	0.4709	0.3346	0.2758	0.2419	0.2195	0.2034	0.1911	0.1815	0.1736	0.1671	0.1429	0.1144	0.0889	0.0667
20	0.3894	0.2705	0.2205	0.1921	0.1735	0.1602	0.1501	0.1422	0.1357	0.1303	0.1108	0.0879	0.0675	0.0500
24	0.3434	0.2354	0.1907	0.1656	0.1493	0.1374	0.1286	0.1216	0.1160	0.1113	0.0942	0.0743	0.0567	0.0417
30	0.2929	0.1980	0.1593	0.1377	0.1237	0.1137	0.1061	0.1002	0.0958	0.0921	0.0771	0.0604	0.0457	0.0333
40	0.2370	0.1576	0.1259	0.1082	0.0968	0.0887	0.0827	0.0780	0.0745	0.0713	0.0595	0.0462	0.0347	0.0250
60	0.1737	0.1131	0.0895	0.0765	0.0682	0.0623	0.0583	0.0552	0.0520	0.0497	0.0411	0.0316	0.0234	0.0167
120	0.0998	0.0632	0.0495	0.0419	0.0371	0.0337	0.0312	0.0292	0.0279	0.0266	0.0218	0.0165	0.0120	0.0083
∞	0	0	0	0	0	0	0	0	0	0	0	0	0	0

$\alpha = 0.01$

k \ v	1	2	3	4	5	6	7	8	9	10	16	36	144	∞
2	0.9999	0.9950	0.9794	0.9586	0.9373	0.9172	0.8988	0.8823	0.8674	0.8539	0.7949	0.7067	0.6062	0.5000
3	0.9933	0.9423	0.8831	0.8335	0.7933	0.7606	0.7335	0.7107	0.6912	0.6743	0.6059	0.5153	0.4230	0.3333
4	0.9676	0.8643	0.7814	0.7212	0.6761	0.6410	0.6129	0.5897	0.5702	0.5536	0.4884	0.4057	0.3251	0.2500
5	0.9279	0.7885	0.6957	0.6329	0.5875	0.5531	0.5259	0.5037	0.4854	0.4697	0.4094	0.3351	0.2644	0.2000
6	0.8828	0.7218	0.6258	0.5635	0.5195	0.4866	0.4608	0.4401	0.4229	0.4084	0.3529	0.2858	0.2229	0.1667
7	0.8376	0.6644	0.5685	0.5080	0.4659	0.4347	0.4105	0.3911	0.3751	0.3616	0.3105	0.2494	0.1929	0.1429
8	0.7945	0.6152	0.5209	0.4627	0.4226	0.3932	0.3704	0.3522	0.3373	0.3248	0.2779	0.2214	0.1700	0.1250
9	0.7544	0.5727	0.4810	0.4251	0.3870	0.3592	0.3378	0.3207	0.3067	0.2950	0.2514	0.1992	0.1521	0.1111
10	0.7175	0.5358	0.4469	0.3934	0.3572	0.3308	0.3106	0.2945	0.2813	0.2704	0.2297	0.1811	0.1376	0.1000
12	0.6528	0.4751	0.3919	0.3428	0.3099	0.2861	0.2680	0.2535	0.2419	0.2320	0.1961	0.1535	0.1157	0.0833
15	0.5747	0.4069	0.3317	0.2882	0.2593	0.2386	0.2228	0.2104	0.2002	0.1918	0.1612	0.1251	0.0934	0.0667
20	0.4799	0.3297	0.2654	0.2288	0.2048	0.1877	0.1748	0.1646	0.1567	0.1501	0.1248	0.0960	0.0709	0.0500
24	0.4247	0.2871	0.2295	0.1970	0.1759	0.1608	0.1495	0.1406	0.1338	0.1283	0.1060	0.0810	0.0595	0.0417
30	0.3632	0.2412	0.1913	0.1635	0.1454	0.1327	0.1232	0.1157	0.1100	0.1054	0.0867	0.0658	0.0480	0.0333
40	0.2940	0.1915	0.1508	0.1281	0.1135	0.1033	0.0957	0.0898	0.0853	0.0816	0.0668	0.0503	0.0363	0.0250
60	0.2151	0.1371	0.1069	0.0902	0.0796	0.0722	0.0668	0.0625	0.0594	0.0567	0.0461	0.0344	0.0245	0.0167
120	0.1225	0.0759	0.0585	0.0489	0.0429	0.0387	0.0357	0.0334	0.0316	0.0302	0.0242	0.0178	0.0125	0.0083
∞	0	0	0	0	0	0	0	0	0	0	0	0	0	0

where n_0 denotes the size of the individual groups, then the null hypothesis of equality of variances must be rejected and the alternative hypothesis $\sigma^2_{\max} \neq \sigma^2$ accepted.

When the sample sizes are not too different [cf., the remark on sample sizes on page 504], **one computes their harmonic mean** $\bar{x}_H$ and interpolates in Table 152 for $v = \bar{x}_H - 1$.

EXAMPLE. Suppose we are given the following 5 variances: $s_1^2 = 26$, $s_2^2 = 51$, $s_3^2 = 40$, $s_4^2 = 24$, and $s_5^2 = 28$, where every variance is based on 9 degrees of freedom. They are to be tested at the 5% level. We have $\hat{G}_{\max} = 51/(26 + 51 + 40 + 24 + 28) = 0.302$. The tabulated value for $\alpha = 0.05, k = 5, v = 9$ is 0.4241. Since $0.302 < 0.4241$, the equality of the variances under consideration cannot be rejected at the 5% level.

A very similar test, which is however based on the ranges of the individual samples, is described by Bliss, Cochran and Tukey (1956); examples and the upper 5% bounds can be found in the original paper.

The tests of Hartley and Cochran lead to the same decisions in most cases. Since the Cochran test utilizes more information, it is somewhat more **sensitive**. Additional suggestions (cf., Section 7.2.3) are contained in the following outline:

Population	Test
slightly skew distributed	Cochran
normally distributed $N(\mu,\sigma)$	$k < 10$: Hartley, Cochran
	$k \geq 10$: Bartlett
less peaked than $N(\mu,\sigma)$	Levene
more peaked than $N(\mu,\sigma)$	$k < 10$: Cochran
	$k \geq 10$: Levene

▶ 7.2.3 Testing the equality of the variances of several samples of the same or different sizes according to Bartlett

The null hypothesis, homogeneity of the variances, can be tested according to Bartlett (1937) when **the data come from normally distributed populations**. The Bartlett test is a combination of a sensitive test of normality, more

7.2 Testing the Equality of Several Variances

precisely the "long-tailedness" of a distribution, with a less sensitive test of equality of the variances:

$$\hat{\chi}^2 = \frac{1}{c}\left[2.3026(v \log s^2 - \sum_{i=1}^{k} v_i \log s_i^2)\right],$$

where

$$c = \frac{\sum_{i=1}^{k}\frac{1}{v_i} - \frac{1}{v}}{3(k-1)} + 1$$

(7.3)

$$s^2 = \frac{\sum_{i=1}^{k} v_i s_i^2}{v} \quad \text{and} \quad DF = k - 1$$

$v = n - k$ = total number of degrees of freedom = $\sum_{i=1}^{k} v_i$,
n = overall sample size,
k = number of groups: (each group must include at least 5 observations),
s^2 = estimate of the common variance = $[\sum (n_i - 1)s_i^2]/[n - k]$,
v_i = number of degrees of freedom in the ith sample = $n_i - 1$,
s_i^2 = estimate of the variance of the ith sample.

The denominator c is always somewhat larger than 1, i.e., c need be computed only if the value in the brackets is expected to give a statistically significant $\hat{\chi}^2$

Given k groups of samples of equal size n_0, where $n_0 \geq 5$, the following simplifications apply:

$$\hat{\chi}^2 = \frac{1}{c}\left[2.3026k(n_0 - 1)\{\log s^2 - \frac{1}{k}\sum_{i=1}^{k} \log s_i^2\}\right]$$

where

$$c = \frac{k+1}{3k(n_0 - 1)} + 1$$

(7.4)

$$s^2 = \frac{1}{k}\sum_{i=1}^{k} s_i^2 \quad (DF = k - 1).$$

If the test statistic $\hat{\chi}^2$ exceeds $\chi^2_{k-1;\alpha}$, then the null hypothesis $\sigma_1^2 = \sigma_2^2 = \cdots = \sigma_i^2 = \cdots = \sigma_k^2 = \sigma^2$ is rejected (alternative hypothesis $\sigma_i^2 \neq \sigma^2$ for some i) at the $100\alpha\%$ significance level.

Harsaae (1969) gives exact critical limits which supplement Table 32 of the Biometrika Tables (Pearson and Hartley 1966 [2], pp. 204, 205). Exact critical values for $k = 3(1)10$; $v_i = 4(1)11, 14, 19, 24, 29, 49, 99$; $\alpha = 0.10$, 0.05, 0.01 are given by Glasser (1976). For unequal sample sizes see M. T. Chao and R. E. Glasser, Journal of the American Statistical Association **73** (1978), 422–426.

EXAMPLE. Given: Three groups of samples of sizes $n_1 = 9$, $n_2 = 6$, and $n_3 = 5$ with the variances specified in Table 153. Test the equality of the variances ($\alpha = 0.05$).

Table 153

No.	s_i^2	$n_i - 1$ v_i	$v_i s_i^2$	$\log s_i^2$	$v_i \log s_i^2$
1	8.00	8	64.00	0.9031	7.2248
2	4.67	5	23.35	0.6693	3.3465
3	4.00	4	16.00	0.6021	2.4084
		17	103.35		12.9797

$$s^2 = \frac{103.35}{17} = 6.079, \qquad \log s^2 = 0.7838,$$

$$\hat{\chi}^2 = \frac{1}{c}[2.3026(17 \cdot 0.7838 - 12.9797)] = \frac{1}{c} \cdot 0.794.$$

Since $\chi^2_{2;0.05} = 5.99$ is substantially larger than 0.794, the null hypothesis is not rejected at the 5% level. With

$$c = \frac{\left[\frac{1}{8} + \frac{1}{5} + \frac{1}{4}\right] - \frac{1}{17}}{3(3-1)} + 1 = 1.086$$

we have $\hat{\chi}^2 = 0.794/1.086 = 0.731 < 5.99$.

> If the number of variances to be tested for equality is large, one can employ a modification of the Bartlett test (cf., Barnett 1962) proposed by Hartley. Since the Bartlett test is **very sensitive to deviations from the normal distribution** (Box 1953, Box and Anderson 1955), apply in case of doubt the procedure suggested by Levene (cf., Section 3.5.1; also Section 3.9.1 and Meyer-Bahlburg 1970) or still better procedures (cf., Games 1972). Several variances can be simultaneously compared by an elegant method due to David (1956; Tietjen and Beckman 1972 give additional tabulated values). See also page 495.

7.3 ONE WAY ANALYSIS OF VARIANCE

▶ 7.3.1 Comparison of several means by analysis of variance

The comparison of the means of two normally distributed populations (Section 3.6) can be broadened into the comparison of an arbitrary number of means. Given are k samples of sizes n_i, $i = 1, \ldots, k$, and combined sample size n, i.e.,

$$\sum_{i=1}^{k} n_i = n.$$

Each sample originates in a normally distributed population.

<small>Since t-test and analysis of variance are relatively robust against skewness but not against too many observations lying outside of the $\bar{x} \pm 2s$ limits (normal distribution: 4.45% of the observations lie outside of $\mu \pm 2\sigma$), the **Nemenyi test** (Section 7.5.2) (p. 546) should be applied when the tails are too heavy.</small>

The k independently normally distributed populations have identical but unknown variances. The sample values x_{ij} have two indices: x_{ij} is the jth value in the ith sample ($1 \le i \le k$; $1 \le j \le n_i$).

The sample means $\bar{x}_{i.}$

$$\bar{x}_{i.} = \frac{1}{n_i} \sum_{j=1}^{n_i} x_{ij}$$

The dot indicates the index over which summation was carried out; thus, e.g., $x_{..} = \sum_{i=1}^{k} \sum_{j=1}^{n_i} x_{ij}$ is the sum of all x-values, the total of all response values. (7.5)

The overall mean $\bar{x}$:

$$\bar{x} = \frac{1}{n} \sum_{i=1}^{k} \sum_{j=1}^{n_i} n_i x_{ij} = \frac{1}{n} \sum_{i=1}^{k} n_i \bar{x}_{i.} \qquad (7.6)$$

or, in simplified notation:

$$\bar{x} = \frac{1}{n} \sum_{i,j} x_{ij} = \frac{1}{n} \sum_i n_i \bar{x}_{i.} \qquad (7.7)$$

An essential aspect of the one way analysis of variance is the **decomposition of the sum of squares of the deviations of the observed values from the overall mean**, SS_{total}, **into two components**:

1. The sum of squares of the deviations of the observed values from the corresponding sample (group) means, called "**within** sample sum of squares" (SS_{within}) or error sum of squares, and

2. The sum of squares of the deviations of the sample (group) means from the overall mean, weighted by the number of elements in the respective samples (groups) (n_i). This sum is called the "**between** samples (groups) sum of squares" ($SS_{between}$).

$$SS_{total} = SS_{within} + SS_{between}.$$

REMARK. The deviation of any observation x_{ij} from the overall mean $\bar{x}$ may be split up into two parts: $x_{ij} - \bar{x} = (x_{ij} - \bar{x}_{i.}) + (\bar{x}_{i.} - \bar{x})$ with the group means $\bar{x}_{i.}$.

Simplified notation: $\bar{x}$ for $\bar{x}_{..}$ and $\bar{x}_i$ for $\bar{x}_{i.}$.

Partition of the total sum of squares:

$$\begin{aligned} \sum_{i,j}(x_{ij} - \bar{x})^2 &= \sum_{i,j}(x_{ij} - \bar{x}_i)^2 + \sum_{i,j}(\bar{x}_i - \bar{x})^2, \\ \sum_{i,j}(x_{ij} - \bar{x})^2 &= \sum_{i,j}(x_{ij} - \bar{x}_i)^2 + \sum_i n_i(\bar{x}_i - \bar{x})^2, \end{aligned} \quad (7.8)$$

with the corresponding degrees of freedom

$$\begin{aligned} n - 1 &= \sum_{i=1}^{k}(n_i - 1) + k - 1 \\ &= (n - k) + (k - 1). \end{aligned} \quad (7.9)$$

The sums of squares divided by the respective degrees of freedom $SS_{total}/(n - 1), \ldots$, i.e., the estimates of the variances are in analysis of variance called the **mean sum of squares** (MS). If all the groups originate in the same population, then the variances, that is, the mean squares

$$s_{between}^2 = MS_{between} = \frac{1}{k - 1} \sum_i n_i(\bar{x}_{i.} - \bar{x})^2, \quad (7.10)$$

and

$$s_{within}^2 = MS_{within} = \frac{1}{n - k} \sum_{i,j}(x_{ij} - \bar{x}_{i.})^2, \quad (7.11)$$

should be of about the same size. If this is not so (i.e., if the quotient of $MS_{between}/MS_{within}$ is larger than the critical value of the F-distribution determined from $v_1 = k - 1$, $v_2 = n - k$, and α), then certain of the groups have different means μ_i. The null hypothesis that the population means of the k treatment groups, classes, categories are all equal, or $\mu_1 = \mu_2 = \cdots = \mu_i = \cdots = \mu_k = \mu$, is then rejected on the basis of the test statistic (7.12) [i.e., (7.13) or (7.14)] if $\hat{F} > F_{(k-1; n-k; \alpha)}$. In this case at least two μ_i's are different, i.e., the alternative hypothesis that $\mu_i \neq \mu_j$ for some (i, j) is accepted. If $MS_{between} < MS_{within}$, the null hypothesis may not be rejected; then (7.6) and (7.11) are estimates of μ and σ^2 with $n - k$ degrees of freedom.

7.3 One Way Analysis of Variance

$MS_{between}$ is also known as the mean square between treatments or categories, or as the "sampling error," and $MS_{within} = s^2_{within} = s^2_{error} =$ Mean square error $= MSE$ is also known as the within group variance, within group mean square, or "experimental error."

By definition

$$\hat{F} = \frac{MS_{between}}{MS_{within}} = \frac{\frac{1}{k-1}\sum_i n_i(\bar{x}_{i.} - \bar{x})^2}{\frac{1}{n-k}\sum_{i,j}(x_{ij} - \bar{x}_{i.})^2} = \frac{\frac{1}{k-1}\sum_i n_i(\bar{x}_{i.} - \bar{x})^2}{\frac{1}{n-k}\sum_i s_i^2(n_i - 1)}.$$

(7.12)

$\hat{F}$ is computed according to

$$\hat{F} = \frac{\frac{1}{k-1}\left[\sum_i \frac{x_{i.}^2}{n_i} - \frac{x_{..}^2}{n}\right]}{\frac{1}{n-k}\left[\sum_{i,j} x_{ij}^2 - \sum_i \frac{x_{i.}^2}{n_i}\right]}.$$

(7.13)

For sample groups of equal size ($n_i = n_0$) the following is preferred:

$$\hat{F} = \frac{\left[k\sum_i x_{i.}^2 - x_{..}^2\right]/(k-1)}{\left[n_0\sum_{i,j} x_{ij}^2 - \sum_i x_{i.}^2\right]/(n_0 - 1)}.$$

(7.14)

For normally distributed observations it is remarkable that mean and variance are **independently** distributed. In our k samples s^2_{within} is independent of $\bar{x}_i$. and consequently independent of $s^2_{between}$. If we assume that the null hypothesis is true, then both the sample variances $s^2_{between}$ and s^2_{within}, that is, the numerator and denominator in (7.12) to (7.14), are independent, usually not too different, unbiased estimates of σ. This holds true for the denominator, even if the population means are not equal. For departures from the null hypothesis the numerator will tend to be greater than the denominator, giving values of F greater than unity. Thus a **one sided test** is adequate. Upper tail values of the F distribution for just this situation are given in Tables 30a to 30f. The **ASSUMPTIONS** for this test are:

1. **Independence** of observations within and between all random samples.
2. Observations from **normally distributed** populations **with equal** population **variances**.

A close look at and a comment on (7.8): We may write $x_{ij} - \bar{x} = (x_{ij} - \bar{x}_i) + (\bar{x}_i - \bar{x})$. Squaring and summing this over i and j gives

$$\sum_{i=1}^{k}\sum_{j=1}^{n_i}(x_{ij} - \bar{x})^2 = \sum_{i=1}^{k}\sum_{j=1}^{n_i}[(x_{ij} - \bar{x}_i) + (\bar{x}_i - \bar{x})]^2.$$

The cross product term $2ab$ vanishes because $\sum (x_{ij} - \bar{x}) = 0$:

$$\sum_{i=1}^{k} \sum_{j=1}^{n_i} (x_{ij} - \bar{x})^2 = \sum_{i=1}^{k} \sum_{j=1}^{n_i} (x_{ij} - \bar{x}_i)^2 + \sum_{i=1}^{k} \sum_{j=1}^{n_i} (\bar{x}_i - \bar{x})^2.$$

The latter term may be written

$$\sum_{i=1}^{k} \sum_{j=1}^{n_i} (\bar{x}_i - \bar{x})^2 = \sum_{i=1}^{k} n_i (\bar{x}_i - \bar{x})^2,$$

since the contribution $(\bar{x}_i - \bar{x})^2$ is the same for all n_i observations in the ith group. Using this, and writing $\bar{x}_i$ instead of $\bar{x}_{i.}$, we have (7.8):

$$\sum_{i=1}^{k} \sum_{j=1}^{n_i} (x_{ij} - \bar{x})^2 = \sum_{i=1}^{k} \sum_{j=1}^{n_i} (x_{ij} - \bar{x}_i)^2 + \sum_{i=1}^{k} n_i (\bar{x}_i - \bar{x})^2.$$

The sum of squares about $\bar{x}$ is partitioned into "within samples" variation (first term) and "between samples" variation (second term). The second term will tend to be large if the means μ_i are not identical, since then the sample means will tend to be more widely dispersed about $\bar{x}$ than if all population means were alike.

The choice of samples of equal size offers several advantages: (1) Deviations from the hypothesis of equality of variances do not carry as much weight, and tests for the equality of the variances are easier. (2) The Type 2 error which occurs with the F-test becomes minimal. (3) Other comparisons of means (see Sections 7.3.2, 7.4.2) are simpler to carry out.

Method of computation

The test statistic (7.13) is computed according to

$$\hat{F} = \frac{MS_{\text{between}}}{MS_{\text{within}}} = \frac{\dfrac{1}{k-1}[SS_{\text{between}}]}{\dfrac{1}{n-k}[SS_{\text{within}}]} = \frac{\dfrac{1}{k-1}[B - K]}{\dfrac{1}{n-k}[A - B]}$$

with a total of n observations from k sample groups or from k samples, and

$$A = \sum (\text{observations})^2 = \sum_{i,j} x_{ij}^2,$$

$$B = \sum \frac{(\text{sample sum})^2}{\text{sample size}} = \sum_i \frac{x_{i.}^2}{n_i},$$

where the sample sum is $x_{i.} = \sum_j x_{ij}$,

$$K = \frac{(\text{sum of all observations})^2}{\text{number of all observations}} = \frac{\left(\sum_i x_{i.}\right)^2}{n} = \frac{x_{..}^2}{n}.$$

(7.15)

7.3 One Way Analysis of Variance

To verify the result, SS_{total} is computed indirectly:

$$SS_{\text{total}} = [SS_{\text{between}}] + [SS_{\text{within}}] = [B - K] + [A - B], \qquad (7.16)$$

and directly:

$$SS_{\text{total}} = \sum_{ij} x_{ij}^2 - \left(\sum_{ij} x_{ij}\right)^2 \Big/ n = A - K. \qquad (7.17)$$

Particularly simple examples
 1. **Samples of unequal sizes** n_i (Table 154): By (7.13) and (7.15),

Table 154

j \ i	1	2	3		
1	3	4	8		
2	7	2	4		
3		7	6		
4		3			
$x_{i\cdot}$	10	16	18	$x_{\cdot\cdot}$	= 44
n_i	2	4	3	n	= 9
$\bar{x}_i$	5	4	6		

$$\hat{F} = \frac{\dfrac{1}{3-1}\left[\left(\dfrac{10^2}{2} + \dfrac{16^2}{4} + \dfrac{18^2}{3}\right) - \dfrac{44^2}{9}\right]}{\dfrac{1}{9-3}\left[(3^2 + 7^2 + 4^2 + 2^2 + 7^2 + 3^2 + 8^2 + 4^2 + 6^2) - \left(\dfrac{10^2}{2} + \dfrac{16^2}{4} + \dfrac{18^2}{3}\right)\right]}$$

$$\hat{F} = \frac{\dfrac{1}{2}[6{,}89]}{\dfrac{1}{6}[30]} = 0.689.$$

Checks of (7.16), (7.17):

$$[6.89] + [30] = 36.89,$$

$$(3^2 + 7^2 + 4^2 + 2^2 + 7^2 + 3^2 + 8^2 + 4^2 + 6^2) - 44^2/9 = 36.89.$$

Since $\hat{F} = 0.689 < 5.14 = F_{(2;6;0.05)}$, the null hypothesis that all three means originated in the same population cannot be rejected at the 5% level; the common mean is (7.6),

$$\bar{x} = [(2)(5) + (4)(4) + (3)(6)]/9 = 4.89,$$

and the variance is (7.11),

$$s^2_{within} = s^2_{error} = MSE = 30/6 = 5.$$

2. **Samples of equal size** $(n_i = \text{const.} = n_0)$ per group (Table 155): By (7.13),

Table 155

j \ i	Sample group 1	2	3		
1	6	5	7		
2	7	6	8		
3	6	4	5		
4	5	5	8		
$x_{i.}$	24	20	28	$x_{..}$	= 72
$n_i = n_0$	4	4	4	n	= 12
$\bar{x}_i$	6	5	7	$\bar{\bar{x}}$	= 6

$$\hat{F} = \frac{\frac{1}{3-1}\left[\frac{1}{4}(24^2 + 20^2 + 28^2) - \frac{72^2}{12}\right]}{\frac{1}{12-3}\left[(6^2 + 7^2 + \cdots + 8^2) - \frac{1}{4}(24^2 + 20^2 + 28^2)\right]} = \frac{\frac{1}{2}[8]}{\frac{1}{9}[10]} = 3.60.$$

Check: $[8] + [10] = 18; (6^2 + 7^2 + \cdots + 5^2 + 8^2) - 72^2/12 = 18.$
By (7.14),

$$\hat{F} = \frac{[3(24^2 + 20^2 + 28^2) - 72^2]/(3-1)}{[4(6^2 + 7^2 + \cdots + 8^2) - (24^2 + 20^2 + 28^2)]/(4-1)} = \frac{96/2}{40/3} = 3.60.$$

Since $\hat{F} = 3.60 < 4.26 = F_{(2;9;0.05)}$, the null hypothesis, equality of the three means ($\bar{\bar{x}} = 6$, $s^2_{within} = s^2_{error} = MSE = 10/9 = 1.11$), cannot be rejected at the 5% level.

Critical bounds for the test $H_A: \mu_1 \leq \mu_2 \leq \mu_3$ ($H_0: \mu_1 = \mu_2 = \mu_3$) with $n_i = $ const. (2 through 240) and $\alpha = 0.005$ through 0.10 are given by Nelson (1976).

7.3 One Way Analysis of Variance

Remarks

1. **Estimating the standard deviation from the range.** If it is assumed that a sample of size n originated in an approximately normally distributed population, then the standard deviation can be estimated from the R:

$$\hat{s} = R(1/d_n). \tag{7.18}$$

The factor $1/d_n$ can, for given n, be read off from Table 156. Usually it is expedient to split up the sample by means of a random process into k groups of 8 (or at least 6 to 10) individual values, and for each group determine the corresponding R and compute the mean range $\bar{R}$:

$$\bar{R} = \frac{1}{k}\sum R_i. \tag{7.19}$$

Using this in

$$\hat{s} = \bar{R}(1/d_n) \tag{7.20}$$

determines the standard deviation ("within the sample") based on the number of effective degrees of freedom, v, given on the right side of Table 156. For $n \geq 5$ and $k > 1$, $v < k(n-1)$ is always true. $\hat{s}^2$ and SS_{within} should be of the same order of magnitude (cf., Table 155 with $\bar{R} = (2 + 2 + 3)/3 = 2.33$, $\hat{s} = (2.33)(0.486) = 1.13$; $\hat{s}^2 = 1.28$ as against $SS_{within} = 10/9 = 1.11$).

Table 156 Factors for estimating the standard deviation of the population from the range of the sample (taken from Patnaik, P. B.: The use of mean range as an estimator of variance in statistical tests, Biometrika **37**, 78–87 (1950))

Size of sample or group n	Factor $1/d_n$	Effective number of degrees of freedom v for k groups of size n				
		k=1	k=2	k=3	k=4	k=5
2	0.8862	1				
3	0.5908	2				
4	0.4857	3				
5	0.4299	4	7	11	15	18
6	0.3946	5	9	14	18	23
7	0.3698	5	11	16	21	27
8	0.3512	6	12	18	24	30
9	0.3367	7	14	21	27	34
10	0.3249	8	15	23	30	38
11	0.3152	9				
12	0.3069	10				
13	0.2998	11				

This table has been extended by Nelson (1975) ($n = 2\text{--}15, k = 1\text{--}15, 20, 30, 50$) and illustrated by additional examples (cf., also the Leslie–Brown test mentioned in Section 7.2.1).

2. A simplified analysis of variance can be carried out with the help of Table 156. We give no example, but refer the reader to the **test of Link and Wallace**, presented in Section 7.5.1, which is also based on the range but which is much more economical thanks to Table 177 (cf., also the graphic procedure of Ott 1967).

3. **The confidence interval of the range** can be estimated using Table 157. Suppose a number of samples of size $n = 6$ are drawn from a population which is at least approximately normally distributed. The mean range $\bar{R}$ equals 3.4 units. A useful

Table 157 Factors for estimating a confidence interval of the range: The product of a standard deviation estimated from the range according to Table 156 and the factors given for the same or some arbitrarily chosen sample size and degree of significance furnishes the upper and lower limits and thus the confidence interval for the range from samples of the size chosen. Column 6 lists a factor v_n for estimating the standard deviation of the mean range. More on this can be found in the text. (Reprinted from Pearson 1941/42 p. 308, Table 2, right part. The values corrected by Harter et al. 1959 have been taken into account.)

n	1% bounds		5% bounds		Factor
	lower	upper	lower	upper	v_n
2	0.018	3.643	0.089	2.772	0.853
3	0.191	4.120	0.431	3.314	0.888
4	0.434	4.403	0.760	3.633	0.880
5	0.665	4.603	1.030	3.858	0.864
6	0.870	4.757	1.253	4.030	0.848
7	1.048	4.882	1.440	4.170	0.833
8	1.205	4.987	1.600	4.286	0.820
9	1.343	5.078	1.740	4.387	0.808
10	1.467	5.157	1.863	4.474	0.797
11	1.578	5.227	1.973	4.552	0.787
12	1.679	5.290	2.071	4.622	0.778

estimate of the standard deviation by (7.20) is then $(3.4)(0.3946) = 1.34$. If the size of future samples is scheduled to be fixed at $n = 4$, we get from Table 157, for the 90% confidence interval, the factors 0.760 and 3.633 and hence the bounds $(1.34)(0.760) = 1.02$ and $(1.34)(3.633) = 4.87$. Assuming we have a normally distributed population with $\sigma = 1.34$, this interval (for future random samples of size $n = 4$) is the exact 90% confidence interval of the range.

7.3 One Way Analysis of Variance

The estimate of the **standard deviation of the mean range**, $s_{\bar{R}}$, is given by

$$\boxed{s_{\bar{R}} = \frac{v_n \cdot (1/d_n)^2 \cdot \bar{R}}{\sqrt{k}}} \qquad (7.21)$$

where

v_n = factor from Table 157,
$1/d_n$ = factor from Table 156,
$\bar{R}$ = mean range,
k = number of samples of size n from which ranges were computed.

For example, for $k = 5$, $n = 6$, $\bar{R} = 7$, $1/d_n = 0.3946$, and $v_n = 0.848$, we get

$$s_{\bar{R}} = \frac{(0.848)(0.3946)^2(7)}{\sqrt{5}} = 0.413.$$

A remark on the factors $1/d_n$ and v_n: For samples of size n from a normally distributed population with standard deviation σ, d_n is the mean and v_n the standard deviation of the standardized range $w = R/\sigma$.

▶ 7.3.2 Assessment of linear contrasts according to Scheffé, and related topics

If the one way analysis of variance leads to a significant result, an effort is then made to determine which of the parameters, $\mu_1, \mu_2, \ldots, \mu_i, \ldots, \mu_k$, or better yet, which two groups of parameters, A and B, with the means μ_A and μ_B, differ from each other. If we have, e.g., estimates of the five parameters $\mu_1, \mu_2, \mu_3, \mu_4, \mu_5$, then we can, among other things, **compare the following means**:

V_1: $\mu_1 = \mu_2 = \mu_A$ with $\mu_3 = \mu_4 = \mu_5 = \mu_B$,

$\mu_A = \frac{1}{2}(\mu_1 + \mu_2)$ with $\mu_B = \frac{1}{3}(\mu_3 + \mu_4 + \mu_5)$,

V_2: $\mu_1 = \mu_A$ with $\mu_2 = \mu_3 = \mu_4 = \mu_5 = \mu_B$,

$\mu_A = \mu_1$ with $\mu_B = \frac{1}{4}(\mu_2 + \mu_3 + \mu_4 + \mu_5)$.

Comparisons of this sort (population contrasts), in the form

V_1: $\boxed{\frac{1}{2}(\mu_1 + \mu_2) - \frac{1}{3}(\mu_3 + \mu_4 + \mu_5)}$

V_2: $\boxed{\mu_1 - \frac{1}{4}(\mu_2 + \mu_3 + \mu_4 + \mu_5)}$

are called **linear contrasts**. They are linear functions of the k means μ_i (7.22) that are determined by the k known constants c_i for which the condition (7.23) holds:

$$\sum_{i=1}^{k} c_i \mu_i \qquad \sum_{i=1}^{k} c_i = 0. \qquad (7.22, 7.23)$$

These constants are as follows:

$$V_1: \quad c_1 = c_2 = \frac{1}{2}; \quad c_3 = c_4 = c_5 = -\frac{1}{3}; \quad \frac{1}{2} + \frac{1}{2} - \frac{1}{3} - \frac{1}{3} - \frac{1}{3} = 0$$

$$V_2: \quad c_1 = 1; \quad c_2 = c_3 = c_4 = c_5 = -\frac{1}{4}; \quad 1 - \frac{1}{4} - \frac{1}{4} - \frac{1}{4} - \frac{1}{4} = 0.$$

If

$$\hat{S} = \frac{|\bar{x}_A - \bar{x}_B|}{s_{\bar{x}_A - \bar{x}_B}} > \sqrt{(k-1)F_{(k-1;n-k;\alpha)}} = S_\alpha \qquad (7.24)$$

with

$$s_{\bar{x}_A - \bar{x}_B} = \sqrt{s_{\text{error}}^2 \sum_{i=1}^{k} \frac{c_i^2}{n_i}},$$

$s_{\text{error}}^2 = s_{\text{within}}^2 = MS_{\text{within}} = MS_{\text{error}} = MSE$, then the parameters underlying the contrasts differ (Scheffé 1953). If we want to compare two means, say μ_3 and μ_5, after the data are collected, and if e.g., $k = 6$, then one sets $c_1 = c_2 = c_4 = c_6 = 0$ and rejects $H_0: \mu_3 = \mu_5$ as soon as

$$\hat{S} = \frac{|\bar{x}_3 - \bar{x}_5|}{\sqrt{s_{\text{error}}^2 \left(\frac{1}{n_3} + \frac{1}{n_5}\right)}} > \sqrt{(k-1)F_{(k-1;n-k;\alpha)}} = S_\alpha. \qquad (7.25)$$

In the case of groups of markedly unequal size, one forms **weighted** linear contrasts, so that, e.g., for V_1 we have

$$\frac{n_1 \mu_1 + n_2 \mu_2}{n_1 + n_2} - \frac{n_3 \mu_3 + n_4 \mu_4 + n_5 \mu_5}{n_3 + n_4 + n_5}$$

estimated by

$$\frac{n_1 \bar{x}_1 + n_2 \bar{x}_2}{n_1 + n_2} - \frac{n_3 \bar{x}_3 + n_4 \bar{x}_4 + n_5 \bar{x}_5}{n_3 + n_4 + n_5}.$$

7.3 One Way Analysis of Variance

EXAMPLE

Table 158

No. (i)	$\bar{x}_i$	s_i^2	n_i I	n_i II
1	10	10	10	15
2	9	8	10	5
3	14	12	10	15
4	13	11	10	10
5	14	7	10	5

$\sum n_I = \sum n_{II} = 50$

Means computed according to (1.47):
$\bar{x}_I = 12.0$,
$\bar{x}_{II} = 12.1$.

Consider the data in Table 158. By (7.15) we have, for the case of equal (I) and unequal (II) sample sizes,

$$\hat{F}_I = \frac{10[(10-12)^2+(9-12)^2+(14-12)^2+(13-32)^2+(14-12)^2]/(5-1)}{9 \cdot 48/(50-5)},$$

$$\hat{F}_I = \frac{55}{9.6} = 5.73,$$

$$\hat{F}_{II} = \frac{(15(10-12.1)^2 + 5(9-12.1)^2 + 15(14-12.1)^2 + 10(13-12.1)^2 + 5(14-12.1)^2)/(5-1)}{(10 \cdot 14 + 8 \cdot 4 + 12 \cdot 14 + 11 \cdot 9 + 7 \cdot 4)/(50-5)},$$

$$\hat{F}_{II} = \frac{48.75}{10.38} = 4.69.$$

Since 5.73 and 4.69 > 3.77 = $F_{(4;45;0.01)}$, we test $\mu_1 = \mu_2 < \mu_3 = \mu_4 = \mu_5$ by (7.24, 7.24a) and form:

for I,

$$|\bar{x}_A - \bar{x}_B| = \frac{1}{2}(\bar{x}_1 + \bar{x}_2) - \frac{1}{3}(\bar{x}_3 + \bar{x}_4 + \bar{x}_5)$$

$$= \frac{1}{2}(10+9) - \frac{1}{3}(14+13+14) = 4.17,$$

$$\sqrt{s_{\text{error}}^2 \sum_{i=1}^{5} c_i^2 \left(\frac{1}{n_i}\right)} = \sqrt{9.6\left[\frac{1}{2^2}\left(\frac{1}{10}+\frac{1}{10}\right) + \frac{1}{3^2}\left(\frac{1}{10}+\frac{1}{10}+\frac{1}{10}\right)\right]}$$

$$= \sqrt{0.8} = 0.894;$$

for II,

$$|\bar{x}_A - \bar{x}_B| = \frac{n_1\bar{x}_1 + n_2\bar{x}_2}{n_1 + n_2} - \frac{n_3\bar{x}_3 + n_4\bar{x}_4 + n_5\bar{x}_5}{n_3 + n_4 + n_5},$$

$$|\bar{x}_A - \bar{x}_B| = \frac{15 \cdot 10 + 5 \cdot 9}{15 + 5} - \frac{15 \cdot 14 + 10 \cdot 13 + 5 \cdot 14}{15 + 10 + 5} = 3.92,$$

and

$$\sqrt{s^2_{error} \sum_{i=1}^{5} c_i^2 \left(\frac{1}{n_i}\right)}$$

$$= \sqrt{10.38 \left(\left\{ \left(\frac{3}{4}\right)^2 \cdot \frac{1}{15} + \left(\frac{1}{4}\right)^2 \cdot \frac{1}{5} \right\} + \left\{ \left(\frac{3}{6}\right)^2 \cdot \frac{1}{15} + \left(\frac{2}{6}\right)^2 \cdot \frac{1}{10} + \left(\frac{1}{6}\right)^2 \cdot \frac{1}{5} \right\} \right)}$$

$$= 0.930$$

$$\left[\text{since } \frac{3}{4} = n_1/(n_1 + n_2) = 15/(15 + 5) \right], \text{ and get}$$

for I	for II
$\frac{4.17}{0.894} = 4.66$	$\frac{3.92}{0.930} = 4.21$

with $F_{(4;45;0.01)} = 3.77$ and $\sqrt{(5-1)3.77} = 3.88$. These are significant differences in both cases:

$$\text{I: } \hat{S}_I = 4.66 > 3.88 = S_{0.01}$$
$$\text{II: } \hat{S}_{II} = 4.21 > 3.88 = S_{0.01}.$$

Remark on the comparison of a large number of means

(p.533) The formula (7.49) in Section 7.4.2 is, for certain problems, more practical than (7.24), (7.24a) ($v_{s^2_{error}} = n - k$). Williams (1970) showed that the effort expended in a one-way analysis with not too small k can be reduced by computing (a) for the smallest n (n_{min}) the greatest nonsignificant difference between $D_{I, below}$ and (b) for the largest n (n_{max}) the least significant difference $D_{I, above}$; D_I in (7.49) then needs to be determined only for the differences lying between $D_{I, below}$ and $D_{I, above}$. One computes $D_{I, below} = \sqrt{W/n_{min}}$ and $D_{I, above} = \sqrt{W/n_{max}}$, where $W = 2s^2_{error}(k - 1) F_{(k-1; n-k; \alpha)}$.

Remark: Forming homogeneous groups of means using the modified LSD test. Whenever the F-test permits the rejection of H_0 ($\mu_1 = \mu_2 = \cdots = \mu_k$) one orders the k means from sample groups of equal size ($n_i = \text{const}; n = \sum n_i = k \cdot n_i$) by decreasing magnitude ($\bar{x}_{(1)} \geq \bar{x}_{(2)} \geq \bar{x}_{(3)} \geq \cdots$) and tests whether

7.3 One Way Analysis of Variance

adjacent means differ by a Δ (delta) which is larger than the least significant difference (LSD)

$$LSD = t_{n-k;\alpha}\sqrt{\frac{2}{n_i}s^2_{error}} = \sqrt{\frac{2}{n_i}s^2_{error}F_{(1;n-k;\alpha)}}. \qquad (7.26)$$

For unequal sample sizes ($n_i \neq$ const., $n = \sum_i n_i$) we have

$$LSD_{(a,b)} = t_{n-k;\alpha}\sqrt{s^2_{error}\left(\frac{n_a+n_b}{n_a n_b}\right)} = \sqrt{s^2_{error}\left(\frac{n_a+n_b}{n_a n_b}\right)F_{(1;n-k;\alpha)}}$$

(7.27)

For $\Delta \leq LSD$ or $\Delta_{(a,b)} \leq LSD_{(a,b)}$, H_0 (equality of adjacent means) cannot be rejected; we mark such means with a common underline.

EXAMPLE

$\bar{x}_i$	Δ
$\bar{x}_{(1)} = 26.8$	0.5
$\bar{x}_{(2)} = 26.3$	1.1
$\bar{x}_{(3)} = 25.2$	5.4
$\bar{x}_{(4)} = 19.8$	5.5
$\bar{x}_{(5)} = 14.3$	2.5
$\bar{x}_{(6)} = 11.8$	

$n_i = 8, k = 6, s^2_{error} = 10.38, v = 48 - 6 = 42$,
$t_{42;0.05} = 2.018, F_{(1;42;0.05)} = 4.07$,

$$LSD = 2.018\sqrt{\frac{2}{8} \cdot 10.38} = 3.25,$$

or

$$LSD = \sqrt{\frac{2}{8} \cdot 10.38 \cdot 4.07} = 3.25.$$

At the 5% level, three regions are apparent: $\underline{\bar{x}_{(1)}\bar{x}_{(2)}\bar{x}_{(3)}}\;\underline{\bar{x}_{(4)}}\;\underline{\bar{x}_{(5)}\bar{x}_{(6)}}$.
[Application of (7.27): $n_1 = 7; n_2 = 9$; other values unchanged;

$$\frac{7+9}{(7)(9)} = 0.254;$$

$LSD_{(1,2)} = 2.018\sqrt{10.38 \cdot 0.254} = 3.28$ or $\sqrt{10.38 \cdot 0.254 \cdot 4.07} = 3.28$;
$\Delta_{(1,2)} = 0.5 < 3.28 = LSD_{(1,2)}$, i.e., $H_0: \mu_1 = \mu_2$ cannot be rejected at the 5% level.]

In the case of equal sample sizes (n_i) one can, following Tukey (1949), do a **further study on groups of 3 or more means each**. Thus one finds for each group the group mean $\bar{x}$ and the largest deviation $d = |\bar{x}_i - \bar{x}|$ within the group, and then tests whether $d\sqrt{n_i/s^2_{\text{error}}}$ exceeds the value in Table 26 (see below). If this is the case, $\bar{x}_i$ is isolated, and a new group mean is formed and tested further (with means split off, if necessary) until every group includes no more than 3 means.

The table cited above is found on pp. 185–186 of the Biometrika Tables (Pearson and Hartley 1966) (n = number of means in the group, v = number of degrees of freedom belonging to s^2_{error}). If this table is unavailable, one can compute for groups of:

---

3 means	> 3 means
$\hat{z} = \dfrac{|d/s_{\text{error}} - 0.5|}{3(0.25 + 1/v)}$	$\hat{z} = \dfrac{|d/s_{\text{error}} - 1.2 \cdot \log n'|}{3(0.25 + 1/v)}$

v = number of degrees of freedom belonging to s^2_{error}.
n' = number of means in the group.

For $\hat{z} < 1.96 = z_{0.05}$ the group can be taken as homogeneous at the 5% level. Other bounds of the standard normal distribution can be read from Tables 14 (Section 1.3.4) and 43 (Section 2.1.6) as needed. For $\hat{z} > z_\alpha$, $\bar{x}_i$ is to be isolated and a new group mean formed, for which d and $\hat{z}$ are again computed.

(p. 62)
(p. 217)

Simultaneous confidence intervals to contain all k population means

A set of exact two sided 95% simultaneous confidence intervals (95% SCI) to contain all of the μ_i ($i = 1, 2, \ldots, k$) is obtained as

$$\bar{x}_i \pm |t_{k; r; \rho = 0.0; 0.05}| \sqrt{s^2_{\text{error}}/n_i} \qquad (95\% \text{ SCI})$$

provided that $\bar{x}_i$ are independent sample means based upon n_i independent observations from k normal populations, $s^2_{\text{error}} = MSE = MS_{\text{within}} = s^2_{\text{within}}$. If the correlations between the sample means are not all zero, the formula (95% SCI) still applies, but is conservative. Hahn and Hendrickson (1971, [8:7a] p. 325, Table 1) give percentage points $|t|_{k; v; \rho = 0.0; \alpha}$ of **the maximum absolute value $|t|$ of the k-variate Student distribution with v degrees of freedom for** $\alpha = 0.01$, 0.05, 0.10, **for many values of $v \leq 60$, and for** $k = 1(1)6(2)12$, 15, and 20. Some values of this **STANDARDIZED**

7.3 One Way Analysis of Variance

MAXIMUM MODULUS DISTRIBUTION for $\alpha = 0.05$ needed to compute several 95% SCIs are given below

v \ k	4	5	6	8
10	2.984	3.103	3.199	3.351
15	2.805	2.910	2.994	3.126
20	2.722	2.819	2.898	3.020
30	2.641	2.732	2.805	2.918
40	2.603	2.690	2.760	2.869
60	2.564	2.649	2.716	2.821

For our example we have $\bar{x}_{(1)} = 26.8, \ldots, \bar{x}_{(6)} = 11.8$; $|t|_{6;42;\rho=0.0;0.05} = 2.760 - 0.004 = 2.756$; $2.756\sqrt{10.38/8} = 3.14$ or $\bar{x}_i \pm 3.1$ and 95% SCI: $23.7 \leq \mu_{(1)} \leq 29.9; \ldots; 8.7 \leq \mu_{(6)} \leq 14.9$.

Hahn and Hendrickson (1971) mention eight further applications of the four tables ($\rho = 0.0; 0.2; 0.4; 0.5$); e.g.: (1) multiple comparisons between k treatment means, and a control mean, and the corresponding SCI for $\mu_i - \mu_{contol}$, and (2) prediction intervals to contain all k further means when the estimate of σ^2 is pooled from several samples.

▶ 7.3.3 Transformations

7.3.3.1 Measured values

Skewed distributions, samples with heterogeneous variances, and frequency data must undergo a transformation aimed at getting **normally distributed values with homogeneous variances** before an analysis of variance is carried out. As an example we compare the ranges of the 4 samples in Table 159, where $9.00 - 5.00 = 4.00$; $\sqrt{9} - \sqrt{5} = 3 - 2.236 = 0.764$, $\log 9 - \log 5 = 0.954 - 0.699 = 0.255$; $\frac{1}{5} - \frac{1}{9} = 0.2 - 0.111 = 0.089$, and the other values are correspondingly determined. The range heterogeneity of the original data is reduced somewhat by the **root transformation**, and even more by the **logarithmic transformation**. The reciprocal transformation is

Table 159

Sample		Range of the samples			
No.	Extreme values	Original data	Square roots	Logarithms (base 10)	Reciprocals
1	5.00 and 9.00	4.00	0.764	0.255	0.089
2	0.20 and 0.30	0.10	0.100	0.176	1.667
3	1.10 and 1.30	0.20	0.091	0.072	0.140
4	4.00 and 12.00	8.00	1.464	0.477	0.168

too powerful, as it enlarges tiny ranges too much. The ranges of the logarithms exhibit no larger heterogeneity than one is to expect on the basis of a random process. If it is assumed further that the standard deviation is proportional to the range, then the logarithmic transformation seems appropriate. Occupying a position midway between the logarithmic transformation and the reciprocal transformation is the transformation based on the **reciprocal of the square root** ($1/\sqrt{x}$). Applying it to our four samples, we obtain $1/\sqrt{5} - 1/\sqrt{9} = 0.114$ and correspondingly 0.410, 0.076, 0.211, a still better homogeneity of the variation spread. The difference from the values of the logarithmic transformation is, however, small, so that in the present case this transformation is preferred for its manageability among other things. Variables with unimodal skewed distributions are frequently mapped by the transformation $x' = \log(x \pm a)$ into variables with (approximately) normal distribution (cf., also Knese and Thews 1960); the constant a (called F in Section 1.3.9) can be rapidly approximated as shown by Lehmann (1970). Other important types of transformations are $x' = (x + a)^c$ with $a = \frac{1}{2}$ or $a = 1$ and $x' = a + bx^c$ with $-3 < c < 6$.

7.3.3.2 Counted values

If the observations consist of **counts**, for example of the number of germs per unit volume of milk, the possible values are 0, 1, 2, 3, etc. In such a case, useful homogeneity is frequently obtained when, in place of 0, 1, 2, 3, ..., the transformed values:

$$\sqrt{\frac{3}{8}},\ \sqrt{1+\frac{3}{8}},\ \sqrt{2+\frac{3}{8}},\ \sqrt{3+\frac{3}{8}}, \ldots,$$

i.e.,

0.61, 1.17, 1.54, 1.84,...,

are used. The same shift by $\frac{3}{8}$ is advantageous when frequencies are subjected to a logarithmic transformation: $\log(x + \frac{3}{8})$ rather than $\log x$. One thereby avoids the logarithm of zero, which is undefined. For the square root transformation of frequencies (Poisson distribution) due to Freeman and Tukey (1950) (the function of $g = \sqrt{x} + \sqrt{x+1}$ maps the interval $0 \le x \le 50$ onto $1.00 \le g \le 14.21$), a suitable table, which also includes the squares of the transformed values, is provided by Mosteller and Youtz (1961). That article contains, moreover, a comprehensive table of the angular transformation (cf., Section 3.6.1) for binomially distributed relative frequencies ($n_i \simeq$ constant and not too small). The angular transformation is not needed if all values lie between 30% and 70%, since then ($\pi \simeq 0.5$) the binomial distribution is a sufficiently good approximation to a normal distribution.

The angular transformation also serves to normalize right-steep distributions, which however are also subjected to the power transformation, $x' = x^a$, where $a = 1.5$ for moderate and $a = 2$ for pronounced right-steepness. Tables are provided by Healy and Taylor (1962).

7.3 One Way Analysis of Variance

Transformation of data: percentages, frequencies and measured observations in order to achieve normality and equality of variances. The type of relation between the parameters (e.g., σ and μ) is decisive.

Data		Suitable transformation
Percentages 0–100%	$\sigma^2 = k\mu(1 - \mu)$	**Angular* transformation:** $$x' = \arcsin \sqrt{x/n} \text{ or } \arcsin \sqrt{\frac{x + 3/8}{n + 3/4}}$$ For percentages between 30% and 70% one can do without the transformation (see text).
Frequencies and measured observations	$\sigma^2 = k\mu$	**Square root transformation:** $$x' = \sqrt{x} \text{ or } \sqrt{x + 3/8}$$ 1. In particular for absolute frequencies of relatively rare events. 2. With small absolute frequencies, including zero: $x' = \sqrt{x + 0.4}$.
Measured observations (frequencies)	$\sigma = k\mu$	**Logarithmic transformation:** $x' = \log x$ 1. Also $x' = \log(x \pm a)$; cf., Section 1.3.9. 2. With measured observations between 0 and 1: $x' = \log(x + 1)$.
	$\sigma = k\mu^2$	**Reciprocal transformation:** $x' = 1/x$ In particular for many time-dependent variables.

* Modifications are discussed by Chanter (1975).
If the choice of an adequate transformation causes difficulties, one can explore visually by means of a diagram (according to appearance) whether in various subgroups of the data set there exist certain relations between the variances or standard deviations and the means, and then choose the logically and formally adequate transformation. If σ is proportional to A, then the transformation B should be used in order to stabilize the variance:

A	const.	$\sqrt{\mu}$	μ	$\sqrt{\mu^3}$	μ^2
B	no transf.	$\sqrt{x}$	$\log x$	$1/\sqrt{x}$	$1/x$

If on plotting the data, e.g., s_i versus $\bar{x}_i$, a scatter diagram or point cloud suggests a linear regression, then the logarithmic transformation $x' = \log x$ or $x' = \log(x \pm a)$ should be used.

Supplementary remarks concerning transformations (cf., also Bartlett 1947, Anscombe 1948, Rives 1960, Box and Cox 1964, David 1981, Chap. 8 [8:1b], and Hoaglin et al. 1983 [8:1]) are contained in Remark 2 in Section 7.4.3.

7.3.3.3 Ranks

The **normal rank transformation** enables us to apply the analysis of variance to rank data. The n ranks are mapped onto the expected values of the corresponding rank order statistic of a sample of size n from a standard normal distribution. They are listed in the tables by Fisher and Yates (1963), Table XX. Other tables are given by Teichroew (1956) and Harter (1961).

> The analysis of variance estimation and test procedures are then applied to the transformed values. Significance statements with respect to the transformed variables hold also for the original data. The means and variances obtained by the inverse transformation are however not always unbiased. More on this can be found in Neyman and Scott (1960).

For the common rank transformation as a bridge between parametric and nonparametric statistics see Conover and Iman (1981, cited on p. 286).

7.4 TWO WAY AND THREE WAY ANALYSIS OF VARIANCE

7.4.1 Analysis of variance for $2ab$ observations

If a classification of the data must be made according to more than one point of view, the use of double or, more generally, multiple indices is very expedient. Here, the first index indicates the row, the second the column, the third the stratum (block, subgroup, or depth). Thus x_{251} denotes the observed value in the second row, fifth column, and first stratum of a three dimensional frequency distribution. In the general formulation, x_{ijk} denotes an observation lying in the ith row, jth column, and kth stratum (cf., Fig. 59).

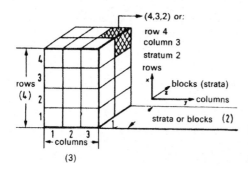

Figure 59 Geometric model of the three way classification: the numbers for a three way analysis of variance are arranged in rows, columns and strata.

The scheme of the three way classification with a groups of the A-classification, $i = 1, 2, \ldots, a$; b groups of the B-classification, $j = 1, 2, \ldots, b$; and 2 groups of the C-classification would look as in Table 160, with a dot always indicating the running index $(1, 2, \ldots, a; 1, 2, \ldots, b; 1$ and $2)$.

7.4.1.1 Analysis of variance for the three way classification with 2ab observations

Experimental variables, here three, are called factors; the intensity setting of a factor is called a level. One observes the outcomes of a trial with the three factors A, B, C at a, b, c ($c = 2$) levels $A_1, \ldots, A_a, B_1, \ldots, B_b, C_1, C_2$ (cf., Table 160 and 162). These levels are chosen systematically and are of particular importance (model I, cf., Section 7.4.1.2). For every possible combination (A_i, B_j, C_k) there is an observation x_{ijk}. The model equation would read:

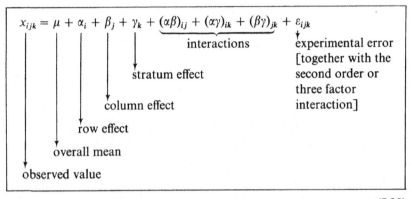

(7.28)

Here α_i are the deviations of the row means from the overall mean μ, the effect of the ith level of the factor A ($i = 1, 2, \ldots, a$); β_j are the deviations of the column means from μ, the effect of the jth level of the factor B ($j = 1, 2, \ldots, b$); γ_k are the deviations of the two means of the strata from μ, the effect of the kth level of the factor C ($k = 1, 2$) (say $k = 1$ is the observed value for the first trial at the instant t_1; $k = 2$ is the observed value for the second trial at the instant t_2; (see below). An **interaction effect** is present if the sum of the isolated effects does not equal the combined effect, i.e., the **effects are not independent** and hence not additive; in comparison with the sum of the individual effects, there is either a diminished or an intensified overall effect. $(\alpha\beta)_{ij}$ is the interaction effect between the ith level of the factor A and the jth level of the factor B ($i = 1, 2, \ldots, a; j = 1, 2, \ldots, b$); $(\alpha\gamma)_{ik}$ is the interaction effect between the ith level of the factor A and the kth level of the factor C ($i = 1, 2, \ldots, a; k = 1, 2$); $(\beta\gamma)_{jk}$ is the interaction effect between the jth level of the factor B and the kth level of the factor C ($j = 1, 2, \ldots, b$; $k = 1, 2$). Let the experimental error ε_{ijk} be independent and normally distributed with mean zero and variance σ^2 for i, j and k.

> Of the three assumptions: (1) equality of variance of the errors, (2) statistical independence of the errors and (3) normality of the errors, (1) is the most critical one for the power of the inference about means.

Table 160

A \ B	B_1	B_2	.	B_j	.	B_b	Σ
A_1	x_{111} x_{121}		.	x_{1j1}	.	x_{1b1}	$S_{1..}$
	x_{112} x_{122}		.	x_{1j2}	.	x_{1b2}	
A_2	x_{211} x_{221}		.	x_{2j1}	.	x_{2b1}	$S_{2..}$
	x_{212} x_{222}		.	x_{2j2}	.	x_{2b2}	
.	.	.	.	.	.	.	.
A_i	x_{i11} x_{i21}		.	x_{ij1}	.	x_{ib1}	$S_{i..}$
	x_{i12} x_{i22}		.	x_{ij2}	.	x_{ib2}	
.	.	.	.	.	.	.	.
A_a	x_{a11} x_{a21}		.	x_{aj1}	.	x_{ab1}	$S_{a..}$
	x_{a12} x_{a22}		.	x_{aj2}	.	x_{ab2}	
Σ	$S_{.1.}$	$S_{.2.}$	.	$S_{.j.}$	.	$S_{.b.}$	S

Here $S_{i..}$ denotes the sum of all values in the ith row, $S_{.j.}$ the sum of all values in the jth column, $S_{..1}$ the sum of all values in the 1st subgroup, and $S_{..2}$ the sum of all values in the 2nd subgroup; S is the sum of all observations (i.e., $S = S_{...}$ = $\sum_i \sum_j \sum_k x_{ijk}$ [with $k = 1, 2$]).

The observations represent random samples from normally distributed populations with a common variance σ^2; for the sample variables, a decomposability of the form (7.28) is assumed. In this model α_i, β_j, γ_k, $(\alpha\beta)_{ij}$, $(\alpha\gamma)_{ik}$, $(\beta\gamma)_{jk}$ are unknown constants which are compared as **systematic portions** of the **random components** ε_{ijk}. In view of the experimental errors ε_{ijk}, hypotheses on the systematic components are tested. Of the following restrictions, those apply that are appropriate to the particular hypotheses to be tested:

$$\sum_i \alpha_i = 0, \qquad \sum_j \beta_j = 0, \qquad \sum_k \gamma_k = 0$$

$$\sum_i (\alpha\beta)_{ij} = 0 \text{ for } j = 1, 2, \ldots, b \qquad \sum_i (\alpha\gamma)_{ik} = 0 \text{ for } k = 1, 2$$

$$\sum_j (\alpha\beta)_{ij} = 0 \text{ for } i = 1, 2, \ldots, a \qquad \sum_k (\alpha\gamma)_{ik} = 0 \text{ for } i = 1, 2, \ldots, a$$

$$\sum_j (\beta\gamma)_{jk} = 0 \text{ for } k = 1, 2 \qquad \sum_k (\beta\gamma)_{jk} = 0 \text{ for } j = 1, 2, \ldots, b.$$

(7.29–7.37)

7.4 Two Way and Three Way Analysis of Variance

We then have the estimates for the parameters

$$\hat{\mu} = (\sum_i\sum_j\sum_k x_{ijk})/2ab = S/2ab \quad (7.38)$$

$$\hat{\alpha}_i = \hat{\mu}_{i..} - \hat{\mu} \qquad (\widehat{\alpha\beta})_{ij} = \hat{\mu}_{ij.} - \hat{\mu}_{i..} - \hat{\mu}_{.j.} + \hat{\mu} \quad (7.39)(7.42)$$
$$\hat{\beta}_j = \hat{\mu}_{.j.} - \hat{\mu} \qquad (\widehat{\alpha\gamma})_{ik} = \hat{\mu}_{i.k} - \hat{\mu}_{i..} - \hat{\mu}_{..k} + \hat{\mu} \quad (7.40)(7.43)$$
$$\hat{\gamma}_k = \hat{\mu}_{..k} - \hat{\mu} \qquad (\widehat{\beta\gamma})_{jk} = \hat{\mu}_{.jk} - \hat{\mu}_{.j.} - \hat{\mu}_{..k} + \hat{\mu}. \quad (7.41)(7.44)$$

Null hypotheses:

$$H_A: \alpha_i = 0 \quad \text{for } i = 1, 2, \ldots, a$$
$$H_B: \beta_j = 0 \quad \text{for } j = 1, 2, \ldots, b$$
$$H_C: \gamma_k = 0 \quad \text{for } k = 1, 2$$
$$H_{AB}: (\alpha\beta)_{ij} = 0 \quad \text{for } i = 1, 2, \ldots, a; j = 1, 2, \ldots, b$$
$$H_{AC}: (\alpha\gamma)_{ik} = 0 \quad \text{for } i = 1, 2, \ldots, a; k = 1, 2$$
$$H_{BC}: (\beta\gamma)_{jk} = 0 \quad \text{for } j = 1, 2, \ldots, b; k = 1, 2.$$

In words:

H_A: There is no row effect of A, or $\alpha_i = 0$ for all i levels; confronted by the alternate hypothesis: not all α_i equal zero, i.e., at least one $\alpha_i \neq 0$.
H_B: The corresponding statement holds for the column effect ($\beta_j = 0$).
H_C: The corresponding statement holds for the stratum effect ($\gamma_k = 0$).
H_{AB}, H_{AC}, H_{BC}: There are no interactions. Alternative hypothesis: at least one $(\alpha\beta)_{ij} \neq 0$; at least one $(\alpha\gamma)_{ik} \neq 0$; at least one $(\beta\gamma)_{jk} \neq 0$.

To reject these hypotheses, we need the associated variances. We recall that the variance, here referred to as the **mean square (MS)**, the **average variation per degree of freedom**,

$$\text{mean square} = \frac{\text{variation}}{\text{degrees of freedom}} = \frac{\text{sum of squares}}{\text{degrees of freedom}} = \frac{SS}{DF} = MS,$$

(7.45)

was estimated by the quotient of the sum of squares SS over the degrees of freedom v, in the case of the variance of a single sample, $v = n - 1$ and

$$s^2 = \frac{\sum(x - \bar{x})^2}{n-1} = \frac{SS}{n-1},$$

when $SS = \sum x^2 - (\sum x)^2/n$ was obtained by subtracting a corrective term from a sum of squares. For the three way classification with $2ab$ observations, the adjustment reads $(1/2ab)S^2$. The sums of squares and the associated DF are found in Table 161. The MS of the 6 effects are then tested by the F-test against the MS of the

Table 161 Analysis of variance for the three dimensional classification with 2ab observations ($c \geq 2$)

	Source of the variation	Sum of squares SS	Degrees of freedom DF	Mean square MS	Computed F $\hat{F}$
Main Effects	Between the levels of the factor A	$SSA = \dfrac{1}{2b}\sum_{i=1}^{a} S_{i..}^{2} - \dfrac{1}{2ab}S^{2}$	$DF_A = a - 1$	$MSA = \dfrac{SSA}{DF_A}$	$\hat{F}_A = \dfrac{MSA}{MSE}$
	Between the levels of the factor B	$SSB = \dfrac{1}{2a}\sum_{j=1}^{b} S_{.j.}^{2} - \dfrac{1}{2ab}S^{2}$	$DF_B = b - 1$	$MSB = \dfrac{SSB}{DF_B}$	$\hat{F}_B = \dfrac{MSB}{MSE}$
	Between the levels of the factor C	$SSC = \dfrac{1}{ab}\sum_{k=1}^{2} S_{..k}^{2} - \dfrac{1}{2ab}S^{2}$	$DF_C = 2 - 1 = 1$	$MSC = \dfrac{SSC}{DF_C}$	$\hat{F}_C = \dfrac{MSC}{MSE}$
Two Factor Interactions	Interaction AB	$SSAB = \dfrac{1}{2}\sum_{i=1}^{a}\sum_{j=1}^{b} S_{ij.}^{2} - \dfrac{1}{2b}\sum_{i=1}^{a} S_{i..}^{2} - \dfrac{1}{2a}\sum_{j=1}^{b} S_{.j.}^{2} + \dfrac{1}{2ab}S^{2}$	$DF_{AB} = (a-1)(b-1)$	$MSAB = \dfrac{SSAB}{DF_{AB}}$	$\hat{F}_{AB} = \dfrac{MSAB}{MSE}$
	Interaction AC	$SSAC = \dfrac{1}{b}\sum_{i=1}^{a}\sum_{k=1}^{2} S_{i.k}^{2} - \dfrac{1}{2b}\sum_{i=1}^{a} S_{i..}^{2} - \dfrac{1}{ab}\sum_{k=1}^{2} S_{..k}^{2} + \dfrac{1}{2ab}S^{2}$	$DF_{AC} = (a-1)(2-1)$	$MSAC = \dfrac{SSAC}{DF_{AC}}$	$\hat{F}_{AC} = \dfrac{MSAC}{MSE}$
	Interaction BC	$SSBC = \dfrac{1}{a}\sum_{j=1}^{b}\sum_{k=1}^{2} S_{.jk}^{2} \times \dfrac{1}{2a}\sum_{j=1}^{b} S_{.j.}^{2} - \dfrac{1}{ab}\sum_{k=1}^{2} S_{..k}^{2} + \dfrac{1}{2ab}S^{2}$	$DF_{BC} = (b-1)(2-1)$	$MSBC = \dfrac{SSBC}{DF_{BC}}$	$\hat{F}_{BC} = \dfrac{MSBC}{MSE}$
	Experimental error	$SSE = \sum\sum\sum x_{ijk}^{2} - \dfrac{1}{2}\sum\sum S_{ij.}^{2} - \dfrac{1}{b}\sum\sum S_{i.k}^{2} - \dfrac{1}{a}\sum\sum S_{.jk}^{2} + \dfrac{1}{ab}\sum_{i} S_{i..}^{2} + \dfrac{1}{2a}\sum_{j} S_{.j.}^{2} + \dfrac{1}{ab}\sum_{k} S_{..k}^{2} - \dfrac{1}{2ab}S^{2}$	$DF_E = (a-1)(b-1)(2-1)$	$MSE = \dfrac{SSE}{DF_E} = \hat{\sigma}^{2}$	
	Total variation	$SS(T) = \sum_{i=1}^{a}\sum_{j=1}^{b}\sum_{k=1}^{2} x_{ijk}^{2} - \dfrac{1}{2ab}S^{2}$	$DF_T = 2ab - 1$		

7.4 Two Way and Three Way Analysis of Variance

experimental error $\hat{\sigma}^2$, referred to as the residual mean square MS(R) or better as error mean square or mean square error MSE. The MSE measures the unexplained variability of a set of data and serves as an estimate of the inherent random variation of the experiment.

These hypotheses H_x, together with the associated mean squares MSX [computed as quotients of the associated sums of squares SSX over the degrees of freedom DF_x (cf., Table 161)] and the mean square of the experimental error $MSE = SSE/[(a-1)(b-1)]$, can be rejected whenever

$$\boxed{\hat{F} = \frac{MSX}{MSE} = \frac{SSX/DF_x}{SSE/[(a-1)(b-1)]} \\ > F_{v_1; v_2; \alpha} \quad \text{with } v_1 = DF_x, \, v_2 = (a-1)(b-1).} \qquad (7.46) \quad \text{(p.145)}$$

Moreover, the following parameters can be estimated:

$$\boxed{\hat{\mu} = \frac{S}{2ab},} \qquad (7.38)$$

$$\boxed{\hat{\alpha}_i = \frac{S_{i..}}{2b} - \hat{\mu},} \qquad (7.39)$$

$$\boxed{\hat{\beta}_j = \frac{S_{.j.}}{2a} - \hat{\mu},} \qquad (7.40)$$

$$\boxed{\hat{\gamma}_k = \frac{S_{..k}}{ab} - \hat{\mu},} \qquad (7.41)$$

$$\boxed{\widehat{(\alpha\beta)}_{ij} = \frac{S_{ij.}}{2} - \frac{S_{i..}}{2b} - \frac{S_{.j.}}{2a} + \hat{\mu},} \qquad (7.42)$$

$$\boxed{\widehat{(\alpha\gamma)}_{ik} = \frac{S_{i.k}}{b} - \frac{S_{i..}}{2b} - \frac{S_{..k}}{ab} + \hat{\mu},} \qquad (7.43)$$

$$\boxed{\widehat{(\beta\gamma)}_{jk} = \frac{S_{.jk}}{a} - \frac{S_{.j.}}{2a} - \frac{S_{..k}}{ab} + \hat{\mu},} \qquad (7.44)$$

the mean row effect

$$\boxed{\hat{\sigma}^2_{\text{row}} = \frac{\sum \alpha_i^2}{a},} \qquad (7.39a)$$

the mean column effect

$$\boxed{\hat{\sigma}^2_{\text{col.}} = \frac{\sum \beta_j^2}{b},} \qquad (7.40a)$$

and the mean stratum effect

$$\hat{\sigma}^2_{\text{str.}} = \frac{\sum y_k^2}{2}. \qquad (7.41a)$$

This is illustrated by a simple numerical example in Table 162.

Table 162

A \ B	B_1	B_2	B_3	Σ
A_1	6 5	6 4	7 6	34
A_2	5 4	5 5	5 5	29
A_3	6 6	7 7	4 4	34
A_4	8 7	6 5	5 2	33
Σ	47	45	38	130

Table 163 (ijk)

C	C_1			C_2			
A \ B	B_1	B_2	B_3	B_1	B_2	B_3	Σ
A_1	6	6	7	5	4	6	34
A_2	5	5	5	4	5	5	29
A_3	6	7	4	6	7	4	34
A_4	8	6	5	7	5	2	33
Σ	25	24	21	22	21	17	130

Table 163a (ij)

A \ B	B_1	B_2	B_3	Σ
A_1	11	10	13	34
A_2	9	10	10	29
A_3	12	14	8	34
A_4	15	11	7	33
Σ	47	45	38	130

Table 163b (ik)

A \ C	C_1	C_2	Σ
A_1	19	15	34
A_2	15	14	29
A_3	17	17	34
A_4	19	14	33
Σ	70	60	130

Table 163c (jk)

B \ C	C_1	C_2	Σ
B_1	25	22	47
B_2	24	21	45
B_3	21	17	38
Σ	70	60	130

7.4 Two Way and Three Way Analysis of Variance

Table 162 (or Table 163) contains the rounded-off yields of a chemical reaction. A_{1-4} are concentration levels, B_{1-3} temperature levels; C_1 and C_2 are instants of time at which the trials were run. Tables 163a, b, c, are auxiliary tables formed by summing.

Table 161 first yields the residual term $(\sum x)^2/n$ for all sums of squares:

$$\frac{1}{2ab}S^2 = \frac{130^2}{(2)(4)(3)} = \frac{16{,}900}{24} = 704.167,$$

and then, using Table 164

Table 164

$$
\begin{aligned}
(34^2 + 29^2 + 34^2 + 33^2)/6 &= 707.000 \\
(47^2 + 45^2 + 38^2)/8 &= 709.750 \\
(70^2 + 60^2)/12 &= 708.333 \\
(11^2 + 10^2 + \ldots + 7^2)/2 &= 735.000 \\
(19^2 + 15^2 + \ldots + 14^2)/3 &= 714.000 \\
(25^2 + 22^2 + \ldots + 17^2)/4 &= 714.000 \\
6^2 + 6^2 + 7^2 + \ldots + 5^2 + 2^2 &= 744.000
\end{aligned}
$$

$$SSA = \frac{1}{2 \cdot 3}(34^2 + 29^2 + 34^2 + 33^2) - 704.167 = 2.833,$$

$$SSB = \frac{1}{2 \cdot 4}(47^2 + 45^2 + 38^2) - 704.167 = 5.583,$$

$$SSC = \frac{1}{4 \cdot 3}(70^2 + 60^2) - 704.167 = 4.166,$$

$$SSAB = \frac{1}{2}(11^2 + 10^2 + 13^2 + 9^2 + \cdots + 7^2) \text{ [see Table 163a]}$$

$$- \frac{1}{2 \cdot 3}(34^2 + 29^2 + 34^2 + 33^2) - \frac{1}{2 \cdot 4}(47^2 + 45^2 + 38^2)$$

$$+ 704.167 = 22.417,$$

$$SSAC = \frac{1}{3}(19^2 + 15^2 + 15^2 + 14^2 + 17^2 + 17^2 + 19^2 + 14^2)$$

[see Table 163b]

$$- \frac{1}{2 \cdot 3}(34^2 + 29^2 + 34^2 + 33^2) - \frac{1}{4 \cdot 3}(70^2 + 60^2) + 704.167$$

$$= 2.834,$$

$$SSBC = \frac{1}{4}(25^2 + 22^2 + 24^2 + 21^2 + 21^2 + 17^2) \text{ [see Table 163c]}$$

$$- \frac{1}{2 \cdot 4}(47^2 + 45^2 + 38^2) - \frac{1}{4 \cdot 3}(70^2 + 60^2) + 704.167 = 0.084,$$

$$SSE = (6^2 + 6^2 + 7^2 + 5^2 + \cdots + 7^2 + 5^2 + 2^2) \text{ [see Table 163]}$$
$$- \frac{1}{2}(11^2 + 10^2 + \cdots + 7^2) - \frac{1}{3}(19^2 + 15^2 + \cdots + 14^2)$$
$$- \frac{1}{4}(25^2 + 22^2 + \cdots + 17^2) + \frac{1}{2 \cdot 3}(34^2 + 29^2 + 34^2 + 33^2)$$
$$+ \frac{1}{2 \cdot 4}(47^2 + 45^2 + 38^2) + \frac{1}{4 \cdot 3}(70^2 + 60^2) - 704.167 = 1.916,$$
$$SST = (6^2 + 6^2 + 7^2 + 5^2 + \cdots + 7^2 + 5^2 + 2^2) - 704.167 = 39.833.$$

These results are summarized in Table 165, which also contains the values for the respective test statistics (7.46) and the critical values. Column 5 compares computed F-values ($\hat{F}$) with tabulated F-values $F_{v_1; v_2; 0.05}$ at a significance level of $\alpha = 0.05$. (If we wish to make all $m = 6$ tests with an overall significance level of $\alpha = 0.05$, we use $\alpha/m = 0.05/6 \simeq 0.0083$ or 0.01 as the significance level for each single F-test).

Table 165 Analysis of variance for Table 163 using Table 161

Source of variation	Sum of squares SS	DF	MS	$\hat{F}$	$F_{0.05}$
(1)	(2)	(3)	(4)	(5)	
Factor A	SSA = 2.833	4 − 1 = 3	0.944	2.96 < 4.76	
Factor B	SSB = 5.583	3 − 1 = 2	2.792	8.75 > 5.14	
Factor C	SSC = 4.166	1	4.166	13.06 > 5.99	
Interaction AB	SSAB = 22.417	6	3.736	11.71 > 4.28	
Interaction AC	SSAC = 2.834	3	0.948	2.97 < 4.76	
Interaction BC	SSBC = 0.084	2	0.042	0.13 < 5.14	
Experimental error	SSE = 1.916	6	0.319 = MSE = $\hat{\sigma}^2$		
Total variation T	SST = 39.833	23			

The null hypotheses
$$\beta_1 = \beta_2 = \beta_3 = 0, \quad \gamma_1 = \gamma_2 = 0, \quad (\alpha\beta)_{11} = \cdots = (\alpha\beta)_{43} = 0$$
can, in accordance with
$$\hat{F}_B = \frac{2.792}{0.319} = 8.75 > 5.14 = F_{2; 6; 0.05},$$
$$\hat{F}_C = \frac{4.166}{0.319} = 13.06 > 5.99 = F_{1; 6; 0.05},$$
$$\hat{F}_{AB} = \frac{3.736}{0.319} = 11.71 > 4.28 = F_{6; 6; 0.05},$$

7.4 Two Way and Three Way Analysis of Variance

be rejected at the 5% level. The corresponding estimates $\hat{\beta}_j, \hat{\gamma}_k, \widehat{(\alpha\beta)}_{ij}$ can be read from Table 166. The mean column effect and the mean stratum effect are

$$\hat{\sigma}^2_{columns} = \frac{1}{abc} SSB = \frac{1}{(4)(3)(2)} 5.583 = 0.233,$$

$$\hat{\sigma}^2_{strata} = \frac{1}{abc} SSC = \frac{1}{(4)(3)(2)} 4.166 = 0.174.$$

Table 166 Note the estimates of the effects of the different levels of a factor add up to zero. The strongest (in absolute value) interactions are positive and belong to the fields $A_1 B_3 \, \widehat{(\alpha\beta)}_{13}$ and $A_4 B_1 \, \widehat{(\alpha\beta)}_{41}$; the strongest negative interactions occur in the combinations $A_3 B_3 \, \widehat{(\alpha\beta)}_{33}$ and $A_4 B_3 \, \widehat{(\alpha\beta)}_{43}$.

$\hat{\mu} = \frac{130}{2 \cdot 4 \cdot 3} = 5.417$	$\widehat{(\alpha\beta)}_{11} = \frac{11}{2} - \frac{34}{2 \cdot 3} - \frac{47}{2 \cdot 4} + 5.42 = -0.63$
$\hat{\alpha}_1 = \frac{34}{2 \cdot 3} - 5.42 = 0.25$	$\widehat{(\alpha\beta)}_{12} = \frac{10}{2} - \frac{34}{2 \cdot 3} - \frac{45}{2 \cdot 4} + 5.42 = -0.87$
$\hat{\alpha}_2 = \frac{29}{2 \cdot 3} - 5.42 = -0.59$	$\widehat{(\alpha\beta)}_{13} = \frac{13}{2} - \frac{34}{2 \cdot 3} - \frac{38}{2 \cdot 4} + 5.42 = 1.50$
$\hat{\alpha}_3 = \hat{\alpha}_1 = 0.25$	$\widehat{(\alpha\beta)}_{21} = \frac{9}{2} - \frac{29}{2 \cdot 3} - \frac{47}{2 \cdot 4} + 5.42 = -0.79$
$\hat{\alpha}_4 = \frac{33}{2 \cdot 3} - 5.42 = 0.08$	$\widehat{(\alpha\beta)}_{22} = \frac{10}{2} - \frac{29}{2 \cdot 3} - \frac{45}{2 \cdot 4} + 5.42 = -0.04$
$\hat{\beta}_1 = \frac{47}{2 \cdot 4} - 5.42 = 0.46$	$\widehat{(\alpha\beta)}_{23} = \frac{10}{2} - \frac{29}{2 \cdot 3} - \frac{38}{2 \cdot 4} + 5.42 = 0.84$
$\hat{\beta}_2 = \frac{45}{2 \cdot 4} - 5.42 = 0.21$	$\widehat{(\alpha\beta)}_{31} = \frac{12}{2} - \frac{34}{2 \cdot 3} - \frac{47}{2 \cdot 4} + 5.42 = -0.13$
$\hat{\beta}_3 = \frac{38}{2 \cdot 4} - 5.42 = -0.67$	$\widehat{(\alpha\beta)}_{32} = \frac{14}{2} - \frac{34}{2 \cdot 3} - \frac{45}{2 \cdot 4} + 5.42 = 1.12$
	$\widehat{(\alpha\beta)}_{33} = \frac{8}{2} - \frac{34}{2 \cdot 3} - \frac{38}{2 \cdot 4} + 5.42 = -1.00$
$\hat{\gamma}_1 = \frac{70}{4 \cdot 3} - 5.42 = 0.42$	$\widehat{(\alpha\beta)}_{41} = \frac{15}{2} - \frac{33}{2 \cdot 3} - \frac{47}{2 \cdot 4} + 5.42 = 1.54$
$\hat{\gamma}_2 = \frac{60}{4 \cdot 3} - 5.42 = -0.42$	$\widehat{(\alpha\beta)}_{42} = \frac{11}{2} - \frac{33}{2 \cdot 3} - \frac{45}{2 \cdot 4} + 5.42 = -0.21$
	$\widehat{(\alpha\beta)}_{43} = \frac{7}{2} - \frac{33}{2 \cdot 3} - \frac{38}{2 \cdot 4} + 5.42 = -1.33$

7.4.1.2 Analysis of variance for the two way classification with 2ab observations

Disregarding the factor C could mean that we have two (c) observations (replications of the experiment) under identical conditions. The model is now

$$x_{ijk} = \mu + \alpha_i + \beta_j + (\alpha\beta)_{ij} + \varepsilon_{ijk} \tag{7.47}$$

—where γ_k, $(\alpha\gamma)_{ik}$, $(\beta\gamma)_{jk}$ are included in the experimental error—with the three constraints

$$\sum_{i=1}^{a}\alpha_i=0, \quad \sum_{j=1}^{b}\beta_j=0, \quad \sum_{i=1}^{a}\sum_{j=1}^{b}(\alpha\beta)_{ij}=0 \qquad (7.48)$$

and the appropriate null hypotheses (cf., Table 167). The experimental error now includes the sources of variation C, AC, BC, and ABC:

$$(SSE \text{ from Table 167}) = (SSC + SSAC + SSBC + SSE \text{ from Table 161}),$$

and (cf., Table 167a; SSC is computed using Table 165; $SSE = 4.166 + 2.834 + 0.084 + 1.916$) only the interaction AB is significant at the 5% level.

Table 167 Two-way analysis of variance with interaction

Source	Sum of squares SS	Degrees of freedom DF	Mean sum of squares MS
Factor A	SSA	a − 1	$MSA = \dfrac{SSA}{a-1}$
Factor B	SSB	b − 1	$MSB = \dfrac{SSB}{b-1}$
Interaction	SSAB	(a − 1)(b − 1)	$MSAB = \dfrac{SSAB}{(a-1)(b-1)}$
Error	SSE	ab	$MSE = \dfrac{SSE}{ab}$
Total	SST	2ab − 1	

Table 167a Analysis of variance for Table 162 using Table 167

Source of the variation	Sum of squares SS	DF	Mean square MS	$\hat{F}$	$F_{0.05}$
Factor A	2.833	3	0.944	1.26 < 3.49	
Factor B	5.583	2	2.792	3.72 < 3.89	
Interaction AB	22.417	6	3.736	4.98 > 3.00	
Experimental error	9.000	12	0.750		
Total	39.833	23			

The analysis of variance for the three way classification with $2ab$ observations [Model (7.28); Table 161] can thus be simplified considerably by ignoring both of the less important interactions [Table 168 with the four constraints $\sum_i \alpha_i = \sum_j \beta_j = \sum_k \gamma_k = \sum_i \sum_j (\alpha\beta)_{ij} = 0$] and the interaction effect of the factor C [Table 168a with (7.48)]. The sets of experiments at the two levels of the factor C (instants t_1 and t_2) are often called randomized blocks, C_1 and C_2, in Table 161 and 168, since all the treatments $A_i B_j$ are assigned randomly to the experimental units within each block. For Table 168a (as for 167, 167a), a so-called completely randomized two variable classification with replication is in effect.

Our example is based, as far as the formulation of the problem and collection of data are concerned, on a model with **fixed effects**; the systematically chosen levels of the factors are of primary interest. In the factors to be tested one can often take all levels (e.g., male and female animals) or only a portion of the possible levels into consideration. In the last case, we distinguish between:

1. **Systematic choice**, e.g., **deliberately chosen** varieties, fertilizers, spacing, sowing times, amount of seed, or the levels of pressure, temperature, time and concentration in a chemical process; and
2. **Random choice**, e.g., soils, localities, and years, test animals, or other test objects, which can be thought of as **random samples** from some imagined population.

According to Eisenhart (1947), two models are distinguished in the analysis of variance:

Model I with **systematic components** or fixed effects, referred to as the "fixed" model (Type 1): Special treatments, medicines, methods, levels of a factor, varieties, test animals, machines are **chosen deliberately** and employed in the trial, since it is precisely they (e.g., the pesticides A, B and C) that are of practical interest, and one would like to learn something about their mean effects and the significance of these effects. The comparisons of means are thus of primary concern.

Model II, with random effects or **random components**, is referred to as the "random" model (Type II): the procedures, methods, test personnel or objects under study are random samples from a population about which we would like to make some statement. The variabilities of the individual factors as portions of the total variability are of interest. The variance components as well as confidence intervals are estimated, and hypotheses on the variance components are tested ("authentic analysis of variance"; cf., e.g., Blischke 1966, Endler 1966, Koch 1967, Wang 1967, Harvey 1970, Searle 1971, Dunn and Clark 1974 [1]). Model II is much more sensitive to nonnormality than Model I.

Table 168 Analysis of variance for the three way classification with 2ab observations. Model: $x_{ijk} = \mu + \alpha_i + \beta_j + \gamma_k + (\alpha\beta)_{ij} + \varepsilon_{ijk}$.

Source of the variation	Sum of squares SS	DF	MS	$\hat{F}$
Between the levels of the factor A	$SSA = \dfrac{1}{2b}\sum_i S_{i..}^2 - \dfrac{1}{2ab}S^2$	$DF_A = a - 1$	$MSA = \dfrac{SSA}{DF_A}$	$\hat{F}_A = \dfrac{MSA}{MSE}$
Between the levels of the factor B	$SSB = \dfrac{1}{2a}\sum_j S_{.j.}^2 - \dfrac{1}{2ab}S^2$	$DF_B = b - 1$	$MSB = \dfrac{SSB}{DF_B}$	$\hat{F}_B = \dfrac{MSB}{MSE}$
Between the levels of the factor C	$SSC = \dfrac{1}{ab}\sum_{k=1}^{2} S_{..k}^2 - \dfrac{1}{2ab}S^2$	$DF_C = 2 - 1 = 1$	$MSC = \dfrac{SSC}{DF_C}$	$\hat{F}_C = \dfrac{MSC}{MSE}$
Interaction AB	$SSAB = SS(T) - (SSA + SSB + SSC + SSE)$	$DF_{AB} = (a-1)(b-1)$	$MSAB = \dfrac{SSAB}{DF_{AB}}$	$\hat{F}_{AB} = \dfrac{MSAB}{MSE}$
Experimental error	$SSE = \sum_i\sum_j\sum_k x_{ijk}^2 - \dfrac{1}{ab}\sum_k S_{..k}^2 - \dfrac{1}{2}\sum_i\sum_j S_{ij.}^2 + \dfrac{1}{2ab}S^2$	$DF_E = (ab-1)(2-1) = ab - 1$	$MSE = \dfrac{SSE}{DF_E} = \hat{\sigma}^2$	
Total variation	$SS(T) = \sum_i\sum_j\sum_k x_{ijk}^2 - \dfrac{1}{2ab}S^2$	$DF_T = 2ab - 1$		

Table 168a Analysis of variance for the three-way classification with 2ab observations. Mode: $x_{ijk} = \mu + \alpha_i + \beta_j + (\alpha\beta)_{ij} + \varepsilon_{ijk}$.

Source of the variation	Sum of squares SS	DF	MS	$\hat{F}$
Between the levels of the factor A	$SSA = \dfrac{1}{2b}\sum_i S_{i..}^2 - \dfrac{1}{2ab}S^2$	$DF_A = a - 1$	$MSA = \dfrac{SSA}{DF_A}$	$\hat{F}_A = \dfrac{MSA}{MSE}$
Between the levels of the factor B	$SSB = \dfrac{1}{2a}\sum_j S_{.j.}^2 - \dfrac{1}{2ab}S^2$	$DF_B = b - 1$	$MSB = \dfrac{SSB}{DF_B}$	$\hat{F}_B = \dfrac{MSB}{MSE}$
Interaction AB	$SSAB = SS(T) - (SSA + SSB + SSE)$	$DF_{AB} = (a-1)(b-1)$	$MSAB = \dfrac{SSAB}{DF_{AB}}$	$\hat{F}_{AB} = \dfrac{MSAB}{MSE}$
Experimental error	$SSE = \sum_i\sum_j\sum_k x_{ijk}^2 - \dfrac{1}{2}\sum_i\sum_j S_{ij.}^2$	$DF_E = ab(2-1) = ab$	$MSE = \dfrac{SSE}{DF_E} = \hat{\sigma}^2$	
Total variation	$SS(T) = \sum_i\sum_j\sum_k x_{ijk}^2 - \dfrac{1}{2ab}S^2$	$DF_T = 2ab - 1$		

Table 169 "Fixed" model

Source	DF	MS	Test	$\hat{F}$	$F_{0.05}$
A	3	0.94		$\hat{F}_A = \dfrac{0.944}{0.75} = 1.26 < 3.49$	
B	2	2.79		$\hat{F}_B = \dfrac{2.79}{0.75} = 3.72 < 3.89$	
AB	6	3.74		$\hat{F}_{AB} = \dfrac{3.74}{0.75} = 4.99 > 3.00$	
E	12	0.75			

Fixed effects are indicated by Greek letters, random ones by Latin letters.

> Only in the "fixed" model can the mean squares (MS) be tested against the MS of the experimental error. In the "random" model, the MS of the row and column effects are tested against the MS of the interaction, which is then tested against the MS of the experimental error.

More on this can be found in Binder (1955), Hartley (1955), Wilk and Kempthorne (1955), Harter (1957), Le Roy (1957–1972), Scheffé (1959), Plackett (1960), Federer (1961), Ahrens (1967), and especially Searle (1971).

Now back to our example. As in the Model I analysis, we obtain only one significant interaction (Table 170). It can also happen that the description "fixed effect" applies to the levels of one characteristic while levels of another characteristic are random (the "mixed" model or Model III). If we assume the levels of the factor A to be "random" and those of the factor B to be

Table 170 The "random" model. This model is less suitable for the example.

Source	DF	MS	Test	$\hat{F}$	$F_{0.05}$
A	3	0.94		$\hat{F}_A = \dfrac{0.944}{3.74} = 0.25 < 4.76$	
B	2	2.79		$\hat{F}_B = \dfrac{2.79}{3.74} = 0.75 < 5.14$	
AB	6	3.74		$\hat{F}_{AB} = \dfrac{3.74}{0.75} = 4.99 > 3.00$	
E	12	0.75			

"fixed" [the interaction effects $(\alpha\beta)_{ij}$ are also random variables due to the random nature of the factor A levels], then the row (A) effect for our example is larger than under the pure model II but still below the 5% level.

Table 171 The "mixed" model

Source	DF	MS	Test	$\hat{F}$	$F_{0.05}$
A	3	0.94		$\hat{F}_A = \dfrac{0.944}{0.75} = 1.26$	< 3.49
B	2	2.79		$\hat{F}_B = \dfrac{2.79}{3.74} = 0.75$	< 5.14
AB	6	3.74		$\hat{F}_{AB} = \dfrac{3.74}{0.75} = 4.99$	> 3.00
E	12	0.75			

The analysis of mixed models is not simple (Wilk and Kempthorne 1955, Scheffé 1959, Searle and Henderson 1961, Hays 1963, Bancroft 1964, Holland 1965, Blischke 1966, Eisen 1966, Endler 1966, Spjøtvoll 1966, Cunningham 1968, Koch and Sen 1968, Harvey 1970, Rasch 1971, Searle 1971).

▶ ### 7.4.2 Multiple comparison of means according to Scheffé, according to Student, Newman and Keuls, and according to Tukey

Multiple comparison procedures like the Scheffé test should only be done after an analysis of variance rejects H_0; otherwise the overall significance level of the multiple comparisons may be much greater than the preselected α and be heavily dependent on the value of k.

Given are k means, ordered by magnitude: $\bar{x}_{(1)} \geq \bar{x}_{(2)} \geq \cdots \geq \bar{x}_{(k)}$. If in the multiple pairwise comparison of means a critical difference D_α is exceeded, then $H_0: \mu_{(i)} = \mu_{(j)}$ is rejected and $H_A: \mu_{(i)} > \mu_{(j)}$, accepted at the $100\alpha\%$ level. The 5% level is preferred.

I. According to Scheffé (1953) we have for sample groups of equal or unequal size and arbitrary pairs of means (cf., also Section 7.3.2): (p. 509)

$$D_I = \sqrt{s_{\text{within}}^2(1/n_i + 1/n_j)(k-1)F_{(k-1;\,\nu_{s_{\text{within}}^2};\,\alpha)}}, \quad (7.49)$$

where $s_{\text{within}}^2 = MSE$ is the mean square of the experimental error, n_i, n_j are sample sizes of the compared means, and $\nu_{s_{\text{within}}^2}$ is the number of degrees of freedom for s_{within}^2.

II. According to Student (1927), Newman (1939), and Keuls (1952), we have for sample groups of equal size n:

$$D_{II} = q\sqrt{\frac{s^2_{within}}{n}} \qquad (7.50)$$

with the approximation for $n_i \neq n_j$:

$$D'_{II} = q\sqrt{s^2_{within}\,0.5\left(\frac{1}{n_i} + \frac{1}{n_j}\right)} \qquad (7.50a)$$

(cf., also Section 7.4.3, Remark 1). Here q is a factor from Table 172 for $P = 0.05$ or $P = 0.01$, depending on the number k of means in the region considered (for $\bar{x}_{(4)} - \bar{x}_{(2)}$ we thus have $k = 3$) and on the number v_2 of degrees of freedom associated with $MSE = s^2_{within}$. A table for $P = 0.10$ is given by Pachares (1959).

Compute $d_k = \bar{x}_{(1)} - \bar{x}_{(k)}$. For $d_k \leq D_{II,k,\alpha}$ all means are taken to be equal. For $d_k > D_{II,k,\alpha}$, $\mu_{(1)}$ and $\mu_{(2)}$ are taken to be unequal and $d'_{k-1} = \bar{x}_{(1)} - \bar{x}_{(k-1)}$ as well as $d''_{k-1} = \bar{x}_{(2)} - \bar{x}_{(k)}$ are computed. For $d'_{k-1} \leq D_{II,k-1,\alpha}$ the means $\mu_{(1)}$ to $\mu_{(k-1)}$ are taken to be equal; for $d'_{k-1} > D_{II,k-1,\alpha}$, $\mu_{(1)} = \mu_{(k-1)}$ is rejected. Corresponding tests are carried out with d''_{k-1}. This procedure is repeated until the h means of a group lead to $d_h \leq D_{II,h,\alpha}$ and are thus considered to be equal.

III. The Tukey procedure: A D_{II} which is based on q with k the total number of means is, according to Tukey (cf., e.g., Scheffé 1953), suitable for testing two arbitrary means $\bar{x}_{(i)} - \bar{x}_{(j)}$ or two arbitrary groups of means, $(\bar{x}_{(1)} + \bar{x}_{(2)} + \bar{x}_{(3)})/3 - (\bar{x}_{(4)} + \bar{x}_{(5)})/2$ say. For the $k(k-1)/2$ differences of means, the 95% confidence intervals can be specified: $\bar{x}_{(i)} - \bar{x}_{(j)} \pm D_{II}$ or D'_{II} with $P = 0.05$.

We use the example in Section 7.3.2: $\alpha = 0.05$; $\bar{x}_{(1)}$ to $\bar{x}_{(6)}$: 26.8, 26.3, 25.2, 19.8, 14.3, 11.8; $n_i = 8$; $s^2_{within} = 10.38$; $v = 48 - 6 = 42$. Then

$$D_{I;0.05} = \sqrt{10.38(1/8 + 1/8)(6-1)2.44} = 5.63,$$

$$D_{II;6;0.05} = 4.22\sqrt{10.38/8} = 4.81,$$

and correspondingly:

$$D_{II;5;0.05} = 4.59, \qquad D_{II;4;0.05} = 4.31,$$

$$D_{II;3;0.05} = 3.91, \qquad D_{II;2;0.05} = 3.25.$$

7.4 Two Way and Three Way Analysis of Variance

Table 172 Upper significance bounds of the Studentized range distribution: $P = 0.05$ (from Documenta Geigy 1968 (cf., Bibliography: Important Tables [2])

v_2 \ k	2	3	4	5	6	7	8	9	10	11	12	13	14	15	16	17	18	19	20
1	17.969	26.98	32.82	37.08	40.41	43.12	45.40	47.36	49.07	50.59	51.96	53.20	54.33	55.36	56.32	57.22	58.04	58.83	59.56
2	6.085	8.33	9.80	10.88	11.74	12.44	13.03	13.54	13.99	14.39	14.75	15.08	15.38	15.65	15.91	16.14	16.37	16.57	16.77
3	4.501	5.91	6.82	7.50	8.04	8.48	8.85	9.18	9.46	9.72	9.95	10.15	10.35	10.52	10.69	10.84	10.98	11.11	11.24
4	3.926	5.04	5.76	6.29	6.71	7.05	7.35	7.60	7.83	8.03	8.21	8.37	8.52	8.66	8.79	8.91	9.03	9.13	9.23
5	3.635	4.60	5.22	5.67	6.03	6.33	6.58	6.80	6.99	7.17	7.32	7.47	7.60	7.72	7.83	7.93	8.03	8.12	8.21
6	3.460	4.34	4.90	5.30	5.63	5.90	6.12	6.32	6.49	6.65	6.79	6.92	7.03	7.14	7.24	7.34	7.43	7.51	7.59
7	3.344	4.16	4.68	5.06	5.36	5.61	5.82	6.00	6.16	6.30	6.43	6.55	6.66	6.76	6.85	6.94	7.02	7.10	7.17
8	3.261	4.04	4.53	4.89	5.17	5.40	5.60	5.77	5.92	6.05	6.18	6.29	6.39	6.48	6.57	6.65	6.73	6.80	6.87
9	3.199	3.95	4.41	4.76	5.02	5.24	5.43	5.59	5.74	5.87	5.98	6.09	6.19	6.28	6.36	6.44	6.51	6.58	6.64
10	3.151	3.88	4.33	4.65	4.91	5.12	5.30	5.46	5.60	5.72	5.83	5.93	6.03	6.11	6.19	6.27	6.34	6.40	6.47
11	3.113	3.82	4.26	4.57	4.82	5.03	5.20	5.35	5.49	5.61	5.71	5.81	5.90	5.98	6.06	6.13	6.20	6.27	6.33
12	3.081	3.77	4.20	4.51	4.75	4.95	5.12	5.27	5.39	5.51	5.61	5.71	5.80	5.88	5.95	6.02	6.09	6.15	6.21
13	3.055	3.73	4.15	4.45	4.69	4.88	5.05	5.19	5.32	5.43	5.53	5.63	5.71	5.79	5.86	5.93	5.99	6.05	6.11
14	3.033	3.70	4.11	4.41	4.64	4.83	4.99	5.13	5.25	5.36	5.46	5.55	5.64	5.71	5.79	5.85	5.91	5.97	6.03
15	3.014	3.67	4.08	4.37	4.59	4.78	4.94	5.08	5.20	5.31	5.40	5.49	5.57	5.65	5.72	5.78	5.85	5.90	5.96
16	2.998	3.65	4.05	4.33	4.56	4.74	4.90	5.03	5.15	5.26	5.35	5.44	5.52	5.59	5.66	5.73	5.79	5.84	5.90
17	2.984	3.63	4.02	4.30	4.52	4.70	4.86	4.99	5.11	5.21	5.31	5.39	5.47	5.54	5.61	5.67	5.73	5.79	5.84
18	2.971	3.61	4.00	4.28	4.49	4.67	4.82	4.96	5.07	5.17	5.27	5.35	5.43	5.50	5.57	5.63	5.69	5.74	5.79
19	2.960	3.59	3.98	4.25	4.47	4.65	4.79	4.92	5.04	5.14	5.23	5.31	5.39	5.46	5.53	5.59	5.65	5.70	5.75
20	2.950	3.58	3.96	4.23	4.45	4.62	4.77	4.90	5.01	5.11	5.20	5.28	5.36	5.43	5.49	5.55	5.61	5.66	5.71
21	2.941	3.56	3.94	4.21	4.43	4.60	4.74	4.87	4.98	5.08	5.17	5.25	5.33	5.40	5.46	5.52	5.58	5.63	5.67
22	2.933	3.55	3.93	4.20	4.41	4.58	4.72	4.85	4.96	5.05	5.15	5.23	5.30	5.37	5.43	5.49	5.55	5.60	5.65
23	2.926	3.54	3.91	4.18	4.39	4.56	4.70	4.83	4.94	5.03	5.12	5.20	5.27	5.34	5.40	5.46	5.52	5.57	5.62
24	2.919	3.53	3.90	4.17	4.37	4.54	4.68	4.81	4.92	5.01	5.10	5.18	5.25	5.32	5.38	5.44	5.49	5.55	5.59
25	2.913	3.52	3.89	4.16	4.36	4.52	4.66	4.79	4.90	4.99	5.08	5.16	5.23	5.30	5.36	5.42	5.48	5.52	5.57
26	2.907	3.51	3.88	4.14	4.34	4.51	4.65	4.78	4.89	4.98	5.07	5.14	5.21	5.28	5.34	5.40	5.46	5.50	5.55
27	2.902	3.51	3.87	4.13	4.33	4.50	4.63	4.76	4.87	4.96	5.06	5.13	5.19	5.26	5.32	5.38	5.43	5.48	5.53
28	2.897	3.50	3.86	4.12	4.32	4.48	4.62	4.75	4.86	4.94	5.04	5.11	5.18	5.24	5.30	5.36	5.42	5.46	5.51
29	2.892	3.49	3.85	4.11	4.31	4.47	4.61	4.73	4.84	4.93	5.01	5.09	5.16	5.22	5.29	5.35	5.40	5.44	5.49
30	2.888	3.49	3.85	4.10	4.30	4.46	4.60	4.72	4.82	4.92	5.00	5.08	5.15	5.21	5.27	5.33	5.38	5.43	5.47
31	2.884	3.48	3.84	4.09	4.29	4.45	4.59	4.71	4.82	4.91	4.99	5.07	5.14	5.20	5.26	5.32	5.37	5.41	5.46
32	2.881	3.48	3.83	4.09	4.28	4.45	4.58	4.70	4.81	4.89	4.98	5.06	5.13	5.19	5.24	5.30	5.35	5.40	5.45
33	2.877	3.47	3.83	4.08	4.28	4.44	4.57	4.69	4.80	4.88	4.97	5.04	5.11	5.17	5.23	5.29	5.34	5.39	5.44
34	2.874	3.47	3.82	4.07	4.27	4.43	4.56	4.68	4.79	4.87	4.96	5.03	5.10	5.16	5.22	5.27	5.33	5.37	5.42
35	2.871	3.46	3.81	4.07	4.26	4.42	4.55	4.67	4.78	4.86	4.95	5.02	5.09	5.15	5.21	5.27	5.32	5.36	5.41
36	2.868	3.46	3.81	4.06	4.25	4.41	4.54	4.66	4.77	4.85	4.94	5.01	5.08	5.14	5.20	5.26	5.31	5.35	5.40
37	2.865	3.45	3.80	4.05	4.24	4.41	4.54	4.65	4.76	4.84	4.93	5.00	5.07	5.13	5.19	5.25	5.30	5.34	5.39
38	2.863	3.45	3.80	4.05	4.24	4.40	4.53	4.64	4.75	4.84	4.92	5.00	5.07	5.13	5.18	5.24	5.29	5.33	5.38
39	2.861	3.44	3.79	4.04	4.24	4.40	4.53	4.64	4.75	4.83	4.92	4.99	5.06	5.12	5.17	5.23	5.28	5.32	5.37
40	2.858	3.44	3.79	4.04	4.23	4.39	4.52	4.63	4.73	4.82	4.90	4.98	5.04	5.11	5.16	5.22	5.27	5.31	5.36
50	2.841	3.41	3.76	4.00	4.19	4.34	4.47	4.58	4.69	4.76	4.85	4.92	4.99	5.05	5.10	5.15	5.20	5.24	5.29
60	2.829	3.40	3.74	3.98	4.16	4.31	4.44	4.55	4.65	4.73	4.81	4.88	4.94	5.00	5.06	5.11	5.15	5.20	5.24
120	2.800	3.36	3.68	3.92	4.10	4.24	4.36	4.47	4.56	4.64	4.71	4.78	4.84	4.90	4.95	5.00	5.04	5.09	5.13
∞	2.772	3.31	3.63	3.86	4.03	4.17	4.29	4.39	4.47	4.55	4.62	4.68	4.74	4.80	4.85	4.89	4.93	4.97	5.01

Remark concerning Table 172: Comprehensive tables can be found in volume 1 of Harter (1970 [2], pp. 623–661 [discussed on pp. 21–231]); bounds for $k > 20$, $P = 0.05$, and $P = 0.01$ are given there on pp. 653 and 657 [see also Applied Statistics **32** (1983), 204–210].

Table 172 (continued). Upper significance bounds of the Studentized range distribution: P = 0.01

ν₂\k	2	3	4	5	6	7	8	9	10	11	12	13	14	15	16	17	18	19	20
1	90.025	135.0	164.3	185.6	202.2	215.8	227.2	237.0	245.6	253.2	260.0	266.2	271.8	277.0	281.8	286.3	290.4	294.3	298.0
2	14.036	19.02	22.29	24.72	26.63	28.20	29.53	30.68	31.69	32.59	33.40	34.13	34.81	35.43	36.00	36.53	37.03	37.50	37.95
3	8.260	10.62	12.17	13.33	14.24	15.00	15.64	16.20	16.69	17.13	17.53	17.89	18.22	18.52	18.81	19.07	19.32	19.55	19.77
4	6.511	8.12	9.17	9.96	10.58	11.10	11.55	11.93	12.27	12.57	12.84	13.09	13.32	13.53	13.73	13.91	14.08	14.24	14.40
5	5.702	6.98	7.80	8.42	8.91	9.32	9.67	9.97	10.24	10.48	10.70	10.89	11.08	11.24	11.40	11.55	11.68	11.81	11.93
6	5.243	6.33	7.03	7.56	7.97	8.32	8.61	8.87	9.10	9.30	9.48	9.65	9.81	9.95	10.08	10.21	10.32	10.43	10.54
7	4.949	5.92	6.54	7.01	7.37	7.68	7.94	8.17	8.37	8.55	8.71	8.86	9.00	9.12	9.24	9.35	9.46	9.55	9.65
8	4.745	5.64	6.20	6.62	6.96	7.24	7.47	7.68	7.86	8.03	8.18	8.31	8.44	8.55	8.66	8.76	8.85	8.94	9.03
9	4.596	5.43	5.96	6.35	6.66	6.91	7.13	7.33	7.49	7.65	7.78	7.91	8.03	8.13	8.23	8.33	8.41	8.49	8.57
10	4.482	5.27	5.77	6.14	6.43	6.67	6.87	7.05	7.21	7.36	7.49	7.60	7.71	7.81	7.91	7.99	8.08	8.15	8.23
11	4.392	5.15	5.62	5.97	6.25	6.48	6.67	6.84	6.99	7.13	7.25	7.36	7.46	7.56	7.65	7.73	7.81	7.88	7.95
12	4.320	5.05	5.50	5.84	6.10	6.32	6.51	6.67	6.81	6.94	7.06	7.17	7.26	7.36	7.44	7.52	7.59	7.66	7.73
13	4.260	4.96	5.40	5.73	5.98	6.19	6.37	6.53	6.67	6.79	6.90	7.01	7.10	7.19	7.27	7.35	7.42	7.48	7.55
14	4.210	4.89	5.32	5.63	5.88	6.08	6.26	6.41	6.54	6.66	6.77	6.87	6.96	7.05	7.13	7.20	7.27	7.33	7.39
15	4.167	4.84	5.25	5.56	5.80	5.99	6.16	6.31	6.44	6.55	6.66	6.76	6.84	6.93	7.00	7.07	7.14	7.20	7.26
16	4.131	4.79	5.19	5.49	5.72	5.92	6.08	6.22	6.35	6.46	6.56	6.66	6.74	6.82	6.90	6.97	7.03	7.09	7.15
17	4.099	4.74	5.14	5.43	5.66	5.85	6.01	6.15	6.27	6.38	6.48	6.57	6.66	6.73	6.81	6.87	6.94	7.00	7.05
18	4.071	4.70	5.09	5.38	5.60	5.79	5.94	6.08	6.20	6.31	6.41	6.50	6.58	6.65	6.73	6.79	6.85	6.91	6.97
19	4.045	4.67	5.05	5.33	5.55	5.73	5.89	6.02	6.14	6.25	6.34	6.43	6.51	6.58	6.65	6.72	6.78	6.84	6.89
20	4.024	4.64	5.02	5.29	5.51	5.69	5.84	5.97	6.09	6.19	6.28	6.37	6.45	6.52	6.59	6.65	6.71	6.77	6.82
21	4.004	4.61	4.99	5.26	5.47	5.65	5.80	5.92	6.04	6.14	6.24	6.32	6.39	6.47	6.53	6.59	6.65	6.70	6.76
22	3.986	4.58	4.96	5.22	5.43	5.61	5.76	5.88	6.00	6.10	6.19	6.27	6.35	6.42	6.48	6.54	6.60	6.65	6.70
23	3.970	4.56	4.93	5.19	5.40	5.57	5.72	5.85	5.96	6.06	6.15	6.23	6.30	6.37	6.43	6.49	6.55	6.60	6.65
24	3.955	4.55	4.91	5.17	5.37	5.54	5.69	5.81	5.92	6.02	6.11	6.19	6.26	6.33	6.39	6.45	6.51	6.56	6.61
25	3.942	4.52	4.89	5.15	5.34	5.51	5.66	5.78	5.89	5.99	6.07	6.15	6.22	6.29	6.35	6.41	6.47	6.52	6.57
26	3.930	4.50	4.85	5.12	5.32	5.49	5.63	5.75	5.86	5.95	6.04	6.12	6.19	6.26	6.32	6.38	6.43	6.48	6.53
27	3.918	4.49	4.83	5.10	5.28	5.46	5.61	5.73	5.83	5.93	6.01	6.09	6.16	6.22	6.28	6.34	6.40	6.45	6.50
28	3.908	4.47	4.82	5.08	5.26	5.44	5.58	5.70	5.80	5.90	5.99	6.06	6.13	6.20	6.25	6.31	6.37	6.42	6.47
29	3.898	4.46	4.82	5.07	5.26	5.44	5.56	5.67	5.78	5.87	5.95	6.03	6.10	6.17	6.23	6.29	6.34	6.39	6.44
30	3.889	4.45	4.80	5.05	5.24	5.40	5.54	5.65	5.76	5.85	5.93	6.01	6.08	6.14	6.20	6.26	6.31	6.36	6.41
31	3.881	4.44	4.79	5.03	5.22	5.37	5.52	5.63	5.74	5.83	5.91	5.99	6.06	6.12	6.18	6.23	6.29	6.34	6.38
32	3.873	4.43	4.78	5.02	5.21	5.35	5.50	5.61	5.72	5.81	5.89	5.97	6.03	6.09	6.15	6.21	6.26	6.31	6.36
33	3.865	4.42	4.76	5.01	5.19	5.35	5.48	5.59	5.70	5.79	5.87	5.95	6.01	6.07	6.13	6.19	6.24	6.29	6.34
34	3.859	4.41	4.75	4.99	5.18	5.35	5.47	5.58	5.68	5.77	5.86	5.93	5.99	6.05	6.12	6.17	6.22	6.27	6.31
35	3.852	4.41	4.74	4.98	5.16	5.33	5.45	5.56	5.67	5.76	5.84	5.91	5.98	6.04	6.10	6.15	6.20	6.25	6.30
36	3.846	4.40	4.73	4.97	5.15	5.31	5.44	5.55	5.65	5.74	5.82	5.90	5.96	6.02	6.08	6.13	6.18	6.23	6.28
37	3.841	4.39	4.72	4.96	5.14	5.30	5.43	5.54	5.64	5.72	5.81	5.88	5.94	6.00	6.06	6.11	6.17	6.22	6.26
38	3.835	4.38	4.72	4.95	5.13	5.29	5.42	5.52	5.62	5.71	5.80	5.87	5.93	5.99	6.05	6.10	6.15	6.20	6.24
39	3.830	4.38	4.71	4.94	5.12	5.28	5.40	5.51	5.61	5.70	5.78	5.85	5.91	5.97	6.03	6.08	6.13	6.18	6.23
40	3.825	4.37	4.70	4.93	5.11	5.26	5.39	5.50	5.60	5.69	5.76	5.74	5.90	5.96	6.02	6.07	6.12	6.16	6.21
50	3.787	4.32	4.64	4.86	5.04	5.19	5.30	5.41	5.51	5.59	5.67	5.74	5.80	5.86	5.84	5.96	6.01	6.06	6.09
60	3.762	4.28	4.59	4.82	4.99	5.13	5.25	5.36	5.45	5.53	5.60	5.67	5.73	5.78	5.84	5.89	5.93	5.97	6.01
120	3.702	4.20	4.50	4.71	4.87	5.01	5.12	5.21	5.30	5.37	5.44	5.50	5.56	5.61	5.66	5.71	5.75	5.79	5.83
∞	3.643	4.12	4.40	4.60	4.76	4.88	4.99	5.08	5.16	5.23	5.29	5.35	5.40	5.45	5.49	5.54	5.57	5.61	5.65

Results:

D_I: $\mu_{(1)} = \mu_{(2)} = \mu_{(3)}$, $\mu_{(1)} > \mu_{(4)-(6)}$, $\mu_{(2)} > \mu_{(5),(6)}$, $\mu_{(3)} = \mu_{(4)}$,
$\mu_{(3)} > \mu_{(5),(6)}$, $\mu_{(4)} = \mu_{(5)}$, $\mu_{(4)} > \mu_{(6)}$, $\mu_{(5)} = \mu_{(6)}$,

D_{II}: $\mu_{(1)} > \mu_{(6)-(4)}$, $\mu_{(2)} > \mu_{(6)-(4)}$, $\mu_{(1)} = \mu_{(2)} = \mu_{(3)}$,
$\mu_{(3)} > \mu_{(6)-(4)}$, $\mu_{(4)} > \mu_{(6),(5)}$, $\mu_{(5)} = \mu_{(6)}$,

Tukey-D_{II}: e.g., $\mu_{(1)} = \mu_{(2)} = \mu_{(3)} > \mu_{(4)} > \mu_{(5)} = \mu_{(6)}$.

One uses D_{II} with sample groups of equal size and D_I when sizes are unequal. D_{II} is more sensitive, more selective, but the experimental type I error rate may be greater than α (in other words: D_{II} is the more liberal approximation whereas D_I is the more conservative approximation); D_I is more robust and suitable in particular when one suspects that the variances are unequal. A very fine multiple comparison procedure for means with equal and unequal n's in the case of equal and unequal variance is presented by P. A. Games and J. F. Howell, Journal of Educational Statistics 1 (1976), 113–125 (cf., H. J. Keselman et al., Journal of the American Statistical Association 74 (1979), 626–627). More on other multiple comparisons of group means (cf., Miller 1966, and Seeger 1966) with a control (Dunnett 1955, 1964) or with the overall mean (Enderlein 1972) can be found in fine survey articles by M. R. Stoline [The American Statisitician 35 (1981), 134–141] and Games et al. (1983).

Simultaneous inference: In Section 6.2.1 we discussed the Bonferroni χ^2 procedure. Tables and charts for the Bonferroni t-statistics are provided by B. J. R. Bailey, Journal of the American Statistical Association 72 (1977), 469–478 and by L. E. Moses, Communications in Statistics—Simulation and Computation B7 (1978), 479–490, respectively, together with examples of multiple comparison problems (cf., also P. A. Games, Journal of the American Statistical Association 72 (1977), 531–534). An interesting sequentially rejective Bonferroni multiple test procedure with a prescribed level of significance protection against error of the first kind for any combination of true hypotheses and with applications is presented by S. Holm, Scandinavian Journal of Statistics 6 (1979), 65–70.

▶ 7.4.3 Two way analysis of variance with a single observation per cell. A model without interaction

If it is known that **no interaction** is present, a single observation per cell suffices. The appropriate scheme involves r rows and c columns (Table 173). The associated model, termed **additive model**, reads

$$\begin{array}{ccccccc}
\text{observed} & & \text{overall} & & \text{row} & & \text{column} & & \text{experimental} \\
\text{value} & = & \text{mean} & + & \text{effect} & + & \text{effect} & + & \text{error} \\
x_{ij} & = & \mu & + & \alpha_i & + & \beta_j & + & \varepsilon_{ij}
\end{array} \quad (7.51)$$

Table 173

A\B	1	2	...	j	...	c	Σ
1	x_{11}	x_{12}	...	x_{1j}	...	x_{1c}	$S_{1.}$
2	x_{21}	x_{22}	...	x_{2j}	...	x_{2c}	$S_{2.}$
.	.	.	.	.	.	.	.
i	x_{i1}	x_{i2}	...	x_{ij}	...	x_{ic}	$S_{i.}$
.	.	.	.	.	.	.	.
r	x_{r1}	x_{r2}	...	x_{rj}	...	x_{rc}	$S_{r.}$
Σ	$S_{.1}$	$S_{.2}$	...	$S_{.j}$	...	$S_{.c}$	S

Let the experimental error ε_{ij} be independent and normally distributed with mean zero and variance σ^2 for all i and j. The scheme of the analysis of variance can be found in Table 174. The variability of an observed value in this table is conditioned by three factors which are mutually independent and which act simultaneously: by the row effect, the column effect, and the experimental error. [Note: (7.51) is a noninteraction or additive model, (7.28) is the corresponding nonadditive or interaction model.]

Table 174 Analysis of variance for a two way classification: 1 observation per class, no interaction

Source of variability	Sum of squares	DF	Mean sum of squares
r Rows (row means)	$SSR = \sum_{i=1}^{r} \dfrac{S_{i.}^2}{c} - \dfrac{S^2}{r \cdot c}$	$r - 1$	$\dfrac{SSR}{r - 1}$
c Columns (column means)	$SSC = \sum_{j=1}^{c} \dfrac{S_{.j}^2}{r} - \dfrac{S^2}{r \cdot c}$	$c - 1$	$\dfrac{SSC}{c - 1}$
Experimental error	$SSE = [SST - SSR - SSC]$	$(c - 1)(r - 1)$	$\dfrac{SSE}{(c - 1)(r - 1)}$
Total variability	$SST = \sum_{i=1}^{r} \sum_{j=1}^{c} x_{ij}^2 - \dfrac{S^2}{r \cdot c}$	$rc - 1$	

1. Null hypotheses:
   H_{01}: The row effects are null (row homogeneity);
   H_{02}: The column effects are null (column homogeneity).
   The two null hypotheses are mutually independent.
2. Choice of significance level: $\alpha = 0.05$.

7.4 Two Way and Three Way Analysis of Variance

3. Decision: Under the usual conditions (cf., Section 7.4.1.1),
   H_{01} is rejected if $\hat{F} > F_{(r-1);(r-1)(c-1);0.05}$;
   H_{02} is rejected if $\hat{F} > F_{(c-1);(r-1)(c-1);0.05}$.

EXAMPLE. Two way analysis of variance: 1 observation per class, no interaction. We take our old example and combine the respective double observations (cf., Table 175).

Table 175

B\A	B_1	B_2	B_3	Σ
A_1	11	10	13	34
A_2	9	10	10	29
A_3	12	14	8	34
A_4	15	11	7	33
Σ	47	45	38	130

We have
r = 4 rows and
c = 3 columns

Method of computation: cf., Table 174

$$SST = \sum_{i=1}^{r=4} \sum_{j=1}^{c=3} x_{ij}^2 - \frac{S^2}{r \cdot c} = 11^2 + 10^2 + 13^2 + \cdots + 7^2 - \frac{130^2}{4 \cdot 3} = 61.667.$$

$$SSR = \sum_{i=1}^{r=4} \frac{S_{i.}^2}{c} - \frac{S^2}{r \cdot c} = \frac{34^2}{3} + \frac{29^2}{3} + \frac{34^2}{3} + \frac{33^2}{3} - \frac{130^2}{12}$$

$$= \frac{1}{3}(34^2 + 29^2 + 34^2 + 33^2) - \frac{130^2}{12} = 5.667.$$

$$SSC = \sum_{j=1}^{c=3} \frac{S_{.j}^2}{r} - \frac{S^2}{r \cdot c} = \frac{47^2}{4} + \frac{45^2}{4} + \frac{38^2}{4} - \frac{130^2}{12}$$

$$= \frac{1}{4}(47^2 + 45^2 + 38^2) - \frac{130^2}{12} = 11.167 \quad \text{(cf., Table 176)}.$$

Decision: Both null hypotheses are retained ($P > 0.05$).

These results are due to the fact that the experimental error is overestimated (blown up), and thus the $\hat{F}$-ratio is underestimated, because of the strong interaction—an indication of the **presence of nonlinear effects**, which we may also call regression effects (compare the opposite trends of columns 1 and 3). Cf., Remark 2 below.

Table 176

Source of variability	Sum of squares	DF	Mean sum of squares	$\hat{F}$	$F_{0.05}$
Rows (row means)	5.667	4 − 1 = 3	1.889	0.253	< 4.76
Columns (column means)	11.167	3 − 1 = 2	5.583	0.747	< 5.14
Experimental error	44.833	(4 − 1)(3 − 1) = 6	7.472		
Total variability	61.667	4 · 3 − 1 = 11			

Remarks

1. More on two way analysis of variance can be found in the books presented in [8:7] and [8:7a] (as well as on pages in [1] and [1a]), which also contain substantially **more complicated** models. The **two way classification with unequal numbers of observations within the cells** is considered by Kramer (1955), Rasch (1960), and Bancroft (1968). Five methods and programs are compared in D. G. Herr and J. Gaebelein, Psychological Bulletin **85** (1978), 207–216. M. B. Brown (1975) comments on **interaction** (J. V. Bradley presents a nonparametric test for interaction of any order, Journal of Quality Technology **11** (1979), 177–184).

2. **The Mandel test for nonadditivity. Additivity** is an important assumption of the analysis of variance. When nonadditive effects appear, they are usually treated as interactions (see Weiling 1972 [8:7]; cf., the reference given there]). In the case of the two way analysis of variance with a single observed value per class, the nonadditive effects can be separated from the interaction (see Weiling 1972) and split up into two components by a method due to Mandel (1961). The first part, to which a single degree of freedom is assigned, can be interpreted as the dispersion of a regression; the second, with $r - 2$ degrees of freedom, as dispersion about the regression. Mandel also introduced the designations "concurrence" and "nonconcurrence" for these two parts. The well-known Tukey test [1949; cf. Journal of the American Statistical Association **71** (1967), 945–948, and Biometrics **34** (1978), 505–513] for testing the "absence of additivity" covers only the first, the regression component. Weiling (1963) showed a possible case in which the nonadditive effects can, in accordance with Mandel, be neatly determined. The interested reader is referred to Weiling's works. The procedure is there demonstrated by means of an example. The **testing for nonadditivity is recommended** if in the case of a two way analysis of variance like the one in Section 7.4.1.2, either no or only weak significance was determined and there is suspicion that nonadditive effects could be present. For, under these conditions, the computed experimental error which enters the test statistic is overestimated, since this quantity contains in addition to the actual experimental error the influence of nonadditive effects as well. Hence this test also yields information on the actual level of the random error. **If possible, one should thus carry out the analysis of variance with at least two observa-**

tions under identical conditions. The nonadditivity test is useful in deciding whether a transformation is recommended, and if so, which transformation is appropriate and to what extent it can be regarded as successful. More on this is given by N. A. C. Cressie, Biometric **34** (1978), 505–513. Introductions to the especially interesting field of **transformations** (cf., Section 7.3.3) can be found in Grimm (1960) and Lienert (1962) (cf., also Tukey 1957 and Taylor 1961); on **outliers** see Barnett and Lewis (1978 [8:1]), Hawkins (1980 [8:1]), and Beckman and Cook (1983 [8:1]). Martin (1962) discusses the particular significance of transformations in clinical–therapeutic research (cf., also Snell 1964).

3. If analyses of variance are planned and if well-founded assumptions on the orders of magnitude of the variances or on the expected mean differences can be made, then tables (Bechhofer 1954, Bratcher et al. 1970, Kastenbaum et al. 1970) permit estimation of the **sample sizes** required to achieve a specified power.

4. The comparison of two similar independent experiments with regard to their **sensitivity** can be conveniently carried out following Bradley and Schumann (1957). It is assumed that the two trials agree in the number of the A and in that of the B classifications (model: two way classification with one observation per class, no interaction). More on this can be found in the original work, which also contains the method of computation, examples, and an important table.

5. For **testing the homogeneity of the profiles of independent samples of response curves** measured at identical points of time, a generalization of the Friedman test is given by W. Lehmacher, Biometrical Journal **21** (1979), 123–130; this multivariate test is illustrated by an example (cf., Lehmacher and Wall (1978) as well as Cole and Grizzle (1966)). A **procedure for comparing groups of time-dependent measurements** by fitting cubic spline functions is presented by H. Prestele et al., Methods of Information in Medicine **18** (1979), 84–88.

6. In many studies the experimenter's goal is **to select the best of several alternatives**. For this problem ranking (ordering) and selection procedures were developed. An overview of how to select the best is given by E. J. Dudewicz, Technometrics **22** (1980), 113–119. See also Gibbons et al. (1979), and Dudewicz and Koo (1982).

Order restrictions; (Note Remark 2 in Section 3.9.5 and in Section 6.1.1, Barlow et al. (1972), Bartholomew (1961) (also Bechhofer 1954), and the Page test in Section 7.6.1). A **test on the order of magnitude of k means** is given by Nelson (1977). Included are tables with critical values for testing order alternatives in a one way analysis of variance, $H_A: \mu_1 \leq \mu_2 \leq \ldots \leq \mu_k$ with not all μ_i equal, that is, the k means, each of n observations, are set (on the basis of intuition and/or prior information) in a monotone increasing rank order for $k = 3(1)10; n = 2(1)20, 24, 30, 40, 60; \alpha = 0.10$, 0.05, 0.025, 0.01, 0.001. Tables and examples for $k = 3$ are given by Nelson (1976). Tests for ordered means are compared by E. A. C. Shirley, Applied Statistics **28** (1979), 144–151.

7. Other important aspects of the analysis of variance are discussed by Anscombe and Tukey (1963; see also Cox and Snell 1968 [5]), Bancroft (1968), and Weiling (1972) (cf., also Dunn 1959, 1961, Green and Tukey 1960, Gabriel 1963, Siotani 1964, Searle 1971).

8. The analysis of two-way layout data with interaction and one observation per cell is discussed by Hegemann and Johnson (1976).

7.5 RAPID TESTS OF ANALYSIS OF VARIANCE

7.5.1 Rapid test of analysis of variance and multiple comparisons of means according to Link and Wallace

We assume we are dealing with at least approximately **normal distribution, identical variances and the same sizes** n of the individual sample groups (Link and Wallace 1952, cf., also Kurtz et al. 1965). This rapid test may also be used in two way classification with a single observation per cell.

The k ranges R_i of the individual groups and the range of the means $R_{(\bar{x}_i)}$ will be needed. The null hypothesis $\mu_1 = \mu_2 = \cdots = \mu_i = \cdots = \mu_k$ is rejected in favor of the alternative hypothesis, not all μ_i are equal, whenever

$$\boxed{\frac{nR_{(\bar{x}_i)}}{\sum R_i} > K.} \qquad (7.52)$$

The critical value of K is taken from Table 177 for given n, k and $\alpha = 0.05$ or $\alpha = 0.01$. Multiple comparisons of means with the mean difference D are significant at the given level if

$$\boxed{\hat{D} > \frac{K \sum R_i}{n}.} \qquad (7.52a)$$

Examples

1. Given three sets of measurements A, B, C with the values in Table 178. Since $1.47 > 1.18 = K_{(8;\,3;\,0.05)}$, the null hypothesis $\mu_A = \mu_B = \mu_C$ is rejected. The corresponding analysis of variance with $\hat{F} = 6.05 > 3.47 = F_{(2;\,21;\,0.05)}$ leads to the same decision. With

$$\begin{array}{l}\bar{x}_C - \bar{x}_B = 3.125 \\ \bar{x}_C - \bar{x}_A = 3.000\end{array} > 2.51 = \frac{(1.18)(17)}{8},$$

the null hypotheses $\mu_A = \mu_C$ and $\mu_B = \mu_C$ can also be rejected; since $\bar{x}_A - \bar{x}_B = 0.125 < 2.51$, it follows that $\mu_A = \mu_B \neq \mu_C$.

2. Given 4 samples with 10 observations each (Table 179). The "triangle" of differences D of means indicates (since $\bar{x}_4 - \bar{x}_1 = 2 > 1.46$) that the special hypothesis $\mu_1 = \mu_4$ must be rejected at the 1% level.

Table 177 Critical values K for the test of Link and Wallace. P = 0.05. k is the number of sample groups; n is the size of the sample groups. (Taken from Kurtz, T. E., Link, R. F., Tukey, J. W., and Wallace, D. L.: Short-cut multiple comparisons for balanced single and double classifications: Part 1, Results. Technometrics **7** (1965), 95–161.)

n \ k	2	3	4	5	6	7	8	9	10	11	12	13	14	15	16	17	18	19	20	30	40	50
2	3.43	2.35	1.74	1.39	1.15	0.99	0.87	0.77	0.70	0.63	0.58	0.54	0.50	0.47	0.443	0.418	0.396	0.376	0.358	0.245	0.187	0.151
3	1.90	1.44	1.14	0.94	0.80	0.70	0.62	0.56	0.51	0.47	0.43	0.40	0.38	0.35	0.335	0.317	0.301	0.287	0.274	0.189	0.146	0.119
4	1.62	1.25	1.01	0.84	0.72	0.63	0.57	0.51	0.47	0.43	0.40	0.37	0.35	0.33	0.310	0.294	0.279	0.266	0.254	0.177	0.136	0.112
5	1.53	1.19	0.96	0.81	0.70	0.61	0.55	0.50	0.45	0.42	0.39	0.36	0.34	0.32	0.303	0.287	0.273	0.260	0.249	0.173	0.134	0.110
6	1.50	1.17	0.95	0.80	0.69	0.61	0.55	0.49	0.45	0.42	0.39	0.36	0.34	0.32	0.302	0.287	0.273	0.260	0.249	0.174	0.135	0.111
7	1.49	1.17	0.95	0.80	0.69	0.61	0.55	0.50	0.46	0.42	0.39	0.36	0.34	0.32	0.304	0.289	0.275	0.262	0.251	0.175	0.136	0.111
8	1.49	1.18	0.96	0.81	0.70	0.62	0.55	0.50	0.46	0.42	0.39	0.37	0.35	0.33	0.308	0.292	0.278	0.265	0.254	0.178	0.138	0.113
9	1.50	1.19	0.97	0.82	0.71	0.62	0.56	0.51	0.47	0.43	0.40	0.37	0.35	0.33	0.312	0.297	0.282	0.269	0.258	0.180	0.140	0.115
10	1.52	1.20	0.98	0.83	0.72	0.63	0.57	0.52	0.47	0.44	0.41	0.38	0.36	0.34	0.317	0.301	0.287	0.274	0.262	0.183	0.142	0.117
11	1.54	1.22	0.99	0.84	0.73	0.64	0.58	0.52	0.48	0.44	0.41	0.38	0.36	0.34	0.322	0.306	0.291	0.278	0.266	0.186	0.145	0.119
12	1.56	1.23	1.01	0.85	0.74	0.65	0.58	0.53	0.49	0.45	0.42	0.39	0.37	0.35	0.327	0.311	0.296	0.282	0.270	0.189	0.147	0.121
13	1.58	1.25	1.02	0.86	0.75	0.66	0.59	0.54	0.49	0.46	0.42	0.40	0.37	0.35	0.332	0.316	0.300	0.287	0.274	0.192	0.149	0.122
14	1.60	1.26	1.03	0.87	0.76	0.67	0.60	0.55	0.50	0.46	0.43	0.40	0.38	0.36	0.337	0.320	0.305	0.291	0.279	0.195	0.152	0.124
15	1.62	1.28	1.05	0.89	0.77	0.68	0.61	0.55	0.51	0.47	0.44	0.41	0.38	0.36	0.342	0.325	0.310	0.295	0.283	0.198	0.154	0.126
16	1.64	1.30	1.06	0.90	0.78	0.69	0.62	0.56	0.52	0.48	0.44	0.41	0.39	0.37	0.348	0.330	0.314	0.300	0.287	0.201	0.156	0.128
17	1.66	1.32	1.08	0.91	0.79	0.70	0.63	0.57	0.52	0.48	0.45	0.42	0.39	0.37	0.352	0.335	0.319	0.304	0.291	0.204	0.158	0.130
18	1.68	1.33	1.09	0.92	0.80	0.71	0.64	0.58	0.53	0.49	0.46	0.43	0.40	0.38	0.357	0.339	0.323	0.308	0.295	0.207	0.161	0.132
19	1.70	1.35	1.10	0.93	0.81	0.72	0.64	0.59	0.54	0.50	0.46	0.43	0.41	0.38	0.362	0.344	0.327	0.312	0.299	0.210	0.163	0.134
20	1.72	1.36	1.12	0.95	0.82	0.73	0.65	0.59	0.54	0.50	0.47	0.44	0.41	0.39	0.367	0.348	0.332	0.317	0.303	0.212	0.165	0.135
30	1.92	1.52	1.24	1.05	0.91	0.81	0.73	0.66	0.60	0.56	0.52	0.49	0.46	0.43	0.408	0.387	0.369	0.352	0.337	0.237	0.184	0.151
40	2.08	1.66	1.35	1.14	0.99	0.88	0.79	0.72	0.66	0.61	0.57	0.53	0.50	0.47	0.444	0.422	0.402	0.384	0.367	0.258	0.201	0.165
50	2.23	1.77	1.45	1.22	1.06	0.94	0.85	0.77	0.71	0.65	0.61	0.57	0.53	0.50	0.476	0.453	0.431	0.412	0.394	0.277	0.216	0.177
200	2.81	2.23	1.83	1.55	1.34	1.19	1.07	0.97	0.89	0.83	0.77	0.72	0.67	0.64	0.60	0.573	0.546	0.521	0.499	0.351	0.273	0.224
200	3.61	2.88	2.35	1.99	1.73	1.53	1.38	1.25	1.15	1.06	0.99	0.93	0.87	0.82	0.78	0.74	0.70	0.67	0.64	0.454	0.353	0.290
500	5.15	4.10	3.35	2.84	2.47	2.19	1.97	1.79	1.64	1.52	1.42	1.32	1.24	1.17	1.11	1.06	1.01	0.96	0.92	0.65	0.504	0.414
1000	6.81	5.43	4.44	3.77	3.28	2.90	2.61	2.37	2.18	2.02	1.88	1.76	1.65	1.56	1.47	1.40	1.33	1.27	1.22	0.85	0.669	0.549

Table 177 (continued). Critical values K for the test of Link and Wallace. P = 0.01

n\k	2	3	4	5	6	7	8	9	10	11	12	13	14	15	16	17	18	19	20	30	40	50
2	7.92	4.32	2.84	2.10	1.66	1.38	1.17	1.02	0.91	0.82	0.74	0.68	0.63	0.58	0.54	0.51	0.480	0.454	0.430	0.285	0.214	0.172
3	3.14	2.12	1.57	1.25	1.04	0.89	0.78	0.69	0.62	0.57	0.52	0.48	0.45	0.42	0.39	0.37	0.352	0.334	0.318	0.217	0.165	0.134
4	2.48	1.74	1.33	1.08	0.91	0.78	0.68	0.62	0.56	0.51	0.47	0.44	0.41	0.38	0.36	0.34	0.323	0.307	0.293	0.200	0.153	0.125
5	2.24	1.60	1.24	1.02	0.86	0.75	0.66	0.59	0.54	0.49	0.46	0.42	0.40	0.37	0.35	0.33	0.314	0.299	0.285	0.196	0.151	0.123
6	2.14	1.55	1.21	0.99	0.85	0.74	0.65	0.59	0.53	0.49	0.45	0.42	0.39	0.37	0.35	0.33	0.313	0.298	0.284	0.196	0.151	0.123
7	2.10	1.53	1.20	0.99	0.84	0.73	0.65	0.59	0.53	0.49	0.45	0.42	0.39	0.37	0.35	0.33	0.314	0.299	0.286	0.198	0.152	0.124
8	2.09	1.54	1.20	0.99	0.85	0.74	0.66	0.59	0.54	0.49	0.46	0.43	0.40	0.37	0.35	0.33	0.318	0.303	0.289	0.200	0.154	0.126
9	2.09	1.55	1.21	1.00	0.85	0.75	0.66	0.60	0.54	0.50	0.46	0.43	0.40	0.38	0.36	0.34	0.318	0.303	0.289	0.200	0.154	0.126
10	2.11	1.55	1.22	1.01	0.86	0.75	0.66	0.60	0.54	0.50	0.46	0.43	0.40	0.38	0.36	0.34	0.322	0.307	0.293	0.203	0.156	0.127
11	2.11	1.56	1.23	1.02	0.87	0.76	0.67	0.61	0.55	0.51	0.47	0.44	0.41	0.38	0.36	0.34	0.327	0.311	0.297	0.206	0.159	0.129
12	2.13	1.58	1.25	1.04	0.89	0.78	0.68	0.62	0.56	0.51	0.48	0.44	0.41	0.39	0.37	0.35	0.332	0.316	0.302	0.209	0.161	0.132
13	2.15	1.60	1.26	1.05	0.90	0.79	0.69	0.62	0.57	0.52	0.48	0.45	0.42	0.40	0.37	0.35	0.337	0.321	0.306	0.213	0.164	0.134
14	2.18	1.62	1.28	1.06	0.91	0.80	0.70	0.63	0.58	0.53	0.49	0.46	0.43	0.40	0.38	0.36	0.342	0.326	0.311	0.216	0.166	0.136
15	2.20	1.63	1.30	1.08	0.92	0.81	0.71	0.64	0.58	0.54	0.50	0.46	0.43	0.41	0.39	0.36	0.347	0.330	0.316	0.219	0.169	0.138
16	2.22	1.65	1.31	1.09	0.93	0.82	0.72	0.65	0.59	0.54	0.50	0.47	0.44	0.41	0.39	0.37	0.352	0.335	0.320	0.222	0.171	0.140
17	2.25	1.67	1.33	1.10	0.95	0.83	0.73	0.66	0.60	0.55	0.51	0.48	0.45	0.42	0.40	0.38	0.357	0.340	0.325	0.226	0.174	0.142
18	2.27	1.69	1.34	1.12	0.96	0.84	0.74	0.67	0.61	0.56	0.52	0.48	0.45	0.43	0.40	0.38	0.362	0.345	0.329	0.229	0.176	0.144
19	2.32	1.71	1.36	1.13	0.97	0.85	0.75	0.68	0.62	0.57	0.53	0.49	0.46	0.43	0.41	0.39	0.367	0.350	0.334	0.232	0.179	0.146
20	2.32	1.73	1.38	1.14	0.98	0.86	0.76	0.69	0.62	0.57	0.53	0.50	0.46	0.44	0.41	0.39	0.372	0.354	0.338	0.235	0.181	0.148
30	2.59	1.95	1.54	1.27	1.09	0.96	0.85	0.77	0.70	0.65	0.60	0.56	0.52	0.49	0.46	0.44	0.419	0.399	0.381	0.266	0.205	0.168
40	2.80	2.11	1.66	1.38	1.18	1.04	0.93	0.84	0.76	0.70	0.65	0.61	0.57	0.54	0.51	0.48	0.456	0.435	0.415	0.289	0.223	0.183
50	2.99	2.25	1.78	1.48	1.27	1.11	0.99	0.90	0.82	0.75	0.70	0.65	0.61	0.57	0.54	0.51	0.489	0.466	0.446	0.310	0.240	0.196
100	3.74	2.83	2.24	1.86	1.60	1.40	1.25	1.13	1.03	0.95	0.88	0.82	0.77	0.73	0.69	0.65	0.62	0.590	0.564	0.393	0.304	0.248
200	4.79	3.63	2.88	2.39	2.06	1.81	1.61	1.46	1.33	1.23	1.14	1.06	0.99	0.94	0.88	0.84	0.80	0.76	0.73	0.507	0.392	0.320
500	6.81	5.16	4.10	3.41	2.93	2.58	2.30	2.08	1.90	1.75	1.62	1.52	1.42	1.34	1.26	1.20	1.14	1.09	1.04	0.73	0.560	0.458
1000	9.01	6.83	5.42	4.52	3.88	3.41	3.05	2.76	2.52	2.32	2.15	2.01	1.88	1.77	1.68	1.59	1.51	1.44	1.38	0.96	0.743	0.608

7.5 Rapid Tests of Analysis of Variance

Table 178

	A	B	C
	3	4	6
	5	4	7
	2	3	8
	4	8	6
	8	7	7
	4	4	9
	3	2	10
	9	5	9
$\bar{x}_i$	4.750	4.625	7.750
R_i	7	6	4

$n=8$
$k=3$

$$\frac{nR_{(\bar{x}_i)}}{\sum R_i} = \frac{8(7.750-4.625)}{7+6+4} = 1.47.$$

Table 179

	$\bar{x}_i$		R_i
$\bar{x}_1$ =	10	R_1 =	3
$\bar{x}_2$ =	11	R_2 =	3
$\bar{x}_3$ =	11	R_3 =	2
$\bar{x}_4$ =	12	R_4 =	4
$R(\bar{x}_i)$ =	2	$\sum R_i$ =	12

	$\bar{x}_1$	$\bar{x}_2$	$\bar{x}_3$	$\bar{x}_4$
$\bar{x}_4$	2	1	1	
$\bar{x}_3$	1			
$\bar{x}_2$	1			
$\bar{x}_1$				

$n=10, \quad k=4, \quad \alpha=0.01,$

$$\frac{nR_{(\bar{x}_i)}}{\sum R_i} = \frac{10 \cdot 2}{12} = 1.67 > 1.22 = K_{(10;\,4;\,0.01)},$$

$$\frac{K\sum R_i}{n} = \frac{1.22 \cdot 12}{10} = 1.46.$$

7.5.2 Distribution-free multiple comparisons of independent samples according to Nemenyi: Pairwise comparisons of all possible pairs of treatments

If several variously treated sample groups of equal size (for unequal sizes see (7.53) at the end of this section) are given and if all these groups or treatment effects are to be compared with each other and tested for possible differences, then a **rank test** proposed by Nemenyi (1963) is used as a **method which is good for "nonnormally" distributed data.** The samples come from k populations with **continuous distributions of the same type**. Two other distribution-free multiple comparison procedures may be found in my booklet (Sachs 1984).

The test in detail: Given k treatment groups of n elements each. Ranks are assigned to the nk observed values of the combined sample; the smallest observation is given the rank 1, the largest the rank nk. Observed values of equal size are given average ranks. If one adds the ranks in the individual

Table 180 Critical differences D for the one way classification: comparison of all possible pairs of treatments according to Nemenyi. P = 0.10 (two sided). (From Wilcoxon, F. and Wilcox, R. A.: Some Rapid Approximate Statistical Procedures, Lederle Laboratories, Pearl River, New York, 1964, 29-31.)

n	k = 3	k = 4	k = 5	k = 6	k = 7	k = 8	k = 9	k = 10
1	2.9	4.2	5.5	6.8	8.2	9.6	11.1	12.5
2	7.6	11.2	14.9	18.7	22.5	26.5	30.5	34.5
3	13.8	20.2	26.9	33.9	40.9	48.1	55.5	63.0
4	20.9	30.9	41.2	51.8	62.6	73.8	85.1	96.5
5	29.0	42.9	57.2	72.1	87.3	102.8	118.6	134.6
6	37.9	56.1	75.0	94.5	114.4	134.8	155.6	176.6
7	47.6	70.5	94.3	118.8	144.0	169.6	195.8	222.3
8	58.0	86.0	115.0	145.0	175.7	207.0	239.0	271.4
9	69.1	102.4	137.0	172.8	209.4	246.8	284.9	323.6
10	80.8	119.8	160.3	202.2	245.1	288.9	333.5	378.8
11	93.1	138.0	184.8	233.1	282.6	333.1	384.6	436.8
12	105.9	157.1	210.4	265.4	321.8	379.3	438.0	497.5
13	119.3	177.0	237.1	299.1	362.7	427.6	493.7	560.8
14	133.2	197.7	264.8	334.1	405.1	477.7	551.6	626.6
15	147.6	219.1	293.6	370.4	449.2	529.6	611.6	694.8
16	162.5	241.3	323.3	407.9	494.7	583.3	673.6	765.2
17	177.9	264.2	353.9	446.6	541.6	638.7	737.6	837.9
18	193.7	287.7	385.5	486.5	590.0	695.7	803.4	912.8
19	210.0	311.9	417.9	527.5	639.7	754.3	871.2	989.7
20	226.7	336.7	451.2	569.5	690.7	814.5	940.7	1068.8
21	243.8	362.2	485.4	612.6	743.0	876.2	1012.0	1149.5
22	261.3	388.2	520.4	656.8	796.6	939.4	1085.0	1232.7
23	279.2	414.9	556.1	702.0	851.4	1004.1	1159.7	1317.6
24	297.5	442.2	592.7	748.1	907.4	1070.2	1236.0	1404.3
25	316.2	470.0	630.0	795.3	964.6	1137.6	1314.0	1492.9

7.5 Rapid Tests of Analysis of Variance

Table 180 (*continued*): P = 0.05 (two sided)

n	k = 3	k = 4	k = 5	k = 6	k = 7	k = 8	k = 9	k = 10
1	3.3	4.7	6.1	7.5	9.0	10.5	12.0	13.5
2	8.8	12.6	16.5	20.5	24.7	28.9	33.1	37.4
3	15.7	22.7	29.9	37.3	44.8	52.5	60.3	68.2
4	23.9	34.6	45.6	57.0	68.6	80.4	92.4	104.6
5	33.1	48.1	63.5	79.3	95.5	112.0	128.8	145.8
6	43.3	62.9	83.2	104.0	125.3	147.0	169.1	191.4
7	54.4	79.1	104.6	130.8	157.6	184.9	212.8	240.9
8	66.3	96.4	127.6	159.6	192.4	225.7	259.7	294.1
9	78.9	114.8	152.0	190.2	229.3	269.1	309.6	350.6
10	92.3	134.3	177.8	222.6	268.4	315.0	362.4	410.5
11	106.3	154.0	205.0	256.6	309.4	363.2	417.9	473.3
12	120.9	176.2	233.4	292.2	352.4	413.6	476.0	539.1
13	136.2	198.5	263.0	329.3	397.1	466.2	536.5	607.7
14	152.1	221.7	293.8	367.8	443.6	520.8	599.4	679.0
15	168.6	245.7	325.7	407.8	491.9	577.4	664.6	752.8
16	185.6	270.6	358.6	449.1	541.7	635.9	732.0	829.2
17	203.1	296.2	392.6	491.7	593.1	696.3	801.5	907.9
18	221.2	322.6	427.6	535.5	646.1	758.5	873.1	989.0
19	239.8	349.7	463.6	580.6	700.5	822.4	946.7	1072.4
20	258.8	377.6	500.5	626.9	756.6	888.1	1022.3	1158.1
21	278.4	406.1	538.4	674.4	813.7	955.4	1099.8	1245.9
22	298.4	435.3	577.2	723.0	872.3	1024.3	1179.1	1335.7
23	318.9	465.2	616.9	772.7	932.4	1094.8	1260.3	1427.7
24	339.8	495.8	657.4	823.5	993.7	1166.8	1343.2	1521.7
25	361.1	527.0	698.8	875.4	1056.3	1240.4	1427.9	1617.6

Table 180 (*continued*): P = 0.01 (two sided)

n	k = 3	k = 4	k = 5	k = 6	k = 7	k = 8	k = 9	k = 10
1	4.1	5.7	7.3	8.9	10.5	12.2	13.9	15.6
2	10.9	15.3	19.7	24.3	28.9	33.6	38.3	43.1
3	19.5	27.5	35.7	44.0	52.5	61.1	69.8	78.6
4	29.7	41.9	54.5	67.3	80.3	93.6	107.0	120.6
5	41.2	58.2	75.8	93.6	111.9	130.4	149.1	168.1
6	53.9	76.3	99.3	122.8	146.7	171.0	195.7	220.6
7	67.6	95.8	124.8	154.4	184.6	215.2	246.3	277.7
8	82.4	116.8	152.2	188.4	225.2	262.6	300.6	339.0
9	98.1	139.2	181.4	224.5	268.5	313.1	358.4	404.2
10	114.7	162.8	212.2	262.7	314.2	366.5	419.5	473.1
11	132.1	187.7	244.6	302.9	362.2	422.6	483.7	545.6
12	150.4	213.5	278.5	344.9	412.5	481.2	551.0	621.4
13	169.4	240.6	313.8	388.7	464.9	542.4	621.0	700.5
14	189.1	268.7	350.5	434.2	519.4	606.0	693.8	782.6
15	209.6	297.8	388.5	481.3	575.8	671.9	769.3	867.7
16	230.7	327.9	427.9	530.1	634.2	740.0	847.3	955.7
17	252.5	359.0	468.4	580.3	694.4	810.2	927.8	1046.5
18	275.0	391.0	510.2	632.1	756.4	882.6	1010.6	1140.0
19	298.1	423.8	553.1	685.4	820.1	957.0	1095.8	1236.2
20	321.8	457.6	597.2	740.0	885.5	1033.3	1183.3	1334.9
21	346.1	492.2	642.4	796.0	952.6	1111.6	1273.0	1436.0
22	371.0	527.6	688.7	853.4	1021.3	1191.8	1364.8	1539.7
23	396.4	563.8	736.0	912.1	1091.5	1273.8	1458.8	1645.7
24	422.4	600.9	784.4	972.1	1163.4	1357.6	1554.8	1754.0
25	449.0	638.7	833.8	1033.3	1236.7	1443.2	1652.8	1864.6

treatment groups and forms all possible absolute differences of the sums, these can then be tested in terms of a critical value D. If the computed difference is equal to or greater than the critical value D, given in Table 180 for a chosen significance level and the values n and k, then there is a genuine difference between the two treatments. If it is less, the two groups are equivalent at the given significance level. More on this can be found in the book by Miller (1966 [8:7a]).

EXAMPLE. In a pilot experiment, 20 rats are partitioned into 4 feeding groups. The weights after 70 days are listed in Table 181, with the ranks as well as their column sums given to the right of the weights (Table 181). The

Table 181

I		II		III		IV	
203	12	213	16	171	5	207	13
184	7 1/2	246	18	208	14	152	2
169	4	184	7 1/2	260	19	176	6
216	17	282	20	193	10	200	11
209	15	190	9	160	3	145	1
	55 1/2		70 1/2		51		33

absolute differences of the rank column sums (Table 182) are then compared with the critical difference D for $n = 5$ and $k = 4$ at the 10% level. Table 180 ($P = 0.10$; $k = 4$; $n = 5$) gives $D = 42.9$. All the differences are smaller than D. A difference between the feeding groups II and IV could perhaps be ascertained by larger sample size.

Table 182

	II (70½)	III (51)	IV (33)
I (55½)	15	4½	22½
II (70½)		19½	37½
III (51)			18

When needed, **additional values of D for $k > 10$ and $n = 1(1)20$** can be computed according to $D = W\sqrt{n(nk)(nk + 1)/12}$, where for $P = 0.05 \, (0.01)$, W is read from the bottom row of Table 172, and for other values of P it is interpolated in Table 23 of the Biometrika Tables (Pearson and Hartley 1966, pp. 178–183). For example, in Table 180, $P = 0.05$, $n = 25$, $k = 10$: 1,617.6; $\sqrt{25(25)(10)((25)(10) + 1)/12} = \sqrt{} = 361.5649$; (1) Table 172, $k = 10$: $W = 4.47$ and $W\sqrt{} = 1{,}616.2$; (2) Table 23, p. 180, column 10; $P' = 0.95$: $W = 4.4745$ and $W\sqrt{} = 1{,}617.8$.

Nemenyi test for unequal sample sizes

This test allows for multiple comparisons of k sample mean ranks. Let $\bar{R}_i = R_i/n_i$ denote the mean of the ranks corresponding to the ith sample, and let $\bar{R}_{i'} = R_{i'}/n_{i'}$ be the analogous mean of the ranks for the i'th sample. The null hypothesis (the expected values of two among k independent sample rank means are equal) is rejected at the $100\alpha\%$ level if

$$|\bar{R}_i - \bar{R}_{i'}| > \sqrt{\chi^2_{k-1;\alpha}\left[\frac{n(n+1)}{12}\right]\left[\frac{n_i + n_{i'}}{n_i n_{i'}}\right]}$$

where 1. $k \geq 4$,
2. $n_i, n_{i'} \geq 6$, and
3. n = total number of observations in all samples; at least 75% of the n observations should be nonidentical, i.e., less than 25% of the observations may be involved in ties.

(7.53)

The samples come from k populations with continuous distributions of the same type. The test can be used to make all $k(k-1)/2$ pairwise comparisons among the k populations with an experimental error rate less than α. For $k = 4$ we have $4(4-1)/2 = 6$ comparisons, and with $\alpha = 0.05$ we get from Table 28a for $k - 1 = 4 - 1 = 3$ degrees of freedom the value $\chi^2_{3;0.05} = 7.81$.

7.6 RANK ANALYSIS OF VARIANCE FOR SEVERAL CORRELATED SAMPLES

▶ 7.6.1 The Friedman test: Double partitioning with a single observation per cell

In Sections 3.9.5 and 7.5.2 we dealt with the distribution-free comparison of several *independent* samples. The rank analysis of variance developed by Friedman (1937), a two way analysis of variance on the ranks, allows a distribution-free **comparison of several correlated samples** of data with respect to their central tendency. n individuals, sample groups or blocks (cf., Section 7.7) are to be studied under k conditions. As an example, see Table 184 with four penicillin preparations on three agar plates, or $k = 4$ conditions [treatments] and $n = 3$ individuals [blocks]. If the experimental units are partitioned into groups of k each, care must be taken that the k elements of a block are as homogeneous as possible with respect to a control characteristic which is correlated as strongly as possible with the characteristic under study. The k individuals in each of the blocks are then assigned randomly to the k conditions. The ranks are written in a scheme such that the columns represent the conditions and the rows the blocks.

Under the hypothesis that the various conditions exert no influence, **the ranks are assigned randomly to the k conditions within each of the n individuals or blocks**. Under the null hypothesis the rank sums deviate from each other only randomly if at all. If however the individual conditions do exert a systematic influence, then the k columns originate in different populations and exhibit different rank sums. Friedman (1937) has provided a test statistic $\hat{\chi}_R^2$ for testing the null hypothesis that there is no treatment effect in a randomized block design with k treatments and n blocks, or to put it more simply, **the k columns originate in the same population**:

$$\hat{\chi}_R^2 = \left[\frac{12}{nk(k+1)} \sum_{i=1}^{k} R_i^2\right] - 3n(k+1), \qquad (7.54)$$

where

n = number of rows (which are assumed independent of each other but not homogeneous among themselves): individuals, replications, sample groups, **blocks**,

k = number of columns (with random ordering of the): conditions, **treatments**, types, factors (to the test units),

$\sum_{i=1}^{k} R_i^2$ = sum of squares of the column rank sums for the k factors, treatments, or conditions to be compared.

When the samples are not too small, the test statistic $\hat{\chi}_R^2$ is distributed almost like χ^2 for $k-1$ degrees of freedom. For small values of n, this approximation is inadequate. Table 183 contains 5% and 1% bounds. Thus a $\hat{\chi}_R^2 = 9.000$ for $k=3$ and $n=8$ is significant at the 1% level. For more tables see R. E. Odeh, Communications in Statistics—Simulation and Computation **B6** (1977), 29–48. For good approximations see R. L. Iman and J. M. Davenport, Communications in Statistics—Theory and Methods **A9** (1980), 571–595.

Ties within a row (i.e., equal data or mean ranks) are, strictly speaking, not allowed; the computation then follows Victor (1972):

$$\hat{\chi}_{R,T}^2 = \left\{ n \Big/ \left[n - \frac{1}{k^3 - k} \left(\sum_{i=1}^{n} \sum_{j=1}^{r_i} (t_{ij}^3 - t_{ij}) \right) \right] \right\} \cdot \hat{\chi}_R^2 \qquad (7.55)$$

with r_i the number of ties within the ith row of the ith block and t_{ij} the multiplicity (see also Section 3.9.4) of the jth tie in the ith block.

If we wish to know whether there are considerable differences among the individuals or groups under investigation, we set up ranks within the individual columns and sum the row ranks. Computationally we merely have to interchange the symbols k and n in the above formula.

7.6 Rank Analysis of Variance for Several Correlated Samples

Table 183 5% and 1% bounds for the Friedman test (from Michaelis, J.: Threshold values of the Friedman test, Biometr. Zeitschr. 13 (1971), pp. 118-129, p. 122 by permission of the author and Akademie-Verlag, Berlin)

Threshold values of χ_R^2 for $P = 0.05$ approximated by the F-distribution; enclosed by line: exact values for $P \leq 0.05$

n/k	3	4	5	6	7	8	9	10	11	12	13	14	15
3	6.000	7.4	8.53	9.86	11.24	12.57	13.88	15.19	16.48	17.76	19.02	20.27	21.53
4	6.500	7.8	8.8	10.24	11.63	12.99	14.34	15.67	16.98	18.3	19.6	20.9	22.1
5	6.400	7.8	8.99	10.43	11.84	13.23	14.59	15.93	17.27	18.6	19.9	21.2	22.4
6	7.000	7.6	9.08	10.54	11.97	13.38	14.76	16.12	17.4	18.8	20.1	21.4	22.7
7	7.143	7.8	9.11	10.62	12.07	13.48	14.87	16.23	17.6	18.9	20.2	21.5	22.8
8	6.250	7.65	9.19	10.68	12.14	13.56	14.95	16.32	17.7	19.0	20.3	21.6	22.9
9	6.222	7.66	9.22	10.73	12.19	13.61	15.02	16.40	17.7	19.1	20.4	21.7	23.0
10	6.200	7.67	9.25	10.76	12.23	13.66	15.07	16.44	17.8	19.2	20.5	21.8	23.1
11	6.545	7.68	9.27	10.79	12.27	13.70	15.11	16.48	17.9	19.2	20.5	21.8	23.1
12	6.167	7.70	9.29	10.81	12.29	13.73	15.15	16.53	17.9	19.3	20.6	21.9	23.2
13	6.000	7.70	9.30	10.83	12.32	13.76	15.17	16.56	17.9	19.3	20.6	21.9	23.2
14	6.143	7.71	9.32	10.85	12.34	13.78	15.19	16.58	17.9	19.3	20.6	21.9	23.2
15	6.400	7.72	9.33	10.87	12.35	13.80	15.20	16.6	18.0	19.3	20.6	21.9	23.2
16	5.99	7.73	9.34	10.88	12.37	13.81	15.23	16.6	18.0	19.3	20.7	22.0	23.2
17	5.99	7.73	9.34	10.89	12.38	13.83	15.2	16.6	18.0	19.3	20.7	22.0	23.3
18	5.99	7.73	9.36	10.90	12.39	13.83	15.2	16.6	18.0	19.4	20.7	22.0	23.3
19	5.99	7.74	9.36	10.91	12.40	13.8	15.3	16.7	18.0	19.4	20.7	22.0	23.3
20	5.99	7.74	9.37	10.92	12.41	13.8	15.3	16.7	18.0	19.4	20.7	22.0	23.3
∞	5.99	7.82	9.49	11.07	12.59	14.07	15.51	16.92	18.31	19.68	21.03	22.36	23.69

Threshold values of χ_R^2 for $P = 0.01$ approximated by the F-distribution; enclosed by line: exact values for $P \leq 0.01$

n/k	3	4	5	6	7	8	9	10	11	12	13	14	15
3	—	9.000	10.13	11.76	13.26	14.78	16.28	17.74	19.19	20.61	22.00	23.38	24.76
4	8.000	9.600	11.20	12.59	14.19	15.75	17.28	18.77	20.24	21.7	23.1	24.5	25.9
5	8.400	9.96	11.43	13.11	14.74	16.32	17.86	19.37	20.86	22.3	23.7	25.2	26.6
6	9.000	10.200	11.75	13.45	15.10	16.69	18.25	19.77	21.3	22.7	24.2	25.6	27.0
7	8.857	10.371	11.97	13.69	15.35	16.95	18.51	20.04	21.5	23.0	24.4	25.9	27.3
8	9.000	10.35	12.14	13.87	15.53	17.15	18.71	20.24	21.8	23.2	24.7	26.1	27.5
9	8.667	10.44	12.27	14.01	15.68	17.29	18.87	20.42	21.9	23.4	24.95	26.3	27.7
10	9.600	10.53	12.38	14.12	15.79	17.41	19.00	20.53	22.0	23.5	25.0	26.4	27.9
11	9.455	10.60	12.46	14.21	15.89	17.52	19.10	20.64	22.1	23.6	25.1	26.6	28.0
12	9.500	10.68	12.53	14.28	15.96	17.59	19.19	20.73	22.2	23.7	25.2	26.7	28.0
13	9.385	10.72	12.58	14.34	16.03	17.67	19.25	20.80	22.3	23.8	25.3	26.7	28.1
14	9.000	10.76	12.64	14.40	16.09	17.72	19.31	20.86	22.4	23.9	25.3	26.8	28.2
15	8.933	10.80	12.68	14.44	16.14	17.78	19.35	20.9	22.4	23.9	25.4	26.8	28.2
16	8.79	10.84	12.72	14.48	16.18	17.81	19.40	20.9	22.5	24.0	25.4	26.9	28.3
17	8.81	10.87	12.74	14.52	16.22	17.85	19.50	21.0	22.5	24.0	25.4	26.9	28.3
18	8.84	10.90	12.78	14.56	16.25	17.87	19.5	21.1	22.6	24.1	25.5	26.9	28.3
19	8.86	10.92	12.81	14.58	16.27	17.90	19.5	21.1	22.6	24.1	25.5	27.0	28.4
20	8.87	10.94	12.83	14.60	16.30	18.00	19.5	21.1	22.6	24.1	25.5	27.0	28.4
∞	9.21	11.35	13.28	15.09	16.81	18.48	20.09	21.67	23.21	24.73	26.22	27.69	29.14

If χ_R^2 equals or exceeds the tabulated values for k, n, and P, then, at the given level, not all k columns originated in a common population.

Several additional bounds for testing at the 10% and 0.1% level with small k and small n:

k	3	3	3	3	3	3	3	3	4	4	4	4	4	5	6
n	4	5	6	7	8	9	10	11	12	3	4	5	6	7	8	4	3
P < 0.10	6.000	5.200	5.333	5.429	5.250	5.556	5.000	5.091	5.167	6.600	6.300	6.360	6.400	6.429	6.450	7.600	8.714
P < 0.001	—	10.000	12.000	12.286	12.250	12.667	12.600	13.273	12.667	—	11.100	12.600	12.800	13.457	13.800	13.200	13.286

The Friedman test is a homogeneity test for k matched samples if normality and common variance cannot be assumed. It is the natural extension of the sign test for $k > 2$. It checks whether the samples dealt with could originate in the same population:

H_0: all k distributions are identical,

H_A: not all k distributions are equal.

One can test which conditions or treatments exhibit significant differences among themselves by pairwise comparisons with (7.56). Statistical significance for

$$\left| \frac{R_i}{n_i} - \frac{R_{i'}}{n_{i'}} \right| > \sqrt{\chi^2_{k-1;\alpha} \left[\frac{k(k+1)}{12} \right] \left[\frac{n_i + n_{i'}}{n_i n_{i'}} \right]} \qquad (7.56)$$

at the $100\alpha\%$ level [corresponding to (7.53)] or by following Wilcoxon and Wilcox (Section 7.6.2), Student, Newman, and Keuls (see below), or Page (see below); cf., also Miller (1966 [8:7a]) as well as Hollander and Wolfe (1973 [8:1b]). Reinach (1965) decomposed $\hat{\chi}^2_R$ into orthogonal components.

The method in detail:

1. The observed values are entered in a two way table—**horizontal**: k treatments or conditions, **vertical**: n individuals, blocks, sample groups, or replications.
2. The values in each row are ordered according to rank; thus each row exhibits the ranks 1 to k.
3. For each column the rank sum R_i (for the ith column) is determined; all rank sums are checked by the equality $\sum_i R_i = \frac{1}{2}nk(k+1)$.
4. $\hat{\chi}^2_R$ is computed according to (7.54) [with ties, $\hat{\chi}^2_{R,T}$ is computed according to (7.55)].
5. $\hat{\chi}^2_R$ (or $\hat{\chi}^2_{R,T}$) is assessed on the basis of Table 183, or for large n on the basis of the χ^2 table (Table 28a, Section 1.5.2).

EXAMPLE. Comparing the effectiveness of $k = 4$ penicillin samples at the 5% level (source: Weber 1964, p. 417). The test is carried out on $r = 3$ plates of agar. From 9 cm diameter agar plates inoculated with *B. subtilis* (hay bacillus) there are cut out 4 small discs, about 0.4 cm in diameter each. Into each cut-out space there is then introduced drop by drop the same amount of one of the several penicillin solutions, so that all 4 penicillin samples are represented on each dish. The penicillin solution diffuses into the layer of agar, inhibiting the growth of *B. subtilis*. This manifests itself by the formation of an apparent region of effectiveness around the cut-out space. The diameter of the inhibition zone is a measure of the concentration of the penicillin solution. The experimental units (cut-out spaces) are assigned randomly to the solutions. The question is raised whether there are differences

7.6 Rank Analysis of Variance for Several Correlated Samples

among the diameters of the inhibition zones; a possible agar plate effect should be taken into consideration. The sizes of the inhibition zones in mm are given in Table 184. A check of the computation of the column sums:

$$\sum_{i=1}^{k} R_i = \frac{nk(k+1)}{2} = \frac{3 \cdot 4(4+1)}{2} = 30,$$

$$\hat{\chi}_R^2 = \left[\frac{12}{3 \cdot 4 \cdot 5}(11^2 + 6^2 + 10^2 + 3^2)\right] - 3 \cdot 3 \cdot 5 = 8.2.$$

Since $\hat{\chi}_R^2 = 8.2 > 7.4 = \chi_R^2$ for $k = 4$, $n = 3$, and $P = 0.05$ (Table 183), H_0 (equality of the four penicillin solutions) must be rejected at the 5% level.

Table 184

Dish No.	Penicillin preparation			
	1	2	3	4
1	27	23	26	21
2	27	23	25	21
3	25	21	26	20

Table 185 Ranks

Dish No.	Penicillin preparation				
	1	2	3	4	
1	4	2	3	1	
2	4	2	3	1	
3	3	2	4	1	
$\sum$	11	6	10	3	30

If we wish to check whether there exist differences among the agar plates, we assign ranks to the columns and form row sums. We obtain

2.5	2.5	2.5	2.5	10.0
2.5	2.5	1	2.5	8.5
1	1	2.5	1	5.5
				24.0

and forgo the test due to the many ties (see above).

Approximate multiple comparisons following Student, Newman, and Keuls (see Section 7.4.2. II)

For $n \geq 6$, (7.50) can be replaced by $q\sqrt{k(k+1)/(12n)}$ with q obtained from Table 172 (let the k of that table be referred to as h) and h equal to the number of ordered rank means in the comparison (in which $h \geq 2$) and $v_2 = \infty$.

Given appropriate prior knowledge, $H_0: \mu_1 = \mu_2 = \cdots = \mu_k$ can be tested against the one sided $H_A: \mu_1 > \mu_2 > \cdots > \mu_k$ **by a method due to Page** (1963); $H_0: \mu_1 = \mu_2 = \cdots = \mu_k$ is rejected if the sum of the products of hypothetical rank and accompanying rank sum equals or exceeds the corresponding tabulated value at the preselected level. If, e.g., the identification numbers of the solutions were identical to the hypothetical ranks in Tables 184 and 185, then [since $(1)(11) + (2)(6) + (3)(10) + (4)(3) = 65 < 84$; cf., Table 186, $k = 4, n = 3, P = 0.05$] H_0 could not have been rejected at the 5% level. Page (1963) gives 5%, 1%, and 0.1% limits for $3 \leq k \leq 10$

Table 186 Some 5% and 1% bounds for the Page test

P	\multicolumn{6}{c	}{0.05}	\multicolumn{6}{c	}{0.01}
n \ k	3	4	5	6	7	8	3	4	5	6	7	8
3	41	84	150	244	370	532	42	87	155	252	382	549
4	54	111	197	321	487	701	55	114	204	331	501	722
5	66	137	244	397	603	869	68	141	251	409	620	893
6	79	163	291	474	719	1,037	81	167	299	486	737	1,063
7	91	189	338	550	835	1,204	93	193	346	563	855	1,232

[called n there] and $2 \leq n \leq 50$ [called m there]. Tables of exact probabilities and critical values for $\alpha = 0.2, 0.1, 0.05, 0.025, 0.01, 0.005$, and 0.001 for $k = 3, 4, \ldots, 8$ and $n = 2, 3, \ldots, 10$ are given by R. E. Odeh, Communications in Statistics—Simulation and Computation **B6** (1977), 49–61. Page (1963) also suggests for the Friedman test the relation (7.57) below, which is quite a bit simpler than (7.54) and which permits a clearer recognition of the χ^2 character of the test statistic:

$$\hat{\chi}_R^2 = \frac{6}{k}\sum \frac{(R_i - E)^2}{E} = \frac{6\sum(R_i - E)^2}{\sum R_i}, \qquad (7.57)$$

where $E = \sum R_i/k$ represents the **mean rank sum**. For our first example we get $E = 30/4 = 7.5$:

$$\hat{\chi}_R^2 = \frac{6\{(11 - 7.5)^2 + (6 - 7.5)^2 + (10 - 7.5)^2 + (3 - 7.5)^2\}}{30} = 8.2.$$

For n individuals and $k = 2$ conditions the statistic $\hat{\chi}_R^2$ is, as was shown by Friedman, connected to the Spearman rank correlation coefficient r_S by way of the following relation:

$$\hat{\chi}_R^2 = (n-1)(1 + r_S) \qquad (7.58)$$

or

$$r_S = \frac{\hat{\chi}_R^2}{n-1} - 1. \quad (7.58a)$$

Thus one can determine, in terms of $\hat{\chi}_R^2$, a statistic for the size of the difference between two data sets.

Remarks

1. If several **rank sequences**—obtained through arrangements by several judges or through transformations from data—are to be assessed as to their degree of agreement (a means, by the way, of objectivizing nonquantifiable biological characteristics), then the Friedman test is to be used. If three ($n = 3$) persons are asked to rank four ($k = 4$) movie stars as to their artistic performance, they could, e.g., end up with Table 185 (with the result: no agreement [$\alpha = 0.05$]).

2. If dichotomous data (but no measured observations or ranks) are available, then the Q-test (Section 6.2.4) replaces the Friedman test.

3. If several products, let us say kinds of cheese, brands of tobacco, or carbon papers, are to be tested in a subjective comparison, then the technique of **paired comparisons** is appropriate: several different samples of a product (e.g., brands A, B, C, D), in every case grouped as pairs ($A - B, A - C, A - D, B - C, B - D, C - D$), are compared. More on this can be found in the monograph by David (1969) (cf., also Trawinski 1965 and Linhart 1966). The variance analytic pairwise comparison proposed by Scheffé (1952) is illustrated by an example due to Mary Fleckenstein et al. (1958), which Starks and David (1961) analyze in great detail by means of further tests. A simple procedure with auxiliary tables and an example is presented by Terry et al. (1952) (cf., also Bose 1956, Jackson and Fleckenstein 1957; Vessereau 1956, and Rao and Kupper 1967). For a survey and bibliography see R. A. Bradley, Biometrics **32** (1976), 213–252.

7.6.2 Multiple comparisons of correlated samples according to Wilcoxon and Wilcox: Pairwise comparisons of several treatments which are repeated under a number of different conditions or in a number of different classes of subjects

The Friedman test is a two way analysis of variance with ranks; the corresponding multiple comparisons [cf. (7.56)] due to Wilcoxon and Wilcox (1964). The test resembles the procedure given by Nemenyi.

The comparison in detail: again k treatments with n replications each are compared. Every treatment is assigned a rank from 1 to k, so that n rank orders result. The ranks of the individual samples are added; their differences are compared with the value of the critical difference from Table 187. If the tabulated critical difference is attained or exceeded then the treatments

Table 187 Critical differences for the two way classification: comparison of all possible pairs of treatments. P = 0.10 (two sided) (taken from Wilcoxon, F. and Wilcox, R. A.: Some Rapid Approximate Statistical Procedures, Lederle Laboratories, Pearl River, New York 1964, 36–38).

n	k = 3	k = 4	k = 5	k = 6	k = 7	k = 8	k = 9	k = 10
1	2.9	4.2	5.5	6.8	8.2	9.6	11.1	12.5
2	4.1	5.9	7.8	9.7	11.6	13.6	15.6	17.7
3	5.0	7.2	9.5	11.9	14.2	16.7	19.1	21.7
4	5.8	8.4	11.0	13.7	16.5	19.3	22.1	25.0
5	6.5	9.4	12.3	15.3	18.4	21.5	24.7	28.0
6	7.1	10.2	13.5	16.8	20.2	23.6	27.1	30.6
7	7.7	11.1	14.5	18.1	21.8	25.5	29.3	33.1
8	8.2	11.8	15.6	19.4	23.3	27.2	31.3	35.4
9	8.7	12.5	16.5	20.5	24.7	28.9	33.2	37.5
10	9.2	13.2	17.4	21.7	26.0	30.4	35.0	39.5
11	9.6	13.9	18.2	22.7	27.3	31.9	36.7	41.5
12	10.1	14.5	19.0	23.7	28.5	33.4	38.3	43.3
13	10.5	15.1	19.8	24.7	29.7	34.7	39.9	45.1
14	10.9	15.7	20.6	25.6	30.8	36.0	41.4	46.8
15	11.2	16.2	21.3	26.5	31.9	37.3	42.8	48.4
16	11.6	16.7	22.0	27.4	32.9	38.5	44.2	50.0
17	12.0	17.2	22.7	28.2	33.9	39.7	45.6	51.5
18	12.3	17.7	23.3	29.1	34.9	40.9	46.9	53.0
19	12.6	18.2	24.0	29.9	35.9	42.0	48.2	54.5
20	13.0	18.7	24.6	30.6	36.8	43.1	49.4	55.9
21	13.3	19.2	25.2	31.4	37.7	44.1	50.7	57.3
22	13.6	19.6	25.8	32.1	38.6	45.2	51.9	58.6
23	13.9	20.1	26.4	32.8	39.5	46.2	53.0	60.0
24	14.2	20.5	26.9	33.6	40.3	47.2	54.2	61.2
25	14.5	20.9	27.5	34.2	41.1	48.1	55.3	62.5

Values for $k \leq 15$ are given by McDonald and Thompson (1967).

involved in the comparison come from different populations. If the computed difference falls below the tabulated D, then the difference can yet be regarded as accidental.

Additional table values of D for $k > 10$ and $n = 1(1)20$ can, when needed, be computed by using the formula $D = W\sqrt{nk(k+1)/12}$, where W for $P = 0.05$ or 0.01 is read from Table 172 (last line), and for other values of P is interpolated in Table 23 of the Biometrika Tables (Pearson and Hartley 1966, pp. 178–183); e.g., $D = 67.7$ [Table 187; $P = 0.05$; $n = 25$; $k = 10$]: for $P' = 0.95$ we get (Biometrika Table 23, p. 180, column 10) $W = 4.4745$ and $4.4745\sqrt{(25)(10)(10+1)/12} = 67.736$; by Table 172, for $k = 10$; $W = 4.47$ and $D = 67.668$.

7.6 Rank Analysis of Variance for Several Correlated Samples

Table 187-1 (*continued*): P = 0.05 (two sided)

n	k = 3	k = 4	k = 5	k = 6	k = 7	k = 8	k = 9	k = 10
1	3.3	4.7	6.1	7.5	9.0	10.5	12.0	13.5
2	4.7	6.6	8.6	10.7	12.7	14.8	17.0	19.2
3	5.7	8.1	10.6	13.1	15.6	18.2	20.8	23.5
4	6.6	9.4	12.2	15.1	18.0	21.0	24.0	27.1
5	7.4	10.5	13.6	16.9	20.1	23.5	26.9	30.3
6	8.1	11.5	14.9	18.5	22.1	25.7	29.4	33.2
7	8.8	12.4	16.1	19.9	23.9	27.8	31.8	35.8
8	9.4	13.3	17.3	21.3	25.5	29.7	34.0	38.3
9	9.9	14.1	18.3	22.6	27.0	31.5	36.0	40.6
10	10.5	14.8	19.3	23.8	28.5	33.2	38.0	42.8
11	11.0	15.6	20.2	25.0	29.9	34.8	39.8	44.9
12	11.5	16.2	21.1	26.1	31.2	36.4	41.6	46.9
13	11.9	16.9	22.0	27.2	32.5	37.9	43.3	48.8
14	12.4	17.5	22.8	28.2	33.7	39.3	45.0	50.7
15	12.8	18.2	23.6	29.2	34.9	40.7	46.5	52.5
16	13.3	18.8	24.4	30.2	36.0	42.0	48.1	54.2
17	13.7	19.3	25.2	31.1	37.1	43.3	49.5	55.9
18	14.1	19.9	25.9	32.0	38.2	44.5	51.0	57.5
19	14.4	20.4	26.6	32.9	39.3	45.8	52.4	59.0
20	14.8	21.0	27.3	33.7	40.3	47.0	53.7	60.6
21	15.2	21.5	28.0	34.6	41.3	48.1	55.1	62.1
22	15.5	22.0	28.6	35.4	42.3	49.2	56.4	63.5
23	15.9	22.5	29.3	36.2	43.2	50.3	57.6	65.0
24	16.2	23.0	29.9	36.9	44.1	51.4	58.9	66.4
25	16.6	23.5	30.5	37.7	45.0	52.5	60.1	67.7

Table 187-2 (*continued*): P = 0.01 (two sided)

n	k = 3	k = 4	k = 5	k = 6	k = 7	k = 8	k = 9	k = 10
1	4.1	5.7	7.3	8.9	10.5	12.2	13.9	15.6
2	5.8	8.0	10.3	12.6	14.9	17.3	19.7	22.1
3	7.1	9.8	12.6	15.4	18.3	21.2	24.1	27.0
4	8.2	11.4	14.6	17.8	21.1	24.4	27.8	31.2
5	9.2	12.7	16.3	19.9	23.6	27.3	31.1	34.9
6	10.1	13.9	17.8	21.8	25.8	29.9	34.1	38.2
7	10.9	15.0	19.3	23.5	27.9	32.3	36.8	41.3
8	11.7	16.1	20.6	25.2	29.8	34.6	39.3	44.2
9	12.4	17.1	21.8	26.7	31.6	36.6	41.7	46.8
10	13.0	18.0	23.0	28.1	33.4	38.6	44.0	49.4
11	13.7	18.9	24.1	29.5	35.0	40.5	46.1	51.8
12	14.3	19.7	25.2	30.8	36.5	42.3	48.2	54.1
13	14.9	20.5	26.2	32.1	38.0	44.0	50.1	56.3
14	15.4	21.3	27.2	33.3	39.5	45.7	52.0	58.4
15	16.0	22.0	28.2	34.5	40.8	47.3	53.9	60.5
16	16.5	22.7	29.1	35.6	42.2	48.9	55.6	62.5
17	17.0	23.4	30.0	36.7	43.5	50.4	57.3	64.4
18	17.5	24.1	30.9	37.8	44.7	51.8	59.0	66.2
19	18.0	24.8	31.7	38.8	46.0	53.2	60.6	68.1
20	18.4	25.4	32.5	39.8	47.2	54.6	62.2	69.8
21	18.9	26.0	33.4	40.9	48.3	56.0	63.7	71.6
22	19.3	26.7	34.1	41.7	49.5	57.3	65.2	73.2
23	19.8	27.3	34.9	42.7	50.6	58.6	66.7	74.9
24	20.2	27.8	35.7	43.6	51.7	59.8	68.1	76.5
25	20.6	28.4	36.4	44.5	52.7	61.1	69.5	78.1

EXAMPLE. Source: Wilcoxon and Wilcox (1964, pp. 11, 12).

Table 188

Person	A		B		C		D		E		F	
1	3.88	1	30.58	5	25.24	3	4.44	2	29.41	4	38.87	6
2	5.64	1	30.14	3	33.52	6	7.94	2	30.72	4	33.12	5
3	5.76	2	16.92	3	25.45	4	4.04	1	32.92	5	39.15	6
4	4.25	1	23.19	4	18.85	3	4.40	2	28.23	6	28.06	5
5	5.91	2	26.74	5	20.45	3	4.23	1	23.35	4	38.23	6
6	4.33	1	10.91	3	26.67	6	4.36	2	12.00	4	26.65	5
		8		23		25		10		27		33

Six persons receive 6 different diuretics each (drugs A to F). Two hours after the treatment the sodium excretion is determined. It is to be decided which diuretics differ from the others on the basis of the sodium excretion. Table 188 contains the data, with the corresponding ranks and the column rank sums on the right. The absolute differences are listed in Table 189.

Table 189

	D	B	C	E	F
	10	23	25	27	33
A 8	2	15	17	19*	25**
D 10		13	15	17	23**
B 23			2	4	10
C 25				2	8
E 27					6

The critical difference for $k = 6$ and $n = 6$ is 18.5 (cf., Table 187) at the 5% level, 21.8 at the 1% level. Each difference significant at the 5% level is marked with a single asterisk (*), while each difference significant at the 1% level is marked with a double asterisk (**). It can thus be established that the preparation F distinguishes itself on the basis of a stronger sodium diuresis at the 1% level from the diuretics A and D. The preparation E differs at the 5% level from the preparation A; other differences are not significant at the 5% level.

▶ 7.7 PRINCIPLES OF EXPERIMENTAL DESIGN

In the design of experiments there are, according to Koller (1964), two opposing viewpoints to be reconciled with each other: The principle of **comparability** and the principle of **generizability**.

Two experiments by which the effects of two types of treatment are to be compared are **comparable** if they differ only in the type of treatment but agree

7.7 Principles of Experimental Design

in all other respects. Agreement should exist with respect to test conditions and sources of variation:

1. the observation and measurement procedures,
2. the performance of the experiment,
3. the individual peculiarities of the objects being tested,
4. the peculiarities of the time, location, equipment, and technicians.

Comparability is seldom attainable with individuals but is attainable for groups of individuals. For a comparison, the specific individual factors of variation must have the same frequency distribution.

If in order to achieve good comparability, e.g., only young male animals of a certain breed, with a certain weight, etc., are used for the test, then the comparability is indeed assured but **generalizability** is impaired. After all, then the tests give no clue to how older or female animals or those of other breeds would behave. This test sequence would furnish only a narrow inductive basis (cf., also Sections 2.1.4–5 and 4.1).

Generalization means identification and description of the collectives and the distribution of their attributes from which the observed values can be viewed as representative samples. Only by examining such collectives of various animals (age, type, hereditary factors, disposition), various test times (time of day, season, weather), various kinds of experiments, various experimenters, various experimental techniques, etc., can we judge to what extent the results are independent of these variability and interference factors, i.e., whether the results may be generalized in this way. In the context of the experiment comparability and generalizability oppose each other, since comparability calls for homogeneous material while on the other hand generalizability requires heterogeneity to obtain a broad inductive basis: **comparisons call for replication collectives, generalizations for variability collectives.** Both principles must interlock in the experimental design. Particularly advantageous are comparisons of various procedures on the same animal. There the comparability is optimal, while at the same time an arbitrarily large extension of the sample range can be carried out.

Herzberg and Cox (1969) give a fine survey of experimental design.

The underlying principles of experimental design are:

1. **Replication:** permits the estimation of the experimental error, at the same time providing for its diminution.
2. **Randomization:** permits—by **elimination** of known and unknown **systematic errors** in particular of trends which are conditioned by time and space—an unbiased estimation of the effects of interest, at the same time bringing about **independence** of the test results. The randomization can be carried out with the help of a table of random numbers.
3. **Block division** (planned grouping): Increases the precision of comparisons within blocks.

Some interesting comments on randomization—pro and con—are presented by H. Bunke and O. Bunke, Statistics, Mathematische Operationsforschung und Statistik **9** (1978), 607–623.

> The smallest subunit of the experimental material receiving a treatment is called an experimental unit; it is the object on which a measurement is made. The idea of randomly assigning the procedures to the experimental units, called **randomization** for short—it originated with R. A. Fisher—can be regarded as the foundation of every experimental design. Through it, one obtains (a) an unbiased estimate of the effects of interest, (b) an unbiased estimate of the **experimental error**, and (c) a more nearly **normal distribution of the data**. Unknown and undesirable correlation systems are removed (by randomization), so that we have uncorrelated and independent experimental errors and our standard significance tests may be applied.

If the **experimental units are very diverse** then the isolation of the effects of interest becomes more difficult. In such cases, it is advisable to group the most similar units before the experiment is even started. Subgroups of comparable experimental units are formed which are internally more uniform than the overall material: **homogeneous** "blocks". Within a block, the randomization principle for the assigning of the treatments to the experimental units again applies.

Examples of blocks are persons or animals or identical twins or paired organs or siblings or leaves from the same plant or the adjacent parcels of a field in an agricultural experiment or other groupings which describe natural or artificial blocks. Blocking criteria are characteristics associated with (1) the experimental units (for persons: sex, age, health condition, income, etc.) or, to maintain a constant experimental environment, (2) the experimental settings (batch of material, observer, measuring instrument, time, etc.). Several blocking criteria may be combined. The individual blocks should always have the same size. The comparisons which are important for the trial objective must be dealt with as fully as possible within the blocks.

Nuisance quantities or nuisance factors (e.g., soil variations) **are eliminated**:

1. By **analysis of covariance** when **quantitatively measurable nuisance factors are known**. Under a covariance model, classifying and influence factors (covariable, as, e.g., weight or blood pressure at the beginning of the test period) act linearly upon the dependent variables. Analysis of covariance helps to eliminate influences which otherwise interfere in the proper evaluation of the experiment through the analysis of variance, and serves to explore regression relations in categorized material (see Winer 1971, Huitema 1980, and Biometrics **38** (1982), 539–753; cf., also Cochran

1957, J. C. R. Li 1964, Enderlein 1965, Harte 1965, Peng 1967, Quade 1967, Rutherford and Stewart 1967, Bancroft 1968, Evans and Anastasio 1968, Reisch and Webster 1969, Sprott 1970). An alternative to the analysis of covariance is given by D. Sörbom, Psychometrika **43** (1978), 381–396.

2. When **nonmeasurable perturbing factors are known,** by the formation of blocks (groups of experimental units which agree as much as possible with respect to the perturbing factor), or by pairing; the experiment is carried out under special conditions (e.g., in a greenhouse).

3. When the **perturbing factors are unknown,** by randomization and replication as well as by consideration of additional characteristics which lead to a subsequent understanding of the perturbing quantities.

> Ambient conditions that can be only hazily identified or are hard to control should be overcome by proper blocking and randomization techniques. Sometimes—as in the case of changing external conditions or unplanned events—measurements or at least qualitative records on these conditions should be taken. Under blocking the effect of a badly controlled variable is removed from the experimental error, while under randomization it usually is not. **If possible, block; otherwise randomize.** Concerning replicate measurements it is important to obtain information about each component of repeatability (e.g., the **same** experimental unit, day, operator, equipment, etc.). It is always useful to include some standard test conditions known as controls.

In comparing surveys and experiments, Kish (1975) gives more hints on the **control of disturbing variables**.

In contrast with absolute experiments, for example, the determination of a natural constant such as the speed of light, the overwhelming majority of experiments belongs to the category of **comparison experiments**: We compare, e.g., the harvest yields realized under fixed conditions (on seeds, fertilizer etc.). The relative values in question are either known as **theoretical values** or are to be determined by **control trials**. Comparison experiments—which can be understood as processes affected by various conditions or "treatments", at the end of which the results are compared and interpreted as "consequences" of the treatments, as specific effects—aim at: (a) testing whether an effect exists and (b) measuring the size of this effect, where errors of Type I and II are avoided if possible, i.e., neither are nonexistent effects to be "detected" in the material, nor are genuine effects to be ignored. Moreover, the smallest effect which will still be regarded as significant should be specified in advance. Genuine effects can be found only when it can be ascertained that (a) neither the heterogeneity of the trial units (e.g., soil differences in the harvest yield experiment) nor (b) random influences **alone** could be responsible for the effect.

Modern experimental design distinguishes itself from the classical or traditional procedure in that at least 2 factors are always considered simultaneously. Previously, if the effect of several factors was to be analyzed, the factors were consecutively tested one factor with respect to its different levels at a time. It can be shown that this procedure is not only **ineffective** but can also yield **incorrect results**. The simultaneous **(optimal) range** of operation of all the factors cannot be found in this way. Moreover, interactions among the factors cannot be recognized with the classical procedure. The principle of modern statistical experimental design consists in combining the factors in such a way that their **effects and interactions** as well as the variability of these effects can be **measured, compared** and **delimited** against the random variability; more on this can be found, e.g., in Natrella 1963 (see C. Daniel there) (cf., also Section 2.4.1 and Table 190). To the three underlying principles of experimental design (replication, randomization, and block division) we add three more: (1) various controls and accompanying control experiments (2) diversity of treatments, any of which could even be encoded to avoid subjective influences and (3) the numbers of replications of a treatment should be proportional to the corresponding deviations $[\sigma_i \neq \text{const.}]$: $n_1/n_2 = \sigma_1/\sigma_2$.

Remark: On experimental designs

1. Arrangement of trials in blocks with random assignment of procedures to the trial units. The test material is partitioned into blocks of the greatest possible homogeneity. Each block contains as many units as there are factors (methods of treatment, procedures) to be tested **(completely randomized blocks)** or an integral multiple of this number. The factors are associated with the experimental units of each block by means of a randomization procedure (e.g., a table of random numbers). The comparison among the factors is made more precise through replication on very different blocks. The two way classification model without interaction is applied in the analysis of variance of these **joint samples**. Here the designations "block" and "factor" are appropriate in place of row and column.

We should perhaps emphasize that the forming of blocks, just like the forming of paired observations, makes sense only if the dispersion between the trial units is clearly greater than that between the individual members of the pairs or between the block units; this is so because correlated samples (paired observations, blocks) exhibit **fewer** degrees of freedom than the corresponding independent samples. If there exists a clear dispersion difference in the sense stated above, then the gain in accuracy through formation of correlated samples is greater than the loss in accuracy due to a decrease in the number of degrees of freedom.

If the number of trial units per block is less than the number of factors to be tested, one speaks of **incompletely randomized blocks**. They are frequently used in case a natural block involves only a small number of elements (e.g., twins, right-left comparisons), when there are technical or temporal limitations on the feasibility of parallel trials on the same day, etc.

2. The Latin square. Whereas only one variation factor is eliminated by means of block division, the experimental design of a so-called **Latin square** serves to eliminate

7.7 Principles of Experimental Design

Table 190 The most important setups for tests on different levels of a factor or of several factors (adapted from Juran, J. M. (Ed.): Quality Control Handbook, 2nd ed., New York, 1962, Table 44, pp. 13-122/123)

Design	Basic approach	Comment
1. Completely randomized	Levels of a single factor are distributed completely at random over the experimental units	The number of trials may vary from level to level; of little sensitivity in detecting significant effects
2. Randomized blocks	Combining of most similar experimental units into blocks to which the levels of a single factor are then assigned	The number of trials may vary from level to level; more sensitive than the completely randomized design
3. Latin squares	Design for testing 3 factors at k levels each consisting of k^2 experimental units which (in accordance with 2 characteristics with k levels each) are so assigned to the rows and columns of a square, that each factor occurs exactly once in each row and in each column	Simultaneous study of two or more factors. It is assumed that the factors act independently of each other (i.e., no interaction)
4. Factorial experiments	Designs for arbitrarily many factors, each with an arbitrary but fixed number of levels. An experiment which, e.g., tests four factors at three levels each and one at two levels, requires $3^4 \times 2 = 162$ trials in a single replication	Exact experiment; encompasses, in particular, all interactions, in addition to the main factors; the experiment may easily become unmanageable if all combinations of factors and levels are tested; moreover, it requires greater homogeneity of material than the other designs
5. Fractional factorial experiments	Only a portion of all combinations of a factorial design, selected so as to allow assessment of the main factors and the most important interactions, is considered in the experiment	More economical experiments. Experimental error larger than in full factorial experiment and estimation less exact. Some interactions cannot be estimated. Interpretation of the result much more complex

two variation factors. It is frequently found that a field being tested clearly exhibits differences in soil conditions along two directions. Through a judicious parceling, the differences along two directions can be successfully eliminated with the help of this model. If k factors (e.g., the fertilizers A and B and the control C) are to be tested, then k^2 trials and hence k^2 (or $3^2 = 9$) trial units (lots) are needed. A simple Latin square is, e.g., the following:

$$A \ B \ C$$
$$B \ C \ A$$
$$C \ A \ B$$

Each factor appears exactly once in each row and each column of this square. In general, with a single replication, only squares with $k \geq 5$ are used, since with smaller squares only a small number of degrees of freedom is available for evaluating the experimental error. With $k = 5$ there are 12. The corresponding experimental designs, which are of course used not only in agriculture but also wherever trial units can be randomly grouped along two directions or characters, are, e.g., found in the tables by Fisher and Yates (1963). With a Greco-Latin square the randomization works in three directions. More on this can be found in Jaech (1969).

3. Factorial experiments. Factorial designs involve running all combinations of conditions or levels of the independent variables. If it is not possible or not practical to apply all combinations, a specially selected fraction is run (fractional factorial experiment).

If n factors are to be compared simultaneously at 2, 3, or k levels each, then experimental designs which enable comparisons of combinations, known as 2^n-, 3^n-, or k^n-designs or experiments, are called for (cf., Box et al. 1978, Chapters 7, 10–13; Davies 1971; also Plackett and Burman 1946, Baker 1957, Daniel 1959, Winer 1971, Addelman 1963, 1969, C. C. Li 1964, J. C. R. Li 1964, Cooper 1967).

4. Hierarchic experimental designs. In hierarchic classification a sample group consists of sample subgroups of, e.g., type 1 and 2 (say: streets, buildings, and apartments). One speaks of "nested design": All levels of a factor always occur in conjunction with a level of some other factor (cf., Gates and Shiue 1962, Gower 1962, Bancroft 1964, Eisen 1966, Ahrens 1967, Kussmaul and Anderson 1967, Tietjen and Moore 1968).

> Several books on experimental design can be found at the end of the bibliography [8 : 7b]. Let us in particular call attention to the comprehensive introduction by Hahn (1977) and to Box and al. (1978, cited in [1] on p. 569), Scheffé (1959), Kempthorne (1960), Davies (1971), Johnson and Leone (1964), C. C. Li (1964), J. C. R. Li (1964), Kendall and Stuart (1968), Peng (1967), Bancroft (1968), Linder (1969), John (1971), Winer (1971), Bätz (1972), and Kirk (1982). Special reference is also made to the works mentioned at the end of Sections 2.4.1.3 and 5.8 and to the survey by Herzberg and Cox (1969), as well as to the bibliography of Federer and Balaam (1973). Surveys of **recent developments** in the design of experiments are provided in the International Statistical Review by W. T. Federer [**48** (1980), 357–368, **49** (1981), 95–109, 185–197] and by A. C. Atkinson [**50** (1982), 161–177].

Scientific investigation: evaluating hypotheses and discovering new knowledge

1. **Formulating the problem and stating the objectives:** It is frequently expedient to subdivide the overall problem into component problems and ask several questions:
   a. Why is the problem posed?
   b. Outlining the initial situation by means of standard questions: what? how? where? when? how much? what is not known? what will be assumed?
   c. Problem type: comparisons? finding optimal conditions? significance of change? association among variables?

2. **Checking all sources of information:** Mainly researching the literature.

3. **Choice of strategy:**
   a. **Developing the model appropriate to the problem.** Number of variables to be taken into consideration. Introduction of simplifying assumptions. Examining whether it is possible to further simplify the problem by modification, e.g., to studies on guinea pigs instead of on men.
   b. **Developing the technique of investigation.** Defining the population (and/or sample units) about which inferences are to be made. Selection of the experimental (and/or sampling) design of the variables, of the auxiliary variables, of the number of replications, and of the form of randomization. Planning for the hypotheses to be tested, for the recording of results, and for the data analysis.
   c. **Developing the statistical model.** Defining the population (and/or sample units) about which inferences are to be made. Selection of the experimental (and/or sampling) design, the number of replications, the form of randomization, and the auxiliary variables. Recording the results and planning for an analysis of all the hypotheses to be tested.

4. **Testing the strategy** by means of exploratory surveys and trials. Examining the method of inquiry and the compatibility of the observed values with the statistical model.

5. **Setting and realizing the strategy** on the basis of the experience gained in items 3 and 4.
   a. **Final specification of all essential points**, e.g., the method of investigation, the objects being studied, the experimental units, the characteristic and influence factors, the controls, the basis of reference; the variables and auxiliary variables; avoiding or

recording uncontrollable variables; blocking and randomization; the sample size or number of replications, taking into account the expenditure of technicians, equipment, material, and time, among other things; setting up of tactical reserves to avoid major shortages; the extent of the overall program; definitive formulation of the statistical analysis model; preparation of special arrangements (computer used?) for recording, checking, and evaluating the data.
   b. **Carrying out the study**, if possible without modification. Analyzing the data, e.g., plotting, giving confidence intervals and testing the hypotheses.

6. **Decisions and conclusions**:
   a. **Result**: Checking the computations. Stating the results in tabulated form and/or graphically.
   b. **Interpretation**: Indications as to the plausibility, practical significance, verifiability, and region of validity of the study. The results of the tests on the hypotheses are scrutinized critically with the simplifying assumptions taken into account; and when it is feasible and of value to do so, they are compared with the findings of other authors. Is a replication of the study necessary with fewer simplifying assumptions, with improved models, newer methods of investigation, etc.? Do there arise new hypotheses, derived from the data, which must be checked by new, independent investigation?
   c. **Report**: Description of the overall program, items 1 to 6b.

Some useful hints for the writing and presentation of reports or papers are given in The American Statistician: (1) by A. S. C. Ehrenberg [**36** (1982), 326–329], and (2) by D. H. Freeman, Jr. and coworkers [**37** (1983), 106–110].

Five periods in the history of probability and statistics

1654 The Chevalier de Méré asked Blaise Pascal (1623–1662) why it would be advantageous in a game of dice to bet on the occurrence of a six in 4 trials but not advantageous in a game involving two dice to bet on the occurrence of a double six in 24 trials. Pascal corresponded with Pierre de Fermat (1601–1665) on the subject. The two probabilities are 0.518 and 0.491. The problem of coming up with assertions which are based on the outcomes of a game and which are determined by underlying probability laws, i.e., the problem of coming up with the probability needed for correct models or hypotheses, was considered by Thomas Bayes (1702–1761).

1713–1718 The texts on probability by Jakob Bernoulli (1654–1705; Ars Conjectandi, opus posthumum, 1713) and Abraham de Moivre (1667–1754; The Doctrine of Chances, 1718) were published. The first contains the notion of statistics, the binomial distribution, and the law of large numbers; the second, the transition from the binomial to the normal distribution.

1812 Pierre Simon de Laplace (1749–1827): Théorie Analytique des Probabilités, the first comprehensive survey of probability.

1901 Founding of the journal Biometrika, around which crystallized the Anglo-Saxon school of statistics, by Karl Pearson (1837–1936), who with Ronald Aylmer Fisher (1890–1962) developed most of the biometrical methods, later extended by Jerzy Neyman (1894–1981) and Egon S. Pearson (1895–1980) to include the confidence interval and general test theory. Fisher also was responsible for pioneering studies in experimental design (The Design of Experiments, 1935), the analysis of variance, and other important subjects. After the axiomatization of probability (1933), Andrei Nikolayevich Kolmogoroff developed the theory of stochastic processes, which originated with Russian mathematicians.

1950 Statistical Decision Functions by Abraham Wald (1902–1950) appeared. Sequential analysis, which was developed during World War II and which can be interpreted as a stochastic process, is a special case of statistical decision theory. The text provides guidelines for procedures in uncertain situations: Statistical inference is understood as a decision problem.

The future of statistics is discussed by Tukey (1962), Kendall (1968), Watts (1968) and Bradley (1982).

BIBLIOGRAPHY AND GENERAL REFERENCES

INDEX

[1] A selection of texts for further reading
[2] Important tables
[3] Dictionaries and directories
[4] Computer programs
[5] Bibliographies and abstracts
    [5:1]   Mathematical-statistical tables
    [5:2]   Articles
    [5:3]   Books
    [5:4]   Abstract journals
    [5:5]   Proceedings
[6] Some periodicals
[7] Sources for technical aids (e.g. function and probability charts)
[8] References and suggestions for further reading for the individual chapters. The authors of material pertaining to the topics specified below are listed separately at the end of the references for the chapter.
    [8:1]   Chapter 1
    [8:1a]   Stochastic processes
    [8:1b]   Distribution-free methods
    [8:2]   Chapter 2
    [8:2a]   Medical statistics
    [8:2b]   Sequential test plans
    [8:2c]   Bioassay
    [8:2d]   Statistics in engineering
    [8:2e]   Linear programming and operations research
    [8:2f]   Game theory and war games

[8:2g]   Monte Carlo method and computer simulation
[8:3]   Chapter 3
[8:3a]   Sampling theory
[8:4]   Chapter 4
[8:5]   Chapter 5
[8:5a]   Factor analysis
[8:5b]   Multiple regression analysis
[8:6]   Chapter 6
[8:7]   Chapter 7
[8:7a]   Multiple comparisons
[8:7b]   Experimental design

[1] A SELECTION OF TEXTS FOR FURTHER READING

Menges, G. (and H. J. Skala): Grundriß der Statistik. 2 volumes (Westdeutscher Verlag, 374 and 475 pages) Opladen 1972, 1973

Sachs, L.: Statistische Methoden, 6th revised ed. (Springer-Verlag, 133 pages) Berlin, Heidelberg, New York 1984

Stange, K.: Angewandte Statistik. Teil I: Eindimensionale Probleme. Teil II: Mehrdimensionale Probleme. (Springer, 592 and 505 pages) Berlin, Heidelberg, New York 1970, 1971

Rasch, D., Herrendorfer, G., Bock, J., and Busch, K.: Verfahrensbibliothek, Versuchsplanung und Auswertung. 2 volumes (VEB Deutscher Landwirtschaftsverlag; 1052 pages) Berlin 1978

Weber, Erna: Grundriß der Biologischen Statistik. Anwendungen der mathematischen Statistik in Naturwissenschaft und Technik. 7th revised edition (Fischer; 706 pages) Stuttgart 1972 [8th revised edition 1980].

Bliss, C. I.: Statistics in Biology, Vol. 1–2 (McGraw-Hill; pp. 558, 639) New York 1967, 1970.

Box, G. E. P., Hunter, W. G., and Hunter, J. S.: Statistics for Experimenters. (Wiley; 653 pages) New York 1978

Bury, K. V.: Statistical Models in Applied Science. (Wiley; 625 pages) New York 1976

Chakravarti, I. M., Laha, R. G. and J. Roy: Handbook of Methods of Applied Statistics. Vol. I and II (Wiley, pp. 460 and 160) New York 1967

Chou, Y.-L.: Statistical Analysis. With Business and Economic Applications. 2nd ed. (Holt, Rinehart and Winston, pp. 912) New York 1974

Dagnelie, P.: (1) Théorie et Méthodes Statistique. Vol. 1, 2. (Duculot; pp. 378, 451) Gembloux, Belgien 1969, 1970. (2) Analyse Statistique a Plusieurs Variables. (Vander; pp. 362) Bruxelles 1975

Daniel, C. and Wood, F. S.: Fitting Equations to Data, Computer Analysis of Multifactor Data. 2nd ed. (Wiley; 458 pages) New York 1980

Dixon, W. J., and F. J. Massey, Jr.: Introduction to Statistical Analysis. 3rd ed. (McGraw-Hill, pp. 638) New York 1969

Dunn, Olive J. and Clark, Virginia, A.: Applied Statistics. Analysis of Variance and Regression. (Wiley; pp. 387) New York 1974

Eisen, M.: Introduction to Mathematical Probability Theory. (Prentice-Hall; pp. 496) Englewood Cliffs, N.J. 1969

Feller, W.: An Introduction to Probability Theory and Its Applications. Vol. 1, 3rd ed., Vol. 2, 2nd ed. (Wiley, pp. 496 and 669) New York 1968 and 1971

Johnson, N. L., and S. Kotz: Distributions in Statistics. (Discr. D.; Cont. Univ. D. I + II; Cont. Multiv. D.) (Wiley; pp. 328, 300, 306, 333) New York 1969/72 [Int. Stat. Rev. **42** (1974), 39–65 and **50** (1982), 71–101]

Johnson, N. L., and F. C. Leone: Statistics and Experimental Design in Engineering and the Physical Sciences. Vol. I and II (Wiley, pp. 523 and 399) New York 1964 (2nd ed. 1977)

Kendall, M. G., and A. Stuart: The Advanced Theory of Statistics. Vol. 1, 4th ed., Vol. 2, 4th ed., Vol. 3, 4th ed. (Griffin, pp. 484, 758, 780) London 1977, 1979, 1983

Kotz, S., Johnson, N. L., and Read, C. B. (Eds.): Encyclopedia of Statistical Sciences. Vol. I–VIII. (Wiley; approx. 5000 pages) New York 1982–1986

Krishnaiah, P. R. (Ed.): Multivariate Analysis, and Multivariate Analysis II, III. (Academic Press; pp. 592 and 696, 450) New York and London 1966 and 1969, 1973

Kruskal, W. H. and Tanur, J. M.: International Encyclopedia of Statistics, Vol. 1 and 2. (The Free Press; 1350 pages) New York 1978

Lindgren, B. W.: Statistical Theory. 2nd ed. (Macmillan and Collier-Macmillan, pp. 521) New York and London 1968 (3rd ed. 1976)

Miller, I. and Freund, J. E.: Probability and Statistics for Engineers, 2nd ed. (Prentice-Hall; 529 pages) Englewood Cliffs 1977

Mood, A. M., Graybill, F. A., and Boes, D. C.: Introduction to the Theory of Statistics. 3rd ed. (McGraw-Hill [Int. Stud. Ed.]; pp. 564) Düsseldorf 1974

Moran, P. A. P.: An Introduction To Probability Theory. (Clarendon Press; pp. 542) Oxford 1968

Ostle, B and R. W. Mensing: Statistics in Research. 3rd ed. (Iowa Univ. Press, pp. 596) Ames 1975

Rahman, N. A.: A Course in Theoretical Statistics. (Griffin; pp. 542) London 1968

Rao, C. R.: Linear Statistical Inference and Its Applications. 2nd ed. (Wiley, pp. 608) New York 1973

Schlaifer, R.: Analysis of Decisions under Uncertainty. (McGraw-Hill; pp. 729) New York 1969

Searle, S. R.: Linear Models. (Wiley, pp. 532) London 1971

Snedecor, G. W., and Cochran, W. G.: Statistical Methods. (Iowa State Univ. Press, pp. 593) Ames, Iowa 1967 [7th ed., pp. 507, 1980]

Tukey, J. W.: Exploratory Data Analysis. (Addison-Wesley; pp. 688) Reading, Mass. 1977

Yamane, T.: Statistics; An Introductory Analysis. 3rd ed. (Harper and Row; pp. 1133) New York 1973

Zacks, S.: The Theory of Statistical Inference. (Wiley; pp. 609) New York 1971

This is a very selective listing. Section 8 contains references to specialized literature.

[2] IMPORTANT TABLES

Documenta Geigy: Wissenschaftliche Tabellen. 7th edition (Geigy AG, pp. 9–199) Basel 1968 (Wissenschaftliche Tabellen Geigy: Statistik, 8th revised and enlarged edition 1980)

Koller, S.: Neue graphische Tafeln zur Beurteilung statistischer Zahlen. 4th revised edition (Dr. Steinkopff, 167 pages) Darmstadt 1969

Wetzel, W., Jöhnk, M.-D. and Naeve, P.: Statistische Tabellen. (de Gruyter, 168 pages) Berlin 1967

Abramowitz, M., and Stegun, Irene A. (Eds.): Handbook of Mathematical Functions with Formulas, Graphs and Mathematical Tables. (National Bureau of Standards Applied Mathematics Series 55, U.S. Government Printing Office; pp. 1046) Washington 1964 (further there are references to additional tables) (7th printing, with corrections, Dover, N.Y. 1968)

Beyer, W. H. (Ed.): CRC Handbook of Tables for Probability and Statistics. 2nd ed. (The Chemical Rubber Co., pp. 642) Cleveland, Ohio 1968

Fisher, R. A., and Yates F.: Statistical Tables for Biological, Agricultural and Medical Research. 6th ed. (Oliver and Boyd, pp. 146) Edinburgh and London 1963

Harter, H. L.: Order Statistics and their Use in Testing and Estimation. Vol. 1: Tests Based on Range and Studentized Range of Samples from a Normal Population. Vol. 2: Estimates Based on Order Statistics of Samples from Various Populations. (ARL, USAF; U.S. Government Printing Office; pp. 761 and 805) Washington 1970

Harter, H. L. and Owen, D. B. (Eds.): Selected Tables in Mathematical Statistics. Vol. I (Markham, pp. 405) Chicago 1970

Isaacs, G. L., Christ, D. E., Novick, M. R. and Jackson, P. H.: Tables for Bayesian Statisticians. (Univ. of Iowa; pp. 377) Iowa City, Iowa 1974 [cf. Appl. Stat. **24** (1975), 360 + 361]

Kres, H.: Statistical Tables for Multivariate Analysis: A Handbook with References to Applications. (Springer; pp. 530) New York, Berlin, Heidelberg, Tokyo 1983

Lienert, G. A.: Verteilungsfreie Methoden in der Biostatistik. Tafelband. (A. Hain; pp. 686) Meisenheim am Glan 1975

Owen, D. B.: Handbook of Statistical Tables. (Addison-Wesley, pp. 580) Reading, Mass. 1962 (Errata: Mathematics of Computation **18**, 87; Mathematical Reviews **28**, 4608) [Selected tables in Mathematical Statistics are edited by D. B. Owen and R. E. Odeh, e.g., Vol. 4 and 5 published in 1977 by the American Mathematical Society, Providence, Rhode Island]

Pearson, E. S., and Hartley, H. O. (Eds.): Biometrika Tables for Statisticians. I, 3rd ed., II (Univ. Press, pp. 264, 385) Cambridge 1966 (with additions 1969), 1972

Rao, C. R., Mitra, S. K. and Matthai, A. (Eds.): Formulae and Tables for Statistical Work. (Statistical Publishing Society, pp. 234) Calcutta 1966 (further there are references to additional tables)

Statistical Tables and Formulas with Computer Applications. (Japanese Standards Association; pp. 750) Tokyo 1972

Section [5:1] contains references to a few further sources of statistical tables.

[3] DICTIONARIES AND DIRECTORIES

1. VEB Deutscher Landwirtschaftsverlag Berlin, H. G. Zschommler (Ed.): Biometrisches Wörterbuch. Erläuterndes biometrisches Wörterbuch in 2 volumes (VEB Deutscher Landwirtschaftsverlag, a total of 1047 pages) Berlin 1968, contents: 1. Illustrated encyclopedia (2712 key words, 795 pages), 2. Foreign language index (French, English, Polish, Hungarian, Czechoslovakian, Russian; 240 pages), 3. Recommendations for standard symbols (9 pages)
2. Müller, P. H. (Ed.): Lexikon, Wahrscheinlichkeitsrechnung und Mathematische Statistik. (Akademie-Vlg., 445 pages) Berlin 1980
3. Kendall, M. G., and Buckland, A.: A Dictionary of Statistical Terms. 4th ed., revised and enlarged (Longman Group, pp. 213) London and New York 1982

4. Freund, J. E., and Williams, F.: Dictionary/Outline of Basic Statistics. (McGraw-Hill, pp. 195) New York 1966
5. Morice, E., and Bertrand, M.: Dictionnaire de statistique. (Dunod, pp. 208) Paris 1968
6. Paenson, I.: Systematic Glossary of the Terminology of Statistical Methods. English, French, Spanish, Russian. (Pergamon Press, pp. 517) Oxford, New York, Braunschweig 1970
7. Fremery, J. D. N., de: Glossary of Terms Used in Quality Control. 3rd ed. (Vol. XII, European Organization for Quality Control; pp. 479) Rotterdam 1972 (400 definitions in 14 languages)

The following directories contain the addresses of most of the authors in the bibliography:

1. Mathematik. Institute, Lehrstühle, Professoren, Dozenten mit Anschriften sowie Fernsprechanschlüssen. Mathematisches Forschungsinstitut Oberwolfach, 762 Oberwolfach-Walke, Lorenzenhof, 1978 Directory
2. World Directory of Mathematicians 1982. International Mathematical Union. (7th ed.; pp. 725) Distrib. by Amer. Math. Soc., P.O. Box 6248, Providence, RI 02940, USA
3. The Biometric Society, 1982 Membership Directory. Edited by Elsie E. Thull, The Biom. Soc., 806 15th Street, N.W., Suite 621, Washington, D.C. 20005, USA
4. 1970 Directory of Statisticians and Others in Allied Professions. (pp. 171) American Statistical Association, 806 15th Street, N. W., Washington (D.C. 20005) 1971
5. Membership Directory 1981–1982: The Institute of Mathematical Statistics. (pp. 219) 3401 Investment Blvd., Suite 6, Hayward, Calif. 94545, USA
6. ISI's Who Is Publishing In Science 1975. International Directory of Research and Development Scientists, Institute for Scientific Information, 325 Chestnut Str., Philadelphia, Pa. 19106
7. Williams, T. I. (Ed.): A Biographical Dictionary of Scientists. (Black, pp. 592), London 1969

[4] COMPUTER PROGRAMS

A few hints to orient the reader. For more details, check with computing centers. For desktop computers see Th. J. Boardman, The American Statistician **36** (1982), 49–58 and [same journal] H. Neffendorf **37** (1983), 83–86.

An introduction is
Afifi, A. A. and Azen, S. P.: Statistical Analysis—A Computer Oriented Approach. 2nd ed. (Academic Press; pp. 442) New York 1979
Kennedy, W. J., Jr. and Gentle, J. E.: Statistical Computing. (M. Dekker; pp. 591) New York 1980

More can be found in, e.g.:
Baker, R. J. and Nelder, J. A.: The GLIM System Release 3 Manual (Numerical Algorithms Group; various paginations) Oxford 1978
Dixon, W. J. and Brown, M. B. (Eds.): BMDP-79: Biomedical Computer Programs P-Series (University of California Press; pp. 880) Berkeley 1979. For more details see BMDP Communications

GENSTAT Manual. User's Reference Manual. (Rothampsted Experimental Station) Harpenden, Herts. 1977, supplemented by GENSTAT Newsletters

Nie, N. H., Hull, C. H., Jenkins, J. G., Steinbrenner, K., and Bent, D. H.: SPSS Statistical Package for the Social Sciences. 2nd ed. (McGraw-Hill) New York 1975, supplemented by updates and newsletters

Survey Research Center—Computer Support Group: OSIRIS IV—Statistical Analysis and Data Management Software System: User's Manual, 4th ed. (Institute for Social Research; pp. 254) Ann Arbor 1979

A good survey is provided by

Statistical Software Newsletter, edited by Hörmann, A. and Victor, N., medis-Institute, z.H. Frau Eder, Arabellastr. 4/III, D-8000 München 81

See, for instance,

Francis, I. and Wood, L.: Evaluating and improving statistical software. Statistical Software Newsletter **6** (1980), No. 1, 12–16 and Francis, I.: Statistical Software. A Comparative Review. (Elsevier, North-Holland; pp. 556) Amsterdam

Many journals now contain computer programs. We mention only four:

Applied Statistics **28** (1979), 94–100 [and **30** (1981), 358–373]
Computer Programs in Biomedicine **10** (1979), 43–47
Journal of Quality Technology **11**, (1979), 95–99
The American Statistician **37** (1983), 169–175

[5] BIBLIOGRAPHIES AND ABSTRACTS

[5:1] Mathematical-statistical tables

Consult for specialized tables:
Greenwood, J. A., and Hartley, H. O.: Guide to Tables in Mathematical Statistics. (University Press, pp. 1014) Princeton, N.J. 1962

For mathematical tables, consult:

1. Fletcher, A., Miller, J. C. P., Rosenhaed, L., and Comrie, L. J.: An Index of Mathematical Tables. 2nd ed., Vol. I and II (Blackwell; pp. 608, pp. 386) Oxford 1962
2. Lebedev, A. V., and Fedorova, R. M. (English edition prepared from the Russian by Fry, D. G.): A Guide to Mathematical Tables. Supplement No. 1 by N. M. Buronova (D. G. Fry, pp. 190) (Pergamon Press, pp. 586) Oxford 1960
3. Schütte, K.: Index mathematischer Tafelwerke und Tabellen aus allen Gebieten der Naturwissenschaften, 2nd edition (Oldenbourg, 239 pages) München and Wien 1966

We specifically mention Mathematical Tables and other Aids to Computation, published by the National Academy of Sciences (National Research Council, Baltimore, Md., **1** [1947] – **13** [1959]) and Mathematics of Computations, published by the American Mathematical Society (Providence, R.I., **14** [1960] – **34** [1980])

These series contain important tables:

1. Applied Mathematics Series. U.S. Govt. Printing Office, National Bureau of Standards, U.S. Department of Commerce, Washington
2. New Statistical Tables. Biometrika Office, University College, London
3. Tracts for Computers. Cambridge University Press, London

[5:2] Articles

1. Revue de l'institut de statistique (La Haye), Review of the International Statistical Institute (The Hague) (e.g., **34** [1966], 93–110 and **40** [1972], 73–81) (since 1972 as International Statistical Review)
2. Allgemeines Statistisches Archiv (e.g., **56** [1972], 276–302)
3. Deming, Lola S., et al.: Selected Bibliography of Literature, 1930 to 1957: in Journal of Research of the National Bureau of Standards
   I Correlation and Regression Theory: **64B** (1960), 55–68
   II Time Series: **64B** (1960), 69–76
   III Limit Theorems: **64B** (1960), 175–192
   IV Markov Chains and Stochastic Processes: **65B** (1961), 61–93
   V Frequency Functions, Moments and Graduation: **66B** (1962), 15–28
   VI Theory of Estimation and Testing of Hypotheses, Sampling Distribution and Theory of Sample Surveys: **66B** (1962), 109–151
   Supplement, 1958–1960: **67B** (1963), 91–133; likewise important
   Haight, F. A.: Index to the distributions of mathematical statistics **65B** (1961), 23–60

[5:3] Books

1. Lancaster, H.: Bibliography of Statistical Bibliographies. (Oliver and Boyd, pp. 103) Edinburgh and London 1968 (with the main sections: personal bibliographies, pp. 1–29, and subject bibliographies, pp. 31–65, as well as the subject and author indexes) (cf.: a second list, Rev. Int. Stat. Inst. 37 [1969], 57–67, . . . , 15th list Int. Stat. Rev. **51** [1983], 207–212) as well as Problems in the bibliography of statistics. With discussion. J. Roy. Statist. Soc. A **133** (1970), 409–441, 450–462 and Gani, J.: On coping with new information in probability and statistics. With discussion. J. Roy. Statist. Soc. A **133** (1970), 442–462 and Int. Stat. Rev. **40** (1972), 201–207 as well as Rubin, E.: Developments in statistical bibliography, 1968–69. The American Statistician **24** (April 1970), 33 + 34
2. Buckland, W. R., and Fox, R. A.: Bibliography of Basic Texts and Monographs on Statistical Methods 1945–1960. 2nd ed. (Oliver and Boyd; pp. 297) Edinburgh and London 1963
3. Kendall, M. G., and Doig, A. G.: Bibliography of Statistical Literature, 3 vol. (Oliver and Boyd, pp. 356, 190, 297) Edinburgh and London 1962/68 (1) Pre-1940, with supplements to (2) and (3), 1968; (2) 1940–49, 1965; (3) 1950–58, 1962. This bibliography, indexed unfortunately only by authors' names, comprises 34082 papers which are characterized per volume by 4-digit numbers. Since 1959 it has been continued by Statistical Theory and Method Abstracts (12 sections with 10–12 subsections) which contains 1000–12000 reviews annually. Publisher: International Statistical Institute, 2 Oostduinlaan, Den Haag, Holland.

4. Kellerer, H.: Bibliography of all foreign language books in statistics and its applications that have been published since 1928 (Deutsche Statistische Gesellschaft, 143 pages) (Nr. 7a) Wiesbaden 1969

Specialized bibliographies:

Menges, G. and Leiner, B. (Eds.): Bibliographie zur statischen Entscheidungstheorie 1950–1967. (Westdeutscher Verlag, 41 pages) Köln and Opladen 1968
Patil, G. P., and Joshi, S. W.: A Dictionary and Bibliography of Discrete Distributions. (Oliver and Boyd, pp. 268) Edinburgh 1968
A bibliography on the foundations of statistics was compiled by L. J. Savage: Reading suggestions for the foundations of statistics. The American Statistician **24** (Oct. 1970), 23–27

In addition to the bibliographies quoted in the text of the book we mention:

Pritchard, A.: Statistical Bibliography. An Interim Bibliography. (North-Western Polytechnic, School of Librarianship, pp. 69) London 1969
Wilkie, J.: Bibliographie Multivariate Statistik und mehrdimensionale Klassifikation. 2 volumes (Akademie Verlag; 1123 pages) Berlin 1978

A source of recent papers in the seven most important journals (up to 1969):

Joiner, B. L., Laubscher, N. F., Brown, Eleanor S., and Levy, B.: An Author and Permuted Title Index to Selected Statistical Journals. (Nat. Bur. Stds. Special Publ. 321, U.S. Government Printing Office, pp. 510) Washington Sept. 1970

The following books and journals are useful in addition to those by Dolby, Tukey, and Ross (cf. end of Section [6]):

Burrington, G. A.: How to Find out about Statistics. (Pergamon, pp. 153) Oxford 1972
Moran, P. A. P.: How to find out in statistical and probability theory. Int. Stat. Rev. **42** (1974), 299–303

Other modern bibliographies are mentioned in the main text and referenced in [8].

[5:4] Abstract journals

1. Statistical Theory and Method Abstracts. International Statistical Institute. Oliver and Boyd, Tweeddale Court, 14 High Street, Edinburgh 1 (cf. above)
2. International Journal of Abstracts on Statistical Methods in Industry. International Statistical Institute. Oliver and Boyd, Tweeddale Court, 14 High Street, Edinburgh 1
3. Quality Control and Applied Statistics. Executive Sciences Institute, Whippany, N.J., Interscience Publ. Inc., 250 Fifth Avenue, New York, N. Y., USA

Mathematical reviewing journals should also be considered: Zentralblatt für Mathematik, Mathematical Reviews and Bulletin Signalétique Mathematiques.

[5:5] Proceedings

Bulletin de l'Institut International de Statistique. Den Haag
Proceedings of the Berkeley Symposium on Mathematical Statistics and Probability. Berkeley, California

[6] SOME PERIODICALS

Allgemeines Statistisches Archiv, Organ der Deutschen Statistischen Gesellschaft, Institut für Statistik und Mathematik, J. W. Goethe-Universität, Mertonstr. 17-19, D-6000 Frankfurt am Main [**64** (1980)]

Applied Statistics, Journal of the Royal Statistical Society (Series C). Royal Statistical Society, 25 Enford Street, London W1H 2BH [**29** (1980)]

Biometrics, Journal of the Biometric Society, Department of Biomathematics, Univ. of Oxford, OXI 2JZ, Pusey Street, England. Biometrics Business Office: 806 15th Street NW, Suite 621, Washington, D.C. 20005, USA [**36** (1980)]

Biometrika, The Biometrika Office, University College London, Gower Street, London WC1E 6BT [**67** (1980)]

Biometrical Journal. Journal of Mathematical Methods in Biosciences. Institut für Mathematik der AdW, DDR-1080 Berlin, Mohrenstr. 39 [**22** (1980)]

Communications in Statistics: Part A—Theory and Methods. M. Dekker, New York, P.O. Box 11305, Church Street Station, N.Y. 10249 (and Basel); Dept. Statist., Southern Methodist Univ., Dallas, Texas 75275 [**A9** (1980)]

Communications in Statistics: Part B—Simulation and Computation. M. Dekker, New York, P.O. Box 11305, Church Street Station, N.Y. 10249 (and Basel); Dept. Statist., Virginia Polytechnic Institute and State University, Blacksburg, VA. 24061 [**B9** (1980)]

International Statistical Review, A Journal of the International Statistical Institute, 428 Prinses Beatrixlaan, Voorburg, The Netherlands [**48** (1980)]

Journal of Multivariate Analysis. Dept. Math. Statist, Univ. of Pittsburgh, Pittsburgh, Pa. 15260 [**10** (1980)]

Journal of Quality Technology. A Quarterly Journal of Methods, Applications, and Related Topics. American Society for Quality Control; Plankinton Building, 161 West Wisconsin Avenue, Milwaukee, WI 53203 [**12** (1980)]

Journal of the American Statistical Association, 806 15th St. N. W., Suite 640, Washington, D.C. 20005, USA [**75** (1980)]

Journal of the Royal Statistical Society, Series A (General), Series B (Methodological), Royal Statistical Society, 25 Enford Street, London W1H 2BH [A **143** (1980); B **42** (1980)]

Metrika, International Journal for Theoretical and Applied Statistics. Seminar für Angewandte Stochastik an der Universität München, Akademiestr. 1/IV, D-8000, München 40 [**27** (1980)]

Psychometrika, A Journal devoted to the Development of Psychology as a Quantitative Rational Science, Journal of the Psychometric Society, Johns Hopkins University, Baltimore, Maryland 21218 [**45** (1980)]

Technometrics, A Journal of Statistics for the Physical, Chemical and Engineering Sciences; published quarterly by the American Society for Quality Control and the American Statistical Association. ASQC: 161 W. Wisconsin Avenue, Milwaukee, Wis., 53203; ASA: 806 15th Street, N.W., Suite 640, Washington, D.C. 20005 [**22** (1980)].

The Annals of Mathematical Statistics, Institute of Mathematical Statistics, Stanford University, Calif. 94305, USA. Since 1973 as The Annals of Probability and as The Annals of Statistics [both **8** (1980)].

For further periodicals see e.g., Journal of the Royal Statistical Society A **139** (1976), 144-155, 284-294.

Beginners and more advanced scientists will find many interesting ideas in Annual Technical Conference Transactions of the American Society for Quality Control and in Journal of Quality Technology (previously: Industrial Quality Control).

Finally we mention the excellent series concerning recent papers up to now:

Dolby, J. L. and J. W. Tukey: The Statistics Cum Index. (The R and D Press, pp. 498), Los Altos, Calif. 1973. Ross, I. C. and J. W. Tukey: Index to Statistics and Probability. Permuted Titles. (pp. 1588); (1975). Locations and Authors. (pp. 1092; 1974).

CURRENT INDEX TO STATISTICS. Applications, Methods and Theory **1** (1975), ..., **6** (1980), ..., jointly published by the American Statistical Association and the Institute of Mathematical Statistics. Editors: B. L. Joiner and J. M. Gwynne.

[7] SOURCES FOR TECHNICAL AIDS (E.G. FUNCTION AND PROBABILITY CHARTS)

Schleicher und Schüll, D-3352 Einbeck/Hannover
Schäfers Feinpapiere, DDR Plauen (Sa.), Bergstraße 4
Rudolf Haufe Verlag, D-7800 Freiburg i. Br.
Keuffel und Esser-Paragon GmbH., D-2000 Hamburg 22, Osterbekstraße 43
Codex Book Company, Norwood, Mass. 02062, 74 Broadway, USA
Technical and Engineering Aids for Management. 104 Belsore Avenue, Lowell, Mass., USA (also RFD, Box 25, Tamworth, New Hampshire 03886)

Statistical work sheets, control cards and further aids:

Arinc Research Corp., Washington D.C., 1700 K Street, USA
Beuth-Vertrieb, D-1000 Berlin 30, Burggrafenstraße 4–7 (Köln and Frankfurt/M.)
Arnold D. Moskowitz, Defense Industrial Supply Center, Philadelphia, Pa. USA
Dyna-Slide Co., 600 S. Michigan Ave., Chicago, Ill., USA
Recorder Charts Ltd., P.O. Box 774, Clyde Vale, London S.E. 23, England
Technical and Engineering Aids for Management. 104 Belrose Avenue, Lowell, Mass., USA (also RFD, Tamworth, New Hampshire 03886)
Howell Enterprizes, Ltd., 4140 West 63rd Street, Los Angeles, Cal. 90043, USA

[8] REFERENCES FOR THE INDIVIDUAL CHAPTERS

[8:1] Chapter 1

Ackoff, R. L.: Scientific Method: Optimizing Applied Research Decisions. (Wiley; pp. 462) New York 1962
Ageno, M., and Frontali, C.: Analysis of frequency distribution curves in overlapping Gaussians. Nature **198** (1963), 1294–1295

Aitchison, J., and Brown, J. A. C.: The Lognormal Distribution. Cambridge 1957 [see Applied Statistics **29** (1980), 58–68; Biometrics **36** (1980), 707–719; and J. Amer. Statist. Assoc. **75** (1980), 399–404]

Alluisi, E. A.: Tables of binary logarithms, uncertainty functions, and binary log functions. Percept. Motor Skills **20** (1965), 1005–1012

Altham, P. M. E.: Two generalizations of the binomial distribution. Applied Statistics **27** (1978), 162–167

Anderson, O.: Probleme der statistischen Methodenlehre in den Sozialwissenschaften, 4th edition. (Physica-Vlg., 358 pages) Würzburg 1963, Chapter IV

Angers, C.: A graphical method to evaluate sample sizes for the multinomial distribution. Technometrics **16** (1974), 469–471

Bachi, R.: Graphical Rational Patterns. A New Approach to Graphical Presentation of Statistics. (Israel Universities Press; pp. 243) Jerusalem 1968

Barnard G. A.: The Bayesian controversy in statistical inference. J. Institute Actuaries **93** (1967), 229–269 [see also J. Amer. Statist. Assoc. **64** (1969), 51–57]

Barnett, V.: Comparative Statistical Inference. 2nd ed. (Wiley; pp. 325) London 1982
—, and Lewis, T.: Outliers in Statistical Data. (Wiley; pp. 384) New York 1978

Bartko, J. J.: (1) Notes approximating the negative binomial. Technometrics **8** (1966), 345–350 (2) Letter to the Editor. Technometrics **9** (1967), 347 + 348 (see also p. 498)

Batschelet, E.: Introduction to Mathematics for Life Scientists. 2nd ed. (Springer; pp. 643) Berlin, Heidelberg, New York 1975

Beckman, R. J., and Cook, R. D.: Outlier . . . . s. With discussion and response. Technometrics **25** (1983), 119–163

Bernard, G.: Optimale Strategien unter Ungewißheit. Statistische Hefte 9 (1968), 82–100

Bertin, J.: Semiology Graphique. Les Diagrammes – Les Reseau – Les Cartes. (Gautier Villars, pp. 431) Paris 1967

Bhattacharya, C. G.: A simple method of resolution of a distribution into Gaussian components. Biometrics **23** (1967), 115–135 [see also **25** (1969), 79–93 and **29** (1973), 781–790]

Birnbaum, A.: Combining independent tests of significance. J. Amer. Statist. Assoc. **49** (1954), 559–574 [see also **66** (1971), 802–806]

Blind, A.: Das harmonische Mittel in der Statistik. Allgem. Statist. Arch. **36** (1952), 231–236

Bliss, C. I.: (1) Fitting the negative binomial distribution to biological data. Biometrics **9** (1953), 176–196 and 199–200. (2) The analysis of insect counts as negative binomial distributions. With discussion. Proc. Tenth Internat. Congr. Entomology 1956, **2** (1958), 1015–1032

Blyth, C. R., and Hutchinson, D. W.: Table of Neyman-shortest unbiased confidence intervals for the binomial parameter. Biometrika **47** (1960), 381–391

Bolch, B. W.: More on unbiased estimation of the standard deviation. The Americar Statistician **22** (June 1968), 27 (see also **25** [April 1971], 40, 41 and [Oct. 1971], 30–32)

Botts, R. R.: Extreme value methods simplified. Agric. Econom. Research **9** (1957), 88–95

Box, G. E. P. and Tiao, G. C.: Bayesian Inference in Statistical Analysis (Addison-Wesley) Reading, Mass. 1973
—, Leonhard, T., and Wu, C.-F. (Eds.): Scientific Inference, Data Analysis, and Robustness. (Academic Press; pp. 320) New York 1983

Boyd, W. C.: A nomogram for chi-square. J. Amer. Statist. Assoc. **60** (1965), 344–346 (cf. **61** [1966] 1246)

Bradley, J. V.: A common situation conducive to bizarre distribution shapes. The American Statistician **31** (1977), 147–150

Bruckmann, G.: Schätzung von Wahlresultaten aus Teilergebnissen. (Physica-Vlg., 148 pages) Wien and Würzburg 1966 [see also P. Mertens (ed.): Prognoserechnung. (Physica-Vlg., 196 pages.) Würzburg and Wien 1972]

Brugger, R. M.: A note on unbiased estimation of the standard deviation. The American Statistician **23** (October 1969), 32 (see also **26** [Dec. 1972], 43)

Bühlmann, H., Loeffel, H. and Nievergelt, E.: Einführung in die Theorie und Praxis der Entscheidung bei Unsicherheit. Heft 1 der Reihe: Lecture Notes in Operations Research and Mathematical Economics. Berlin-Heidelberg-New York 1967 (122 pages) (2nd edition 1969, 125 pages)

Calot, G.: Signicatif ou non signicatif? Réflexions à propos de la théorie et de la pratique des tests statistiques. Revue de Statistique Appliquée **15** (No. 1, 1967), 7–69 (see **16** [No. 3, 1968], 99–111 and Cox, D. R.: The role of significance tests. Scand. J. Statist. **4** [1977] 49–70)

Campbell, S. K.: Flaws and Fallacies in Statistical Thinking. (Prentice-Hall; pp. 200) Englewood Cliffs, N.J. 1974 [see also I. J. Good, Technometrics **4** (1962), 125–132 and D. J. Ingle, Perspect. Biol. Med. **15**, 2 (Winter 1972), 254–281]

Cetron, M. J.: Technological Forecasting: A Practical Approach. (Gordon and Breach, pp. 448) New York 1969

Chambers, J., Cleveland, W., Kleiner, B., and Tukey, P.: Graphical Methods for Data Analysis. (Wadsworth; pp. 330) Belmont, Calif. 1983

Chernoff, H., and Moses, L. E.: Elementary Decision Theory. New York 1959

Chissom, B. S.: Interpretation of the kurtosis statistic. The American Statistician **24** (Oct. 1970), 19–22

Cleary, T. A., and Linn, R. L.: Error of measurement and the power of a statistical test. Brit. J. Math. Statist. Psychol. **22** (1969), 49–55

Cochran, W. G.: Note on an approximate formula for the significance levels of z. Ann. Math. Statist. **11** (1940), 93–95

Cohen, A. C. jr.: (1) On the solution of estimating equations for truncated and censored samples from normal populations. Biometrika **44** (1957), 225–236. (2) Simplified estimators for the normal distribution when samples are singly censored or truncated. Technometrics **1** (1959), 217–237. (3) Tables for maximum likelihood estimates: singly truncated and singly censored samples. Technometrics **3** (1961), 535–541 [see also **18** (1976), 99–103 and Applied Statistics **25** (1976), 8–11]

Cohen, J.: Statistical Power Analysis for the Behavioral Sciences. (Academic Press, pp. 496) New York 1977 [see also, p. 581, Guenther 1973, and p. 611, Odeh and Fox 1975]

Cornfield, J.: (1) Bayes theorem. Rev. Internat. Statist. Inst. **35** (1967), 34–49. (2) The Bayesian outlook and its application. With discussion. Biometrics **25** (1969), 617–642 and 643–657 [see also J. Roy. Statist. Soc. A **145** (1982), 250–258]

Cox, D. R.: (1) Some simple approximate tests for Poisson variates. Biometrika **40** (1953), 354–360. (2) Some problems connected with statistical inference. Ann. Math. Statist. **29** (1958), 357–372. (3) The role of significance tests. Scand. J. Statist. **4** (1977), 49–70. (4) Some remarks on the role in statistics of graphical methods. Applied Statistics **27** (1978), 4–9 [cf. Biometrika **66** (1979), 188–190]

Craig, I.: On the elementary treatment of index numbers. Applied Statistics **18** (1969), 141–152 [see also Econometrica **41** (1973), 1017–1025]

Crowe, W. R.: Index Numbers, Theory and Applications. London 1965

D'Agostino, R. B.: Linear estimation of the normal distribution standard deviation. The American Statistician **24** (June 1970), 14 + 15 [see also J. Amer. Statist. Assoc. **68** (1973), 207–210]

Dalenius, T.: The mode—a neglected statistical parameter. J. Roy. Statist. Soc. A **128** (1965), 110–117 [see also Ann. Math. Statist. **36** (1965), 131–138 and **38** (1967), 1446–1455]

Darlington, R. B.: Is kurtosis really "peakedness"? The American Statistician **24** (April 1970), 19–22 (see also **24** [Dec. 1970], 41, **25** [Febr. 1971], 42, 43, 60 and **30** [1976], 8–12)

David, Florence N.: (1) A Statistical Primer, Ch. Griffin, London 1953. (2) Games, Gods and Gambling. New York 1963

Day, N. E.: Estimating the components of a mixture of normal distributions. Biometrika **56** (1969), 463–474 (see also **59** [1972], 639–648 and Technometrics **12** [1970], 823–833)

Defense Systems Department, General Electric Company: Tables of the Individual and Cumulative Terms of Poisson Distribution. Princeton, N.J. 1962

DeLury, D. B. and Chung, J. H.: Confidence Limits for the Hypergeometric Distribution. Toronto 1950

Dickinson, G. C.: Statistical Mapping and the Presentation of Statistics. (E. Arnold, pp. 160) London 1963

Dietz, K.: Epidemics and rumours: a survey. J. Roy. Statist. Soc. A **130** (1967), 505–528

Documenta Geigy: Wissenschaftliche Tabellen. (6th and) 7th editions., Basel (1960 and) 1968, pages 85–103, 107, 108, 128 [8th revised edition 1980]

Dubey, S. D.: Graphical tests for discrete distributions. The American Statistician **20** (June 1966), 23 + 24 [see also D. I. Holmes, The Statistician **23** (1974), 129–134]

Dudewicz, E. J. and Dalal, S. R.: On approximations to the t-distribution. J. Qual. Technol. **4** (1972). 196–198 [see also Ann. Math. Statist. **34** (1963), 335–337 and Biometrika **61** (1974), 177–180]

Ehrenberg, A. S. C. Graphs or tables? The Statistician **27** (1978), 87–96.

Elderton, W. P., and Johnson, N. L.: Systems of Frequency Curves. (Cambridge University Press, pp. 214) Cambridge 1969

Faulkner, E. J.: A new look at the probability of coincidence of birthdays in a group. Mathematical Gazette **53** (1969), 407–409 (see also **55** [1971], 70–72)

Federighi, E. T.: Extended tables of the percentage points of Student's t-distribution. J. Amer. Statist. Assoc. **54** (1959), 683–688

Fenner, G.: Das Genauigkeitsmaß von Summen, Produkten und Quotienten der Beobachtungsreihen. Die Naturwissenschaften **19** (1931), 310

Ferris, C. D., Grubbs, F. E., and Weaver, C. L.: Operating characteristics for the common statistical tests of significance. Ann. Math. Statist. **17** (1946), 178–197

Fienberg, S. E.: Graphical methods in statistics. The American Statistician **33** (1979), 165–178 [see also **32** (1978), 12–16]

Finucan, H. M.: A note on kurtosis. J. Roy. Statist. Soc., Ser. B **26** (1964), 111 + 112, p. 112

Fishburn, P. C.: (1) Decision and Value Theory. (Wiley) New York 1964. (2) Decision under uncertainty: an introductory exposition. Industrial Engineering **17** (July 1966), 341–353

Fisher, N. I.: Graphical methods in nonparametric statistics: or review and annotated bibliography. International Statistical Review **51** (1983), 25–58

Fisher, R. A.: (1) The negative binomial distribution. Ann. Eugenics **11** (1941), 182–187. (2) Theory of statistical estimation. Proc. Cambr. Phil. Soc. **22** (1925), 700–725. (3) Note on the efficient fitting of the negative binomial. Biometrics **9** (1953), 197–200. (4) The Design of Experiments, 7th ed. (1st ed. 1935), Edinburgh 1960, Chapter II [see also J. Roy. Statist. Soc. B **37** (1975), 49–53]

Fisher, R. A. and Yates, F.: Statistical Tables for Biological, Agricultural and Medical Research. (Oliver and Boyd; pp. 146) Edinburgh and London 1963

Fraser, C. O.: Measurement in psychology. Brit. J. Psychol. **71** (1980), 23–34

Freudenthal, H., and Steiner, H.-G.: Aus der Geschichte der Wahrscheinlichkeitstheorie und der mathematischen Statistik. In H. Behnke, G. Bertram and R. Sauer

(Eds.): Grundzüge der Mathematik. Vol. IV: Praktische Methoden und Anwendungen der Mathematik. Göttingen 1966, Chapter 3, pp. 149-195, cf. p. 168

Garland, L. H.: Studies on the accuracy of diagnostic procedures. Amer. J. Roentg. **82** (1959), 25-38 (particularly p. 28) [see also M. Jacobsen, Applied Statistics **24** (1975), 229-249]

Gawronski, W., and Stadtmüller, U.: Smoothing histograms by means of lattice and continuous distributions. Metrika **28** (1981), 155-164

Gbur, E. E.: On the Poisson index of dispersion. Commun. Statist.—Simul. Comput. B **10** (1981), 531-535

Gebhardt, F.: (1) On the effect of stragglers on the risk of some mean estimators in small samples. Ann. Math. Statist. **37** (1966), 441-450. (2) Some numerical comparisons of several approximations to the binomial distribution. J. Amer. Statist. Assoc. **64** (1969), 1638-1646 (see also **66** [1971], 189-191)

Gehan, E. A.: Note on the "Birthday Problem". The American Statistician **22** (April 1968), 28 [see also Commun. Statist.—Simul. Comp. **11** (1982), 361-370]

Gini, C.: Logic in statistics. Metron **19** (1958), 1-77

Glick, N.: Hijacking planes to Cuba: an up-dated version of the birthday problem. The American Statistician **24** (Febr. 1970), 41-44 [see also **30** (1976), 197 + 198]

Good. I. J.: How random are random numbers? The American Statistician **23** (October 1969), 42-45 [see Applied Statistics **29** (1980), 164-171, Biometrical Journal **22** (1980), 447-461 and J. Amer. Statist. Assoc. **77** (1982), 129-136]

Gridgeman, N. T.: The lady tasting tea, and allied topics. J. Amer. Statist. Assoc. **54** (1959), 776-783

Griffiths, D. A.: Interval estimation for the three-parameter lognormal distribution via the likelihood function. Applied Statistics **29** (1980), 58-68

Grimm, H.: (1) Tafeln der negativen Binomialverteilung. Biometrische Zeitschr. **4** (1962), 239-262. (2) Tafeln der Neyman-Verteilung Typ A. Biometrische Zeitschr. **6** (1964), 10-23. (3) Graphical methods for the determination of type and parameters of some discrete distributions. In G. P. Patil (Ed.): Random Counts in Scientific Work. Vol. I: Random Counts in Models and Structures. (Pennsylvania State University Press, pp. 268) University Park and London 1970, pp. 193-206 (see also J. J. Gart: 171-191)

Groot, M. H. de: Optimal Statistical Decisions. (McGraw-Hill, pp. 489) New York 1970

Guenther, W. C.: (1) Concepts of Statistical Inference. (McGraw-Hill, pp. 553) 2nd ed. New York 1973 [see also Statistica Neelandica **27** (1973), 103-110]. (2) The inverse hypergeometric—a useful model. Statist. Neerl. **29** (1975), 129-144

Gumbel, E. J.: (1) Probability Tables for the Analysis of Extreme-Value Data. National Bureau of Standards, Appl. Mathem. Ser. 22, Washington, D.C., July 1953. (2) Statistics of Extremes. New York 1958 [see also Biometrics **23** (1967), 79-103 and J. Qual. Technol. **1** (Oct. 1969), 233-236] (3) Technische Anwendungen der statistischen Theorie der Extremwerte. Schweiz. Arch. angew. Wissenschaft Technik **30** (1964), 33-47

Gurland, J.: Some applications of the negative binomial and other contagious distributions. Amer. J. Public Health **49** (1959), 1388-1399 [see also Biometrika **49** (1962), 215-226 and Biometrics **18** (1962), 42-51]

Guterman, H. E.: An upper bound for the sample standard deviation. Technometrics **4** (1962), 134 + 135

Hahn, G. J., and Chandra, R.: Tolerance intervals for Poisson and binomial variables. Journal of Quality Technology **13** (1981), 100-110

Haight, F. A.: (1) Index to the distributions of mathematical statistics. J. Res. Nat. Bur. Stds. **65B** (1961), 23-60. (2) Handbook of the Poisson Distribution. New York 1967

Hald, A.: (1) Statistical Tables and Formulas. New York 1952, pp. 47–59. (2) Statistical Theory with Engineering Applications. New York 1960, Chapter 7

Hall, A. D.: A Methodology for Systems Engineering. Princeton, N.J. 1962

Hamaker, H. C.: Approximating the cumulative normal distribution and its inverse. Applied Statistics **27** (1978), 76–77 [see **26** (1978), 75–76 and **28** (1979), 175–176 as well as P. A. P. Moran, Biometrika **67** (1980), 675–677]

Hamblin, C. L.: Fallacies. (Methuen; pp. 326) London 1970

Hampel, F.: Robuste Schätzungen. Ein anwendungsorientierter Überblick. Biometrical Journal **22** (1980), 3–21

Harris, D.: A method of separating two superimposed normal distributions using arithmetic probability paper. J. Animal Ecol. **37** (1968), 315–319

Harter, H. L.: (1) A new table of percentage points of the chisquare distribution. Biometrika **51** (1964), 231–239. (2) The use of order statistics in estimation. Operations Research **16** (1968), 783–798. (3) The method of least squares and some alternatives—part I/VI. Int. Stat. Rev. **42** (1974), 147–174, 235–264, 282, **43** (1975), 1–44. 125–190, 269–272, 273–278, **44** (1976), 113–159. (4) A bibliography on extreme-value theory. Int. Stat. Rev. **46** (1978). 279–306

Harvard University, Computation Laboratory: Tables of the Cumulative Binomial Probability Distribution; Annals of the Computation Laboratory of Harvard University, Cambridge, Mass. 1955

Hasselblad, V., Stead, A. G., and Galke, W.: Analysis of coarsely grouped data from the lognormal distribution. J. Amer. Statist. Assoc. **75** (1980), 771–778

Hawkins, D. M.: Identifications of Outliers. (Chapman and Hall; pp. 224) London 1980

Hemelrijk, J.: Back to the Laplace definition. Statistica Neerlandica **22** (1968), 13–21

Herdan, G.: (1) The relation between the dictionary distribution and the occurrence distribution of word length and its importance for the study of quantitative linguistics. Biometrika **45** (1958), 222–228. (2) The Advanced Theory of Language as Choice and Chance. Berlin-Heidelberg-New York 1966, pp. 201–206

Herold, W.: Ein Verfahren der Dekomposition einer Mischverteilung in zwei normale Komponenten mit unterschiedlichen Varianzen. Biometrische Zeitschr. **13** (1971), 314–328

Hill, G. W.: Reference table. "Student's" t-distribution quantiles to 20 D. CSIRO Div. Math. Statist. Tech. Paper **35** (1972), 1–24

Hoaglin, D. C., Mosteller, F., and Tukey, J. W. (Eds.): Understanding Robust and Exploratory Data Analysis. (Wiley; pp. 448) New York 1983

Hodges, J. L., Jr. and E. L. Lehmann: A compact table for power of the t-test. Ann. Math. Statist. **39** (1968), 1629–1637

Hotelling, H.: The statistical method and the philosophy of science. The American Statistician **12** (December 1958), 9–14

Huber, P. J.: (1) Robust statistics: a review. Ann. Math. Statist. **43** (1972), 1041–1967 (cf. F. Hampel, J. Amer. Statist. Assoc. **69** (1974), 383–393). (2) Robust Statistics. (Wiley; pp. 300) New York 1981

Huddleston, H. F.: Use of order statistics in estimating standard deviations. Agric. Econom. Research **8** (1956), 95–99

Johnson, E. E.: (1) Nomograph for binomial and Poisson significance tests. Industrial Quality Control **15** (March 1959), 22 + 24. (2) Empirical equations for approximating Tabular F values. Technometrics **15** (1973), 379–384

Johnson, N. L. and Kotz, S.: Urn Models and their Application. An Approach to Modern Discrete Probability Theory. (Wiley, pp. 402) New York 1977

Jolly, G. M.: Estimates of population parameters from multiple recapture data with both death and dilution—deterministic model. Biometrika **50** (1963), 113–126

Kane, V. E.: Standard and goodness-of-fit parameter estimation methods for the three-parameter lognormal distribution. Commun. Statist.—Theory Meth. **11** (1982), 1935–1957 [see also E. Mohn: Biometrika **66** (1979), 567–575]

Keeney, R. L.: Decision analysis: an overview. Operations Research **30** (1982), 803–835
King, A. C., and Read, C. B.: Pathways to Probability. History of the Mathematics of Certainty and Chance. (Holt, Rinehart and Winston, pp. 139) New York 1963
King, J. R.: Probability Charts for Decision Making. (Industrial Press; pp. 290) New York 1971
Kitagawa, T.: Tables of Poisson Distribution. (Baifukan) Tokyo 1952
Koehler, K. J.: A simple approximation for the percentiles of the t distribution. Technometrics **25** (1983), 103–105
Kolmogoroff, A. N.: Grundbegriffe der Wahrscheinlichkeitsrechnung. Berlin 1933
Kramer, G.: Entscheidungsproblem, Entscheidungskriterien bei völliger Ungewißheit und Chernoffsches Axiomensystem. Metrika **11** (1966), 15–38 (cf. Table 1, pp. 22–23)
Kübler, H.: On the parameter of the three-parameter distributions: lognormal, gamma and Weibull. Statistische Hefte **20** (1979), 68–125
Lancaster, H. O.: (1) The combination of probabilities. (Query 237) Biometrics **23** (1967) 840–842 [cf. **31** (1975), 987–992 and Ann. Math. Statist. **38** (1967), 659–680]. (2) The Chi-Squared Distribution. (Wiley, pp. 356), New York 1969
Larson, H. R.: A nomograph of the cumulative binomial distribution. Industrial Quality Control **23** (Dec. 1966), 270–278 [see also Qualität u. Zuverlässigkeit **17** (1972) 231–242 and 247–254]
Laubscher, N. F.: Interpolation in F-tables. The American Statistician **19** (February 1965), 28 + 40
Lee, Elisa T.: Statistical Methods for Survival Data Analysis. (Lifetime Learning Publications; pp. 557) Belmont, Calif. 1980
Lehmann, E. L.: Significance level and power. Ann. Math. Statist. **29** (1958), 1167–1176
Lesky, Erna: Ignaz Philipp Semmelweis und die Wiener medizinische Schule. Österr. Akad. Wiss., Philos.-histor. Kl. **245** (1964), 3. Abh. (93 pages) [cf. Dtsch. Med. Wschr. **97** (1972), 627–632]
Lieberman, G. J., and Owen, D. B.: Tables of the Hypergeometric Probability Distribution. Stanford, Calif. 1961 [see also A. M. Mathai and R. K. Saxena, Metrika **14** (1969), 21–39]
Lienert, G. A.: Die zufallskritische Beurteilung psychologischer Variablen mittels verteilungsfreier Schnelltests. Psycholog. Beiträge **7** (1962), 183–217
Ling, R. F.: Study of the accuracy of some approximations for t, χ^2, and F tail probabilities. J. Amer. Stat. Assoc. **73** (1978), 274–283
Linstone, H. A. and Turoff, M.: The Delphi Method. Techniques and Applications. (Addison-Wesley; pp. 620) London 1975
Lockwood, A.: Diagrams. A Survey of Graphs, Maps, Charts and Diagrams for the Graphic Designer. (Studio Vista; pp. 144) London 1969
Lubin, A.: Statistics. Annual Review of Psychology **13** (1962), 345–370
Mahalanobis, P. C.: A method of fractile graphical analysis. Econometrica **28** (1960), 325–351
Mallows, C. L.: Generalizations of Tchebycheff's inequalities. With discussion. J. Roy. Statist. Soc., Ser. B **18** (1956), 139–176
Manly, B. F. J., and Parr, M. J.: A new method of estimating population size, survivorship, and birth rate from capture-recapture data. Trans. Soc. Brit. Ent. **18** (1968), 81–89 [cf. Biometrika **56** (1969), 407–410 and Biometrics **27** (1971), 415–424, **28** (1972), 337–343, **29** (1973), 487–500]
Marascuilo, L. A. and Levin, J. R.: Appropriate post hoc comparisons for interaction and nested hypotheses in analysis of variance designs. The elimination of type IV errors. Amer. Educat. Res. J. **7** (1970), 397–421 [see also Psychol. Bull. **78** (1972), 368–374 and **81** (1974), 608–609]
Maritz, J. S.: Empirical Bayes Methods. (Methuen, pp. 192) London 1970

Martin, J. D., and Gray, L. N.: Measurement of relative variation: sociological examples. Amer. Sociol. Rev. **36** (1971), 496–502

Martino. J. P.: The precision of Delphi estimates. Technol. Forecastg. (USA) **1** (1970), 293–299

Matthijssen, C., and Goldzieher, J. W.: Precision and reliability in liquid scintillation counting. Analyt. Biochem. **10** (1965), 401–408

McLaughlin, D. H., and Tukey, J. W.: The Variance of Means of Symmetrically Trimmed Samples from Normal Populations and its Estimation from such Trimmed Samples. Techn. Report No. 42, Statist. Techn. Res. Group, Princeton University, July 1961

McNemar, Q.: Psychological Statistics. 4th ed. (Wiley; pp. 529) New York 1969, p. 75 [see also The American Statistician **29** (1975), 20–25]

Miller, J. C. P. (Ed.): Table of Binomial Coefficients. Royal Soc. Math. Tables Vol. III, Cambridge (University Press) 1954

Molenaar, W.: Approximations to the Poisson, Binomial, and Hypergeometric Distribution Functions. (Mathematisch Centrum, pp. 160) Amsterdam 1970

Molina, E. C.: Poisson's Exponential Binomial Limit. (Van Nostrand) New York 1945

Montgomery, D. C.: An introduction to short-term forecasting. J. Ind. Engg. **19** (1968), 500–504

Morice, E.: Puissance de quelques tests classiques effectif d'échantillon pour des risques α, β fixes. Revue de Statistique Appliquée **16** (No. 1, 1968), 77–126

Moses, L. E.: Statistical theory and research design. Annual Review of Psychology **7** (1956), 233–258

Moses, L. E., and Oakford, R. V.: Tables of Random Permutations. (Allen and Unwin, pp. 233) London 1963

Moshman, J.: Critical values of the log-normal distribution. J. Amer. Statist. Assoc. **48** (1953), 600–605

Mosteller, F., and Tukey, J. W.: The uses and usefulness of binomial probability paper. J. Amer. Statist. Assoc. **44** (1949), 174–212

Mudgett, B. D.: Index Numbers. (Wiley) New York 1951

National Bureau of Standards: Tables of the Binomial Probability Distribution. Applied Math. Series No. 6, Washington 1950

Natrella, Mary, G.: Experimental Statistics. NBS Handbook 91. (U.S. Govt. Print. Office) Washington 1963, Chapters 3 + 4

Naus, J. I.: An extension of the birthday problem. The American Statistician **22** (Feb. 1968), 227–229.

Nelson, W.: The truncated normal distribution—with applications to component sorting. Industrial Quality Control **24** (1967), 261–271

Neumann von, J.: Zur Theorie der Gesellschaftsspiele. Math. Ann. **100** (1928), 295–320

Neyman, J.: (1) On a new class of "contagious" distributions, applicable in entomology and bacteriology. Ann. Math. Statist. **10** (1939), 35–57. (2) Basic ideas and some recent results of the theory of testing statistical hypotheses. J. Roy. Statist. Soc. **105** (1942), 292–327. (3) First Course in Probability and Statistics. New York 1950, Chapter V: Elements of the Theory of Testing Statistical Hypotheses; Part 5 · 2 · 2: Problem of the Lady tasting tea. (4) Lectures and Conferences on Mathematical Statistics and Probability, 2nd rev. and enlarged ed., Washington 1952

Neyman, J., and Pearson, E. S.: (1) On the use and interpretation of certain test criteria for purposes of statistical inference. Part I and II. Biometrika **20A** (1928), 175–240 and 263–294. (2) On the problem of the most efficient type of statistical hypotheses. Philosophical Transactions of the Royal Society A **231** (1933), 289–337

Noether, G. E.: Use of the range instead of the standard deviation. J. Amer. Statist. Assoc. **50** (1955), 1040–1055

Norden, R. H.: A survey of maximum likelihood estimation. Int. Stat. Rev. **40** (1972), 329–354; (Part 2) **41** (1973), 39–58

Novik, M. R. and Jackson, P. H.: Statistical Methods for Educational and Psychological Research (McGraw-Hill) New York 1974

Odeh, R. E , and Owen, D. B.: Tables for Normal Tolerance Limits, Sampling Plans, and Screening. (M. Dekker; pp. 316) New York 1980

Ord, J. K.: Graphical methods for a class of discrete distributions. J. Roy. Statist. Soc. A **130** (1967), 232–238

Owen, D. B.: Handbook of Statistical Tables. (Addison-Wesley) London 1962 (Errata: Mathematics of Computation **18**, 87; Mathematical Reviews **28**, 4608)

Pachares, J.: Table of confidence limits for the binomial distribution. J. Amer. Statist. Assoc. **55**, (1960), 521–533 [see also **70** (1975), 67–69]

Page, E.: Approximations to the cumulative normal function and its inverse for use on a pocket calculator. Applied Statistics **26** (1977), 75–76 [cf. **28** (1979), 175–176 and **27** (1978), 76–77]

Parks, G. M.: Extreme value statistics in time study. Industrial Engineering **16** (1965), 351–355

Parratt, L. G.: Probability and Experimental Errors in Science. London 1961

Paulson, E.: An approximate normalization of the analysis of variance distribution. Ann. Math. Statist. **13** (1942), 233–235 [see also J. Statist. Comput. Simul. **3** (1974), 81–93 as well as Biometrika **66** (1979), 681–683]

Pearson, E. S.: The History of Statistics in the 17th and 18th Centuries. (Griffin; pp. 711) London 1978

Pearson, E. S. and Hartley, H. O. (Eds.): Biometrika Tables for Statisticians, Vol. I, Cambridge University Press, Cambridge, 1958

Pearson, E. S. and Kendall, M. G. (Eds.): Studies in the History of Statistics and Probability. (Griffin, pp. 481) London 1970

Pearson, E. S., and Tukey, J. W.: Approximate means and standard deviations based on distances between percentage points of frequency curves. Biometrika **52** (1965), 533–546

Pettitt, A. N.: (1) A non-parametric approach to the change point problem. Applied Statistics **28** (1979), 126–135. (2) A simple cumulative sum type statistic for the change point problem with zero-one observations. Biometrika **67** (1980), 79–84

Pfanzagl, J.: (1) Verteilungsunabhängige statistische Methoden. Zschr. angew. Math. Mech. **42** (1962), T71–T77. (2) Allgemeine Methodenlehre der Statistik, Vols. I, II (page 63) (Sammlung Göschen), Berlin 1964, 1966

Pitman, E. J. G.: (1) Lecture Notes on Nonparametric Statistics. Columbia University, New York 1949. (2) Statistics and science. J. Amer. Statist. Assoc. **52** (1957), 322–330

Plackett, R. L.: Random permutations. J. Roy. Statist. Soc., B **30** (1968), 517–534

Polak, F. L.: Prognostics. (Elsevier, pp. 450) Amsterdam 1970

Popper, K. R.: (1) Science: problems, aims, responsibilities. Fed. Proc. **22** (1963), 961–972. (2) Logik der Forschung, 2nd enlarged edition, Tübingen 1966 [see also Nature **241** (1973), 293–294 and Social Studies of Science **8** (1978), 287–307]

Pratt, J. W., Raiffa, H., and Schlaifer, R.: The foundations of decision under uncertainty: an elementary exposition. J. Amer. Statist. Assoc. **59** (1964), 353–375

Prescott, P.: A simple method of estimating dispersion from normal samples. Applied Statistics **17** (1968), 70–74 [see also **24** (1975), 210–217 and Biometrika **58** (1971), 333–340]

Pressat, R.: Demographic Analysis. Methods, Results, Applications. (Transl. by J. Matras). (Aldine-Atherton, pp. 498) London 1972

Preston, E. J.: A graphical method for the analysis of statistical distributions into two normal components. Biometrika **40** (1953), 460–464 [see also J. Amer. Statist. Assoc. **70** (1975), 47–55 and **78** (1983), 228–237]

Price de, D. J. S.: (1) Science Since Babylon. New Haven, Connecticut 1961. (2) Little Science, Big Science. New York 1963. (3) Research on Research. In Arm, D. L. (Ed.), Journeys in Science: Small Steps – Great Strides. The University of New Mexico Press, Albuquerque 1967. (4) Measuring the size of science. Proc. Israel Acad. Sci. Humanities **4** (1969), No. 6, 98–111

Quandt, R. E.: Old and new methods of estimation and the Pareto distribution. Metrika **10** (1966), 55–82 [see also **30** (1983), 15–19]

Raiffa, H., and Schlaifer, R.: Applied Statistical Decision Theory. Division of Research, Harvard Business School, Boston, Mass. 1961

Rao, C. R. and Chakravarti, I. M.: Some small sample tests of significance for a Poisson distribution. Biometrics **12** (1956), 264–282

Rasch, D.: Zur Problematik statistischer Schlußweisen. Wiss. Z. Humboldt-Univ. Berlin, Math.-Nat. R. **18** (1969) (2), 371–383

Rider, P. R.: The distribution of the quotient of ranges in samples from a rectangular population. J. Amer. Statist. Assoc. **46** (1951), 502–507

Rigas, D. A.: A nomogram for radioactivity counting statistics. International Journal of Applied Radiation and Isotopes **19** (1968), 453–457

Riordan, J.: (1) An Introduction to Combinatorial Analysis. New York 1958. (2) Combinatorial Identities. (Wiley, pp. 256) New York 1968

Roberts, H. V.: Informative stopping rules and inferences about population size. J. Amer. Statist. Assoc. **62** (1967), 763–775

Robson, D. S.: Mark-Recapture Methods of Population Estimation. In N. L. Johnson and H. Smith, Jr. (Eds.): New Developments in Survey Sampling. (Wiley-Interscience, pp. 732) New York 1969, pp. 120–146 [see also Biometrics **30** (1974), 77–87]

Romig, H. G.: 50–100 Binomial Tables. (Wiley) New York 1953

Rosenthal, R.: Combining results of independent studies. Psychological Bulletin **85** (1978), 185–193 [see **88** (1980), 494–495 and **92** (1982), 500–504]

Rusch, E., and Deixler, A.: Praktische und theoretische Gesichtspunkte für die Wahl des Zentralwertes als statistische Kenngröße für die Lage eines Verteilungszentrums. Qualitätskontrolle **7** (1962), 128–134

Saaty, L.: Seven more years of queues: a lament and a bibliography. Naval Res. Logist. Quart. **13** (1966), 447–476

Saaty, T. L. and Alexander, J. M.: Thinking With Models. Math. Models in the Phys., Biol., and Social Sci. (Pergamon; pp. 181) Oxford and New York 1981

Sachs, L.: (1) Statistische Methoden. 6th revised edition (Springer, 133 pages) Berlin, Heidelberg, New York 1984, pages 3, 7, 12–18, 23–26, 100–105 (2) Graphische Methoden in der Datenanalyse. Klin. Wschr. **55** (1977), 973–983

Sarhan, A. E. and Greenberg, B. G. (Eds.): Contributions to Order Statistics. New York 1962

Savage, I. R.: Probability inequalities of the Tchebycheff type. J. Res. Nat. Bur. Stds. **65B** (1961), 211–222 [see also Trab. Estadist. **33** (1982), 125–132]

Schmid, C. F., and Schmid, S. E.: Handbook of Graphic Presentation. 2nd ed. (Ronald Press, Wiley; pp. 308) New York 1979 [see also Stat. Graphics, Wiley 1983]

Schmitt, S. A.: Measuring Uncertainty. An Elementary Introduction to Bayesian Statistics. (Addison-Wesley; pp. 400) Reading, Mass. 1969

Schneeweiss, H.: Entscheidungskriterien bei Risiko. Berlin-Heidelberg 1967

Severo, N. C. and Zelen, M.: Normal approximation to the chi-square and non-central F probability functions. Biometrika **47** (1960), 411–416 [see also J. Amer. Statist. Assoc. **66** (1971), 577–582]

Sheynin, O. B.: C. F. Gauss and the theory of errors. Archive for History of Exact Sciences **20** (1979), 21–72 [see also **16** (1976), 137–187 and **17** (1977), 1–61]

Smirnov, N. V. (Ed.): Tables for the Distribution and Density Function of t-Distribution. London, Oxford 1961

Smith, J. H.: Some properties of the median as an average. The American Statistician **12** (October 1958), 24, 25, 41 [see also J. Amer. Statist. Assoc. **63** (1968), 627–635]
Snyder, R. M.: Measuring Business Changes. New York 1955
Sonquist, J. A., and Dunkelberg, W. C.: Survey and Opinion Research: Procedures for Processing and Analysis. (Prentice International; pp. 502) Englewood Cliffs, N.J. 1977
Southwood, T. R. E.: Ecological Methods with Particular Reference to the Study of Insect Populations. (Methuen, pp. 391), London 1966 [see also Science **162** (1968), 675 + 676]
Spear, Mary, E.: Practical Charting Techniques. (McGraw-Hill, pp. 394) New York 1969
Spinchorn, E.: The odds on Hamlet. American Statistician **24** (Dec. 1970), 14–17
Stange, K.: Eine Verallgemeinerung des zeichnerischen Verfahrens zum Testen von Hypothesen im Wurzelnetz (Mosteller-Tukey-Netz) auf drei Dimensionen. Qualitätskontrolle **10** (1965), 45–52
Stegmüller, W.: Personelle und Statistische Wahrscheinlichkeit. Semivolume 1 and 2 (Vol. 4 of Probleme und Resultate der Wissenschaftstheorie und Analytischen Philosophie) (Springer, 560 and 420 pages) Berlin, Heidelberg, New York 1972 [see also Statist. Neerl. **28** (1974), 225–227, **29** (1975), 29]
Stephenson, C. E.: Letter to the editor. The American Statistician **24** (April 1970), 37 + 38
Stevens, S. S.: On the theory of scales of measurement. Science **103** (1946), 677–680
Student: The probable error of a mean. Biometrika **6** (1908), 1–25 [see also **30** (1939), 210–250, Metron **5** (1925), 105–120 as well as The American Statistician **26** (Dec. 1972), 43 + 44]
Sturges, H. A.: The choice of a class interval. J. Amer. Statist. Assoc. **21** (1926), 65 + 66
Szameitat, K., and Deininger, R.: Some remarks on the problem of errors in statistical results. Bull. Int. Statist. Inst. **42**, I (1969), 66–91 [see also Allgem. Statist. Arch. **55** (1971), 290–303]
Teichroew, D.: A history of distribution sampling prior to the era of the computer and its relevance to simulation. J. Amer. Statist. Assoc. **60** (1965), 27–49
Theil, H.: (1) Optimal Decision Rules for Government and Industry. Amsterdam 1964. (2) Applied Economic Forecasting. (Vol. 4 of Studies in Mathematical and Managerial Economics; North-Holland Publ. Co., pp. 474) Amsterdam 1966
Thöni, H.: A table for estimating the mean of a lognormal distribution. J. Amer. Statist. Assoc. **64** (1969), 632–636. Corrigenda **65** (1970), 1011–1012
Thorndike, Frances: Applications of Poisson's probability summation. Bell System Techn. J. **5** (1926), 604–624
Troughton, F.: The rule of seventy. Mathematical Gazette **52** (1968), 52 + 53
Tukey, J. W.: (1) Some sampling simplified. J. Amer. Statist. Assoc. **45** (1950), 501–519. (2) Unsolved problems of experimental statistics. J. Amer. Statist. Assoc. **49** (1954), 706–731. (3) Conclusions vs. decisions. Technometrics **2** (1960), 423–433. (4) A survey of sampling from contaminated distributions. In I. Olkin and others (Eds.): Contributions to Probability and Statistics. Essays in Honor of Harold Hotelling, pp. 448–485, Stanford 1960. (5) The future of data analysis. Ann. Math. Statist. **33** (1962), 1–67 [Trimming and Winsorization: Statistische Hefte **15** (1974), 157–170]. (6) Data analysis, computation and mathematics. Quart. Appl. Math. **30** (1972), 51–65
—, and McLaughlin, D. H.: Less vulnerable confidence and significance procedures for location based on a single sample: Trimming/Winsorization I. Sankhya Ser. A **25** (1963), 331–352
Vahle, H. and Tews, G.: Wahrscheinlichkeiten einer χ^2-Verteilung. Biometrische Zeitschr. **11** (1969), 175–202 [see also **18** (1976), 13–22]

Waerden, B. L., van der: Der Begriff Wahrscheinlichkeit. Studium Generale **4** (1951), 65–68; p. 67, left column

Wagle, B.: Some techniques of short-term sales forecasting. The Statistician **16** (1966), 253–273

Wainer, H.: (1) Robust statistics and some prescriptions. J. Educational Statistics. (2) —, and Thissen, D.: Graphical Data Analysis. Annual Review of Psychology **32** (1981), 191–241

Wald, A.: Statistical Decision Functions. New York 1950

Wallis, W. A., and Roberts, H. V.: Methoden der Statistik, 2nd edition, Freiburg/Br. 1962 [Transl. from English: Statistics, A New Approach. The Free Press, Glencoe, Ill.]

Walter, E.: (1) Review of the book Verteilungsfreie Methoden in der Biostatistik by G. Lienert. Biometrische Zeitschr. **6** (1964), 61 + 62. (2) Personal communication, 1966.

Wasserman, P., and Silander, F. S.: Decision-Making: An Annotated Bibliography Supplement, 1958–1963. Ithaca, N.Y. 1964

Weber, Erna: Grundriß der Biologischen Statistik. 7th revised edition, Stuttgart 1972, pp. 143–160 (8th revised edition 1980)

Weibull, W.: Fatigue Testing and Analysis of Results. New York 1961

Weichselberger, K.: Über ein graphisches Verfahren zur Trennung von Mischverteilungen und zur Identifikation kupierter Normalverteilungen bei großem Stichprobenumfang. Metrika **4** (1961), 178–229 [see also J. Amer. Statist. Assoc. **70** (1975), 47–55]

Weintraub, S.: Tables of the Cumulative Binomial Probability Distribution for Small Values of p. (The Free Press of Glencoe, Collier-Macmillan) London 1963

Weiss, L. L.: (1) A nomogram based on the theory of extreme values for determining values for various return periods. Monthly Weather Rev. **83** (1955), 69–71. (2) A nomogram for log-normal frequency analysis. Trans. Amer. Geophys. Union **38** (1957), 33–37

Wellnitz, K.: Kombinatorik, 6th edition (Bagel/Hirt, 56 pages) Düsseldorf/Kiel 1971

Westergaard, H.: Contributions to the History of Statistics. (P. S. King, pp. 280) London 1932

Wilk, M. B., and Gnanadesikan, R.: Probability plotting methods for the analysis of data. Biometrika **55** (1968), 1–17 [see also **70** (1983), 11–17 and V. Barnett, Applied Statistics **24** (1975), 95–108]

Williams, C. B.: (1) A note on the statistical analysis of sequence length as a criterion of literary style. Biometrika **31** (1940), 356–361. (2) Patterns in the Balance of Nature. New York 1964

Williamson, E., and Bretherton, M. H.: Tables of the Negative Binomial Distribution. London 1963

Wilson, E. B., and Hilferty, M. M.: The distribution of chi-square. Proc. Nat. Acad. Sci. **17** (1931), 684–688 (see also Pachares, J.: Letter to the Editor. The American Statistician **22** [Oct. 1968], 50)

Winkler, R. L.: An Introduction to Bayesian Inference and Decisions (Holt, Rinehart and Winston) New York 1972

Wold, H. O. A.: Time as the realm of forecasting. Annals of the New York Academy of Sciences **138** (1967), 525–560

Yamane, T.: Statistics. An Intoductory Analysis. Chapter 8, pp. 168–226. New York 1964

Yasukawa, K.: On the probable error of the mode of skew frequency distributions. Biometrika **18** (1926), 263–292, see pages 290, 291

Zahlen, J. P.: Über die Grundlagen der Theorie der parametrischen Hypothesentests. Statistische Hefte **7** (1966), 148–174

Zar, J. H.: Approximations for the percentage points of the chi-squared distribution. Applied Statistics **27** (1978), 280–290

Zinger, A.: On interpolation in tables of the F-distribution. Applied Statistics **13** (1964), 51–53

[8:1a] Stochastic processes

Bailey, N. T. J.: The Elements of Stochastic Processes with Applications to the Natural Sciences. (Wiley; pp. 249) New York 1964

Bartholomew, D. J.: Stochastic Models for Social Processes. 3rd ed. (Wiley; pp. 384) New York 1982

Bartlett, M. S.: (1) An Introduction to Stochastic Processes. Cambridge 1955 (2nd ed. 1966). (2) Stochastic Population Models. London 1960. (3) Essays on Probability and Statistics. London 1962

Basawa, I. V., and Rao, B. L. S. P.: Statistical Inference for Stochastic Processes. (Academic Press; pp. 438) London 1980

Billingsley, P.: (1) Statistical Methods in Markov chains. Ann. Math. Statist. **32** (1961), 12–40. (2) Statistical Inference for Markov Processes. Chicago 1961

Chiang, C. L.: (1) Introduction to Stochastic Processes in Biostatistics. (Wiley, pp. 312) New York 1968. (2) Chiang, C. L.: An Introduction to Stochastic Processes and their Applications. (R. E. Krieger Publ. Co.; pp. 517) Huntington, N.Y. 1980

Cox, D. R., and Lewis, P. A. W.: The Statistical Analysis of Series of Events. London 1966

—, and Miller, H. D.: The Theory of Stochastic Processes. (Wiley; pp. 337) London 1965

—, and Smith, W. L.: Queues. London 1961

Cramér, H.: Model building with the aid of stochastic processes. Technometrics **6** (1964), 133–159

Cramér, H., and Leadbetter, M. R.: Stationary and Related Stochastic Processes: Sample Function Properties and Their Applications. (Wiley, pp. 348) New York and London 1967

Doig, A.: A bibliography on the theory of queues. Biometrika **44** (1957), 490–514

Feller, W.: An Introduction to Probability Theory and Its Applications. Vol. 1, 3rd ed. (Wiley; pp. 509) New York 1968; Vol. 2, 2nd ed. (Wiley; pp. 669) New York 1971

Gold, R. Z.: Tests auxiliary to χ^2 tests in a Markov chain. Ann. Math. Statist. **34** (1963), 56–74

Grimmet, G. R., and Stirzacker, D. R.: Probability and Random Processes. (Clarendon Press; pp. 354) Oxford 1982

Gurland, J. (Ed.): Stochastic Models in Medicine and Biology. Madison (Univ. of Wisconsin) 1964

Karlin, S., and Taylor, H.: (1) A First Course in Stochastic Processes. 2nd ed. (Acad. Press; pp. 560) New York 1975. (2) A Second Course in Stochastic Processes. (Acad. Press; pp. 576) New York and London 1980

Kemeny, J. G., and Snell, J. L.: Finite Markov Chains. Princeton, N.J. 1960; reprinted by Springer-Verlag, New York.

—, Snell, J. L., and Knapp, A. W.: Denumerable Markov Chains. Princeton, N.J. 1966; reprinted by Springer-Verlag, New York

Kullback, S., Kupperman, M., and Ku, H. H.: Tests for contingency tables and Markov chains. Technometrics **4** (1962), 573–608 [cf. Applied Statistics **22** (1973), 7–20]

Lee, A. M.: Applied Queueing Theory. London 1966

Parzen, E.: (1) Modern Probability Theory and Its Applications. (Wiley; pp. 464) New York 1960. (2) Stochastic Processes. San Francisco 1962

Prabhu, N. U.: (1) Stochastic Processes. Basic Theory and Its Applications. New York 1965. (2) Queues and Inventories: a Study of Their Basic Stochastic Processes. New York 1965

Saaty, T. L.: Elements of Queueing Theory. New York 1961

Takacs, L.: (1) Introduction to the Theory of Queues. Oxford Univ. Press 1962. (2) Stochastische Prozesse. München 1966. (3) Combinatorial Methods in the Theory of Stochastic Processes. (Wiley, pp. 262) New York 1967

Wold, H. O. A.: The I.S.I. Bibliography on Time Series and Stochastic Processes. London 1965
Zahl, S.: A Markov process model for follow-up studies. Human Biology **27** (1955), 90–120

[8:1b] Distribution-free methods

Bradley, J. V.: Distribution-Free Statistical Tests. (Prentice-Hall; pp. 388) Englewood Cliffs, N.J. 1968
Büning, H. and Trenkler, G.: Nichtparametrische Statistische Methoden. (de Gruyter, 435 pages) Berlin and New York 1978
Conover, W. J.: Practical Nonparametric Statistics. 2nd ed. (Wiley; pp. 510) London 1980
David, H. A.: Order Statistics. 2nd ed. (Wiley, pp. 360) New York 1981
Gibbons, Jean D.: Nonparametric Statistical Inference. (McGraw-Hill, pp. 306) New York 1971 [see also Biometrics **28** (1972), 1148–1149]
Hàjek, J., and Sidàk, Z.: Theory of Rank-Tests (Academic Press, pp. 297) New York and London 1967
Hollander, M., and Wolfe, D. A.: Nonparametric Statistical Methods. (Wiley, pp. 503) New York 1973
Kraft, C. H., and Eeden, Constance Van: A Nonparametric Introduction to Statistics. (Macmillan, pp. 304) New York 1968
Lehmann, E. L.: Nonparametrics. Statistical Methods Based on Ranks. (McGraw-Hill; pp. 457) New York 1975
Lienert, G. A.: Verteilungsfreie Methoden in der Biostatistik. 2nd revised edition (A. Hain, Meisenheim am Glan) Vol. I (736 pages), 1973, Vol. II (1246 pages), 1978, Book of tables (686 pages) 1975
Milton, R. C.: Rank Order Probabilities. Two-Sample Normal Shift Alternatives. (Wiley, pp. 320) New York 1970
Noether, G. E.: Elements of Nonparametric Statistics. (Wiley, pp. 104) New York 1967
Pratt, J. W., and Gibbons, J. D.: Concepts of Nonparametric Theory. (Springer; pp. 462) New York, Heidelberg, Berlin
Puri, M. L. and Sen, P. K.: Nonparametric Methods in Multivariate Analysis. (Wiley, pp. 450) London 1971
Savage, I. R.: Bibliography of Nonparametric Statistics. (Harvard University Press, pp. 284) Cambridge, Mass. 1962
Walsh, J. E.: Handbook of Nonparametric Statistics. Vol. I–III (Van Nostrand, pp. 549, 686, 747) Princeton, N.J. 1962, 1965, 1968

[8:2] Chapter 2

[8:2a] Medical statistics

Abdel-Hamid, A. R., Bather, J. A., and Trustrum, G. B.: The secretary problem with an unknown number of candidates. J. Appl. Probab. **19** (1982), 619–630
Armitage, P.: Statistical Methods in Medical Research. (Blackwell; pp. 504) Oxford 1971, pages 8–18, 176–184, 384–391, 408–414, 426–441
Barnard, G. A.: Control charts and stochastic processes. J. Roy. Statist. Soc., Ser. B **21** (1959), 239–257 [cf. A **132** (1969), 205–228; V-mask: see Technometrics **15** (1973), 833–847]

Barnett, R. N., and Youden, W. J.: A revised scheme for the comparison of quantitative methods. Amer. J. Clin. Pathol. **54** (1970), 454–462 [see also **43** (1965), 562–569 and **50** (1968), 671–676 as well as Biometrics **34** (1978), 39–45; Technometrics **21** (1979), 397–409; and Applied Statistics **29** (1980), 135–141]

Benjamin, B.: Demographic Analysis. (Allen and Unwin; pp. 160) London 1968

Bissel, A. F.: Cusum techniques for quality control (with discussion). Applied Statistics **18** (1969), 1–30 [see also **23** (1974), 420–433 and Clin. Chem. **21** (1975), 1396–1405]

Bogue, D. J.: Principles of Demography (Wiley, pp. 917) New York 1969

Brown, B. W., Jr.: (1) Statistical aspects of clinical trials. Proc. 6th Berkeley Symp. Math. Stat. Prob. 1970/71, Vol. IV, 1–13. Univ. of Calif. Press, Berkeley 1972. (2) The crossover experiment for clinical trials. Biometrics **36** (1980), 69–79

Burch, P. R. J.: Smoking and lung cancer. Tests of a causal hypothesis. J. Chronic Diseases **33** (1980), 221–238

Burdette, W. J. and Gehan, E. A.: Planning and Analysis of Clinical Studies. (C. Thomas; pp. 104) Springfield 1970 [see Int. J. Cancer **13** (1974), 16–36]

Burr, I. W.: The effect of non-normality on constants for $\bar{x}$ and R charts. Industrial Quality Control **23** (May 1967) 563–569 [see also J. Qual. Technol. **1** (1969), 163–167]

Castleman, B., and McNeely, Betty U. (Eds.): Normal laboratory values. New Engl. J. Med. **283** (1970), 1276–1285 [see also Clin. Chem. **21** (1975), 1457–1467 and 1873–1877]

Caulcutt, R., and Boddy, R.: Statistics for Analytical Chemists. (Chapman and Hall; pp. 265) London 1983

Chow, Y. S., Moriguti, S., Robbins, H., and Samuels, S. M.: Optimal selection based on relative rank (the "secretary problem"). Israel Journal of Mathematics **2** (1964), 81–90 [see also Operations Research **29** (1981), 130–145]

Chun, D.: Interlaboratory tests – short cuts. Annu. Tech. Conf. Trans., Amer. Soc. Qual. Contr. **20** (1966), 147–151 [see also Clin. Chem. **19** (1973), 49–57]

Cochran, W. G.: (1) The planning of observational studies of human populations. With discussion. J. Roy. Statist. Soc., Ser. A **128** (1965), 234–265 [see also C **18** (1969), 270–275; C **23** (1974), 51–59 and Biometrics **24** (1968), 295–313, see also Sonja M. McKinlay, J. Amer. Statist. Assoc. **70** (1975), 503–520 (and 521/3)]. (2) Errors of measurement in statistics. Technometrics **10** (1968), 637–666

Cox, P. R.: Demography. 4th ed. (University Press, pp. 470) Cambridge 1970

Dobben de Bruyn, C. S., van: Cumulative Sum Tests, Theory and Practice. (Griffin, pp. 82) London 1968 [see Applied Statistics **29** (1980), 252–258]

Documenta Geigy: (1) Placebos und Schmerz. In Schmerz, pages 3 and 4, 1965 [see also Selecta **10** (1968), 2386–2390]. (2) Wissenschaftliche Tabellen, 6th edition, Basel 1960, 7th edition, Basel 1968 (8th revised edition 1980)

Dorfman, R.: The detection of defective members of large populations. Ann. Math. Statist. **14** (1943), 436–440 [see also Biometrika **63** (1976), 671–673]

Duncan, A. J.: Quality Control and Industrial Statistics. 4th ed. (Irwin; pp. 1047) Homewood, Ill. 1974

Eilers, R. J. (Chairman): Total quality control for the medical laboratory (14 papers). Amer. J. Clin. Path. **54** (1970), 435–530 [see also Fed. Proc. **34** (1975), 2125–2165]

Eisenhart, Ch.: Realistic evaluation of the precision and accuracy of instrument calibration systems. J. Res. Nat. Bur. Stds. C**67** (1963), 161–187 [see also Science **160** (1968), 1201–1204 and Technometrics **15** (1973), 53–66]

Elandt-Johnson, R. C., and Johnson, N. L.: Survival Models and Data Analysis. (Wiley; pp. 450) New York 1980

Elveback, Lila, R., Guillier, C. L., and Keating, F. R.: Health, normality, and the ghost of Gauss. J. Amer. Med. Ass. **211** (1970), 69–75 [see also Ann. N.Y. Acad. Sci. **161** (1969), 538–548]

Ewan, W. D.: When and how to use cu-sum charts. Technometrics **5** (1963), 1–22 [see **16** (1974), 65–71 and **24** (1982), 199–205]

Federer, W. T.: Procedures and Designs useful for screening material in selection and allocation, with a bibliography. Biometrics **19** (1963), 553–587 [see also Postgraduate Medicine **48** (Oct. 1970), 57–61, Fed. Proc. **34** (1975), 2157–2161 and Am. J. Publ. Health **66** (1976), 145–150]

Feinstein, A. R.: Clinical Biostatistics. (Mosby; pp. 468) St. Louis 1977

Fienberg, S. E., and Straf, M. L.: Statistical assessments as evidence. J. Roy. Statist. Society A **145** (1982), 410–421 [see also "law": 395–409 and 422–438]

Fleiss, J. L.: Statistical Methods for Rates and Proportions. 2nd ed. (Wiley; pp. 321) New York 1981

Gabriels, R.: A general method for calculating the detection limit in chemical analysis. Analytical Chemistry **42** (1970), 1434 [see also Clin. Chem. **21** (1975), 1542]

Gehan, E. A., and Freireich, E. J.: Non-randomized controls in cancer clinical trials. New Engl. J. Med. **290** (1974), 198–203 [see also **291** (1974), 1278–1285, 1305–1306 and Int. J. Cancer **13** (1974), 16–36]

Gilbert, J. P., and Mosteller, F.: Recognizing the maximum of a sequence. J. Amer. Statist. Assoc. **61** (1966), 35–73 [see also **70** (1975), 357–361 and Metrika **29** (1982), 87–93]

Glazebrook, K. D.: On the optimal allocation of two or more treatments in a controlled clinical trial. Biometrika **65** (1978), 335–340

Goldberg, J. D., and Wittes, J. T.: The evaluation of medical screening procedures. The American Statistician **35** (1981), 4–11

Graff. L. E. and Roeloffs, R.: Group testing in the presence of test error; an extension of the Dorfman procedure. Technometrics **14** (1972), 113–122 [see also J. Amer. Statist Assoc. **67** (1972), 605–608]

Hill, A. B.: Principles of Medical Statistics. 9th ed. (The Lancet; pp. 390) London 1971

Hinkelmann, K.: Statistische Modelle und Versuchspläne in der Medizin. Method. Inform. Med. **6** (1967), 116–124 [see also Biometrics **17** (1961), 405–414 and **21** (1966), 467–480]

Hosmer, D. W., and Hartz, S. C.: Methods for analyzing odds ratios in a $2 \times c$ contingency table. Biometrical Journal **23** (1981), 741–748

Hwang, F. K.: Group testing with a dilution effect. Biometrika **63** (1976) 671–673

Jellinek, E. M.: Clinical tests on comparative effectiveness of analgesic drugs. Biometrics **2** (1946), 87–91 [clinical trial: see also Amer. J. Med. **55** (1973), 727–732 and **58** (1975), 295–299]

Johnson, N. L., and Leone, F. C.: Statistics and Experimental Design in Engineering and the Physical Sciences, Vol. I, pp. 320–339, New York 1964

Kemp, K. W.: The average run length of the cumulative sum chart when a V-mask is used. J. Roy. Statist. Soc., Ser. B **23** (1961), 149–153 [see also B **29** (1967), 263–265 and Applied Statistics **11** (1962), 16–31]

Keyfitz, N., and Beekman, J. A.: Demography Through Problems. (Springer; pp. 240) New York, Berlin, Heidelberg, Tokyo 1983

Koller, S.: (1) Die Aufgaben der Statistik und Dokumentation in der Medizin. Dtsch. med. Wschr. **88** (1963). 1917–1924. (2) Einführung in die Methoden der ätiologischen Forschung – Statistik und Dokumentation. Method. Inform. Med. **2** (1963). 1–13. (3) Systematik der statistischen Schlußfehler. Method. Inform. Med. **3** (1964), 113–117. (4) Problems in defining normal values. Bibliotheca Haematologica **21** (1965), 125–128. (5) Mathematisch-statistische Grundlagen der Diagnostik. Klin. Wschr. **45** (1967), 1065–1072. (6) Mögliche Aussagen bei Fragen der statistischen Ursachenforschung. Metrika **17** (1971), 30–42

Kramer, K. H.: Use of mean deviation in the analysis of interlaboratory tests. Technometrics **9** (1967), 149–153

Lange, H.-J.: Syntropie von Krankheiten. Method. Inform. Med. **4** (1965), 141–145 [see also **14** (1975), 144–149 and Internist **11** (1970), 216–222]

Lasagna, L.: Controlled trials: nuisance or necessity. Method. Inform. Med. **1** (1962), 79–82 [For clinical trials see, e.g., Biometrics **36** (1980), 677–706]

Lawless, J. F.: Statistical Models and Methods for Lifetime Data. (Wiley-Interscience; pp. 736) New York 1982

Lawton, W. H., Sylvestre, E. A., and Young-Ferraro, B. J.: Statistical comparison of multiple analytic procedures: application to clinical chemistry. Technometrics **21** (1979), 397–409 [see Anal. Chem. **47** (1975), 1824–1829]

Loyer, M. W.: Bad probability, good statistics, and group testing for binomial estimation. The American Statistician **37** (1983), 57–59

Lilienfeld, A. M., and Lilienfeld, D. E.: Foundations of Epidemiology. 2nd rev. ed. (Oxford Univ. Press; pp. 400) New York 1980

Mainland, D.: (1) The clinical trial – some difficulties and suggestions. J. Chronic Diseases **11** (1960), 484–496 [see **35** (1982), 413–417]. (2) Experiences in the development of multiclinic trials. J. New Drugs **1** (1961), 197–205. (3) Elementary Medical Statistics. 2nd ed. (W. B. Saunders; pp. 381), Philadelphia and London 1963.

Mandel, J.: The Statistical Analysis of Experimental Data. (Interscience-Wiley, pp. 410) New York 1964, Chapter 14

Mandel, J., and Lashof, T. W.: The interlaboratory evaluation of testing methods. Amer. Soc. for Testing Materials Bulletin No. 239 (July 1959), 53–61 [see J. Qual. Technol. **6** (1974), 22–36]

Mandel, J., and Stiehler, R. D.: Sensitivity – a criterion for the comparison of methods of test. J. Res. Nat. Bur. Stds. **53** (Sept. 1954), 155–159 [see also J. Qual Technol. **4** (1972), 74–85]

Martini, P.: (1) Methodenlehre der therapeutisch-klinischen Forschung. 3rd edition, Berlin 1953 (4th edition see Martini-Oberhoffer-Welte). (2) Die unwissentliche Versuchsanordnung und der sogenannte doppelte Blindversuch. Dtsch. med. Wschr. **82** (1957), 597–602. (3) Grundsätzliches zur therapeutisch-klinischen Versuchsplanung. Method. Inform. Med. **1** (1962), 1–5

—, Oberhoffer, G. and Welte, E.: Methodenlehre der therapeutisch-klinischen Forschung. 4th rev. ed. (Springer, 495 pages) Berlin-Heidelberg-New York 1968

McFarren, E. F., Lishka, T. R. J., and Parker, J. H.: Criterion for judging acceptability of analytical methods. Analytical Chemistry **42** (1970), 358–365

Mendoza, G. and Iglewicz, B.: A three-phase sequential model for clinical trials. Biometrika **64** (1977), 201–205

Morgenstern, H., Kleinbaum, D. G., and Kupper, L. L.: Measures of disease incidence used in epidemiologic research. International Journal of Epidemiology **9** (1980), 97–104 [for summary estimators of relative risk see J. Chronic Diseases **34** (1981), 463–468]

Page, E. S.: (1) Cumulative sum charts. Technometrics **3** (1961), 1–9. (2) Controlling the standard deviation by cusums and warning lines. Technometrics **5** (1963), 307–315

Peto, R., Pike, M. C., Armitage, P., Breslow, N. E., Cox, D. R., Howard, S. V., Mantel, N., McPherson, K., Peto, J., and Smith, P. G.: Design and analysis of randomized clinical trials requiring prolonged observation of each patient. British Journal of Cancer **34** (1976), 585–612, **35** (1977), 1–39 [cf., The Statistician **28** (1979), 199–208 and Controlled Clinical Trials **1** (1980), 37–58, **2** (1981), 15–29, 31–49 and **3** (1982), 29–46, 311–324.

Reed, A. H., Henry, R. J., and Mason, W. B.: Influence of statistical method used on the resulting estimate of normal range. Clin. Chem. **17** (1971), 275–284 (page 281) [cf. **20** (1974), 576–581]

Reynolds, J. H.: The run sum control chart procedure. J. Qual. Technol. **3** (Jan. 1971), 23–27 [cf. **5** (1973), 166]

Roos, J. B.: The limit of detection of analytical methods. Analyst. **87** (1962), 832

Rümke, Chr. L.: Über die Gefahr falscher Schlußfolgerungen aus Krankenblattdaten (Berkson's Fallacy). Method. Inform. Med. **9** (1970), 249–254

Rümke, Chr. L. en Bezemer, P.D.: Methoden voor de bepaling van normale waarden. I, II. Nederl. Tijdschr. Geneesk. **116** (1972), 1124–1130, 1559–1568 (pages 1561–1565)

Ryan, J. C., and Fisher, J. W.: Management of clinical research "architecture". J. Clin. Pharmacol. **14** (1974), 233–248

Sachs, L.: (1) Statistische Methoden in der Medizin: In Unterricht, Beratung und Forschung. Klin. Wschr. **55** (1977), 767–775. (2) Statistische Methoden. 6th revised edition (Springer, 133 pages) Berlin, Heidelberg, New York 1984

Schindel, L. E.: (1) Placebo in theory and practice. Antibiotica et Chemotherapia, Advances **10** (1962), 398–430. (2) Die Bedeutung des Placebos für die klinischtherapeutische Forschung. Arzneim.-Forsch. **15** (1965), 936–940. (3) Placebo und Placeboeffekte in Klinik und Forschung. Arzneim.-Forschg. **17** (1967), 892–918 [see also J. Amer. Med. Assoc. **232** (1975), 1225–1227]

Schlesselman, J. J.: Case Control Studies. Design, Conduct, Analysis. (Oxford Univ. Press; pp. 354) New York 1982

Schneiderman, M. A.: The proper size of a clinical trial: "Grandma's Strudel" method. J. New Drugs **4** (1964), 3–11 [see also J. Chronic Diseases **21** (1968), 13–24; **25** (1972), 673–681; **26** (1973), 535–560]

Sobel, M., and Groll, P. A.: (1) Group testing to eliminate efficiently all defectives in a binomial sample. Bell System Technical Journal **38** (1959), 1179–1252. (2) Binomial group-testing with an unknown proportion of defectives. Technometrics **8** (1966), 631–656 [see also J. Qual. Technol. **1** (1969), 10–16, J. Amer. Statist. Assoc. **70** (1975), 923–926 as well as Biometrika **62** (1975), 181–193]

Svodoba, V., and Gerbatsch, R.: Zur Definition von Grenzwerten für das Nachweisvermögen. Z. analyt. Chem. **242** (1968), 1–12

Taylor, H. M.: The economic design of cumulative sum control charts. Technometrics **10** (1968), 479–488 [see also **16** (1974), 65–71, 73–80]

Tygstrup, N., Lachin, J. M., and Juhl, E. (Eds.): The Randomized Clinical Trial and Therapeutic Decisions. (M. Dekker; pp. 320) New York 1982

Vessereau, A.: Efficacité et gestion des cartes de contrôle. Revue Statistique Appliquée **20**, 1 (1970), 21–64 [see also **22**, 1 (1974), 5–48]

Williams, G. Z., Harris, E. K., Cotlove, E., Young, D. S., Stein, M. R., Kanofsky, P., and Shakarji, G.: Biological and analytic components of variation in long-term studies of serum constituents in normal subjects. I, II, III. Clinical Chemistry **16** (1970), 1016–1032 [see also **18** (1972), 605–612]

Wilson, A. L.: The precision and limit of detection of analytical methods. Analyst **86** (1961), 72–74

Woodward, R. H. and Goldsmith, P. L.: Cumulative Sum Techniques. (I.C.I. Monograph No. 3) Edinburgh 1964 [V-mask: see J. Qual. Technol. **8** (1976), 1–12]

Yates, F.: The analysis of surveys on computers—features of the Rothamsted survey program. Applied Statistics **22** (1973), 161–171 [cf. **23** (1974), 51–59; Biometrics **31** (1975), 573–584, and especially **34** (1978), 299–304]

Youden, W. J.: (1) Graphical diagnosis of interlaboratory test results. Industrial Quality Control **15** (May 1959), 1–5. [see also J. Qual. Technol. **6** (1974), 22–36]. (2) The sample, the procedure, and the laboratory. Anal. Chem. **13** (December 1960), 23 A–37 A. (3) Accuracy of analytical procedures. J. Assoc. Offic. Agricult. Chemists **45** (1962), 169–173. (4) Systematic errors in physical constants. Technometrics **4** (1962). 111–123 [see also **14** (1972), 1–11.] (5) The collaborative test. J. Assoc. Offic. Agricult. Chemists **46** (1963), 55–62. (6) Ranking laboratories by round-robin tests. Materials Research and Standards **3** (January 1963), 9–13. (7) Statistical Techniques for Collaborative Tests. (Association of Official Analytical Chemists, pp. 60) Washington 1967 [see D. M. Rocke, Biometrika **70** (1983), 421–431]

[8:2b] Sequential test plans

Alling, D. W.: Closed sequential tests for binomial probabilities. Biometrika **53** (1966), 73–84

Armitage, P.: (1) Sequential methods in clinical trials. Amer. J. Public Health **48** (1958), 1395–1402. (2) Sequential Medical Trials. 2nd ed. (Wiley; pp. 194) N.Y. 1975. (3) Sequential analysis in medicine. Statistica Neerlandica **15** (1961), 73–82. (4) Some developments in the theory and practice of sequential medical trials. In Proc. Fifth Berkeley Symp. Mathem. Statist. Probab., Univ. of Calif. 1965/66. Univ. of Calif. Press, Berkeley and Los Angeles 1967, Vol. 4: Biology and Problems of Health, pp. 791–804 (see also 805–829) [see also Biometrika **62** (1975), 195–200]

Beightler, C. S., and Shamblin, J. E.: Sequential process control. Industrial Engineering **16** (March–April 1965), 101–108

Bertram, G.: Sequenzanalyse für zwei Alternativfolgen. Zschr. Angew. Math. Mechanik **40** (1960), 185–189

Billewicz, W. Z.: (1) Matched pairs in sequential trials for significance of a difference between proportions. Biometrics **12** (1956), 283–300. (2) Some practical problems in a sequential medical trial. Bull. Intern. Statist. Inst. **36** (1958), 165–171

Bross, I. D. J.: (1) Sequential medical plans. Biometrics **8** (1952), 188–205. (2) Sequential clinical trials. J. Chronic Diseases **8** (1958), 349–365

Chilton, N. W., Fertig, J. W., and Kutscher, A. H.: Studies in the design and analysis of dental experiments. III. Sequential analysis (double dichotomy). J. Dental Research **40** (1961), 331–340

Cole, L. M. C.: A closed sequential test design for toleration experiments. Ecology **43** (1962), 749–753

Davies, O. L.: Design and Analysis of Industrial Experiments. London 1956, Chapter 3

Fertig, J. W., Chilton, N. W., and Varma, A. O.: Studies in the design of dental experiments. 9–11. Sequential analysis. J. Oral Therapeutics and Pharmacol. **1** (1964), 45–56, 175–182, **2** (1965), 44–51

Freeman, H.: Sequential analysis of statistical data: Applications. Columbia University Press, New York 1957

Fülgraff, G.: Sequentielle statistische Prüfverfahren in der Pharmakologie. Arzneim.-Forschg. **15** (1965), 382–387

Greb, D. J.: Sequential sampling plans. Industrial Quality Control **19** (May 1963), 24–28, 47–48

Jackson, J. E.: Bibliography on sequential analysis. J. Amer. Statist. Assoc. **55** (1960), 561–580

Johnson, N. L.: Sequential analysis: a survey. J. Roy. Statist. Soc. A **124** (1961), 372–411

Lienert, G. A. and Sarris, V.: (1) Eine sequentielle Modifikation eines nicht-parametrischen Trendtests. Biometrische Zeitschr. **10** (1967), 133–147. (2) Testing monotonicity of dosage-effect relationship by Mosteller's test and its sequential modification. Method. Inform. Med. **7** (1968), 236–239

Litchfield, J. T.: Sequential analysis, screening and serendipity. J. Med. Pharm. Chem. **2** (1960), 469–492

Maly, V.: Sequenzprobleme mit mehreren Entscheidungen und Sequenzschätzung. I and II. Biometr. Zeitschr. **2** (1960), 45–64 and **3** (1961), 149–177 [see also **5** (1963), 24–31 and **8** (1966), 162–178]

Sachs, V.: Die Sequenzanalyse als statistische Prüfmethode im Rahmen medizinischer experimenteller, insbesondere klinischer Untersuchungen. Ärztl. Forschg. **14** (1962), 331–345

Schneiderman, M. A.: A family of closed sequential procedures. Biometrika **49** (1962), 41–56

—, and Armitage, P.: Closed sequential t-tests. Biometrika **49** (1962), 359–366 (see also 41–56) Corrections **56** (1969), 457

Spicer, C. C.: Some new closed sequential designs for clinical trials. Biometrics **18** (1962), 203–211
Vogel, W.: Sequentielle Versuchspläne. Metrika **4** (1961), 140–157 [see also Unternehmensforschung **8** (1964), 65–74]
Wald, A.: Sequential Analysis. (Wiley) New York 1947
Weber, Erna: Grundriß der Biologischen Statistik. 7th edition (G. Fischer, 706 pages) Stuttgart 1972, pp. 412–499 (8th revised edition 1980)
Wetherill, G. B.: (1) Sequential estimation of quantal response curves. With discussion. J. Roy. Statist. Soc., Ser. B **25** (1963), 1–48. (2) Sequential Methods in Statistics. 2nd ed. (Chapman and Hall; pp. 242) London 1975
Whitehead, J.: The Design and Analysis of Sequential Clinical Trials. (Wiley; pp. 315) New York 1983
Winne, D.: Die sequentiellen statistischen Verfahren in der Medizin. Arzneim.-Forschg. **15** (1965), 1088–1091 [see also J. Amer. Statist. Assoc. **61** (1966), 577–594]
Wohlzogen, F. X. and Wohlzogen-Bukovics, E.: Sequentielle Parameterschätzung bei biologischen Alles-oder-Nichts-Reaktionen. Biometr. Zeitschr. **8** (1966), 84–120

[8:2c] Bioassay

Armitage, P., and Allen, Irene: Methods of estimating the LD 50 in quantal response data. J. Hygiene **48** (1950), 298–322
Ashford, J. R.: An approach to the analysis of data for semiquantal responses in biological assay. Biometrics **15** (1959), 573–581
Ashton, W. D.: The Logit Transformation with Special Reference to Its Uses in Bioassay. Griffin, pp. 88) London 1972
Axtell, Lilian M.: Computing survival rates for chronic disease patients. A simple procedure. J. Amer. Med. Assoc. **186** (1963), 1125–1128
Bennett, B. M.: Use of distribution-free methods in bioassay. Biometr. Zeitschr. **11** (1969), 92–104
Bliss, C. I.: The Statistics of Bioassay. (Acad. Press) New York 1952
Borth, R., Diczfalusy, E., and Heinrichs, H. D.: Grundlagen der statistischen Auswertung biologischer Bestimmungen. Arch. Gynäk. **188** (1957), 497–538 (see also Borth et al.: Acta endocr. **60** [1969], 216–220)
Brock, N., and Schneider, B.: Pharmakologische Charakterisierung von Arzneimitteln mit Hilfe des Therapeutischen Index. Arzneim.-Forschg. **11** (1961), 1–7
Bross, I.: Estimates of the LD_{50}: A Critique. Biometrics **6** (1950), 413–423
Brown, B. W.: Some properties of the Spearman estimator in bioassay. Biometrika **48** (1961), 293–302 [see also **60** (1973), 535–542, **63** (1976), 621–626 and Biometrics **28** (1972), 882–889]
Brown, B. W., Jr.: Planning a quantal assay of potency. Biometrics **22** (1966), 322–329
Buckland, W. R.: Statistical Assessment of the Life Characteristic: A Bibliographic Guide. (Griffin; pp. 125) London 1964
Cavalli-Sforza, L. (revised by R. J. Lorenz): Biometrie. Grundzüge biologisch-medizinischer Statistik, 4th edition (G. Fischer; 212 pages) Stuttgart 1974 [Dilution Assay: cf. Biometrics **22** (1966), 610–619 and **31** (1975), 619–632]
Cochran, W. G., and Davis, M.: The Robbins-Monro method for estimating the median lethal dose, J. Roy. Statist. Soc., Ser. B **27** (1965), 28–44
Cornfield, J., and Mantel, N.: Some new aspects of the application of maximum likelihood to the calculation of the dosage response curve. J. Amer. Statist. Assoc. **45** (1950), 181–209
—, Gordon, T., and Smith, W. W.: Quantal response curves for experimentally uncontrolled variables. Bull. Intern. Statist. Inst. **38** (1961), 97–115

Cox, C. P.: Statistical analysis of log-dose response bioassay experiments with experiments with unequal dose ratios for the standard and unknown preparations. J. Pharmaceut. Sci. **56** (1967), 359–364

Cox, C. P., and Ruhl, Donna, J.: Simplified computation of confidence intervals for relative potencies using Fieller's theorem. J. Pharmaceutical Sci. **55** (1966), 368–379

Das, M. N., and Kulkarni, G. A.: Incomplete block designs for bio-assays. Biometrics **22** (1966), 706–729

Davis, M.: Comparison of sequential bioassays in small samples. J. Roy. Statist. Soc. **33** (1971), 78–87

Dixon, W. J.: The up-and-down method for small samples. J. Amer. Statist. Assoc. **60** (1965), 967–978 [see also R. E. Little, **69** (1974), 202–206]

—, and Mood, A. M.: A method for obtaining and analyzing sensitivity data. J. Amer. Statist. Assoc. **43** (1948), 109–126

Fink, H., and Hund, G.: Probitanalyse mittels programmgesteuerter Rechenanlagen. Arzneim.-Forschg. **15** (1965), 624–630

—, Hund, G., and Meysing, D.: Vergleich biologischer Wirkungen mittels programmierter Probitanalyse. Method. Inform. Med. **5** (1966), 19–25

Finney, D. J.: (1) Probit Analysis. 2nd ed. London 1952, 3rd ed. (Cambridge Univ. Press, pp. 334) Cambridge and London 1971. (2) Statistical Methods in Biological Assay. 2nd ed. London 1964 (3rd edition Macmillan 1979)

Gaddum, J. H.: (1) Simplified mathematics for bioassay. J. Pharmacy and Pharmacology **6** (1953), 345–358. (2) Bioassay and mathematics. Pharmacol. Rev. **5** (1953), 87–134

Golub, A., and Grubbs, F. E.: Analysis of sensitivity experiments when the levels of stimulus cannot be controlled. J. Amer. Statist. Assoc. **51** (1956), 257–265

Hubert, J. J.: Bioassay. (Kendall/Hunt Publ. Co.; pp. 164) Dubuque, Iowa 1980

International Symposium on Biological Assay Methods. (Ed.: R. H. Regamey) (Karger, pp. 262) Basel, New York 1969

Kärber, G.: Ein Beitrag zur kollektiven Behandlung pharmakologischer Reihenversuche. Archiv für experimentelle Pathologie und Pharmakologie **162** (1931), 480–483

Kaufmann, H.: Ein einfaches Verfahren zur Auswertung von Überlebenskurven bei tödlich verlaufenden Erkrankungen. Strahlentherapie **130** (1966), 509–527

Kimball, A. W., Burnett, W. T., Jr., and Doherty, D. G.: Chemical protection against ionizing radiation. I. Sampling methods for screening compounds in radiation protection studies with mice. Radiation Research **7** (1957), 1–12

King, E. P.: A statistical design for drug screening. Biometrics **19** (1963), 429–440

Lazar, Ph.: Les essais biologiques. Revue de Statistique Appliquée **16** (No. 3, 1968), 5–35

Litchfield, J. T., Jr., and Wilcoxon, F.: A simplified method of evaluating dose-effect experiments. J. Pharmacol. Exptl. Therap. **96** (1949), 99–113

Little, R. E.: The up-and-down method for small samples with extreme value response distributions. Journal of the American Statistical Association **69** (1974), 803–806

McArthur, J. W., and Colton, T. (Eds.): Statistics in Endocrinology. Proc. Conf., Dedham, Mass., Dec. 1967. (MIT Press, pp. 476) Cambridge, Mass. 1970

Oberzill, W.: Mikrobiologische Analytik. Grundlagen der quantitativen Erfassung von Umwelteinwirkungen auf Mikroorganismen. (Carl, 519 pages) Nürnberg 1967

Olechnowitz, A. F.: Ein graphisches Verfahren zur Bestimmung von Mittelwert und Streuung aus Dosis-Wirkungs-Kurven. Arch. exp. Veterinärmed. **12** (1958), 696–701

Petrusz, P., Diczfalusy, E., and Finney, D. J.: Bioimmunoassay of gonadotrophins. Acta endocrinologica **67** (1971), 40–62

Schneider, B.: Probitmodell und Logitmodell in ihrer Bedeutung für die experimentelle Prüfung von Arzneimitteln. Antibiot. et Chemother. **12** (1964), 271–286

Stammberger, A.: Über ein nomographisches Verfahren zur Lösung der Probleme des Bio-Assay. Biometr. Zeitschr. **12** (1970), 35–53 (see also pages 351–361)

Ther, L.: Grundlagen der experimentellen Arzneimittelforschung. (Wiss. Verlagsges., 439 pages) Stuttgart 1965. pp. 74–112

Vølund, A.: Multivariate bioassay. Biometrics **36** (1980), 225–236

Warner, B. T.: Method of graphical analysis of 2 + 2 and 3 + 3 biological assays with graded responses. J. Pharm. Pharmacol. **16** (1964), 220–233

Waud, D. R.: On biological assays involving quantal responses. J. Pharmacol. Exptl. Therap. **183** (1972), 577–607

[8:2d] Statistics in engineering

Amstadter, B. L.: Reliability Mathematics. (McGraw-Hill, pp. 320) New York 1970

Bain, L. J., and Thoman, D. R.: Some tests of hypotheses concerning the three-parameter Weibull distribution. J. Amer. Statist. Assoc. **63** (1968), 853–860 [see also Technometrics **14** (1972), 831–840, **16** (1974), 49–56, **17** (1975), 369–374, 375–380 **19** (1977), 323–331, **20** (1978), 167–169, and **21** (1979), 233–237]

Barlow, R. E., Fussell, J. B., and Singpurwalla, N. D. (Eds.): Reliability and Fault Tree Analysis; Theoretical and Applied Aspects of System Reliability and Safety Assessment, Society for Industrial and Applied Mathematics, Philadelphia, 1975, pp. 927

Barlow, R. E., and Proschan, F.: Mathematical Theory of Reliability. (Wiley; pp. 265) New York 1965

Beard, R. E., Pentikäinen, T., and Pesonen, E.: Risk Theory. The Stochastic Basis of Insurance. 3rd ed. (Chapman and Hall; pp. 230) London 1984

Beightler, C. S., and Shamblin, J. E.: Sequential process control. Industrial Engineering **16** (March–April 1965), 101–108

Berrettoni, J. N.: Practical applications of the Weibull distribution. ASQC Convention (Cincinnati, Ohio, USA) Transactions 1962, pp. 303–323

Bingham, R. S., Jr.: EVOP for systematic process improvement. Industrial Quality Control **20** (Sept. 1963), 17–23

Bowker, A. H., and Lieberman, G. J.: Engineering Statistics. Englewood Cliffs, N.J. 1961

Box, G. E. P.: (1) Multi-factor designs of first order. Biometrika **39** (1952), 49–57. (2) The exploration and exploitation of response surfaces: some general considerations and examples. Biometrics **10** (1954), 16–60. (3) Evolutionary operation: a method for increasing industrial productivity. Applied Statistics **6** (1957), 3–23. (4) A simple system of evolutionary operation subject to empirical feedback. Technometrics **8** (1966), 19–26

—, and Draper, N. R.: (1) A basis for the selection of a response surface design. J. Amer. Statist. Assoc. **54** (1959), 622–654 [see also **70** (1975), 613–617]. (2) Evolutionary Operation. A Statistical Method for Process Improvement. (Wiley, pp. 237) New York 1969; Übers., Das EVOP-Verfahren. (Oldenbourg; 268 pages) München 1975

—, and Hunter, J. S.: (1) Multifactor experimental designs for exploring response surfaces. Ann. Math. Statist. **28** (1957), 195–241. (2) Experimental designs for the exploration and exploitation of response surfaces. In V. Chew (Ed.), Experimental Designs in Industry. New York 1958, pp. 138–190. (3) Condensed calculations for evolutionary operations programs. Technometrics **1** (1959), 77–95. (4) A useful method for model-building. Technometrics **4** (1962), 301–318

—, and Lucas, H. L.: Design of experiments in non-linear situations. Biometrika **46** (1959), 77–90

—, and Wilson, K. B.: On the experimental attainment of optimum conditions. J. Roy. Statist. Soc., Ser. B **13** (1951), 1–45 [see also Biometrika **62** (1975), 347–352]

—, and Youle, P. V.: The exploration and exploitation of response surfaces: an example of the link between the fitted surface and the basic mechanism of the system. Biometrics **11** (1955), 287–323 [see also Technometrics **18** (1976), 411–417]

Brewerton, F. J.: Minimizing average failure detection time. J. Qual. Technol. **2** (1970), 72–77

Brooks, S. H.: A comparison of maximum seeking methods. Operations Research **7** (1959), 430–457

—, and Mickey, M. R.: Optimum estimation of gradient direction in steepest ascent experiments. Biometrics **17** (1961), 48–56

Burdick, D. S., and Naylor, T. H.: Response surface methods in economics. Rev. Internat. Statist. Inst. **37** (1969), 18–35

Cohen, A. C., Jr.: Maximum likelihood estimation in the Weibull distribution based on complete and on censored samples. Technometrics **7** (1965), 579–588

Cox, D. R., and Oakes, D. O.: Analysis of Survival Data. (Chapman and Hall; pp. 200) London 1984

D'Agostino, R. B.: Linear estimation of the Weibull parameters. Technometrics **13** (1971), 171–182

Davies, O. L.: Design and Analysis of Industrial Experiments, London 1956, Chapter 11

Dean, B. V., and Marks, E. S.: Optimal design of optimization experiments. Operations Research **13** (1965), 647–673

Dhillon, B. S., and Singh, C.: Engineering Reliability. New Techniques and Applications. (Wiley; pp. 339) New York 1981

Drnas, T. M.: Methods of estimating reliability. Industrial Quality Control **23** (1966), 118–122

Dubey, S. D.: (1) On some statistical inferences for Weibull laws. Naval Res. Logist. Quart. **13** (1966), 227–251. (2) Normal and Weibull distribution. Naval Res. Logist. Quart. **14** (1967), 69–79. (3) Some simple estimators for the shape parameter of the Weibull laws. Naval Res. Logist. Quart. **14** (1967), 489–512. (4) Some percentile estimators for Weibull parameters. Technometrics **9** (1967), 119–129. (5) On some permissible estimators of the location parameter of the Weibull and certain other distributions. Technometrics **9** (1967), 293–307

Duckworth, W. E.: Statistical Techniques in Technological Research: An Aid to Research Productivity. (Methuen, pp. 303) London 1968

Duncan, A. J.: Quality Control and Industrial Statistics. 4th ed. (Irwin; pp. 1047) Homewood, Ill. 1974

Durr, A. C.: Accurate reliability prediction. Reliability Engineering **3** (1982), 475–485

Eagle, E. L.: Reliability sequential testing. Industrial Quality Control **20** (May 1964), 48–52

Ferrell, E. B.: Control charts for log-normal universes. Industrial Quality Control **15** (Aug. 1958), 4–6

Freudenthal, H. M., and Gumbel, E. J.: On the statistical interpretation of fatigue tests. Proc. Roy. Soc., Ser A **216** (1953), 309–332 [see also Technometrics **19** (1977); 87–93]

Fussell, J. B. and Burdick, G. R.: Nuclear Systems Reliability Engineering and Risk Assessment, Society for Industrial and Applied Mathematics, Philadelphia, 1977, pp. 849

Gheorghe, A.: Applied Systems Engineering. (Wiley; pp. 342) New York 1983

Goldberg, H.: Extending the Limits of Reliability Theory. (Wiley; pp. 263) New York 1981

Goldman, A. S., and Slattery, T. B.: Maintainability. A Major Element of System Effectiveness. (Wiley; pp. 282) New York 1964

Gottfried, P., and Roberts, H. R.: Some pitfalls of the Weibull distribution. Ninth Symp. on Reliability and Quality Control, pp. 372–379, San Francisco, Calif. (Jan. 1963)

Gross, A. J., and Clark, Virginia, A.: Survival Distributions. Reliability Applications in the Biomedical Sciences. (Wiley; pp. 331) New York 1975

Gryna, F. M., Jr., McAfee, N. J., Ryerson, C. M., and Zwerling, S.: Reliability Training Text. (Institute of Radio Engineers) New York 1960

Guild, R. D. and Chipps, J. D.: High reliability systems by multiplexing. Journal of Quality Technology **9** (1977), 62–69

Hahn, G. J.: Some things engineers should know about experimental design. Journal of Quality Technology **9** (1977), 13–20

Harter, H. L.: A bibliography of extreme-value theory. International Statistical Review **46** (1978), 279–306

Harter, H. L., and Dubey, S. D.: Theory and tables for tests of hypotheses concerning the mean and the variance of a Weibull population. ARL Tech. Rep. No. 67–0059 (Aerospace Research Laboratories. pp. 393), Wright-Patterson Air Force Base, Ohio 1967

Henley, E. J., and Kumamoto, H.: Reliability Engineering and Risk Assessment. (Prentice-Hall; pp. 568) New York 1980

Hill, W. J., and Hunter, W. G.: A review of response surface methodology: a literature survey. Technometrics **8** (1966), 571–590 [see also **15** (1973), 113–123 and 301–317 as well as Biometrics **31** (1975), 803–851]

Hillier, F. S.: (1) Small sample probability limits for the range chart. J. Amer. Statist. Assoc. **62** (1967), 1488–1493. (2) X- and R-chart control limits based on a small number of subgroups. J. Qual. Technol. **1** (1969), 17–26 [see also N. I. Gibra, J. Qual. Technol. **7** (1975), 183–192]

Honeychurch, J.: Lambda and the question of confidence. Microelectronics Reliability **4** (1965), 123–130

Hunter, W. G., and Kittrel, J. R.: Evolutionary operation: a review. Technometrics **8** (1966), 389–397

Johns, M. V., Jr., and Lieberman, G. J.: An exact asymptotically efficient confidence bound for reliability in the case of the Weibull distribution. Technometrics **8** (1966), 135–175

Johnson, L. G.: Theory and Technique of Variation Research. (Elsevier, pp. 105) Amsterdam 1964

Kabe, D. G.: Testing outliers from an exponential population. Metrika **15** (1970), 15–18 [see also J. Likes, **11** [1966], 46–54 and L. R. Lamberson, AIIE Transact. **6** (1974), 327–337]

Kao, J. H. K.: A graphical estimation of mixed parameters in life testing of electron tubes. Technometrics **1** (1959), 389–407 [see U. Hjorth **22** (1980), 99–107]

Kapur, K. C., and Lamberson, L. R.: Reliability in Engineering Design. (Wiley; pp. 586) New York 1977

Kaufman, A., Grouchko, D., and Cruon, R.: Mathematical Models for the Study of the Reliability of Systems, (Academic Press, pp. 221) New York 1977

Kenworthy, I. C.: Some examples of simplex evolutionary operation in the paper industry. Applied Statistics **16** (1967), 211–224

Kiefer, J. C.: Optimum experimental designs. J. Roy. Statist. Soc., Ser. B **21** (1959), 272–319

Knowler, L. A., Howell, J. M., Gold, B. K., Coleman, E. P., Moan, O. B., and Knowler, W. C.: Quality Control by Statistical Methods. (McGraw-Hill, pp. 139) New York 1969

Kumar, S., and Patel, H. I.: A test for the comparison of two exponential distributions. Technometrics **13** (1971), 183–189 [see also **15** (1973), 177–182, 183–186]

Lawless, J. F.: Confidence interval estimation for the Weibull and extreme value distributions. Technometrics **20** (1978), 355–364 [see also 365–368]

Lie, C. H., Hwang, C. L., and Tillman, F. A.: Availability of maintained systems. A. state-of-the-art survey. AIIE Transactions **9**, No. 3 (1977), 247–259

Lieblein, J., and Zelen, M.: Statistical investigation of the fatigue life of deep-groove ball bearings. J. Res. Nat. Bur. Stds. **57** (Nov. 1956), 273–319 [see also Technometrics **14** (1972), 693–702]

Little, R. E., and Jebe, E. H.: Statistical Design of Fatigue Experiments. (Applied Science Publishers; pp. 280) London 1975

Lloyd, D. K., and Lipow, M.. Reliability: Management, Methods and Mathematics. (Prentice-Hall; pp. 528) Englewood Cliffs, N.J. 1962 (2nd edition 1977)

Lowe, C. W.: Industrial Statistics. Vol. 2. (Business Books Ltd., pp. 294) London 1970, Chapter 12

Mann, Nancy R.: Tables for obtaining the best linear invariant estimates of parameters of the Weibull distribution. Technometrics **9** (1967), 629–645 [see also **10** (1968), 231–256, **15** (1973), 87–101, **16** (1974), 49–56, **17** (1975), 361–368, **22** (1980), 83–97 and 567–573]

—, Schafer, R. E., and Singpurwalla, N. D.: Methods for Statistical Analysis of Reliability and Life Data. (Wiley; pp. 564) New York 1974

Martz, H. F., and Waller, R. A.: Bayesian Reliability Analysis. (Wiley; pp. 745) New York 1982

McCall, J. J.: Maintenance policies for stochastically failing equipment: a survey. Management Science **11** (1965), 493–524

Morice, E.: (1) Quelques modèles mathématiques de durée de vie. Revue de Statistique Appliquée **14** (1966), No. 1, 45–126; Errata **14** (1966), No. 2, 99–101. (2) Quelques problèmes d'estimation relatifs à la loi de Weibull. Revue de Statistique Appliquée **16** (1968), No. 3, 43–63

Morrison, J.: The lognormal distribution in quality control. Applied Statistics **7** (1958), 160–172

Myhre, J. M., and Saunders, S. C.: Comparison of two methods of obtaining approximate confidence intervals for system reliability. Technometrics **10** (1968), 37–49

Nelson, L. S.: (1) Tables for a precedence life test. Technometrics **5** (1963), 491–499. (2) Weibull probability paper. Industrial Quality Control **23** (1967), 452–455

Nelson, W.: (1) A statistical test for equality of two availabilities. Technometrics **10** (1968), 594–596 [see also pages 883 and 884 as well as **15** (1973), 889–896 and J. Qual. Technol. **4** (1972), 190–195]. (2) Applied Life Data Analysis. (Wiley-Interscience, pp. 634) New York 1982

—, and Thompson, V. C.: Weibull probability papers. J. Qual. Technol. **3** (1971), 45–50

Oakes, D.: Survival analysis. European Journal of Operational Research **12** (1983), 3–14

Ostle, B.: Industry use of statistical test design. Industrial Quality Control **24** (July 1967), 24–34

Pearson, E. S.: Comments on the assumption of normality involved in the use of some simple statistical techniques. Rev. belge Statist. Rech. opérat. **9** (1969), Nr. 4, 2–18

Peng, K. C.: The Design and Analysis of Scientific Experiments. (Addison-Wesley, pp. 252) Reading, Mass. 1967, Chapter 8

Plait, A.: The Weibull distribution – with tables. Industrial Quality Control **19** (Nov. 1962), 17–26 [see also **15** (1973), 87–101]

Prairie, R. R.: Probit analysis as a technique for estimating the reliability of a simple system. Technometrics **9** (1967), 197–203

Proschan, F., and Serfling, R. J. (Eds.): Reliability and Biometry, Statistical Analysis of Lifelength. Society for Industrial and Applied Mathematics, Philadelphia, 1974, pp. 815

Qureishi, A. S., Nabavian, K. J., and Alanen, J. D.: Sampling inspection plans for discriminating between two Weibull processes. Technometrics **7** (1965), 589–601

Ravenis, J. V. J.: Estimating Weibull-distribution parameters. Electro-Technology, March 1964, 46–54 [see also Naval Res. Logist. Quart. **29** (1982), 419–428]

Rice, W. B.: Control Charts in Factory Management, 3rd ed. New York 1955

Roberts, N. H.: Mathematical Methods in Reliability Engineering. New York 1964

Sen, I., and Prabhashanker, V.: A nomogram for estimating the three parameters of the Weibull distribution. Journal of Quality Technology **12** (1980), 138–143

Sherif, Y. S.: Reliability analysis. Optimal inspection and maintenance schedules for failing systems. Microelectron. Reliab. **22** (1982), 59–115

—, and Smith, M. L.: Optimal maintenance models for systems subject to failure—a review. Naval Research Logistics Quarterly **28** (1981), 47–74 [for availability see **29** (1982), 411–418]

Shooman, M. L.: Probabilistic Reliability: An Engineering Approach. (McGraw-Hill, pp. 524) New York 1968 [see also J. Roy. Statist. Soc. A **136** (1973), 395–420]

Simonds, T. A.: MTBF confidence limits. Industrial Quality Control **20** (Dec. 1963), 21–27 [see also Technometrics **24** (1982), 67–72]

Sinha, S. K., and Kale, B. K.: Life Testing and Reliability Estimation. (Wiley; pp. 196) New York 1980

Stange, K.: (1) Ermittlung der Abgangslinie für wirtschaftliche und technische Gesamtheiten. Mitteilungsbl. Mathem. Statistik **7** (1955), 113–151. (2) Stichprobenpläne für messende Prüfung: Aufstellung und Handhabung mit Hilfe des doppelten Wahrscheinlichkeitsnetzes. Deutsche Arbeitsgemeinschaft für statistische Qualitätskontrolle beim Ausschuß für wirtschaftliche Fertigung. (ASQ/AWF). Beuth-Vertrieb, Berlin 1962. (3) Optimalprobleme in der Statistik. Ablauf- und Planungsforschung (Operational Research) **5** (1964), 171–190. (4) Die Berechnung wirtschaftlicher Pläne für messende Prüfung. Metrika **8** (1964), 48–82. (5) Statistische Verfahren im Betrieb zur Überwachung, Prüfung und Verbesserung der Qualität. Allgem. Statist. Arch. **49** (1965), 14–46. (6) Die zeichnerische Ermittlung von Folgeplänen für messende Prüfung bei bekannter Varianz der Fertigung. Biometrische Zeitschr. **8** (1966), 55–74. (7) Ein Näherungsverfahren zur Berechnung optimaler Pläne für messende Prüfung bei bekannten Kosten und bekannter Verteilung der Schlechtanteile in den vorgelegten Liefermengen. Metrika **10** (1966), 92–136. (8) Die Wirksamkeit von Kontrollkarten. I. Die $\bar{x}$- und $\tilde{x}$-Karte. Qualitätskontrolle **11** (1966), 129–137. (9) Die Wirksamkeit von Kontrollkarten. II. Die s- und R-Karte zur Überwachung der Fertigungsstreuung. Qualitätskontrolle **12** (1967), 13–20 (see also 73–75). (10) Die Bestimmung von Toleranzgrenzen mit Hilfe statistischer Überlegungen. Qualitätskontrolle **14** (1969), 57–63. (11) Folgepläne für messende Prüfung bei bekannter Varianz der Fertigung und einem nach oben und unten abgegrenzten Toleranzbereich für die Merkmalwerte. Biometrische Zeitschr. **11** (1969), 1–24. (12) Kontrollkarten für meßbare Merkmale. (Springer; 158 pages) 1975

—, and Henning, H.-J.: Formeln und Tabellen der mathematischen Statistik. 2nd revised edition, (Springer; pp. 362) Berlin 1966, pages 189–220

Störmer, H.: Mathematische Theorie der Zuverlässigkeit. Einführung und Anwendung. (Oldenbourg; 329 pages) München-Wien 1970 [see also Statist. Neerl. **28** (1974), 1–10]

Thoman, D. R., and Bain, L. J.: Two sample tests in the Weibull distribution. Technometrics **11** (1969), 805–815 [see Bain and Thoman]

Thoman, D. R., Bain, L. J., and Antle, C. E.: Inferences on the parameters of the Weibull distribution. Technometrics **11** (1969), 445–460 [see Bain and Thoman as well as Biometrika **61** (1974), 123–129 and H. Kübler, Statistische Hefte **20** (1979), 68–125]

Tillman, F. A., Hwang, C. L., and Kuo, W.: Optimization of Systems Reliability. (M. Dekker; pp. 311) New York 1980

—, Kuo, W., Hwang, C. L., and Grosh, D. L.: Bayesian reliability and availability. A review. IEEE Trans. Reliab. **R-31** (1982), 362–372

Tsokos, C. P. and Shimi, I. N. (Eds.): The Theory and Applications of Reliability with Emphasis on Bayesian and Nonparametric Methods, Academic Press, New York, 1977, Vols. I and II, pp. 549, 582

Uhlmann, W.: Kostenoptimale Prüfpläne. Tabellen, Praxis und Theorie eines Verfahrens der statistischen Qualitätskontrolle. (Physica-Vlg., 129 S.) Würzburg-Wien 1969

Watson, G. S., and Leadbetter, M. R.: Hazard analysis. I. Biometrika **51** (1964), 175–184

Weibull, W.: (1) A statistical distribution function of wide applicability. J. Applied Mechanics **18** (1951), 293–297. (2) Fatigue Testing and Analysis of Results. Oxford 1961

Wilde, D. J.: Optimum Seeking Methods. (Prentice-Hall; pp. 202) Englewood Cliffs, N.J. 1964

Yang, C.-H., and Hillier, F. S.: Mean and variance control chart limits based on a small number of subgroups. J. Qual. Technol. **2** (1970), 9–16

Zaludova, Agnes H.: Problèmes de durée de vie. Applications à l'industrie automobile (1). Revue de Statistique Appliquée **13**, (1965), No. 4, 75–98 [see also H. Vogt, Metrika **14** (1969), 117–131]

[8:2e] Linear programming and operations research

Anderson, M. Q.: Quantitative Management Decision Making. (Wadsworth; pp. 622) Belmont, Calif. 1982

Dantzig, G. B. (translated and revised by A. Jaeger): Lineare Programmierung und Erweiterungen. Berlin-Heidelberg 1966 [Linear Programming and Extensions. Princeton University Press, Princeton, New Jersey]

Fabrycky, W. J., and Torgersen, P. E.: Operations Economy: Industrial Applications of Operations Research. Englewood Cliffs, N.J. 1966, Chapter 16

Flagle, C. D., Huggins, W. H., and Roy, R. H. (Eds.): Operations Research and Systems Engineering. Baltimore 1960

McCormick, G.: Nonlinear Programming: Theory, Algorithms and Applications. (Wiley; pp. 464) New York 1983

Gass, S. I.: Linear Programming. Methods and Applications. 3rd ed. (McGraw-Hill, pp. 325), New York 1969

Harnes, A., and Cooper, W. W.: Management Models and Industrial Applications of Linear Programming. New York 1961

Harper, W. M., and Lim, H. C.: Operational Research. 2nd ed. (Macdonald and Evans; pp. 310) Plymouth 1982

Hertz, D. B. (Ed.): Progress in Operations Research II. (Wiley; pp. 455) New York 1964

Hillier, F. S., and Lieberman, G. J.: Introduction to Operations Research. (Holden-Day, pp. 639) San Francisco 1967

Holzman, A. G. (Ed.): Operations Research Support Methodology. (M. Dekker; pp. 647) New York 1979

Kohlas, J.: Stochastic Methods of Operations Research. (Cambridge Univ. Press; pp. 224) Cambridge 1982

Moore, P. G.: A survey of operational research. J. Roy. Statist. Soc. A **129** (1966), 399–447

Müller-Merbach, H.: Operations Research. 3rd edition (Vahlen; 565 pages) München 1973

Philipson, C.: A review of the collective theory of risk. Skand. Aktuarietidskr. **51** (1968), 45–68 and 117–133 (see also H. Bühlmann: 174–177)

Saaty, T. L.: Operations research. Some contributions to mathematics. Science **178** (1972), 1061–1070

Sakarovitch, M.: Linear Programming. (Springer; pp. 210) New York, Berlin, Heidelberg, Tokyo 1983

Sasieni, M., Yaspan, A., and Friedman, L. Operations Research: Methods and Problems. (Wiley) New York 1959

Stoller, D. S.: Operations Research: Process and Strategy. Univ. of Calif. Press, Berkeley 1965

Theil, H., Boot, J. C. G., and Kloek, T.: Operations Research and Quantitative Economics, an Elementary Introduction. New York 1965

[8:2f] Game theory and war games

Bauknecht, K.: Panzersimulationsmodell „Kompaß". Industrielle Organisation **36** (1967), 62–70

Berlekamp, E. R., Conway, J. H., and Guy, R. K.: Winning Ways. Vol. 1: Games in General, Vol. 2: Games in Particular (Academic Press; pp. 472 and 480) New York 1982

Bowen, K. C.: Research Games. An Approach to the Study of Decision Processes. (Halsted Press; pp. 126) New York 1978

Brams, S. J., Schotter, A., and Schwödiauer, G. (Eds): Applied Game Theory. (Physica-Vlg.; pp. 447) Würzburg 1979

Charnes, A., and Cooper, W. W.: Management Models and Industrial Applications of Linear Programming. Vol. I, II (Wiley; pp. 467, 393) New York 1961

Dresher, M.: Games of Strategy: Theory and Applications, Englewood Cliffs, N.J. 1961

—, Shapley, L. S., and Tucker, A. W. (Eds.): Advances in Game Theory. Princeton (Univ. Press), N.J. 1964 [see also the bibliography on gaming in Simulation **23** (1974), 90–95, 115–116]

Eckler, A. R.: A survey of coverage problems associated with point and area targets. Technometrics **11** (1969), 561–589

Edwards, W.: The theory of decision making. Psychological Bulletin **51** (1954), 380–417 (see also Psychol. Rev. **69** [1962], 109)

Fain, W. W., Fain, J. B., and Karr, H. W.: A tactical warfare simulation program. Naval Res. Logist. Quart. **13** (1966), 413–436

Horvath, W. J.: A statistical model for the duration of wars and strikes. Behavioral Science **13** (1968), 18–28

Isaacs, R.: Differential Games. A Mathematical Theory with Applications to Warfare and Pursuit, Control and Optimization. (Wiley; pp. 384) New York 1965

Jones, A. J.: Game Theory. Mathematical Models of Conflict. (Halsted Press, Wiley; pp. 309) New York 1979

Kaplan, E. L.: Mathematical Programming and Games. Vol. 1 (Wiley; pp. 588) New York 1982

Kemeny, J. G., Schleifer, A., Jr., Snell, J. L., and Thompson, G. L.: Mathematik für die Wirtschaftspraxis. Translated by H.-J. Zimmermann. Berlin 1966, pp. 410–475
Luce, R. D., and Raiffa, H.: Games and Decisions. New York 1957
Neumann, J. von: Zur Theorie der Gesellschaftsspiele. Math. Annalen **100** (1928), 295–320
—, and Morgenstern, O.: Theory of Games and Economic Behavior. Princeton 1944, 3rd ed. 1953
Owen, G.: Game Theory. 2nd ed. (Academic Press; pp. 368) New York 1982
Packel, E. W.: The Mathematics of Games and Gambling. (The Mathematical Association of America, and Wiley; pp. 141) Washington 1981
Rapoport, A., and Orwant, C.: Experimental games: a review. Behavioral Science **7** (1962), 1–37 (see also 38–80) [see also D. Martin, Social Studies of Science **8** (1978), 85–110]
Riley, V., and Young, R. P.: Bibliography on War Gaming. (Johns Hopkins Univ. Press, pp. 94) Baltimore 1957
Vogelsang, R.: Die mathematische Theorie der Spiele. (Dümmler, 254 pages) Bonn 1963
Williams, J. D.: The Compleat Strategyst. Rev. ed., London 1966
Wilson, A. (translated by W. Höck): Strategie und moderne Führung (List, 240 pages) München 1969
Young, J. P.: A Survey of Historical Development in War Games. Operations Research Office. (The Johns Hopkins Univ.) Bethesda. Md. August 1959

[8:2g] Monte Carlo method and computer simulation

Adler, H., and Neidhold, G.: Elektronische Analog- und Hybridrechner. (VEB Dt. Vlg. d. Wissenschaften, 415 pages) Berlin 1974
Anke, K., Kaltenecker, H., and Oetker, R.: Prozeßrechner. Wirkungsweise und Einsatz. (Oldenbourg, 602 pages) München and Wien 1970 (2nd edition 1971)
Anke, K., and Sartorius. H.: Industrielle Automatisierung mit Prozeßrechnern. Elektrotechnische Zeitschrift A **89** (1968), 540–544
Barney, G. C., and Hambury, J. H.: The components of hybrid computation. Computer Bulletin **14** (1970), 31–36
Bekey, G. A., and Karplus, W. J.: Hybrid Computation. (Wiley, pp. 464) New York 1969
Böttger, R.: Die Leistungsfähigkeit von Simulationsverfahren bei der Behandlung von Straßenverkehrsproblemen. Ablauf und Planungsforschung **8** (1967), 355–369
Bratley, P., Fox, B. L., and Schrage, L. E.: A Guide to Simulation. (Springer; pp. 385) New York, Heidelberg, Berlin, Tokyo 1983
Buslenko, N. P., and Schreider, J. A.: Monte-Carlo-Methode und ihre Verwirklichung mit elektronischen Digitalrechnern. Leipzig 1964
Cellier, F. E. (Ed.): Progress in Modelling and Simulation. (Academic Press; pp. 466) New York 1982
Chambers, J. M.: Computers in statistical research. Simulation and computer-aided mathematics. Technometrics **12** (1970), 1–15
Conway, R. W.: Some tactical problems in simulation. Management Science **10** (Oct. 1963), 47–61
Dutter, R., and Ganster, I.: Monte Carlo investigation of robust methods. Prob. Statist. Infer. 1982, 59–72
Ehrenfeld, S., and Ben-Tuvia, S.: The efficiency of statistical simulation procedures. Technometrics **4** (1962), 257–275
Eilon, S., and Deziel, D. P.: The use of an analogue computer in some operational research problems. Operations Research Quarterly **16** (1965), 341–365

Fernbach, S., and Taub, A. H. (Eds.): Computers and their Role in the Physical Sciences. (Gordon and Breach, pp. 638) London 1970

Gibbons, Diane I., and Vance, L. C.: A simulation study of estimators for the 2-parameter Weibull distribution. IEEE Transactions on Reliability R-30, (1981), No. 1, 61–66.

Gorenflo, R.: Über Pseudozufallszahlengeneratoren und ihre statistischen Eigenschaften. Biometrische Zeitschr. **7** (1965), 90–93

Guetzkow, H. (Ed.): Simulation in Social Sciences: Readings. Englewood Cliffs, N.J. 1962

Halton, J. H.: A retrospective and prospective survey of the Monte Carlo Method. SIAM Review **12** (Jan. 1970), 1–63 [see also J. Econometrics **1** (1973), 377–395]

Hammersley, J. M., and Handscomb, D. C.: Monte Carlo Methods. London 1964

James, M. L., Smith, G. M., and Wolford, J. C.: Analog Computer Simulation of Engineering Systems. New York 1966

Karplus, W. J.: Analog Simulation. New York 1958

—, and Soroka, W. J.: Analog Methods, Computation and Simulation. New York 1959

Klerer, M., and Korn, G. A. (Eds.): Digital Computer User's Handbook. (McGraw-Hill, pp. 922) New York 1967

Kohlas, J.: Monte Carlo Simulation im Operations Research. (Springer, 162 pages) Berlin, Heidelberg, New York 1972

Kowalski, C. J.: On the effect of non-normality on the distribution of the sample product-moment correlation coefficient. Applied Statistics **21** (1972), 1–12

Lehmann, F.: Allgemeiner Bericht über Monte-Carlo-Methoden. Blätter Dtsch. Ges. Versich.-math. **8** (1967), 431–456

Martin, F. F.: Computer Modeling and Simulation. (Wiley, pp. 331) New York 1968

Maryanski, F. J.: Digital Computer Simulations. (Wiley; pp. 304) New York 1981

Mize, J. H., and Cox, J. G.: Essentials of Simulation. (Prentice-Hall International, pp. 234) London 1968 [see also the bibliography in Simulation **23** (1974), 90–95, 115–116]

Morgenthaler, G. W.: The Theory and Application of Simulation in Operations Research. In Ackoff, R. L. (Ed.): Progress in Operations Research I. New York 1961, Chapter 9

Newman, T. G., and Odell, P. L.: The Generation of Random Variates. (No. 29 of Griffin's Stat. Monogr. and Courses) (Griffin, pp. 88) London 1971

Namneck, P.: Vergleich von Zufallszahlen-Generatoren. Elektronische Rechenanlagen **8** (1966), 28–32 [see also J. Amer. Statist. Assoc. **77** (1982), 129–136]

Naylor, Th. H., Balintfy, J. L., Burdick, D. S., and Chu, K.: Computer Simulation Techniques. New York 1966

Naylor, Th. H., Burdick, D. S., and Sasser, W. E.: Computer simulation experiments with economic systems: the problem of experimental design. J. Amer. Statist. Assoc. **62** (1967), 1315–1337

Page, W. F.: A simulation study of two sequential designs by I. D. Bross. Biometrical Journal **20** (1978), 285–298

Payne, J. A.: Introduction to Simulation. Programming Techniques and Methods of Analysis. (McGraw-Hill; pp. 324) New York 1982

Pritsker, A. A. B., and Pegden, C. D.: Introduction to Simulation and SLAM. (Wiley; 588) New York 1979

Rechenberg, P.: Grundzüge digitaler Rechenautomaten. 2nd edition (Oldenbourg; 220 pages) München 1968

Richards, R. K.: Electronic Digital Systems. (Wiley; pp. 637) New York 1966

Röpke, H., and Riemann, J.: Analogcomputer in Chemie und Biologie. Eine Einführung. (Springer, 184 pages) Berlin, Heidelberg, New York 1969

Rogers, A. E., and Connolly, T. W.: Analog Computation in Engineering Design. New York 1960

Rubinstein, R. Y.: Simulation and the Monte Carlo Method. (Wiley; pp. 300) New York 1981
Schreider, Y. A. (Ed.): Method of Statistical Testing (Monte Carlo Method). Amsterdam 1964
Shah, M. J.: Engineering Simulation Using Small Scientific Computers. (Prentice-Hall; pp. 401) Englewood Cliffs, N.J. 1976
Shubik, M.: Bibliography on simulation, gaming, artificial intelligence and applied topics. J. Amer. Statist. Assoc. 55 (1960), 736–751
Smith, J. M.: Mathematical Modeling and Digital Simulation for Engineers and Scientists. (Wiley; pp. 332) New York 1977
Smith, J. U. M.: Computer Simulation Models. (Griffin, pp. 112) London 1968
Sowey, E. R.: (1) A chronological and classified bibliography on random number generation and testing. Internat. Statist. Rev. 40 (1972), 355–371 [see also J. Econometrics 1 (1973), 377–395]. (2) A second classified bibliography on random number generation and testing. International Statistical Review 46 (1978), 89–102
Thomas, D. A. H.: Error rates in multiple comparisons among means—results of a simulation exercise. Applied Statistics 23 (1974) 284–294
Tocher, K. D.: (1) The Art of Simulation. London 1963. (2) Review of simulation languages. Operations Research Quarterly 16 (1965), 189–217
Van der Laan, P., and Oosterhoff, J.: Monte Carlo estimation of the powers of the distribution-free two-sample tests of Wilcoxon, van der Waerden and Terry and comparison of these powers. Statistica Neerlandica 19 (1965), 265–275 and 21 (1967), 55–68
Wilkins, B. R.: Analogue and Iterative Methods in Computation, Simulation and Control (Chapman and Hall; pp. 276) London 1970

[8:3] Chapter 3

Alling, D. W.: Early decision in the Wilcoxon two-sample test. J. Amer. Statist. Assoc. 58 (1963), 713–720 [see also 69 (1974), 414–422]
Anscombe, F. J.: Rejection of outliers. Technometrics 2 (1960), 123–166 [see also 11 (1969) 527–550, 13 (1971), 110–112, 15 (1973), 385–404, 723–737]
Bailey, B. J. R.: Tables of the Bonferroni t statistics. Journal of the American Statistical Association 72 (1977), 469–478
Banerji, S. K.: Approximate confidence interval for linear functions of means of k populations when the population variances are not equal. Sankhya 22 (1960), 357 + 358
Bauer, R. K.: Der „Median-Quartile-Test": Ein Verfahren zur nichtparametrischen Prüfung zweier unabhängiger Stichproben auf unspezifizierte Verteilungsunterschiede. Metrika 5 (1962), 1–16
Behrens, W.-V.: Ein Beitrag zur Fehlerberechnung bei wenigen Beobachtungen. Landwirtschaftliche Jahrbücher 68 (1929), 807–837
Belson, I., and Nakano, K.: Using single-sided non-parametric tolerance limits and percentiles. Industrial Quality Control 21 (May 1965), 566–569
Bhapkar, V. P., and Deshpande, J. V.: Some nonparametric tests for multisample problems. Technometrics 10 (1968), 578–585
Birnbaum, Z. W., and Hall, R. A.: Small sample distribution for multisample statistics of the Smirnov type. Ann. Math. Stat. 31 (1960), 710–720 [see also 40 (1969), 1449–1466 as well as J. Amer. Statist. Assoc. 71 (1976), 757–762]
Bowker, A. H.: Tolerance Factors for Normal Distributions, in (Statistical Research Group, Columbia University), Techniques of Statistical Analysis (edited by Eisenhart, C., Hastay, M. W., and Wallis, W. A.) (McGraw-Hill) New York and London 1947

Bowker, A. H., and Lieberman, G. J.: Engineering Statistics. (Prentice-Hall) Englewood Cliffs, N.J. 1959
Box, G. E. P.: Non-normality and tests on variances. Biometrika **40** (1953), 318–335
—, and Andersen, S. L.: Permutation theory in the derivation of robust criteria and the study of departures from assumption. With discussion. J. Roy. Statist. Soc., Ser. B **17** (1955), 1–34
Boyd, W. C.: A nomogram for the "Student"-Fisher t test. J. Amer. Statist. Assoc. **64** (1969), 1664–1667
Bradley, J. V.: Distribution-Free Statistical Tests. (Prentice-Hall, pp. 388) Englewood Cliffs, N.J. 1968, Chapters 5 and 6
Bradley, R. A., Martin, D. C., and Wilcoxon, F.: Sequential rank-tests. I. Monte Carlo studies of the two-sample procedure. Technometrics **7** (1965), 463–483
—, Merchant, S. D., and Wilcoxon, F.: Sequential rank tests II. Modified two-sample procedures. Technometrics **8** (1966), 615–623
Breny, H.: L'état actuel du problème de Behrens-Fisher. Trabajos Estadist. **6** (1955), 111–131
Burrows, G. L.: (1) Statistical tolerance limits – what are they? Applied Statistics **12** (1963), 133–144. (2) One-sided normal tolerance factors. New tables and extended use of tables. Mimeograph, Knolls Atomic Power Lab., General Electric Company, USA 1964
Cacoullos, T.: A relation between t and F-distributions. J. Amer. Statist. Assoc. **60** (1965), 528–531
Cadwell, J. H.: (1) Approximating to the distributions of measures of dispersion by a power of chi-square. Biometrika **40** (1953), 336–346. (2) The statistical treatment of mean deviation. Biometrika **41** (1954), 12–18
Carnal, H., and Riedwyl, H.: On a one-sample distribution-free test statistic V. Biometrika **59** (1972), 465–467 [see also Statistische Hefte **14** (1973), 193–202]
Chacko, V. J.: Testing homogeneity against ordered alternatives. Ann. Math. Statist. **34** (1963), 945–956 [see also **38** (1967), 1740–1752]
Chakravarti, I. M.: Confidence set for the ratio of means of two normal distributions when the ratio of variances is unknown. Biometrische Zeitschr. **13** (1971), 89–94
Chun, D.: On an extreme rank sum test with early decision. J. Amer. Statist. Assoc. **60** (1965), 859–863
Cochran, W. G.: (1) Some consequences when the assumptions for the analysis of variance are not satisfied. Biometrics **3** (1947), 22–38. (2) Modern methods in the sampling of human populations. Amer. J. Publ. Health **41** (1951), 647–653. (3) Query 12, Testing two correlated variances. Technometrics **7** (1965), 447–449
—, Mosteller, F., and Tukey, J. W.: Principles of sampling. J. Amer. Statist. Assoc. **49** (1954), 13–35
Cohen, J.: Statistical Power Analysis for the Behavioral Sciences. (Academic Press, pp. 474) New York 1977; see p. 579, Cohen 1977 and the [ ] note
Conover, W. J.: Two k-sample slippage tests. J. Amer. Statist. Assoc. **63** (1968), 614–626
Croarkin, Mary C.: Graphs for determining the power of Student's t-test. J. Res. Nat. Bur. Stand. **66 B** (1962), 59–70 (cf. Errata: Mathematics of Computation **17** (1963), 83 [334])
D'Agostino, R. B.: (1) Simple compact portable test of normality: Geary's test revisited. Psychol. Bull. **74** (1970), 138–140 [see also **78** (1972), 262–265]. (2) An omnibus test of normality for moderate and large size samples. Biometrika **58** (1971), 341–348 [see also **63** (1976), 143–147]. (3) Small sample probability points for the D test of normality. Biometrika **59** (1972), 219–221 [see also **60** (1973), 169–173, 613–622, 623–628, **61** (1974), 181–184, 185–189, **64** (1977), 638–640]
Danziger, L., and Davis, S. A.: Tables of distribution-free tolerance limits. Ann. Math. Statist. **35** (1964), 1361–1365 [see also J. Qual. Technol. **7** (1975), 109–114]

Darling, D. A.: The Kolmogorov-Smirnov, Cramér-von Mises tests. Ann. Math. Statist. **28** (1957), 823–838

Davies, O. L.: The Design and Analysis of Industrial Experiments. (Oliver and Boyd), Edinburgh 1956, pp. 614

Dietze, Doris: t for more than two. Perceptual and Motor Skills **25** (1967), 589–602

Dixon, W. J.: (1) Analysis of extreme values. Ann. Math. Statist. **21** (1950), 488–506. (2) Processing data for outliers. Biometrics **9** (1953), 74–89. (3) Rejection of Observations. In Sarhan, A. E., and Greenberg, B. G. (Eds.): Contributions to Order Statistics. New York 1962, pp. 299–342

Dixon, W. J., and Tukey, J. W.: Approximate behavior of the distribution of Winsorized t (trimming/Winsorization 2). Technometrics **10** (1968), 83–98 [see Statistische Hefte **15** (1974), 157–170]

Edington, E. S.: The assumption of homogeneity of variance for the t-test and non-parametric tests. Journal of Psychology **59** (1965), 177–179

Faulkenberry, G. D., and Daly, J. C.: Sample size for tolerance limits on a normal distribution. Technometrics **12** (1970), 813–821

Fisher, R. A.: (1) The comparison of samples with possibly unequal variances. Ann. Eugen. **9** (1939), 174–180. (2) The asymptotic approach to Behrens's integral, with further tables for the d test of significance. Ann. Eugen. **11** (1941), 141–172

Fisher, R. A., and Yates, F.: Statistical Tables for Biological, Agricultural and Medical Research. 6th ed. London 1963

Ford, B. L., and Tortora, R. D.: A consulting aid to sample design. Biometrics **34** (1978), pp. 299–304

Geary, R. C.: (1) Moments of the ratio of the mean deviation to the standard deviation for normal samples. Biometrika **28** (1936), 295–305 (see also **27**, 310–332, **34**, 209–242, **60**, 613–622 as well as **61**, 181–184 and **66**, 400–401). (2) Tests de la normalité. Ann. Inst. Poincaré **15** (1956), 35–65 [cf., Biometrika **64** (1977), 135–139]

Gibbons, J. D.: On the power of two-sample rank tests on the equality of two distribution functions. J. Roy. Statist. Soc. B **26** (1964), 292–304

Glasser, G. J.: A distribution-free test of independence with a sample of paired observations. J. Amer. Statist. Assoc. **57** (1962), 116–133

Goldman, A.: On the Determination of Sample Size. (Los Alamos Sci. Lab.; LA-2520; 1961) U.S. Dept. Commerce, Washington, D.C. 1961 [see also Biometrics **19** (1963), 465–477]

Granger, C. W. J., and Neave, H. R.: A quick test for slippage. Rev. Inst. Internat. Statist. **36** (1968), 309–312

Graybill, F. A., and Connell, T. L.: Sample size required to estimate the ratio of variances with bounded relative error. J. Amer. Statist. Assoc. **58** (1963), 1044–1047

Grubbs, F. E.: Procedures for detecting outlying observations in samples. Technometrics **11** (1969), 1–21 [see also 527–550 and **14** (1972), 847–854; **15** (1973), 429]

Guenther, W. C.: Determination of sample size for distribution-free tolerance limits. The American Statistician **24** (Feb. 1970), 44–46

Gurland, J., and McCullough, R. S.: Testing equality of means after a preliminary test of equality of variances. Biometrika **49** (1962), 403–417

Guttmann, I.: Statistical Tolerance Regions. Classical and Bayesian. (Griffin, pp. 150) London 1970

Haga, T.: A two-sample rank test on location. Annals of the Institute of Statistical Mathematics **11** (1960), 211–219

Hahn, G. J.: Statistical intervals for a normal population. Part I and II. J. Qual. Technol. **2** (1970), 115–125 and 195–206 [see also **9** (1977), 6–12, **5** (1973), 178–188, Biometrika **58** (1971), 323–332, J. Amer. Statist. Assoc. **67** (1972), 938–942 as well as Technometrics **15** (1973), 897–914]

Hall, I. J., Prairie, R. R., and Motlagh, C. K.: Non-parametric prediction intervals. Journal of Quality Technology **7** (1975), 103–114

Halperin, M.: Extension of the Wilcoxon-Mann-Whitney test to samples censored at the same fixed point. J. Amer. Statist. Assoc. **55** (1960), 125–138 [cf., Biometrika **52** (1965), 650–653]

Harmann, A. J.: Wilks' tolerance limit sample sizes. Sankhya A **29** (1967), 215–218

Harter, H. L.: Percentage points of the ratio of two ranges and power of the associated test. Biometrika **50** (1963), 187–194

Herrey, Erna, M. J.: Confidence intervals based on the mean absolute deviation of a normal sample. J. Amer. Statist. Assoc. **60** (1965), 257–269 (see also **66** [1971], 187–188)

Hodges, J. L., Jr., and Lehmann, E. L.: (1) The efficiency of some nonparametric competitors of the t-test. Ann. Math. Statist. **27** (1956), 324–335. (2) A compact table for power of the t-test. Ann. Math. Statist. **39** (1968), 1629–1637. (3) Basic Concepts of Probability and Statistics. 2nd ed. (Holden-Day, pp. 401) San Francisco 1970

Hsiao, F. S. T.: The diagrammatical representation of confidence-interval estimation and hypothesis testing. The American Statistician **26** (Dec. 1972), 28–29

Jacobson, J. E.: The Wilcoxon two-sample statistic: tables and bibliography. J. Amer. Statist. Assoc. **58** (1963), 1086–1103

Johnson, N. L., and Welch, B. L.: Applications of the noncentral t-distribution. Biometrika **31** (1940), 362–389

Kendall, M. G.: The treatment of ties in ranking problems. Biometrika **33** (1945), 239–251

Kim, P. J.: On the exact and approximate sampling distribution of the two sample Kolmogorov-Smirnov criterion D_{mn}, $m \leq n$. J. Amer. Statist. Assoc. **64** (1969), 1625–1637 [see also **68** (1973), 994–997 and Ann. Math. Statist. **40** (1969), 1449–1466]

Kolmogoroff, A. N.: Sulla determinazione empirica di una legge di distribuzione. Giornale Istituto Italiano Attuari **4** (1933), 83–91

Krishnan, M.: Series representations of the doubly noncentral t-distribution. J. Amer. Statist. Assoc. **63** (1968), 1004–1012

Kruskal, W. H.: A nonparametric test for the several sampling problem. Ann. Math. Statist. **23** (1952), 525–540

Kruskal, W. H., and Wallis, W. A.: Use of ranks in one-criterion variance analysis. J. Amer. Statist. Assoc. **47** (1952), 583–621 [see **48** (1953), 907–911 and **70** (1975), 794]

Krutchkoff, R. G.: The correct use of the sample mean absolute deviation in confidence intervals for a normal variate. Technometrics **8** (1966), 663–674

Laan, P. van der: Simple distribution-free confidence intervals for a difference in location. Philips Res. Repts. Suppl. 1970, No. 5, pp. 158

Lachenbruch, P. A.: Analysis of data with clumping at zero. Biom. Z. **18** (1976), 351–356

Levene, H.: Robust tests for equality of variances. In I. Olkin and others (Eds.): Contributions to Probability and Statistics. Essays in Honor of Harold Hotelling, pp. 278–292. Stanford 1960 [cf. J. Statist. Comput. Simul. **1** (1972), 183–194 and J. Amer. Statist. Assoc. **69** (1974), 364–367]

Lieberman, G. J.: Tables for one-sided statistical tolerance limits. Industrial Quality Control **14** (Apr. 1958), 7–9

Lienert, G. A., and Schulz, H.: Zum Nachweis von Behandlungswirkungen bei heterogenen Patientenstichproben. Ärztliche Forschung **21** (1967), 448–455

Lindgren, B. W.: Statistical Theory. (Macmillan; pp. 427) New York 1960, p. 401, Table VI

Lindley, D. V., East, D. A., and Hamilton, P. A.: Tables for making inferences about the variance of a normal distribution. Biometrika **47** (1960), 433–437

Linnik, Y. V.: Latest investigation on Behrens-Fisher-problem. Sankhya **28** A (1966), 15–24

Lohrding, R. K.: A two sample test of equality of coefficients of variation or relative errors. J. Statist. Comp. Simul. **4** (1975), 31–36

Lord, E.: (1) The use of range in place of standard deviation in the t-test. Biometrika **34** (1947), 41–67. (2) Power of the modified t-test (u-test) based on range. Biometrika **37** (1950), 64–77

Mace, A. E.: Sample-Size Determination. (Reinhold; pp. 226) New York 1964

MacKinnon, W. J.: Table for both the sign test and distribution-free confidence intervals of the median for sample sizes to 1,000. J. Amer. Statist. Assoc. **59** (1964), 935–956

Mann, H. B., and Whitney, D. R.: On a test of whether one of two random variables is stochastically larger than the other. Ann. Math. Statist. **18** (1947), 50–60

Massey, F. J., Jr.: (1) The distribution of the maximum deviation between two sample cumulative step functions. Ann. Math. Statist. **22** (1951), 125–128. (2) Distribution table for the deviation between two sample cumulatives. Ann. Math. Statist. **23** (1952), 435–441

McCullough, R. S., Gurland, J., and Rosenberg, L.: Small sample behaviour of certain tests of the hypothesis of equal means under variance heterogeneity. Biometrika **47** (1960), 345–353

McHugh, R. B.: Confidence interval inference and sample size determination. The American Statistician **15** (April 1961), 14–17

Mehta, J. S. and Srinivasan, R.: On the Behrens-Fisher problem. Biometrika **57** (1970), 649–655

Meyer-Bahlburg, H. F. L.: A nonparametric test for relative spread in k unpaired samples. Metrika **15** (1970), 23–29

Miller, L. H.: Table of percentage points of Kolmogorov statistics. J. Amer. Statist. Assoc. **51** (1956), 113–115

Milton, R. C.: An extended table of critical values for the Mann-Whitney (Wilcoxon) two-sample statistic. J. Amer. Statist. Assoc. **59** (1964), 925–934

Minton, G.: (1) Inspection and correction error in data processing. J. Amer. Statist. Assoc. **64** (1969), 1256–1275 [see also **71** (1976), 17–35 and particularly Maria E. Gonzalez et al., J. Amer. Statist. Assoc. **70** (Sept. 1975), No. 351, Part II, 1–23]. (2) Some decision rules for administrative applications of quality control. J. Qual. Technol. **2** (1970), 86–98 [see also **3** (1971), 6–17]

Mitra, S. K.: Tables for tolerance limits for a normal population based on sample mean and range or mean range. J. Amer. Statist. Assoc. **52** (1957), 88–94

Moore, P. G.: The two sample t-test based on range. Biometrika **44** (1957), 482–489

Mosteller, F.: A k-sample slippage test for an extreme population. Ann. Math. Stat. **19** (1948), 58–65 [see also **21** (1950), 120–123]

Neave, H. R.: (1) A development of Tukey's quick test of location. J. Amer. Statist. Assoc. **61** (1966), 949–964. (2) Some quick tests for slippage. The Statistician **21** (1972), 197–208 [cf. **22** (1973) 269–280]

Neave, H. R., and Granger, C. W. J.: A Monte Carlo study comparing various two-sample tests for differences in mean. Technometrics **10** (1968), 509–522

Nelson, L. S.: (1) Nomograph for two-sided distribution-free tolerance intervals. Industrial Quality Control **19** (June 1963), 11–13. (2) Tables for Wilcoxon's rank sum test in randomized blocks. J. Qual. Technol. **2** (Oct. 1970), 207–218

Neyman, J.: First Course in Probability and Statistics. New York 1950

Odeh, R. E., and Fox, M.: Sample Size Choice (M. Dekker, pp. 190) New York 1975 [cf. Guenther, W. C., Statistica Neerlandica **27** (1973), 103–113]

Owen, D. B.: (1) Factors for one-sided tolerance limits and for variables sampling plans. Sandia Corporation, Monograph 607, Albuquerque, New Mexico, March 1963. (2) The power of Student's t-test. J. Amer. Statist. Assoc. **60** (1965), 320–333 and 1251. (3) A survey of properties and applications of the noncentral t-distribution. Technometrics **10** (1968), 445–478

—, and Frawley, W. H.: Factors for tolerance limits which control both tails of the normal distribution. J. Qual. Technol. **3** (1971), 69–79

Parren, J. L. Van der: Tables for distribution-free confidence limits for the median. Biometrika **57** (1970), 613–617 [cf., **60** (1973), 433–434]

Pearson, E. S., and Hartley, H. O.: Biometrika Tables for Statisticians, Cambridge University Press, 1954

Pearson, E. S., and Stephens, M. A.: The ratio of range to standard deviation in the same normal sample. Biometrika **51** (1964), 484–487

Penfield, D. A., and McSweeney, Maryellen: The normal scores test for the two-sample problem. Psychological Bull. **69** (1968), 183–191

Peters, C. A. F.: Über die Bestimmung des wahrscheinlichen Fehlers einer Beobachtung aus den Abweichungen der Beobachtungen von ihrem arithmetischen Mittel. Astronomische Nachrichten **44** (1856), 30–31

Pierson, R. H.: Confidence interval lengths for small numbers of replicates. U.S. Naval Ordnance Test Station, China Lake, Calif. 1963

Pillai, K. C. S., and Buenaventura, A. R.: Upper percentage points of a substitute F-ratio using ranges. Biometrika **48** (1961), 195–196

Potthoff, R. F.: Use of the Wilcoxon statistic for a generalized Behrens-Fisher problem. Ann. Math. Stat. **34** (1963), 1596–1599 [see Biometrika **66** (1979) 645–653]

Pratt, J. W.: Robustness of some procedures for the two-sample location problem. J. Amer. Statist. Assoc. **59** (1964), 665–680

Proschan, F.: Confidence and tolerance intervals for the normal distribution. J. Amer. Statist. Assoc. **48** (1953), 550–564

Quesenberry, C. P., and David, H. A.: Some tests for outliers. Biometrika **48** (1961), 379–390

Raatz, U.: Eine Modifikation des White-Tests bei großen Stichproben. Biometrische Zeitschr. **8** (1966), 42–54 [see also Arch. ges. Psychol. **118** (1966), 86–92]

Reiter, S.: Estimates of bounded relative error for the ratio of variances of normal distributions. J. Amer. Statist. Assoc. **51** (1956), 481–488

Rosenbaum, S.: (1) Tables for a nonparametric test of dispersion. Ann. Math. Statist. **24** (1953), 663–668. (2) Tables for a nonparametric test of location. Ann. Math. Statist. **25** (1954), 146–150. (3) On some two-sample non-parametric tests. J. Amer. Statist. Assoc. **60** (1965), 1118–1126

Rytz, C.: Ausgewählte parameterfreie Prüfverfahren im 2- und k-Stichproben-Fall. Metrika **12** (1967), 189–204 and **13** (1968), 17–71

Sachs, L.: Statistische Methoden. 6th revised edition (Springer, 133 pages) Berlin, Heidelberg, New York 1984, pages 39 and big table, column 7; 51; 52–54; 95, 96

Sandelius, M.: A graphical version of Tukey's confidence interval for slippage. Technometrics **10** (1968), 193–194

Saw, J. G.: A non-parametric comparison of two samples one of which is censored. Biometrika **53** (1966), 599–602 [see also **52** (1965), 203–223 and **56** (1969), 127–132]

Scheffé, H.: Practical solutions of the Behrens-Fisher problem, J. Amer. Statist. Assoc. **65** (1970), 1501–1508 [see also **66** (1971), 605–608 and J. Pfanzagl, Biometrika **61** (1974), 39–47, 647]

Scheffé, H., and Tukey, J. W.: Another Beta-Function Approximation. Memorandum Report 28, Statistical Research Group, Princeton University 1949

Sheesley, J. H.: Tests for outlying observations. Journal of Quality Technology **9** (1977), 38–41 and 208

Shorak, G. R.: Testing and estimating ratios of scale parameters. J. Amer. Statist. Assoc. **64** (1969), 999–1013 [see also Commun. Statist.—Theor. Meth. A **5** (1976), 1287–1312 and A **6** (1977), 649–655]

Siegel, S.: Nonparametric Statistics for the Behavioral Sciences. New York 1956, p. 278

Siegel, S., and Tukey, J. W.: A nonparametric sum of ranks procedure for relative spread in unpaired samples, J. Amer. Statist. Assoc. **55** (1960), 429–445

Smirnoff, N. W.: (1) On the estimation of the discrepancy between empirical curves of distribution for two independent samples. Bull. Université Moskov. Ser. Internat.,

Sect. A 2. (2) (1939), 3–8. (2) Tables for estimating the goodness of fit of empirical distributions. Ann. Math. Statist. **19** (1948), 279–281

Stammberger, A.: Über einige Nomogramme zur Statistik. [Fertigungstechnik und Betrieb **16** (1966). 260–263 or] Wiss. Z. Humboldt-Univ. Berlin, Math.-Nat. R. **16** (1967), 86–93

Sukhatme, P. V.: On Fisher and Behrens's test of significance for the difference in means of two normal samples. Sankhya **4** (1938), 39–48

Szameitat, K., und Koller, S.: Über den Umfang und die Genauigkeit von Stichproben. Wirtschaft u. Statistik **10** NF (1958), 10–16

—, and K.-A. Schäffer: (1) Fehlerhaftes Ausgangsmaterial in der Statistik und seine Konsequenzen für die Anwendung des Stichprobenverfahrens. Allgemein. Statist. Arch. **48** (1964), 1–22. (2) Kosten und Wirtschaftlichkeit von Stichprobenstatistiken. Allgem. Statist. Arch. **48** (1964), 123–146

—, and Deininger, R.: Some remarks on the problem of errors in statistical results. Bull. Int. Statist. Inst. **42** I (1969), 66–91 [cf. **41** II (1966), 395–417 and Allgem. Statist. Arch. **55** (1971), 290–303]

Thompson, W. A., Jr., and Endriss, J.: The required sample size when estimating variances. The American Statistician **15** (June 1961), 22–23

Thompson, W. A., and Willke, T. A.: On an extreme rank sum test for outliers. Biometrika **50** (1963), 375–383 [see also J. Qual. Technol. **9** (1977), 38–41, 208]

Thöni, H. P.: Die nomographische Lösung des t-Tests. Biometrische Zeitschr. **5** (1963), 31–50

Tiku, M. L.: Tables of the power of the F-test. J. Amer. Statist. Assoc. **62** (1967), 525–539 [see also **63** (1968), 1551 and **66** (1971), 913–916 as well as **67** (1972), 709–710]

Trickett, W. H., Welch, B. L., and James, G. S.: Further critical values for the two-means problem. Biometrika **43** (1956), 203–205

Tukey, J. W.: (1) A quick, compact, two-sample test to Duckworth's specifications. Technometrics **1** (1959), 31–48. (2) A survey of sampling from contaminated distributions. In I. Olkin and others (Eds.): Contributions to Probability and Statistics. Essays in Honor of Harold Hotelling. pp. 448–485, Stanford 1960. (3) The future of data analysis. Ann. Math. Statist. **33** (1962), 1–67, 812

Waerden, B. L., van der: Mathematische Statistik. 2nd edition, Berlin-Heidelberg-New York 1965, pages 285–295, 334–335, 348–349 [cf. X-Test Schranken: Math. Operatforsch, u. Statist. **3** (1972), 389–400]

Walter, E.: Über einige nichtparametrische Testverfahren. Mitteilungsbl. Mathem. Statist. **3** (1951), 31–44 and 73–92

Weiler, H.: A significance test for simultaneous quantal and quantitative responses. Technometrics **6** (1964), 273–285

Weiling, F.: Die Mendelschen Erbversuche in biometrischer Sicht. Biometrische Zeitschr. **7** (1965), 230–262

Weir, J. B. de V.: Significance of the difference between two means when the population variances may be unequal. Nature **187** (1960), 438

Weissberg, A., and Beatty, G. H.: Tables of tolerance-limit factors for normal distributions. Technometrics **2** (1960), 483–500 [see also J. Amer. Statist. Assoc. **52** (1957), 88–94 and **64** (1969), 610–620 as well as Industrial Quality Control **19** (Nov. 1962), 27–28]

Welch, B. L.: (1) The significance of the difference between two means when the population variances are unequal. Biometrika **29** (1937), 350–361. (2) The generalization of "Student's" problem when several different population variances are involved. Biometrika **34** (1947), 28–35

Wenger, A.: Nomographische Darstellung statistischer Prüfverfahren. Mitt. Vereinig. Schweizer. Versicherungsmathematiker **63** (1963), 125–153

Westlake, W. J.: A one-sided version of the Tukey-Duckworth test. Technometrics **13** (1971), 901–903 [see also D. J. Gans, **23** (1981), 193–195]

Wetzel, W.: Elementare Statistische Tabellen, Kiel 1965; (De Gruyter) Berlin 1966
Wilcoxon. F.: Individual comparisons by ranking methods. Biometrics **1** (1945), 80–83
—, Katti, S. K., and Wilcox, Roberta, A.: Critical Values and Probability Levels for the Wilcoxon Rank Sum Test and the Wilcoxon Signed Rank Test. Lederle Laboratories, Division Amer. Cyanamid Company, Pearl River, New York, August 1963
—, Rhodes, L. J., and Bradley, R. A.: Two sequential two-sample grouped rank tests with applications to screening experiments. Biometrics **19** (1963), 58–84 (see also **20** [1964], 892)
—, and Wilcox, Roberta A.: Some Rapid Approximate Statistical Procedures. Lederle Laboratories, Pearl River, New York 1964
Wilks, S. S.: (1) Determination of sample sizes for setting tolerance limits. Ann. Math. Statist. **12** (1941), 91–96 [see also The American Statistician **26** (Dec. 1972), 21]. (2) Statistical prediction with special reference to the problem of tolerance limits. Ann. Math. Statist. **13** (1942), 400–409
Winne, D.: (1) Zur Auswertung von Versuchsergebnissen: Der Nachweis der Übereinstimmung zweier Versuchsreihen. Arzneim.-Forschg. **13** (1963), 1001–1006. (2) Zur Planung von Versuchen: Wieviel Versuchseinheiten? Arzneim.-Forschg. **18** (1968), 1611–1618

[8:3a] Sampling theory

Cochran, W. G.: Sampling Techniques. 3rd ed. (Wiley; pp. 428) New York 1977
Conway, Freda: Sampling: An Introduction for Social Scientists. (Allen and Unwin, pp. 154) London 1967
Deming, W. E.: Sampling Design in Business Research. (Wiley; pp. 517) New York 1960
Raj, D.: (1) Sampling Theory. (McGraw-Hill, pp. 225) New York 1968. (2) The Design of Sample Surveys. (McGraw-Hill, pp. 416) New York 1972
Hansen, M. H., Hurwitz, W. N., and Madow, W. G.: Sample Survey Methods and Theory. Vol. I and II (Wiley, pp. 638, 332) New York 1964
Hogarth, R. M. (Ed.): Question Framing and Response Consistency. (Jossey-Bass; pp. 109) San Francisco, Calif. 1982
Jessen, R. J.: Statistical Survey Techniques. (Wiley; pp. 528) New York 1978
Kalton, G. and Schuman, H.: The effect of the question on survey responses; a review. With discussion. J. Roy. Statist. Society A **145** (1982), 42–73
Kish, L.: Survey Sampling. New York 1965 [see also J. Roy. Statist. Soc. B **36** (1974), 1–37, A **139** (1976), 183–204 and Dalenius, T., Int. Stat. Rev. **45** (1977), 71–89, 181–197, 303–317, **47** (1979) 99–109 as well as Kalton, G. **51** (1983), 175–188]
Konijn, H. S.: Statistical Theory of Sample Survey Design and Analysis. (North-Holland; pp. 429) Amsterdam 1973
Krewski, D., Platek, R., and Rao, J. N. K.: Current Topics in Survey Sampling. (Academic Press; pp. 509) New York 1981
Kruskal, W., and Mosteller, F.: Representative sampling III: the current statistical literature. Internat. Statist. Rev. **47** (1979), 245–265 [cf. 13–24, 111–127, and **48** (1980), 169–195]
Murthy, M. N.: Sampling Theory and Methods. (Statistical Publ. Soc., pp. 684) Calcutta 1967
Parten, Mildred: Surveys, Polls, and Samples: Practical Procedures. (Harper and Brothers, pp. 624) New York 1969 (Bibliography pp. 537–602 [see also Struening, E. L. and Marcia Guttentag (Eds.): Handbook of Evaluation Research. (Sage; pp. 696) London 1975])
Sampford, M. R.: An Introduction to Sampling Theory with Applications to Agriculture. London 1962

Strecker, H.: Model for the decomposition of errors in statistical data into components, and the ascertainment of respondent errors by means of accuracy checks. Jahrbücher für Nationalökonomie und Statistik **195** (Sept. 1980), 385–420

Sukhatme, P. V., and Sukhatme, B. V.: Sampling Theory of Surveys With Applications. 2nd rev. ed. (Iowa State Univ. Press; pp. 452) Ames, Iowa 1970

United Nations: A short Manual on Sampling. Vol. I. Elements of Sample Survey Theory. Studies in Methods Ser. F No. 9, rev. 1, New York 1972

Weisberg, H. F., and Bowen, B. D.: An Introduction to Survey Research and Data Analysis. (W. H. Freeman; pp. 243) Reading and San Francisco 1977

Yamane, T.: Elementary Sampling Theory. (Prentice-Hall, pp. 405) Englewood Cliffs, N.J. 1967

Yates, F.: Sampling Methods for Censuses and Surveys. 4th rev. enlarged ed. (Griffin; pp. 458) London 1981 [see the bibliography, pp. 416–450]

[8:4] Chapter 4

Adler, F.: Yates correction and the statisticians. J. Amer. Statist. Assoc. **46** (1951), 490–501 [see also **47** (1952), 303 and American Statistician **30** (1976), 103–104]

Bateman, G.: On the power function of the longest run as a test for randomness in a sequence of alternatives. Biometrika **35** (1948), 97–112 [see also **34** (1947), 335–339; **44** (1957), 168–178; **45** (1958), 253–256; **48** (1961), 461–465]

Bennett, B. M.: (1) Tests of hypotheses concerning matched samples. J. Roy. Statist. Soc. B **29** (1967), 468–474. (2) On tests for order and treatment differences in a matched 2×2. Biometrische Zeitschr. **13** (1971), 95–99

—, and Horst, C.: Supplement to Tables for Testing Significance in a 2×2 Contingency Table. New York 1966

—, and Hsu, P.: On the power function of the exact test for the 2×2 contingency table. Biometrika **47** (1960), 393–397 [editorial note 397, 398, correction **48** (1961), 475]

—, and Underwood, R. E.: On McNemar's test for the 2×2 table and its power function. Biometrics **26** (1970), 339–343 [see also **27** (1971), 945–959 and Psychometrika **47** (1982), 115–118]

Berchtold, W.: Die Irrtumswahrscheinlichkeiten des χ^2-Kriteriums für kleine Versuchszahlen. Z. angew. Math. Mech. **49** (1969), 634–636

Berkson, J.: In dispraise of the exact test. Do the marginal totals of the 2×2 table contain relevant information respecting the table proportions? Journal of Statistical Planning and Inference **2** (1978), 27–42

Bihn, W. R.: Wandlungen in der statistischen Zeitreihenanalyse und deren Bedeutung für die ökonomische Forschung. Jahrb. Nationalök. Statistik **180** (1967), 132–146 (see also Parzen 1967 and Nullau 1968)

Birnbaum, Z. W.: Numerical tabulation of the distribution of Kolmogorov's statistic for finite sample size. J. Amer. Statist. Assoc. **47** (1952), 425–441

Blyth, C. R., and Hutchinson, D. W.: Table of Neyman-shortest unbiased confidence intervals for the binomial parameter. Biometrika **47** (1960), 381–391

—, and Still, H. A.: Binomial confidence intervals. J. Amer. Statist. Assoc. **78** (1983), 108–116

Bogartz, R. S.: A least squares method for fitting intercepting line segments to a set of data points. Psychol. Bull. **70** (1968), 749–755 (see also **75** [1971], 294–296)

Box, G. E. P., and Jenkins, G. M.: Time Series Analysis, Forecasting and Control. (Holden-Day, pp. 575) San Francisco 1976

Bradley, J. V.: A survey of sign tests based on the binomial distribution. J. Qual. Technol. **1** (1969), 89–101

Bredenkamp, J.: F-Tests zur Prüfung von Trends und Trendunterschieden. Z. exper. angew. Psychologie **15** (1968), 239–272

Bross, I. D. J.: Taking a covariable into account. J. Amer. Statist. Assoc. **59** (1964), 725–736

Casagrande, J. T., Pike, M. C., and Smith, P. G.: The power function of the "exact" test for comparing two binomial distributions. Applied Statistics **27** (1978), 176–180

Clopper, C. J., and Pearson, E. S.: The use of confidence or fiducial limits illustrated in the case of the binomial. Biometrika **26** (1934), 404–413

Cochran, W. G.: (1) The comparison of percentages in matched samples. Biometrika **37** (1950), 256–266. (2) The χ^2-test of goodness of fit. Ann. Math. Statist. **23** (1952), 315–345 [see Applied Statistics **29** (1980), 292–298 and Biometrika **67** (1980), 447–453]. (3) Some methods for strengthening the common chi-square tests. Biometrics **10** (1954), 417–451

Cochran, W. G.: Sampling Techniques, 2nd edition, J. Wiley, New York, 1963

Conover, W. J.: A Kolmogorov goodness-of-fit test for discontinuous distributions. J. Amer. Statist. Assoc. **67** (1972), 591–596

Cox, D. R., and Stuart, A.: Some quick sign tests for trend in location and dispersion. Biometrika **42** (1955), 80–95 [cf., **55** (1968), 381–386; **67** (1980), 375–379]

Crow, E. L.: Confidence intervals for a proportion. Biometrika **43** (1956), 423–435

Crow, E. L., and Gardner, R. S.: Confidence intervals for the expectation of a Poisson variable. Biometrika **46** (1959), 441–453

Croxton, F. E., and Cowden, D. J.: Applied General Statistics. 2nd ed. (Prentice-Hall) New York 1955

Csorgo, M., and Guttman, I.: On the empty cell test. Technometrics **4** (1962), 235–247

Cureton, E. E.: The normal approximation to the signed-rank sampling distribution when zero differences are present. J. Amer. Statist. Assoc. **62** (1967), 1068–1069 [see also **69** (1974), 368–373]

Darling, D. A.: The Kolmogorov-Smirnov, Cramér-von Mises tests. Ann. Math. Statist. **28** (1957), 823–838

David, F. N.: (1) A χ^2 'smooth' test for goodness of fit. Biometrika **34** (1947), 299–310. (2) Two combinatorial tests of whether a sample has come from a given population. Biometrika **37** (1950), 97–110

David, H. A., Hartley, H. O., and Pearson, E. S.: The distribution of the ratio, in a single normal sample, of range to standard deviation. Biometrika **41** (1954), 482–493

Davis, H. T.: The Analysis of Economic Time Series. San Antonio, Texas 1963

Dixon, W. J., and Mood, A. M.: The statistical sign test. J. Amer. Statist. Assoc. **41** (1946), 557–566 [see Int. Statist. Rev. **48** (1980), 19–28]

Documenta Geigy: Wissenschaftliche Tabellen, 7th edition, Basel 1968, pages 85–103 and 109–123 (8th revised edition 1980)

Duckworth, W. E., and Wyatt, J. K.: Rapid statistical techniques for operations research workers. Oper. Res. Quarterly **9** (1958), 218–233

Dunn, J. E.: A compounded multiple runs distribution. J. Amer. Statist. Assoc. **64** (1969), 1415–1423

Eisenhart, C., Hastay, M. W., and Wallis, W. A.: Techniques of Statistical Analysis. New York 1947

Feldman, S. E., and Klinger, E.: Short cut calculation of the Fisher-Yates "exact test". Psychometrika **28** (1963), 289–291

Finkelstein, J. M., and Schafer, R. E.: Improved goodness-of-fit tests. Biometrika **58** (1971), 641–645

Finney, D. J., Latscha, R., Bennett, B. M., and Hsu, P.: Tables for Testing Significance in a 2×2 Contingency Table. Cambridge 1963

Gail, M., and Gart, J. J.: The determination of sample sizes for use with the exact conditional test in 2 × 2 comparative trials. Biometrics **29** (1973), 441–448 [see Haseman (1978)]

Gart, J. J.: (1) Approximate confidence limits for the relative risk. Journal of the Royal Statistical Society B **24** (1962), 454–463, 458 [for odds ratio and relative risk see Amer. J. Epidemiology **115** (1982), 453–470 and **118** (1983), 396–407]. (2) An exact test for comparing matched proportions in crossover designs. Biometrika **56** (1969), 75–80 [see also Biometrics **27** (1971), 945–959 and Rev. Int. Stat. Inst. **39** (1971), 148–169]. (3) The analysis of ratios and cross product ratios of Poisson variates with application to incidence rates. Commun. Statist.—Theory and Methods A **7** (1978), 917–937

Gebhardt, F.: Verteilung und Signifikanzschranken des 3. und 4. Stichprobenmomentes bei normal-verteilten Variablen. Biometrische Zeitschr. **8** (1966), 219–241

Gildemeister, M., and Van der Waerden, B. L.: Die Zulässigkeit des χ^2-Kriteriums für kleine Versuchszahlen. Ber. Verh. Sächs. Akad. Wiss. Leipzig, Math.-Nat. Kl. **95** (1944), 145–150

Glasser, G. J.: A distribution-free test of independence with a sample of paired observations. J. Amer. Statist. Assoc. **57** (1962), 116–133

Good, I. J.: Significance tests in parallel and in series J. Amer. Statist. Assoc. **53** (1958), 799–813 [see also Biometrics **31** (1975), 987–992]

Grizzle, J. E.: Continuity correction in the χ^2-test for 2 × 2 tables. The American Statistician **21** (Oct. 1967), 28–32 [as well as **23** (April 1969), 35; cf. J. Amer. Statist. Assoc. **69** (1974), 374–382]

Harris, B. (Ed.): Spectral Analysis of Time Series. (Wiley, pp. 319) New York 1967

Hart, B. I.: Significance levels for the ratio of the mean square successive difference to the variance. Ann. Math. Statist. **13** (1942), 445–447

Haseman, J. K.: Exact sample sizes for use with the Fisher-Irwin test for 2 × 2 tables. Biometrics **34** (1978), 106–109 [see Fleiss, J. L., Tytun, A., and Ury, H. K.: Biometrics **36** (1980), 343–346 and 347–351]

Jenkins, G. M.: Spectral Analysis and Its Applications. (Holden-Day, pp. 520) San Francisco 1968

—, and Watts, D. E.: Spectrum Analysis and Its Applications. (Holden-Day, pp. 350) San Francisco 1968

Jesdinsky, H. J.: Orthogonale Kontraste zur Prüfung von Trends. Biometrische Zeitschrift **11** (1969), 252–264

Johnson, E. M.: The Fisher-Yates exact test and unequal sample sizes. Psychometrika **37** (1972), 103–106 [see also Applied Statistics **28** (1979), 302]

Kincaid, W. M: The combination of tests based on discrete distributions. J. Amer. Statist. Assoc. **57** (1962), 10–19 [see also **66** (1971), 802–806 and **68** (1973), 193–194]

Klemm, P. G.: Neue Diagramme für die Berechnung von Vierfelderkorrelationen. Biometrische Zeitschr. **6** (1964), 103–109

Kolmogorov, A.: Confidence limits for an unknown distribution function. Ann. Math. Statist. **12** (1941), 461–463

Koziol, J. A., and Perlman, M. D.: Combining independent chi-squared tests. Journal of the American Statistical Association **73** (1978), 753–763

Kruskal, W. H.: A nonparametric test for the several sample problem. Ann. Math. Statist. **23** (1952), 525–540

Kullback, S., Kupperman, M., and Ku, H. H.: An application of information theory to the analysis of contingency tables, with a table of $2n \ln n$, $n = 1(1)10,000$. J. Res. Nat. Bur. Stds. B **66** (1962), 217–243

Le Roy, H. L.: Ein einfacher χ^2-Test für den Simultanvergleich der inneren Struktur von zwei analogen 2 × 2 - Häufigkeitstabellen mit freien Kolonnen- und Zeilentotalen. Schweizer. landw. Forschg. **1** (1962), 451–454

Levene, H.: On the power function of tests of randomness based on runs up and down. Ann. Math. Statist. **23** (1952), 34–56

Li, J. C. R.: Statistical Inference. Vol. I (Edwards Brothers, pp. 658) Ann Arbor, Mich. 1964, p. 466

Lienert, G. A.: Die zufallskritische Beurteilung psychologischer Variablen mittels verteilungsfreier Schnelltests. Psychol. Beiträge **7** (1962), 183–215

Lillefors, H. W.: (1) On the Kolmogorov-Smirnov test for normality with mean and variance unknown. J. Amer. Statist. Assoc. **62** (1967), 399–402, Corrigenda **64** (1969), 1702. (2) On the Kolmogorov-Smirnov test for the exponential distribution with mean unknown. J. Amer. Statist. Assoc. **64** (1969), 387–389 [see also Biometrika **63** (1976), 149–160]

Ludwig, O.: Über die stochastische Theorie der Merkmalsiterationen. Mitteilungsbl. math. Statistik **8** (1956), 49–82

MacKinnon, W. J.: Table for both the sign test and distribution-free confidence intervals of the median for sample sizes to 1,000. J. Amer. Statist. Assoc. **59** (1964), 935–956

Makridakis, S.: (1) A survey of time series. International Statistical Review **44** (1976), 29–70. (2) Time-series analysis and forecasting: an update and evaluation. International Statistical Review **46** (1978), 255–278 [see J. Roy. Statist. Soc. A **142** (1979), 97–145]

Marascuilo, L. A., and McSweeney, Maryellen: Nonparametric post hoc comparisons for trend. Psychological Bulletin **67** (1967), 401–412 [see also **92** (1982), 517–525]

Massey, F. J. Jr.: The Kolmogorov-Smirnov test for goodness of fit. J. Amer. Statist. Assoc. **46** (1951). 68–78 [see also Allgem. Stat. Arch. **59** (1975), 228–250]

Maxwell, A. E.: Analysing Qualitative Data. 2nd edition. (Methuen) London 1970

McCornack, R. L.: Extended tables of the Wilcoxon matched pair rank statistic. J. Amer. Statist. Assoc. **60** (1965), 864–871 [see also **65** (1970), 974–975, **69** (1974), 255–258, 368–373 and Method. Inform. Med. **14** (1975), 224–230]

McNemar, Q.: Note on sampling error of the differences between correlated proportions or percentages. Psychometrika **12** (1947), 153–154

Miller, L. H.: Table of percentage points of Kolmogorov statistics. J. Amer. Statist. Assoc. **51** (1956), 111–121

Moore, P. G.: The properties of the mean square successive difference in samples from various populations. J. Amer. Statist. Assoc. **50** (1955), 434–456

Neumann, J. von, Kent, R. H., Bellinson, H. B., and Hart, B. I.: The mean square successive difference. Ann. Math. Statist. **12** (1941), 153–162

Nicholson, W. I.: Occupancy probability distribution critical points. Biometrika **48** (1961), 175–180

Nullau, B.: Verfahren zur Zeitreihenanalyse. Vierteljahreshefte zur Wirtschaftsforschung, Berlin 1968, 1, 58–82 (see DIW-Beitr. z. Strukturf., H. 7/1969; Wirtsch. u. Stat. H. 1/1973, H. 2 and 5/1975)

Olmstead P. S.: Runs determined in a sample by an arbitrary cut. Bell Syst. Techn. J. **37** (1958), 55–82

Ott, R. L., and Free, S. M.: A short-cut rule for a one-sided test of hypothesis for qualitative data. Technometrics **11** (1969), 197–200

Parzen, E.: The role of spectral analysis in time series analysis. Rev. Int. Statist. Inst. **35** (1967), 125–141 (cf. Empirical Time Series Analysis. Holden-Day, San Francisco, Calif. 1969)

Patnaik, P. B.: The power function of the test for the difference between two proportions in a 2×2 table. Biometrika **35** (1948), 157–175

Paulson, E., and Wallis, W. A.: Planning and analyzing experiments for comparing two percentages. In Eisenhart, Ch., M. W. Hastay and W. A. Wallis (Eds.), Selected Techniques of Statistical Analysis, McGraw-Hill, New York and London 1947, Chapter 7

Pearson, E. S.: Table of percentage points of $\sqrt{b_1}$ and b_2 in normal samples; a rounding off. Biometrika **52** (1965), 282–285

Pearson, E. S., and Hartley, H. O.: Biometrika Tables for Statisticians. Vol. I, 3rd ed., Cambridge 1966, 1970
Pearson, E. S., and Stephens, M. A.: The ratio of range to standard deviation in the same normal sample. Biometrika **51** (1964), 484–487
Plackett, R. L.: The continuity correction in 2×2 tables. Biometrika **51** (1964), 327–337 [see Biometrical Journal **22** (1980), 241–248]
Quandt, R. E.: (1) Statistical discrimination among alternative hypotheses and some economic regularities. J. Regional Sci. **5** (1964), 1–23. (2) Old and new methods of estimation and the Pareto distribution. Metrika **10** (1966), 55–82
Radhakrishna, S.: Combination of results from several 2×2 contingency tables. Biometrics **21** (1965), 86–98
Rao, C. R.: Linear Statistical Inference and Its Applications. 2nd ed. (Wiley) New York 1973, pp. 404–10
Rehse, E.: Zur Analyse biologischer Zeitreihen. Elektromedizin **15** (1970), 167–180
Rhoades, H. M., and Overall, J. E.: A sample size correction for Pearson chi-square in 2×2 contingency tables. Psychological Bulletin **91** (1982), 418–428
Runyon, R. P., and Haber, A.: Fundamentals of Behavioral Statistics. (Addison-Wesley, pp. 304) Reading, Mass. 1967, p. 258
Sachs, L.: (1) Statistische Methoden. 6th revised edition (Springer, 133 pages) Berlin, Heidelberg, New York 1984, pages 69–70, 72–75. (2) Numerischer Vergleich von 11 Konkurrenten des klassischen Vierfelder-χ^2-Tests bei kleinem Stichprobenumfang. Habilitationsschrift, Kiel 1974
Sandler, J.: A test of the significance of the difference between the means of correlated measures, based on a simplification of Student's t. Brit. J. Psychol. **46** (1955), 225–226
Sarris, V.: Nichtparametrische Trendanalysen in der klinisch-psychologischen Forschung. Z. exper. angew. Psychologie **15** (1968), 291–316
Seeger, P.: Variance analysis of complete designs: Some practical aspects. (Almqvist and Wiksell, pp. 225) Uppsala 1966, pp. 166–190
Seeger, P., and Gabrielsson, A.: Applicability of the Cochran Q test and the F test for statistical analysis of dichotomous data for dependent samples. Psychol. Bull. **69** (1968), 269–277
Shapiro, S. S., and Wilk, M. B.: (1) An analysis of variance test for normality (complete samples). Biometrika **52** (1965), 591–611. (2) Approximations for the null distribution of the W statistic Technometrics **10** (1968), 861–866 [cf. Statist. Neerl. **22** (1968), 241–248 and **27** (1973), 163–169]
Shapiro, S. S., Wilk, M. B., and Chen, H. J.: A comparative study of various tests for normality. J. Amer. Statist. Assoc. **63** (1968), 1343–1372 [cf. **66** (1971), 760–762 and **67** (1972), 215–216]
Slakter, M. J.: A comparison of the Pearson chi-square and Kolmogorov goodness-of-fit tests with respect to validity. J. Amer. Statist. Assoc. **60** (1965), 854–858; Corrigenda: **61** (1966), 1249 [cf. **69** (1974), 730–737 and **71** (1976), 204–209]
Smirnov, N.: Tables for estimating the goodness of fit of empirical distributions. Ann. Math. Statist. **19** (1948), 279–281 [cf. J. Roy. Statist. Soc. **38** (1976), 152–156]
Stephens, M. A.: Use of the Kolmogorov-Smirnov, Cramér-Von Mises and related statistics without extensive tables. J. Roy. Statist. Soc. **B32** (1970), 115–122
Stevens, W. L.: (1) Distribution of groups in a sequence of alternatives. Ann. Eugenics **9** (1939), 10–17. (2) Accuracy of mutation rates. J. Genetics **43** (1942), 301–307
"Student" (W. S. Gosset): The probable error of a mean. Biometrika **6** (1908), 1–25
Suits, D. B.: Statistics: An Introduction to Quantitative Economic Research. Chicago, Ill. 1963, Chapter 4
Swed, Frieda, S., and Eisenhart, C.: Tables for testing randomness of grouping in a sequence of alternatives. Ann. Math. Statist. **14** (1943), 83–86

Tate, M. W., and Brown, Sara, M.: Note on the Cochran Q-test. J. Amer. Statist. Assoc. **65** (1970), 155–160 [see also **68** (1973), 989–993; **70** (1975), 186–189; **72** (1977), 658–661; Biometrics **21** (1965), 1008–1010 and **36** (1980), 665–670]

Thomson, G. W.: Bounds for the ratio of range to standard deviation. Biometrika **42** (1955), 268–269

Tukey, J. W., and McLaughlin, D. H.: Less vulnerable confidence and significance procedures for location based on a single sample: Trimming/Winsorization. Sankhya Ser. A **25** (1963), 331–352

Ury, H. K.: A note on taking a covariable into account. J. Amer. Statist. Assoc. **61** (1966), 490–495

Vessereau, A.: Sur les conditions d'application du criterium χ^2 de Pearson. Bull. Inst. Int. Statistique **36** (3) (1958), 87–101

Waerden, B. L. van der: Mathematical Statistics. Springer-Verlag, New York 1969

Wallis, W. A.: Rough-and-ready statistical tests. Industrial Quality Control **8** (1952) (5), 35–40

—, and Moore, G. H.: A significance test for time series analysis. J. Amer. Statist. Assoc. **36** (1941), 401–409

Walter, E.: (1) Über einige nichtparametrische Testverfahren. I, II. Mitteilungsbl. Mathemat. Statistik **3** (1951), 31–44, 73–92. (2) χ^2-Test zur Prüfung der Symmetrie bezüglich Null. Mitteilungsbl. Mathemat. Statistik **6** (1954), 92–104. (3) Einige einfache nichtparametrische überall wirksame Tests zur Prüfung der Zweistichprobenhypothese mit paarigen Beobachtungen. Metrika **1** (1958), 81–88

Weichselberger, K.: Über eine Theorie der gleitenden Durchschnitte und verschiedene Anwendungen dieser Theorie. Metrika **8** (1964), 185–230

Wilcoxon, F., Katti, S. K., and Wilcox, Roberta A.: Critical Values and Probability Levels for the Wilcoxon Rank Sum Test and the Wilcoxon Signed Rank Test. Lederle Laboratories, Division Amer. Cyanamid Company, Pearl River, New York, August 1963

—, and Wilcox, Roberta A.: Some Rapid Approximate Statistical Procedures. Lederle Laboratories, Pearl River, New York 1964

Wilk, M. B., and Shapiro, S. S.: The joint assessment of normality of several independent samples. Technometrics **10** (1968), 825–839 [see also J. Amer. Statist. Assoc. **67** (1972), 215–216]

Woolf, B.: The log likelihood ratio test (the G-test). Methods and tables for tests of heterogeneity in contingency tables. Ann. Human Genetics **21** (1957), 397–409

Yamane, T.: Statistics: An Introductory Analysis. 2nd ed. (Harper and Row, pp. 919) New York 1967, pp. 330–367, 845–873

Yates, F.: Contingency tables involving small numbers and the χ^2-text. Supplement to the Journal of the Royal Statistical Society **1** (1934), 217–235 [cf. Biometrika **42** (1955), 404–411 and Biometrical Journal **22** (1980), 241–248]

[8:5]   Chapter 5

Abbas, S.: Serial correlation coefficient. Bull. Inst. Statist. Res. Tr. **1** (1967), 65–76

Acton, F. S.: Analysis of Straight-Line Data. New York 1959

Anderson, R. L., and Houseman, E. E.: Tables of Orthogonal Polynomial Values Extended to $N = 104$. Res. Bull. 297, Agricultural Experiment Station, Ames, Iowa 1942 (Reprinted March 1963)

Anderson, T. W.: An Introduction to Multivariate Statistical Analysis. New York 1958

—, Gupta, S. D., and Styan, G. P. H.: A Bibliography of Multivariate Statistical Analysis. (Oliver and Boyd; pp. 654) Edinburgh and London 1973 [see also Subrahmaniam, K. and K.: Multivariate Analysis, A Selected and Abstracted Bibliography 1957–1972. (M. Dekker; pp. 265) New York 1973]

Bancroft, T. A.: Topics in Intermediate Statistical Methods. (Iowa State Univ.) Ames, Iowa 1968

Bartlett, M. S.: Fitting a straight line when both variables are subject to error. Biometrics **5** (1949), 207–212

Barton, D. E., and Casley, D. J.: A quick estimate of the regression coefficient. Biometrika **45** (1958), 431–435

Blomqvist, N.: (1) On a measure of dependence between two random variables. Ann. Math. Statist. **21** (1950), 593–601. (2) Some tests based on dichotomization. Ann. Math. Statist. **22** (1951), 362–371

Brown, R. G.: Smoothing, Forecasting and Prediction of Discrete Time Series. (Prentice-Hall, pp. 468) London 1962 [cf. E. McKenzie, The Statistician **25** (1976), 3–14]

Carlson, F. D., Sobel, E., and Watson, G. S.: Linear relationships between variables affected by errors. Biometrics **22** (1966), 252–267

Chambers, J. M.: Fitting nonlinear models: numerical techniques. Biometrika **60** (1973), 1–13 [see also Murray, W. (Ed.): Numerical Methods for Unconstrained Optimization. (Acad. Press) London 1972]

Cohen, J.: A coefficient of agreement for nominal scales. Educational and Psychological Measurement **20** (1960), 37–46 [see Biometrics **36** (1980), 207–216]

Cole, La M. C.: On simplified computations. The American Statistician **13** (February 1959), 20

Cooley, W. W., and Lohnes, P. R.: Multivariate Data Analysis. (Wiley, pp. 400) London 1971

Cornfield, J.: Discriminant functions. Rev. Internat. Statist. Inst. **35** (1967), 142–153 [see also J. Amer. Statist. Assoc. **63** (1968), 1399–1412, Biometrics **35** (1979), 69–85 and **38** (1982), 191–200 and Biometrical Journal **22** (1980), 639–649]

Cowden, D. J., and Rucker, N. L.: Tables for Fitting an Exponential Trend by the Method of Least Squares. Techn. Paper 6, University of North Carolina, Chapel Hill 1965

Cox, D. R., and Snell, E. J.: A general definition of residuals. J. Roy. Statist. Soc. B **30** (1968), 248–275 [see also J. Qual. Technol. **1** (1969), 171–188, 294; Biometrika **58** (1971), 589–594; Biometrics **31** (1975), 387–410; Technometrics **14** (1972), 101–111, 781–790; **15** (1973), 677–695, 697–715; **17** (1975), 1–14]

Cureton, E. E.: Quick fits for the lines $y = bx$ and $y = a + bx$ when errors of observation are present in both variables. The American Statistician **20** (June 1966), 49

Daniel, C., and Wood, F. S. (with J. W. Gorman): Fitting Equations to Data. Computer Analysis of Multifactor Data for Scientists and Engineers. (Wiley-Interscience, pp. 342) New York 1971 [2nd edition, pp. 458, 1980] [see also Applied Statistics **23** (1974), 51–59 and Technometrics **16** (1974), 523–531]

Dempster, A. P.: Elements of Continuous Multivariate Analysis. (Addison-Wesley, pp. 400) Reading, Mass. 1968

Draper, N. R., and Smith, H.: Applied Regression Analysis. 2nd ed. (Wiley; pp. 709) New York 1981

Duncan, D. B.: Multiple comparison methods for comparing regression coefficients. Biometrics **26** (1970), 141–143 (see also B. W. Brown, 143–144)

Dunn, O. J.: A note on confidence bands for a regression line over a finite range. J. Amer. Statist. Assoc. **63** (1968), 1028–1033

Ehrenberg, A. S. C.: Bivariate regression is useless. Applied Statistics **12** (1963), 161–179

Elandt, Regina, C.: Exact and approximate power function of the non-parametric test of tendency. Ann. Math. Statist. **33** (1962), 471–481

Emerson, Ph. L.: Numerical construction of orthogonal polynomials for a general recurrence formula. Biometrics **24** (1968), 695–701

Enderlein, G.: Die Schätzung des Produktmoment-Korrelationsparameters mittels Rangkorrelation. Biometrische Zeitschr. **3** (1961), 199–212

Ferguson, G. A.: Nonparametric Trend Analysis. Montreal 1965

Fisher, R. A.: Statistical Methods for Research Workers, 12th ed. Edinburgh 1954, pp. 197–204

Friedrich, H.: Nomographische Bestimmung und Beurteilung von Regressions- und Korrelationskoeffizienten. Biometrische Zeitschr. **12** (1970), 163–187

Gallant, A. R.: Nonlinear regression. The American Statistician **29** (1975), 73–81, 175 [see also **30** (1976), 44–45]

Gebelein, H., and Ruhenstroth-Bauer, G.: Über den statistischen Vergleich einer Normalkurve und einer Prüfkurve. Die Naturwissenschaften **39** (1952), 457–461

Gibson, Wendy, M., and Jowett, G. H.: "Three-group" regression analysis. Part I. Simple regression analysis. Part II. Multiple regression analysis. Applied Statistics **6** (1957), 114–122 and 189–197

Glasser, G. J., and Winter, R. F.: Critical values of the coefficient of rank correlation for testing the hypothesis of independence. Biometrika **48** (1961), 444–448

Gregg, I. V., Hossel, C. H., and Richardson, J. T.: Mathematical Trend Curves – An Aid to Forecasting. (I.C.I. Monograph No. 1), Edinburgh 1964

Griffin, H. D.: Graphic calculation of Kendall's tau coefficient. Educ. Psychol. Msmt. **17** (1957), 281–285

Hahn, G. J.: Simultaneous prediction intervals for a regression model. Technometrics **14** (1972), 203–214

Hahn, G. J., and Hendrickson, R. W.: A table of percentage points of the distribution of the largest absolute value of k student t variates and its applications. Biometrika **58** (1971), 323–332

Hocking, R. R., and Pendleton, O. J.: The regression dilemma. Commun. Statist.— Theor. Meth. **12** (1983), 497–527

Hotelling, H.: (1) The selection of variates for use in prediction with some comments on the general problem of nuisance parameters. Ann. Math. Statist. **11** (1940), 271–283 [cf. O. J. Dunn et al., J. Amer. Statist. Assoc. **66** (1971), 904–908, Biometrics **31** (1975), 531–543 and Biometrika **63** (1976), 214–215]. (2) New light on the correlation coefficient and its transforms. J. Roy. Statist. Soc. B **15** (1953), 193–232

Hiorns, R. W.: The Fitting of Growth and Allied Curves of the Asymptotic Regression Type by Stevens Method. Tracts for Computers No. 28. Cambridge Univ. Press 1965

Hoerl, A. E., Jr.: Fitting Curves to Data. In J. H. Perry (Ed.): Chemical Business Handbook. (McGraw-Hill) London 1954, 20–55/20–77 (see also 20–16)

Kendall, M. G.: (1) A new measure of rank correlation. Biometrika **30** (1938), 81–93. (2) Multivariate Analysis. (Griffin; pp. 210) London 1975. (3) Rank Correlation Methods, 3rd ed. London 1962, pp. 38–41 (4th ed. 1970). (4) Ronald Aylmer Fisher, 1890–1962. Biometrika **50** (1963), 1–15. (5) Time Series. (Griffin; pp. 197) London 1973

Kerrich, J. E.: Fitting the line $y = ax$ when errors of observation are present in both variables. The American Statistician **20** (February 1966), 24

Koller, S.: (1) Statistische Auswertung der Versuchsergebnisse. In Hoppe-Seyler/Thierfelder's Handb. d. physiologisch- und pathologisch-chemischen Analyse, 10th edition, vol. II, pp. 931–1036, Berlin-Göttingen-Heidelberg 1955, pp. 1002–1004. (2) Typisierung korrelativer Zusammenhänge. Metrika **6** (1963), 65–75 [see also **17** (1971), 30–42]. (3) Systematik der statistischen Schlußfehler. Method. Inform. Med. **3** (1964), 113–117. (4) Graphische Tafeln zur Beurteilung statistischer Zahlen. 3rd edition. Darmstadt 1953 (4th edition 1969)

Konijn, H. S.: On the power of certain tests for independence in bivariate populations. Ann. Math. Statist. **27** (1956), 300–323

Kramer, C. Y.: A First Course in Methods of Multivariate Analysis. (Virginia Polytech. Inst.; pp. 353) Blacksburg, Virginia 1972

—, and Jensen, D. R.: Fundamentals of multivariate analysis. Part I–IV. Journal of Quality Technology **1** (1969), 120–133, 189–204, 264–276, **2** (1970), 32–40 and **4** (1972), 177–180

Kres, H.: Statistische Tafeln zur Multivariaten Analysis. (Springer; pp. 431) New York 1975

Krishnaiah, P. R. (Ed.): Multivariate Analysis and Multivariate Analysis II, III. (Academic Press; pp. 592 and 696, 450), New York and London 1966 and 1969, 1973

Kymn, K. O.: The distribution of the sample correlation coefficient under the null hypothesis. Econometrica 36 (1968), 187–189

Lees, Ruth, W., and Lord, F. M.: (1) Nomograph for computing partial correlation coefficients. J. Amer. Statist. Assoc. 56 (1961), 995–997. (2) Corrigenda 57 (1962), 917–918

Lieberson, S.: Non-graphic computation of Kendall's tau. Amer. Statist. 17 (Oct. 1961), 20–21

Linder, A.: (1) Statistische Methoden für Naturwissenschaftler, Mediziner und Ingenieure. 3rd edition. Basel 1960, page 172. (2) Anschauliche Deutung und Begründung des Trennverfahrens. Method. Inform. Med. 2 (1963), 30–33. (3) Trennverfahren bei qualitativen Merkmalen. Metrika 6 (1963), 76–83

Lord, F. M.: Nomograph for computing multiple correlation coefficients. J. Amer. Statist. Assoc. 50 (1955), 1073–1077 [see also Biometrika 59 (1972), 175–189]

Ludwig, R.: Nomogramm zur Prüfung des Produkt-Moment-Korrelationskoeffizienten r. Biometrische Zeitschr. 7 (1965), 94–95

Madansky, A.: The fitting of straight lines when both variables are subject to error. J. Amer. Statist. Assoc. 54 (1959), 173–205 [see also 66 (1971), 587–589 and 77 (1982), 71–79]

Mandel, J.: (1) Fitting a straight line to certain types of cumulative data. J. Amer. Statist. Assoc. 52 (1957), 552–566. (2) Estimation of weighting factors in linear regression and analysis of variance. Technometrics 6 (1964), 1–25

—, and Linning, F. J.: Study of accuracy in chemical analysis using linear calibration curves. Analyt. Chem. 29 (1957), 743–749

Meyer-Bahlburg, H. F. L.: Spearmans rho als punktbiserialer Korrelationskoeffizient. Biometrische Zeitschr. 11 (1969), 60–66

Miller, R. G.: Simultaneous Statistical Inference. (McGraw-Hill, pp. 272), New York 1966 (Chapter 5, pp. 189–210)

Morrison, D. F.: Multivariate Statistical Methods. 2nd ed. (McGraw-Hill, pp. 425), New York 1979

Natrella, M. G.: Experimental Statistics, National Bureau of Standards Handbook 91, U.S. Govt. Printing Office, Washington, D.C., 1963, pp. 5–31

Neter, J., and Wasserman, W.: Applied Linear Statistical Models. R. D. Irwin, Homewood, IL, 1974

Nowak, S.: in Blalock, H. M., et al.: Quantitative Sociology. Academic Press, New York, 1975, Chapter 3 (pp. 79–132)

Olkin, I., and Pratt, J. W.: Unbiased estimation of certain correlation coefficients. Ann. Math. Statist. 29 (1958), 201–211

Olmstead, P. S., and Tukey, J. W.: A corner test of association. Ann. Math. Statist. 18 (1947), 495–513

Ostle, B., and Mensing, R. W.: Statistics in Research. 3rd edition. (Iowa Univ. Press; pp. 596), Ames, Iowa 1975

Pfanzagl, J.: Über die Parallelität von Zeitreihen. Metrika 6 (1963), 100–113

Plackett, R. L.: Principles of Regression Analysis. Oxford 1960

Potthoff, R. F.: Some Scheffé-type tests for some Behrens-Fisher type regression problems. J. Amer. Statist. Assoc. 60 (1965), 1163–1190

Press, S. J.: Applied Multivariate Analysis. (Holt, Rinehart and Winston; pp. 521) New York 1972

Prince, B. M., and Tate, R. F.: The accuracy of maximum likelihood estimates of correlation for a biserial model. Psychometrika 31 (1966), 85–92

Puri, M. L., and Sen, P. K.: Nonparametric Methods in Multivariate Analysis. (Wiley, pp. 450) London 1971

Quenouille, M. H.: Rapid Statistical Calculations. Griffin, London 1959
Raatz, U.: Die Berechnung des SPEARMANschen Rangkorrelationskoeffizienten aus einer bivariaten Häufigkeitstabelle. Biom. Z. **13** (1971), 208–214
Radhakrishna, S.: Discrimination analysis in medicine. The Statistician **14** (1964), 147–167
Rao, C. R.: (1) Multivariate analysis: an indispensable aid in applied research (with an 81 reference bibliography). Sankhya **22** (1960), 317–338. (2) Linear Statistical Inference and Its Applications. New York 1965 (2nd ed. 1973). (3) Recent trends of research work in multivariate analysis. Biometrics **28** (1972), 3–22
Robson, D. S.: A simple method for constructing orthogonal polynomials when the independent variable is unequally spaced. Biometrics **15** (1959), 187–191 [see Int. Statist. Rev. **47** (1979), 31–36]
Roos, C. F.: Survey of economic forecasting techniques. Econometrica **23** (1955), 363–395
Roy, S. N.: Some Aspects of Multivariate Analysis. New York and Calcutta 1957
Sachs, L.: Statistische Methoden. 6th revised edition. (Springer, 133 pages) Berlin, Heidelberg, New York 1984, pages 92–94
Sahai, H.: A bibliography on variance components. Int. Statist. Rev. **47** (1979), 177–222.
Salzer, H. E., Richards, Ch. H., and Arsham, Isabelle: Table for the Solution of Cubic Equations. New York 1958
Samiuddin, M.: On a test for an assigned value of correlation in a bivariate normal distribution. Biometrika **57** (1970), 461–464 [cf., **65** (1978), 654–656 and K. Stange: Statist. Hefte **14** (1973), 206–236]
Saxena, A. K.: Complex multivariate statistical analysis: an annotated bibliography. International Statistical Review **46** (1978), 209–214
Saxena, H. C., and Surendran, P. U.: Statistical Inference. (Chand, pp. 396), Delhi, Bombay, Calcutta 1967 (Chapter 6, 258–342), (2nd ed. 1973)
Schaeffer, M. S., and Levitt, E. E.: Concerning Kendall's tau, a nonparametric correlation coefficient. Psychol. Bull. **53** (1956), 338–346
Scharf, J.-H.: Was ist Wachstum? Nova Acta Leopoldina NF (Nr. 214) **40** (1974), 9–75 [see also Biom. Z. **16** (1974), 383–399 **23** (1981), 41–54; Kowalski, Ch. J. and K. E. Guire, Growth **38** (1974), 131–169 as well as Peil, J., Gegenbaurs morph. Jb. **120** (1974), 832–853, 862–880; **121** (1975), 163–173, 389–420; **122** (1976), 344–390; **123** (1977), 236–259; **124** (1978), 525–545, 690–714; **125** (1979), 625–660 and Biometrics **35** (1979), 255–271, 835–848; **37** (1981), 383–390]
Seal, H.: Multivariate Statistical Analysis for Biologists. London 1964
Searle, S. R.: Linear Models. (Wiley, pp. 532) New York 1971
Spearman, C.: (1) The proof and measurement of association between two things. Amer. J. Psychol. **15** (1904), 72–101. (2) The method "of right and wrong cases" ("constant stimuli") without Gauss' formulae. Brit. J. Phychol. **2** (1908), 227–242
Stammberger, A.: Ein Nomogramm zur Beurteilung von Korrelationskoeffizienten. Biometrische Zeitschr. **10** (1968), 80–83
Stilson, D. W., and Campbell, V. N.: A note on calculating tau and average tau on the sampling distribution of average tau with a criterion ranking. J. Amer. Statist. Assoc. **57** (1962), 567–571
Stuart, A.: Calculation of Spearman's rho for ordered two-way classifications. American Statistician **17** (Oct. 1963), 23–24
Student: Probable error of a correlation coefficient. Biometrika **6** (1908), 302–310
Swanson, P., Leverton, R., Gram, M. R., Roberts, H., and Pesek, I.: Blood values of women: cholesterol. Journal of Gerontology **10** (1955) 41–47, cited by Snedecor, G. W., Statistical Methods, 5th ed., Ames 1959, p. 430
Tate, R. F.: (1) Correlation between a discrete and a continuous variable. Pointbiserial correlation. Ann. Math. Statist. **25** (1954), 603–607. (2) The theory of correlation between two continuous variables when one is dichotomized. Biometrika **42** (1955), 205–216. (3) Applications of correlation models for biserial data. J. Amer. Statist.

Assoc. **50** (1955), 1078–1095. (4) Conditional-normal regression models. J. Amer. Statist. Assoc. **61** (1966), 477–489

Thöni, H.: Die nomographische Bestimmung des logarithmischen Durchschnittes von Versuchsdaten und die graphische Ermittlung von Regressionswerten. Experientia **19** (1963), 1–4

Tukey, J. W.: Components in regression. Biometrics **7** (1951), 33–70

Waerden, B. L. van der: Mathematische Statistik. 2nd edition. (Springer, 360 pages), Berlin 1965, page 324

Wagner, G.: Zur Methodik des Vergleichs altersabhängiger Dermatosen. (Zugleich korrelationsstatistische Kritik am sogenannten „Status varicosus"). Zschr. menschl. Vererb.-Konstit.-Lehre **53** (1955), 57–84

Walter, E.: Rangkorrelation und Quadrantenkorrelation. Züchter Sonderh. 6, Die Frühdiagnose in der Züchtung und Züchtungsforschung II (1963), 7–11

Weber, Erna: Grundriß der biologischen Statistik. 7th revised edition. (Fischer, 706 pages), Stuttgart 1972, pages 550–578 [Discr. Anal.: see also Technometrics **17** (1975), 103–109] (8th revised edition 1980)

Williams, E. J.: Regression Analysis. New York 1959

Yule, G. U., and Kendall, M. G.: Introduction to the Theory of Statistics. London 1965, pp. 264–266

[8:5a] Factor analysis

Adam, J., and Enke, H.: Zur Anwendung der Faktorenanalyse als Trennverfahren. Biometr. Zeitschr. **12** (1970), 395–411

Bartholomew, D. J.: Factor analysis for categorical data. J. Roy. Statist. Soc. B **42** (1980), 293–321

Browne, M. W.: A comparison of factor analytic techniques. Psychometrika **33** (1968), 267–334

Corballis, M. C., and Traub. R. E.: Longitudinal factor analysis. Psychometrika **35** (1970), 79–98 [see also **36** (1971), 243–249 and Brit. J. Math. Statist. Psychol. **26** (1973), 90–97]

Derflinger, G.: Neue Iterationsmethoden in der Faktorenanalyse. Biometr. Z. **10** (1968), 58–75

Gollob, H. F.: A statistical model which combines features of factor analytic and analysis of variance techniques. Psychometrika **33** (1968), 73–115

Harman, H. H.: Modern Factor Analysis. 2nd rev. ed. (Univ. of Chicago, pp. 474), Chicago 1967

Jöreskog, K. G.: A general approach to confirmatory maximum likelihood factor analysis. Psychometrika **34** (1969), 183–202 [see also **36** (1971), 109–133, 409–426 and **37** (1972), 243–260, 425–440 as well as Psychol. Bull. **75** (1971), 416–423]

Lawley, D. N., and Maxwell, A. E.: Factor Analysis as a Statistical Method. 2nd ed. (Butterworths; pp. 153) London 1971 [see also Biometrika **60** (1973), 331–338]

McDonald, R. P.: Three common factor models for groups of variables. Psychometrika **35** (1970), 111–128 [see also 401–415 and **39** (1974), 429–444]

Rummel, R. J.: Applied Factor Analysis. (Northwestern Univ. Press, pp. 617) Evanston, Ill. 1970

Sheth, J. N.: Using factor analysis to estimate parameters. J. Amer. Statist. Assoc. **64** (1969), 808–822

Überla, K.: Faktorenanalyse. Eine systematische Einführung in Theorie und Praxis für Psychologen, Mediziner, Wirtschafts- und Sozialwissenschaftler. 2nd edition. (Springer, 399 pages), Berlin-Heidelberg-New York 1971 (see in particular pages 355–363)

Weber, Erna: Einführung in die Faktorenanalyse. (Fischer, 224 pages), Stuttgart 1974

[8:5b] Multiple regression analysis

Abt. K.: On the identification of the significant independent variables in linear models. Metrika **12** (1967), 1–15, 81–96

Anscombe, F. J.: Topics in the investigation of linear relations fitted by the method of least squares. With discussion. J. Roy. Statist. Soc. B **29** (1967), 1–52 [see also A **131** (1968), 265–329]

Beale, E. M. L.: Note on procedures for variable selection in multiple regression. Technometrics **12** (1970), 909–914 [see also **16** (1974), 221–227, 317–320 and Biometrika **54** (1967), 357–366 (see J. Amer. Statist. Assoc. **71** (1976), 249)]

Bliss, C. I.: Statistics in Biology. Vol. 2. (McGraw-Hill, pp. 639), New York 1970, Chapter 18

Cochran, W. G.: Some effects of errors of measurement on multiple correlation. J. Amer. Statist. Assoc. **65** (1970), 22–34

Cramer, E. M.: Significance tests and tests of models in multiple regression. The American Statistician **26** (Oct. 1972), 26–30 [see also **25** (Oct. 1971), 32–34, **25** (Dec. 1971), 37–39 and **26** (April 1972), 31–33 as well as **30** (1976), 85–87]

Darlington, R. B.: Multiple regression in psychological research and practice. Psychological Bulletin **69** (1968), 161–182 [see also **75** (1971), 430–431]

Donner, A.: The relative effectiveness of procedures commonly used in multiple regression analysis for dealing with missing values. Amer. Statist. **36** (1982), 378–381

Draper, N. R., and Smith, H.: Applied Regression Analysis. (Wiley, pp. 407), New York 1966 [2nd edition, pp. 709, 1981]

Dubois, P. H.: Multivariate Correlational Analysis. (Harper and Brothers, pp. 202), New York 1957

Enderlein, G.: Kriterien zur Wahl des Modellansatzes in der Regressionsanalyse mit dem Ziel der optimalen Vorhersage. Biometr. Zeitschr. **12** (1970), 285–308 [see also **13** (1971), 130–156]

Enderlein, G., Reiher, W., and Trommer, R.: Mehrfache lineare Regression, polynomiale Regression und Nichtlinearitätstests. In: Regressionsanalyse und ihre Anwendungen in der Agrarwissenschaft. Vorträge des 2. Biometr. Seminars d. Deutsch. Akad. d. Landwirtschaftswissensch. Berlin, März 1965. Tagungsber. Nr. 87, Berlin 1967, pages 49–78

Folks, J. L., and Antle, C. E.: Straight line confidence regions for linear models. J. Amer. Statist. Assoc. **62** (1967), 1365–1374

Goldberger, A. S.: Topics in Regression Analysis. (Macmillan, pp. 144), New York 1968

Graybill, F. A., and Bowden, D. C.: Linear segment confidence bands for simple linear models. J. Amer. Statist. Assoc. **62** (1967), 403–408

Hahn, G. J., and Shapiro. S. S.: The use and misuse of multiple regression. Industrial Quality Control **23** (1966), 184–189 [see also Applied Statistics **14** (1965), 196–200; **16** (1967), 51–64, 165–172; **23** (1974), 51–59]

Herne, H.: How to cook relationships. The Statistician **17** (1967), 357–370

Hinchen, J. D.: Multiple regression with unbalanced data. J. Qual. Technol. **2** (1970), 1, 22–29

Hocking, R. R.: The analysis and selection of variables in linear regression. Biometrics **32** (1976), 1–49

Huang, D. S.: Regression and Econometric Methods. (Wiley, pp. 274), New York 1970

La Motte, L. R., and Hocking, R. R.: Computational efficiency in the selection of regression variables. Technometrics **12** (1970), 83–93 [see also **13** (1971), 403–408 and **14** (1972), 317–325, 326–340]

Madansky, A.: The fitting of straight lines when both variables are subject to error. J. Amer. Statist. Assoc. **54** (1959), 173–205

Robinson, E. A.: Applied Regression Analysis. (Holden-Day, pp. 250), San Francisco 1969

Rutemiller, H. C., and Bowers, D. A.: Estimation in a heteroscedastic regression model. J. Amer. Statist. Assoc. **63** (1968), 552–557

Schatzoff, M., Tsao, R., and Fienberg, S.: Efficient calculation of all possible regressions. Technometrics **10** (1968), 769–779 [see also **14** (1972), 317–325]

Seber, G. A. F.: The Linear Hypothesis. A General Theory. (No. 19 of Griffin's Statistical Monographs and Courses. Ch. Griffin, pp. 120), London 1966

Smillie, K. W.: An Introduction to Regression and Correlation. (Acad. Pr., pp. 168), N.Y. 1966

Thompson, M. L.: Selection of variables in multiple regression. Part I. A review and evaluation. Part II. Chosen procedures, computations and examples. International Statistical Review **46** (1978), 1–19 and 129–146

Toro Vizcarrondo, C., and Wallace, T. D.: A test of the mean square error criterion for restrictions in linear regression. J. Amer. Statist. Assoc. **63** (1968), 558–572

Ulmo, J.: Problèmes et programmes de regression. Revue de Statistique Appliquée **19** (1971), No. 1, 27–39

Väliaho, H.: A synthetic approach to stepwise regression analysis. Commentationes Physico-Mathematicae **34** (1969), 91–131 [supplemented by **41** (1971), 9–18 and 63–72]

Wiezorke, B.: Auswahlverfahren in der Regressionsanalyse. Metrika **12** (1967), 68–79

Wiorkowski, J. J.: Estimation of the proportion of the variance explained by regression, when the number of parameters in the model may depend on the sample size. Technometrics **12** (1970), 915–919

[8:6] Chapter 6

Altham, Patricia, M. E.: The measurement of association of rows and columns for an $r \cdot s$ contingency table. J. Roy. Statist. Soc. B **32** (1970), 63–73

Armitage, P.: Tests for linear trends in proportions and frequencies. Biometrics **11** (1955), 375–386

Bartholomew, D. J.: A test of homogeneity for ordered alternatives. I and II. Biometrika **46** (1959), 36–48 and 328–335 [see also **63** (1976), 177–183, 647–654; J. Roy. Statist. Soc. B **23** (1961), 239–281. J. Amer. Statist. Assoc. **67** (1972), 55–63, **75** (1980), 454–459 as well as Biometrics **30** (1974), 589–597]

Benedetti, J. K., and Brown, M. B.: Strategies for the selection of log linear models. Biometrics **34** (1978), 680–686

Bennett, B. M.: Tests for marginal symmetry in contingency tables. Metrika **19** (1972), 23–26

—, and Hsu, P.: Sampling studies on a test against trend in binomial data. Metrika **5** (1962), 96–104

—, and E. Nakamura: (1) Tables for testing significance in a 2×3 contingency table. Technometrics **5** (1963), 501–511. (2) The power function of the exact test for the 2×3 contingency table. Technometrics **6** (1964), 439–458

Bennett, B. M., and Kaneshiro, C.: Small sample distribution and power of the binomial index of dispersion and log likelihood ratio tests. Biometrical Journal **20** (1978) 485–493

Berg, Dorothy, Leyton, M., and Maloney, C. J.: Exact contingency table calculations. Ninth Conf. Design Exper. in Army Research Developments and Testing (1965), (N.I.H., Bethesda Md.)

Bhapkar, V. P.: On the analysis of contingency tables with a quantitative response. Biometrics **24** (1968), 329–338

—, and Koch, G. G.: (1) Hypotheses of "no interaction" in multidimensional contingency tables. Technometrics **10** (1968), 107–123. (2) On the hypotheses of "no interaction" in contingency tables. Biometrics **24** (1968), 567–594

Bishop, Yvonne, M. M.: Full contingency tables, logits, and split contingency tables. Biometrics **25** (1969), 383–399 (see also 119–128)
—, Fienberg, S. E., and Holland, P. W.: Discrete Multivariate Analysis. Theory and Practice. (MIT Press: pp. 557) Cambridge, Mass. 1975 [cf. Biometrical Journal **22** (1980), 159–167, 779–793, and 795–789]
Bowker, A. H.: A test for symmetry in contingency tables. J. Amer. Statist. Assoc. **43** (1948), 572–574 [see also Biometrics **27** (1971), 1074–1078]
Bresnahan, J. L., and Shapiro, M. M.: A general equation and technique for the exact partitioning of chi-square contingency tables. Psychol. Bull. **66** (1966), 252–262
Castellan, N. J., Jr.: On the partitioning of contingency tables. Psychol. Bull. **64** (1965), 330–338
Caussinus, H.: Contribution à l'analyse statistique des tableaux de corrélation. Ann. Fac. Sci. Univ. Toulouse. Math., 4. Ser., **29** (1965), 77–183
Cochran, W. G.: (1) Some methods of strengthening the common χ^2 tests. Biometrics **10** (1954), 417–451. (2) Analyse des classifications d'ordre. Revue Statistique Appliquée **14**, 2 (1966), 5–17
Dunn, O. J.: Multiple comparisons using rank sums. Technometrics **6** (1964), 241–252
Eberhard, K.: $\overline{FM}$ – Ein Maß für die Qualität einer Vorhersage aufgrund einer mehrklassigen Variablen in einer $k \cdot 2$-Felder-Tafel. Z. exp. angew. Psychol. **17** (1970), 592–599
Enke, H.: On the analysis of incomplete two-dimensional contingency tables. Biometrical Journal **19** (1977), 561–573 [and **20** (1978), 229–242, **22** (1980), 779–793]
Everitt, B. S.: The Analysis of Contingency Tables. (Chapman and Hall; pp. 128) London 1977
Fairfield Smith, H: On comparing tables. The Philippine Statistician **6** (1957), 71–81
Fienberg, S. E.: The Analysis of Cross-Classified Categorical Data. (MIT Press; pp. 151) Cambridge, Mass. 1978
Gabriel, K. R.: Simultaneous test procedures for multiple comparisons on categorical data. J. Amer. Statist. Assoc. **61** (1966), 1080–1096
Gart, J. J.: Alternative analyses of contingency tables. J. Roy. Statist. Soc. B **28** (1966), 164–179 [see also Biometrika **59** (1972), 309–316 and **65** (1978), 669–672]
Gokhale, D. V., and Kullback, S.: The Information in Contingency Tables. M. Dekker, New York, 1978
Goodman, L. A.: (1) On methods for comparing contingency tables. J. Roy. Statist. Soc., Ser. A **126** (1963), 94–108. (2) Simple methods for analyzing three-factor interaction in contingency tables. J. Amer. Statist. Assoc. **59** (1964), 319–352. (3) On partitioning χ^2 and detecting partial association in three-way contingency tables. J. Roy. Statist. Soc. B **31** (1969), 486–498. (4) The multivariate analysis of qualitative data: interactions among multiple classifications. J. Amer. Statist. Assoc. **65** (1970), 226–256. (5) The analysis of multidimensional contingency tables. Stepwise procedures and direct estimation methods for building models for multiple classifications. Technometrics **13** (1971), 33–61 [see also J. Amer. Statist. Assoc. **68** (1973), 165–175 and **74** (1979), 537–552 as well as Biometrika **68** (1981), 347–355]
Goodman, L. A., and Kruskal, W. H.: Measures of association for cross classifications, IV. Simplification of asymptotic variances. J. Amer. Statist. Assoc. **67** (1972), 415–421 [see also **49** (1954), 732–764; **52** (1957), 578; **54** (1959), 123–163; **58** (1963), 310–364; **74** (1979), 537–552 as well as A. Agresti **71** (1976), 49–55]
Grizzle, J. E., Starmer, C. F., and Koch, G. G.: Analysis of categorical data by linear models. Biometrics **25** (1969), 489–504 [see also **26** (1970), 860; **28** (1972), 137–156, J. Amer. Statist. Assoc. **67** (1972), 55–63 and Method. Inform. Med. **12** (1973), 123–128 as well as Int. Stat. Rev. **48** (1980), 249–265]
Hamdan, M. A.: Optimum choice of classes for contingency tables. J. Amer. Statist. Assoc. **63** (1968), 291–297 [see also Psychometrika **36** (1971), 253–259]
Hartwig, F.: Statistical significance of the lambda coefficients. Behavioral Science **18** (1973), 307–310

Hettmansperger, T. P., and McKean, J. W.: On testing for significant change in c × c tables. Commun. Statist. **2** (1973), 551–560

Ireland, C. T., Ku, H. H., and Kullback, S.: Symmetry and marginal homogeneity of an r · r contingency table. J. Amer. Statist. Assoc. **64** (1969), 1323–1341

Ireland, C. T., and Kullback, S.: Minimum discrimination information estimation. Biometrics **24** (1968), 707–713 [see also **27** (1971), 175–182 and Biometrika **55** (1968), 179–188]

Kastenbaum, M. A.: (1) A note on the additive partitioning of chi-square in contingency tables. Biometrics **16** (1960), 416–422. (2) Analysis of categorical data: some well-known analogues and some new concepts. Commun. Statist. **3** (1974), 401–417

Killion, R. A., and Zahn, D. A.: A bibliography of contingency table literature: 1900 to 1974. International Statistical Review **44** (1976), 71–112

Kincaid, W. M.: The combination of $2 \times m$ contingency tables. Biometrics **18** (1962), 224–228

Kramer, C. Y.: A First Course in Methods of Multivariate Analysis. Virginia Polytechnic Institute and State University, Blacksburg 1972

Ku, H. H.: A note on contingency tables involving zero frequencies and the 2I test. Technometrics **5** (1963), 398–400

Ku, H. H., and Kullback, S.: Interaction in multidimensional contingency tables: an information theoretic approach. J. Res. Nat. Bur. Stds. **72B** (1968), 159–199

Ku, H. H., Varner, R. N., and Kullback, S.: On the analysis of multidimensional contingency tables. J. Amer. Statist. Assoc. **66** (1971), 55–64 [see also **70** (1975), 503–523 and 624–625]

Kullback, S.: Information Theory and Statistics. (Wiley, pp. 395) New York 1959 [cf. J. Adam, H. Enke and G. Enderlein: Biometrische Zeitschr. **14** (1972), 305–323, **15** (1973), 53–78, **17** (1975), 513–523, **22** (1980), 779–793]

—, Kupperman, M., and Ku, H. H.: (1) An application of information theory to the analysis of contingency tables, with a table of $2n \ln n$, $n = 1(1)10{,}000$. J. Res. Nat. Bur. Stds. B **66** (1962), 217–243. (2) Tests for contingency tables and Markov chains. Technometrics **4** (1962), 573–608

—, and Leibler, R. A.: On information and sufficiency. Ann. Math. Statist. **22** (1951), 79–86

Lancaster, H. O.: The Chi-Squared Distribution. (Wiley, pp. 356), New York 1969

Lewontin, R. C., and Felsenstein, J.: The robustness of homogeneity tests in $2 \times n$ tables. Biometrics **21** (1965), 19–33

Mantel, N.: (1) Chi-square tests with one degree of freedom; extensions of the Mantel-Haenszel procedure. J. Amer. Statist. Assoc. **58** (1963), 690–700 [see also Biometrics **29** (1973), 479–486]. (2) Tests and limits for the common odds ratio of several 2×2 contingency tables: methods in analogy with the Mantel-Haenszel procedure. Journal of Statistical Planning and Inference **1** (1977), 179–189

—, and Haenszel, W.: Statistical aspects of the analysis of data from retrospective studies of disease. J. Natl. Cancer Institute **22** (1959), 719–748 [see also W. J. Youden: Cancer **3** (1950), 32–35 and Biometrika **66** (1979), 181–183, 419–427; Biometrics **35** (1979), 385–391, **36** (1980), 355–356, 381–399]

Martini, P.: Methodenlehre der theraupeutisch-klinischen Forschung. Springer-Verlag, Berlin-Göttingen-Heidelberg, 1953

Maxwell, A. E.: Analysing Qualitative Data. London 1961

Meng, R. C., and Chapman, D. G.: The power of chi-square tests for contingency tables. J. Amer. Statist. Assoc. **61** (1966), 965–975 [see also Amer. Statist. **31** (1977), 83–85]

Mosteller, F.: Association and estimation in contingency tables. J. Amer. Statist. Assoc. **63** (1968), 1–2 [see also L. A. Goodman, **74** (1979), 537–552, J. R. Landis and G. G. Koch, Statistica Neerlandica **29** (1975), 101–123, 151–161 as well as L. A. Goodman, Biometrika **66** (1979), 413–418 and J. Wahrendorf, Biometrika **67** (1980), 15–21]

Nass, C. A. G.: The χ^2 test for small expectations in contingency tables with special reference to accidents and absenteeism. Biometrika **46** (1959), 365–385

Nelder, J. A.: Log linear models for contingency tables. Applied Statistics **23** (1974), 323–329 [cf. **25** (1976), 37–46 and The Statistician **25** (1976), 51–58]

Odoroff, C. L.: A comparison of minimum logit chi-square estimation and maximum likelihood estimation in $2 \times 2 \times 2$ and $3 \times 2 \times 2$ contingency tables: tests for interaction. J. Amer. Statist. Assoc. **65** (1970), 1617–1631 [see also **69** (1974), 164–168]

Pawlik, K.: Der maximale Kontingenzkoeffizient im Falle nichtquadratischer Kontingenztafeln. Metrika **2** (1959), 150–166

Plackett, R. L.: Analysis of Categorical Data. 2nd rev. ed. (Griffin; pp. 215) London 1981

Ryan, T.: Significance tests for multiple comparison of proportions, variances and other statistics. Psychological Bull. **57** (1960), 318–328

Sachs, L.: Statistische Methoden. 6th ed. (Springer; pp. 133) New York, Berlin, Heidelberg, Tokyo 1984

Shaffer, J. P.: Defining and testing hypotheses in multidimensional contingency tables. Psychol. Bull. **79** (1973), 127–141

Upton, G. J. G.: (1) The Analysis of Cross Tabulated Data. (Wiley; pp. 148) New York 1978. (2) A comparison of alternative tests for the 2×2 comparative trial. J. Roy. Statist. Society A **145** (1982), 86–105

Wisniewski, T. K. M.: Power of tests of homogeneity of a binomial series. Journal of the American Statistical Association **67** (1972), 680–683

Woolf, B.: The log likelihood ratio test (the G-Test). Methods and tables for tests of heterogeneity in contingency tables. Ann. Human Genetics **21** (1957), 397–409

Yates, F.: The analysis of contingency tables with groupings based on quantitative characters. Biometrika **35** (1948), 176–181 [see also **39** (1952), 274–289]

[8:7] Chapter 7

Addelman, S.: (1) Techniques for constructing fractional replicate plans. J. Amer. Statist. Assoc. **58** (1963), 45–71 [see also **67** (1972), 103–111]. (2) Sequences of two-level fractional factorial plans. Technometrics **11** (1969), 477–509 [see also Davies-Hay, Biometrics **6** (1950), 233–249]

Ahrens, H.: Varianzanalyse. (WTB, Akademie-Vlg, 198 pages), Berlin 1967

Anscombe, F. J.: The transformation of Poisson, binomial and negative-binomial data. Biometrika **35** (1948), 246–254 [see also Applied Statistics **24** (1975), 354–359 and **29** (1980), 190–197]

—, and Tukey, J. W.: The examination and analysis of residuals. Technometrics **5** (1963), 141–160 [see also **17** (1975), 1–14 and Applied Statistics **29** (1980), 190–197]

Baker, A. G.: Analysis and presentation of the results of factorial experiments. Applied Statistics **6** (1957), 45–55 [see also **3** (1954), 184–195 and **22** (1973), 141–160]

Bancroft, T. A.: (1) Analysis and inference for incompletely specified models involving the use of preliminary test(s) of significance. Biometrics **20** (1964), 427–442. [mixed model: see also Industrial Quality Control **13** (1956), 5–8]. (2) Topics in Intermediate Statistical Methods. Vol. I. (Iowa State University Press; pp. 129) Ames, Iowa 1968, Chapters 1 and 6 [Nonorthog. ANCOVA: see W. J. Hemmerle, J. Amer. Statist. Assoc. **71** (1976), 195–199]

Barlow, R. E., Bartholomew, D. J., Brenner, J. M., and Brunk, H. D.: Statistical Inference Under Order Restrictions. (Wiley; pp. 388) New York 1972

Barnett, V. D.: Large sample tables of percentage points for Hartley's correction to Bartlett's criterion for testing the homogeneity of a set of variances. Biometrika **49** (1962), 487–494

Bartholomew, D. J.: Ordered tests in the analysis of variance. Biometrika **48** (1961), 325–332

Bartlett, M. S.: (1) Properties of sufficiency and statistical tests. Proc. Roy. Soc. A **160** (1937), 268–282. (2) Some examples of statistical methods of research in agriculture and applied biology. J. Roy. Statist. Soc. Suppl. **4** (1937), 137–170. (3) The use of transformations. Biometrics **3** (1947), 39–52 [see also Technometrics **11** (1969), 23–40]

Bechhofer, R. E.: A single-sample multiple decision procedure for ranking means of normal populations with known variances. Ann. Math. Statist. **25** (1954), 16–39 [cf. Biometrika **54** (1967), 305–308 and **59** (1972), 217–219 as well as Biometr. Z. **16** (1974), 401–430 and J. Amer. Statist. Assoc. **71** (1976), 140–142]

Binder, A.: The choice of an error term in analysis of variance designs. Psychometrika **20** (1955), 29–50

Bishop, T. A., and Dudewicz, E. J.: Exact analysis of variance with unequal variances: test procedures and tables. Technometrics **20** (1978), 419–430

Blischke, W. R.: Variances of estimates of variance components in a three-way classification. Biometrics **22** (1966), 553–565 [see also **2** (1946), 110–114 and **26** (1970), 243–254, 677–686]

Bliss, C. I., Cochran, W. G., and Tukey, J. W.: A rejection criterion based upon the range. Biometrika **43** (1956), 418–422

Bose, R. C.: Paired comparison designs for testing concordance between judges. Biometrika **43** (1956), 113–121

Box, G. E. P.: (1) Non-normality and tests on variances. Biometrika **40** (1953), 318–335. (2) The exploration and exploitation of response surfaces. Biometrics **10** (1954), 16–60

—, and Andersen, S. L.: Permutation theory in the derivation of robust criteria and the study of departures from assumption. With discussion. J. Roy. Statist. Soc., Ser. B **17** (1955), 1–34

—, and Cox, D. R.: An analysis of transformations. J. Roy. Statist. Soc., Ser. B **26** (1964), 211–252 [cf. B **35** (1973), 473–479, B **42** (1980), 71–78 and J. J. Schlesselman: J. Amer. Statist. Assoc. **68** (1973), 369–378, **78** (1983), 411–417 as well as Applied Statistics **29** (1980), 190–197]

Bradley, R. A.: The Future of statistics as a discipline. J. Amer. Statist. Assoc. **77** (1982), 1–10

—, and Schumann, D. E. W.: The comparison of the sensitivities of similar experiments: applications. Biometrics **13** (1957), 496–510

Bratcher, T. L., Moran, M. A., and Zimmer, W. J.: Tables of sample sizes in the analysis of variance. J. Qual. Technol. **2** (1970), 156–164 [cf. Technometrics **15** (1973), 915–921; **16** (1974), 193–201]

Chanter, D. O.: Modifications of the angular transformation. Applied Statistics **24** (1975), 354–359

Cochran, W. G.: (1) The distribution of the largest of a set of estimated variances as a fraction of their total. Ann. Eugen. (Lond.) **11** (1941). 47–61. (2) Some consequences when assumptions for the analysis of variance are not satisfied. Biometrics **3** (1947), 22–38. (3) Testing a linear relation among variances. Biometrics **7** (1951), 17–32. (4) Analysis of covariance: its nature and use. Biometrics **13** (1957), 261–281 [cf. Industr. Qual. Contr. **22** (1965), 282–286]. (5) The Design of Experiments. In Flagle, C. D., Huggins, W. H., and Roy, R. H. (Eds.): Operations Research and Systems Engineering, pp. 508–553. Baltimore 1960

Cole, J. W. L., and Grizzle, J. E.: Applications of multivariate analysis of variance to repeated measurements experiments. Biometrics **22** (1966), 810–828 [see also **28** (1972), 39–53 and 55–71]

Conover, W. J.: Two k-sample slippage tests. J. Amer. Statist. Assoc. **63** (1968), 614–626

Cooper, B. E.: A unifying computational method for the analysis of complete factorial experiments. Communications of the ACM **10** (Jan. 1967), 27–34

Cunningham, E. P.: An iterative procedure for estimating fixed effects and variance components in mixed model situations. Biometrics **24** (1968), 13–25

Daniel, C.: Use of half-normal plots in interpreting factorial two-level experiments. Technometrics **1** (1959), 311–341 [see also **2** (1960), 149–156, **8** (1966), 259–278, **17** (1975), 189–211, **24** (1982), 213–222; J. Qual. Technol. **6** (1974), 2–21]

David, H. A.: (1) Further applications of range to the analysis of variance. Biometrika **38** (1951), 393–409. (2) The ranking of variances in normal populations. J. Amer. Statist. Assoc. **51** (1956), 621–626. (3) The Method of Paired Comparisons (Griffin, pp. 124) London 1969 [cf. N. Wrigley, Environment and Planning A **12** (1980), 21–40]

—, Hartley, H. O., and Pearson, E. S.: The distribution of the ratio, in a single normal sample of range to standard deviation. Biometrika **41** (1954), 482–493

Dudewicz, E. J., and Koo, J. O.: The Complete Categorized Guide to Statistical Selection and Ranking Procedures. (American Sciences Press; pp. 627) Columbus, Ohio 1982

Duncan, D. B.: (1) Multiple range and multiple F tests. Biometrics **11** (1955), 1–42 [see also for $n_i \neq$ const. C. Y. Kramer, **12** (1956), 307–310], [see also Technometrics **11** (1969), 321–329]. (2) Multiple range tests for correlated and heteroscedastic means. Biometrics **13** (1957), 164–176. (3) A Bayesian approach to multiple comparisons. Technometrics **7** (1965), 171–222

Dunn, Olive, J.: (1) Confidence intervals for the means of dependent, normally distributed variables. J. Amer. Statistic. Assoc. **54** (1959), 613–621. (2) Multiple comparisons among means. J. Amer. Statist. Assoc. **56** (1961), 52–64 [see also Technometrics **6** (1964), 241–252 and Biometrical Journal **23** (1981), 29–40]

Eisen, E. J.: The quasi-F test for an unnested fixed factor in an unbalanced hierarchal design with a mixed model. Biometrics **22** (1966), 937–942 [cf. J. Amer. Statist. Assoc. **69** (1974), 765–771]

Eisenhart, C.: The assumptions underlying the analysis of variance. Biometrics **3** (1947), 1–21

Eisenhart, C., Hastay, M. W., and Wallis, W. A.: Techniques of Statistical Analysis. McGraw-Hill, New York 1947

Enderlein, G.: Die Kovarianzanalyse. In: Regressionsanalyse und ihre Anwendungen in der Agrarwissenschaft. Vorträge des 2. Biometrischen Seminars der Deutschen Akademie der Landwirtschaftswissenschaften zu Berlin, im März 1965. Tagungsberichte Nr. 87, Berlin 1967, pp. 101–132

Endler, N. S.: Estimating variance components from mean squares for random and mixed effects analysis of variance models. Perceptual and Motor Skills **22** (1966), 559–570 [see also the corrections given by Whimbey et al. **25** (1967), 668]

Enrick, N. L.: An analysis of means in a three-way factorial. J. Qual. Technology **8** (1976), 189–196

Evans, S. H., and Anastasio, E. J.: Misuse of analysis of covariance when treatment effect and covariate are confounded. Psychol. Bull. **69** (1968), 225–234 [see also **75** (1971), 220–222]

Federer, W. T.: Experimental error rates. Proc. Amer. Soc. Hort. Sci. **78** (1961), 605–615

Fisher, L., and McDonald, J.: Fixed Effects Analysis of Variance. (Academic Press; pp. 177) New York 1978

Fisher, R. A., and Yates, F.: Statistical Tables for Biological, Agricultural and Medical Research. 6th ed. London 1963

Fleckenstein, Mary, Freund, R. A., and Jackson, J. E.: A paired comparison test of typewriter carbon papers. Tappi **41** (1958), 128–130

Freeman, M. F., and Tukey, J. W.: Transformations related to the angular and the square root. Ann. Math. Statist. **21** (1950), 607–611

Friedman, M.: (1) The use of ranks to avoid the assumption of normality implicit in the analysis of variance. J. Amer. Statist. Assoc. **32** (1937), 675–701. (2) A comparison of alternative tests of significance for the problem of m rankings. Ann. Math. Statist. **11** (1940), 86–92

Gabriel, K. R.: Analysis of variance of proportions with unequal frequencies. J. Amer. Statist. Assoc. **58** (1963). 1133–1157 [cf. W. Berchtold and A. Linder, EDV Med. Biol. **4** (1973), 99–108]

Games, P. A.. (1) Robust tests for homogeneity of variance. Educat. Psychol. Msmt. **32** (1972), 887–909 [see also J. Amer. Statist. Assoc. **68** (1973), 195–198]. (2) —, Keselman, H. J., and Rogan, J. C.: A review of simultaneous pairwise multiple comparisons. Statistica Neerlandica **37** (1983), 53–58

Gates, Ch. E., and Shiue, Ch.-J.: The analysis of variance of the s-stage hierarchal classification. Biometrics **18** (1962), 529–536

Ghosh, M. N., and Sharma, D.: Power of Tukey's test for non-additivity. J. Roy. Statist. Soc. **B 25** (1963), 213–219

Gibbons, J. D., Olkin, I., and Sobel, M.: An introduction to ranking and selection. American Statistician **33** (1979), 185–195 [see also J. Qual. Technol. **14** (1982), 80–88]

Glasser, R. E.: Exact critical values for Bartlett's test for homogeneity of variances. Journal of the American Statistical Association **71** (1976), 488–490 [see D. D. Dyer and J. P. Keating, J. Amer. Statist. Assoc. **75** (1980), 313–319]

Gower, J. C.: Variance component estimation for unbalanced hierarchical classifications. Biometrics **18** (1962), 537–542

Green, B. F., Jr., and Tukey, J. W.: Complex analyses of variance: general problems. Psychometrika **25** (1960), 127–152

Grimm, H.: Transformation von Zufallsvariablen. Biometr. Z. **2** (1960), 164–182

Hahn, G. J.: Some things engineers should know about experimental design. Journal of Quality Technology **9** (1977), 13–22

Harsaae, E.: On the computation and use of a table of percentage points of Bartlett's M. Biometrika **56** (1969), 273–281 [see also R. E. Glaser, J. Amer. Statist. Assoc. **71** (1976), 488–490]

Harte, Cornelia: Anwendung der Covarianzanalyze beim Vergleich von Regressionskoeffizienten. Biometrische Zeitschr. **7** (1965), 151–164

Harter, H. L.: (1) Error rates and sample sizes for range tests in multiple comparisons. Biometrics **13** (1957), 511–536. (2) Tables of range and Studentized range. Ann. Math. Statist. **31** (1960), 1122–1147. (3) Expected values of normal order statistics. Biometrika **48** (1961), 151–165. (3) Order Statistics and their Use in Testing and Estimation. Vol. 1: Tests based on Range and Studentized Range of Samples from a Normal Population. (ARL, USAF; U.S. Government Printing Office; pp. 761) Washington 1970, 21–23, 623–661, especially 653 and 657

Harter, H. L., Clemm, D. S., and Guthrie, E. H.: The Probability Integrals of the Range and of the Studentized Range. Vol. I. Wright Air Development Center Technical Report 58-484, 1959

Hartley, H. O.: (1) The use of range in analysis of variance. Biometrika **37** (1950), 271–280. (2) The maximum F-ratio as a short cut test for heterogeneity of variance. Biometrika **37** (1950), 308–312 [see also **60** (1973), 213–214 and J. Amer. Statist. Assoc. **70** (1975), 180–183]. (3) Some recent developments in the analysis of variance. Comm. Pure and Applied Math. **8** (1955), 47–72

—, and Pearson, E. S.: Moments constants for the distribution of range in normal samples. I. Foreword and tables. Biometrika **38** (1951), 463–464

Harvey, W. R.: Estimation of variance and covariance components in the mixed model. Biometrics **26** (1970), 485–504 [see also **30** (1974), 157–169 and Technometrics **15** (1973), 819–831]

Hays, W. L.: Statistics for Psychologists. (Holt, Rinehart and Winston, pp. 719), New York 1963, pp. 439–455

Healy, M. J. R., and Taylor, L. R.: Tables for power-law transformations. Biometrika **49** (1962), 557–559

Herzberg, Agnes, M., and Cox, D. R.: Recent work on the design of experiments: a bibliography and a review. J. Roy. Statist. Soc. A **132** (1969), 29–67

Holland, D. A.: Sampling errors in an orchard survey involving unequal numbers of orchards of distinct type. Biometrics **21** (1965), 55–62

Huitema, B. E.: The Analysis of Covariance and Alternatives. (Wiley; pp. 440) New York 1980 [see also Biometrics **38** (1982), 540–753]

Jackson, J. E., and Fleckenstein, Mary: An evaluation of some statistical techniques used in the analysis of paired comparison data. Biometrics **13** (1957), 51–64

Jaech, J. L.: The latin square. J. Qual. Technol. **1** (1969), 242–255 [cf. Biom. Z. **17** (1975), 447–454]

Kastenbaum, M. A., Hoel, D. G., and Bowman, K. O.: (1) Sample size requirements: one-way analysis of variance. Biometrika **57** (1970), 421–430. (2) Sample size requirements: randomized block designs. Biometrika **57** (1970), 573–577 [see also Biometrika **59** (1972), 234, and Amer. Statistician **31** (1977), 117–118; **33** (1979) 209–210]

Kempthorne, V.: The randomization theory of experimental inference. J. Amer. Statist. Assoc. **50** (1955), 946–967

—, and Barclay, W. D.: The partition of error in randomized blocks. J. Amer. Statist. Assoc. **48** (1953), 610–614

Kendall, M. G.: On the future of statistics – a second look. J. Roy. Statist. Soc. A **131** (1968), 182–294 [see also Amer. Statist. **24** (Dec. 1970), 10–13]

Kiefer, J. C.: Optimum experimental designs. J. Roy. Statist. Soc., Ser. B **21** (1959), 272–319

Knese, K. H., and Thews, G.: Zur Beurteilung graphisch formulierter Häufigkeitsverteilungen bei biologischen Objekten. Biometrische Zeitschr. **2** (1960), 183–193

Koch, G. G.: A general approach to the estimation of variance components. Technometrics **9** (1967), 93–118 [cf. **5** (1963), 421–440, 441–450, **10** (1968), 551–558 and **13** (1971), 635–650]

—, and Sen, K. P.: Some aspects of the statistical analysis of the "mixed model." Biometrics **24** (1968), 27–48 [see also Industr. Qual. Contr. **13** (1956), 5–8 and Amer. Statist. **27** (1973), 148–152]

Koller, S.: (1) Statistische Auswertung der Versuchsergebnisse. In Hoppe-Seyler/Thierfelder's Handb. d. physiologisch- und pathologisch-chemischen Analyse, 10th edition, Vol. II, pp. 931–1036. Berlin-Göttingen-Heidelberg 1955, pp. 1011–1016 [see also Arzneim.-Forschg. **18** (1968), 71–77; **24** (1974), 1001–1004]. (2) Statistische Auswertungsmethoden. In H. M. Rauen (Eds.), Biochemisches Taschenbuch, part II, pp. 959–1046, Berlin-Göttingen-Heidelberg-New York 1964

Kramer, C. Y.: On the analysis of variance of a two-way classification with unequal sub-class numbers. Biometrics **11** (1955), 441–452 [see also Psychometrika **36** (1971), 31–34]

Kurtz, T. E., Link, R. F., Tukey, J. W., and Wallace, D. L.: (1) Short-cut multiple comparisons for balanced single and double classifications: Part 1, Results. Technometrics **7** (1965), 95–161. (2) Short-cut multiple comparisons for balanced single and double classifications: Part 2. Derivations and approximations. Biometrika **52** (1965), 485–498

Kussmaul, K., and Anderson, R. L.: Estimation of variance components in two-stage nested designs with composite samples. Technometrics **9** (1967), 373–389

Lehmacher, W. and Wall, K. D.: A new nonparametric approach to the comparison of k independent samples of response curves. Biometrical Journal 20 (1978), 261–273 [see also 21 (1979), 123–130 and 24 (1982), 717–722]

Lehmann, W.: Einige Probleme der varianzanalytischen Auswertung von Einzelpflanzenergebnissen. Biometrische Zeitschr. 12 (1970), 54–61

Le Roy, H. L.: (1) Wie finde ich den richtigen F-Test? Mitteilungsbl. f. math. Statistik 9 (1957), 182–195. (2) Testverhältnisse bei der doppelten Streuungszerlegung (Zweiwegklassifikation). Schweiz. Landw. Forschg. 2 (1963), 329–340. (3) Testverhältnisse beim $a \cdot b \cdot c$- und $a \cdot b \cdot c \cdot d$- Faktorenversuch. Schweiz. Landw. Forschg. 3 (1964), 223–234. (4) Vereinfachte Regel zur Bestimmung des korrekten F-Tests beim Faktorenversuch. Schweiz. Landw. Forschg. 4 (1965), 277–283. (5) Verbale und bildliche Interpretation der Testverhältnisse beim Faktorenversuch. Biometrische Zeitschr. 14 (1972), 419–427 [see also Metrika 17 (1971), 233–242]

Leslie, R. T., and Brown, B. M.: Use of range in testing heterogeneity of variance. Biometrika 53 (1966), 221–227 [cf. L. S. Nelson: J. Qual. Technol. 7 (1975), 99–100]

Li, C. C.: Introduction to Experimental Statistics. (McGraw-Hill, pp. 460), New York 1964, pp. 258–334 [Faktorial Exper.: cf. J. Roy. Statist. Soc. B 27 (1965), 251–263 and J. Qual. Technol. 6 (1974), 2–21]

Li, J. C. R.: Statistical Inference I, II. (Edwards Brothers, pp. 658, 575), Ann Arbor, Mich. 1964

Lienert, G. A.: Über die Anwendung von Variablen-Transformationen in der Psychologie. Biometrische Zeitschr. 4 (1962), 145–181

Linhart, H.: Streuungszerlegung für Paar-Vergleiche. Metrika 10 (1966), 16–38

Link, R. F., and Wallace, D. L.: Some Short Cuts to Allowances. Princeton University, March 1952

Mandel, J.: Non-additivity in two-way analysis of variance. J. Amer. Statist. Assoc. 56 (1961), 878–888

Martin, L.: Transformations of variables in clinical-therapeutical research. Method. Inform. Med. 1 (1962), 1938–1950

McDonald, B. J., and Thompson, W. A., Jr.: Rank sum multiple comparisons in one- and two-way classifications. Biometrika 54 (1967), 487–497

Michaelis, J.: Schwellenwerte des Friedman-Tests. Biometr. Zeitschr. 13 (1971), 118–129

Mosteller, F., and Youtz, C.: Tables of the Freeman-Tukey transformations for the binomial and Poisson distributions. Biometrika 48 (1961), 433–440

Natrella, Mary G.: Experimental Statistics. NBS Handbook 91. (U.S. Gov't Printing Office) Washington 1963, Chapters 11–14 [see C. Daniel, J. Amer. Statist. Assoc. 68 (1973), 353–360]

Nelson, L. S.: Ordered tests for a three-level factor. Journal of Quality Technology 8 (1976), 241–243 [see 7 (1975), 46–48]

Newman, D.: The distribution of the range in samples from normal population, expressed in terms of an independent estimate of standard deviation. Biometrika 31 (1939), 20–30

Neyman, J. and Scott. E. L.: Correction for bias introduced by a transformation of variables. Ann. Math. Statist. 31 (1960), 643–655 [see also Int. Statist. Rev. 41 (1973), 203–223]

Ott, E. R.: Analysis of means – a graphical procedure. Industrial Quality Control 24 (August 1967), 101–109 [see J. Qual. Technol. 5 (1973), 47–57, 93–108, 147–159; 6 (1974), 2–21, 175–181; 13 (1981), 115–119; J. Amer. Statist. Assoc. 73 (1978), 724–729 and American Statistician 34 (1980), 195–199]

Pachares, J.: Table of the upper 10% points of the Studentized range. Biometrika 46 (1959), 461–466

Page, E. B.: Ordered hypotheses for multiple treatments: A significance test for linear ranks. J. Amer. Statist. Assoc. **58** (1963), 216–230 [see also **67** (1972), 850–854 and Psychological Review **71** (1964), 505–513 as well as Psychometrika **38** (1973), 249–258]

Patnaik, B.: The use of mean range as an estimator of variance in statistical tests. Biometrika **37** (1950), 78–87

Pearson, E. S.: The probability integral of the range in samples of n observations from a normal population. Biometrika **32** (1941/42), 301–308

—, and Stephens, M. A.: The ratio of range to standard deviation in the same normal sample. Biometrika **51** (1964), 484–487

Peng, K. C.: The Design and Analysis of Scientific Experiments. (Addison-Wesley, pp. 252) Reading, Mass. 1967, Chapter 10

Plackett, R. L.: Models in the analysis of variance. J. Roy. Statist. Soc. B **22** (1960), 195–217

Plackett, R. L., and Burman, J. P.: The design of optimum multifactorial experiments. Biometrika **33** (1946), 305–325 [see also J. Amer. Statist. Assoc. **68** (1973), 353–360]

Quade, D.: Rank analysis of covariance. J. Amer. Statist. Assoc. **62** (1967), 1187–1200

Rao, P. V., and Kupper, L. L.: Ties in paired-comparison experiments: a generalization of the Bradley-Terry model. J. Amer. Statist. Assoc. **62** (1967), 194–204 (see **63** (1968) 1550)

Rasch, D.: (1) Probleme der Varianzanalyse bei ungleicher Klassenbesetzung. Biometrische Zeitschr. **2** (1960), 194–203. (2) Gemischte Klassifikationen der dreifachen Varianzanalyse. Biometrische Zeitschr. **13** (1971), 1–20 [see **16** (1974), 2–14 and J. Qual. Technol. **6** (1974), 98–106, 187]

Reinach, S. G.: A nonparametric analysis for a multiway classification with one element per cell. South Africa J. Agric. Sci. (Pretoria) **8** (1965), 941–960 [see also S. Afr. Statist. J. **2** (1968), 9–32]

Reisch, J. S., and Webster, J. T.: The power of a test in covariance analysis. Biometrics **25** (1969), 701–714 [see also Biometrika **61** (1974), 479–484; more on covariance analysis may be found in the special issue of Communications in Statistics—Theory and Methods **A8** (1979), 719–854]

Rives, M.: Sur l'analyse de la variance. I. Emploi de transformations. Ann. Inst. nat. Rech. agronom., Ser. B **3** (1960), 309–331 [see also Technometrics **4** (1962), 531–550, **5** (1963), 317–325 and **11** (1969), 23–40]

Rutherford, A. A., and Stewart, D. A.: The use of subsidiary information in the improvement of the precision of experimental estimation. Record of Agricultural Research **16**, Part 1 (1967), 19–24

Ryan, T. A.: (1) Multiple comparisons in psychological research. Psychol. Bull. **56** (1959), 26–47. (2) Comments on orthogonal components. Psychol. Bull. **56** (1959), 394–396

Sachs, L.: Statistische Methoden. 6th revised edition. (Springer, 133 pages), Berlin, Heidelberg, New York 1984, pp. 95–98

Scheffé, H.: (1) An analysis of variance for paired comparisons. J. Amer. Statist. Assoc. **47** (1952), 381–400. (2) A method for judging all contrasts in the analysis of variance. Biometrika **40** (1953), 87–104, Corrections **56** (1969), 299. (3) The Analysis of Variance. (Wiley, pp. 477), New York 1959

Searle, S. R.: (1) Linear Models. (Wiley, pp. 532) London 1971. (2) Topics in variance component estimation. Biometrics **27** (1971), 1–76 [see also **30** (1974), 157–169]

Searle, S. R., and Henderson, C. R.: Computing procedures for estimating components of variance in the two-way classification, mixed model. Biometrics **17** (1961), 607–616

Siotani, M.: Internal estimation for linear combinations of means. J. Amer. Statist. Assoc. **59** (1964), 1141–1164 [see also **60** (1965), 573–583]

Snell, E. J.: A scaling procedure for ordered categorical data. Biometrics **20** (1964), 592–607

Spjøtvoll, E.: A mixed model in the analysis of variance. Optimal properties. Skand. Aktuarietidskr. **49** (1966), 1–38 [see also Technometrics **15** (1973), 819–831; **18** (1976), 31–38 and Metrika **30** (1983), 85–91]

Sprott, D. A.: Note on Evans and Anastasio on the analysis of covariance. Psychol. Bull. **73** (1970), 303–306 [see also **69** (1968), 225–234 and **79** (1973), 180]

Starks, T. H., and David, H. A.: Significance tests for paired-comparison experiments. Biometrika **48** (1961), 95–108, Corrigenda 475

Student: Errors of routine analysis. Biometrika **19** (1927), 151–164

Taylor, L. R.: Aggregation, variance and the mean. Nature **189** (1961), 723–735 [see also Biometrika **49** (1962), 557–559]

Teichroew, D.: Tables of expected values of order statistics and products of order statistics for samples of size twenty and less from the normal distribution. Ann. Math. Statist. **27** (1956), 410–426

Terry, M. E., Bradley, R. A., and Davis, L. L.: New designs and techniques for organoleptic testing. Food Technology **6** (1952), 250–254 [see also Biometrics **20** (1964), 608–625]

Tietjen, G. L., and Beckman, R. J.: Tables for use of the maximum F-ratio in multiple comparison procedures. J. Amer. Statist. Assoc. **67** (1972), 581–583 [see also Biometrika **60** (1973), 213–214]

Tietjen, G. L., and Moore, R. H.: On testing significance of components of variance in the unbalanced nested analysis of variance. Biometrics **24** (1968), 423–429

Trawinski, B. J.: An exact probability distribution over sample spaces of paired comparisons. Biometrics **21** (1965), 986–1000

Tukey, J. W.: (1) Comparing individual means in the analysis of variance. Biometrics **5** (1949), 99–114. (2) One degree of freedom for non-additivity. Biometrics **5** (1949), 232–242 [cf. **10** (1954), 562–568], (see also Ghosh and Sharma 1963). (3) Some selected quick and easy methods of statistical analysis. Trans. N.Y. Acad. Sciences (II) **16** (1953), 88–97. (4) Answer to query 113. Biometrics **11** (1955), 111–113. (5) On the comparative anatomy of transformations. Ann. Math. Statist. **28** (1957), 602–632. (6) The future of data analysis. Ann. Math. Statist. **33** (1962), 1–67

Vessereau, A.: Les méthodes statistiques appliquées au test des caractères organoleptiques. Revue de Statistique Appliquée **13** (1965, No. 3), 7–38

Victor, N.: Beschreibung und Benutzeranleitung der interaktiven Statistikprogramme ISTAP. Gesellschaft für Strahlen- und Umweltforschung mbH; 52 pages) München 1972, page 39

Wang, Y. Y.: A comparison of several variance component estimators. Biometrika **54** (1967), 301–305

Watts, D. G. (Ed.): The Future of Statistics. (Proc. Conf. Madison, Wisc., June 1967; Academic Press, pp. 315), New York 1968

Weiling, F.: (1) Weitere Hinweise zur Prüfung der Additivität bei Streuungszerlegungen (Varianzanalysen). Der Züchter **33** (1963), 74–77. (2) Möglichkeiten der Analyse von Haupt- und Wechselwirkungen bei Varianzanalysen mit Hilfe programmgesteuerter Rechner. Biometrische Zeitschr. **14** (1972), 398–408 [see also EDV in Med. u. Biol. **4** (1973), 88–98 and **10** (1979), 31]

Wilcoxon, F., and Wilcox, Roberta, A.: Some Rapid Approximate Statistical Procedures. Lederle Laboratories, Pearl River, New York 1964

Wilk, M. B., and Kempthorne, O.: Fixed, mixed, and random models. J. Amer. Statist. Assoc. **50** (1955), 1144–1167 [see also J. Qual. Technol. **6** (1974), 98–106, 187]

Williams, J. D.: (Letter) Quick calculations of critical differences for Scheffé's test for unequal sample sizes. The American Statistician **24** (April 1970), 38–39

Winer, B. J.: Statistical Principles in Experimental Design. (McGraw-Hill, pp. 672), New York 1962, pp. 140–455 [2nd ed., pp. 907, N.Y. 1971]

[8:7a] Multiple comparisons

Bancroft, T. A.: Topics in Intermediate Statistical Methods. Vol. I. (Iowa State University Press; pp. 129) Ames, Iowa 1968, Chapter 8

Bechhofer, R. E.: Multiple comparisons with a control for multiply-classified variances of normal populations. Technometrics **10** (1968), 715–718 (as well as 693–714)

Box, G. E. P., Hunter, W. G., and Hunter, J. S.: Statistics for Experimenters. (Wiley; pp. 653) New York 1978

Brown, M. B.: Exploring interaction effects in the ANOVA. Applied Statistics **24** (1975), 288–298

Crouse, C. F.: A multiple comparison of rank procedure for a one-way analysis of variance. S. Afr. Statist. J. **3** (1969), 35–48

Dudewicz, E. J., and Ramberg, J. S.: Multiple comparisons with a control; unknown variances. Annual Techn. Conf. Transact. Am. Soc. Qual. Contr. **26** (1972), 483–488 [see also Biometr. Z. **17** (1975), 13–26]

Duncan, D. B.: A Bayesian approach to multiple comparisons. Technometrics **7** (1965), 171–222 [see also Biometrics **11** (1955), 1–42 and **31** (1975), 339–359]

Dunn, Olive J.: (1) Confidence intervals for the means of dependent, normally distributed variables. J. Amer. Statist. Assoc. **54** (1959), 613–621. (2) Multiple comparisons among means. J. Amer. Statist. Assoc. **56** (1961), 52–64. (3) Multiple comparisons using rank sums. Technometrics **6** (1964), 241–252 [see also Communications in Statistics **3** (1974), 101–103]

—, and Massey, F. J., Jr.: Estimating of multiple contrasts using t-distributions. J. Amer. Statist. Assoc. **60** (1965), 573–583

Dunnett, C. W.: (1) A multiple comparison procedure for comparing several treatments with a control. J. Amer. Statist. Assoc. **50** (1955), 1096–1121 [see also Dudewicz and Ramberg (1972)]. (2) New tables for multiple comparisons with a control. Biometrics **20** (1964), 482–491. (3) Multiple comparison tests. Biometrics **26** (1970), 139–141 [see also **33** (1977), 293–303]

Enderlein, G.: Die Maximum-Modulus-Methode zum multiplen Vergleich von Gruppenmitteln mit dem Gesamtmittel. Biometrische Zeitschr. **14** (1972), 85–94

Gabriel, K. R.: (1) A procedure for testing the homogeneity of all sets of means in analysis of variance. Biometrics **20** (1964), 458–477. (2) Simultaneous test procedures for multiple comparisons on categorical data. J. Amer. Statist. Assoc. **61** (1966), 1081–1096

Games, P. A.: Inverse relation between the risks of type I and type II errors and suggestions for the unequal n case in multiple comparisons. Psychol. Bull. **75** (1971), 97–102 [see also **71** (1969), 43–54, **81** (1974), 130–131 and Amer. Educat. Res. J. **8** (1971), 531–565]

Hahn, G. J. and Hendrickson, R. W.: A table of percentage points of the distribution of the largest absolute value of k Student t variates and its applications. Biometrika **58** (1971), 323–332

Hegemann, V., and Johnson, D. E.: On analyzing two-way AoV data with interaction. Technometrics **18** (1976), 273–281

Hollander, M.: An asymptotically distribution-free multiple comparison procedure treatments vs. control. Ann. Math. Statist. **37** (1966), 735–738

Keuls, M.: The use of the Studentized range in connection with an analysis of variance. Euphytica **1** (1952), 112–122

Kramer, C. Y: Extension of multiple range tests to group correlated adjusted means. Biometrics **13** (1957), 13–18 [see also Psychometrika **36** (1971), 31–34]

Kurtz, T. E., Link, R. F., Tukey, J. W., and Wallace, D. L.: Short-cut multiple comparisons for balanced single and double classifications: Part 1, Results. Technometrics **7** (1965), 95–161

—, Link, R. F., Tukey, J. W., and Wallace, D. L.: Short-cut multiple comparisons for balanced single and double classifications: Part 2. Derivations and approximations. Biometrika **52** (1965), 485–498

Lehmacher, W., and Wall, K. D.: A new nonparametric approach to the comparison of k independent samples of response curves. Biometrical Journal **20** (1978), 261–273

Link, R. F.: On the ratio of two ranges. Ann. Math. Statist. **21** (1950), 112–116

—, and Wallace, D. L.: Some Short Cuts to Allowances. Princeton University, March 1952

Marascuilo, L. A.: Large-sample multiple comparisons. Psychol. Bull. **65** (1966), 280–290 [see also J. Cohen. **67** (1967), 199–201]

McDonald, B. J., and Thompson, W. A., Jr.: Rank sum multiple comparisons in one- and two-way classifications. Biometrika **54** (1967), 487–497; Correction **59** (1972), 699

Miller, R. G.: Simultaneous Statistical Inference. (McGraw-Hill, pp. 272), New York 1966 (Chapter 2, pp. 37–109) [2nd ed. (Springer; pp. 299) New York 1981]

Morrison, D. F.: Multivariate Statistical Methods. (McGraw-Hill, pp. 338), New York 1967

Nelson, L. S.: Ordered tests for a three-level factor. Journal of Quality Technology **8** (1976), 241–243

Nemenyi, P.: Distribution-Free Multiple Comparisons. New York, State University of New York, Downstate Medical Center 1963

O'Neill, R., and Wetherill, G. B.: The present state of multiple comparison methods. With discussion. J. Roy. Statist. Soc. **B33** (1971), 218–250 [see also The Statistician **22** (1973), 16–42, 236]

Perlmutter, J., and Myers, J. L.: A comparison of two procedures for testing multiple contrasts. Psychol. Bull. **79** (1973), 181–184 [see also **81** (1974), 130–131]

Petrinovich, L. F., and Hardyck, C. D.: Error rates for multiple comparison methods. Some evidence concerning the frequency of erroneous conclusions. Psychol. Bull. **71** (1969), 43–54 [see also **75** (1971), 97–102 and **80** (1973), 31–32, 480, **81** (1974), 130–131 as well as H. J. Keselman et al., J. Amer. Statist. Assoc. **70** (1975), 584–587]

Rhyne, A. L., and Steel, R. G. D.: A multiple comparisons sign test: all pairs of treatments. Biometrics **23** (1967), 539–549

Rhyne, A. L., Jr., and Steel, R. G. D.: Tables for a treatment versus control multiple comparisons sign test. Technometrics **7** (1965), 293–306

Ryan, T. A.: Significance tests for multiple comparison of proportions, variances and other statistics. Psychol. Bull. **57** (1960), 318–328

Scheffé, H.: A method for judging all contrasts in the analysis of variance. Biometrika **40** (1953), 87–104, Corrections **56** (1969), 229

Seeger, P.: Variance Analysis of Complete Designs. Some Practical Aspects. (Almqvist and Wiksell, pp. 225) Uppsala 1966, pp. 111–160

Siotani, M.: Interval estimation for linear combinations of means. J. Amer. Statist. Assoc. **59** (1964), 1141–1164

Slivka, J.: A one-sided nonparametric multiple comparison control percentile test: treatment versus control. Biometrika **57** (1970), 431–438

Steel, R. G. D.: (1) A multiple comparison rank sum test: treatment versus control. Biometrics **15** (1959), 560–572. (2) A rank sum test for comparing all pairs of treatments. Technometrics **2** (1960), 197–208. (3) Answer to Query: Error rates in multiple comparisons. Biometrics **17** (1961), 326–328. (4) Some rank sum multiple comparisons tests. Biometrics **17** (1961), 539–552

Thöni, H.: A nomogram for testing multiple comparisons. Biometrische Z. **10** (1968), 219–221

Tobach, E. Smith, M., Rose, G., and Richter, D.: A table for making rank sum multiple paired comparisons. Technometrics **9** (1967), 561–567

[8:7b] Experimental design

Bätz, G. et al.: Biometrische Versuchsplanung. (VEB Dtsch. Landwirtschaftsvlg., 355 pages) Berlin 1972

Bancroft, T. A.: Topics in Intermediate Statistical Methods (Iowa State Univ. Press) Ames, Iowa 1968

Bandemer, H., Bellmann, A., Jung, W., and Richter, K.: Optimale Versuchsplanung. (WTB Vol. 131) (Akademie-Vlg., 180 pages) Berlin 1973

Bose, R. C.: The design of experiments. Proc. Indian Sci. Congr. **34**(II) (1947), 1–25 [see also Ann. Eugenics **9** (1939), 353–399, Sankhya **4** (1939), 337–372, J. Amer. Statist. Assoc. **75** (1952), 151–181 and K. D. Tocher: J. Roy. Statist. Soc. B **14** (1952), 45–100]

Brownlee, K. A.: Statistical Theory and Methodology in Science and Engineering. (Wiley, pp. 570), New York 1960

Chew, V. (Ed.): Experimental Designs in Industry. (Wiley) New York 1958

Cochran, W. G., and Cox, G. M.: Experimental Designs. 2nd ed. (Wiley) New York 1962

Das, M. N., and Giri, N. C.: Design and Analysis of Experiments. (Wiley; pp. 308) New York 1980

Davies, O. L.: The Design and Analysis of Industrial Experiments. 2nd ed. (Oliver and Boyd; pp. 636) Edinburgh 1971

Dugue, D., and Girault, M.: Analyse de Variance et Plans d'Expérience. Paris 1959

Edwards, A. L.: Versuchsplanung in der Psychologischen Forschung. (Belts, 501 pages) Weinheim, Berlin, Basel 1971

Federer, W. T.: Experimental Design. (Macmillan) New York 1963

—, and Balaam, L. N.: Bibliography on Experiment and Treatment Design Pre 1968. (Oliver and Boyd; pp. 765) Edinburgh and London 1973 [see also Int. Statist. Rev. **48** (1980), 357–368; **49** (1981), 95–109, 185–197]

Fisher, R. A.: The Design of Experiments. (Oliver and Boyd) Edinburgh 1935 (7th ed. 1960)

Gill, J. L.: Design and Analysis of Experiments in the Animal and Medical Sciences. Vol. 1–3 (Iowa State Univ. Press) Ames, Iowa 1979

Hahn, G. J.: Some things engineers should know about experimental design. Journal of Quality Technology **9** (1977), 13–20

Hall, M., Jr.: Combinatorial Theory. (Blaisdell, pp. 310) Waltham, Mass. 1967

Hedayat, A.: Book Review. Books on experimental design. (Lists 43 books). Biometrics **26** (1970), 590–593

Herzberg, Agnes M., and Cox, D. R.: Recent work on the design of experiments: a bibliography and a review. J. Roy. Statist. Soc. **132 A** (1969), 29–67

Hicks, C. R.: Fundamental Concepts in the Design of Experiments. 2nd ed. (Holt, Rinehart and Winston; pp. 368) New York 1973

John, P. W. M.: Statistical Design and Analysis of Experiments. (Macmillan; pp. 356) New York 1971

Johnson, N. L., and Leone, F. C.: Statistics and Experimental Design in Engineering and the Physical Sciences. Vol. II, (Wiley) New York 1964 (2nd edition 1977)

Kempthorne, O.: The Design and Analysis of Experiments. 2nd ed. (Wiley) New York 1960

Kendall, M. G., and Stuart, A.: The Advanced Theory of Statistics. Vol. 3, Design and Analysis, and Time Series. 2nd ed. (Griffin; pp. 557) London 1968, Chapters 35–38 (3rd edition 1976)

Kirk, R. E.: Experimental Design. Procedures for the Behavioral Sciences. 2nd ed. (Wadsworth; pp. 840), Belmont, Calif. 1982

Kish, L., in Blalock, N. M., et al. (Eds.): Quantitative Sociology, Academic Press, New York 1975, Chapter 9, pp. 261–284

Li, C. C.: Introduction to Experimental Statistics. (McGraw-Hill; pp. 460) New York 1964

Li, J. C. R.: Statistical Inference. Vol. I, II. (Edwards Brothers; pp. 658, 575) Ann Arbor, Mich. 1964

Linder, A.: Planen und Auswerten von Versuchen. 3rd enlarged edition. (Birkhäuser, 344 pages) Basel and Stuttgart 1969

Mendenhall, W.: Introduction to Linear Models and the Design and Analysis of Experiments (Wadsworth Publ. Comp., pp. 465), Belmont, Calif. 1968

Montgomery, D. C.: Design and Analysis of Experiments. (Wiley, pp. 418) New York 1976

Myers, J. L.: Fundamentals of Experimental Design. 2nd ed. (Allyn and Bacon, pp. 407) Boston 1972

Peng, K. C.: The Design and Analysis of Scientific Experiments. (Addison-Wesley; pp. 252) Reading, Mass. 1967

Quenouille, M. H., and John, J. A.: Experiments: Design and Analysis. 2nd ed. (Macmillan; pp. 264) New York 1977

Rasch, D., Herrendörfer, G., Bock, J., and Busch, K.: Verfahrensbibliothek. Versuchsplanung und Auswertung. 2 volumes. (VEB Deutscher Landwirtschaftsverlag; 1052 pages) Berlin 1978

Scheffé, H.: The Analysis of Variance. (Wiley; pp. 477) New York 1959

Winer, B. J.: Statistical Principles in Experimental Design. 2nd ed. (McGraw-Hill, pp. 907) New York 1971

Yates, F.: Experimental Design. Selected Papers. (Hafner, pp. 296) Darien, Conn. (USA) 1970

EXERCISES

CHAPTER 1

Probability calculus

1. Two dice are tossed. What is the probability that the sum of the dots on the faces is 7 or 11?
2. Three guns are fired simultaneously. The probabilities of hitting the mark are 0.1, 0.2, and 0.3 respectively. What is the total probability that a hit is made?
3. The sex ratio among newborn (male:female) drawn from observations taken over many years is 514:486. The relative frequency of individuals with blond hair is known to be 0.15. Both attributes, sex and hair color, are stochastically independent. What is the relative frequency of blond males?
4. What is the probability of obtaining at least one 6 in four tosses of a die?
5. How many tosses are needed for the probability of getting at least one 6 to be 50%?
6. What is the probability of getting (a), 5, (b) 6, (c) 7, (d) 10 heads in 5, 6, 7, 10 tosses of a coin?

Mean and standard deviation

7. Compute the mean and standard deviation of the frequency distribution

x	5	6	7	8	9	10	11	12	13	14	15	16
n	10	9	94	318	253	153	92	40	26	4	0	1

Chapter 1

8. Compute the mean and standard deviation of the following 45 values:

40, 43, 43, 46, 46, 46, 54, 56, 59,
62, 64, 64, 66, 66, 67, 67, 68, 68,
69, 69, 69, 71, 75, 75, 76, 76, 78,
80, 82, 82, 82, 82, 82, 83, 84, 86,
88, 90, 90, 91, 91, 92, 95, 102, 127.

(a) directly, (b) by using the class limits 40 but less than 45, 45 but less than 50, etc., (c) by using the class limits 40 but less than 50, 50 but less than 60, etc.

9. Compute the median, the mean, the standard deviation, the skewness II, and the coefficient of excess of the sampling distribution:

62, 49, 63, 80, 48, 67, 53, 70, 57, 55, 39, 60, 65, 56, 61, 37,
63, 58, 37, 74, 53, 27, 94, 61, 46, 63, 62, 58, 75, 69, 47, 71,
38, 61, 74, 62, 58, 64, 76, 56, 67, 45, 41, 38, 35, 40.

10. Sketch the frequency distribution and compute the mean, median, mode, first and third quartile, first and ninth decile, standard deviation, skewness I–III, and coefficient of excess.

Class limits	Frequencies
72.0 – 73.9	7
74.0 – 75.9	31
76.0 – 77.9	42
78.0 – 79.9	54
80.0 – 81.9	33
82.0 – 83.9	24
84.0 – 85.9	22
86.0 – 87.9	8
88.0 – 89.9	4
Total	225

F-distribution

11. Given $F = 3.84$ with $v_1 = 4$ and $v_2 = 8$ degrees of freedom. Find the level of significance corresponding to the F-value.

Binomial coefficients

12. Suppose 8 insecticides are to be tested in pairs as to their effect on mosquitoes. How many tests must be run?

13. Of those afflicted with a certain disease, 10% die on the average. What is the probability that out of 5 patients stricken with the disease (a) all get well, (b) exactly 3 fail to survive, (c) at least 3 fail to survive?
14. What is the probability that 5 cards drawn from a well-shuffled deck (52 cards) all turn out to be diamonds?
15. A die is tossed 12 times. What is the probability that the 4 shows up exactly twice?
16. Of the students registered in a certain department, 13 are female and 18 are male. How many possible ways are there of forming a committee consisting of 2 female and 3 male students?

Binomial distribution

17. What is the probability of getting heads five times in 10 flips of a coin?
18. The probability that a thirty year old person will live another year is 99% according to life tables ($p = 0.99$). What is the probability that out of 10 thirty year olds, 9 survive for another year?
19. What is the probability that among 100 tosses of a die the 6 comes up exactly 25 times?
20. Twenty days are singled out at random. What is the probability that 5 of them fall on a certain day of the week—say a Sunday?
21. Suppose that on the average 33% of the ships involved in battle are sunk. What is the probability that out of 6 ships (a) exactly 4, (b) at least 4 manage to return?
22. One hundred fair coins are flipped. What is the probability that exactly 50 come up heads? Use Stirling's formula.
23. An urn contains 2 white and 3 black balls. What is the probability that in 50 consecutive drawings with replacement a white ball is drawn exactly 20 times? Use Stirling's formula.

Poisson distribution

24. A hungry frog devours 3 flies per hour on the average. What is the probability that an hour passes without it devouring any flies?
25. Suppose the probability of hitting the target is $p = 0.002$ for each shot. What is the probability of making exactly 5 hits when all together $n = 1000$ shots are fired?
26. Assume the probability of a manufacturer producing a defective article is $p = 0.005$. The articles are packed in crates of 200 units each. What is the probability that a crate contains exactly 4 defective articles?

27. In a warehouse a certain article is seldom asked for, on the average only 5 times a week, let us say. What is the probability that in a given week the article is requested k times?
28. Suppose 5% of all schoolchildren wear glasses. What is the probability that in a class of 30 children (a) no, (b) one, (c) two, (d) three children wear glasses?

CHAPTER 2

Formulate and solve a few problems on the basis of Figures 33 through 37.

CHAPTER 3

1. By means of a random process 16 sample elements with $\bar{x} = 41.5$ and $s = 2.795$ are drawn from a normally distributed population. Are there grounds for rejecting the hypothesis that the population mean is 43 ($\alpha = 0.05$)?
2. Test the equality of the variances of the two samples, A and B, at the 5% level using the F-test:

    A: 2.33 4.64 3.59 3.45 3.64 3.00 3.41 2.03 2.80 3.04
    B: 2.08 1.72 0.71 1.65 2.56 3.27 1.21 1.58 2.13 2.92

3. Test at the 5% level the equality of the central tendency (H_0) of the two independent samples, A and B, using (a) the Tukey rapid test, (b) the U-test:

    A: 2.33 4.64 3.59 3.45 3.64 3.00 3.41 2.03 2.80 3.04
    B: 2.08 1.72 0.71 1.65 2.56 3.27 1.21 1.58 2.13 2.92

CHAPTER 4

1. Two sleep-inducing preparations, A and B, were tested on each of 10 persons suffering from insomnia (Student 1908, Biometrika 6, p. 20). The resulting additional sleep, in hours, was as follows:

Patient	1	2	3	4	5	6	7	8	9	10
A	1.9	0.8	1.1	0.1	−0.1	4.4	5.5	1.6	4.6	3.4
B	0.7	−1.6	−0.2	−1.2	−0.1	3.4	3.7	0.8	0.0	2.0
Diff.	1.2	2.4	1.3	1.3	0.0	1.0	1.8	0.8	4.6	1.4

Can A and B be distinguished at the 1% level? Formulate the null hypothesis and apply (a) the t-test for paired observations and (b) the maximum test.

2. Test the equality of the central tendencies (H_0) of two dependent samples, A and B, at the 5%-level by means of the following tests for paired observations: (a) t-test, (b) Wilcoxon test, (c) maximum test.

No.	1	2	3	4	5	6	7	8	9
A	34	48	33	37	4	36	35	43	33
B	47	57	28	37	18	48	38	36	42

3. Gregor Mendel, as the result of an experiment involving peas, ended up with 315 round yellow peas, 108 round green ones, 101 with edges and yellow, and 32 with edges and green. Do these values agree with the theory according to which the four frequencies are related as $9:3:3:1$ ($\alpha = 0.05; S = 95\%$)?

4. Does the following frequency distribution represent a random sample which could have originated in a Poisson distributed population with parameter $\lambda = 10.44$? Test the fit at the 5% level by means of the χ^2-test.
Number of events, E: 0 1 2 3 4   5   6   7   8
Observed frequency, O: 0 5 14 24 57 111 197 278 378
E: 9   10   11   12   13   14   15   16 17 18 19 20 21 22
O: 418 461 433 413 358 219 145 109 57 43 16 7 8 3

5. The frequencies of a fourfold table are: $a = 140, b = 60, c = 85, d = 90$. Apply the test for independence at the 0.1% level.

6. The frequencies of a fourfold table are: $a = 605, b = 135, c = 195, d = 65$. Apply the test for independence at the 5% level.

7. The frequencies of a fourfold table are: $a = 620, b = 380, c = 550, d = 450$. Apply the test for independence at the 1% level.

CHAPTER 5

1. Test the significance of $r = 0.5$ at the 5% level ($n = 16$).

2. How large at least must r be to be statistically significant at the 5% level for $n = 16$?

3. Estimate the regression lines and the correlation coefficient for the following pairs of values:

x: | 22 24 26 26 27 27 28 28 29 30 30 30 31 32 33 34 35 35 36 37

y: | 10 20 20 24 22 24 27 24 21 25 29 32 27 27 30 27 30 31 30 32

Should we reject the H_0 hypothesis that $\rho = 0$ at the 0.1% level?

Chapter 5

4. Given the following two dimensional frequency distribution:

y \ x	42	47	52	57	62	67	72	77	82	Total
52	3	9	19	4						35
57	9	26	37	25	6					103
62	10	38	74	45	19	6				192
67	4	20	59	96	54	23	7			263
72		4	30	54	74	43	9			214
77			7	18	31	50	19	5		130
82				2	5	13	15	8	3	46
87						2	5	8	2	17
Total	26	97	226	244	189	137	55	21	5	1000

Estimate the correlation coefficient, the standard deviations s_x, s_y, the sample covariance s_{xy}, the regression line of y on x, and the correlation ratio. Test the correlation and the linearity of the regression ($\alpha = 0.05$).

5. A correlation based on 19 paired observations has the value 0.65. (a) Can this sample originate in a population with parameter $\rho = 0.35$ ($\alpha = 0.05$)? (b) Estimate the 95% confidence interval for ρ on the basis of the sample. (c) If a second sample, also consisting of 19 paired observations, has a correlation coefficient $r = 0.30$, could both samples have originated in a common population ($\alpha = 0.05$)?

6. Fit a function of the type $y = ab^x$ to the following values:

x	0	1	2	3	4	5	6
y	125	209	340	561	924	1525	2512

7. Fit a function of the type $y = ab^x$ to the following values:

x	273	283	288	293	313	333	353	373
y	29.4	33.3	35.2	37.2	45.8	55.2	65.6	77.3

8. Fit a function of the type $y = ax^b$ to the following values:

x	19	58	114	140	181	229
y	3	7	13.2	17.9	24.5	33

9. Fit a second degree parabola to the following values:

x	7.5	10.0	12.5	15.0	17.5	20.0	22.5
y	1.9	4.5	10.1	17.6	27.8	40.8	56.9

10. Fit a second degree parabola to the following values:

x	1.0	1.5	2.0	2.5	3.0	3.5	4.0
y	1.1	1.3	1.6	2.1	2.7	3.4	4.1

CHAPTER 6

1. Test the 2 × 6 table

13	10	10	5	7	0
2	4	9	8	14	7

for homogeneity ($\alpha = 0.01$).

2. Test the independence and the symmetry of the 3 × 3 contingency table

102	41	57
126	38	36
161	28	11

at the 1% level.

3. Test whether both sampling distributions I and II could have originated in the same population ($\alpha = 0.05$). Use (a) the formula (6.1) to test the homogeneity of the two samples, and (b) the information statistic $2I$ to test the homogeneity of a two way table consisting of $k \times 2$ cells.

Category	Frequencies		Total
	I	II	
1	160	150	310
2	137	142	279
3	106	125	231
4	74	89	163
5	35	39	74
6	29	30	59
7	28	35	63
8	29	41	70
9	19	22	41
10	6	11	17
11	8	11	19
12	13	4	17
Total	644	699	1343

4. Test the homogeneity of this table at the 5% level:

23	5	12
20	13	10
22	20	17
26	26	29

CHAPTER 7

1. Test the homogeneity of the following three variances at the 5% level:

   $s_A^2 = 76.84$ ($n_A = 45$), $s_B^2 = 58.57$ ($n_B = 82$), $s_C^2 = 79.64$ ($n_C = 14$).

2. Test the independent samples A, B, C for equality of the means ($\alpha = 0.05$) (a) by analysis of variance, (b) by means of the H-test:

   A: 40,   34,   84,   46,   47,   60
   B: 59,   92,   117,   86,   60,   67,   95,   40,   98,   108
   C: 92,   93,   40,   100,   92

3. Given

A \ B	B_1	B_2	B_3	B_4	B_5	B_6	Σ
A_1	9.5	11.5	11.0	12.0	9.3	11.5	64.8
A_2	9.6	12.0	11.1	10.8	9.7	11.4	64.6
A_3	12.4	12.5	11.4	13.2	10.4	13.1	73.0
A_4	11.5	14.0	12.3	14.0	9.5	14.0	75.3
A_5	13.7	14.2	14.3	14.6	12.0	13.2	82.0
Σ	56.7	64.2	60.1	64.6	50.9	63.2	359.7

   Test possible column and row effects at the 1% level.

4. Three methods of determination are compared on 10 samples. Test by means of the Friedman test (a) the equivalence of the methods ($\alpha = 0.001$), (b) the equivalence of the samples ($\alpha = 0.05$).

Sample	Method of determination		
	A	B	C
1	15	18	9
2	22	25	20
3	44	43	25
4	75	80	58
5	34	33	31
6	15	16	11
7	66	64	45
8	56	57	40
9	39	40	27
10	30	34	21

SOLUTIONS TO THE EXERCISES

CHAPTER 1

Probability calculus

1. There are six different ways to get the sum 7, and only two ways to get the sum 11, so that

$$P = \frac{6}{36} + \frac{2}{36} = \frac{2}{9} = 0.222.$$

2. The total probability of hitting the target is not quite 50%:

$$P(A+B+C) = P(A)+P(B)+P(C)-P(AB)-P(AC)-P(BC)+P(ABC)$$
$$P(A+B+C) = 0.1+0.2+0.3-0.02-0.03-0.06+0.006 = 0.496.$$

3. $P = 0.514 \cdot 0.15 = 0.0771$: blond males can be expected in about 8% of all births.

4. $1 - (5/6)^4 = 0.5177$: in a long sequence of tosses one can count on getting this event in about 52% of all cases.

5.
$$P = \left(\frac{5}{6}\right)^n = \frac{1}{2}; \quad n = \frac{\log 2}{\log 6 - \log 5} = \frac{0.3010}{0.7782 - 0.6990} \simeq 4.$$

6. The probabilities are (a) $(\frac{1}{2})^5$, (b) $(\frac{1}{2})^6$, (c) $(\frac{1}{2})^7$, (d) $(\frac{1}{2})^{10}$, or approximately 0.031, 0.016, 0.008, 0.001.

Chapter 1

Mean and standard deviation

7. $\bar{x} = 9.015$, $s = 1.543$.

8. For a: $\bar{x} = 73.2$, $s = 17.3$.
   For b: $\bar{x} = 73.2$, $s = 17.5$.
   For c: $\bar{x} = 73.2$, $s = 18.0$.

   With increasing class size the standard deviation also gets larger (cf., Sheppard's correction).

9. Statistics   Rough estimates

   $\tilde{x} = 59.5$ $\bar{x} \simeq 56.3$
   $\bar{x} = 57.3$ $s \simeq 14.1$
   $s = 13.6$

   Skewness II $= -0.214$,

   Coefficient of excess $= 0.250$.

10.
   $\bar{x} = 79.608$,
   $s = 3.675$,
   $\tilde{x} = 79.15$,
   $Q_1 = 76.82$,   first decile $= 74.95$,
   $Q_3 = 82.10$,   ninth decile $= 84.99$,
   Mode $= 78.68$.

   Skewness I $= -2.07$,
   Skewness II $= 0.163$,
   Skewness III $= 0.117$.

   Coefficient of excess $= 0.263$.

11.
$$\hat{z} = \frac{\left(1 - \dfrac{2}{9 \cdot 8}\right)3.84^{1/3} - \left(1 - \dfrac{2}{9 \cdot 4}\right)}{\sqrt{\dfrac{2}{9 \cdot 8} \cdot 3.84^{2/3} + \dfrac{2}{9 \cdot 4}}} = 1.644, \text{ i.e., } P_{\hat{z}} = 0.05.$$

   For $v_1 = 4$ and $v_2 = 8$ the exact 5% bound is 3.8378.

Binomial coefficients

12. $P = {}_8C_2 = \dfrac{8!}{6! \cdot 2!} = \dfrac{8 \cdot 7}{2} = 28$.

13. For (a): $P = 0.90^5 = 0.59049$.
   For (b): $_5C_3 = 5!/(3! \cdot 2!) = 5 \cdot 4/2 \cdot 1 = 10$; thus
   $$P = 10 \cdot 0.90^2 \cdot 0.10^3 = 0.00810.$$
   For (c): $_5C_3 = 10$, $_5C_4 = 5$; thus
   $$P = 10 \cdot 0.90^2 \cdot 0.10^3 + 5 \cdot 0.90 \cdot 0.10^4 + 0.10^5,$$
   $$P = 0.00810 + 0.00045 + 0.00001 = 0.00856.$$

14. $P = \dfrac{_{13}C_5}{_{52}C_5} = \dfrac{13! \cdot 47! \cdot 5!}{8! \cdot 5! \cdot 52!} = \dfrac{13 \cdot 12 \cdot 11 \cdot 10 \cdot 9}{52 \cdot 51 \cdot 50 \cdot 49 \cdot 48}$,
   $$P = \dfrac{11 \cdot 3}{17 \cdot 5 \cdot 49 \cdot 16} = \dfrac{33}{66,640} = 0.0004952,$$
   $P \simeq 0.0005$ or $1 : 2,000$.

15. There are $_{12}C_2 = 12!/(10! \cdot 2!) = 12 \cdot 11/(2 \cdot 1)$ ways of choosing two objects from a collection of twelve. The probability of tossing 2 fours and 10 nonfours equals $(1/6)^2(5/6)^{10} = 5^{10}/6^{12}$. The probability that four occurs exactly twice in 12 tosses is thus
   $$P = \dfrac{12 \cdot 11 \cdot 5^{10}}{2 \cdot 1 \cdot 6^{12}} = \dfrac{11 \cdot 5^{10}}{6^{11}} = 0.296.$$
   In a long series of tosses in aggregates of twelve with a fair die, one can count on the double occurrence of a four in about 30% of all cases.

16. The answer is the product of the numbers of possible ways of choosing representatives for each of the two groups, i.e.,
   $$P = {}_{13}C_2 \cdot {}_{18}C_3 = \dfrac{13!}{11! \cdot 2!} \cdot \dfrac{18!}{15! \cdot 3!} = \dfrac{13 \cdot 12}{2 \cdot 1} \cdot \dfrac{18 \cdot 17 \cdot 16}{3 \cdot 2 \cdot 1},$$
   $$P = 13 \cdot 18 \cdot 17 \cdot 16 = 63,648.$$

Binomial distribution

17. $P = {}_{10}C_5 \left(\dfrac{1}{2}\right)^5 \left(\dfrac{1}{2}\right)^5 = \dfrac{10!}{5! \cdot 5!} \cdot \dfrac{1}{2^{10}} = \dfrac{10 \cdot 9 \cdot 8 \cdot 7 \cdot 6}{5 \cdot 4 \cdot 3 \cdot 2 \cdot 1} \cdot \dfrac{1}{1024} = \dfrac{252}{1024}$,
   $P = 0.2461$.
   In a long series of tosses in aggregates of ten, one can count on the occurrence of this event in almost 25% of all cases.

18. $P = {}_{10}C_9 \cdot 0.99^9 \cdot 0.01^1 = 10 \cdot 0.9135 \cdot 0.01 = 0.09135$.

19. $P = \binom{100}{25}(\frac{1}{6})^{25}(\frac{5}{6})^{75} = 0.0098$. In a large number of tosses, this event can be expected in about 1% of all cases.

Chapter 1

20. $P(X = 5) = \dfrac{20!}{15! \cdot 5!} \left(\dfrac{6}{7}\right)^{15} \left(\dfrac{1}{7}\right)^5 = \dfrac{20 \cdot 19 \cdot 18 \cdot 17 \cdot 16}{5 \cdot 4 \cdot 3 \cdot 2 \cdot 1} \cdot \dfrac{6^{15}}{7^{20}},$

$P = 0.0914.$

21. For a: $P = {}_6C_4 \cdot 0.67^4 \cdot 0.33^2 = 15 \cdot 0.2015 \cdot 0.1089 = 0.3292.$

For b: $P = \sum_{x=4}^{6} {}_6C_4 0.67^x 0.33^{6-x} = 0.3292 + 6 \cdot 0.1350 \cdot 0.33 + 0.0905,$

$P = 0.6870.$

22. $P = \dfrac{100!}{50! \cdot 50!} \cdot \left(\dfrac{1}{2}\right)^{50} \left(\dfrac{1}{2}\right)^{50} = 0.0796.$

23. $P = {}_{50}C_{20} \left(\dfrac{2}{5}\right)^{20} \left(\dfrac{3}{5}\right)^{30} = \dfrac{50!}{20! \cdot 30!} \left(\dfrac{2}{5}\right)^{20} \left(\dfrac{3}{5}\right)^{30}.$

Applying Stirling's formula,

$P = \dfrac{\sqrt{2\pi 50} \cdot 50^{50} \cdot e^{-50} \cdot 2^{20} 3^{30}}{\sqrt{2\pi 20} \cdot 20^{20} \cdot e^{-20} \cdot \sqrt{2\pi 30} \cdot 30^{30} \cdot e^{-30} \cdot 5^{20} \cdot 5^{30}},$

$P = \dfrac{\sqrt{5} \cdot 5^{50} \cdot 10^{50} \cdot 2^{20} \cdot 3^{30}}{\sqrt{2}\sqrt{2\pi 30} \cdot 2^{20} \cdot 10^{20} \cdot 3^{30} \cdot 10^{30} \cdot 5^{20} \cdot 5^{30}} = \dfrac{\sqrt{5}}{20\sqrt{3\pi}} = 0.0364.$

Poisson distribution

24. $P = \dfrac{\lambda^x \cdot e^{-\lambda}}{x!} = \dfrac{3^0 \cdot e^{-3}}{0!} = \dfrac{1 \cdot e^{-3}}{1} = \dfrac{1}{e^3} = \dfrac{1}{20.086} \approx 0.05.$

25. $\lambda = n \cdot \hat{p} = 1{,}000 \cdot 0.002 = 2,$

$P = \dfrac{\lambda^x \cdot e^{-\lambda}}{x!} = \dfrac{2^5 \cdot e^{-2}}{5!} = 0.0361.$

26. $\lambda = n \cdot \hat{p} = 200 \cdot 0.005 = 1,$

$P = \dfrac{\lambda^x \cdot e^{-\lambda}}{x!} = \dfrac{1^4 \cdot e^{-1}}{4!} = \dfrac{0.3679}{24} = 0.0153.$

27. $P(k, 5) = 5^k e^{-5}/k!.$

28. $\lambda = n \cdot \hat{p} = 30 \cdot 0.05 = 1.5, \qquad P = \dfrac{\lambda^x \cdot e^{-\lambda}}{x!}.$

(a) No children: $P = \dfrac{1.5^0 \cdot e^{-1.5}}{0!} = 0.2231,$

(b) One child: $P = \dfrac{1.5^1 \cdot e^{-1.5}}{1!} = 0.3346,$

(c) Two children: $P = \dfrac{1.5^2 \cdot e^{-1.5}}{2!} = 0.2509,$

(d) Three children: $P = \dfrac{1.5^3 \cdot e^{-1.5}}{3!} = 0.1254.$

CHAPTER 3

1. Yes: $\hat{t} = \dfrac{|41.5 - 43|}{2.795} \cdot \sqrt{16} = 2.15 > t_{15;\,0.05} = 2.13.$

2. $\hat{F} = \dfrac{s_B^2}{s_A^2} = \dfrac{0.607}{0.542} = 1.12 < F_{9;\,9;\,0.05} = 3.18.$

3. For a: $\hat{T} = 10 > 7$; H_0 is rejected at the 5% level.
   For b: $\hat{U} = 12 < U_{10,10;\,0.05} = 27$; H_0 is likewise rejected.

CHAPTER 4

1. For a: $\hat{t} = 4.06 > t_{9;\,0.01} = 3.25.$

   The null hypothesis: both sleep-inducing medications A and B have the same effect is rejected; it must be assumed that A is more effective than B.
   For b: Same conclusion as in a.

2. For a: $\hat{t} = 2.03 < t_{8;\,0.05} = 2.31.$
   For b: $\hat{R}_p = 7 > R_{8;\,0.10} = 6.$

   For c: The difference is assured at only the 10% level.
   The H_0 is retained in all three cases.

3. Yes: $\hat{\chi}^2 = 0.47 < \chi^2_{3;\,0.05} = 7.815.$

4. No: $\hat{\chi}^2 = 43.43 > \chi^2_{20;\,0.05} = 31.4.$

5. As $\hat{\chi}^2 = 17.86 > \chi^2_{1;\,0.001} = 10.83$; the independence hypothesis is rejected.

6. As $\hat{\chi}^2 = 5.49 > \chi^2_{1;\,0.05} = 3.84$; the independence hypothesis is rejected.

7. As $\hat{\chi}^2 = 10.09 > \chi^2_{1;\,0.01} = 6.635$; the independence hypothesis is rejected.

CHAPTER 5

1. $\hat{t} = 2.16 > t_{14;0.05} = 2.14,$
   $\hat{F} = 4.67 > F_{1;14;0.05} = 4.60.$

2. $r^2 \cdot \dfrac{16-2}{1-r^2} = 4.60; |r| \geq 0.497.$

3. $\hat{y} = 1.08x - 6.90,$
   $\hat{x} = 0.654y + 13.26,$
   $r = 0.842,$
   $\hat{t} = 6.62 > t_{18;0.001} = 3.92.$

4. $r = 0.6805,$
   $s_x = 7.880; s_y = 7.595; s_{xy} = 40.725,$
   $E_{yx}^2 = 0.4705 \simeq 0.47; E_{yx} = 0.686,$
   $\hat{F}_{\text{Corr.}} = 860.5 > F_{1;998;0.05} \simeq F_{1;\infty;0.05} = 3.84.$ The correlation coefficient differs considerably from zero.
   $F_{\text{Lin.}} = 2.005 < F_{(7;991;0.05)} \simeq F_{7;\infty;0.05} = 2.01.$ As $F_{7;1,000;0.05} = 2.02,$ which is larger than
   $\hat{F}_{\text{Lin.}} = 2.005,$ the deviations from linearity cannot be assured at the 5% level.

5. For a: $\hat{z} = 1.639 < 1.96;$ yes.
   For b: $0.278 \leq \varrho \leq 0.852.$
   For c: $\hat{z} = 1.159 < 1.96;$ yes.

6. $\hat{y} = 125 \cdot 1.649^x.$

7. $\hat{y} = 2.2043 \cdot 1.0097^x.$

8. $\hat{y} = 0.1627 \cdot x^{0.9556}.$

9. $\hat{y} = 0.2093x^2 - 2.633x + 10.$

10. $\hat{y} = 0.950 - 0.098x + 0.224x^2.$

CHAPTER 6

1. As $\hat{\chi}^2 = 20.7082$ $(2\hat{I}_{\text{Corr.}} = 23.4935)$ is larger than $\chi_{5;0.01}^2 = 15.086,$ the hypothesis of homogeneity is rejected.

2. As $\hat{\chi}_{\text{indep.}}^2 = 48.8 > \chi_{4;0.01}^2 = 13.3,$ the hypothesis of independence must be rejected. As $\hat{\chi}_{\text{sym.}}^2 = 135.97 > \chi_{3;0.01}^2 = 11.345,$ the hypothesis of symmetry is also to be rejected.

3. (a) $\hat{\chi}^2 = 11.12$.
   (b) $2\hat{I} = 11.39$.

   In neither case is $\chi^2_{11;0.05} = 19.675$ attained. There is thus no reason to doubt the hypothesis of homogeneity.

4. As $\hat{\chi}^2 = 10.88 < \chi^2_{6;0.05} = 12.59$, the hypothesis of homogeneity is retained.

CHAPTER 7

1. $\hat{\chi}^2 = 1.33 < \chi^2_{2;0.05} = 5.99$ (c not yet taken into account). We can spare ourselves further computation; H_0 is retained.
2. For a: $\hat{F} = 4.197 > F_{2;18;0.05} = 3.55$.
   For b: $\hat{H} = 6.423 > \chi^2_{2;0.05} = 5.99$.
3.

Source of variability	Sum of squares	DF	Mean square	$\hat{F}$	$F_{0.01}$
Among the A's	36.41	4	9.102	19.12 > 4.43	
Among the B's	28.55	5	5.710	12.00 > 4.10	
Experimental error	9.53	20	0.476		
Total variability	74.49	29			

Multiple comparisons of the row as well as the column means at the 1% level in accordance with Scheffé or Student, Newman, and Keuls is recommended (cf. $D_{I,\text{row means}} = 1.80$ and $D_{I,\text{column means}} = 1.84$).

4. For a: $\hat{\chi}^2_R = 15.8 > \chi^2_{2;0.001} = 13.82$.
   For b: $\hat{\chi}^2_R = 26.0 > \chi^2_{9;0.01} = 21.67$.

FEW NAMES AND SOME PAGE NUMBERS

Abbas, S. 408, 620
Abdel-Hamid, A.R. 219, 590
Abramowitz, M. 571
Abt, K. 626
Ackoff, R.L. 134
Acton, F.S. 390
Addelman, S. 564
Adler, F. 351
Adler, H. 605
Ageno, M. 83
Agresti, A. 493, 628
Ahrens, H. 532
Aitchinson, J. 107
Alanen, J.D. 602
Allen, I. 596
Allin, D.W. 222, 301, 595
Alluisi, E.A. 16
Altham, P.M.I. 482
Amstadter, B.L. 237
Anastasio, E.J. 561
Andersen, S.L. 67
Anderson, M.Q. 238, 603
Anderson, R.L. 460, 461
Anderson, T.W. 461
Anke, K. 244
Anscombe, F.J. 517
Antle, C.E. 626
Armitage, P. 473
Arsham, I. 624

Ashford, J.R. 596
Ashton, J.R. 228, 596
Ashton, W.D. 222
Atkinson, A.C. 564
Axtell, L.M. 235, 596

Bachi, R. 578
Bätz, G. 640
Bailey, J.R. 537
Bailey, N.T.J. 589
Bain, L.J. 598
Baker, A.G. 564
Balaam, L.N. 564
Balintfy, J.L. 606
Bancroft, T.A. 460, 540, 553, 561, 564
Bandemer, H. 640
Banerji, S.K. 271
Barclay, W.D. 634
Barlow, R.I. 237
Barnard, G.A. 41, 202
Barnett, R.N. 200
Barnett, V. 28, 41, 82, 280, 447, 578
Barnett, V.D. 500
Barney, G.C. 244
Barron, B.A. 200
Bartholomew, D.J. 464
Bartko, J.J. 190
Bartlett, M.S. 390, 392
Barton, D.E. 621
Bateman, G. 378, 615

Bather, J.A. 590
Batschelet, E. 455, 578
Bauer, R.K. 302
Bauknecht, K. 240
Bayes, T. 41, 124, 134, 567
Bazovsky, I. 598
Beale, E.M.L. 626
Beard, R.E. 134, 598
Beatty, G.H. 283
Bechhofer, R.E. 541
Beckman, R.J. 280, 496, 578
Behnke, H. 580
Behrens, W.-V. 271, 272
Beightler, C.S. 230
Bekey, G.A. 244, 605
Bellinson, H.B. 618
Bellman, A. 640
Belson, I. 285
Benjamin, B. 197
Bennett, B.M. 364, 365, 371, 373, 464, 465, 473
Bent, D.H. 573
Ben-Tuvia, S. 605
Berchtold, W. 348
Berettoni, G.N. 598
Berg, D. 627
Berkson, J. 246
Bernard, G. 134

657

Bernoulli, J. 27, 35, 163, 567
Bertram, G. 595
Beyer, W.H. 129
Bezemer, P.D. 197
Bhapkar, V.P. 306
Bhattacharya, C.G. 83
Bienaymé, I.J. 64
Bihn, W.R. 379, 615
Billewicz, W.Z. 595
Billingsley, P. 589
Binder, A. 532
Bindgham, R.S. 233
Birnbaum, Z.W. 293, 331
Bishop, T.A. 495, 631
Bishop, Y.M.M. 482, 484, 628
Bissel, A.F. 202
Blishke, W.R. 533
Bliss, C.I. 177, 190
Blomqvist, N. 403
Blyth, C.R. 333
Boddy, R. 197, 591
Bogartz, R.S. 381
Bogue, D.J. 197
Bolch, B.W. 71
Boomsma, A. 428
Boot, J.C.G. 604
Borth, R. 596
Bose, R.C. 555, 640
Böttger, R. 605
Botts, R.R. 111
Bowden, D.C. 626
Bowers, D.A. 627
Bowker, A.H. 211, 282, 488
Bowman, K.O. 634
Box, G.E.P. 55, 66, 232, 233, 261, 381, 453, 500, 578, 598
Boyd, W.C. 148, 274
Bradley, J.V. 77, 293, 578, 608
Bradley, R.A. 26, 301, 541, 555, 567, 608, 614, 631, 637
Bratcher, T.L. 541
Bredenkamp, J. 381, 616
Breny, H. 271
Bresnahan, J.L. 480
Bretherton, M.H. 190
Brewerton, F.J. 237

Brock, N. 596
Brooks, S.H. 232
Broos, I.D.S. 220, 221, 225, 236
Brown, B.M. 496
Brown, B.W. 225
Brown, B.W., Jr. 591
Brown, E.S. 575
Brown, J.A.C. 107
Brown, M.B. 493
Brown, R.G. 394
Brown, S.M. 365, 620
Browne, M.W. 625
Brownlee, K.A. 640
Bruckmann, G. 53
Brugger, R.M. 71
Buckland, A. 571
Buckland, W.R. 574, 596
Buenaventura, A.R. 275
Bühlmann, H. 134
Bunke, H. 560
Bunke, O. 560
Bunt, L. 129
Burdette, W.J. 197, 210, 214
Burdick, G.R. 238, 599
Burman, J.P. 564
Burnett, W.T., Jr. 597
Burr, I.W. 202, 591
Burrington, G.A. 575
Burrows, G.L. 282
Buslenko, N.P. 241

Cacoullos, T. 261
Cadwell, T. 252
Calot, G. 122
Campbell, S.K. 26, 579
Campbell, V.N. 403, 624
Carlson, F.D. 390
Carnal, H. 300
Casella, G. 392
Casley, D.J. 621
Castellan, N.J. 480
Castelman, B. 197, 591
Caulcutt, R. 197, 591
Caussinos, H. 482
Cavalli-Sforza, L. 596
Cetron, M.J. 53
Chacko, V.J. 306
Chakravarti, I.M. 190,

192, 213, 272, 569, 586, 608
Chambers, J. 6
Chambers, J.M. 241, 453, 621
Chapman, D.G. 282
Charnes, A. 240
Chebyshev, P.L. 64
Chernoff, H. 134
Chew, V. 598, 640
Chiang, C.L. 589
Chilton, N.W. 595
Chissom, B.S. 103
Chou, Y.-L. 569
Chu, K. 606
Chun, D. 200
Clark, V. 235
Cleary, T.A. 127
Clopper, C.J. 340
Cochran, W.G. 146, 197, 200, 206, 209, 261, 336, 351, 365, 381, 453, 466, 473, 497, 498, 579, 591, 608, 614, 616, 626, 628, 631, 640
Cohen, J. 127, 129, 274, 579, 608
Cole, J.W.L. 541
Cole, L.M.C. 222, 223, 460
Coleman, E.P. 601
Colton, T. 597
Comrie, L.J. 573
Connell, T.L. 263
Connolly, T.W. 243
Conover, W.J. 285, 286, 306, 330, 518, 590, 608
Conway, F. 614
Conway, R.W. 605
Cook, R.D. 280, 578
Cooley, W.W. 461
Cooper, B.E. 564
Cooper, W.W. 240
Corballis, M.C. 685
Cornfield, J. 41, 160, 178
Cotlove, E. 594
Cowden, D.J. 323, 461
Cox, C.P. 597
Cox, D.R. 123, 187, 381, 476, 517, 559,

564, 579, 616, 621, 634
Cox, G.M. 640
Cox, J.G. 241
Cox, P.R. 197
Craig, I. 579
Cramér, E.M. 460
Cramer, H. 589
Cressie, N.A.C. 541
Croacrin, M.C. 274
Crouse, C.F. 638
Crow, E.L. 333
Crowe, W.R. 579
Croxton, F.E. 323
Csorgo, M. 332
Cunningham, E.P. 533
Cureton, E.E. 312, 391, 392

Dagnelie, P. 569
D'Agostino, R.B. 71, 252, 329, 484, 493, 579, 599, 608
Dalal, S.R. 137
Dalenius, T. 91
Daly, J.C. 283
Daniel, C. 460
Dantzig, G.B. 239
Danziger, L. 285
Darling, D.A. 291
Darlington, R.B. 103
Das, M.N. 597
Davenport, J.M. 550
David, F.N. 35, 330, 350, 424
David, H.A. 66, 517, 555, 590, 632
Davies, O.L. 219, 232, 262, 609, 640
Davis, H.T. 616
Davis, L.L. 637
Davis, M. 228, 597
Davis, S.A. 285
Day, N.E. 83, 580
Dean, B.V. 233
Deininger, R. 67, 246
Deixler, A. 90
Delucchi, K.L. 462
Delury, D.B. 174
Deming, Lola S. 574
Deming, W.E. 614
Dempster, A.P. 461
Derflinger, G. 625

Despande, J.W. 306
Deziel, D.P. 605
Dhillon, B.S. 237, 238, 599
Dickinson, G.C. 580
Diczfalusy, E. 596, 597
Dietz, K. 580
Dietze, D. 274
Dixon, W.J. 161, 226, 274, 277, 279, 280, 569, 597, 609, 616
Dobben de Bruyn, C.S., van 202, 591
Doherty, D.G. 597
Doig, A.G. 574, 589
Doornbos, R. 447
Donner, A. 626
Dorfman, R. 218, 591
Draper, N.R. 460, 621
Dresher, M. 240
Drnas, T.M. 237
Dubey, S.D. 166, 181, 234
Dubois, P.H. 626
Duckworth, W.E. 319, 616
Dudewicz, E.J. 137, 495
Dugue, D. 640
Duncan, A.J. 230, 232
Duncan, D.B. 621, 632, 638
Dunn, J.E. 378
Dunn, O.J. 632, 638
Dunnet, C.W. 495, 537, 638
Durr, A.C. 238, 599

Eagle, E.L. 237, 599
Eberhard, K. 628
Eckler, A.R. 240
Edington, E.S. 293, 609
Edwards, A.L. 640
Edwards, W. 604
Eeden, Constance, Van 305
Ehrenberg, A.S.C. 566, 621
Ehrenfeld, S. 605
Eilers, R.J. 197, 591
Eilon, S. 605

Eisen, E.J. 533, 564, 632
Eisen, M. 569
Eisenhart, Ch. 200, 282, 376, 497
Elandt, R.C. 210, 403, 591
Elderton, W.P. 580
Elveback, L.R. 197, 591
Emerson, Ph.L. 460
Enderlein, G. 537, 621, 626, 632, 638
Endler, N.S. 533
Endriss, J. 250, 613
Enke, H. 625, 629
Evans, S.H. 561, 632
Everitt, B.S. 482, 628
Ewan, W.D. 202, 591

Fabrycky, W.J. 603
Fain, J.B. 604
Fain, W.W. 604
Fairfield, Smith H. 487, 628
Faulkenberry, G.D. 283, 609
Faulkner, E.J. 37, 580
Federer, W.T. 218, 532, 564, 592, 632, 640
Federighi, E.T. 138
Federova, R.M. 573
Feinstein, A.R. 209
Feldman, S.E. 372, 616
Feller, W. 570, 589
Fellsenstein, J. 464
Fenner, G. 96
Ferguson, G.A. 394, 461
Fermat, P., de 35, 567
Fernbach, S. 606
Ferrel, E.B. 230, 599
Ferris, C.D. 129
Fertig, J.W. 595
Fienberg, S.E. 6, 26, 209, 482, 580, 592, 628
Fifer, S. 244
Fink, H. 597
Finkelstein, J.M. 231
Finney, D.J. 228, 371, 597, 616

Finucan, H.M. 103
Fishburn, P.C. 134
Fisher, N.I. 6, 131, 580
Fisher, R.A. 67, 121, 136, 137, 143, 166, 190, 269, 271, 272, 302, 351, 370, 372, 424, 427, 428, 460, 494, 518, 560, 564, 567, 622, 640
Flagle, C.D. 238
Fleckenstein, M. 553
Fleiss, J.L. 206, 209, 351, 592
Fletcher, A. 573
Folks, J.L. 626
Fox, M. 250, 579, 611
Frawley, 282
Free, S.M. 373
Freeman, D.H., Jr. 566
Freeman, H. 595
Freeman, M.F. 516
Freiman, Jennie A. 218
Fremery, J.D.N., de 572
Freudenthal, H. 35, 50, 59, 235, 580
Freudenthal, H.M. 235
Freund, J.E. 572
Freund, R.A. 633
Friedman, L. 604
Friedman, M. 306, 549, 550, 551, 552, 555
Friedrich, H. 427
Frontali, C. 83
Fülgraff, G. 595
Fussel, J.B. 238, 599

Gabriel, K.R. 480, 490, 541, 633
Gabriels, R. 592
Gabrielsson, A. 285
Gaddum, J.H. 597
Gaede, K.-W. 600
Games, P.A. 427, 495, 500, 537, 633, 638
Gani, J. 574
Gans, D.J. 613
Gardner, R.S. 344
Garland, L.H. 41

Gart, J.J. 351, 364, 482
Gass, S.I. 603
Gates, Ch.E. 564
Gauss, C.F. 58, 64
Geary, R.C. 252
Gebelein, H. 622
Gebhart, F. 66, 165
Gehan, E.A. 37, 197, 210, 214
Gerbatsh, R. 200
Ghosh, M.N. 633
Gibbons, J.D. 293, 590, 609
Gibra, N.I. 600
Gibson, W. 391, 460, 461
Gilbert, J.P. 219
Gildemeister, M. 617
Gini, C. 26, 581
Girault, M. 640
Glasser, G.J. 306, 316
Glick, N. 37
Gokhale, D.V. 482
Gold, B.K. 601
Gold, R.Z. 589
Goldberg, J.D. 218, 592
Goldberger, A.S. 626
Goldman, A. 250, 609
Goldman, A.S. 238, 600
Goldsmith, P.L. 202
Goldzieher, J.W. 177
Gollob, H.F. 625
Golub, A. 597
Good, I.J. 367
Goodman, L.A. 482, 484, 492, 628
Gordon, T. 596
Gorenflo, R. 606
Gosset, W.S. (s. Student) 135
Gottfried, P. 600
Gower, J.C. 569
Graff, L.E. 219
Granger, C.W.J. 290, 609
Graybill, F.A. 163
Greb, D.J. 595
Green, B.F., Jr. 541, 633
Greenberg, B.G. 66
Greenwood, J.A. 573

Gregg, I.V. 196, 461
Gridgeman, N.T. 121
Griffin, H.D. 622
Griffiths, D. 389
Griffiths, D.A. 111, 581
Grimm, H. 185, 190, 540, 633
Grizzle, J.E. 351, 492, 541
Groll, P.A. 218
Groot, M.H., de 41
Gross, A.J. 234
Grubs, F.E. 279
Gryna, F.M., Jr. 237, 600
Guenther, W.C. 129, 285
Guetzkow, H. 241
Gumbel, E.J. 111, 235
Gupta, S.D. 461
Gurland, J. 190, 581, 589, 609
Guterman, H.E. 98
Guttman, I. 332

Haber, A. 310
Haenszel, W. 351
Haga, T. 290
Hahn, G.J. 55, 166, 183, 232, 249, 250, 392, 444, 445, 514, 515, 564, 609, 622, 626, 633, 638, 640
Haight, F.A. 134, 179, 574
Hàjek, J. 590
Hald, A. 196
Hall, A.D. 134
Hall, M., Jr. 640
Hall, R.A. 299
Halperin, M. 301
Halton, J.H. 241
Hambury, J.H. 243
Hamdan, M.A. 482
Hammersley, J.M. 241
Hampel, F. 66, 253, 582
Hanscomb, D.S. 191
Hansen, M.H. 614
Hanson, W.R. 172
Hardyck, C.D. 639
Harman, H.H. 625

Harmann, A.J. 285, 610
Harnes, A. 603
Harris, B. 379
Harris, D. 83
Harris, E.K. 594
Harsaae, E. 499
Hart, B.I. 375
Harte, C. 561
Harter, H.L. 67, 97, 142, 291, 294, 508, 518, 532, 535, 582, 610, 633
Hartley, H.O. 150, 228, 281, 326, 329, 341, 460, 461, 496, 498, 515, 532
Harvey, W.R. 533
Hasselblad, V. 109, 582
Hastay, M.W. 282, 497
Hawkins, D.M. 280, 582
Hays, W.L. 533
Healy, M.J.R. 516, 634
Hedayat, A. 640
Heinhold, J. 600
Heinrichs, H.D. 596
Helmert, F.R. 139
Hemelrijk, J. 582
Henderson, C.R. 533
Henley, E.J. 237, 238, 600
Henning, H.-J. 602
Henry, R.J. 593
Herdan, G. 108
Herne, H. 626
Herold, W. 83
Herrey, E.M.J. 252
Hertz, D.B. 238
Herzberg, A.M. 559, 564, 634
Hibon, M. 381
Hicks, C.R. 640
Hilferty, M.M. 588
Hill, A.B. 210
Hill, G.W. 138
Hill, W.J. 232, 600
Hillier, F.S. 229, 600, 603
Hinchen, J.D. 626
Hinkelmann, K. 210
Hiorns, R.W. 450, 455, 460

Hoaglin, D.C. 6, 66, 517, 582
Hocking, R.R. 390, 447, 622, 626
Hodges, J.L., Jr. 129, 274, 293, 610
Hoel, D.G. 634
Hoerl, A.E., Jr. 455
Höhndorf, K. 600
Hogg, R.V. 66, 447
Holland, D.A. 533
Hollander, M. 305, 552
Holm, S. 537
Honeychurch, J. 235
Horn, M. 464
Horst, C. 371
Horvath, W.J. 604
Hossel, C.H. 460, 461
Hotelling, H. 26, 156, 426, 427, 622
Houseman, E.E. 461
Howell, J.F. 537
Howel, J.M. 601
Hsu, P. 371, 373, 473
Huang, D.S. 626
Huber, P.J. 66, 582
Huddleston, H.F. 97
Huggins, W.H. 603, 631
Huitema, B.E. 560, 634
Hull, C.H. 573
Hund, G. 597
Hunter, J.S. 233
Hunter, W.G. 233
Hurwitz, W.N. 614
Hutchinson, D.W. 333
Hwang, F.K. 218

Iman, R.L. 286, 518, 550
Ireland, C.T. 482, 488, 629
Isaacs, R. 240

Jaech, J.L. 564
Jaeger, A. 603
James, G.S. 271
James, M.L. 606
Jackson, J.E. 219, 555, 301
Jacobson, J.E. 300

Jellinek, E.M. 214
Jenkins, T.N. 381
Jensen, D.R. 461
Jesdinsky, H.J. 381, 617
Jöhnk, M.-D. 570
Johns, M.V., Jr. 600
Johnson, E.E. 147, 188
Johnson, E.M. 371, 617
Johnson, L.G. 600
Johnson, M.E. 283
Johnson, N.L. 166, 194, 202, 210, 219, 259, 564, 591
Joiner, B.L. 575
Jolly, G.M. 172
Jöreskog, K.G. 625
Joshi, S.W. 166, 194, 575
Jowett, D. 391, 622
Jung, W. 640
Juran, J.M. 563

Kabe, D.G. 234
Kaltenecker, H. 605
Kane, V.E. 111, 583
Kanofsky, P. 594
Kao, J.H.K. 600
Kapur, K.C. 237, 238, 600
Kärber, G. 225
Karlin, S. 589
Karplus, W.J. 243, 244
Karr, H.W. 604
Kastenbaum, M.A. 482, 541
Katti, S.K. 614, 620
Kaufmann, H. 235, 597
Keeney, R.L. 134, 583
Kemeny, J.G. 589, 605
Kemp, K.W. 202, 592
Kempthorne, O. 532, 533, 564
Kempthorne, V. 634
Kendall, M.G. 35, 123, 349, 373, 401, 461, 567, 570, 574, 625, 640
Kent, R.H. 618
Kenworthy, I.C. 233
Kerrich, J.E. 390, 392

Keselman, H.J. 537, 633, 639
Keuls, M. 534
Keyfitz, N. 197, 592
Kiefer, J.C. 634
Kim, P.J. 291
Kimball, A.W. 597
Kincaid, W.M. 367, 617, 629
King, A.C. 35, 583
King, E.P. 597
King, J.R. 235
Kirk, R.E. 564, 641
Kirkpatrick, R.L. 283
Kish, L. 614, 641
Kitagawa, T. 179
Kittrel, J.R. 233
Klemm, P.G. 288
Klerer, M. 606
Klinger, E. 372, 616
Kloek, T. 604
Knese, K.H. 516, 634
Knowler, L.A. 230
Knowler, W.C. 230
Koch, G.G. 482, 533
Koehler, K.J. 137, 583
Kohlas, J. 241
Koller, S. 26, 197, 206, 209, 393, 394, 558, 570, 592, 622
Kolmogoroff, A.N. 28, 291, 293, 330, 331, 567
Konijn, H.S. 403
Korn, G.A. 606
Kotz, S. 166, 194
Kraft, C.H. 240
Kramer, C.W. 461, 540
Kramer, G. 134
Kramer, K.H. 200
Kres, H. 461, 571, 623
Krishaiah, P.R. 461
Krishnan, M. 274, 610
Kruskal, W.H. 35, 300, 302, 306, 484, 570, 610, 614, 617
Krutchkoff, R.G. 253
Ku, H.H. 482, 490, 492
Kübler, H. 111, 583
Kulkarni, G.A. 597
Kullback, S. 359, 360, 482, 490, 493, 617, 629

Kumar, S. 234
Kupper, L.L. 555
Kupperman, M. 589, 617, 629
Kurz, T.E. 542, 543
Kussmaul, K. 564
Kutscher, A.H. 595
Kymn, K.O. 426, 623

Laan, P., van der 303
LaMotte, L.R. 626
Lancaster, H.O. 119, 574
Lange, H.-J. 592
Laplace, P.S., de 27, 134, 567
Lasagna, L. 213
Larson, H.R. 165
Lashof, T.W. 200
Latscha, R. 371
Laubscher, N.F. 152
Lawless, J.F. 235, 593
Lawley, D.N. 457
Lazar, Ph. 597
Laedbetter, M.R. 210, 589, 603
Lebedev, A.V. 573
Lee, A.M. 589
Lee, Elisa T. 109, 214, 218, 234, 235, 583
Lees, R.W. 456, 623
Lehmaher, W. 541, 635
Lehmann, E.L. 127, 129, 274, 293, 583, 610
Lehmann, F. 241, 606
Lehmann, W. 516, 635
Leibler, R.A. 490
Leiner, B. 490
Leone, F.C. 202, 564
Leonhard 578
Le Roy, H.L. 360, 532
Lesky, E. 23, 583
Leslie, R.T. 496
Levene, H. 261, 378, 500
Levin, J.R. 119
Levitt, E.E. 403
Levy, B. 575
Levy, K.L. 427
Lewis, P.A.W. 589
Lewis, T. 280, 447, 578

Lewontin, R.C. 464
Leyton, M. 627
Li, C.C. 564
Li, J.C.R. 564
Lieberman, G.J. 174, 231, 282, 373, 583, 610
Lieberson, S. 403
Lieblein, J. 235
Lilienfeld, A.M. 206, 593
Lilienfeld, D.E. 206, 593
Lienert, G.A. 131, 204, 306, 541, 583, 590, 595, 610, 618, 635
Lilliefors, H.W. 331, 618
Linder, A. 623, 641
Lindgren, B.W. 291
Lindley, D.V. 610
Ling, R.F. 137
Linhart, H. 555
Link, R.F. 542, 543, 635, 639
Linn, R.L. 127
Linnik, Y.V. 271
Linstone, H.A. 53, 583
Lipow, M. 237
Lishka, T.R.J. 593
Litchfield, J.T. 595
Little, R.E. 226, 597
Lloyd, D.K. 237
Lockwood, A. 583
Loeffel, H. 579
Lohnes, P.R. 461
Lord, E. 265, 276, 277
Lord, F.M. 456, 458
Lorenz, R.J. 596
Lowe, C.W. 233
Loyer, M.W. 218, 593
Lubin, A. 130, 583
Lucas, H.L. 598
Luce, R.D. 605
Ludwig, O. 378, 618
Ludwig, R. 427, 623

Mace, A.E. 250, 611
MacKinnon, W.J. 254, 318, 618
Makridakis, S. 381
Martz, H.F. 41, 238, 601

Minton, G. 197, 246, 611
Mitra, S.K. 611
Mize, J.H. 241
Moan, O.B. 601
Moivre, A., de 58, 567
Molenaar, W. 165, 175, 185, 334
Molina, E.C. 179
Montgomery, D.C. 53
Mood, A.M. 315, 316
Moore, G.H. 378, 379
Moore, P.G. 238, 277
Moore, R.H. 564
Moran, P.A.P. 570, 575
Morgerstern, H. 206, 593
Morgenstern, O. 240, 605
Morgenthaler, G.W. 606
Morice, E. 129, 186
Morrison, D.F. 461
Morrison, J. 230, 601
Moses, L.E. 51, 119, 134, 537, 584
Moshman, J. 111
Mosteller, F. 166, 219, 230, 285, 516, 584, 611, 614, 629, 635
Mudgett, B. 584
Müller, P.H. 571
Murthy, M.N. 614
Myers, J.L. 641
Myhre, J.M. 601

Nabavian, K.J. 602
Naeve, P. 570
Nakamura, E. 464
Nakano, K. 285, 607
Namneck, P. 606
Narula, S.C. 457, 461
Nass, C.A.G. 475, 630
Natrella, Mary G. 129, 454, 562, 584, 623, 635
Nauss, J.I. 37
Neave, H.R. 285, 611
Neidhold, G. 244, 605
Neill, J.J. 426
Nelder, J.A. 482, 492, 630

Nelson, L.S. 232, 235, 282, 541, 601
Nelson, W. 234, 238, 249, 250, 601
Neter, J. 446, 460, 461, 623
Neumann, J., von 133, 605, 618
Newmann, D. 534, 635
Newman, T.G. 241
Newton, I. 159
Neyman, J. 123, 246, 518, 567, 635
Nicholson, W.I. 332, 618
Novik, M.R. 41, 585
Nowak, S. 394, 623
Nullau, B. 381, 618

Oakes 235, 599, 601
Odeh, R.E. 250, 282, 554, 579, 585, 611
Oosterhoff, J. 607
Ostle, B. 453, 623
Overall, J.E. 348, 619
Owen, D.B. 282, 283, 585, 611

Pachares, J. 534, 635
Page, E.B. 553, 554, 636
Page, E.S. 202, 593
Page, W.F. 241, 606
Parzen, E. 381
Patefield, W.M. 493
Patel, H.I. 234
Patil, G.P. 166, 194, 575
Pawlik 483, 630
Pearson, E.S. 35, 107, 123, 241, 460, 567
Pearson, K. 77, 106, 139, 567
Pendelton, O.J. 390, 447, 622
Peng, K.C. 564, 601
Perlman, M.D. 367, 617
Perng, S.K. 234
Pesek, I. 353, 624
Peters, C.A.F. 2, 53, 612

Peto, R. 210, 214, 593
Pettit, A.N. 166, 585
Pfanzagl, J. 166, 585
Pfeifer, C.G. 218
Pillai, K.C.S. 275, 276, 612
Plakett, R.L. 81, 482, 532, 564, 630, 636
Plait, A. 235, 601
Please, N.W. 107, 241
Pitman, E.J.G. 130, 585
Poisson, S.D. 175, 516
Popper, K.R. 26, 585
Potthoff, R.F. 441, 623
Prabhashanker, V. 235
Prairie, R.R. 237
Pratt, J.W. 293, 407, 585, 590, 612, 623
Prestele, H. 541
Price, B. 461
Prince, B.M. 408, 623
Proschan, F. 237, 602

Quetlet, L.A.J. 58

Raatz, U. 300
Radhakrishna, S. 367, 460
Rahman, N.A. 570
Raiffa, H. 134, 586
Raj, D. 614
Ramberg, J.S. 323, 638
Rao, C.R. 192, 363, 461
Rao, P.V. 555, 635
Rapoport, A. 605
Rasch, D. 28, 533, 540
Ravenis, J.V.J. 35, 482
Read, C.B. 35, 480
Rechenberg, P. 242
Reed, A.H. 197
Regamey, R.H. 597
Reshe, E. 381
Reiher, W. 626
Reinach, S.G. 552
Reisch, J.S. 561
Reiter, S. 263
Reynolds, J.H. 201, 202
Rhoades, H.M. 619
Rhodes, L.J. 614

Rhyne, A.L. 639
Rhyne, A.L., Jr. 639
Rice, W.B. 230, 602
Richards, Ch. H. 624
Richards, R.K. 242
Richardson, J.T. 461
Richter, K. 640
Ricker, W.E. 446
Rider, P.R. 85
Riedwyl, H. 300
Rigas, D.A. 177
Riley, V. 605
Riordan, J. 156, 162
Rives, M. 317
Roberts, H.R. 600
Roberts, H.V. 166
Robinson, E.A. 626
Robson, D.S. 172, 460
Rocke, D.M. 200, 594
Roeloffs, R. 218
Rogers, A.E. 243
Roming, H.G. 164
Roos, C.F. 460, 624
Roos, J.B. 199
Röpke, H. 606
Rose, G. 640
Rosenbaum, S. 285, 286, 612
Rosenberg, L. 611
Rosenhead, L. 573
Rosman, B. 484, 493
Roy, J. 569
Roy, R.H. 603, 631
Roy, S.N. 461
Rubin, E. 574
Rucker, N.L. 461
Ruhenstroth-Bauer, G. 451, 622
Ruhl, D.J. 597
Rümke, Chr. L. 197, 209, 593, 594
Rummel, R.J. 457
Runyon, R.P. 310
Rusch, E. 90
Rutemiller, H.C. 627
Rutheford, A.A. 561
Ryan, T.A. 464
Ryerson, C.M. 600
Rytz, C. 293

Saaty, L. 48, 586
Saaty, T.L. 55, 238, 586, 589, 604

Sachs, L. 26, 51, 56, 81, 83, 107, 129, 209, 259, 265, 272, 306, 348, 362, 364, 367, 373, 447, 546
Sachs, V. 595
Salzer, H.E. 461
Samiuddin, M. 426
Sampford, M.R. 614
Sandelius, M. 290
Sandler, J. 310
Sarhan, A.E. 66
Sarris, V. 381, 619
Sasieni, M. 238
Sauer, R. 580
Saunders, S.C. 601
Savage, I.R. 65
Savage, L.J. 575
Saxena, A.K. 461, 624
Saxena, H.C. 461, 624
Schaeffer, M.S. 403
Schafer, R.E. 234, 331
Schäffer, K.-A. 613
Scharf, J.-H. 455, 624
Schatzoff, M. 627
Selten, E. 605
Sheffe, H. 271, 509, 532, 555, 564
Schindel, L.E. 212, 213
Schlaifer, R. 134
Schleifer, A., Jr. 605
Schlesselman, J.J. 206, 351, 594, 631
Schneeweiss, H. 134, 586
Schneider, B. 597
Schneiderman, M.H. 215, 216
Schreider, J.A. (Y.A.) 241, 607
Schultz, H. 294, 306
Schumann, D.E.W. 541, 631
Scott, D.W. 107
Scott, E.L. 518, 635
Seal, H. 461
Searle, S.R. 460, 461, 529, 532
Seber, G.A.F. 627
Seeger, P. 365, 537
Semmelweis, I. 23
Sen, K.P. 533
Sen, P.K. 461

Severo, N.C. 142
Shaffer, J.P. 73, 480, 492, 630
Shakarji, G. 594
Shamblin, J.E. 230
Shapiro, M.M. 480
Shapiro, S.S. 329, 460
Shapley, L.S. 604
Sharma, D. 633
Shaw, G.B. 108
Sherif, Y.S. 238, 602
Shet, J.N. 625
Sheynin, O.B. 64, 67, 586
Shiue, Ch.-J. 564
Shooman, M.L. 602
Shorack, G.R. 262
Shubik, M. 241, 604
Sidàk, Z. 590
Siegel, S. 260, 286, 288, 291
Silander, F.S. 134, 588
Simonds, T.A. 235
Singpurwalla, N.D. 234
Sinha, S.K. 235
Siotani, M. 636, 639
Skala, H.J. 569
Skillings, J.H. 306
Slakter, M.J. 332
Slattery, T.B. 238
Slivka, J. 639
Smillie, K.W. 627
Smirnoff (Smirnov), K.W.V. 138, 291, 293, 330, 331
Smith, G.M. 606
Smith, H. 460
Smith, J.H. 90
Smith, J.U.M. 241, 607
Smith, M. 640
Smith, W.L. 589
Smith, W.W. 596
Snedecor, G.W. 453, 457
Snee, R.D. 493
Snell, E.J. 541, 637
Snell, J.L. 589, 605
Snyder, R.M. 587
Sobel, M. 218
Sonquist, J.A. 55, 393, 587
Soroka, W.J. 241
Southwood, T.R.E. 172

Sowey, E.R. 51, 241, 607
Spear, M.E. 587
Spearman, C. 225, 384, 395, 531, 402, 554, 624
Spicer, C.C. 222
Spjøtvoll, E. 533
Sprott, D.A. 561
Stammberger, A. 228, 275, 598, 613
Stange, K. 166, 229, 230, 233, 446, 460
Starks, T.H. 637
Steel, R.G.D. 639
Stegmüller, W. 26, 587
Stegun, I.A. 571
Stein, M.R. 594
Steiner, H.G. 35, 59, 580
Stephens, M.A. 323, 332, 378
Stephenson, C.E. 71
Stevens, S.S. 132
Stevens, W.L. 378
Stewart, D.A. 561
Stiehler, R.D. 200
Stilson, D.W. 403
Stirling, D. 81
Stoline, M.R. 537
Stoller, D.S. 238
Störmer, H. 237
Strecker, H. 96, 197
Stuart, A. 349, 373, 380, 403, 564, 570, 616, 624, 640
Student (W.S. Gosset) 130, 534, 552, 624, 645
Sturges, H.A. 587
Styan, G.P.H. 461
Suits, D.B. 380
Sukhatme, P.V. 613
Surendran, P.U. 461
Swanson, P.P. 457
Swed, F.S. 376
Swoboda, V. 199
Szameitat, K. 67, 197, 246, 613

Takacs, L. 589
Tamhane, A.C. 495
Tate, M.W. 365, 620
Tate, R.F. 408, 623
Taub, A.H. 606
Taylor, H.M. 202
Taylor, L.R. 516, 541
Teichroew, D. 51, 518, 587, 637
Terry, M.E. 555
Tews, G. 144
Theil, H. 53, 134
Ther, L. 598
Thews, G. 516, 634
Thissen, D. 6, 588
Thoman, D.R. 598, 602, 603
Thompson, G.L. 605
Thompson, V.C. 601
Thompson, W.A. 279, 613
Thompson, W.A., Jr. 250, 556, 613
Thomson, G.W. 620
Thöni, H.P. 109, 274
Thorndike, F. 183
Tietjen, G.L. 283, 496, 500, 564
Tiku, M.L. 262
Tillman, F.A. 41, 238, 603
Tobach, E. 640
Tocher, K.D. 241
Torgersen, P.E. 603
Toro-Vizcarrondo, C. 627
Traub, R.E. 625
Trawinski, B.J. 555
Trickett, W.H. 613
Trommer, R. 626
Troughton, F. 587
Trustrum, G.B. 590
Tsao, R. 627
Tucker, A.W. 604
Turoff, M. 53, 583
Tukey, J.W. 6, 26, 66, 134, 166, 230, 252, 260, 280, 286–290, 390, 514, 516, 534, 540, 541, 567, 570, 575, 577, 584, 587, 612, 613, 620, 623, 625, 630, 633, 637
Tygstrup, N. 214, 218, 594

Übrela, K. 625
Uhlmann, W. 603
Ulmo, J. 627
Underwood, R.E. 363
Upton, G.J.G. 348, 482, 630
Ury, H.K. 363

Vahle, H. 142
Väliaho, H. 460
Varma, A.O. 595
Varner, R.N. 492
Vessereau, A. 202, 351, 555
Victor, N. 550
Vogel, W. 496
Vogelsang, R. 240
Vogt, H. 603
Vølund, A. 228, 598

Waerden, B.L., van der 28, 293, 317, 344
Wagle, B. 53
Wagner, G. 625
Wainer, H. 6, 66, 588
Wald, A. 24, 118, 133, 219, 567, 588, 596
Wallace, T.D. 627
Wallis, W.A. 282, 300, 302, 303, 306, 370, 378, 497
Walsh, J.E. 590
Walter, E. 28, 129, 130, 297, 316, 395, 588, 613, 620, 625
Warner, B.T. 598
Wasserman, P. 134, 588
Watts, D.E. 380
Watson, G.S. 210, 603
Watts, D.E. 380
Watts, D.G. 567
Waud, D.R. 228, 598
Weaver, C.L. 580
Weber, Erna 190, 219, 569, 588, 625
Weber, Ernst 625
Webster, J.T. 561
Weibull, W. 111, 234
Weichselberger, K. 83, 381
Weiler, H. 275

Weiling, F.   257, 540, 541, 613, 637
Weintraub, S.   164
Weir, J.B., de V.   272, 613
Weiss, L.L.   111, 588
Weissberg, A.   282, 283
Welch, B.L.   253, 271
Wellnitz, K.   162
Welte, E.   593
Wenger, A.   274
Westergaard, H.   35, 588
Westlake   290
Wetherill, G.B.   219, 596
Wetzel, W.   284, 570, 614
Whitehead, J.   214, 223, 596
Whitney, D.R.   293, 296, 298
Widdra, W.   181
Wiezorke, B.   627
Wilcox, R.A.   546, 555, 556, 558
Wilcox, R.R.   495
Wilcoxon, F.   286, 293, 294, 296, 297, 298, 299, 312, 313, 316, 546, 555, 556, 558
Wilde, D.J.   232
Wilk, M.B.   329, 532, 533
Wilkie, J.   575
Wilkins, B.R.   241

Wilks, S.S.   283, 614
Williams, C.B.   108, 588
Williams, E.J.   625
Williams, F.   572
Williams, G.Z.   197, 594
Williams, J.D.   240, 512, 605, 637
Williams, T.I.   572
Williamson, E.   190
Willke, T.A.   279
Wilson, A.   240
Wilson, A.L.   199
Wilson, E.B.   588
Wilson, K.B.   232
Winer, B.J.   564, 637, 641
Winne, D.   250, 274
Winsor, C.P.   65, 280
Winter, R.F.   622
Wiorkowski, J.J.   627
Wohlzogen, F.X.   596
Wohlzogen-Bukovics, E.   596
Wold, H.O.A.   53, 588, 590
Wolfe, D.A.   305, 552
Wolford, J.C.   606
Wolfowitz, J.   24
Wood, F.S.   460
Woodward, R.H.   202
Woolf, B.   351, 358, 359, 360
Wu, C.-F.   578

Wyatt, J.K.   319, 616

Yamane, T.   380, 588, 620
Yang, C.-H.   229
Yaspan, A.   604
Yasukawa, K.   93, 588
Yates, F.   136, 197, 272, 351, 366, 367, 380, 428, 518, 571, 580, 594, 609, 615, 620, 630, 632, 641
Youden, W.J.   197, 198, 200, 590, 594
Young, D.S.   594
Young, J.P.   240, 605
Young, R.P.   605
Youtz, C.   516
Yule, G.U.   401, 625

Zacks, S.   570
Zahl, S.   590
Zahlen, J.P.   122, 588
Zaludova, A.H.   235
Zar, J.H.   142, 588
Zelen, M.   142, 235
Zimmer, W.J.   631
Zimmermann, H.-J.   605
Zinger, A.   153, 588
Zschommler, H.G.   571
Zwerling, S.   600
Zwick, Rebecca   488

SUBJECT INDEX

Abbreviations, mathematical 7
Aberrant observations (*see* Outlier)
Abscissa of a normal
  distribution 60
Absolute deviations 262
Absolute experiments 561
Absolute frequencies 54
Absolute value 9
Abstract models 5, 242
Acceptance inspection 129, 231, 232
  sampling plans 231
Acceptance line 129
Acceptance region 256, 257
Acceptance sampling 231, 232
  number 231
Accidental rate 403
Accuracy of a method 198–202
Action limit 200
Active placebo 213
Addition theorem 28–31, 42
Addition tests (in clinical chemistry) 198
Addition theorem of the χ^2
  distribution 139, 366
Additive effects 59, 108, 537, 538
Additive model 537, 538
Additive partitioning of the degrees of
  freedom of a contingency table 468–474, 480, 481
Additivity (noninteraction; analysis of
  variance) 538
  absence of 540
  of chi-squared 139, 366, 367

Actiological, etiological studies 204–209
Alpha error 112–123
Alphabet, Greek endpaper at front of book
Alternate hypothesis 125, 128, 187, 295, 302, 303, 309, 320, 330, 347, 376, 488, 521 (see also alternative hypothesis)
  and rank order 464
  one sided 553
Alternating pattern 211
  test sequence with equalization 211
Alternative hypothesis 114–129, 215, 258, 260, 264, 266, 292, 293, 338, 339, 374, 425, 438, 441, 465, 471, 477, 496, 498, 499, 502, 542
Alternatives, mutually exclusive 464
Analog computers 243
Analysis
  of fourfold tables 346–373
  of information 351–360, 490–493
  of sums of squares 494, 502, 503
  of 2 × 2 tables 346–373
  of two way tables 462–493
Analysis of variance 494–567
  authentic 529
  distribution-free 303–306, 493, 546–558
  for multiple comparison of k relative
    frequencies 464
  for three way classification 519

667

668 Subject Index

Analysis of variance (*cont.*)
   for two way classification 537
   Model 1 with systematic components 529
   Model 2 with random components 529
   of joint samples 526
   one way 501–515
   rapid tests of 542–545
   robust 495, 537
   with ranks, two way 549–558
Analytic chemistry 108, 197–202
Angular transformation 269, 339, 516, 517
   for binomially distributed alternative frequencies 516
Antilogarithm (or numerous) 16
Antilogarithm table 14, 15
Antilogarithms, table of natural 177
Approximating empirical functions using analog computers 243
Arcsine transformation, approximation based on 339
Areas under the standard normal distribution 61, 62, 217
Argument 255, 256
Arithmetic mean 46–49, 68–77, 132, 199, 281, 470
   combined 76, 77
   of deviations
      differences of means, assessing of 452
   weighted 76, 77
Arithmetical computations 11–19, 20
Arithmetical operations 7–20
Artists and scientists 3
Artificial associations 203, 246, 393–395
Associated mean squares (MSX) 523
Associated sum of squares (SSX) 523
Associated variances 521
Association 202–204, 224, 369–370, 379–462, 472–474, 476, 482–487
Assumed imaginary population 49, 107
Assumption-free applicability 131
Assumptions
   of regression analysis 437
   of t-test 130, 264–267
Asymetry (*see* Skewness)
Asymptotic efficiency 127, 130, 286, 303, 397, 403, 405
   of median test 302
Asymptotic relative efficiency 127
Attribute 47

distribution of classes of 475
Attributes 47, 53, 195, 196
   alternative 53
   assigning scores to levels or categories of 484, 485
   categorically joined 53
   dichotomous 53
   orderable or ordinal 53
   plans 231
   qualitative 53
   quantitative 53
Attribute superiority comparisons, minimum sample size for 232
Authentic analysis of variance 529
Authentic effects 210
Autosuggestion 212
Availability 238
Average (*see* Mean)
Average rate of change (86–87), 449
Axioms in probability 29, 30
Axis intercept or intercept 386, 439
   confidence limits for 439, 440
   hypothetical 439
   testing difference between estimated and hypothetical 439, 440

Back transformation 518
Bartlett procedure 390, 391
Bartlett test 498–500
   modification to as proposed by Hartley 500
Basic tasks of statistics 3–6, 23–26, 116, 123–124
Bayesian method 40, 41
Bayes's theorem 38–41, 43
Bayes, T. 38, 134, 567
Behrens-Fisher problem 241, 271–273
   (*see also* Fisher–Behrens Problem)
Bell-Doksum test 293
Bell-shaped curves 55–60
   operating characteristic (OC) 129
Bernoulli distribution 162–170, 333–343
Bernoulli, Jakob 27, 35, 163, 567
Bernoulli ("either-or", "success–failure") trial 162–170, 333–343
Beta error 117–119, 125–129, 504
Bias 5, 66, 67, 196–203, 206, 208, 212, 246, 559
   due to selection 203, 204, 246
Bibliometry 196
Bienaymé, I.J. 64
Binary numbers 242

Subject Index

Binary representation 243
Binomial 157
  coefficients 155–159, 373
  distribution 132, 162, 163–166, 170, 171, 174, 175, 177, 185, 188, 190, 215, 218, 316, 333, 347, 365, 516, 566
    approximation by normal 164
    confidence limits of 165
    control limits computed with the help of 230
    mean of 163
    variance of 163
  homogeneity test 462
  populations 164, 269 465
  probabilities 164, 167
  proportion 269
Binomial to normal distribution, transition from 566
Bioassay 224–228, 596–598
Biological assay (see Bioassay)
Biological attributes 108
Biometrika 567, 576
  biometrical methods 567
Biometrics (or biometry) 196
Birthday problems 36, 37, 180
Birthday selection 246
Biserial correlation, point 407
Bivariate frequency table
  determining r_s in accordance with Raatz from 403
  examination for correlation of two variables 484
Bivariate independence, test of 476
Bivariate normal distribution 242, 383–385, 390, 397, 447
Bivariate normality, test for 447
Blind studies 212, 213
Block division (planned grouping) 559–562
  as underlying principle of experimental design 562
Block formation 561
Blocking 559–562, 565
Blocks 562, 563
  completely randomized 562
  homogeneous 562
Bonferroni
  χ^2 procedure 537
  inequality 117, 267, 360, 426, 478–479, 537
  multiple test procedure, sequentially rejective 537
  procedure 478

χ^2-table 478
χ^2-statistics, upper bounds of 479
  one-stage and multistage 426
t-Statistic 267, 537
Bowker test 488, 489
  generalization of 489
Bravais and Pearson product moment correlation coefficient 383, 384, 406, 407
Breakdown probability 237
Bross
  sequential test plan 220–222, 241
Brownian motion 48

Calculators 19, 20, 572, 575
Calibration 461
Canonical correlation analysis 461
Capture–recapture study 172
Card playing 30, 31, 36, 47
Caret 362
Carrying out study 565
Case-control study 205, 206, 209
Categorically itemized characteristics, testing strength of relation between 482
Causal correlation 393, 394
Causal interpretations 394
Causal relations 206, 224, 393
  topology of 394
Causal relationships 206, 393, 394
Causal statement 394
Causation and experimental design 232, 233, 558–564
Causing and correlation analysis 206, 207, 393–395
Cell frequency, expected 321, 348, 464
Censoring 66, 195
Centile 92
Central limit theorem 46, 59, 98, 155, 279, 704
Central tendency of a distribution 100
Central tendency of a population (sign test) 319
Chain sampling plans 231
Change point problem 166
Characteristic
  and influence factors 565
  fixed levels of 532
  random levels of 532, 533
Chebyshev, P.L. 64
Chebyshev's inequality 64
Checking sources of information 565
Chi-square 139–143, 154, 155

χ^2 distribution (chi-square distribution) 139–143, 154, 155, 321, 345, 431, 464, 488
χ^2 goodness of fit test 320–324, 329, 330, 332
χ^2 statistic (Fisher's combination procedure) 366
χ^2 table 140, 141, 349
χ^2 test 320–324, 329, 330, 332, 346–351, 462–482
    simple, for comparing two fourfold tables 362
CI, confidence interval 248
Classed data 53–57, 73
Class frequency 54–56
Class number 73, 76
Class width 73, 76, 92, 97, 106, 420, 422
Clinical statistics 37, 195–224
Clinical testing of drugs for side effects 223, 224, 337
Clinical trial (*see also* Therapeutic comparison) 209–218, 223
Clinical trials, planning of 214, 218
Clinically normal values 197
Clopper and Pearson's quick estimation of confidence intervals of relative frequency 340–341
Cluster 209, 246
    effect 376
        hypothesis 378
    sample 209, 246
Cochran test, homogeneity of several variances 497, 498
Codeviance 413
Coefficient
    of correlation 309, 382–384, 388, 390, 406, 407
    of determination 389, 449
        nonlinear 449
    of kurtosis 102–106, 325–327
    of skewness 102–106, 164, 325–327
    of variability 77, 78
    of variation 77, 78, 85, 107, 110, 133, 198, 234, 259, 260, 275, 392
        comparison of two 275
Cohort studies 204, 205, 208–210
Coin tossing 27, 47, 115, 157, 166
    problem 157
Cole sequential plan 223
Collinearity 390
Column effect 519, 521, 532, 538
    mean 523, 527, 538, 540

Combination of kth order 161
Combinations
    and permutations 155–162, 192, 193
    of means
        with corresponding standard errors of independent samples 96
        with corresponding standard errors of stochastically dependent samples 96
    with or without regard for order 161, 162
    with or without replication 161, 162
Combinatorics 155–162, 192, 193
Combined two tailed rank correlation statistics 403
Combining comparable test results 119, 366, 367
Combining errors 21, 22, 96
Combining evidence from fourfold tables 367–369
Comparability 207, 208, 558–561
    principle 558, 559
Comparing groups of Time-dependent measurements 541
Comparing surveys and experiments 561
Comparing therapies 210–224
Comparing two
    fourfold tables 362
    r-c tables 484
Comparison
    experiments 561
    groups 207–214
    of a mean with standard value 121–123, 255, 256, 274
    of central tendency of empirical log-normal distributions 111
    of means 264, 494, 529
    of observed frequencies with expectation 321
    of regression lines 442
    of survival distributions 195, 210
    of two percentages 364
    tests 198–200
    two sided 124–129
    within blocks, increasing precision of 560
Complement 42
Complementary date 196
Complete association, functional dependence 383
Completely randomized design 563
Completely randomized two variable

classification with replication 529
Comparability principle 558, 559
Computational
 aids 19, 20, 572, 573
 economy of tests 131
Computer 241-244, 572
Computer programs 572, 573
Computer simulation 241
 examples 241, 242
Conclusion 24, 25, 134, 246, 566
Concurrence and nonconcurrence 540
Conditional probability 31-43, 266
Conditions of comparability 207, 208, 558-561
Confidence belt 443-446
Confidence bounds for the median 254, 319
Confidence coefficient (S) 112-119, 123, 220, 249, 250, 255, 260, 280, 293, 341, 343
Confidence ellipse for estimated parameters 392
Confidence interval 70, 112, 113, 116, 183, 198, 234, 246, 247, 249, 250, 252, 253, 255, 258, 259, 261, 267, 268, 273, 311, 334, 335, 392, 419, 423, 428-430, 440, 443-446, 534, 567
 and test 116, 256, 257, 268, 273
 for difference between two means 267, 268, 271
 distribution-free 254
 for axis intercept 439, 440
 for coefficient of correlation 423, 424, 427, 429
 for coefficient of variation 259
 for difference between medians 303
 for difference between two means 267, 268, 271
 for expectation of Poisson variable 182, 183, 343-345
 for failure rate 235, 236
 for lambda (mean of Poisson distributions) 182, 183, 343-345
 for mean 198, 247, 248, 252, 253
 for mean (lognormal distribution) 111
 for mean difference of paired observations 310, 311
 for mean lethal dose 225-227
 for means (k means simultaneously) 514
 for mean times between failures 235
 for median 95, 254, 319
 for median differences 254
 for parameters of the binomial distribution 333-337
 for quantiles 197, 198, 254
 for rare events 182, 183, 343-345
 for ratio of two means 272
 for regression coefficient 392, 439, 440
 for regression line 443, 446
 for relative frequency 333-337, 340, 341
 for residual variance 440
 for standard deviation 259
 for true mean difference of paired observations 310, 311
 for variance 440
 of binomial distribution 333
 of mean 246-248
 of median 254, 319
 of parameters of multinominal distribution 194
 of range 508
 of relative frequency, estimation of 166, 333
 one sided 247
 simultaneous, to contain all k population means 514
 two sided 247
Confidence level, confidence coefficient (S) 112-119, 123, 220, 249, 255, 260, 280, 293, 341, 343
Confidence limits 247, 248, 268, 281, 334, 341, 432, 443
 for the mean 253
 of correlation coefficients 423, 424
 of observed frequency 333
 of Poisson distribution 182, 183, 343-345
Confidence probability 247
Confidence region 445, 446
Configurations of interactions 495
Conservative decision/test 131
Conservative statistical decision 131
Consistent estimator 67
Constant growth rate 87
Consumer risk 231
Contingencies, systematic study of 238
Contingency coefficients 369, 370, 384, 482
 according to Cramér 483
 according to Pawlik 482, 483
 according to Pearson 369, 370, 482, 483
 analysis 384

Contingency coefficients (*cont.*)
  corrected   462, 483
  maximal, for nonsquare contingency table   483
Contingency die   493
Contingency table   370, 476, 482
  examining for independence   475
  square   475, 482, 488, 489
  weakly occupied   464, 475
Contingency tables   139, 462, 482, 490, 492, 493, 627–630
  analysis of incomplete two and three way   493
  of type $(r \times c)$, testing for homogeneity or independence   194, 474–493
  ordered categories in   472–474, 484–487, 493
Continuity assumption   131
Continuity correction   165, 215, 217, 302, 335, 345, 350–352, 364, 365, 379
Continuous quantities   53
Continuous random variable   45, 46, 57
Continuous sampling plans   231
  acceptance or rejection decided on unit-by-unit basis in   231
Continuous uniform distribution
  mean of   84
  probability density of   84
  variance of   84, 85
Contrasts
  population   509–512
  weighted linear   509–512
Control charts   200–202, 229, 230
  for countable properties (error numbers and fractions defective)   230
Control correlation   395
Control group   203, 205–209, 562–565
  matching   50, 206
Controlled error   4–6, 50–52, 58, 196–198, 245, 246, 273, 558–562
Controlled experiments   494–566
Controlled trial involving random allocation   49–51, 195–224
Control limits   200, 230
Controls (standard test conditions)   561, 565
  and accompanying control experiments, as underlying principle of experimental design   562
Control sequence   207
Control trials   561
Control variable   349

Convergence statements   68
Corner-n (or corner sum)   464, 474, 475
Corner test   384, 395, 405, 406
  of Olmstead and Tukey   405, 406
Correction according to Yates   351
  testing for independence or homogeneity   351
  fourfold table   352
Correlated samples   307–320, 518–541, 549–558, 562
  rank analysis of variance for   549–558
Correlation   382–385, 387–389, 393–436
  coefficient   383, 388, 390, 406, 407
  control   395
  multiple   457–460
  partial   395, 456–458
  testing for   424–432
  true   393–395
Correlation analysis   382–461
  canonical   461
  corner test (Olmstead–Tukey)   395–396, 405–406
  multiple correlation   457–460
  normal   242
  partial correlation   456–458
  point biserial correlation   507, 508
  product-moment correlation   309, 382–384, 388–390, 406, 407, 419–432
  quadrant correlation   395, 396, 403–405
  Spearman correlation   395–403, 554, 555
Correlation coefficient (ρ)   133, 383, 402, 406, 407, 410, 414, 416, 419, 420, 423–425, 427–431, 436, 449
  common   430, 432
  computing and assesing, nomograms for   427
  estimated   406–410
  for fourfold tables
    estimation of   370
    exact computation of developed by Pearson   370
  hypothetical   430, 432
  multiple   384, 457–460
  nonlinear   449
  other tests for equality   426
  partial   384, 456–458
  point biserial   407, 408
  ρ 383, 389, 395

r, testing for significance against zero 425
significance of 425
test of homogeneity among 431
testing two estimated for equality according to Hotelling 426
two sided comparison of two estimated 430
Correlation ratio 435, 436
Correlation table 419–423
Correlations
multiple tests of 426
pairwise comparisons among k independent 427
partial and multiple 456–460
Counted observations 53, 57, 333, 516, 517
Counting process 57
Covariance 413
Cox and Stuart sign test 379–381
Critical bounds, critical limits, critical values 112–155, 255–257, 362
Critical difference 533, 556, 558
for one way classification 547
for two way classification 556
Critical limit 112–155, 255–257, 362
Critical probability 118
Critical region of test 256, 257, 266
Critical time 87
Critical value 112–155, 255–257, 362
Cross-sectional time series, analytical procedures for 381
Cubes 10
Cubic equation 447–453
Cumulative binomial probability 165
Cumulative distribution function (of the population) 44, 56, 68, 291–298
empirical 68
sample 68
Cumulative empirical distribution 68, 82
Cumulative frequency 56
Cumulatively added relative portions of reacting individuals 225
Cumulative percentage curve 224
frequency distribution 224
Cumulative percentage line 82
Cumulative percentages 56, 82
Cumulative sampling results, criteria for acceptance or rejection applied to 231
Cumulative sum chart 201, 229
Cumulative sum line of normal distribution 83

Curve fitting 447–461
by orthogonal polynomial regression 460
Curve forms 453
Curvilinearities 495

D'Agostino test for nonnormality 329
Data 4–6, 25
analyzing 5, 6, 565
binary 363
collection 55
dichotomous 53, 463, 469, 489, 555
disclosure 55
evaluation 55
grouped 53–57
into two classes 72–76, 92, 106
hard 210
preparation 55
preparation of arrangements for recording, checking and evaluation of 566
soft 210
ungrouped 57, 72
Data analysis 5, 6, 565
Data compilation 54
Data management 55
Data pairs 307–320
Data processing 55, 246, 572, 573
David method for comparing several variances simultaneously 500
Decile 92, 98–102, 132, 325
Decimal numbers 242
Decimal representation 242
Decision 24, 134, 239, 566
Decision criteria 133, 134
Decision principles 133
Decision problem 238
Decisions
and conclusions 566
and strategies 240
optimal 24
Deduction 124
Deductive inference 116, 124, 257
Deductive procedure 124
Defective units 230–232
Definitions of statistics 3, 24, 25, 46, 123
Definitive assessment of significance 132
Degree of association, Pearson contingency coefficient statistic for 370
Degree of connection 309

Degrees of freedom (DF)   135, 137–140, 153, 260, 262, 266, 271, 272, 303, 308, 310, 321, 323, 345, 348, 349, 352, 366, 381, 424, 425, 431, 433, 441, 444, 457, 464, 467, 476, 477, 480, 481, 484, 488–490, 492, 495–499, 502, 523, 528, 533, 534, 540, 549, 550, 562, 564
  of a $k \times 2$ table, partitioning   468
Delphi technique   53
de Méré, C.   35, 567
Demographic evolution   48
de Moivre, A.   58, 59, 567
de Moivre distribution (normal distribution)   59
Density function   45, 59, 79, 138, 139, 143, 166, 267
Departure from normality   65, 66, 82, 83, 90–111, 327
Dependence   395–396
Dependent samples   363
Depletion function   233, 234
Descriptive significance level   119, 120, 266
Descriptive statistics   4, 24–25
Design, experimental   558–566
Detection limit   199
Developing model appropriate to problem   565
Developing statistical model   5, 6, 47–49, 55, 81, 116, 124, 161, 238, 240–242, 257
Developing technique of investigation   565
Deviation (*see also* Dispersion, Variance, Variation)
  from normal distribution   65, 66, 82, 83, 90–111, 261, 327
  mean absolute (MD)   251, 252, 281
  measure of   122
  median   253, 254
  of column means from overall mean   519
  of ratios   345, 346
  of row means from overall mean   519
  of strata means from overall mean   519
  probable   60
Diagonal product   370
Diagonal sum   486
Dice
  games with   169, 566
  tossing   27-35, 43, 44, 47, 169–194, 239, 322, 332, 375, 567
Dichotomous attributes   53, 463
Dichotomous data   333–373, 375, 462, 463, 555
  comparison of several samples of   462
Dichotomous result   224
Dichotomy   407
Die game   47, 322, 332
Difference between regression limits   440–442
Difference between sample means   246–275, 501–518
Difference between sample medians   293–306
Difference-sign run test   379
Diffusion   48
Digital computers   242, 243
Digital devices   242
Direct inference   249
Directories   572
Discordant observations (*see* Outliers)
Discoverer and critic   119
Discrete distributions   162–194
Discrete frequency distribution   44, 57
Discrete quantities   53
Discrete random variable   44–46, 57
Discrete uniform distribution, definition of   84
Discrete random   44–46, 155–194
Discriminant analysis   460
Disease   37, 40, 50, 195–218
Dispersion   24, 286, 303, 388, 389
  about the regression   540
  between individual members   562
  between trial units   562
  between units   562
  difference   287–289, 562
  index   189–191
  interdecile range   98–101
  kurtosis   100–107, 252, 325–327
  measure of   251
  median deviation   253, 254, 281
  multiple correlation   457–459
  of a process   229
  of a regression, interpretation as   540
  of observations   389
  of placebo-reaction   213
  of predicted values   389
  parameters   46
  partial correlation   456–458
  range   97–100, 507–509
  standard deviation   69–76, 97, 98
  standard error

Subject Index

of the mean 94
of the median 95
statistics 99, 100
test (of Poisson frequencies) 190, 465
within groups 310
Distribution 44–48, 53–57
 arbitrary 64
 asymmetric 65
 bell-shaped 55–60
 Bernoulli 162–170
 bimodal 91
 binomial 162–170, 215, 230, 566, 704
 bivariate normal 242, 383–385, 390
 bizarre 111
 χ^2 (chi squared) 139–143, 154, 155, 362, 366, 476, 704
 de Moivre 59
 discrete 162–194
 dispersion of 291
 exponential 234, 252
 extreme value response 226
 extreme values 111
 F 143–155, 362, 704
 F' (Pillai–Buenaventura) 276
 frequency 53–66
 function 44–46, 61, 68, 291, 295
 Gaussian (see Normal distribution, Standard normal distribution)
 hypergeometric 171–175, 371, 704
 inverse hypergeometric 172
 J-shaped 391
 left-steep 99, 100
 lognormal 83, 107–111, 225, 230, 235, 323, 515, 517
 L-shaped 111, 139, 143
 mixed distributions 83, 234
 multimodal 65, 91, 93, 107
 multinomial 193, 194, 475
 negative binomial 177, 190
 Neyman 190
 normal 57–66, 78–83, 94, 95, 98, 111, 155, 225–227, 234, 235, 241, 252, 322, 329, 331, 391, 566 (see also Distribution, standard normal)
 odd 100
 of Cochran statistic for several variances 497
 of confidence limits for the mean of a Poisson distribution 343–345
 of cumulative standard normal 62
 of differences 310
 of estimator 98

of extreme values 111, 235
of failure indices 236
of Friedman statistic 551
of Hart statistic (mean square successive difference) 375
of Hartley statistic, several variances 496
of life span 233–236
of Link–Wallace statistic 544–547
of Nemenyi statistic 546, 547
of number of runs 376, 377
of ordinate of standard normal curve 79
of r 242, 384, 395, 425
of R/s (Pearson–Stephens) 325, 328, 329
of rank correlation coefficient 398, 399
of rank sum statistic H 305
of rank sum statistic U 296–301
of ratio
 largest variance/smallest variance 495, 496
 largest variance/sum of variances 497, 498
 range/standard deviation 325, 328, 329
of signed rank statistic (Wilcoxon) 313
of signed test statistic 317
of standardized extreme deviation 281
of standardized 3rd and 4th moments 326
of Studentized range 535, 536
of tolerance factors for normal distribution 282
of Wilcoxon–Wilcox statistic 556, 557
overdispersed 190
Pareto 111
Poisson 175–192, 230, 236, 241, 329, 343–345, 516–517
positively skewed 93, 100, 107
rectangular 83–85, 98, 391
right-steep 107
sample 53–60
Skewed (skew) 90–93, 100, 111, 391
skewness of 91–93, 100–107, 291, 293, 391
special cases $p = 1$ 336, 337
standard Gaussian (see Distribution, standard normal)
standard (standardized) normal 60–64, 78–81, 154, 155, 217, 704

Distribution (*cont.*)
  Student or *t*-distribution (*see t*)
  t 135–138, 154, 155, 255–257, 264–275, 309–312, 424, 426, 437–442, 704
  theoretical 61–63, 135–155, 362
  triangular 98, 190, 252
  trimodal 91
  two dimensional 107, 383, 384, 390
  underdispersed 190
  uniform 83–85, 98, 241, 252, 322
  unimodal 64, 91, 99, 107
  unsymmetric 90, 91, 99, 103, 107, 234
  U-shaped 93, 94, 98, 99, 391
  Weibull (generalized exponential) 234, 235, 323
Distribution-free comparison of several correlated samples with respect to their central tendency 549–558
Distribution-free comparison of several independent samples 303–306, 546–549
Distribution-free measures of association 395–406
Distribution-free methods 123, 130–133, 590
Distribution-free multiple comparisons of correlated samples 553–558
Distribution-free multiple comparisons of independent samples according to Nemenyi 546–549
Distribution-free 90% confidence interval for quantiles 197
Distribution-free procedures 130, 131, 254, 261, 283–306, 312–320, 395–406, 546–558
Distribution-free tests 123, 128, 130, 131, 133, 285–306, 312–320, 395–406, 546–558
Distribution function 44–46, 56, 61, 68, 291, 299, 396
Diversity of treatments
  as underlying principle of experimental design 562
  encoding to avoid subjective influences 562
Dixon and Mood sign test 315, 316
Dixon comparison 277, 278
Documentation 55, 208, 209, 566
Dosage 226
Dosage–dichotomous effect curves 224
Dosage–effect curve 224, 225
  mean and deviation from 225

Dosage scale
  linear 226
  logarithmic 226
Dose 224, 225, 227, 228
  bringing on narcosis 224
  causing impairment 224
  lethal 224, 225
  mean effective 224, 225, 228
  mean lethal 225, 227
  median effective 225
  median lethal 225
  symmetrically grouped about mean 225
Double blind study 213
Double blind trial 195, 213
Double logarithmic paper 455
Double partitioning with a single observation per cell 549
Doubling period 87
Drawing inferences on population parameter from portion in sample (indirect inference) 340, 341
Drug side effects 223, 224, 337
Duckworth and Wyatt modification 319
Dunn test 477
Durability curve 233

Econometrics 196
Economic statistics 108
Effect
  additive 59, 108, 537, 538
  column 532
  fixed 532
  interaction 519, 538, 540, 541
  main 518–533
  multiplicative 107
  nonlinear 539
  random 532
  row 532
Effect curve
  dosage–dichotomous 224
Effect error secondary cause 211
Effect relation
  dosage–dichotomous 224
  dosage–quantitative 224
Effects between strata, comparison of 205
Effects within strata, comparison of 205
Efficiency
  of asymptotic test 130
  of nonparametric test 130
Efficient estimator 67
Election 51, 53, 245

Elementary events, space of 28
Empirical cumulative distribution function 44, 46, 56, 61, 68, 291, 295
Empirical data 53–58, 333, 346
Empirical distribution
 comparison with
  normal distribution 322
  Poisson distribution 329
  uniform distribution 322
 left-steep 91, 100, 101, 107, 111
 multimodal 65, 91
 right-steep 516
 symmetric 58, 100
 unimodal 64, 91, 99, 107
Empirical facts and scientific inquiry 116
Empirical frequency distribution and normal distribution 82
Empirical generalizations 116
Empirical mean in comparison 255, 264
Empirical regression curve 446
Empirical variance in comparison 258, 260
Empty set 29
Engineering and industrial statistics 107–111, 175–189, 228–238, 241–244, 343–345, 382–461, 494–566, 598–603 (*see also* Quality control)
Engineering design 232
Epidemiological studies 206
Equality of variance of errors 519
Equality of variance of several samples, testing according to Bartlett 498
Equality or homogeneity hypothesis concerning variances 495–500
Equalizing alternation 211
 principle 220, 222
Equalizing line of point cloud 382
Equivalence of two contingency tables 362, 484
Error 3–6, 26, 53, 55, 66–68, 96, 195–203
 mean relative 96
 of the first kind 112–113
 of the second kind 117–119, 125–129, 504
 sources of 26, 66–67, 195–197
 systematic, bias 5, 66, 67, 196, 197, 199, 201–203, 212, 246, 559
 Type I 117, 118, 119, 120, 125, 251, 561
 Type II 117, 118, 119, 125, 128, 251, 504, 564

Type III 119
Type IV 119
 unconscious and unintentional 212
Error law distribution 58
Error numbers 230
 number of defects per test unit as 230
Error of the first kind, prescribed level of significance protection against 537
Error rates in multiple comparisons among means 242
Errors of substitution 195
Error sum of squares 414
Estimate 47
 biased 66, 67, 71
 of kurtosis 105
 of mean 104
 of parameter 66, 67
 of skewness 105
 of variance 105
 unbiased 70
Estimates in analysis of variance 527
Estimating linear regression portion of total variation 462, 472–474, 484–487
Estimating sample size, counted data 342, 350
Estimation of the correlation coefficient 406–407
Estimation of minimum size of sample 49, 214–218, 249–251, 274, 341–343, 350, 351, 541
Estimation of parameters 66–67
Estimation of the regression line 408–413
Estimation of sample size 49, 214, 218, 249–251, 274, 341–343, 350, 351, 541
Estimation procedure 66–68, 124
Estimation theory 25, 66, 67, 124
Estimator 66, 67, 98
 consistent 67
 efficient 67
 sufficient 67
 unbiased 66, 70
 Estimator of a parameter 66
  biased 66
  consistent 67
  efficient 67
  sufficient 67
  unbiased 66, 70
Eta, correlation ratio 436
Ethical aspects in clinical research 195
Etiological studies 204–209

Etiology 205, 206, 208
Euler's circles 30
Evaluating hypotheses 565
Evaluation forms 111
Evaluation of biologically active substances 224
Event
 arbitrary 30
 certain 28, 29
 complementary 29
 elementary 28
 impossible 29
 independent 5, 32, 38
 intersection of 29, 30
 mutually exclusive 5, 29, 30, 38
 mutually independent 33, 34, 42
 pairwise independent 42
 random 28
 stochastic 43
 union of 28
Evidence 3–6, 134
Evolutionary operation, optimal increase in performance through an 233
Exact test according to Fisher 302, 370–373
Excess (*see* Kurtosis)
Expected cell frequencies 320–330, 348, 351, 464–466, 475
Expected value 46
Experimental design 55, 559, 562, 564, 567, 640, 641
 modern 562
 principles 558–564
 underlying principles of 559
Experimental error 503, 519–523, 526, 528, 532, 538–540, 559, 560–563, 564
Experimental strategy for quality improvement 232, 494
Experimental Type I error rate 537
Experimental units 560, 563, 565
Experimentation on the model 240
Experimenting with a model 242
Experiments, planning and evaluation of 494–564
Experiments without replicates 537–540
Explanatory variable 384
Explicit parameters 166
Exponential distribution 234, 252
Exponential growth 87 (*see also* Growth curves)
Exponential regression 451
Exposition and risk 205, 209
Extension of sample range 559

Extreme value probability paper 111
Extreme value response distribution 226
Extreme value theory 111

Factor 526–529
Factor analysis 457, 625
Factorial 155, 160
Factorial design 563, 564
Factorial experiments 563, 564
Factorials 160
Factor levels 533
Factors (or experimental variables) 494, 519, 558–564
 comparison among 562
 effects of 562
 interaction among 562
 systematically chosen levels of 529
 variability of 562
Failure index 236
Failure rate 234–236, 238
Failure time 235, 236
Fallacies in statistics 26
False negative (*see* Beta error)
 in medical diagnosis 223
False positive (*see* Alpha error)
Faulty decisions 118
F-distribution 143–155, 261, 426, 502, 503, 551
 relationship to other test distribution 154
 upper significance levels 144–149
Fermat, P. de 35, 567
Field trials 494
Filter function of statistics 196
Finite population correction 248, 336, 340
Finney tables for one sided test 372
Fisher–Behrens problem 271 (*see also* Behrens-Fisher problem)
Fisher-Irwin test for 2×2 tables 350, 370–373
Fisher, R.A. 143, 166, 427, 494, 567
Fisher's combination procedure 366, 367
Fisher's exact test 348, 351, 370–373
Fisher's r to z transformation 427–432
Fisher test
 of independence, exact binomially distributed populations, approximation for comparison of two 370–373
 two sided 372

Subject Index

Fitting distribution to set of data 320, 382, 383
Fixed effects 532
  model of 501–541
Fixed model 532
Forecast feedback 53
Forecasting, empirical investigation on accuracy of 381
Formal correlation 393, 394
Formulating problem and stating objectives 565
$4 \times k$ median quartile test 306
Fourfold χ^2 test 346–351, 360
Fourfold scheme 346–349
Fourfold table 346–349, 361, 363, 366–370, 405, 462, 471, 483, 490
  analysis of 346–373
  combination of 367–369
  evaluation of 346–373
  regarded as simplest two way table 474
  standardization 351
  test for comparing association in two independent 370
  to general case, extension of 474
  values, adjustment according to Yates 352
Fourfold test 346–373
Four parameter probability function 323
4-sigma region (6ŝ) 279
Fractile 92
Fractional factorial experiment 564
Free hand line 409
Freeman–Halton test for $r \times c$ tables 475
Frequency
  absolute 55
  comparison of two 236, 338–373
  cumulative 56
  information content of 462
  inner field 350
  marginal 350
  relative 55
Frequency data 515–517
  comparison of two independent empirical distributions of 467–468
Frequency distribution 53–66, 200, 463, 469, 475, 486, 492
  cumulative 44
  three dimensional 518
Frequency function 44
Frequency profile 107
Frequency, relative 26, 333

Frequency table 467
Friedman rank analysis of variance 549–555
Friedman rank test 306, 549–555
Friedman test 306, 315, 495, 549–555
  generalization 541
F-test 133, 260–264, 501–506
  due to Gart 360, 361
Function
  defined as allocation rule 255, 256
  distribution 44
  frequency 44
  probability 44
Functional parameters 166
Functional relation 382, 393
Functional relationship 385
Function value 256
Futurology 51
F-value 143–155

Gain and loss 133, 134
Games
  evasion 240
  pursuit 240
  supply 240
  war 239, 240
Game theory 239, 240, 604, 605 (see also War games)
Gart approximation 348, 351, 360, 361
Gart's F-test 348, 351, 360, 361
Gauss, C.F. 58, 64, 67
Gaussian (or normal) distribution 57–66, 68, 78–83
Gauss test 121–123
Generalizability principle 4, 47–49, 95, 116, 202, 209, 210, 558, 559
Generalized median test 302
Generalized sign test 363
Genetic ratio test 321, 322
Glivenko and Cantelli theorem 68
Goodness of a test, estimation of 241
Goodness of fit test 123, 320–332, 449
  chi-square 320–324, 329, 330
  due to David 332
  due to Quandt 332
  for singly truncated bivariate normal distribution 447
  nonnormality due to skewness and kurtosis 325–327
  of Kolmogoroff and Smirnoff 330–332
Gosset, W.S. 135
Grandma's strudel method 216

Graphical methods 6, 55–57, 78–81, 83, 86, 200, 201, 229, 230, 383, 409, 444, 449, 455, 493, 495, 517
Greco-Latin square 564
Grouped data 53–57, 72, 73
Group homogeneity 207–210
Grouping into classes 72, 73
Group means 501–515
Groups of time-dependent measurements, comparing by fitting cubic spline functions 541
Growth 86, 87, 214, 455
Growth curves 455
G-test 302, 349, 351–360, 370 (*see also* Woolf's G-test, Log likelihood ratio test)
g-values 351

Hamlet 27
Hard data 210
Harmonic interpolation 151
Harmonic mean 88, 498
Hartley's test 495–498
Hay bacillus 552
Hazen line 83
Heterogeneity
 hypothesis, general 474
 of population 204
 of sample units 65, 83, 196, 197, 204, 209, 245, 273, 393–395, 559–561
 of trial units 561
Heterogeneous samples 65, 83, 234, 470
Hierarchic experimental designs 564
 classification 564
Histogram 55, 56, 63, 107, 204, 449
Historigram 55
History of probability theory and statistic 27, 35, 59, 64, 123, 567
Homogeneity
 hypothesis 260, 348, 462, 476, 495
 obtaining of 516
 of sample 471
 table 348
 test 192, 291, 328, 346–363, 474–482, 490, 491
  for a $k \times 2$ table, applying, to compare two frequency distributions 462–468
  for matched samples 552

 for s correlated samples of dichotomous data 489
 of binomial samples 338, 339, 346–362, 462–468
 of groups of time-dependent measurements 541
 of multinomial samples 474–478, 490, 491
 of Poisson samples 186–189, 236, 474–477
 of profiles of independent samples of response curves 541
 of variances 260–263, 311, 312
 testing for 192, 291, 328, 346–363, 431, 432, 462–468, 474–482, 490, 491
Homogeneous group 207–211
 using modified LSD test to form 512, 513
Homomer 130
Homoscedasticity 126, 437
H-test 132, 293, 303–306, 495
 for the comparison of variances 262
 modified to test paired observations for independence 306
 of Kruskal and Wallis 262, 300, 302–306
 power of 306
H-values 305
Hybrid computer 243
Hypergeometric distribution 171, 174, 175, 371
 cumulative probability of 174
 generalized (or polyhypergeometric distribution) 171, 193
 inverse 172
 symmetric 371
Hypergeometric distribution tables of Lieberman and Owen 373
Hypotheses
 on causation 393
 testing partial 480
Hypothesis 3–6, 23–26, 114–129, 256, 257, 266
 alternative (or alternate) 114, 115, 116, 129
 deductive relations among 116
 null 114, 115, 116, 129, 131
 of independence (or homogeneity) 350, 475
 ordered by rank 116
 substantiated 125
 testing 24, 112–133, 255–257
Hypothetical population 49, 107

Illness 37, 93, 209–214, 246
Imperfections in implementations (*see* Bias, Error, and Fallacies)
Incidence and relative risk 206
Incidence probabilities 206
Inclusion inference 249
Incompletely randomized blocks 562
Independence 32–37, 46, 204
Independence
 lack of 476
 statistical 31
 stochastic 33, 34, 42, 46
 testing for 462, 474
 testing of 480
 test of 346–348, 476
Independence condition 204
Independence hypothesis 402
Independence of attributes 347, 476
Independence of observations within and between samples 503
Independence of test results 559
Independent data samples 245
Independent observations 46, 47, 121, 130, 264
Independent probabilities 366, 367
Independent sample groups 495
Independent (uncorrelated) samples 254, 260, 266, 562
 events 32–34
 variables 46
Independent unbiased estimate 503
Indexing critical values of the χ^2 distribution 142
Index number 77
Indirect inference 340
Induction 124
Inductive basis 559
Inductive inference 24, 124, 257, 559
Inductive procedure 124
Industrial planning 232, 238
Inequalities 7, 64, 65
Inequality
 Bienayme's 64
 Chebyschev 64
 Gauss' 64
Inference
 deductive 124, 257
 inductive 124, 257
 nonparametric 130–133, 283–306, 312–320, 546–558
Inference about defined population 5, 25, 46–49, 65, 107, 196, 565
Infinite population 46, 48, 248

Influence curve 253
Influence factors 195, 494, 560
Influence of different variances on comparison of two means 241
Influence quantities (factors) 457
Influence quantity (regressor) 389, 458, 459
Influence variable 385, 408, 457, 458
Influencing factors 232, 233
Information analysis 462, 492, 493
Information content of frequencies 462
Information loss in grouping 76
Information statistic 462
 for testing two way tables for independence or homogeneity 462, 490–493
Inherent uncertainty 224, 239
Inhibition zone 552, 553
Inhomogeneities, existence of 65, 83, 207, 208, 234
Inhomogeneity correlation 393, 394
Insurance 133, 134
Interaction 492, 518–533, 539–541
 effect 519, 529, 538
Interactions 519, 521, 538–540, 563
 nonadditive effects separated from 540
Intercept 386
 confidence limits of 439
 determination of 391, 408, 417, 419–423
 standard deviation of 415
 testing the difference between an estimated and a hypothetical value of 439
Interdecile range 98, 99, 101
Interdecile region 85
Interpolation of intermediate values 152
Interpolation of probabilities 153
Interpretation 120, 134, 195–197, 393–395, 566
Interquartile range 100
Intersection 29
Interval estimation 112, 113, 116, 246, 247, 249, 254, 258, 268, 273
Interval scale 132, 133
Intervals (open and closed) 165
Interview 55, 196, 197, 202, 208
Interviewer's bias 196, 197, 208
Inverse inference 249
Inverse transformation 518
 means and variances are not always unbiased when obtained under 518

Investigations 3–6, 55, 107, 565, 566

Joint correlation 394, 395
J-shaped distribution 391

k designs or experiments 564
$k \times 2$ χ^2 test 164, 462–474
$k \times 2$ contingency table 465, 467
$k \times 2$ table 463, 464, 468, 473
  two way analysis of 462
Kendall's tau 403
Kerrich procedure 392
Knapsack problem 239
Kolmogorov (Kolmogoroff), A.N. 28, 567
Kolmogoroff and Smirnoff
  comparison of two independent samples 291–293
  goodness of fit test 330–332
Ku correction 490
Kullback tables 360
Kurtosis 85, 100–107, 252, 255, 299, 325–327
  related quick test for nonnormality 252

Laboratory control chart 200–202, 229, 230
Laplace, P.S. de 27, 58, 134, 567
Latin square 562–564
  experimental design of a 562
Law of large numbers 49, 67, 68, 566
  strong 49, 67, 68
  weak 49, 67, 68
Least significant difference (LSD) 512, 513
Least square estimation of regression coefficients 67, 385, 461
Least square line 386, 387, 408–413
Least square method 67, 385
Leslie–Brown test 496, 508
Lethality 37
Level of significance 112, 123, 138, 260, 261, 263, 264, 283, 285, 286, 291, 349, 403, 425, 497
Level (or intensity) setting of a factor 519
Levels of a factor 533, 563, 564
Levels of risk II 117–119, 125–129
Levene procedure 498, 500
Levene test 498, 500

Life span 206, 210, 233–236
Likelihood ratio test (*see* G- and 2*I*-test)
Liliefors method with the Kolmogoroff–Smirnoff goodness of fit test 331
Limits
  action 200, 229
  warning 200, 229
Linear approximation of nonlinear problems 239
Linear contrasts 509–512
Linear correlation 456
Linear inequalities 239
Linearity hypothesis 435, 436
Linearity test 435
Linearizing transformations 453–455
Linear programming (or linear optimization) 238, 239
  and operations research 238, 239, 603, 604
Linear regression 382–392, 408–423, 433, 437–446, 462, 472–474, 487, 517
  χ^2 473, 484
  departure from 433–437, 472, 487
  estimation of 408–423
  in overall variation
    component due to 484
    share of 472
  multiple 459, 460
  outlier and robust regression 447
  test for 433–437, 472, 487
  testing significance of 433–437, 472
Linear trend 472, 484
Link and Wallace multiple comparisons of means 542–545
Link and Wallace test 508, 542–545
Location measures 85, 106
Location parameter 132, 286
Location test 285
Logarithm 11–19
  binary (or logarithmus dualis) 19
  Briggs (or common) 11–19
  characteristic of 17
  common 11–19
  decadic (or common) 11–19
  mantissa of 16, 17
  table 12, 13
  to base *e* (or natural) 13, 19, 142, 177
  to base 10 11–19
  to base 2 19
Logarithmic interpolation 142
Logarithmic probability grid 108

Logarithmic transformation 65, 107–111, 515–518
Logarithms
  base ten factorials 160
  calculation with 11–19
Logarithmus dualis 19
Logit transformation 228, 269
Log-likelihood ratio for contingency tables (see G- and 2I-Tests)
Log likelihood ratio test (G-test) 353, 359
Lognormal distribution 83, 107–111, 225, 230, 235, 323, 515–517
  comparison with empirical distribution 323
  three parameter 111
Logrank test 195
Longitudinal studies 210 (see also Cohort studies)
Lord comparison 276, 277
Lottery procedure 49
Lower and upper percentiles of G_1 and b_2 326
LSD test, forming homogeneous groups by means of modified 512

Maintainability 238
Mandel test for nonadditivity 540
Mann and Whitney rank test 293–303
Marginal (sum) frequencies 352, 474, 476
Marginal sums 346, 348, 351, 463, 464, 466, 475, 485, 490
Mark-recapture estimation 172
Markov, A.A. 48
Markov chains 48
Marriage problem 218, 219
Matched pairs 50, 206, 254, 318, 550
Matching 50, 206, 307, 308 (see also Blocking)
Mathematical abbreviations 7
Mathematical models for clinical trials 214
Mathematical preliminaries 7–22
Mathematical statistics 123
Matrix algebra 461
Maverick, outlier 74, 131, 252, 279, 280, 328, 541
Maximal contingency coefficient of fourfold table 370
Maximax criterion 133
Maximaxer 133
Maximum likelihood method 67, 385

Maximum test 315, 316
  for pair differences 315
McNemar χ^2 test 363–365
McNemar's test for correlated proportions in a 2 × 2 table 363–365
McNemar test 363–365, 489
  extended to a multivariate distribution of dichotomous variables 489
MD, mean absolute deviation from the mean 100, 251–253, 281
Mean 48, 59, 60, 102, 106, 229, 234, 392, 407, 503, 517, 518
  and standard deviation 68–77, 81
  arithmetic 46–49, 68–77, 86, 88, 90, 91, 93
  combined harmonic 90
  comparison of 251, 264
  estimation of sample sizes 249, 251
  for nearly normally distributed values, estimate based on deciles 100
  geometric 85, 86, 90, 133
  harmonic 85, 88–90, 133
  overall 519
  predicted 443, 444
  variance of 46
  weighted harmonic 88
Mean$_L$ 109, 110
Mean absolute deviation from the mean 100, 251–253, 281
Mean breakdown frequency 236
  time 236
Mean chart 200, 201, 229
Mean column effect 523, 527
Mean deviation 100, 251, 252, 281
Mean effective dose 224, 225
Mean effects 523–529
Mean growth rate 86
Meaning of test results 119–123, 266
Mean line 200, 201
Mean loss, total 236
Mean of pair differences, testing for zero 309–311
Mean range 509
  of all samples 229
Means
  comparison of 121–123, 255–257, 264–279, 494–545
  comparison of a large number of 512
  comparison of several by analysis of variance 494–545
Mean square (MS) 502, 503, 521, 523
Mean square error (MSE) 503, 523
Mean square experimental error 533, 534

Mean square of experimental error *MSE* 523, 533
Mean square successive difference Δ^2 373, 374
Mean stratum effect 524, 527
Mean sum of squares (MS) 502, 528
Mean survival time 236
Mean susceptibility to breakdown 236
Mean time between failures 235
  confidence limits of 235
Mean value
  chart ($\bar{X}$-chart) 201, 229
    to control variability between samples 229
    computation for large sample sizes 72–76
    computation for small sample sizes 69–72
Measurement scales 132, 133
Measures of association 382, 472, 484
  distribution-free 395–406
Median 70, 71, 85, 90–95, 100, 106, 132, 254, 293, 299, 319
  chart ($\tilde{X}$-chart) 229
  class 92
  deviation 253, 254, 281
  estimation of 91, 92
  lethal concentrations in toxicity bioassays 228
  quartile test 302
  test of independence 405
  tests 301, 302
Median$_L$ 109, 110
Medical records 204, 209, 210
Medical statistics 37, 40, 195–218, 246, 590–594
Method of least squares 67, 385, 386
Method of steepest ascent 232
Methods research 238
Migration effect 203
Minimal samples size (*see* Sample size)
Minimax criterion 133
Minimaxer 133
Minimax principle for decisions 133
Minimum discrimination information statistic 490, 491
  application of 490
  computed 490
  of a multiway table 492
  of a three way table 492
Minimum effective doses, symmetrization of distribution 226
Minimum sample size for attribute superiority comparisons 232
Mixed model 533

analysis 533
Mixtures of distributions 83, 234
Mixture tests in clinical chemistry 198
Mode 90–93, 102, 132
  estimate of 93
Model 5, 6, 25, 47–49, 55, 81, 116, 124, 161, 238, 240–242, 257, 279, 348, 408, 437, 476, 519, 565
  in analysis of variance 519, 529, 532, 533, 537, 538
  in contingency tables 476
  in experimental design 563
  in fourfold tables 348
  in regression analysis 408, 437
Model$_L$ 109, 110
Model I analysis 532
Model II 533
Model III (or mixed model) 532
Model building for multiway contingency table analyses 493
  with respect to final choice of model 55, 493
  with respect to intermediate information available 493
Models
  linear 408, 437
  nonlinear 447–455
  with unspecified requirements 239
  with variable costs taken into account 239
Model without interaction 537, 538
Moivre, A. de 58, 59
Moment coefficients 85, 103–107
  for kurtosis 105, 106, 325, 326
  for skewness 105, 106, 325, 326
Moments
  about mean 102–106, 325, 326
  Sheppard's modification of 106
Monotone trend 402
Monotonic decreasing sequence 396
Monotonic increasing sequence 396
Monotonic trend 380
Monte Carlo method 68, 241
  and computer simulation 241–244, 605–607
Morbidity 37, 210, 214, 368
Mosteller–Tukey–Kayser tester (MTK sample tester) 230μ 46, 48, 49, 59, 60, 66–68, 94, 247–249
Multiclinic trial 214
Multicollinearity 461
Multicomparative plotting of means 495
Multidimensional analysis 461
Multilevel random selections 246

Multimodal distribution 65, 91, 93, 107
Multinomial coefficient 192
Multinomial distributions 161, 193, 194, 475
  comparison of population probabilities of 475
Multinomial probability 193
Multiple comparison of correlated samples after Wilcoxon and Wilcox 555–558
Multiple comparison of group means 537
Multiple comparisons 242, 306, 533–537, 542–558, 638–640
  of means 533–537
Multiple correlation 457–460
  and regressions 456
  coefficient 457, 458
  population 459
Multiple determination measure 458
Multiple linear regression 459–460
Multiple pairwise comparisons of means 533
Multiple regression 460
  analysis 459–461, 626, 627
  Multiple test of correlations 426
Multiplication rule multiplication theorem 33, 42
Multiplication theorem 32, 33, 42
Multiplicative effects 107
Multisampling plans 231
Multivariate analysis 461
Multivariate statistical procedures 461
Multivariate statistics 461
Multi-way tables 489, 493
Murphy's law 214
Mutually exclusive alternatives 464
Mutually exclusive events 29
  addition rule for 29

$N(0,1)$ Distribution 61–63, 79, 138
Natural logarithm 19
Natural variability 23, 24, 85, 279
Negative binomial distribution 177, 190
Negative correlation 397, 406
Nemenyi
  procedure 546–549, 555
  test 501, 546–549
Nested design 564
Newton, I. 159
Neyman distribution 190
Neyman's rule 119
No intercept model, considerations in fitting 392

Nominal scale 132, 133
Nominal significance level 119, 120, 266
Nomogram 183, 340, 424
  for comparison of two relative frequencies 188
  for computing and assessing correlation and regression coefficients 427
  for determining multiple correlation coefficient 458
  for determining partial correlation coefficient 456
  for flow scintillation spectrometry 177
  for tolerance limits 285
Nonadditive effects, suspicion of presence of 540
Nonadditive model 538
Nonadditivity test 541
Noncausal correlation 393
Noncentral t-distribution 256
Noninteraction model 538
Nonlinear coefficient of determination 449
Nonlinear correlation coefficient 449
Nonlinear monotone regression 395
Nonlinear regression 447, 449, 453
Nonnormality 65, 66, 131, 242, 322–329
  due to skewness and kurtosis 324
  sensitivity of distribution against 242
  test for 323
Nonparametric 131, 518
  hypothesis (or parameter-free methods) 130
  tests 130–133, 285–306, 312–320, 546–558
  tolerance limits 283–285
Nonrepresentativeness 203, 204, 208
Nonresponse 203
Nonsampling error 196, 197
  in surveys 197
Nonuniform distribution, testing for 322, 492
Normal distribution 47, 57–66, 78–83, 85, 114, 130, 166, 169, 175, 179, 186, 200, 225, 227, 252–255, 258–260, 263–266, 279, 280, 282, 283, 286, 293, 302, 303, 322, 323, 325–328, 330, 331, 334, 341, 365, 367, 374, 391, 395, 427, 437, 494, 516, 538
  bivariate 242, 383, 384, 389, 395, 397, 427
  chi-square goodness of fit test for 323

Normal distribution (*cont.*)
  quantiles of 325
  two dimensional 383, 384, 390, 395, 397, 403, 427
Normal equations 447–452, 454
  exact and approximate 452
Normality of errors 519
Normal law of error 58
Normally distributed data 58–60, 198, 503
Normally distributed experimental error 519
Normally distributed observations 58–60, 198, 503
Normally distributed population 57–67, 129, 250, 251, 267, 269, 281, 282, 310, 328, 453, 495, 499, 501, 509
  with equal variances, observations from 503
Normally distributed values with homogeneous variances 515
Normal rank transformation 517
Null hypothesis 114–129, 169, 183, 186, 187, 215, 255–258, 264–266, 268, 271, 272, 277, 280, 285–287, 292, 294, 295, 302, 303, 305, 306, 309, 310, 312, 316, 319, 320–322, 330, 332, 338–340, 343, 345, 347, 348, 372, 374, 375, 378, 393, 400, 424–426, 430–433, 436, 438, 439, 441, 442, 459, 463, 465, 466, 469, 471, 477, 488, 489, 499, 500, 502, 503, 521, 526, 538, 539, 542
  for homogeneity or independence 348, 476
  of equality of adjacent means 513
  of equality of population means 502
  of equality of regression coefficients 438–442, 485–487
  of equality of three means 506
  of equality of variances 495–500
  of homogeneity of k independent samples from a common binomial population 462–467
  of homogeneity of variances 495–500
  of independence (or homogeneity) 348, 476, 480
  on homogeneity of means 264–275
  on independence between two attributes 346–363
  on independence or homogeneity 349
  on two regression coefficients 438–442, 485–487
Number of years for amount to double 87
Numbers 43

Observation, predicted 51, 249, 443, 446
Observational study 55, 196–197, 202–204, 207, 208, 214, 245–246, 565, 566
Observations 47
Observation unit 343
Observed frequency 466, 467
Occupancy numbers 346
Occupation number 55, 420, 421, 475
  observed 490
  of class 321
Odds 27
Olmstead and Tukey corner test 405, 406
One dimensional frequency distribution 85
One sample Gauss test 121–123
One sample test of Carnal and Riedwyl 300
One sample *t*-test 255
One sided confidence limits 311, 345
  lower 337, 345
  upper 345
One sided problem 124, 127, 138, 222, 260, 272, 274, 278, 285, 295, 297–301, 312, 315, 338, 425, 441
One sided procedures 258
One sided question 186, 187, 215, 256, 294, 371, 376
One sided statistical test 199
One sided test 124, 127, 128, 137–139, 217, 222, 251, 257, 261, 277, 287, 309, 349, 350, 364, 378, 381, 397–399, 401, 402, 437, 438, 471, 503
  generalization of 306
1–9 decile coefficient of skewness 101
Operating characteristic (OC) 129
  curve for sampling plan 231
Operation "Sea Lion" 240
Operations research (or management science) 238–240
Opinion polls 246
Optimal decision making 40, 133, 134

Optimal range of operation of all factors 562
Optimal solutions for compound systems, organizations and processes 238
Optimal test 123
 for comparison of paired observations 312
 parametric or nonparametric 131
Optimization of processes using hybrid computer 244
Ordered categories in contingency tables 472–474, 484–487, 493
Order of depletion 233
Order of magnitude of k means, test on 541
Order restrictions 541
 in comparing means/medians 306, 464, 506, 541
 in comparing regression lines 442
Order statistics (see also Median) 286
Ordinates of a normal distribution 59, 60
Ordinates of the standard normal distribution 79
Orthogonal polynomials 460, 461
Ott and Free test 362, 373
Ott procedure 495, 508
Outliers 74, 131, 279, 280, 395, 447, 541
Outlier test 278–280, 447
Overall error 198, 199
Overall mean 519
Overall significance level 267

Page method 553
Page test 541, 554
Paired comparison 555
Paired data 307–309
Paired measurements 307–319
Paired observations 254, 307–320, 562
 testing equality of variance of 311
Paired sample procedure 310
Paired samples (or correlated samples) 309
Paired (connected) samples 307, 308, 311
 distribution-free methods for 312–319
 t-tests 307–312
Pairing 561
 by randomization 561
 by replication 561
Pairwise comparisons 308, 548, 549
 among k independent correlations 427
 of all possible pairs of treatments according to Nemenyi 546–549
 of mean ranks 305
 of means, multiple 533–537
Pairwise differences 316
Paper
 double logarithmic (log–log paper) 455
 exponential 455
 function 455
 graph 455
 power function 455
 semi-logarithmic (exponential paper) 455
 sine 455
Parameters 25, 48, 66–68, 112–116, 120, 125, 166, 246–248, 268, 273, 320, 362
 estimates for 521
 of normal distributions 60
 test 123
Parametric methods 123–133, 255–278, 494–545
Partial correlation 395, 456–458
 and regression 456
 coefficient 457
 of second order 458
Partial independence 492
Partition 42
Partitioning chi-square 468–474, 480, 482, 484–487, 492
Pascal, B. 35, 156, 567
Pascal's triangle 156, 157
Path 39
Patients of a clinic 209, 210
Peakedness (see Kurtosis)
Pearson contingency coefficient 369, 370, 482
Pearson, Karl 106, 567
Pearson product moment coefficient of correlation 309, 382–384, 388–390, 406, 407, 419–432
Pearson's contingency coefficient 369, 370, 482
Penicillin 552, 553
Percentage frequency 333, 335
Percentile 92
Periodicals 576
Permutations and combinations 155–162
Permutation test 131
Perturbing factors 560, 561
 nonmeasurable 561
 unknown 561

Perturbing quantities 561
Phase 379
Phase frequency test of Wallis and Moore 378
Phenotype 321
Philosophical roots of statistics 26
Pillai and Buenaventura comparison 275, 276
Placebo 195, 212, 213, 364, 365
 -dependent results 213
 effect 365
 reactors 213
Planned grouping (block division) 559, 560
Planning hypotheses to be tested 565, 566
Planning of investigations 4–6, 55, 195–197, 202–213, 558–566
Plausibility considerations 224
Plotting 6, 55–57, 78–81, 83, 86, 200, 201, 229, 230, 383, 409, 444, 449, 455, 495, 517
Point cloud 382–384, 387, 392, 403–406, 409, 517
Poisson distribution 132, 170, 175–192, 230, 235, 236, 241, 329, 330, 343–345, 516–517
 comparison of several 188
 compound 170, 190
 confidence intervals for the mean 182, 183, 343–345
 confidence limits of 343–345
 control limits computed with the help of 230
 mean and variance of 176
Poisson homogeneity test 190, 465
Poisson probability 183–185
Poisson, S.D. 175
Polling 342
Polynomials, orthogonal 460
Population 4, 5, 24, 25, 46, 47, 65, 107, 196, 203, 209
 at risk 37, 205–208, 218
 checking a set of samples for a common underlying 475
 contrasts 509
 correction, simplified 342
 correlation coefficient 383, 384
 covariance 413
 error of the mean 46, 94
 mean 46, 48, 49, 59, 60, 66–68, 94, 247–249, 266
 median 254, 319

 multiple correlation 459
 multivariate normal 456
 size 51, 172
 standard deviation 46, 48, 59, 60, 71, 94
 variance 46, 67, 70, 71
Positive correlation 397
Positive probability contagion 190
Positive reagents 225
Positive skewness 91, 93
 index 101
Power 10
Power of a test 125–131, 215, 216, 251, 350, 351, 519
Power transformation for right-steep distributions 516
Practical long-range consideration regarding a method 199
Practical significance 119
Precision in chemistry 198, 201
Precision of data 198–201
Precision (or reproducibility) of a method 198
Predictant 384, 386, 446
Predicted mean 443, 446
Predicted observation 443, 446
Prediction 3–6, 24, 51–53, 249, 443
Prediction interval 249, 285, 443, 444, 515
 for a future observation 444
 for a future sample mean 444
Predictions 51
Predictor 384
Preliminary trial 131, 350
Preparation
 specific biological activity of 228
 standard 228
Principle of equalizing alternation 220
Probabilistic statements 6
Probabilities of two binomial distributions, comparison of 338–341, 346–362
Probability 4–6, 26–49
 and statistics, periods in history of 567
 a posteriori 27
 a priori 27
 at least one hit 36, 218
 axioms of 29, 42
 computational 26
 conditional 31, 32, 42
 contagion 190
 cumulative 234

density 45, 59, 234
function 45
distribution 48, 228
  hypothetical 241
element 45
estimates 47
function 44
linkage 48
mathematical 27
of at least one success 36, 218
of combinations 157
path 39
plot 81–83
plotting 78–83, 108–111, 166, 446
sample 49, 50
statistical 26, 27
theory 26–46, 123, 257
Probit analysis 225
Probit transformation 269
Problem of traveling salesman 239
Problem solving 565, 566
Process computer 244
  process automatization using 244
  technology 244
Process control 197–202, 228–238, 243, 244
Producer's risk 231
Production planning 239
Product moment correlation 395
  coefficient 383, 384, 388, 389, 406, 407, 419–432
  of Bravis and Pearson 383, 402
Profitability studies 238
Prognosis models 48
Programs, computer 572, 573
Propagation of errors 21, 22, 96
  power product law of 96
Proportionality in two way tables, testing of 480
Proportions 269, 333
Prospective etiological studies 204–209
Prospective studies 204–209
Prospective study (see also Cohort study) 204–209
Pseudorandom numbers 51, 241
Psychometrics 196
Pursuit or evasion games 240
P-value 119–120, 266

Quadrant correlation 384, 396, 403–405
  of Quenouille (r_Q) 395

Quality control 39, 172, 196–202, 204, 219, 228–236, 281
  charts 200–202, 228–230
  in industry 228
    continuous 219, 229
    graphical 229
Quality improvement 232
  experimental strategy for 494
  guidance toward more favorable working conditions as special case of 232
  technological economic complexity of questions connected with 232
Quality of conformance 229
Quality response 224–228
Quantifiable qualitative outcomes, counting 210
Quantiles 92, 100–102, 106, 197, 254, 325
  confidence interval 197, 198, 254, 319
  of the normal distribution 60, 325
  of the standard normal distribution 62, 63, 217
Quantitative methods, comparison of 200
Quantities
  continuous 53
  discrete 53
Quartiles 101, 302, 325
Quenouille quadrant correlation (r_Q) 395, 403–405
Questionnaire 55, 202
Questions to put 3–6, 107, 204, 565–566
Quetelet, A. 58
Queuing theory 48
Quick distribution-free procedures for evaluating the differences of paired observations 315, 316
Quick estimation
  differences between two samples 285, 286
  mean 81, 100
  standard deviation 71, 81, 97–100
Quick test of Ott and Free 362, 373
Q-symbol 264, 417, 459
Q-test 489, 555
  due to Cochran 489
  of Cochran for comparison of several percentages in matched samples 365

Quota procedure 245
  in sampling 245, 246

r, coefficient of correlation 383, 384, 388, 389, 406, 407, 419–432, 449
$r \times c$ contingency table 474–491
$r \times c$ table 476, 490
  test statistic χ^2 for independence or homogeneity in 475
Raatz, computation according to 477
Radioactive desintegration 48
Radioimmunoassay 228
Random
  allocation 4, 5, 49–51, 211, 212
  assignments of procedures to trial units 558–562
  choice 529
  components model (or random effects model) (Model II) 529, 532
  connections 224, 393–395
  effects 532
  error 5, 58, 67, 196, 198, 560–562
  influences 561
  model 519, 532
  number generator 243
    generation and testing 243
  numbers 49–52, 84, 241, 243, 559
    drawing samples from theoretical population with the help of 241
  procedures (or processes) 5, 49–51, 123, 170, 230
  sample 4, 24, 44, 47–51, 70, 95, 107, 121, 123, 124, 130, 204, 209, 218, 219, 230, 231, 264, 266, 273, 339, 389, 390, 529
  selection 50, 196, 203, 245
  statistically significant result 116
  variability 198, 273, 562
  variable 43–46, 57, 134, 139, 219, 408, 443, 457, 459, 523, 533
    continuous 45, 46, 57
    discrete 44–48, 163, 175
    normally distributed 57–64
    standard normally distributed 61–64
    variation of experiment, inherent 523
Randomization 559, 560, 565
  as underlying principle of experimental design 562
  procedure 562
  techniques 561

Randomized blocks 529, 562, 563
Randomness
  of observations 373–381
  of the fit 332, 449–451
  principle 49–51, 246
Range 85, 97–100, 198, 230, 328, 507–509
  chart (R-chart) 201, 229
    in localizing and removing excessive dispersions in controlling variability within samples 229
    preparation of, for upper limits 229
  confidence interval of 508
  interdecile 98–101
  of a sample 97, 229, 515
    heterogeneity 515, 516
  of individual groups and means 542
  tables and/or tests 275–277, 328, 507–509, 542–545
Rank 286, 396, 517
Rank analysis of variance
  for several correlated samples 549
  for several independent samples 303–306, 546–549
  of Friedman 549–558
Rank correlation 132, 395–403, 554
  coefficient 395, 396, 400
    multiple 403
    of Kendall 403
    of Spearman 395–403
    partial 403
    τ (Kendall's tau) 403
  statistics, combined two tailed 403
Rank (or ranked) data 130–133, 286
Rank differences 400
Rank dispersion test of Siegel and Tukey 286–289
Ranking and selection procedures 541
Ranking of medians/means 306, 464, 541
Rank means, ordered 553
Rank, normal rank transformation 517, 518
Rank order 464, 541
Rank (or ordinal) scale 132, 133
Rank sum 293–295
Rank tests 131, 286–306, 312–315, 546–558
  of Nemenyi 546–549
Rank transformation 286, 518
Rapid test 131, 132, 285, 286, 315–320
Rare events 175–192, 236, 241, 329, 330, 343–345, 516–517

Ratio scale 132, 133
Reaction threshold 226
Reagents, positive 225
Realization 43, 256
Real numbers 43
Reciprocal of the square root transformation 516
Reciprocal transformation 515–517
Recovery duration 210, 465, 477
Recovery percentage 215, 347, 465, 477
Rectangular distribution 83–85, 98, 391
  definition of probability 84
  density of 84
Recursion formula (Feldman and Klinger) 372
References 568–641
Reference value 197
Regressand 384, 458
Regression 382–388, 408–413, 417, 419–423, 437–446, 459, 460, 484
  analysis of multiple 460
  Bartlett and Kerrich's quick estimates of 390–392
  estimated from correlation table, testing linearity of 435, 436
  exponential 451
  fitted 384, 389, 415, 449
  linear 382, 389, 408–423, 437–446, 462
  multiple 385, 459, 460
  nonlinear (curvilinear) 385, 447
    monotone 395
  robust 390–392, 447
  testing linearity of 433–437
  variance analytic testing of 418
Regression analysis 382–461, 484
  and variance analysis 418, 433, 442
  assumption of linearity in 437
  Bartlett procedure 390–392
  Model I, II 408
  models for 408
  multiple linear 459, 460
  no intercept model 392
  nonlinear models 447–455
Regression coefficient 107, 133, 384–390, 392, 408–423, 425, 437–446, 453, 462, 484, 485
  comparing two 440–442
  comparing two way tables with respect to 462, 472–474, 484–487
  computing and assessing 408–423
  confidence interval for 439, 440

confidence limits for 439, 440
least squares estimation of 67, 385, 386
of corresponding two way tables, comparison of 484–487
testing against zero 425, 437, 438
testing difference between estimated and hypothetical 438, 439
Regression equation 384
Regression function 433, 458
Regression line 386–391, 408–423, 437–455
  component of the χ^2 value 484
  confidence interval for 443–446
  deviation from 472–474
  estimation of 408, 410, 437
  of two corresponding or matching tables, comparing 487
  testing equality of 440–442
  total 442
Regressions, partial and multiple 456
Regressor 384, 386, 458
Rejection of observations (see Outliers)
Rejection probability 127
Rejection region 256, 257
Relation
  causal 393, 394, 480
  formal 393, 394, 480
Relationship (see Association, Correlation, Regression)
Relative coefficient of variation 78
Relative frequency 26, 48, 218, 269, 330, 333, 338, 340, 469
  comparison with underlying parameter 333, 338
  in the k classes 464
Relative modes 91
Relative portion 226
  of reacting individuals, sum of 227
Relative risk 206
Relative variation coefficient 70, 78
Relevant statistic 124
Reliability 236–238
  analysis 238
  criteria for a method 198
  of a device 236–238
  of an estimate 95
  of data 196–202, 279
  of laboratory methods 197
  of measurements 96, 197–202, 279
  theory 108, 237, 238
Repeatability
  component of 561
  of observations (23), 47, 212

Replacement 36, 47, 48, 94, 247, 248
Replicability of a sample 47
Replicate measurements 559, 561
Replication 559, 562, 565, 566
  as underlying principle of experimental design 562
Replication collectives 559
Replications 563, 564
  number of 565
Report writing 566
Reporting uncertainty 95
Representative inference 3–6, 47–51, 245, 246
Representative random sample 4, 5, 47–51
Representative samples 49, 207, 208
Representative symptoms and characteristics 212
Reproducibility 23
  of a method 199
  of observations 211
Research 24, 116, 565, 566
Residual mean square (MSR) 523
Residuals 418, 419, 449
  examination of 449
Residual sum of squares 414
Residual variance (or dispersion about regression line) 415, 418, 437, 439, 441, 474, 487
  confidence interval for 440
Response error 196, 197, 202, 203
Response surface experimentation 232, 494
Response variable 384
Results 95, 128, 129, 566
  interpretation of 120, 134, 195–197, 393–395, 566
  stating in tabulated form and/or graphically 566
Retrospective interrogation 202, 206, 208
Retrospective etiological studies (see also Case-control study) 202–210
Rho 383, 384
Ridge regression 461
Right-left comparison 307
Right-steep distribution 107, 516
Risk 134, 205–208, 233–238
  I 117–119, 127, 129, 199, 215, 565
  II 117–119, 126, 129, 199, 216, 315, 565
  levels of 215, 216
Risk exposition 37, 205–208, 233–238
Risk, relative 206

Robust estimate for dispersion 253, 254
Robust multiple comparisons 495
Robustness 66, 100, 123, 131, 241, 253, 258, 262, 279, 501
  of intervals and tests 258
  of test 123, 265
  of the t-test 241, 265, 501
Robust statistics (with robust estimation) 66, 81, 90–95, 97–102, 123, 130–133, 251–254, 258, 280, 390–392, 395, 447, 495, 501, 537
Root transformation 515, 516
Rosenbaum tests 285, 286
Roulette 27, 166
Rounding off 20, 21
Row effect 519, 521, 532, 533, 538
Row homogeneity 538
Row mean 538, 540
rth sample moment 103
Run 375
Run test 132, 329, 332, 375–378
  critical values for 376
  for testing whether sequence of data is random 375

S_3 sign test of Cox and Stuart for monotone trend 379, 380
S test (see Scheffé test)
Sample 4, 24, 25, 43, 44, 51, 257
  random 43, 44, 47, 49, 70, 95, 107, 121, 123, 204, 218
  simulated 241
Sample accuracy 246
Sample correlation coefficient r 383, 384, 388, 389, 406, 407, 419–432
Sample cumulative distribution function 68
Sample function 134
Sample heterogeneity 65, 209, 234
  selection 203, 209
  testing 469
Sample mean 68–78, 134, 266, 269, 273
Sample moments 102–106, 325
Samples
  correlated 307–309, 382–384, 549–558
  in pairs 309
  having zero defectives, nomograph for 232

independent 260, 266, 459, 501, 503, 562
random 4, 24, 49, 95, 121, 130, 264
Sample size 49, 207, 214–218, 246, 249–251, 262, 263, 274, 283–285, 341–343, 350, 351, 466, 504, 533, 541, 562, 565
Sample sizes
approximate choice of, for clinical trial 214–218
Sample space 47
Sample survey 55, 202
Samples with heterogeneous variances 515
Sample variance 68–78, 103, 134, 139, 260, 283, 503
Sampling 48, 203, 245, 246, 425
without replacement 36, 48
with replacement 36, 48
Sampling distribution 266
n-variate 461
three dimensional 461
two dimensional 461
Sampling error 49, 196, 203, 502
protecting against 231
Sampling experiments 241, 242
Sampling in clusters 246
Sampling plans 231
Sampling schemes for quality control 129
Sampling technique 48, 65, 245, 246
Sampling theory 25, 48–51, 196, 197, 245, 246, 614, 615
Sampling variation 114, 266, 273
Sampling without/with replacement 47, 48, 94, 155–175, 247, 248, 335, 336
Sandbox exercise 240
Scales of measurement 132, 133
Scatter diagram 382, 383, 405, 409, 411, 517
Scheffé assessment of linear contrasts 509–512
Scheffé test 509–512, 533–537
Shewhart control chart 200–201, 229
determination of sample size 214–217
SCI (simultaneous confidence intervals) 514, 515
Scientific inquiry 116, 565, 566
Scientific investigation 3–6, 116, 565, 566
Scientific method 3–6, 23–26, 114–116, 123, 124, 128, 129, 134, 257, 565, 566
Scientific model of the world 116
Scientometry 196
Scores 473
assigned to levels or categories of attributes 484, 485
Scoring 472–474, 484–487
Screening 218
Secondary cause, spontaneous tendency towards recovery as 211
Second degree equation, correspondence between acutal relation and 447
Second order partial correlation coefficient 458
Second order regression 447–450
Secretary problem 218, 219
Selecting the best of several alternatives 541
Selection
bias 203, 204, 246
probabilities 203
Selection correlation 246, 394
Selection procedures 246, 541
Semi-logarithmic plotting paper 455
Sensitivity
comparison of two similar independent experiments with regard to 541
of a chemical determination method 199
of a method 199
of experiments 541
Sensitivity ratio of methods 199
Sensitivity test 226
Sequence of trials 220
Sequential analysis 219–224, 567
Sequential analytic method of quality control 230
Sequential charts, test characteristics of 241
Sequential clinical trials, analysis of 223
Sequential medical plan 220–224
Sequential test, efficiency of 221
Sequential test plans 219–223, 595, 596
closed 220
design for toleration experiments 223
due to Bross 220–222
due to Cole 223
due to Spicer 222
maximum sample size for 222
one sided 222
open 220
quick 222

Service theory 48
Set
  complement of 29, 30, 42
  partition 42
Setting and realizing strategy 565
Shaw, G.B. 108
Sheppard correction 76
Shortcut statistics 285, 286
Side effects 223, 224, 337
Siegel and Tukey rank dispersion test 286–289
Siegel-Tukey test 260, 288, 293
Sigma, Σ (see Summation sign)
Sigma, σ (see Standard deviation) 46, 48, 59, 60, 71, 94
Sigma squared, σ^2 (see Variance) 46, 67, 70, 71
Significance, localizing cause of 480
Significance level 112, 120–123, 126, 127–130, 136, 137, 153, 199, 214, 273, 274, 299, 330, 339, 350, 437, 465, 470, 471, 474, 477, 478, 487, 495, 497, 499, 526, 538
  nominal (or descriptive) 119, 120, 266
Significance probabilities, combination of independent 119, 366
Significance test 123
Significant, statistically 118, 266
Significant difference 266, 273, 274, 305, 306
Significant digits (figures) 20, 21, 107
  of characteristic values 107
Sign test 132, 316–320, 365, 552 (see also McNemar test)
  applications for orientation 319
  efficiency of 318
  modified by McNemar 363
  of Dixon and Moot 315, 316
Similarity of samples 273
Simple blind trial 213
Simplifying assumptions 565
Simulated sample 241
Simulation 238, 241–243
  analogue 242
  computers in 241
  studies 241–244
Simultaneous confidence intervals (SCI) 514
Simultaneous correlation 393
Simultaneous inference 514, 515, 537
  (see Bonferroui inequality)

Simultaneous nonparametric comparison 300, 303–306, 546–558
Single blind trial 212
Single-sampling plans 231
Size of population 51, 172 (see also Tolerance interval)
Skewed distribution 90–93, 100, 111, 291, 293, 391, 515
Skewness 85, 90, 100–107, 252, 255, 293, 299, 325–327, 501
  and kurtosis, measures for 106
  determination from moments 101
  negative 100
  I 101, 102
  positive 100
  III 101, 102
  II 101, 102
Skip-lot sampling plans 231
Slide rule 242
Slope 386
Smoking and lung cancer 204, 366
Software 572, 573
Spearman–Kärber method 225–228
  a rapid distribution-free method 225
  for estimating mean and standard deviation 225
  for estimating mean effective or lethal dose 225
  trimmed 228
Spearman rank correlation 395–403, 554
  coefficient 395, 396
  critical values of 398, 399
  with ties 396, 401, 402
Specificity of a method 198
Spicer sequential plan 222
Spurious association 203, 393–395
Spurious correlation 393, 394
Square root transformation 517
  of frequencies 516
Squares and square roots 10, 11
Square tables
  analysis of 475
  testing for symmetry of 487
Stabilizing variance 269
Standard deviation 48, 60, 68–76, 85, 97, 98, 133, 164, 166, 198, 199, 225, 228, 264, 279, 281, 283, 328, 388, 392, 407, 413–417, 422, 427, 431, 432, 443, 453, 507–509, 516, 517
  chart (S-chart) 229
  computation for large sample sizes 72–76

Subject Index

computation for small sample sizes 69–72
estimating from range 507
estimation of 249, 250, 413
for nearly normally distributed values, estimate based on deciles 100
from grouped data 72–75
from ungrouped data 69
of axis intercept 415–417
of difference 309
of mean range estimate of 509
of regression coefficient 415–417
of standardized range 509
using range to estimate 97, 507–509
using range to estimate maximum 98
within samples 76
Standard error
of arithmetic mean 94–96
of axis intercept 415–417
of coefficient of variance 275
of difference 207
of difference in means 309
of estimate 415
of mean 94–96
  range 509
of median 95
of prediction 414, 415
Standardization of fourfold tables 351
Standardized deviation 273, 274
Standardized extreme deviation 281
Standardized maximum modulus distribution 514–515
Standardized (or standard) normal distribution 60–64, 78–81, 217
Standardized 3rd and 4th moments 325
  percentiles of 326
Standard normal density function 79
Standard normal distribution 60–64, 78–81, 135, 174, 217, 267, 272, 275, 289, 294, 315, 338, 378, 379, 397, 487, 515, 517, 518
  bounds for 62, 63, 217
Standard normal variable 61–63, 79–82, 139, 187, 247, 335, 374, 427, 429, 452, 473
Standard preparation 228
Standard treatment, expected recovery rate for 216
Statistic 66, 85, 98, 106, 256, 257
Statistical analysis model, definitive formulation of 565, 566
Statistical decision 112–134
  functions 567
  theory 567

Statistical experimental design, principles of modern 562
Statistical independence of errors 519
Statistical inference 3–6, 23–26, 116, 123, 124, 195, 196, 559–562
  future of 567
Statistical investigation of causes 203, 393, 394
Statistically significant 115
Statistical method 3–6, 23–26, 116, 123, 124, 195, 196, 559–562
Statistical model 5, 81, 257, 348, 408, 437, 476, 519, 565
Statistical procedures, multivariate 461
Statistical significance 112–119, 195, 199, 236, 266, 289, 292, 310, 400, 425, 467, 474, 552
Statistical software 572, 573
Statistical source material gathering 54, 55
Statistical symmetry 27
Statistical test 112–132
Statistics 3–6, 23–26, 46, 48, 53, 98, 116, 123, 124, 195, 196, 559–562, 566
  biometrical methods of 567
  descriptive 24
  dispersion 85
  future of 567
  inductive or analytic 24, 124, 257
  inference 25, 46, 257
  location 85
  methods 23–26
  robust 66, 95, 99, 100, 123, 130–133, 253, 254, 280, 390, 395, 447, 495, 501, 537
  shape 85
Steepest ascent method 232
Stirling's formula 159, 161, 179
Stochastic
  dependence 393
  event 43
  experiments 4, 43
  independence 31–34, 42, 204, 383, 385, 393
  inductive inference 257
  model 241, 257
    artificial 241
  processes 48, 567, 589, 590
    simulation and analysis of 241
  relation 383–385, 393–395
Stochastically independent
  events 33, 34
  observations 188

Stochastically independent (*cont.*)
  variables 204
Straight line, equation of 385, 386, 408–419
Stratification 205, 206, 245 (*see also* Blocking)
Stratified sample 205, 245
Stratum 205
Stratum effect 519, 521
  mean 524, 527
Structural homogeneity 207, 208
Student distribution 135–139
  k-variate 514
Studentized range 534–537
Student's comparison of two means 70, 264–275
Student's distribution 135–139, 266, 267
Student's t-test
  one sample 255–257
  two independent samples 264–275
  data arranged in pairs 309–312
Subgroups, formation of 83
Subjective criteria 213, 223
Subjective symptoms 212–214
Substitute F ratio 275, 276
Substitute t ratio 276, 277
Substitution error 195
Success-failure situation 215, 216, 346
Success ratio 212, 346
Sufficient estimator 67
Suggestibility 213
Summation procedure 74, 75
Summation sign 9, 10
Summation symbol 9, 10
Sum of squares (SS) 264, 413, 417, 501–505, 521, 528, 538, 540
  between samples (groups) ($SS_{between}$) 502
  mean 538-540
  within sample (SS_{within}) 501
Survey (50), 55, 196, 197, 202–204, 207, 208, 214, 245, 246, 393–395, 565, 566, 614, 615
Surveys and experiment, comparing 561
Survival distributions, equality of 195, 206, 210, 235
Survival model 107–111, 206–210, 233–236
Survival probabilities 206, 237
Survival time 210, 236
  analysis 108–111, 233–238
  distribution of 206
Symmetric distribution 58–64, 99, 100

Symmetry 27
  testing square tables for 363–365, 462, 488, 489
  with respect to diagonal in an $r \times r$ table 488
  Bowker test for 488
Systematic components 529
Systematic error (or bias) 5, 66, 67, 196, 197, 199, 201–203, 212, 246, 559
Systematic sample 245

Tables, sixfold or larger 366
Tabulated breakdown 209
Tactical reserves, setting up 565
Target function 239
Target population 209
Target quantity (regressand) 232, 389, 408, 458, 459
Target variable 385, 457, 458
Tasks of statistic 3–6, 23, 26, 116, 123, 124, 558–562
t-distribution 135–139, 154, 155, 256–258, 260, 273, 321, 424, 438, 441
Technometrics 196
Terry–Hoeffding test 293
Test
  and confidence interval 116, 256, 257, 268
  asymptotic efficiency 130, 293, 302
  based on standardized 3rd and 4th moments 323, 325–327
  conservative 131
  critical region 256, 257, 266, 267
  difference in means 264–275, 494–545
  dispersion difference 258, 260–264, 286–289
  distribution 53–60
  distribution-free 130–133, 261, 262, 285–306
  efficiency 130, 276, 293
    of paired samples 307, 308
  for a single mean 121–123, 255, 256
  for bivariate normality 447
  for departure from normality 322–329
  for nonequivalence of underlying populations of two samples 467, 468
  for ordered means, comparison of 541

homogeneity 186–189, 236, 260–263, 311, 312, 338, 339, 346–362, 462–468, 474–478, 490
independence in contingency tables 346–362, 474–478
nonparametric 130–133, 261, 262, 285–306, 312–320, 546–558
of dispersion differences 258, 260–264, 286–289
of equal means 264–275, 494–545
of equal medians ($n_1 = n_2$) 293–306
of equal regression lines 440–442
of equal slopes of regression lines 440–442
of goodness of fit 123, 320–332
of location differences 121–123, 264–275, 494–545
of nonnormality 322–329
of randomness 373–381
of significance 123
of slope of regression line 437–439
of symmetry in contingency tables 363–365, 488, 489
of the significance of a mean 121–123, 255, 256
on different levels
of a factor 563
of several factors 563
power of 125–131, 215, 216, 251, 350, 351, 519
procedure 124
randomness 373–381
theory 25, 123
general 567
trend 379–381
in contingency tables 468–470, 472–474, 484–487
two sided and one sided 124–129
Testing equality of variances after Cochran 497, 498
Testing homogeneity of profiles of independent samples of response curves 541
Testing hypotheses 114–131
Testing in groups 218
Testing $k \times 2$ table for trend 472–474
Testing of an empirical distribution for uniform distribution 492
Testing of two distributions of frequency data for homogeneity 67, 468
Testing order alternatives in a one way analysis of variance, critical values for 541
Testing randomness of sequence of dichotomous or measured data 373–375
Testing sample for nonnormality, method due to David et al. 235
Testing significance of linear regression 425, 433–442, 472–474, 484–487
Testing square tables for symmetry 363–365, 488, 489
Test on ranks 285–306, 395–403, 549–558
one sample 121–123, 255, 256
one way layout (see Analysis of variance, one way)
optimal 123
ordered alternatives 306, 464, 541
parameter 123
parametric (see t-test or Analysis of variance)
power of 125–129, 131, 215, 251, 350, 351, 519
randomness 373–379
relative efficiency 130, 293
robust 132, 241, 262, 501
sensitive 123, 258, 279, 500
sequential 219
significance 123
statistic 121
statistical 121
"tea" 120, 121
theory 25, 112–132
two sample 260–275, 285–303
two sided 125
Tests (by author)
Bartlett 498–500
Bauer 302
Behrens-Fisher 271–273
Bowker 488, 489
Bross 220–222, 241
Cochran 497, 498
Cox-Stuart 379–381
Dixon 277–279
Dixon-Mood 316–320
Fisher 370–373
Friedman 549–555
Gart 361
Gauss 121–123, 268
Hartley 495, 496
Hutchinson 403
Kolmogoroff-Smirnoff 291–293, 330–332
Kruskal-Wallis 132, 303–306
Kullback-Leibler 490–493
Le Roy 362
Levene 262

Tests (by author) (*cont.*)
  Link-Wallace 542–545
  Lord 276, 277
  Mandel 540
  McNemar 363–365
  Mosteller 285
  Nemenyi 546–549
  Pillai–Buenaventura 275, 276
  Rao–Chakravarti 192
  Rosenbaum 285, 286
  Scheffé 509–512, 533–537
  Shapiro–Wilk 329
  Siegel–Tukey 286–289
  Student 130, 255, 256, 264–275, 309–312
  Student–Newman–Keuls 533–537
  Tukey 289–290, 514, 534, 537
  Wallis-Moore 378, 379
  Weir 272
  Welch 271
  Wilcoxon 293–302, 312–315, 555–558
  Wilcoxon–Mann–Whitney 132
  Wilcoxon–Wilcox 555–558
  Williams 512
  Woolf 351–360, 370
Tests (by letter)
  $\sqrt{b_1}$ 325–327
  b_2 325–327
  χ^2(chi^2, chi–squared)
    approximate 320–324, 345–349, 462–479
    exact 258, 259
  F 132, 260–264, 360, 361, 494, 501–506
  F' 275, 276
  G 351–360, 370
  H 132, 262, 303–306
  $2I$ 490–493
  Q 365, 489, 555
  S 509–512, 533–537
  t 130, 255, 256, 264–275, 309–312
  U 132, 241, 262, 293–303
  $\hat{u}$ 276, 277
  W (Shapiro-Wilk) 329
  X 293
  e.g. 121–123
Test statistic 121, 122, 126, 134, 138, 155, 256, 278, 280, 294, 309, 311, 312, 430, 437, 475, 495–497, 502, 504, 526, 550, 554
χ^2 139–143, 154, 155, 320–324, 329, 330, 332, 346–351, 462–482
  critical value of 125, 126

F 143–155, 260–264, 360, 361, 494, 501–506
for homogeneity 432
for weakly occupied tables 363
of Kruskal and Wallis 306
t 135–138, 154, 155, 255–257, 264–275, 309–312, 424, 426, 437–442, 494, 501
U 299
Theorem of Bayes 38–43
Theorem of Glivenko and Cantelli 68
Theorem of total probabilities 39, 42
Theoretical distributions 61–63, 135–155, 362
Theoretical values 561
Theory 5, 6, 24, 25, 116, 124
Theory of optimal design 232, 233, 494, 558–564
Therapeutic assessment, uncontested 212
Therapeutic comparison 195–197, 205–214, 220–224, 347
Therapeutic results 210, 347
Therapies, comparison of two 214, 347
Therapy difference 216
Thorndike nomogram 183, 184
3 × 3 × 3 table (or contingency die) 493
Threefold tables 490
Three sigma limit 201
Three sigma rule 64, 114
Three-variable analyses, topology of 394
Three way classification 518, 521
Three-way tables 489
Tied observations 288, 296, 297
Ties 287, 396, 401, 550, 552
  in ranking 287, 288, 296–299, 304, 550
Time-dependent population 320
Time sequence 381
Time series 379–381, 394, 395, 461
  analysis and forecasting 381
  correlation among 394
  testing of 379
Tolerance factors 282
Tolerance interval 113, 114, 281, 283–285, 444
Tolerance limits 113, 114, 281–285
  for distribution-free case 283, 284
Tolerance region 282, 446
Total probabilities theorem 33, 38, 39, 42
Total variation 462, 474, 526

Toxicity bioassay 228
Transformation 65, 107–111, 269, 395, 437, 515–518, 541
 angular 65, 269
 logarithmic 65, 107, 515–518
 r to $\dot{z}$ 427–432
 square root or cube root 65, 515–518
Transformations 515–518
 in clinical-therapeutic research, particular significance of 541
Translation law 84
Transposition inference 249
 prediction intervals for 249
Traveling salesman problem 239
Treatment
 particular 211
 specific 465, 466, 478
 symptomatic 211, 465, 466, 478
Treatment means, multiple comparisons between a control mean, corresponding SCI and k 514
Treatments, pairwise comparison of all pairs of 547
Trend 374, 379–381, 394, 395, 461
 early detection of 201, 229
 testing for 472, 484
Trend analysis 53, 379–381, 391–395, 461
Trend recognition 201, 229, 379–381, 394, 395, 461
Trend test in terms of dispersion of sample values 373, 374
Trial (*see* Bernoulli trial, or Clinical trial)
Trial in medicine
 between patients 211
 multiclinic 214
 sample sizes 214–218, 340–343, 350, 351
 sequential analysis 219–223
 simple blind 212
 single blind 212
 within patient 211
Trials 558–564
Trial units per block, number of 562
Triangular distribution 98, 190, 252
Truncated 95% probability ellipse 447
t-statistics 135–138, 154, 155, 241, 255–257, 264–275, 309–312, 424, 426, 437–442
t-test (*see also* Student's t-test) 70, 133, 255–257, 309–312, 424–426, 437–442, 494, 501
 based on the MD 252
 for data arranged in pairs 309

 for normally distributed differences 309
 nomographic presentation of 275
 of the product-moment correlation coefficient 424, 426
Tukey-D_{II} 534, 537
Tukey procedures 514, 534, 537
Tukey's quick and compact test 289, 290
Tukey test 540
27% rule (*see also* Secretary problem, Marriage problem) 218, 219
Two by two table 346, 347, 462, 476, 488
Two way classification 538
Two factor interaction 518–522
Two sample rank-sequential test 301
Two sample test of Wilcoxon, Mann, and Whitney 293–303
Two sample t-test for independent random samples 264–275
Two sided alternative 125
Two sided Fisher test 372
Two sided problem 125, 127, 138, 222, 260, 274, 277, 285, 289, 292, 296–301, 312, 316, 318, 376, 425, 441, 452, 464
Two sided procedures 258
Two-sided question 137, 169, 186, 188, 215, 255, 257, 294, 350, 380, 381, 463
Two sided tests 124, 125, 127, 128, 137, 215, 217, 251, 261, 266, 276, 277, 287, 309, 311, 312, 315, 318, 319, 349, 371, 378, 397–399, 401, 429, 430, 437, 439
 bounds for 318
Two way analysis of variance
 with a single observation per cell 537–541
 with a single observed value per class 537–541
 with ranks 549–558
Two way classification 541
 analysis of variance for 527
 model without interaction 562
 with one observation per class, no interaction 541
 with unequal numbers of observations with cells 540
Two way layout data with interaction and one observation per cell, analysis of 541

700   Subject Index

Two way table 462, 474, 484, 490
   testing for independence or homogeneity 490
Type I error 112, 123, 219, 251, 273
   rate, experimental 537
Type II error 117–119, 125–129, 186, 219, 223, 251, 260, 265, 273, 504
Typical cases 246, 280

Unbiased estimates 433
Unbiased estimator 66
Unbiased information 196, 197, 202–204
Uncertainty 23–26, 113, 224, 239, 246, 249, 561
Unconscious and unintentional error 212–214
Undirected retrospective analysis 208
Uniform distribution 83–85, 98, 241, 252, 322
Uniformity (consistency) in observing 207, 208
Union 28
Unit of observation, experimental unit 47, 560
Unit sequential sampling inspection 231
Universe (*see* Population)
Upper confidence limit, one sided 259, 337
Upper significance bounds of Studentized range distribution 535, 536
Urn 37, 39, 47, 48
   model 47, 48
U-shaped distribution 93, 94, 98, 99, 391
U-test 130, 132, 262, 294, 295, 297, 299, 303, 366
   asymptotic efficiency of 293
   for the comparison of 2 variances 262
   of Wilcoxon, Mann and Whitney, one sided 293
   power of 241
   with tied ranks 296

V-mask 202
Variability 5, 23–26, 68, 85, 559
Variability collectives 559
Variability difference 289
Variability test 285, 286
Variability of individual factors as components of total variability 519, 529
Variable, random 43–46, 134
   continuous 46
   discrete 44, 45
   range of 43
   realization of 43
Variable plans 231
Variance 46, 68–77, 85, 106, 139, 232, 234, 251, 264, 266, 267, 269, 271, 293, 308, 389, 495, 497–499, 503, 517, 518, 521
   analysis of 232, 494–545
   between samples 502–504
   between two flanks of an indivuduum 308
   combined 76, 77
   estimate of 68–77, 501, 502
   of measured points 446
   of sample rank sums 303
   of the mean 46
   partial 232
   within samples 501–504
Variance analysis 232, 494–545
Variance analysis in regression analysis 418, 433, 442
Variance analytic pairwise comparison 555
Variance components
   estimation of 529
   testing hypotheses on 529
Variance components model (Model II) 529, 532
Variances
   additivity of 46
   deviations from the hypothesis of equality of 504
   identical, of individual sample groups 542
   testing equality or homogeneity of 495–500
   tests for equality of 495–500
Variation
   between levels of factor 522
   between sample groups 470
   between samples 502–504
   coefficient of 77, 78, 85, 107, 110, 133, 198, 259, 260, 275
   per degree of freedom, average 521
   total 462, 474, 487, 502, 522, 526
   within sample 502–504
   within samples 501–504
Venn diagram 30–33

Subject Index

Wald, A. 24, 133, 567
Wallis and Moore phase frequency test 378, 379
War game 240
Warning limit 200, 230
Wasserman test 218
Weakly skewed distributions and MD 281
Weak significance determined by two way analysis of variance 540
Weibull distribution 234, 235, 323
Weighted
  arithmetic mean 77
  geometric mean 86, 90
  harmonic mean 88
  linear contrasts 510
Wilcoxon and Wilcox multiple comparison of correlated samples 555–558
Wilcoxon–Mann–Whitney test 293–303
Wilcoxon matched pair signed rank test 312–315
Wilcoxon test
  for independent samples 293–303
  for pair differences 312–315
  for paired data 312–315
Wilcoxon two sample sequential test scheme 301
Wild animal population size 172
Wilks equation 283
Winsorization 280
Within group mean square 503
With/without replacement 36, 47, 48, 94, 161, 162, 171
Woolf's G-test 351–360
Work force 245
Writing a report 566
W-test of Shapiro and Wilk 329

X-test of van der Waerden 293

Yates correction 351, 366, 367, 380

Zero correlation, random deviation from 424–426
Zero-one observations 166
z-transformation 427–432

Remarks and examples concerning the two sample Kolmogoroff–Smirnoff test for nonclassified data

K–S tests use the maximum vertical distance between two empirical distribution functions F_1 and F_2 (the two sample test) or between an empirical distribution function F_e and a hypothesized distribution function F_0 (the goodness of fit test, p. 330). For the two sample K–S test with $H_0: F_1 = F_2$ the assumptions are: Both samples are mutually independent random samples from continuous populations. If the random variables are not continuous but discrete, the test is still valid but becomes conservative. For the one sample goodness of fit test with $H_0: F_e = F_0$ the assumption is: The sample is a random sample.

For the two sample K–S test we use

$$\hat{D} = \max |F_1 - F_2|$$

instead of (3.46) on page 291.

Example 1 Two sided K–S test, $\alpha = 0.05$; $n_1 = n_2 = 10$

x_1	x_2	$F_1 - F_2$	x_1	x_2	$F_1 - F_2$
0.6		1/10 − 0/10 = 1/10	3.0	3.0	9/10 − 3/10 = **6/10**
1.2		2/10 − 0/10 = 2/10		3.1	9/10 − 4/10 = 5/10
1.6		3/10 − 0/10 = 3/10	3.2	3.2	
1.7				3.2	1 − 6/10 = 4/10
1.7		5/10 − 0/10 = 5/10		3.5	1 − 7/10 = 3/10
2.1	2.1	6/10 − 1/10 = 5/10		3.8	1 − 8/10 = 2/10
	2.3	6/10 − 2/10 = 4/10		4.6	1 − 9/10 = 1/10
2.8		7/10 − 2/10 = 5/10		7.2	1 − 1 = 0
2.9		8/10 − 2/10 = **6/10**			$\hat{D} = 6/10 < 7/10 = D_{10(0.05)}$: H_0 is retained

The value $D_{10(0.05); \text{two sided}} = 7/10$ is from Table 61 on page 292.

Example 2 Two sided K–S test, $\alpha = 0.05$; $n_1 = 12$, $n_2 = 8$

x_1	x_2	$F_1 - F_2$	x_1	x_2	$F_1 - F_2$
0.6		1/12 − 0/8 = 1/12	3.0	3.0	9/12 − 3/8 = 3/8
1.2		2/12 − 0/8 = 2/12		3.1	9/12 − 4/8 = 2/8
1.6		3/12 − 0/8 = 3/12	3.2	3.2	
1.7				3.2	10/12 − 6/8 = 1/12
1.7		5/12 − 0/8 = **5/12**	3.5		11/12 − 6/8 = 2/12
2.1	2.1	6/12 − 1/8 = 3/8	3.8		1 − 6/8 = 2/8
	2.3	6/12 − 2/8 = 2/8		4.6	1 − 7/8 = 1/8
2.8		7/12 − 2/8 = 1/8		7.2	1 − 1 = 0
2.9		8/12 − 2/8 = **5/12**			$\hat{D} = 5/12 < 7/12 = D_{0.05}$: H_0 is retained

For the two sided test at the 5% level H_0 is rejected if $\hat{D} > D_{n_1;n_2;0,05;\text{two sided}}$. Some values of this D from Massey (1952) for small sample sizes are given below. More values give Kim (1969) and nearly all books on nonparametric statistics.

$$D_{n_1;n_2;0,05 \text{ two sided}}$$

n_2	n_1	$D_{0,05}$	n_2	n_1	$D_{0,05}$	n_2	n_1	$D_{0,05}$
6	7	29/42	7	8	5/8	8	9	5/8
	8	2/3		9	40/63		10	23/40
	9	2/3		10	43/70		12	7/12
	10	19/30		14	4/7		16	9/16
9	10	26/45	10	15	1/2	12	15	1/2
	12	5/9		20	1/2		16	23/48
	15	8/15	15	20	13/30		18	17/36
	18	1/2	16	20	17/40		20	7/15

Selected 95% confidence intervals for π (binomial distribution) (n = sample size, x = number of hits; e.g. $\hat{p} = x/n = 10/300$ or 3.33%, 95% CI: $1.60\% \leq \pi \leq 6.07\%$)

x	n: 25	50	75	100	200	300	400	500	1000	x
0	0.00– 13.72	0.00– 7.11	0.00– 4.80	0.00– 3.62	0.00– 1.83	0.00– 1.22	0.00– 0.92	0.00– 0.74	0.00– 0.37	0
1	0.10– 20.35	0.05–10.65	0.03– 7.21	0.03– 5.45	0.03– 2.75	0.01– 1.84	0.01– 1.38	0.01– 1.11	0.00– 0.56	1
2	0.98– 26.03	0.49–13.71	0.32– 9.30	0.24– 7.04	0.12– 3.57	0.08– 2.39	0.06– 1.79	0.05– 1.44	0.02– 0.72	2
3	2.55– 31.22	1.25–16.55	0.83–11.25	0.62– 8.52	0.31– 4.32	0.21– 2.89	0.16– 2.18	0.12– 1.74	0.06– 0.87	3
4	4.54– 36.08	2.22–19.23	1.47–13.10	1.10– 9.93	0.55– 5.04	0.36– 3.38	0.27– 2.54	0.22– 2.04	0.11– 1.02	4
5	6.83– 40.70	3.33–21.81	2.20–14.88	1.64–11.28	0.80– 5.78	0.53– 3.88	0.40– 2.92	0.32– 2.34	0.16– 1.17	5
6	9.36– 45.13	4.53–24.31	2.99–16.60	2.23–12.60	1.09– 6.46	0.73– 4.33	0.54– 3.26	0.43– 2.61	0.22– 1.31	6
7	12.07– 49.39	5.82–26.74	3.84–18.29	2.86–13.89	1.40– 7.12	0.93– 4.77	0.70– 3.59	0.56– 2.88	0.28– 1.44	7
8	14.95– 53.50	7.17–29.11	4.72–19.94	3.52–15.16	1.73– 7.76	1.15– 5.21	0.86– 3.92	0.69– 3.14	0.34– 1.58	8
9	17.97– 57.48	8.58–31.44	5.64–21.56	4.20–16.40	2.07– 8.40	1.37– 5.64	1.03– 4.25	0.82– 3.40	0.41– 1.71	9
10	21.13– 61.33	10.03–33.72	6.58–23.16	4.90–17.62	2.41– 9.03	1.60– 6.07	1.20– 4.57	0.96– 3.66	0.48– 1.84	10
11	24.40– 65.07	11.53–35.96	7.56–24.73	5.62–18.83	2.77– 9.66	1.84– 6.49	1.37– 4.88	1.10– 3.92	0.55– 1.97	11
12	27.80– 68.69	13.06–38.17	8.55–26.28	6.36–20.02	3.13–10.28	2.08– 6.90	1.55– 5.20	1.24– 4.17	0.62– 2.09	12
13	31.31– 72.20	14.63–40.34	9.57–27.81	7.11–21.20	3.50–10.89	2.32– 7.32	1.74– 5.51	1.39– 4.42	0.69– 2.22	13
14	34.93– 75.60	16.23–42.49	10.60–29.33	7.87–22.37	3.88–11.49	2.57– 7.73	1.92– 5.82	1.54– 4.67	0.77– 2.34	14
15	38.67– 78.87	17.86–44.61	11.65–30.83	8.65–23.53	4.26–12.09	2.82– 8.13	2.11– 6.12	1.69– 4.91	0.84– 2.47	15
16	42.52– 82.03	19.52–46.70	12.71–32.32	9.43–24.68	4.64–12.69	3.08– 8.53	2.30– 6.43	1.84– 5.16	0.92– 2.59	16
17	46.50– 85.05	21.21–48.77	13.79–33.79	10.23–25.82	5.03–13.29	3.33– 8.94	2.49– 6.73	1.99– 5.40	0.99– 2.71	17
18	50.61– 87.93	22.92–50.81	14.89–35.25	11.03–26.95	5.42–13.88	3.59– 9.33	2.69– 7.03	2.14– 5.64	1.07– 2.84	18
19	54.87– 90.64	24.65–52.83	15.99–36.70	11.84–28.07	5.82–14.46	3.85– 9.73	2.88– 7.33	2.30– 5.88	1.15– 2.96	19
20	59.30– 93.17	26.41–54.82	17.11–38.14	12.67–29.18	6.22–15.04	4.12–10.12	3.08– 7.63	2.46– 6.12	1.22– 3.08	20
21	63.92– 95.46	28.19–56.79	18.24–39.56	13.49–30.29	6.62–15.62	4.38–10.52	3.28– 7.93	2.62– 6.36	1.30– 3.20	21
22	68.78– 97.45	29.99–58.75	19.38–40.98	14.33–31.39	7.03–16.20	4.65–10.91	3.48– 8.22	2.78– 6.60	1.38– 3.32	22
23	73.97– 99.02	31.81–60.68	20.53–42.38	15.17–32.49	7.44–16.78	4.92–11.30	3.68– 8.51	2.94– 6.83	1.46– 3.44	23
24	79.65– 99.90	33.66–62.58	21.69–43.78	16.02–33.57	7.85–17.35	5.19–11.68	3.88– 8.81	3.10– 7.07	1.54– 3.55	24
25	86.28–100.00	35.53–64.47	22.86–45.17	16.88–34.66	8.26–17.92	5.47–12.07	4.08– 9.10	3.26– 7.30	1.62– 3.67	25
30			28.85–51.96	21.24–39.98	10.37–20.73	6.85–13.98	5.12–10.54	4.08– 8.46	2.03– 4.26	30
35			35.05–58.55	25.73–45.18	12.52–23.51	8.27–15.86	6.17–11.97	4.92– 9.61	2.45– 4.84	35
40				30.33–50.28	14.71–26.24	9.71–17.72	7.24–13.38	5.78–10.74	2.87– 5.41	40
45				35.03–55.27	16.93–28.94	11.16–19.56	8.33–14.77	6.64–11.86	3.30– 5.98	45
50				39.83–60.17	19.18–31.61	12.64–21.39	9.43–16.15	7.52–12.98	3.73– 6.54	50
60					23.77–36.88	15.63–24.99	11.65–18.89	9.29–15.18	4.61– 7.66	60
70					28.44–42.06	18.68–28.55	13.91–21.59	11.08–17.36	5.50– 8.76	70
80					33.19–47.16	21.76–32.06	16.20–24.27	12.90–19.52	6.40– 9.86	80
90					38.02–52.18	24.89–35.54	18.51–26.92	14.74–21.66	7.30–10.95	90
100					42.89–57.11	28.04–38.99	20.84–29.55	16.59–23.78	8.21–12.03	100
150						44.21–55.79	32.75–42.45	26.02–34.23	12.84–17.37	150
200							45.00–55.00	35.69–44.45	17.56–22.64	200
250								45.54–54.46	22.35–27.81	250
500									46.85–53.15	500

For $x/n > 0.5$ read off the 95% CI for $(1 - x/n)$ and subtract both limits from 100: e.g. $\hat{p} = x/n = 20/25$, = 5/25 read off 6.83 to 40.70 and obtain the 95% CI: $100 - 40.70 = 59.30$ to $93.17 = 100 - 6.83$, i.e. $59.30\% \leq \pi \leq 93.17\%$.

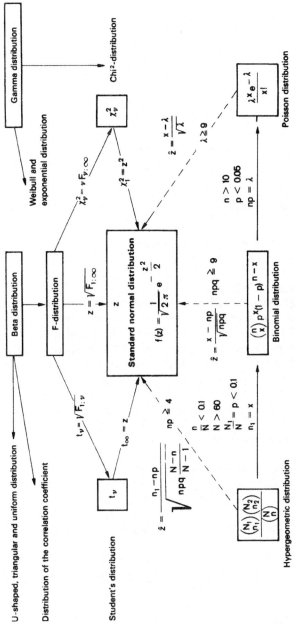

The central limit theorem

If the elements X_i of a set of n independent random variables have the same distribution function with mean μ and variance σ^2, then the larger n is, the more nearly will the random variable

$$z_n = \frac{\frac{1}{n}\sum_{i=1}^{n} X_i - \mu}{\sigma} \sqrt{n} = \frac{\sum_{i=1}^{n} X_i - n\mu}{\sigma \sqrt{n}}$$

be standard normally distributed.

It follows that the means $\bar{X}_i$ of a sequence of independent random variables which are identically distributed (with the same mean and the same variance), this original distribution of the X_i's being arbitrary to a great extent, are asymptotically normally distributed, i.e., the larger the sample size, the better is the approximation to the normal distribution. Analogous theorems apply to sequences of many other sample functions.

The standard techniques of statistics are given in the following table: they are employed in testing e.g.:

1. **The randomness of a sequence of data:** run test, phase test of Wallis and Moore, trend test of Cox and Stuart, mean square successive difference

2. **Distribution type, the agreement of an empirical distribution with a theoretical one** so-called fit tests: the χ^2-test, the Kolmogoroff-Smirnoff test, and especially the tests for
   (a) log-normal distribution: logarithmic probability chart;
   (b) normal distribution: probability chart, Lilliefors test, $\sqrt{b_1}$ and b_2 test for departure from normality;
   (c) simple and compound Poisson distribution: Poisson probability paper (or Thorndike nomogram).

3. **Equivalence of two or more independent populations:**
   (a) **Dispersion** of two or several populations on the basis of two or several independent samples: Siegel-Tukey test, Pillai-Buenaventura test, F-test or Levene test, Cochran test, Hartley test, Bartlett test;
   (b) **Central tendency:** median or mean of two (or several) populations on the basis of two (or several) independent samples: Median test, Mosteller test, Tukey test; U-test of Wilcoxon, Mann, and Whitney; Lord test, t-test and extended median tests, H-test of Kruskal and Wallis, Link-Wallace test, Nemenyi comparisons, analysis of variance, Scheffé test, Student-Newman-Keuls test.

4. **Equivalence of two or more correlated populations:** Sign tests, maximum test, Wilcoxon test, t-test or Q-test, Friedman test, Wilcoxon-Wilcox comparisons, analysis of variance.

5. **Independence or dependence of two characteristics:**
   (a) Fourfold and other two way tables: Fisher test, χ^2-tests with McNemar test, G-test, $2I$-test, coefficients of contingency;
   (b) Ranks or data sequences: quadrant correlation, corner test, Spearman rank correlation, product-moment correlation, linear regression.

Important statistical tests. One distinguishes three levels, in the order of increasing information content of the primary data: (1) Frequencies, (2) Ranks (e.g. school grades) and (3) Measurement data (derived from a scale with constant intervals). The information content of the data and the question posed suggest the appropriate statistical tests. Tests which are valid for data with low information content — they are listed in the upper part of the table — can also be used on data rich in information

DATA		1 sample	TESTS FOR 2 samples independent	2 samples correlated	more than 2 samples independent	more than 2 samples correlated
Frequencies (discrete)	with special discrete distributions	Computations in terms of the: Poisson distribution, binomial distribution, hypergeometric distribution χ^2-fit test 2I-fit test	Fisher test χ^2-tests G-test Median test Contingency coefficients Sequential test plans	Sign tests	χ^2-tests 2I-test Contingency coefficients	Q-test
Ranks (discrete)	not normally distributed	Run tests	Siegel-Tukey test Mosteller test Tukey test U-test	Maximum test Wilcoxon test Quadrant correlation Corner test Spearman rank correlation	Median tests H-test Nemenyi comparisons	Friedman test Multiple comparisons of Wilcoxon and Wilcox
Measurement data (continuous)	approximately normally distributed	Kolmogoroff-Smirnoff test Cox-Stuart trend test Mean square successive difference test Probability chart Tests for departure from normality Lilliefors test χ^2-test	Kolmogoroff-Smirnoff test Levene test Pillai–Buenaventura test F-test	t-test Product-moment correlation	Levene test Cochran, Hartley, and Bartlett test Link-Wallace test Analysis of variance	Analysis of variance
Measurement data (continuous)	possibly after a transformation	t-test outlier tests	Lord test t-tests	Linear regression	Multivariate methods	Multivariate methods

Printed in Great Britain
by Amazon